THE MOLECULAR BIOLOGY OF THE YEAST SACCHAROMYCES

METABOLISM
AND
GENE EXPRESSION

THE MOLECULAR BIOLOGY OF THE YEAST SACCHAROMYCES

METABOLISM AND GENE EXPRESSION

Edited by

Jeffrey N. Strathern
Cold Spring Harbor Laboratory

Elizabeth W. Jones
Carnegie-Mellon University

James R. Broach
State University of New York at Stony Brook

Cold Spring Harbor Laboratory
1982

**COLD SPRING HARBOR
MONOGRAPH SERIES**

The Lactose Operon
The Bacteriophage Lambda
The Molecular Biology of Tumour Viruses
Ribosomes
RNA Phages
RNA Polymerase
The Operon
The Single-Stranded DNA Phages
Transfer RNA:
 Structure, Properties, and Recognition
 Biological Aspects
Molecular Biology of Tumor Viruses, Second Edition:
 DNA Tumor Viruses
 RNA Tumor Viruses
The Molecular Biology of the Yeast Saccharomyces:
 Life Cycle and Inheritance
 Metabolism and Gene Expression
Mitochondrial Genes
Nucleases
Lambda II

**THE MOLECULAR BIOLOGY OF THE YEAST SACCHAROMYCES
METABOLISM AND GENE EXPRESSION**

© 1982 by Cold Spring Harbor Laboratory
Printed in the United States of America
Book design by Emily Harste

Library of Congress Cataloging in Publication Data
Main entry under title:

The Molecular biology of the yeast Saccharomyces.

   (Cold Spring Harbor monograph series ; 11B)
   Includes index.
   1. Saccharomyces—Genetics. 2. Saccharomyces—
Physiology. 3. Fungi—Genetics. 4. Fungi—Physiology.
5. Molecular biology. I. Strathern, Jeffrey N.
II. Jones, Elizabeth W. III. Broach, James R.
IV. Series [DNLM: 1. Molecular biology.
2. Saccharomyces cerevisiae. 3. Saccharomyces
cerevisiae—Metabolism. WI CO133C v.11B / QW 180.5.A8
M718]
QK623.S23M643   1982   589.2'33   81-68203
ISBN 0-87969-149-2

All Cold Spring Harbor Laboratory publications are available through booksellers or may be ordered directly from Cold Spring Harbor Laboratory, Box 100, Cold Spring Harbor, New York 11724.

SAN 203-6185

Contents

Preface, ix

Carbohydrate Metabolism
D. G. Fraenkel, 1

Nitrogen Metabolism in *Saccharomyces cerevisiae*
T. G. Cooper, 39

Membrane Lipids of Yeast: Biochemical and Genetic Studies
S. A. Henry, 101

Regulatory Circuits for Gene Expression: The Metabolism of Galactose and Phosphate
Y. Oshima, 159

Regulation of Amino Acid and Nucleotide Biosynthesis in Yeast
E. W. Jones and G. R. Fink, 181

Mutations Altering Initiation of Translation of Yeast Iso-1-cytochrome *c*; Contrasts between the Eukaryotic and Prokaryotic Initiation Process
F. Sherman and J. W. Stewart, 301

Yeast Cell Wall and Cell Surface
C. E. Ballou, 335

The Secretory Process and Yeast Cell-surface Assembly
R. Schekman and P. Novick, 361

Transport in *Saccharomyces cerevisiae*
T. G. Cooper, 399

Suppression in the Yeast *Saccharomyces cerevisiae*
F. Sherman, 463

Organization and Expression of tRNA Genes in *Saccharomyces cerevisiae*
C. Guthrie and J. Abelson, 487

The Yeast Ribosome: Structure, Function, and Synthesis
J. R. Warner, 529

Yeast Nuclear RNA Polymerases and Their Role in Transcription
A. Sentenac and B. Hall, 561

Principles and Practice of Recombinant DNA Research with Yeast
D. Botstein and R. W. Davis, 607

Appendices

I. Genetic Map of *Saccharomyces cerevisiae*, 639
   R. K. Mortimer and D. Schild
II. Biochemical Markers for Yeast Organelles, 651
   R. Schekman

Subject Index, 653
Gene Index, 675

Contents of companion volume

THE MOLECULAR BIOLOGY OF THE YEAST SACCHAROMYCES
LIFE CYCLE AND INHERITANCE

Development of Yeast as an Experimental Organism
H. Roman

Genetic Mapping in *Saccharomyces cerevisiae*
R. K. Mortimer and D. Schild

Genome Structure and Replication
W. L. Fangman and V. A. Zakian

Cytology of the Yeast Life Cycle
B. Byers

The *Saccharomyces cerevisiae* Cell Cycle
J. R. Pringle and L. H. Hartwell

Pheromonal Regulation of Development in *Saccharomyces cerevisiae*
J. Thorner

Control of Cell Type in *Saccharomyces cerevisiae*. Mating Type and Mating-type Interconversion
I. Herskowitz and Y. Oshima

Meiosis and Ascospore Development
R. E. Esposito and S. Klapholz

Mechanisms of Meiotic Gene Conversion, or "Wanderings on a Foreign Strand"
S. Fogel, R. K. Mortimer, and K. Lusnak

Mechanisms of Mitotic Recombination
M. S. Esposito and J. E. Wagstaff

DNA Repair and Mutagenesis in Yeast
R. H. Haynes and B. A. Kunz

Killer Systems in *Saccharomyces cerevisiae*
R. B. Wickner

The Yeast Plasmid 2μ Circle
J. R. Broach

Mitochondrial Structure
B. Stevens

Mitochondrial Genetics and Functions
B. Dujon

Appendices
  I. Genetic Nomenclature
    F. Sherman
 II. Genetic Map of *Saccharomyces cerevisiae*
    R. K. Mortimer and D. Schild
III. Genes of *Saccharomyces cerevisiae*
    J. R. Broach
    A. Yeast Genes Listed Alphabetically by Gene Designation
    B. Alphabetical Listing of Gene Products
    C. Genes Grouped by Common Function

Preface

With relief we note that the publication of this volume marks the completion of a project we naively began 3 years ago. Inspired by the enthusiasm and encouragement of our colleagues at the Molecular Biology of Yeast meeting at Cold Spring Harbor Laboratory in 1979 and by the excitement engendered by the results presented at the meeting, we set out to assemble a series of reviews of the multifarious topics that in sum constitute the molecular biology of the yeast *Saccharomyces.* The goal of this project was to produce a monograph that covered the field in sufficient depth and breadth so that it could serve as the definitive guide for the current workers in the field, as well as for those scientists who wished to enter the field or who merely had a passing interest in some aspect of yeast molecular biology. Clearly, though, yeast molecular biology, like any dynamic and rapidly expanding field of science, did not cease to grow during the course of our compilation of the reviews. Each of the articles was designed and edited to provide a comprehensive and insightful review of one particular aspect of the field. Many of the papers represent an updated view of an area that has been reviewed before. In other papers the authors have brought together observations and results not previously juxtaposed and evaluated. In total, therefore, the compilation of all of these reviews in a single monograph places the complete molecular biology of yeast at one's fingertips. The end result of this process has been, in several cases, the revelation of novel and significant insights into an area of research and a focusing of direction that should be invaluable to workers in the field.

Although our goal in this monograph was a comprehensive review of the molecular biology of yeast, certain topics have received less attention than some might like. For instance, we have not included a separate and prolonged analysis of transposable elements in yeast, nor have we delved into the topic of mRNA splicing in yeast. Such omissions resulted from the fact that at the time specific reviews were written, these phenomena were not sufficiently developed to warrant extensive discussion. Obviously, both of these areas represent exciting new directions of investigation in the field of yeast molecular biology. We expect that the productive combination of classical genetic analysis, genetic engineering, and enzymology to which yeast is so amenable will yield such significant insights into these issues that they will be essential topics in any future monograph on yeast. In other cases our failure to include certain topics in this monograph was due to the rapidity with which certain areas of molecular biology are progressing. Thus, anyone who sits down to write a review of cloning technology in yeast, approaches to DNA and genomic manipulations, or mapping techniques does so with the knowledge that new techniques will undoubtedly have supplanted those described even before the galleys are corrected. Finally, certain topics have not been covered simply because even the most exhaustive review of a field must be limited. For these omissions we invoke editorial prerogative and hope that they have been few.

Like many editors before us, we have been caught between the Scylla of attempting to provide as thorough a monograph as possible and the Charybdis of producing as timely a monograph as possible. Our decision at the outset, which was continually reaffirmed during the course of the editorial process, was to compile as complete a review as could be realistically expected. Thus, in many cases the timeliness of certain reviews has been sacrificed. For instance, the manuscripts by Dan Fraenkel, Yasuji Oshima, Fred Sherman and John Stewart, Clint Ballou, Randy Schekman and Peter Novick, and Jon Warner have been in hand since 1980. Consequently, the extent to which these reviews appear dated is due to our editorial decision and not to any shortcomings on the part of these authors. It is a tribute to the integrity and diligence of these authors that they abided by the deadlines we invoked. It is also a tribute to their intellect and insights that these papers are certainly still valuable and timely resources. We appreciate the patience of all of the authors whom our editorial policy might have provoked. We are especially grateful to Bob Mortimer and David Schild, who insured that the genetic map was current by providing us with an updated version just prior to publication.

We thank Nancy Ford and Nadine Dumser for their inimitable and valuable assistance in editing this monograph and Mary Cozza, Gail Anderson, and Michaela Cooney for helping to prepare it for publication. We also thank Drs. J. Abraham and A. Sutton for indexing the first and second volumes of this monograph, respectively. Finally, we thank each of the authors of the papers in this monograph. It is their efforts that have made this monograph the valuable resource we believe it to be.

Jeffrey N. Strathern
Elizabeth W. Jones
James R. Broach

THE MOLECULAR BIOLOGY OF THE YEAST SACCHAROMYCES

METABOLISM AND GENE EXPRESSION

Carbohydrate Metabolism

Dan G. Fraenkel
Department of Microbiology and Molecular Genetics
Harvard Medical School, Boston, Massachusetts 02115

1. Introduction
   A. Scope and Organization
   B. An Overall View of Yeast Metabolism
2. **Glycolysis**
   A. Glycolysis Mutants: General Comments
   B. Transport of Glucose
   C. Glucose Phosphorylation
   D. Phosphoglucose Isomerase
   E. Phosphofructokinase
   F. Role of Anomers in Glucose Metabolism
   G. From Fructose-1, 6-P_2 to Phosphoenolpyruvate
   H. Pyruvate Kinase
   I. Fermentation Products
   J. Regulation of the Amount of Glycolysis Enzymes
3. **Pathways into Glycolysis**
   A. Fructose, Mannose, and Glucosamine
   B. Galactose
   C. Glycosides
   D. Glycerol (Utilization)
4. **Hexose-monophosphate Oxidation Pathway**
5. **Pathways of Carboxylic Acid Metabolism**
   A. Functions
   B. Routes to the TCA Cycle
   C. TCA Cycle
   D. Routes out of the TCA Cycle and Gluconeogenic Growth
   E. Other Mutants
6. **Induction and Repression of Respiratory Enzymes**
   A. Oxygen Effects
   B. Catabolite Repression
7. **Catabolite Inactivation**
8. **Enzyme Location and Isoenzymes**
9. **Pasteur Effect**
10. **Reserves**
    A. Glycogen and Trehalose
    B. Polyphosphate
    C. Reserve Materials, ATPase, and the Control of Glycolysis

INTRODUCTION

Scope and Organization

This paper, which I dedicate to B. L. Horecker, is a review of intermediary carbohydrate metabolism in *Saccharomyces*. It emphasizes the central pathways, including glycolysis and the tricarboxylic-acid (TCA) cycle, and also treats some specific catabolic routes; biosynthetic pathways are not included. It covers some of the same material as the review by Sols et al. (1971), in addition to mutant studies.

Many of the reactions with their gene symbols (some already used, others suggested here) are shown in Figure 1 and are also listed in Table 1 with references to structural and genetic studies. The gene symbols are not intended as abbreviations for the enzymes and appear in the text mainly for convenience in referring to Figure 1 and Table 1.

An Overall View of Yeast Metabolism

Strains of *Saccharomyces* in common use for genetics and biochemistry grow best by fermentation, even aerobically; this is illustrated in Table 2. Data from Lagunas (1976) show aerobic growth in batch culture. It was fastest on glucose (k, the first-order growth rate constant, of 0.34 hr^{-1}), but it was inefficient, with only 13% assimilation to cell material and most of the remaining carbon accumulating as fermentation products. Batch aerobic growth on galactose was a little slower than on glucose (k of 0.26) and there was 26% assimilation, but fermentation products still accounted for 55% of the sugar used. In contrast, aerobic growth on ethanol was slow (k of 0.087) but efficient (39% assimilation).

The accumulation of fermentation products from sugars in aerobic batch culture depends on catabolite repression, for in aerobic growth in a glucose-limited chemostat (Table 2, line 4, from Oura 1974), they do not accumulate and cell yield was very high (55% assimilation). For comparison, characteristics of anaerobic growth in a glucose-limited chemostat are given in line 5 of Table 2. As expected, cell yield was very low (9% assimilation), and most glucose formed ethanol.

The energetics of these situations are as follows. In aerobic growth on ethanol, almost all ATP must come by mitochondrial oxidative phosphorylation, and the latter process must also be very active in the aerobic glucose-limited chemostat. In anaerobic growth on glucose, on the other hand, ATP may come entirely from substrate level phosphorylations in glycolysis. For batch culture growth on sugars, the situation is less clear. Considerations of respiratory rate led to the suggestion that even in batch growth on glucose, respiration might make an important contribution (Lagunas 1976; see also Barford and Hall 1979), but Lagunas (1979) showed similar growth rates and yields aerobically and anaerobically. On galactose, however, respiration was important.

GLYCOLYSIS

The glycolysis enzymes are an appreciable fraction of "soluble" protein—65% according to Hess et al. (1969) and about 30% according to Table 1. These

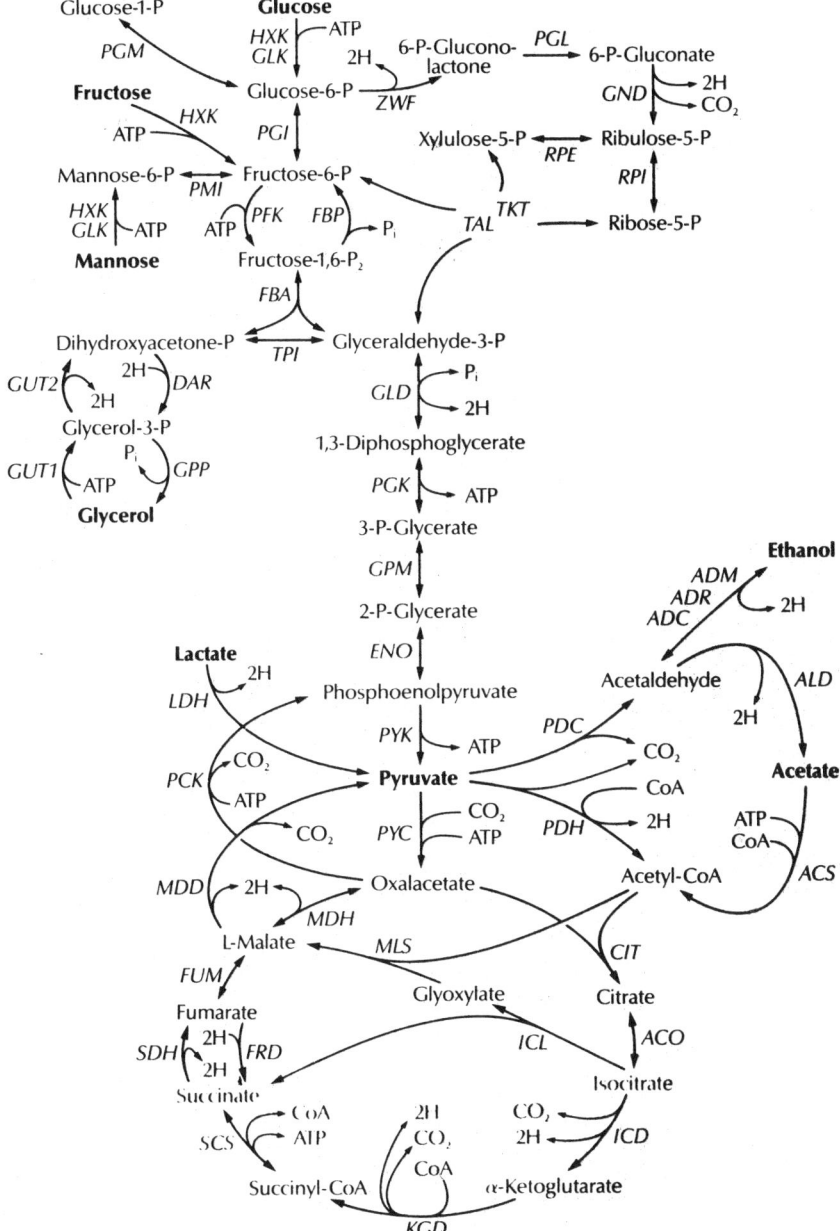

Figure 1 Pathways of yeast intermediary metabolism. Gene symbols are italicized (see footnote to Table 2). All sugars and sugar derivatives are of D configuration. Substances in heavy type are commonly used as carbon sources for growth. Many of the reactions are incompletely described. For example, for kinases, ATP is indicated as a reactant, but ADP is not shown as a product. The existence, function, or direction of several of the indicated reactions is speculative (see text). (P_i) Inorganic phosphate; 2H indicates a dehydrogenase reaction, whether pyridine nucleotide (e.g., glyceraldehyde-3-P dehydrogenase)- or flavoprotein (e.g., succinate dehydrogenase)-dependent.

Table 1 Enzymes and genes of yeast intermediary metabolism

ACO^a (ACS)	aconitate hydratase 4.2.1.3[b]; R: Boyer (1971, V, 413–439)[c] acetyl-CoA synthetase 6.2.1.1; Frenkel and Kitchens (1977); 2×78, 0.5%[d]
ADC ADM ADR	alcohol dehydrogenase 1.1.1.1; Wills and Jörnvall (1979); Wenger and Bernofsky (1971); 4×36, $\sim 1\%$; R: Boyer (1975, XI, 103–190)
(ALD)	aldehyde dehydrogenase (NADP') 1.2.1.5; Bostian and Betts (1978); 4×62
CIT	citrate synthase 4.1.3.7; Parvin (1969); 0.2%; R: Weitzman and Dawson (1976)
(DAR)	dihydroxyacetone-P reductase (glycerol-3-P dehydrogenase) 1.1.1.94
(ENO)	enolase 4.2.1.11; Brewer and Fairwell (1970); 2×44; R: Boyer (1971, V, 499–538)
(FBA)	fructose-bisphosphate aldolase 4.1.2.13; Harris et al. (1969); Lubini et al. (1977); 2×40, 4%; R: Boyer (1972, VII, 213–258)
(FBP)	fructose bisphosphatase 3.1.3.11; Funayama et al. (1979); 2×56, 0.1%; R: Boyer (1971, VI, 612–646)
(FRD)	fumarate reductase 1.3.99.1; Tisdale et al. (1968)
(FUM)	fumarase 4.2.1.2; Cataldi and Stoppani (1966); R: Boyer (1971, V, 539–571)
(GLD)	glyceraldehyde-3-P dehydrogenase 1.2.1.12; Kirschner and Voigt (1968); Butler and Jones (1970); Stallcup et al. (1972); Holland and Westhead (1973); 4×36, $2-20\%$; R: Boyer (1976, XIII, 1–49)
GLK	glucokinase 2.7.1.2; Maitra and Lobo (1977c); $? \times 51$, 0.2%

(GND)	6-phosphogluconate dehydrogenase 1.1.1.44
GPM	phosphoglycerate mutase 2.7.5.3; Grisolia and Carreras (1975); Sasaki et al. (1976); 4×28, 0.9%; R: Boyer (1972, VI, 407–477)
(GPP)	glycerol-1-phosphatase 3.1.3.21
GUT1	glycerol kinase 2.7.1.30; R: Boyer (1973, VIII, 487–508)
GUT2	glycerol-3-P dehydrogenase 1.1.99.5; R: Boyer (1976, XIII, 175–298)
HXK1	hexokinase; Bernard (1975); McDonald et al. (1979); 2×52, 0.5%; R: Boyer (1973, IX, 1–48); $HXK1$ on $6R^c$, Lobo and Maitra (1977a); $HXK2$ on 7L, P.K. Maitra (pers. comm.)
HXK2	
(ICD1)	isocitrate dehydrogenase (NAD$^+$) 1.1.1.42; Barnes et al. (1971); Illingworth (1972); 8×40, $\alpha_4\beta_4$.
(ICD2)	isocitrate dehydrogenase (NADP$^+$) 1.1.1.43; Kornberg (1955)
ICL	isocitrate lyase 4.1.3.1; Olson (1959)
KGD	α-ketoglutarate dehydrogenase 1.2.4.2; Hirabayashi and Harada (1971); R: Boyer (1970, I, 213–240)
(LDH)	lactate dehydrogenase 1.1.2.3 (cytochrome b_2); Labeyrie et al. (1978); 4×57, 1.1.2.4, 1.1.96.6; R: Boyer (1976, XIII, 175–298)
(MDD)	malate dehydrogenase (decarboxylating) or the "malic enzyme" 1.1.1.40; Temperli et al. (1965); Fuck et al. (1973); R: Frenkel (1975)
MDH	malate dehydrogenase 1.1.1.37; Hägele et al. (1978); 2×37 (mitochondrial), 2×34 (cytoplasmic); R: Boyer (1975, XI, 369–396)
(MLS)	malate synthase 1.1.3.2; Dixon et al. (1960); Zipper and Durschlag (1978); 3×63
(PCK)	phosphoenolpyruvate carboxykinase 4.1.1.49

Table 1 *(Continued)*

PDC	pyruvate decarboxylase 4.1.1.1; Gounaris et al. (1975); 4×60, 1%
(PDH)	pyruvate dehydrogenase 1.2.4.1; Wais et al. (1973); 0.1%; R: Boyer (1970, I, 213–240)
PFK	phosphofructokinase 2.7.1.11; Tamaki and Hess (1975); Kopperschläger et al. (1977); $\alpha_4\beta_4$, $8 \times \sim 100$, 0.3%; R: Boyer (1973, VIII, 240–278); Uyeda (1979)
PGI	phosphoglucose isomerase 5.3.1.9; Kempe et al. (1974); 2×60, 1%; R: Boyer (1972, VI, 272–354); on 2R, Maitra and Lobo (1977d); Herrera and Pascual (1978)
PGK	phosphoglycerate kinase 2.7.2.3; Scopes (1975); Spragg et al. (1976); 1×40, 5%; R: Boyer (1973, VIII, 335–351)
(PGL)	6-phosphogluconolactonase 3.1.1.31
PGM1	phosphoglucomutase 2.7.5.1; Daugherty et al. (1975);
PGM2	2×32, 0.04%; R: Boyer (1972, VI, 407–477)
PMI	phosphomannose isomerase 5.3.1.8; Gracy and Noltmann (1968); $1 \times 45?$, 0.06%; R: Boyer (1972, VI, 272–354)
(PYC)	pyruvate carboxylase 6.4.1.1; Cohen et al. (1979); 4×120, 0.3%; R: Boyer (1972, VI, 1–36)
PYK	pyruvate kinase 2.7.1.40; Aust and Suelter (1978); 4×57, 3–6%; R: Boyer (1973, VIII, 353–382); on 1L, Maitra and Lobo (1977a); Sprague (1977)
(RPE)	ribulose-5-P 3-epimerase; Williamson and Wood (1966)
(RPI)	ribose-5-phosphate isomerase 5.3.1.6; R: Boyer (1972, VI, 272–354)
SDH	succinate dehydrogenase 1.3.99.1; R: Boyer (1976, XIII, 175–298)
(SCS)	succinyl-CoA synthetase 6.2.1.4

(TAL)	transaldolase 2.2.1.2; Vankataraman and Racker (1961); 0.4%; R: Boyer (1972, VII, 259–280)
(TKT)	transketolase 2.2.1.1; Cavalieri et al. (1975); Belyaeva et al. (1978); 2 × 80
TPI	triosephosphate isomerase 5.3.1.1; Alber et al. (1980); 2 × 27, 2%; R: Boyer (1972, VI, 272–354)
(ZWF)	glucose-6-phosphate dehydrogenase 1.1.1.49; Yue et al. (1969); 4 × 51, 0.04%; R: Levy (1979)

[a] Gene symbol. Ones already used are written without parentheses. For the reactions without assigned gene symbols, suggested symbols are indicated in parentheses. Many of the latter symbols are already in use in prokaryotic genetics. Some have been chosen so as not to conflict with other usage in yeast. Thus, *PGM* is kept for phosphoglucomutase (Bevan and Douglas 1969), and *GPM* is suggested for phosphoglycerate mutase instead; *GLD* is used for glyceraldehyde-3-P dehydrogenase, because *GAP* has been used for permeability. Two TCA-cycle gene symbols have been changed (from *GLU*) to fit the custom of naming genes in these pathways after their reactions rather than after their mutant phenotypes. In general, gene symbols are only given numbers in those cases where more than one gene has been identified (e.g., *HXK1*, *HXK2*).

[b] Name and probable enzyme commission number (*Enzyme Nomenclature*, 1973, Elsevier, New York).

[c] Reference to a recent review on the reaction (R); most are from the series edited by P.D. Boyer (*The Enzymes*), with year, volume, and pages of relevant chapter given.

[d] References to enzyme purification or structure, followed by an indication of structure (e.g., 4 × 25 means a tetramer of subunits of 25,000) and an estimate, in percent, of the amount of the protein in a crude extract (specific activity [sp.act.] of "pure" enzyme/sp. act. in crude extract). The data on structure should be used critically and the original papers consulted. Many estimates of subunit size have increased since the recognition of proteolytic artifacts (see Pringle 1975). The value of the amount of protein is only meant to indicate whether the enzyme is a major or minor cell component. (In most cases there is no knowledge of whether growth conditions were optimal; extracts were made in different ways and commercial yeasts have been used often.)

[e] Genetic map position, if known, with references.

Table 2 Growth of yeast in minimal medium

Substrate and growth condition	k (hr^{-1})	Yield (g substrate/g cell)	Yield (% carbon assimilated)	O_2 used[a]	Percentage substrate oxidized	Products[a] ethanol	Products[a] glycerol	Rate of utilization[b] substrate	Rate of utilization[b] O_2
1. Glucose, O_2, log phase[c]	0.34	8.1	13	0.19	3	1.42	0.22	15.3	2.4
2. Galactose, O_2, log phase[c]	0.26	4.2	26	0.88	15	1.06	0.05	6.1	6.0
3. Ethanol, O_2, log phase[c]	0.087	2.2	39	1.66	55	0.0	0.0	4.2	7.5
4. Glucose, O_2, chemostat[d]	0.10	1.9	55	2.0	33	0.0	—	1.2	2.5
5. Glucose, AnO_2, chemostat[d]	0.10	11.2	9	—	—	1.75	—	6.2	—

[a] Given in moles per mole substrate.
[b] Rate of utilization is given in moles per hour per gram dry weight.
[c] Data adapted from Lagunas (1976) (haploid strain S-13-Gal).
[d] Data adapted from Oura (1974) (commercial strain 306); AnO_2 indicates anaerobic growth.

values reflect the quantitative importance of the fermentation pathway. Intracellular enzymes were once called ferments, with the word enzyme used first by Kuhne in 1877 and meaning "in yeast," referring to easily extractable activities. When the ferment of glucose fermentation was obtained in cell-free form by Buchner in 1897, it was called zymase, i.e., the enzyme of yeast itself (Oppenheimer and Mitchell 1901).

Hess and his colleagues have studied the dynamics of glycolysis in *S. carlsbergensis* in several ways (see Hess et al. 1969; Hess 1973). The concentrations of the individual enzymes are in the 10^{-5}–10^{-4}-M range; most are oligomers (phosphoglycerate kinase [*PGK*] being the exception), so the concentration of substrate and ligand-binding sites is high and diffusion time between reactions should be negligible. The principal reactions showing cooperative kinetics and allosteric control are phosphofructokinase (*PFK*) and pyruvate kinase (*PYK*), and they are usually thought of as important for controlling the pathway. Computations based on in vitro properties of the enzymes and in vivo substrate concentrations give good fit to the in vivo fluxes and even account for oscillations in NADH concentration (see also Pyruvate Kinase).

Glycolysis Mutants: General Comments

Mutant studies of yeast glycolysis began with the work of Lutstorf and Megnet (1968) on alcohol dehydrogenase and of Maitra (1971) on phosphoglucose isomerase. Mutants now have been reported for 9 of the 12 reactions between glucose and ethanol. The recent paper of Ciriacy and Breitenbach (1979) is a useful introduction. A mutagenized culture was screened for survivors unable to grow on enriched medium with glucose but able to grow on a gluconeogenic medium. Of 45 mutants, 25 were assigned single glycolytic enzyme defects by assay. The set included mutants for six different reactions. Incubations with glucose and assay of the soluble pool showed the metabolite accumulations that were generally expected, and Navon et al. (1979) have used ^{31}P nuclear magnetic resonance spectroscopy with whole cells to the same purpose. Enrichment techniques have also been employed to obtain glycolysis mutants (e.g., nystatin [Sprague 1977], netropsin, or inositol starvation [Clifton et al. 1978]).

In general, the growth description of glycolysis mutants is incomplete. Some are known or were selected as unable to grow on glucose in minimal medium. But most studies have used enriched medium such as YP (1% yeast extract, 2% peptone), which is by itself adequate for slow growth but is usually supplemented with pyruvate, glycerol plus lactate, glycerol plus ethanol, etc. (as permissive media), or with glucose (as nonpermissive medium). The mutants are usually inhibited by glucose, so their phenotype is clearly negative on YP plus glucose. Other nonpermissive sugars are not necessarily equally toxic. For example, the pyruvate kinase (*PYK*) mutant

DFY3 (Clifton et al. 1978) does not grow on maltose or glucose in minimal medium (the parental strain grows on both) or on YP plus glucose, but it is not inhibited on YP plus maltose. The reasons for toxicity or indifference to sugars are not understood and may relate partly to catabolite repression. One general complication is that few sugar preparations are completely free of glucose, and yeast extract itself contains about 5 $\mu g/mg$.

Transport of Glucose

The mechanism of uptake of glucose, or its analogs, is not understood. Many studies agree on the existence of a membrane carrier (an isolation was reported by Horak and Kotyk [1973]), which may deliver free glucose internally (Kotyk and Michaljanicova 1974). However, the internal steady-state concentration of glucose is very low (Becker and Betz 1972), and the phosphorylated sugar may even be found inside before the free sugar, implying a group translocation mechanism (Jaspers and van Steveninck 1975; Meredith and Romano 1977) or some other close linkage between entrance and phosphorylation. Thus, affinity for uptake differs between anaerobic, aerobic, and poisoned cells (Serrano and Delafuente 1974), and there have been suggestions of metabolic control of carrier affinity (e.g., by glucose-6-P; but see Perea and Gancedo 1978). A mechanism of cotransport of glucose with protons, as in *Neurospora* (Bowman and Slayman 1977), might also account for apparent changes in affinity; but in *Saccharomyces,* Seaston et al. (1973) found cotransport of protons with maltose but not with glucose. Glucose uptake by plasma-membrane vesicles was reported by Fuhrmann et al. (1976).

Glucose Phosphorylation

There are three enzymes: hexokinase A or P-I (*HXK1*), hexokinase B or P-II (*HXK2*), and glucokinase (*GLK*). The hexokinases are distinguished from glucokinase in being active with fructose as well as glucose and mannose; so mutants lacking both hexokinases are unable to grow on fructose, but they still grow on glucose. They were obtained first (Maitra 1970) by repetitive selection from the wild type for resistance to 2-deoxyglucose and later by selection for moderate glucose resistance starting with a pyruvate kinase mutant (Gancedo et al. 1977; Lobo and Maitra 1977a). Reversion studies show the loci *HXK1* and *HXK2* to be the structural genes for hexokinase P-I and P-II, respectively (Lobo and Maitra 1977a).

Glucose-negative mutants obtained from the *hxk1* and *hxk2* strains also lacked glucokinase and failed to grow on maltose, sucrose, or trehalose. These results show that there probably are no other major enzymes for initial glucose metabolism in yeast and that the disaccharides are used via conversion to glucose. The triple mutant grew on galactose, as expected for a separate pathway to glucose-6-P. Mutants containing just one of the three enzymes were constructed, and their sugar metabolism was assessed. Each grew on glucose or mannose, although somewhat more slowly than wild type, and glucose phosphorylation activities ranged from 25–40% of wild type, which was judged to be enough to account for the in vivo glucose utilization rates (Lobo and Maitra 1977b,c).

The individual functions of the enzymes in the wild-type strain are not understood. Special kinetic properties of enzyme P-II, such as activation by citrate and 3-phosphoglycerate at low pH, have now been ascribed to aluminum ion chelation, and their physiological significance is uncertain (Womack and Colowick 1979). The amounts of individual isozymes in the wild-type strain seem to be regulated with P-II predominant in glucose growth and P-I predominant in ethanol growth (Kopperschläger and Hofmann 1969; Gancedo et al. 1977; Muratsubaki and Katsume 1979).

Phosphoglucose Isomerase

This reaction (*PGI*) should be needed for growth on sugars whose metabolism proceeds via glucose-6-P and, in growth on other substances, for the synthesis of glucose-6-P. The first mutant was isolated as fructose-positive, glucose-negative, in minimal medium (Maitra 1971); Herrera and Pascual (1978) used rich medium to obtain a similar phenotype, and others have identified them among glucose-negative mutants in rich medium (Clifton et al. 1978; Ciriacy and Breitenbach 1979; Navon et al. 1979). Glucose-6-P accumulates in such strains (see also Maitra and Lobo 1977d; Gancedo and Gancedo 1979), and the pentose-P pathway via 6-phosphogluconate is evidently inadequate for growth although it may be used to a limited degree (Maitra 1971). Growth on galactose and maltose is less affected than growth on glucose (Maitra 1971; Herrera and Pascual 1978), but enriched media were used, so it is not clear whether the difference represents less toxicity or more metabolism.

Herrera and Pascual (1978) showed that mutants with 0.25% residual activity only grew on fructose if supplemented with a low concentration of glucose, and mutants that did not need a supplementation were "leakier." Studies of glucan content would be of interest.

Reversion studies show that *PGI* is the structural gene (Maitra and Lobo 1977d). Thus, the catabolic and biosynthetic functions of the phosphoglucose isomerase reaction in yeast are probably carried by the same enzyme, and the significance of apparent isozymes (Kempe et al. 1974) is unclear.

Phosphofructokinase

Phosphofructokinase is thought to play a role in regulation of glycolytic flux because of its sensitivity to a variety of effectors (Hess 1973; Uyeda 1979). The yeast enzyme is typical in this regard, with the list of apparent control ligands including fructose-6-P, ATP (inhibitor as well as substrate), AMP (activator), H^+, K^+, and phosphate. Because of proteolysis, early kinetic studies should be viewed with reservation, but recent purifications are believed to give a more native product. Nonetheless, there are still interesting uncertainties about the enzyme structure. Many types of studies (see Laurent et al. 1979) favor an $\alpha_4\beta_4$ composition, with the two subunits antigenically different (Herrmann et al. 1973) and possibly heterogeneous or modified (Tamaki and Hess 1975; Kopperschläger et al. 1977).

Phosphofructokinase mutants grow fairly well on glucose. They have been obtained indirectly by selection for glucose resistance in a pyruvate kinase mutant (Clifton et al. 1978), or as apparent glucose-negative mutants (Ciriacy and Breitenbach 1979; Navon et al. 1979), in which case other mutations were present. The *pfk* mutants accumulate hexose monophosphates from glucose (Ciriacy and Breitenbach 1979; Gancedo and Gancedo 1979; Navon et al. 1979), but somehow have metabolism adequate for growth. Two genes govern the activity (Clifton and Fraenkel (1982).

Role of Anomers in Glucose Metabolism

This subject was reviewed by Benkovic and Schray (1976). Hexoses and their phosphorylated derivatives exist mainly as rings—pyranose or furanose, α or β. Glucose in solution is 37% α. Some reactions are specific for anomers. Thus, both α- and β-glucose compete for the same site for entry (den Hollander et al. 1979), both are phosphorylated, but glucose-6-P dehydrogenase (*ZWF*) is specific for β-glucose-6-P, whereas phosphoglucose isomerase can use either anomer. Anomerization can occur spontaneously, with half-lives of about 2 hours for D-glucose, and about 1 second for D-fructose-1,6-P_2, but it may also be enzyme-catalyzed. Phosphoglucose isomerase has anomerase activity for glucose-6-P (Wurster and Hess 1975a), and also there is a separate enzyme, glucose-6-P-1-epimerase (*GPE*), catalyzing the reaction (Wurster and Hess 1975b). Models have been proposed for how anomeric specificity might affect metabolism (see Benkovic and Schray 1976; Koerner et al. 1977).

From Fructose-1,6-P_2 to Phosphoenolpyruvate

Mutants have been reported for triose-P isomerase (*TPI*) (Ciriacy and Breitenbach 1979; K.-B. Lam et al., pers. comm.), phosphoglycerate kinase, and phosphoglycerate mutase (*GPM*) ("*PGM*"; Lam and Marmur 1977; Ciriacy and Breitenbach 1979). *PGK* has been cloned and is the structural gene (Hitzeman et al. 1979). Lam and Marmur (1977) found that phosphoglycerate kinase and phosphoglycerate mutase mutants did not grow unless both glycerol and ethanol were supplied and both glucose and galactose were inhibitory. Growth in minimal medium was not studied.

Glycerate-2,3-P_2 is a cofactor of phosphoglycerate mutase. Its synthesis has been studied in other systems (e.g., Rose and Dube 1976).

The fact that, to date, aldolase (*FBA*), glyceraldehyde-3-P dehydrogenase, and enolase mutants have not been obtained suggests that such mutants may have unexpected properties or that there are multiple functional enzymes. There is abundant biochemical evidence for apparent isozymes of aldolase (e.g., Harris et al. 1969), glyceraldehyde-3-P dehydrogenase (e.g., Kirschner and Voigt 1968), and enolase (e.g., Bloomquist 1976), and cloning has revealed three different genes for glyceraldehyde-3-P dehydrogenase (Holland and Holland 1978, 1979, 1980, and pers. comm.) and two for enolase (Holland et al. 1981).

Pyruvate Kinase

More pyruvate kinase mutants have been identified among glucose-negative mutants than any other glycolysis lesion (Lam and Marmur 1977; Maitra and Lobo 1977a; Sprague 1977; Clifton et al. 1978; Ciriacy and Breitenbach 1979). The locus was mapped by Sprague (1977) and by Sinha and Maitra (1977), with the latter work showing it to be the structural gene on the basis of mutants having antigenically related protein and allelic revertants having labile enzyme. Some recessive alleles may be regulatory (Maitra and Lobo 1977a). The *cdc19* mutation is an allele of *PYK1* and has a temperature-sensitive pyruvate kinase (G. Kawasaki, pers. comm.).

The pyruvate kinase mutants fail to grow on glucose and are inhibited by it. The inhibition is accompanied by accumulation of phosphoenolpyruvate and earlier glycolysis intermediates and by loss of ATP (Maitra and Lobo 1977a; Sprague 1977; Ciriacy and Breitenbach 1979). The retention of ability to grow on ethanol or lactate shows that pyruvate kinase is not needed for gluconeogenesis.

Pyruvate kinase is activated in vitro by fructose-1,6-P_2, and a physiological role for such activation is supported by comparisons of glycolytic flux, in vivo concentrations of fructose-1,6-P_2 and phosphoenolypyruvate, and kinetic properties of the enzyme (Barwell et al. 1971; Barwell and Hess 1972; Hess 1973). A new way to study the regulation has been opened up by the report of a revertant containing a pyruvate kinase active without fructose-1,6-P_2 (Maitra and Lobo 1977b).

Fermentation Products

In fermentation, pyruvate is decarboxylated by pyruvate decarboxylase (*PDC*). Mutants were obtained by Lam and Marmur (1977). Acetaldehyde is reduced to ethanol by alcohol dehydrogenase. Three isoenzymes are known. ADHI (*ADC*, constitutive) is cytoplasmic and assumed to function in fermentation, and there are two repressible enzymes: ADHII (*ADR*) cytoplasmic and m-ADH (*ADM*) mitochondrial. The first mutants were obtained by Lutstorf and Megnet (1968), who selected for resistance to allyl alcohol (which is converted to a toxic compound, acrolein, by alcohol dehydrogenases). Selection for resistance on glucose-containing medium gave ADHI mutants, whereas resistance on glycerol medium gave mutants in the repressible enzymes. The method has been employed extensively by Ciriacy (e.g., 1975, 1979), and by Wills and Phelps (1975). *ADC* is known to be the structural gene for ADHI (Williamson et al. 1980). There are two genes for ADHII: *adr2* mutations affect structure, and *adr1* mutations are regulatory.

The mutants lacking ADHI have been reported variously as unable to grow on glucose or as slow growers, inhibited by high glucose concentrations or not; the differences may depend on leakiness of the mutation, the degree of repression of the other two enzymes, and the extent to which NADH might be oxidized by repressed cells. Many studies have used mutants lacking all three enzymes. Selection for reversion on glucose has yielded strains with derepressed ADHII by *cis*-acting mutations (*ADR3^c*, close to *ADR2* [Ciriacy 1979], and the system has also been used for selection of mutants derepressed in several enzymes (see below, Catabolite Repression). Thus, ADHII can serve the fermentative function if available.

Glycerol is a minor fermentation product (Table 1). Its amount increases if acetaldehyde metabolism is blocked with sulfite, and in *adc1* mutants (Wills and Phelps 1975). The pathway is probably reduction of dihydroxyacetone-P (*DAR*) with NADH, followed by a phosphatase (*GPP*) cleavage (Gancedo et al. 1968). These two steps are probably different from the kinase and dehydrogenase used for glycerol catabolism, and it is not known whether the dihydroxyacetone-P reductase activity is the same one used in synthesis of glycerol-3-P for phospholipids.

Succinic acid is made in even lower amount. Its origin is of interest partly because it is formed by a reductive pathway in prokaryotes, giving ATP (see Thauer et al. 1977). Although there is some evidence for its reductive formation in yeast (Lupianez et al. 1974), it comes mainly by oxidation of glutamate (Harden 1923; Heerde and Radler 1978).

Regulation of the Amount of Glycolysis Enzymes

Maitra and Lobo (1971a) showed that the levels of many glycolysis enzymes were very low after growth in an acetate-supplemented rich medium, and addition of glucose caused 3- to 100-fold increases in their differential rates of synthesis over several hours. The increases were prevented by cycloheximide. Other sugars had similar effects. Experiments on loss of the enzyme activities after sugar removal were also reported. The induction was studied with analogs and mutants (Maitra and Lobo 1971b). 2-Deoxyglucose, which is phosphorylated but not used in glycolysis, induced, whereas 6-deoxyglucose, which is not phosphorylated, did not. In a mutant unable to phosphorylate fructose (*hxk1, hxk2*), glucose induced but fructose did not; whereas in a phosphoglucose isomerase mutant, glucose induced better than in the wild type, and fructose induced worse, as if glucose-6-P were the inducer. Experiments were also done with a phosphomannose isomerase mutant. The induction effects were most marked in a hybrid strain of *S. fragilis* × *S. dobzhanski*, but were also found in a haploid of *S. cerevisiae*.

In this laboratory comparison of glycolysis enzyme levels in the steady state in glycolytically versus gluconeogenically grown cultures do not reveal large differences for most of them. However, there have been few, if any, systematic studies, and there are many reports of apparent variations for individual enzymes. (For example, Foy and Bhattacharjee (1978) reported conditions where phosphofructokinase is almost absent; see also Glucose Phosphorylation.) There is a mutation, *gcr1*, which somehow causes most glycolysis enzyme levels to be in the 1–10% range as compared with wild type in gluconeogenic conditions (Clifton et al. 1978); the affected enzyme levels are substantially higher in cultures grown with sugars, and the mutation affects mRNA levels (Clifton and Fraenkel 1981).

PATHWAYS INTO GLYCOLYSIS

Fructose, Mannose, and Glucosamine

The phosphorylation of fructose and mannose has been discussed (Glucose Phosphorylation). Mannose-6-P is converted to fructose-6-P by phosphomannose isomerase (*PMI*). Mutants were isolated by Herrera et al. (1976) as unable to grow on minimal medium with mannose (see also Maitra and Lobo 1971b). Mannose-6-P was accumulated from mannose, and mannose was inhibitory to growth on glycerol. Several mutations were allelic, and all showed a few percent or more of phosphomannose isomerase, in some cases, more labile than in the wild type. Mannan content was not reported.

Glucosamine is not known to be used as carbon source by *S. cerevisiae*. However, it can be used to supplement a glucosamine auxotroph (Whelan and Ballou 1975), probably being converted first to glucosamine-6-P by one of the glucose-phosphorylating enzymes. Growth on *N*-acetylglucosamine was reported by Singh and Datta (1978), who also found a deaminase for glucosamine-6-P (giving fructose-6-P). Glucosamine causes catabolite repression.

Galactose

Galactokinase (*GAL1*) converts galactose (see also Oshima, this volume) to galactose-1-P, and uridyl transferase (*GAL7*) converts galactose-1-P and UDP-glucose into glucose-1-P and UDP-galactose; the latter compound is reconverted to UDP-glucose by an epimerase (*GAL10*). Transferase and epimerase mutants are sensitive to galactose even in enriched medium; kinase mutants are indifferent. Douglas and Hawthorne (1964) used a vegetative petite strain for mutant selection, since all three types would be unable to grow. They also showed that resistance to galactose in an epimerase mutant is acquired by loss of the kinase.

Two other reactions are sometimes thought of as belonging to the galactose pathway, but have a broader role: phosphoglucomutase (*PGM*), which interconverts glucose-6-P and glucose-1-P, and UDP-glucose pyrophosphorylase, which forms UDP-glucose from glucose-1-P and UTP. Both reactions are needed in growth on other carbon sources for synthesis of UDP-glucose and, at least, glucan. In growth on galactose, on the other hand, phosphoglucomutase is on the primary catabolic pathway, whereas the pyrophosphorylase reaction is needed to make the cofactor UDP-glucose. One class of galactose-negative mutants (originally *gal5*, now *pgm2*) are deficient in phosphoglucomutase and lack the main isozyme but still contain the minor one (*PGM1*), whose role is presumably glucose-1-P synthesis (Bevan and Douglas 1969).

Glucose-1,6-P_2 is a cofactor for phosphoglucomutase; its synthesis has been studied in other systems (see Rose et al. 1975).

The genetics of UDP-glucose pyrophosphorylase is also incomplete. Studies of the *gal3* mutants (long-term adaptation) led Tsuyumu and Adams (1974) to speculate that UDP-glucose might be involved in induction.

Glycosides

Barnett (1976) has reviewed this subject. Disaccharides and oligosaccharides are abundant. Disaccharides, whose metabolism has been studied in *Saccharomyces*, include sucrose, the nonreducing double glycoside of glucose and fructose (glucosyl-α-[β-fructofuranoside]), maltose (glucosyl-α-1-4-glucose), isomaltose (glucosyl-α-1-6-glucose), and melibiose (galactosyl-α-1-6-glucose). Their utilization has several unusual features. First, the enzymes often have limited specificity for the particular glucoside and anomer. Thus, sucrose is generally metabolized by a β-D-fructofuranoside, called sucrase or invertase. But it can also be split by α-D-glucosidase, called maltase (Needleman et al. 1978), which hydrolyzes maltose (α-1-4-) but not isomaltose (α-1-6-). An oligo-α-1-6-glucosidase splits isomaltose but not maltose. Both of the latter enzymes split the chromogenic substrate *p*-nitrophenyl-α-D-glucopyranoside and may act on α-methylglucoside too. Considering that specificity of induction and hydrolysis are not necessarily the same, one may appreciate that in some cases it is not known just which enzyme is being used.

Second, some glycoside-hydrolyzing enzymes are mannoproteins, principally located external to the cell membrane. Invertase is the best studied case (Schekman and Novick, this volume); another is melibiase. Many other glucoside-utilizing enzymes, including maltase, are cytoplasmic, but it should be remembered that information on these pathways is incomplete.

Third, glycoside utilization often involves several dominant genes, any one of which alone suffices (polymeric genes). For example, Winge and Roberts (1950) identified four genes (*MAL1-MAL4*) for maltose metabolism in *S. cerevisiae* by analyzing a cross with a strain unable to grow on maltose, *S. chevalieri*; *S. carlsbergensis* has only one (*MAL6*). A similar multiplicity of genes applies to sucrose metabolism (*SUC*), and other examples are in Lindegren and Lindegren (1956) and Mortimer and Hawthorne (1969).

Much genetic analysis of the *MAL* loci has been consistent with a regulatory function (see Mowshowitz 1979). For example, ten Berge et al. (1973) showed that segregational maltose-negative mutants (strains lacking any *MAL* genes), as well as maltose-negative mutants, e.g., *mal6*, obtained from *MAL6* strains, had low basal levels of maltase indistinguishable from the enzyme in the parental strain; temperature-sensitive *mal6* mutants also had normal enzyme. Constitutive mutations at the *MAL* loci have also been reported (e.g., ten Berge et al. 1974). And Halvorson et al. (1963) showed that induced maltase activities from strains with five different *MAL* genes were not readily distinguishable. However, Naumov (e.g., Naumov 1976) has reported genetic evidence for the *MAL* loci containing both regulatory and structural genes. Evidence for the *SUC* loci containing structural gene information for invertase has been reported (Grossman and Zimmermann 1979).

Mutations called *dsf* (disaccharide fermentation), unlinked to *MAL* and not affecting maltase, have also been obtained and suggested to affect transport, trehalose metabolism, or induction (Zimmermann et al. 1973; Khan 1975).

α-Methylglucoside utilization in *S. cerevisiae* depends on pairs of genes: *MGL1 MGL2*, *MGL2 MGL3*, or *MGL1 MGL4*. In this case, the function of one locus of each pair may be transport (*MGL2*; Okada and Halvorson 1964) and the other, to regulate or specify the hydrolytic enzyme (Khan 1974). Isomaltose is probably the natural substrate.

Melibiose utilization involves another polymeric gene system. The α-galactosidase is external to the cell membrane (Lazo et al. 1977). Kew and Douglas (1976) have shown involvement of several galactose regulatory genes in melibiose and maltose utilization.

Trehalose is the nonreducing disaccharide composed of two glucose residues, linked α,α. Many strains can use it as carbon source. In some conditions it is formed endogenously (see below, Glycogen and Trehalose).

Lactose (galactosyl-β-1-4-glucose) and cellobiose (glucosyl-β-1-4-glucose) hydrolases are present in *Kluyveromyces lactis* (Tingle and Halvorson 1972; Dickson and Markin 1980).

Raffinose is galactosyl-α-1-6-(sucrose), with the linkage to position 6 of the glucose residue; melizitose is glucosyl-α-1-3-(sucrose), with the linkage to position 3 of the fructose residue. Metabolism of such oligosaccharides seems to depend primarily on the individual linkages and sugars; therefore, raffinose may be cleaved by invertase, giving melibiose and fructose, or by α-galactosidase, giving galactose and sucrose, and the products are accumulated or used according to the availability of routes for their further metabolism. Barnett (1976) gives many examples. Melizitose genetics was analyzed by Lindegren and colleagues (see Mortimer and Hawthorne 1969; Barnett 1976).

Glycerol (Utilization)

Gancedo et al. (1968) showed *S. cerevisiae* to be relatively impermeable to glycerol. However, some strains use it aerobically as sole carbon source, and Sprague and Cronan (1977) isolated mutants. *gut1* lacked glycerol kinase and *gut2* lacked glycerol-P dehydrogenase, presumably a mitochondrial enzyme that normally uses oxygen as terminal electron acceptor. The kinase mutant still incorporated glycerol to macromolecules at one-fifth the rate of wild type. The two enzymes do not need added glycerol for induction and are repressed by glucose. The mutations are nuclear and unlinked.

HEXOSE-MONOPHOSPHATE OXIDATION PATHWAY

This pathway, also named the hexose-monophosphate shunt or pentose-P pathway, has two branches. The oxidative branch gives pentose-P and reduced $NADP^+$ from glucose-6-P. It has two dehydrogenases, for glucose-6-P and 6-phosphogluconate (*GND*). The intermediate reaction, hydrolysis of 6-phosphogluconalactone, occurs at appreciable rate spontaneously, but indications for enzymatic catalysis were presented by Brodie and Lipmann (1955). 6-Phosphogluconolactonase (*PGL*) has been studied little (see Kupor and Fraenkel 1972).

The nonoxidative branch allows interconversion of three pentose-P with two fructose-6-P and one glyceraldehyde-3-P. Not all of the reactions are indicated in Figure 1. The usual formulation is a sequence of three: the first, catalyzed by transketolase, gives sedoheptulose-7-P and glyceraldehyde-3-P from the two pentose-P; second, a transaldolase-catalyzed reaction gives fructose-6-P and erythrose-4-P; and third, a transketolase reaction between erythrose-4-P and another xylulose-5-P gives fructose-6-P and glyceraldehyde-3-P. Other sequences and reactions (Williams et al. 1978) have not been evaluated yet in yeast.

Neither the functions nor the regulation of the hexose-monophosphate pathway are understood. In *Saccharomyces*, it probably should be considered a biosynthetic route to erythrose-4-P for aromatic biosynthesis, ribose-5-P and NADPH for reductive steps in biosynthesis and, perhaps, also hydroxylations (see Horecker 1978). Ribose-5-P might arise oxidatively from glucose-6-P or nonoxidatively from fructose-6-P and glyceraldehyde-3-P. Methods of evaluating the routes have been discussed by Katz and Rognstad (1967), but data for *Saccharomyces* are not available. There are some indications that the oxidative reactions do supply NADPH. Calculations based on $^{14}CO_2$ production from 1- and 6-labeled glucose give estimates of fluxes that are sufficient to match need in the two conditions of growth in minimal or enriched medium (Lagunas and Gancedo 1973; see also van de Poll 1973). A method for more direct assessment of whether NADPH arises from these reactions has been reported (Csonka and Fraenkel 1977).

In some other microbes the pentose-P pathway is used for catabolism of pentoses and pentitols, but *Saccharomyces* sp. are not known to grow on these compounds (Barnett 1976). A role in normal glucose-6-P metabolism apart from biosynthesis is also conceivable, i.e., if the flux to pentose is faster than its use in biosynthesis (for discussion of this question, see Katz and Rognstad [1967]). But Gilvarg (1952) showed that the radioactivity of 1-^{14}C glucose was little diluted in acetate, which suggests no major flux back into glycolysis from the shunt. Limited estimates from data on $^{14}CO_2$ formation (e.g., Wang et al. 1958; Gancedo and Langunas 1973) give values of fractional use of the oxidative pathway of 10% or less. However, with rapid metabolism a 10% value is considerable, and data from a glucose-limited culture (Chen 1959) indicated a higher fractional value.

In phosphoglucose isomerase mutants of *Escherichia coli*, the pentose-P pathway allows slow growth on glucose (Fraenkel and Vinopal 1973). Although yeast phosphoglucose isomerase mutants do not grow on glucose, perhaps their growth on other sugars does use the shunt.

PATHWAYS OF CARBOXYLIC ACID METABOLISM

Functions

Growth on pyruvate, lactate, acetate, and ethanol occurs oxidatively by the TCA cycle and electron transport phosphorylation in mitochondria. The same pathways are used for respiration of ethanol and glycerol made in fermentation, and some of the reactions have an essential biosynthetic role even anaerobically. Labeling data in accord with the TCA cycle are in the experiments of Wang et al. (1956), which showed formation of radioactively labeled CO_2 from the original position 6 of glucose after glucose exhaustion (i.e., from the ethanol), and of DeMoss and Swim (1957), who performed degradations of citrate and succinate obtained from cells using methyl-labeled acetate.

Routes to the TCA Cycle

The key compounds are pyruvate, acetyl-CoA, and oxalacetate. Pyruvate may be made by the glycolytic pathway, it may be used directly as carbon source, and it may come from lactate by flavoprotein dehydrogenases (*LDH*).

Acetyl-CoA may be made aerobically by the pyruvate dehydrogenase complex (*PDH*) (Ullrich and Wais 1975). In growth on acetate, it is made by the acetyl-CoA synthetase reaction (*ACS*); there may be two enzymes, one of which is mitochondrial (see Klein and Jahnke 1979). Acetate comes from ethanol by alcohol dehydrogenase (three possible enzymes; see above, Fermentation Products) and acetaldehyde dehydrogenase (*ALD*) (one mitochondrial and one cytoplasmic enzyme; Jacobson and Bernofsky 1974). In growth on glucose, acetyl-CoA also might be formed from acetaldehyde.

Oxalacetate and other dicarboxylic acids are formed by carboxylation in growth on glucose or pyruvate (Stoppani et al. 1958). The enzyme is probably pyruvate carboxylase (*PYC*), which is present in growth on glucose but repressed by aspartate (Haarasilta and Oura 1975a,b). Pyruvate carboxylase is a biotin enzyme (Haarasilta et al. 1979), and biotin starvation causes an aspartate requirement (Cazzulo et al. 1968).

In growth on acetate or ethanol, dicarboxylic acids are formed by the glyoxylate cycle, which uses several reactions of the TCA cycle, plus isocitrate lyase (*ICL*) and malate synthase (*MLS*) (Duntze et al. 1969).

TCA Cycle

α-Ketoglutarate comes by the citrate synthase (*CIT*), aconitase (*ACO*), and isocitrate dehydrogenase (*ICD*) reactions. Strong evidence for the dual biosynthetic and respiratory role of the first two comes from mutants obtained as glutamate auxotrophs on glucose, which are unable to grow on nonfermentable carbon sources even when supplemented. They lack citrate synthase (the *glu3* mutant of Burand et al. 1975), or aconitase (*glu1* [Ogur et al. 1964], and *glu2* [Ogur et al. 1965], which also requires lysine). An aconitase mutant was shown to accumulate citrate from acetate (Crocker and Bhattacharjee 1973). Weitzman and Hewson (1973) have discussed kinetic regulation of citrate synthase in extracts and permeabilized cells, and conditions for its induction have also been reported (Nuñez de Castro et al. 1976).

There are two isocitrate dehydrogenases. The NAD^+ enzyme is activated by AMP and has been studied extensively. Less is known about the $NADP^+$ enzyme, for which Machado et al. (1975) have proposed a biosynthetic role in anaerobiosis.

The α-ketoglutarate dehydrogenase (*KGD*) complex is analogous to pyruvate dehydrogenase and forms succinyl-CoA. A mutant was unable to grow on glycerol, lactate, pyruvate, or acetate, but grew slowly on glucose or ethanol (Subik et al. 1972). The exact lesion is not known.

Succinyl-CoA synthase (*SCS*) is the single TCA-cycle enzyme that gives a substrate-level phosphorylation. There have been almost no studies of the

Carbohydrate Metabolism 21

yeast enzyme, and there have been few recent studies of yeast succinate dehydrogenase (*SDH*); apparent nuclear mutants were obtained in a selection for strains unable to grow by respiration (de Kok and Slater 1975). Fumarate reductase (*FRD*) seems to exist as several isoenzymes, distinct from succinate dehydrogenase (Tisdale et al. 1968).

Malate is formed from fumarate by fumarase (*FUM*) and oxalacetate, from malate by malic dehydrogenase (*MDH*). There are two malate dehydrogenases in yeast, one mitochondrial and the other cytoplasmic (Hägele et al. 1978).

Routes Out of the TCA Cycle and Gluconeogenic Growth

In growth on substances like pyruvate or acetate, key intermediates of glycolysis are needed for biosynthesis. Pyruvate itself, in growth on acetate, might be made by the malic enzyme reaction (*MDD*). Phosphoenolpyruvate probably comes from oxalacetate by the phosphoenolpyruvate carboxykinase (*PCK*) reaction (Hansen et al. 1976), a role supported by studies of its inducibility (Haarasilta and Oura 1975a).

Above phosphoenolpyruvate, most other intermediates come by the same reactions used in glycolysis, which are reversible. The exception is fructose-6-P, probably formed from fructose-1,6-P_2 by a specific phosphatase (*FBP*) inhibited by AMP.

The high potential rate of gluconeogenesis was measured directly by Maitra and Lobo (1978), who showed starved ethanol-grown cells to form fructose-1,6-P_2 and hexose monophosphates within a few seconds of ethanol addition to the culture, presumably from their pools of phosphoenolpyruvate and 2- and 3-phosphoglycerates. Several mutants were also employed and, in glucose-grown cells, which lacked fructose diphosphatase, hexose monophosphates were not formed.

Other Mutants

Understanding the functions of many enzymes discussed in this section might be aided by mutant studies. But few mutants have been described, perhaps because they would not grow on nonfermentable carbon sources and thus have a phenotype similar to ρ mutants. Ciriacy (1977) devised a selection for defective growth on ethanol with retention of some mitochondrial functions and obtained mutants lacking isocitrate lyase (*icl*), succinate dehydrogenase (*sdh*), and malate dehydrogenase (*mdh1* and *mdh2*, two loci each apparently affecting both isoenzymes) or mutants unable to derepress several enzymes (*ccr*, for carbon catabolite repression). Other collections of nuclear mutants affecting aerobic growth (Parker and Mattoon 1969; Tzagoloff et al. 1975) might also contain mutants of interest in the present context. In *Aspergillus*, selection for acetate auxotrophy or fluoroacetate resistance has been useful (Armitt et al. 1976).

INDUCTION AND REPRESSION OF RESPIRATORY ENZYMES

Oxygen Effects

Many of the enzymes of respiratory metabolism depend on oxygen for their synthesis, and chemostat studies (Rogers and Stewart 1973) have shown affinity constants for induction of 1 μM or less. The mechanism of oxygen induction is not understood and is not known to be related to the need for oxygen in biosynthetic reactions (a requirement met in anaerobic minimal medium cultures by yeast extract [e.g., Rogers and Stewart 1973] or ergosterol and Tween 80 [e.g., Oura 1974]). Levels of the affected enzymes are often even lower anaerobically than in aerobic cultures in conditions of catabolite repression. However, potential respiratory ability (Q_{O_2}) is not zero in anaerobic cultures (Rogers and Stewart 1973; Oura 1974), and the levels of respiratory enzymes are not zero either (Heerde and Radler 1978; Wales et al. 1980), even when efforts are made to remove all traces of oxygen. The presence of certain TCA-cycle enzymes anaerobically might be related to their role in biosynthesis. But it is not known in most cases whether the enzymes in anaerobic and aerobic cultures are the same.

Catabolite Repression

Besides oxygen induction, respiratory capacity and the levels of individual respiratory components and enzymes depend on the carbon source used for growth. Thus, levels are "repressed" in batch growth on glucose, which is fermentative until glucose exhaustion and "derepression" occur. Particularly extensive data have been reported by Polakis and Bartley (1965), Duntze et al. (1969), Perlman and Mahler (1974), and Wales et al. (1980). The actual differences between repressed and derepressed levels vary greatly for different enzymes, conditions, and strains. The phenomenon of lowered enzyme levels in glucose growth also applies to specific sugar degradative pathways, even when constitutive (e.g., $MAL6^c$; ten Berge et al. 1974).

There are many glucose effects in microorganisms. For the *lac* operon of *E. coli*, at least two mechanisms operate: inducer exclusion (Magasanik 1970) and cAMP control (Perlman and Pastan 1971). The term catabolite repression was used for bacteria (Magasanik 1962) to indicate that enzyme synthesis is affected rather than activity and that it is catabolism rather than glucose itself that is effective. In yeast, with regard to the first point, it is now known, for some cases, that glucose affects the amount of mRNA (e.g., St. John and Davis 1979; Zitomer et al. 1979).

The idea that catabolism rather than glucose itself signals the control fits the fact that sugars like galactose have analogous effect (Table 1). Best repression is with unrestricted growth on glucose; and with its limitation in a chemostat, respiratory enzymes are derepressed (Rogers and Stewart 1973). There have been many searches for the putative controlling metabolite (e.g., van Wijk et al. 1973), and incompletely metabolized substances like glucosamine (Witt et al. 1966; Furst and Michels 1977) as well as glycolysis mutants (Ciriacy and Breitenbach 1979) have been employed. A possible role for cyclic (c) AMP has been proposed (see Mahler and Lin 1978).

There are mutations, such as $ACR3^c$ (Ciriacy 1979), that are thought to specifically affect catabolite repression of a single enzyme. There are also mutations affecting catabolite repression in several pathways. These include *ccr1, ccr2, ccr3* (see above, Other Mutants; Ciriacy 1977), and *cat1* and *cat2* (Zimmermann et al. 1977), which do not derepress certain enzymes; and $CCR80^r$ (Ciriacy 1978), *glr1* (Michels and Romanowski 1980), *cat80, hex1*, and *hex2* (see Entian and Zimmermann 1980), where several enzymes are insensitive to glucose repression. The ranges of affected enzymes differ among these mutants, but they have not yet been systematically compared, and the number of genes is not known. Some are altered in growth or glucose fermentation, in which case the effect on enzyme synthesis might be an indirect consequence of altered metabolism. But for *hex1*, shown to be allelic to mutants (*hxk2*) in the main glucose-phosphorylating enzyme, a direct role in repression for the kinase has been suggested (Entian 1980).

CATABOLITE INACTIVATION

In yeast, glucose does not seem to inhibit respiration directly (i.e., show a Crabtree effect). Thus, aerobic cells from a glucose-limited chemostat (Oura 1974) showed similar Q_{O_2} with 5 mM glucose and ethanol, and a lack of respiratory inhibition by higher glucose concentrations may be seen in Lagunas' work (1976, 1979). However, Chapman and Bartley (1968) showed that after aerobically adapted cells were shifted to a medium with glucose, although specific activities of many derepressed enzymes decreased in growth as expected from dilution, some decreased more rapidly. Gancedo (1971), e.g., showed fructose diphosphatase activity to be completely lost an hour after glucose addition to acetate-grown cells. The phenomenon was reviewed by Holzer (1976) and named catabolite inactivation. It also occurs with the other gluconeogenic enzymes: phosphoenolpyruvate carboxykinase and cytoplasmic malate dehydrogenase. Glucose was known early to provoke loss of maltose- and galactose-fermentative capacity; in those two cases, transport is affected (Holzer 1976; Matern and Holzer 1977).

The mechanism of catabolite inactivation is not known. In general, inactivation still occurs in conditions of limited protein synthesis or inhibited energy metabolism. Holzer (1976) has suggested that a structural or conformational alteration in the target protein causes its entry into the vacuole and subsequent breakdown, and evidence is emerging for an intermediate modified form of enzyme (e.g., Tortora et al. 1981). Irreversible loss of activity is accompanied by loss of antigenicity (Neeff et al. 1978; Funayama et al. 1980). A general role for a proteinase B in inactivation has been ruled out by the finding of near normal inactivation of several enzymes in mutants lacking this enzyme (Wolf and Ehmann 1979; Zubenko and Jones 1979).

As with catabolic repression, there have been many attempts to observe correlation of metabolite levels with inactivation, and there has also been some use of glycolysis mutants for this purpose (Ciriacy and Breitenbach 1979; Gancedo and Gancedo 1979).

Mutants specifically affected in catabolite inactivation would be of much interest. Often, it has been suggested (see Holzer 1976) that the inability to perform catabolite inactivation might result in wasteful ATP splitting cycles. Accordingly, mutants affected in the inactivation might be defective in growth on glucose and other sugars. Two candidates have been proposed: *fdp* (van de Poll et al. 1974) and *cif* (Navon et al. 1979). They grow much more slowly than wild type on glucose but seem to contain all the glycolysis enzymes and do not inactivate fructose diphosphatase rapidly upon glucose additon; however, they do lose ATP and accumulate fructose-1-6-P_2. The detailed phenotypes are complex, and for the *fdp* mutant, van de Poll and Schamhart (1977) and Schamhart et al. (1977) have presented data arguing against catabolite inactivation as the primary lesion; a similar conclusion has been drawn from assessment of futile cycling (Bañuelos and Fraenkel 1982). *cif* and *fdp* have similar positions on chromosome 2R.

ENZYME LOCATION AND ISOENZYMES

Yeast has several organelles: nucleus, mitochondria, vacuole (Wiemken et al. 1979), and peroxisomes (Parish 1975). Most enzymes mentioned in this paper are thought to have a "soluble" cytoplasmic location. But TCA-cycle enzymes are at least partly associated with mitochondria (see Duntze et al. 1969; Perlman and Mahler 1970; Wales et al. 1980); the same studies find the enzymes of the glyoxylate cycle to be soluble, and earlier work suggesting a peroxisomal location is uncertain (see also Susani et al. 1976). The enzymes of gluconeogenesis, phosphoenolpyruvate carboxykinase, and fructose diphosphatase, as well as pyruvate carboxylase, are soluble (Haarasilta and Taskinen 1977).

It usually is not known whether the soluble and particle-associated enzymes are different or, if different, whether they are different gene products or modified products of the same gene. Several examples have also been given of multiple soluble enzymes for the same reaction. Differences in electrophoretic or chromatographic characteristics are difficult to interpret and sometimes depend partly on preparative artifacts (Pringle 1975; Porcelli et al. 1980). But, in some cases, studies of regulation, mutants, sequencing, or cloning strongly suggest or prove that the isozymes derive from different genes (e.g., hexokinase, glyceraldehyde-3-P dehydrogenase, and alcohol dehydrogenase). Ureta (1978) has suggested that isozymes reflect the existence of enzyme complexes for whole pathways, with the different isozymes functioning in different complexes.

PASTEUR EFFECT

Several examples have been given of ways that glucose metabolism affects respiration. The converse question would be, how does respiration affect glucose metabolism? Oura (1974) reported similar fermentation rates with cells taken from anaerobic and aerobic chemostat cultures. Thus, the ability to ferment is not repressed by respiration. However, respiration inhibits glycolysis. In the same report, cells from the aerobic chemostat used 5 mM glucose at a rate of 0.4, aerobically (Q_O of 2.5), but 4.0, anaerobically (Q_{CO} of 8). The tenfold difference is an expression of the Pasteur effect.

Recent explanations of the Pasteur effect have focused on the availability of key intermediates, such as ADP, needed in both glycolysis and respiration (Racker 1974), and on the sensitivity of certain steps, particularly phosphofructokinase and glucose transport, to their effectors (Krebs 1972; Sols 1976). High glucose concentration (as is usual in batch culture) overcomes the Pasteur effect (see Serrano and DelaFuente 1974), and Lagunas (1979) has provided new data to this point.

RESERVES

Glycogen and Trehalose

The subject of storage polymers in microbes was reviewed by Dawes and Senior (1973). Yeast glycogen and trehalose were discussed by Manners (1971), and there has been considerable speculation about their function. Roles as utilizable energy reserves in starvation, respiratory adaptation, sporulation, germination, and the cell cycle have been considered (see Lillie and Pringle 1980). In batch cultures on glucose, both glycogen and trehalose are less than 1% dry weight, but values after logarithmic growth may be much higher. Glycogen and trehalose do not vary coordinately, and there may be two metabolically different glycogen fractions (Gunja-Smith et al. 1977).

Trevelyan and Harrison (1956) showed accumulation of glycogen and trehalose from glucose in cells starved of NH_4^+. It might be thought reasonable that cells accumulate glycogen if glucose and ATP are available and growth is restricted, for glycogen phosphorolysis later conserves the energy used in its synthesis and makes the carbon available again. However, if there is a general signal for carbohydrate accumulation, it is not excess catabolism but restricted growth. The accumulation is seen in carbon, nitrogen, or phosphorous starvation in batch culture (Lillie and Pringle 1980); in nitrogen or carbon-limited chemostats (Küenzi and Fiechter 1972); and on maltose (Panek et al. 1978).

There have been many assessments of glycogen or trehalose synthesis or degradation with respect to the properties or amounts of the enzymes involved. Glycogen synthetase (UDP-glucose:glycogen α-4-glucosyl transferase) is found in two forms, D and I (for glucose-6-P-*d*ependent or -*i*ndependent). The I form predominates in stationary-phase cells and is convertible to D by phosphorylation (Huang and Cabib 1974).

Glycogen degradation presumably uses glycogen phosphorylase (α-1,4-glucan:orthophosphate glucosyl transferase); glucosidases have also been found in sporulating yeast (Colonna and Magee 1978). Glycogen phosphorylase also occurs in two interconvertible forms, a and b, differing by phosphorylation (Fosset et al. 1971; Lerch and Fischer 1975). cAMP is not known to be involved in the modification system of either glycogen synthetase or glycogen phosphorylase.

Trehalose is made by the enzymes trehalose-6-P synthetase (UDP-glucose + glucose-6-P $\rightarrow$ trehalose-6-P + UDP) and trehalose-P phosphatase (Manners 1971). Trehalase (Kelly and Catley 1976), which hydrolyzes trehalose to glucose, has two forms; the active one apparently formed from the inactive by cAMP-dependent modification (van Solingen and van der Plaat 1975).

Glycogen-deficient mutants can be obtained by an iodine-staining technique (Chester 1968). A *glc1* strain has been found to contain only the D form of glycogen synthetase in vivo (Rothman-Denes and Cabib 1970) and also does not accumulate trehalose (Panek et al. 1978), but the exact lesion is not known.

Polyphosphate

Polyphosphate was reviewed by Harold (1966). It is inorganic phosphate in pyrophosphate linkage, with chain lengths from three to several hundred, and it forms the metachromatic or volutin granules, which are identified cytologically. The quantity in growing cells may be relatively low but can increase in resting cells or after phosphate addition to phosphate-starved cells. Low-molecular-weight polyphosphates are a prominent component of the acid-soluble pool of yeast (Navon et al. 1979; Solimene et al. 1980).

Polyphosphate is made by the polyphosphate kinase reaction: ATP + $(P_i)_n \rightarrow$ ADP + $(P_i)_{n+1}$, and the enzyme from yeast has been purified (Felter and Stahl 1973). Polyphosphate can be broken down, and the phosphate is found later in nucleic acid.

Polyphosphate is found mainly in the vacuole (Urech et al. 1978) and is correlated with the presence of basic amino acids in the medium (Ludwig et al. 1977) and their accumulation also in the vacuole (Durr et al. 1979). Accordingly, these workers suggest that polyphosphate may act as a cation trap. Exopolyphosphatase and endopolyphosphatase activities are also found in the vacuole (Wiemken et al. 1979).

Reserve Materials, ATPase, and the Control of Glycolysis

Although the term reserve implies that the material may be used later, another function might be served merely by the consumption of ATP in its synthesis.

In the Harden-Young effect, cell-free extracts performing glycolysis accumulate fructose-1,6-P_2 (2 glucose + 2 P_i → 1 fructose-1,6-P_2 + 2 ethanol + 2 CO_2). Meyerhof (1949) showed that ATPase activity prevented the fructose-1,6-P_2 accumulation, and polyphosphate formation itself has been suggested as another way to use the ATP (Harold 1966; Bañuelos and Gancedo 1978). The yeast cytoplasmic membrane ATPase is the subject of considerable recent work (Ahlers et al. 1978; Dufour and Goffeau 1978; Wiemken et al. 1979; Willsky 1979; Serrano 1980). Regulation in vivo of this ATPase activity by K^+ or NH_4^+ has long been speculated (see Peña et al. 1972), and perhaps the stimulation of glycolysis and prevention of glycogen accumulation by dinitrophenol (Berke and Rothstein 1957) reflects its uncoupling. Proton translocation during glycolysis has been reported (Riemersma and Alsbach 1974: see also Navon et al. 1979; Serrano 1980).

ACKNOWLEDGMENTS

Work from this laboratory has been supported by the National Science Foundation and the National Institutes of Health.

REFERENCES

Ahlers, J., E. Ahr, and A. Seyfrath. 1978. Kinetic characterization of plasma membrane ATPase from *Saccharomyces cerevisiae*. *Mol. Cell. Biochem.* **22:** 39.

Alber, T., F.C. Hartman, R.M. Johnson, G.A. Petsko, and D. Tsernoglau. 1980. Crystallization of yeast triose phosphate isomerase from polyethylene glycol: Protein crystal formation following phase separation. *J. Biol. Chem.* **256:** 1356.

Armitt, S., W. McCullough, and C.F. Roberts. 1976. Analysis of acetate non-utilizing (*acu*) mutants in *Aspergillus nidulans*. *J. Gen. Microbiol.* **92:** 263.

Aust, Q.E. and C.H. Suelter. 1978. Homogeneous pyruvate kinase isolated from yeast by two different methods is indistinguishable from pyruvate kinase in cell free extract. *J. Biol. Chem.* **253:** 7508.

Bañuelos, M. and D.G. Fraenkel. 1982. Futile cycling of fructose-6-P, and the "*fdP*" mutant of *Saccharomyces carlsbergensis*. *Mol. Cell. Biol.* (in press).

Bañuelos, M. and C. Gancedo. 1978. In situ study of the glycolytic pathway in *Saccharomyces cerevisiae*. *Arch. Microbiol.* **117:** 197.

Barford, J.B. and R.J. Hall. 1979. An examination of the Crabtree effect in *Saccharomyces cerevisiae:* The role of respiratory adaptation. *J. Gen. Microbiol.* **114:** 267.

Barnard, E.A. 1975. Hexokinases from yeast. *Methods Enzymol.* **42:** 6.

Barnes, L.D., G.D. Kuehn, and D.E. Atkinson. 1971. Yeast diphosphopyridine nucleotide specific isocitrate dehydrogenase. Purification and some properties. *Biochemistry* **10:** 3939.

Barnett, J.S. 1976. The utilization of sugars by yeasts. *Adv. Carbohydr. Chem. Biochem.* **32:** 125.

Barwell, C.J. and B. Hess. 1972. Application of kinetics of yeast pyruvate kinase in vitro to calculation of glycolytic flux in the anaerobic yeast cell. *Hoppe-Seyler's Z. Physiol. Chem.* **353:** 1178.

Barwell, C.J., B. Woodward, and R.V. Brunt. 1971. Regulation of pyruvate kinase by fructose 1,6-diphosphate in *Saccharomyces cerevisiae*. *Eur. J. Biochem.* **18:** 59.

Becker, J.-U. and A. Betz. 1972. Membrane transport as controlling pacemaker of glycolysis in *Saccharomyces carlsbergensis*. *Biochim. Biophys. Acta* **274:** 584.

Belyaeva, R.K., V.Y. Chernyak, N.N. Magretova, and G.A. Kochetov. 1978. Molecular weight and quaternary structure of transketolase of baker's yeast. *Biochemistry* (Biorak) **43:** 435.

Benkovic, S.J. and K.J. Schray. 1976. The anomeric specificity of glycolytic enzymes. *Adv. Enzymol.* **44:** 139.

Berke, H.L. and A. Rothstein. 1957. The metabolism of storage carbohydrates in yeast, studied with glucose-1-C^{14} and dinitrophenol. *Arch. Biochem. Biophys.* **72:** 380.

Bevan, P. and G.C. Douglas. 1969. Genetic control of phosphoglucomutase variants in *Saccharomyces cerevisiae*. *J. Bacteriol.* **98:** 532.

Bloomquist, G. 1976. Cross partition and determination of net charge of the isoenzymes of enolase. *Biochim. Biophys. Acta* **420:** 81.

Bostian, K.A. and G.F. Betts. 1978. Rapid purification and properties of a potassium activated aldehyde dehydrogenase from *S. cerevisiae*. *Biochem. J.* **173:** 773.

Bowman, B.J. and C.W. Slayman. 1977. Characterization of plasma membrane adenosine triphosphatase of *Neurospora crassa*. *J. Biol. Chem.* **252:** 3357.

Boyer, P.D., ed. 1970–1976. *The enzymes*, vol. I–XIII (3rd ed.). Academic Press, New York.

Brewer, J.M. and T. Fairwell. 1970. An investigation of the subunit structure of yeast enolase. *Biochemistry* **9:** 1011.

Brodie, A.F. and F. Lipmann. 1955. Identification of a gluconolactonase. *J. Biol. Chem.* **212:** 677.

Burand, J.P., R. Drillien, and J.K. Bhattacharjee. 1975. Citrate synthaseless glutamic acid auxotroph of *Saccharomyces cerevisiae*. *Mol. Gen. Genet.* **139:** 303.

Butler, P.J.G. and G.M.T. Jones. 1970. The purification of alcohol dehydrogenase and glyceraldehyde-3-P dehydrogenase from baker's yeast. *Biochem. J.* **118:** 375.

Cataldi, M.A. and A.O.M. Stoppani. 1966. Purification and properties of fumarate hydratase from baker's yeast. *Biochim. Biophys. Acta* **118:** 631.

Cavalieri, S.W., K.E. Neet, and H.Z. Sable. 1975. Enzymes of pentose biosynthesis. The quaternary structure and reactive form of transketolase from baker's yeast. *Arch. Biochem. Biophys.* **171:** 527.

Cazzulo, J.J., I.M. Claisse, and A.O.M. Stoppani. 1968. Carboxylase levels and carbon dioxide fixation in baker's yeast. *J. Bacteriol.* **96:** 623.

Chapman, C. and W. Bartley. 1968. The kinetics of enzyme changes in yeast under conditions that cause the loss of mitochondria. *Biochem. J.* **107:** 455.

Chen, S.L. 1959. Carbohydrate assimilation in actively growing yeast, *Saccharomyces cerevisiae*. I. Metabolic pathways for ^{14}C glucose utilization by yeast during aerobic fermentation. *Biochim. Biophys. Acta* **32:** 470.

Chester, V.E. 1968. Heritable glycogen-storage deficiency in yeast and its induction by ultraviolet light. *J. Gen. Microbiol.* **51:** 49.

Ciriacy, M. 1975. Genetics of alcohol dehydrogenase in *Saccharomyces cerevisiae*. *Mutat. Res.* **29:** 315.

———. 1977. Isolation and characterization of yeast mutants defective in intermediary carbon metabolism and in carbon catabolite derepression. *Mol. Gen. Genet.* **154:** 213.

———. 1978. A yeast mutant with glucose-resistant formation of mitochondrial enzymes. *Mol. Gen. Genet.* **159:** 329.

———. 1979. Isolation and characterization of further *cis*- and *trans*-acting regulatory elements involved in the synthesis of glucose-repressible alcohol dehydrogenase (ADHII) in *Saccharomyces cerevisiae*. *Mol. Gen. Genet.* **176:** 427.

Ciriacy, M. and I. Breitenbach. 1979. Physiological effects of seven different blocks in glycolysis in *Saccharomyces cerevisiae*. *J. Bacteriol.* **139:** 152.

Clifton, D. and D.G. Fraenkel. 1981. The *gcr* (glycolysis regulation) mutation of *Saccharomyces cerevisiae*. *J. Biol. Chem.* **256:** 13074.

Clifton, D. and D.G. Fraenkel. 1982. Mutant studies of yeast phosphofructokinase. *Biochemistry* **21:** 1935.

Clifton, D., S.B. Weinstock, and D.G. Fraenkel. 1978. Glycolysis mutants in *Saccharomyces cerevisiae*. *Genetics* **88:** 1.

Cohen, N.D., M.F. Utter, N.G. Wrigley, and A.N. Barrett. 1979. Quaternary structure of yeast pyruvate carboxylase: Biochemical and electron microscope studies. *Biochemistry* **18:** 2197.

Colonna, W.J. and P.T. Magee. 1978. Glycogenolytic enzymes in sporulating yeast. *J. Bacteriol.* **134:** 844.
Crocker, W.H., Jr. and J.K. Bhattacharjee. 1973. Biosynthesis of glutamic acid in *Saccharomyces*: Accumulation of tricarboxylic acid cycle intermediates in a glutamate auxotroph. *Appl. Microbiol.* **26:** 303.
Csonka, L.N. and D.G. Fraenkel. 1977. Pathways of NADPH formation in *Escherichia coli. J. Biol. Chem.* **252:** 3383.
Daugherty, J.P., W.F. Kraemern, and J.G. Joshi. 1975. Purification and properties of phosphoglucomutase from Fleischmann's yeast. *Eur. J. Biochem.* **57:** 115.
Dawes, E.A. and P.J. Senior. 1973. The role and regulation of energy reserve polymers in microorganisms. *Adv. Microb. Physiol.* **10:** 135.
De Kok, J., J.L.M. Muller, and E.C. Slater. 1975. EPR studies on the respiratory chain of wild-type *Saccharomyces cerevisiae* and mutants with a deficiency in succinate dehydrogenase. *Biochim. Biophys. Acta* **387:** 441.
DeMoss, J.A. and H.E. Swim. 1957. Quantitative aspects of the tricarboxylic acid cycle in baker's yeast. *J. Bacteriol.* **74::** 445.
den Hollander, J.A., T.R. Brown, K. Ugurbil, and R.G. Shulman. 1979. ^{13}C nuclear magnetic resonance studies of anaerobic glycolysis in suspensions of yeast cells. *Proc. Natl. Acad. Sci.* **76:** 6096.
Dickson, R.C. and J.S. Markin. 1980. Physiological studies of β-galactosidase induction in *Kluyveromyces lactis. J. Bacteriol.* **142:** 777.
Dixon, G.H., H.L. Kornberg, and P. Lund. 1960. Purification and properties of malate synthetase. *Biochim. Biophys. Acta* **41:** 217.
Douglas, H.C. and D.C. Hawthorne. 1964. Enzymatic expression and genetic linkage of genes controlling galactose utilization in *Saccharomyces. Genetics.* **49:** 837.
Dufour, J.-P. and A. Goffeau. 1980. Molecular and kinetic properties of the purified plasma membrane ATPase of the yeast *Schizosaccharomyces pombe. Eur. J. Biochem.* **105:** 145.
Duntze, W., D. Neumann, J.M. Gancedo, W. Atzpodien, and H. Holzer. 1969. Studies on the regulation and localization of the glyoxylate cycle enzymes in *Saccharomyces cerevisiae. Eur. J. Biochem.* **10:** 83.
Durr, M., K. Urech, T. Boller, A. Wiemken, J. Schwencke, and M. Nagy. 1979. Sequestration of arginine by polyphosphate in vacuoles of yeast (*Saccharomyces cerevisiae*). *Arch. Mikrobiol.* **121:** 169.
Entian, K.D. 1980. Genetic and biochemical evidence for hexokinase PII as a key enzyme involved in carbon catabolite repression in yeast. *Mol. Gen. Genet.* **178:** 633.
Entian, K.D. and F.K. Zimmermann. 1980. Glycolytic enzymes and intermediates in carbon catabolite repression mutants of *Saccharomyces cerevisiae. Mol. Gen. Genet.* **177:** 345.
Felter, S. and A.J.C. Stahl. 1973. Enzymes du metabolisme des polyphosphates dans le levure. III. Purification et proprietés de la polyphosphate-ADP-phosphotransferase. *Biochimie* **55:** 245.
Fossett, M., L.W. Muir, L.D. Nielsen, and E.H. Fischer. 1971. Purification and properties of yeast glucogen phosphorylase *a* and *b. Biochemistry* **10:** 4105.
Foy, J.J. and J.K. Bhattacharjee. 1978. Biosynthesis and regulation of fructose-1,6-bisphosphatase and phosphofructokinase in *Saccharomyces cerevisiae* grown in the presence of glucose and gluconeogenic carbon sources. *J. Bacteriol.* **136:** 647.
Fraenkel, D.G. and R.T. Vinopal. 1973. Carbohydrate metabolism in bacteria. *Annu. Rev. Microbiol.* **27:** 69.
Frenkel, E.P. and R.L. Kitchens. 1977. Purification and properties of acetyl coenzyme A synthetase from baker's yeast. *J. Biol. Chem.* **252:** 504.
Frenkel, R. 1975. Malic enzyme. *Curr. Top. Cell. Regul.* **9:** 157.
Fuck, E., G. Stark, and F. Radler. 1973. Apfelsaurestoffwechsel bei *Saccharomyces.* II. Anreicherung und Eigenschaften eines Malatenzymes. *Arch. Mikrobiol.* **89:** 223.
Fuhrmann, G.F., C. Boehm, and A.P.R. Theuvenet. 1976. Sugar transport and potassium permeability in yeast plasma membrane vesicles. *Biochim. Biophys. Acta* **433:** 583.

Funayama, S., J.-M. Gancedo, and C. Gancedo. 1980. Turnover of yeast fructose bisphosphatase in different metabolic conditions. *Eur. J. Biochem.* **109:** 61.

Funayama, S., J. Molano, and C. Gancedo. 1979. Purification and properties of a D-fructose 1,6-bisphosphatase from *Saccharomyces cerevisiae*. *Arch. Biochem. Biophys.* **197:** 170.

Furst, A. and C.S. Michels. 1977. An evaluation of D-glucosamine as a gratuitous catabolite repressor of *Saccharomyces carlsbergensis*. *Mol. Gen. Genet.* **155:** 309.

Gancedo, C. 1971. Inactivation of fructose-1,6-diphosphatase by glucose in yeast. *J. Bacteriol.* **107:** 401.

Gancedo, C., J.-M. Gancedo, and A. Sols. 1968. Glycerol metabolism in yeasts. Pathways of utilization and production. *Eur. J. Biochem.* **5:** 165.

Gancedo, J.-M. and C. Gancedo. 1979. Inactivation of gluconeogenic enzymes in glycolytic mutants of *Saccharomyces cerevisiae*. *Eur. J. Biochem.* **101:** 455.

Gancedo, J.-M. and R. Lagunas. 1973. Contribution of pentose-phosphate pathway to glucose metabolism in *Saccharomyces cerevisiae*: A critical analysis of the use of labelled glucose. *Plant Sci. Lett.* **1:** 193.

Gancedo, J.-M., D. Clifton, and D.G. Fraenkel. 1977. Yeast hexokinase mutants. *J. Biol. Chem.* **252:** 4443.

Gilvarg, C. 1952. Utilization of glucose-1-C^{14} by yeast. *J. Biol. Chem.* **199:** 57.

Gounaris, A.D., I. Turkenkopf, L.L. Civerchia, and J. Greelie. 1975. Pyruvate decarboxylase. III. Specificity restrictions for thiamine pyrophosphate in the protein association step, subunit structure. *Biochim. Biophys. Acta* **405:** 492.

Gracy, R.W. and E.A. Noltmann. 1968. Studies on phosphomannose isomerase. I. Isolation, homogeneity measurements, and determination of some physical properties. *J. Biol. Chem.* **243:** 3161.

Grisolia, S. and J. Carreras. 1975. Phosphoglycerate mutase from yeast, chicken breast, muscle, and kidney (2,3-PGA-dependent). *Methods Enzymol.* **42:** 435.

Grossman, M.K. and F.K. Zimmermann. 1979. The structural genes of internal invertases in *Saccharomyces cerevisiae*. *Mol. Gen. Genet.* **175:** 223.

Gunja-Smith, Z., N.B. Patil, and E.E. Smith. 1977. Two pools of glycogen in *Saccharomyces*. *J. Bacteriol.* **130:** 818.

Haarasilta, S. and E. Oura. 1975a. Effect of aeration on the activity of gluconeogenetic enzymes in *Saccharomyces cerevisiae* growing under glucose limitation. *Arch. Microbiol.* **106:** 271.

———. 1975b. On the activity and regulation of anaplerotic and gluconeogenic enzymes during the growth process of baker's yeast. *Eur. J. Biochem.* **52:** 1.

Haarasilta, S. and L. Taskinen. 1977. Location of three key enzymes of gluconeogenesis in baker's yeast. *Arch. Microbiol.* **113:** 159.

Haarasilta, S., E. Oura, and H. Suomalainen. 1979. Pyruvate holo- and apocarboxylase content of biotin-deficient baker's yeast and the characteristics of the holoenzyme formation in permeabilized cells. *Arch. Microbiol.* **122:** 121.

Hägele, E., J. Neeff, and D. Mecke. 1978. The malate dehydrogenase isoenzymes of *Saccharomyces cerevisiae*. Purification, characterization and studies on their regulation. *Eur. J. Biochem.* **83:** 67.

Halvorson, H.O., S. Winderman, and J. Gorman. 1963. Comparison of the α-glucosidases of *Saccharomyces* produced in response to five non-allelic maltose genes. *Biochim. Biophys. Acta* **67:** 42.

Hansen, R.J., H. Hinze, and H. Holzer. 1976. Assay of phosphoenolpyruvate carboxykinase in crude yeast extracts. *Anal. Biochem.* **74:** 576.

Harden, A. 1923. *Alcoholic fermentation*. Longmans and Green, London.

Harold, F.H. 1966. Inorganic polyphosphates in biology: Structure, metabolism, and function. *Bacteriol. Rev.* **30:** 772.

Harris, C.E., R.D. Kobes, D.C. Teller, and W.J. Rutter. 1969. The molecular characteristics of yeast aldolase. *Biochemistry* **8:** 2442.

Heerde, E. and F. Radler. 1978. Metabolism of the anaerobic formation of succinic acid by *Saccharomyces cerevisiae*. *Arch. Microbiol.* **117:** 269.

Herrera, L.S. and C. Pascual. 1978. Genetical and biochemical studies of glucosephosphate isomerase deficient mutants in *Saccharomyces cerevisiae*. *J. Gen. Microbiol.* **108:** 305.

Herrera, L.S., C. Pascual, and X. Alvarez. 1976. Genetic and biochemical studies of phosphomannose isomerase deficient mutants of *Saccharomyces cerevisiae*. *Mol. Gen. Genet.* **144:** 223.

Herrmann, K., W. Diezel, G. Kopperschläger, and E. Hofmann. 1973. Immunological evidence for non-identical subunits in yeast phosphofructokinase. *FEBS Lett.* **36:** 190.

Hess, B. 1973. Organization of glycolysis: Oscillatory and stationary control. *SEB Symp.* **27:** 105.

Hess, B., A. Boiteux, and J. Kruger. 1969. Cooperation of glycolytic enzymes. *Adv. Enzyme Regul.* **7:** 149.

Hirabayashi, T. and T. Harada. 1971. Isolation and properties of α-ketoglutarate dehydrogenase complex from baker's yeast (*Saccharomyces cerevisiae*). *Biochem. Biophys. Res. Commun.* **45:** 1369.

Hitzeman, R.A., A.C. Chinault, A.J. Kingsman, and J. Carbon. 1979. Detection of *E. coli* clones containing specific yeast genes by immunological screening. *ICN-UCLA Symp. Mol. Cell Biol.* **14:** 57.

Holland, J.P. and M.J. Holland. 1980. Structural comparison of two nontandemly repeated yeast glyceraldehyde-3-P dehydrogenase genes. *J. Biol. Chem.* **255:** 2596.

Holland, M.J. and J.P. Holland. 1978. Isolation and identification of yeast messenger ribonucleic acids coding for enolase, glyceraldehyde-3-phosphate dehydrogenase, and phosphoglycerate kinase. *Biochemistry* **17:** 4900.

———. 1979. Isolation and characterization of a gene coding for glyceraldehyde-3-phosphate dehydrogenase from *Saccharomyces*. *J. Biol. Chem.* **254:** 5466.

Holland, M.J. and E.W. Westhead. 1973. Purification and characterization of aspartic β-semialdehyde dehydrogenase from yeast and purification of an isozyme of glyceraldehyde-3-phosphate dehydrogenase. *Biochemistry* **12:** 2264.

Holland, M.J., J.P. Holland, G.P. Thill, and K.A. Jackson. 1981. The primary structure of two yeast enolase genes. Homology between the 5' noncoding flanking regions of yeast enolase and glyceraldehyde-3-phosphate dehydrogenase genes. *J. Biol. Chem.* **256:** 1385.

Holzer, H. 1976. Catabolite inactivation in yeast. *Trends Biochem. Sci.* **1:** 178.

Horak, J. and A. Kotyk. 1973. Isolation of a glucose-binding lipoprotein from yeast plasma membrane. *Eur. J. Biochem.* **32:** 36.

Horecker, B.L. 1978. Yeast enzymology: Retrospectives and perspectives. In *Biochemistry and genetics of yeasts. Pure and applied aspects* (ed. M. Bacila et al.), p. 1. Academic Press, New York.

Huang, K.-P. and E. Cabib. 1974. Yeast glycogen synthetase in the glucose-6-phosphate dependent form. 1. Purification and properties. *J. Biol. Chem.* **249:** 3851.

Illingworth, J.A. 1972. Purification of yeast isocitrate dehydrogenase. *Biochem. J.* **129:** 1119.

Jacobson, M.K. and C. Bernofsky. 1974. Mitochondrial acetaldehyde dehydrogenase from *Saccharomyces cerevisiae*. *Biochim. Biophys. Acta* **350:** 277.

Jaspers, H.T.A. and J. van Steveninck. 1975. Transport-associated phosphorylation of 2-deoxy-D-glucose in *Saccharomyces fragilis*. *Biochim. Biophys. Acta* **406:** 370.

Katz, J. and R. Rognstad. 1967. The labeling of pentose phosphate from glucose-^{14}C and estimation of the rates of transaldolase, transketolase, and contribution of the pentose cycle, and ribose phosphate synthesis. *Biochemistry* **6:** 2227.

Kelly, P.J. and B.J. Catley. 1976. A purification of trehalase from *Saccharomyces cerevisiae*. *Anal. Biochem.* **72:** 353.

Kempe, T.D., D.M. Gee, G.M. Hathaway, and E.A. Noltmann. 1974. Subunit and peptide composition of yeast phosphoglucose isomerase isoenzymes. *J. Biol. Chem.* **249:** 4625.

Kew, O.M. and H.C. Douglas. 1976. Genetic co-regulation of galactose and melibiose utilization in *Saccharomyces*. *J. Bacteriol.* **125:** 33.

Khan, N.A. 1974. Constitutive alpha-methyl-glucosidase synthesis in yeast. *Mol. Gen. Genet.* **133:** 363.

———. 1975. Genetic control of maltase formation in yeast. III. Isolation and characterization of temperature sensitive mutants affecting maltase induction and maltose utilization. *Mol. Gen. Genet.* **136:** 55.

Kirschner, K. and B. Voigt. 1968. Reinheitskriterien für das kristallisierbare Isoenzym der D-Glycerinaldehyd-3-Phosphat-Dehydrogenase aus Bäckerhefe. *Hoppe-Seyler's Z. Physiol. Chem.* **349:** 632.

Klein, H.P. and L. Jahnke. 1979. Effects of aeration on formation and localization of the acetyl coenzyme A synthetases of *Saccharomyces cerevisiae*. *J. Bacteriol.* **137:** 179.

Koerner, T.A.W., Jr., R.J. Voll, and S. Younathan. 1977. A proposed model for the regulation of phosphofructokinase and fructose 1,6-bisphosphatase based on their reciprocal anomeric specificities. *FEBS Lett.* **84:** 207.

Kopperschläger, G. and E. Hofmann. 1969. Über multiple formen der hexokinase in hefe. *Eur. J. Biochem.* **9:** 419.

Kopperschläger, G., J. Bär, K. Nissler, and E. Hofmann. 1977. Physicochemical parameters and subunit composition of yeast phosphofructokinase. *Eur. J. Biochem.* **81:** 317.

Kornberg, A. 1955. Isocitric dehydrogenase of yeast (TPN). *Methods Enzymol.* **1:** 705.

Kotyk, A. and D. Michaljanicova. 1974. Nature of the uptake of D-galactose, D-glucose and α-methyl-D-glucoside by *Saccharomyces cerevisiae*. *Biochim. Biophys. Acta* **332:** 104.

Krebs, H.A. 1972. The Pasteur effect and the relations between respiration and fermentation. *Essays Biochem.* **8:** 1.

Kuenzi, M.T. and A. Fiechter. 1972. Regulation of carbohydrate composition of *Saccharomyces cerevisiae* under growth limitation. *Arch. Mikrobiol.* **84:** 254.

Kupor, S.R. and D.G. Fraenkel. 1972. Glucose metabolism in 6-phosphogluconolactonase mutants of *Escherichia coli*. *J. Biol. Chem.* **247:** 1904.

Labeyrie, F., A. Baudras, and F. Lederer. 1978. Flavocytochrome b_2 or L-lactate cytochrome c reductase from yeast. *Methods Enzymol.* **53:** 238.

Lagunas, R. 1976. Energy metabolism of *Saccharomyces cerevisiae*. Discrepancy between ATP balance and known metabolic functions. *Biochim. Biophys. Acta* **440:** 661.

———. 1979. Energetic irrelevance of aerobiosis for *S. cerevisiae* growing on sugars. *Mol. Cell. Biochem.* **27:** 139.

Lagunas, R. and J.-M. Gancedo. 1973. Reduced pyridine-nucleotides balance in glucose-growing *Saccharomyces cerevisiae*. *Eur. J. Biochem.* **37:** 90.

Lam, K.B. and J. Marmur. 1977. Isolation and characterization of *Saccharomyces cerevisiae* glycolysis mutants. *J. Bacteriol.* **130:** 746.

Laurent, M., F.J. Seydoux, and P. Dessen. 1979. Allosteric regulation of yeast phosphofructokinase. Correlation between equilibrium binding, spectroscopic and kinetic data. *J. Biol. Chem.* **254:** 7515.

Lazo, P.D., A.G. Ochoa, and S. Gascon. 1977. α-galactosidase from *Saccharomyces carlsbergensis*. Cellular localization, and purification of the external enzyme. *Eur. J. Biochem.* **77:** 375.

Lerch, K. and E.H. Fischer. 1975. Amino acid sequence of two functional sites in yeast glycogen phosphorylase. *Biochemistry* **14:** 2009.

Levy, H.R. 1979. Glucose-6-phosphate dehydrogenases. *Adv. Enzymol.* **48:** 97.

Lillie, S.H. and J.R. Pringle. 1980. Reserve carbohydrate metabolism in *Saccharomyces cerevisiae*: Responses to nutrient limitation. *J. Bacteriol.* **143:** 1384.

Lindegren, D.D. and G. Lindegren. 1956. Eight genes controlling the presence or absence of carbohydrate fermentation in *Saccharomyces*. *J. Gen. Microbiol.* **15:** 19.

Lobo, Z. and P.K. Maitra. 1977a. Genetics of yeast hexokinase. *Genetics* **86:** 727.

———. 1977b. Physiological role of glucose-phosphorylating enzymes in *Saccharomyces cerevisiae*. *Arch. Biochem. Biophys.* **182:** 639.

———. 1977c. Resistance to 2-deoxyglucose in yeast: A direct selection of mutants lacking glucose phosphorylating enzymes. *Mol. Gen. Genet.* **157:** 297.

Lubini, D., M.J. Healy, and P. Christen. 1977. Coupling of fructose-1,6-P_2 to aminated agarose by Schiff base reduction. Affinity chromatography of yeast aldolase. *Experientia* **33:** 709.

Ludwig, J.R., II, S.G. Oliver, and C.S. McLaughlin. 1977. The effect of amino acids on growth and phosphate metabolism in a prototrophic yeast strain. *Biochem. Biophys. Res. Commun.* **79:** 16.

Lupianez, J.A., A. Machado, I. Nuñez de Castro, and F. Mayor. 1974. Succinic acid production by yeasts grown under different hypoxic conditions. *Mol. Cell. Biochem.* **3:** 113.

Lutstorf, U. and R. Megnet. 1968. Multiple forms of alcohol dehydrogenase in *Saccharomyces cerevisiae*. *Arch. Biochem. Biophys.* **126:** 933.

Machado, A., I. Nuñez de Castro, and F. Mayor. 1975. Isocitrate dehydrogenases and oxoglutarate dehydrogenase activities of baker's yeast grown in a variety of hypoxic conditions. *Mol. Cell. Biochem.* **6:** 93.

Magasanik, B. 1962. Catabolite repression. *Cold Spring Harbor Symp. Quant. Biol.* **26:** 249.

———. 1970. Glucose effects: Inducer exclusion and repression. In *The lac operon* (ed. J.R. Beckwith and D. Zipser), p. 189. Cold Spring Harbor Laboratory, Cold Spring Harbor, New York.

Mahler, H.R. and C.C. Lin. 1978. Exogenous adenosine 3':5'-monophosphate can release yeast from catabolite repression. *Biochem. Biophys. Res. Commun.* **83:** 1039.

Maitra, P.K. 1970. A glucokinase from *Saccharomyces cerevisiae*. *J. Biol. Chem.* **245:** 2423.

———. 1971. Glucose and fructose metabolism in a phosphogluco-isomeraseless mutant of *Saccharomyces cerevisiae*. *J. Bacteriol.* **107:** 759.

Maitra, P.K. and Z. Lobo. 1971a. A kinetic study of glycolytic enzyme synthesis in yeast. *J. Biol. Chem.* **246:** 475.

———. 1971b. Control of glycolytic enzyme synthesis in yeast by products of the hexokinase reaction. *J. Biol. Chem.* **246:** 489.

———. 1977a. Pyruvate kinase mutants of *Saccharomyces cerevisiae*: Biochemical and genetic characterization. *Mol. Gen. Genet.* **152:** 193.

———. 1977b. Yeast pyruvate kinase: A mutant form catalytically insensitive to fructose 1,6-bisphosphate. *Eur. J. Biochem.* **78:** 353.

———. 1977c. Molecular properties of yeast glucokinase. *Mol. Cell. Biochem.* **18:** 21.

———. 1977d. Genetic studies with a phosphoglucose isomerase mutant of *Saccharomyces cerevisiae*. *Mol. Gen. Genet.* **156:** 55.

———. 1978. Reversal of glycolysis in yeast. *Arch. Biochem. Blophys.* **185:** 535.

Manners, D.J. 1971. Structure and biosynthesis of storage polymers in yeast. In *The yeasts* (ed. A.H. Rose and J.S. Harrison), vol. 2, p. 418. Academic Press, New York.

Matern, H. and H. Holzer. 1977. Catabolite inactivation of the galactose uptake system in yeast. *J. Biol. Chem.* **252:** 6399.

McDonald, R.C., T.A. Steitz, and D.M. Engelman. 1979. Yeast hexokinase in solution exhibits a large conformational change upon binding glucose or glucose-6-phosphate. *Biochemistry* **18:** 338.

Meredith, S.A. and A.H. Romano. 1977. Uptake and phosphorylation of 2-deoxyglucose by wild-type and respiration deficient baker's yeast. *Biochim. Biophys. Acta* **497:** 745.

Meyerhof, O. 1949. Further studies of the Harden-Young effect in alcoholic fermentation of yeast preparations. *J. Biol. Chem.* **180:** 575.

Michels, C.A. and A. Romanowski. 1980. Pleiotropic glucose repression-resistant mutation in *Saccharomyces carlsbergensis*. *J. Bacteriol.* **143:** 674.

Mortimer, R.K. and D.C. Hawthorne. 1969. Yeast genetics. In *The yeasts* (ed. A.H. Rose and J.S. Harrison), vol. 1, p. 385. Academic Press, New York.

Mowshowitz, D.B. 1979. Gene dosage effects on the synthesis of maltase in yeast. *J. Bacteriol.* **137:** 1200.

Muratsubaki, H. and T. Katsume. 1979. Distribution of hexokinase isoenzymes depending on a carbon source in *Saccharomyces cerevisiae*. *Biochem. Biophys. Res. Commun.* **86:** 1030.

Naumov, G.I. 1976. Comparative genetics of yeast. XVI. Genes for maltose fermentation in *Saccharomyces carlsbergensis*. N.C.Y. C.74 *Genetica* **12:** 87.

Navon, G., R.G. Shulman, T. Yamane, T.R. Eccleshall, K.-B. Lam, J.J. Baronofsky, and J. Marmur. 1979. Phosphorus-31 nuclear magnetic resonance studies of wild type and glycolytic pathway mutants of *Saccharomyces cerevisiae*. *Biochemistry* **18:** 4487.

Needleman, R.B., H.J. Federoff, T.R. Eccleshall, B. Buchferer, and J. Marmur. 1978. Purification and characterization of an α-glucosidase from *Saccharomyces carlsbergensis*. *Biochemistry* **17**: 4657.

Neeff, J., E. Hägele, J. Neuhaus, U. Heer, and D. Mecke. 1978. Application of an immuno-assay to the study of yeast malate dehydrogenase inactivation. *Biochem. Biophys. Res. Commun.* **80**: 276.

Nuñez de Castro, I., J.M. Arias de Saavedra, A. Machado, and F. Mayor. 1976. Regulation of the level of yeasts citrate synthase by oxygen availability. *Mol. Cell. Biochem.* **12**: 161.

Ogur, M., L. Coker, and S. Ogur. 1964. Glutamate auxotrophs in *Saccharomyces*. I. The biochemical lesion in the glt_1 mutants. *Biochem. Biophys. Res. Commun.* **14**: 193.

Ogur, M., A. Roshanmanesh, and S. Ogur. 1965. Tricarboxylic acid cycle mutants in *Saccharomyces*: Comparison of independently derived mutants. *Science* **147**: 1590.

Okada, H. and H.O. Halvorson. 1964. Uptake of alpha-thioethyl-glucopyranoside by *Saccharomyces cerevisiae*. 1. The genetic control of facilitated diffusion and active transport. *Biochim. Biophys. Acta* **82**: 538.

Olson, J.A. 1959. Purification and properties of yeast isocitrate lyase. *J. Biol. Chem.* **234**: 5.

Oppenheimer, C. and C.A. Mitchell. 1901. *Ferments and their actions*. Charles Griffin, London.

Oura, E. 1974. Effect of aeration intensity on the biochemical composition of baker's yeast. I. Factors affecting the type of metabolism. *Biotechnol. Bioeng.* **16**: 1197.

Panek, A.D., A.L. Sampaie, G.C. Braz, and J.R. Mattoon. 1978. Regulation of energy metabolism in yeast. Relationship between carbohydrate reserves, catabolite repression, and maltose utilization. In *Biochemistry and genetics of yeasts. Pure and applied aspects* (ed. M. Bacila et al.), p. 145. Academic Press, New York.

Parish, W.W. 1975. The isolation and characterization of peroxisomes (microbodies) from baker's yeast, *Saccharomyces cerevisiae*. *Arch Microbiol.* **105**: 187.

Parker, J.H. and J.R. Mattoon. 1969. Mutants of yeast with altered oxidative energy metabolism: Selection and genetic characterization. *J. Bacteriol.* **100**: 647.

Parvin, R. 1969. Citrate synthase from yeast. *Methods Enzymol.* **13**: 16.

Peña, A., G. Cinco, A. Gómez-Puyou, and M. Tuena. 1972. Effect of the pH of the incubation medium on glycolysis and respiration in *Saccharomyces cerevisiae*. *Arch. Biochem. Biophys.* **153**: 413.

Perea, J. and C. Gancedo. 1978. Glucose transport in a glucosephosphate isomeraseless mutant of *Saccharomyces cerevisiae*. *Curr. Microbiol.* **1**: 209.

Perlman, P. and H.R. Mahler. 1970. Intracellular localization of enzymes in yeast. *Arch. Biochem. Biophys.* **136**: 245.

———. 1974. Derepression of mitochondria and their enzymes in yeast: Regulatory aspects. *Arch. Biochem. Biophys.* **162**: 248.

Perlman, R.L. and I. Pastan. 1971. The role of cyclic AMP in bacteria. *Curr. Top. Cell. Regul.* **3**: 117.

Polakis, E.S. and W. Bartley. 1965. Changes in the enzyme activities of *Saccharomyces cerevisiae* during aerobic growth on different carbon sources. *Biochem. J.* **97**: 284.

Porcelli, L.J., Jr., E.D. Small, and J.M. Brewer. 1978. Origin of multiple species of yeast enolase A on isoelectric focusing. *Biochem. Biophys. Res. Commun.* **82**: 316.

Pringle, J.R. 1975. Methods for avoiding proteolytic artefacts in studies of enzymes and other proteins from yeasts. *Methods Cell Biol.* **12**: 149.

Racker, E. 1974. History of the Pasteur effect and its pathobiology. *Mol. Cell. Biochem.* **5**: 17.

Riemersma, J.C. and E.J.J. Alsbach. 1974. Proton translocation during anaerobic energy production in *Saccharomyces cerevisiae*. *Biochim. Biophys. Acta* **339**: 274.

Rogers, P.J. and P.R. Stewart. 1973. Mitochondrial and peroxisomal contributions to the energy metabolism of *Saccharomyces cerevisiae* in continuous culture. *J. Gen. Microbiol.* **79**: 205.

Rose, Z.B. and S. Dube. 1976. Rates of phosphorylation and dephosphorylation of phosphoglycerate mutase and bisphosphoglycerate synthase. *J. Biol. Chem.* **251**: 4817.

Rose, I.A., J.V.B. Warms, and G. Kaklij. 1975. A specific enzyme for glucose 1,6-bisphosphate synthesis. *J. Biol. Chem.* **250:** 3466.

Rothman-Denes, L.B. and E. Cabib. 1970. Two forms of yeast glycogen synthetase and their role in glycogen accumulation. *Proc. Natl. Acad. Sci.* **66:** 967.

Sasaki, R., S. Utsumi, E. Sugimoto, and H. Chiba. 1976. Subunit structure and multifunctional properties of yeast phosphoglyceromutase. *Eur. J. Biochem.* **66:** 523.

Schamhart, D.H.J., M.P.M. van de Heijkart, and K.W. van de Poll. 1977. Inactivation of fructose diphosphatase by sucrose in yeast. *J. Bacteriol.* **130:** 526.

Scopes, R.K. 1975. 3-Phosphoglycerate kinase. *Methods Enzymol.* **42:** 134.

Seaston, A., C. Inkson, and A.A. Eddy. 1973. The absorption of protons with specific amino acids and carbohydrates by yeast. *Biochem. J.* **134:** 1031.

Serrano, R. 1980. Effect of ATPase inhibitors on the proton pump of respiratory-deficient yeast. *Eur. J. Biochem.* **105:** 419.

Serrano, R. and G. DelaFuente. 1974. Regulatory properties of the constitutive hexose transport in *Saccharomyces cerevisiae. Mol. Cell. Biochem.* **3:** 161.

Singh, B.R. and A. Datta. 1978. Glucose repression of the inducible catabolic pathway for N-acetylglucosamine in yeast. *Biochem. Biophys. Res. Commun.* **84:** 58.

Sinha, P. and P.K. Maitra. 1977. Mutants of *Saccharomyces cerevisiae* having structurally altered pyruvate kinase. *Mol. Gen. Genet.* **158:** 171.

Solimene, R., A.M. Guerrini, and P. Donini. 1980. Levels of acid-soluble polyphosphate in growing cultures of *Saccharomyces cerevisiae. J. Bacteriol.* **143:** 710.

Sols, A. 1976. The Pasteur effect in the allosteric era. In *Reflections in biochemistry. In honor of Severo Ochoa* (ed. A. Kornberg et al.), p. 199. Academic Press, New York.

Sols, A., C. Gancedo, and G. DelaFuente. 1971. Energy yielding metabolism in yeasts. In *The yeasts* (ed. A.H. Rose and J.S. Harrison), vol. 2, p. 271. Academic Press, New York.

Spragg, S.P., J.K. Wilcox, J.J. Roche, and W.A. Barnett. 1976. The association of yeast phosphoglycerate kinase. *Biochem. J.* **153:** 423.

Sprague, G.F., Jr. 1977. Isolation and characterization of a *Saccharomyces cerevisiae* mutant deficient in pyruvate kinase activity. *J. Bacteriol.* **130:** 232.

Sprague, G.F., Jr. and J.E. Cronan, Jr. 1977. Isolation and characterization of *Saccharomyces cerevisiae* mutants defective in glycerol catabolism. *J. Bacteriol.* **129:** 1335.

St. John, T.P. and R.W. Davis. 1979. Isolation of galactose-inducible DNA sequences from *Saccharomyces cerevisiae* by differential plaque filter hybridization. *Cell* **16:** 443.

Stallcup, W.B., S.C. Mockrin, and D.E. Koshland, Jr. 1972. A rapid purification procedure for glyceraldehyde 3-phosphate dehydrogenase from baker's yeast. *J. Biol. Chem.* **247:** 6277.

Stoppani, A.O.M., L. Conches, S.L.S. de Favelukes, and F.L. Sacerdote. 1958. Assimilation of carbon dioxide by yeasts. *Biochem. J.* **70:** 438.

Subik, J., J. Kolarov, and T.M. Lachowicz. 1972. A mutant of *Saccharomyces cerevisiae* lacking α-ketoglutarate dehydrogenase activity. *FEBS Lett.* **27:** 81.

Susani, M., P. Zimniak, F. Fessl, and H. Ruis. 1976. Localization of catalase A in vacuoles of *Saccharomyces cerevisiae*: Evidence for the vacuolar nature of isolated "yeast peroxisomes." *Hoppe-Seyler's Z. Physiol. Chem.* **357:** 961.

Tamaki, N. and B. Hess. 1975. Subunit structure of 6-phosphofructokinase from brewer's yeast. *Hoppe-Seyler's Z. Physiol. Chem.* **356:** 1663.

Temperli, A., V. Kunsch, K. Mayer, and F. Bush. 1965. Reinigung und Eigenschaften der Malatdehydrogenase (decarboxyliert) aus Hefe. *Biochim. Biophys. Acta* **110:** 630.

ten Berge, A.M.A., G. Zoutewelle, and R.D. Needleman. 1974. Regulation of maltose fermentation in *Saccharomyces carlsbergensis*. II. Constitutive mutations at the *MAL6*-locus and suppressors changing a constitutive phenotype into a maltose negative phenotype. *Mol. Gen. Genet.* **131:** 113.

ten Berge, A.M.A., G. Zoutewelle, and K.W. van de Poll. 1973. Regulation of maltose fermentation in *Saccharomyces carlsbergensis*. I. The function of the gene *MAL6*, as recognized by *mal6*-mutants. *Mol. Gen. Genet.* **123:** 233.

Thauer, R.K., K. Jungermann, and R. Decker. 1977. Energy conservation in chemotrophic anaerobic bacteria. *Bacteriol. Rev.* **41:** 100.

Tingle, M. and H.O. Halvorson. 1972. Biochemical and genetic characterization of β-glucosidase mutants in *Saccharomyces lactis*. *J. Bacteriol.* **110:** 196.

Tisdale, H., J. Hauber, G. Prager, P. Turini, and T.P. Singer. 1968. Studies on succinate dehydrogenase. 15. Isolation, molecular properties, and isoenzymes of fumarate reductase. *Eur. J. Biochem.* **4:** 472.

Tortora, P., M. Birtel, A.-G. Lenz, and H. Holzer. 1981. Glucose-dependent metabolic interconversion of fructose-1,6-bisphosphatase in yeast. *Biochem. Biophys. Res. Commun.* **100:** 688.

Trevelyan, W.E. and J.S. Harrison. 1956. Studies on yeast metabolism. Yeast carbohydrate fractions. Separation from nucleic acid, analysis, and behavior during anaerobic fermentation. *Biochem. J.* **63:** 23.

Tsuyumu, S. and B.G. Adams. 1974. Dilution kinetic studies of yeast populations; *in vivo* aggregation of galactose utilizing enzymes and positive regulator molecules. *Genetics* **77:** 491.

Tzagoloff, A., A. Akal, and R.B. Needleman. 1975. Assembly of the mitochondrial membrane system: Isolation of nuclear and cytoplasmic mutants of *Saccharomyces cerevisiae* with specific defects in mitochondrial functions. *J. Bacteriol.* **122:** 826.

Ullrich, J. and U. Wais. 1975. Pyruvate dehydrogenase complex from brewer's yeast: Regulation by the carbon sources. *Biochem. Soc. Trans.* **3:** 920.

Urech, K., M. Dürr, T. Boller, and A. Wiemken. 1978. Localization of polyphosphate in vacuoles of *Saccharomyces cerevisiae*. *Arch Microbiol.* **116:** 275.

Ureta, T. 1978. The role of isozymes in metabolism: A model of metabolic pathways as the basis for the biological role of isozymes. *Curr. Top. Cell. Regul.* **13:** 233.

Uyeda, K. 1979. Phosphofructokinase. *Adv. Enzymol.* **48:** 193.

van de Poll, K.W. 1973. Activity of the hexose monophosphate shunt in a mutant of *Saccharomyces carlsbergensis* lacking NADP dependent glutamate dehydrogenase activity. *FEBS Lett.* **32:** 33.

van de Poll, K.W.. and D.H.J. Schamhart. 1977. Characterization of a regulatory mutant of fructose-1,6-biphosphatase in *Saccharomyces carlsbergensis*. *Mol. Gen. Genet.* **154:** 61.

van de Poll, K.W., A. Kerkenaar, and H.J. Schamhart. 1974. Isolation of a regulatory mutant of fructose-1,6-diphosphatase in *Saccharomyces carlsbergensis*. *J. Bacteriol.* **117:** 965.

Vankataraman, R. and E. Racker. 1961. Mechanism of action of transaldolase. 1. Crystallization and properties of the yeast enzyme. *J. Biol. Chem.* **236:** 1876.

van Solingen, P. and J.B. van der Plaat. 1975. Partial purification of the protein system controlling the breakdown of trehalose in baker's yeast. *Biochem. Biophys. Res. Commun.* **62:** 553.

van Wijk, R., J.H. Slavenburg, and K.W. van de Poll. 1973. Levels of glycolytic intermediates, adenosine phosphates and polyphosphates in *Saccharomyces carlsbergensis* under various degrees of catabolite repression. *Proc. K. Ned. Akad. Wet. C* **76:** 512.

Wais, U., U. Gillmann, and J. Ullrich. 1973. Isolation and characterization of pyruvate dehydrogenase complex from brewer's yeast. *Hoppe-Seyler's Z. Physiol. Chem.* **354:** 1378.

Wales, D.S., T.G. Cartledge, and D. Lloyd. 1980. Effects of glucose repression and anaerobiosis on the activities and subcellular distribution of tricarboxylic acid and associated enzymes in *Saccharomyces carlsbergensis*. *J. Gen. Microbiol.* **116:** 93.

Wang, C.H., C.T. Gregg, I.A. Forbusch, B.E. Christensen, and V.H. Cheldelin. 1956. Carbohydrate metabolism in baker's yeast. I. Time course study of glucose utilization. *J. Am. Chem. Soc.* **78:** 1869.

Wang, C.H., I. Stern, C.M. Gilmour, S. Klungsoyr, D.J. Reed, J.J. Bialy, B.E. Christensen, and V.H. Cheldelin. 1958. Comparative study of glucose catabolism by the radiorespirometric method. *J. Bacteriol.* **76:** 207.

Weitzman, P.D.J. and M.J. Dawson. 1976. Citrate synthase. *Curr. Top. Cell. Regul.* **10:** 161.

Weitzman, P.D.J. and J.K. Hewson. 1973. In situ regulation of yeast citrate synthase. Absence of ATP inhibition observed *in vitro. FEBS Lett.* **36:** 227.
Wenger, J.I. and C. Bernofsky. 1971. Mitochondrial alcohol dehydrogenase from *Saccharomyces cerevisiae. Biochim. Biophys. Acta.* **227:** 479.
Whelan, W.L. and C.E. Ballou. 1975. Sporulation in D-glucosamine auxotrophs of *Saccharomyces cerevisiae*: Meiosis with defective ascospore wall formation. *J. Bacteriol.* **124:** 1545.
Wiemken, A., M. Schellenberg, and K. Urech. 1979. Vacuoles: The sole compartments of digestive enzymes in yeast (*Saccharomyces cerevisiae*)? *Arch. Microbiol.* **123:** 23.
Williams, J.F., P.F. Blackmore, and M.G. Clark. 1978. New reaction sequences for the non-oxidative pentose phosphate pathway. *Biochem. J.* **176:** 257.
Williamson, V.M., J. Bennetzen, E.T. Young, K. Nasmyth, and B.J. Hall. 1980. Isolation of the structural gene for alcohol dehydrogenase by genetic complementation in yeast. *Nature* **283:** 214.
Williamson, W.T. and W.A. Wood. 1966. D-Ribulose 5-phosphate 3-epimerase. *Methods Enzymol.* **9:** 605.
Wills, C. and H. Jörnvall. 1979. The two major isozymes of yeast alcohol dehydrogenase. *Eur. J. Biochem.* **99:** 323.
Wills, C. and J. Phelps. 1975. A technique for the isolation of yeast alcohol dehydrogenase mutants with altered substrate specificity. *Arch. Biochem. Biophys.* **167:** 627.
Willsky, G.R. 1979. Characterization of the plasma membrane Mg^{2+}-ATPase from the yeast, *Saccharomyces cerevisiae. J. Biol. Chem.* **254:** 3326.
Winge, O. and C. Roberts. 1950. The polymeric genes for maltose fermentation in yeasts and their mutability. *C.R. Trav. Lab.* **25:** 36.
Witt, I., R. Kronau, and H. Holzer. 1966. Repression von alkoholdehydrogenase, isocitratlyase und malatsynthase in hefe durch glucose. *Biochim. Biophys. Acta* **118:** 522.
Wolf, D.H. and C. Ehmann. 1979. Studies on a proteinase B mutant of yeast. *Eur. J. Biochem.* **98:** 375.
Womack, F.C. and S.P. Colowick. 1979. Proton-dependent inhibition of yeast and brain hexikinases by aluminum in ATP preparations. *Proc. Natl. Acad. Sci.* **76:** 5080.
Wurster, B. and B. Hess. 1975a. Quantitative assay of the anomerase activity of glucosephosphate isomerase from baker's yeast. *Methods Enzymol.* **41:** 57.
———. 1975b. Glucose-6-phosphate-1-epimerase from baker's yeast. *Methods Enzymol.* **41:** 488.
Yue, R.H., E.A. Noltmann, and S.A. Kuby. 1969. Glucose 6-phosphate dehydrogenase from brewer's yeast (*Zwischenferment*). III. Studies of the subunit structure and on the molecular association phenomenon induced by triphosphopyridine nucleotide. *J. Biol. Chem.* **244:** 1353.
Zimmermann, F.K., N.A. Khan, and N.R. Eaton. 1973. Identification of new genes involved in disaccharide fermentation in yeast. *Mol. Gen. Genet.* **123:** 29.
Zimmermann, F.K., I. Kaufmann, H. Rasenberger, and P. Haussmann. 1977. Genetics of carbon catabolite repression in *Saccharomyces cerevisiae*: Genes involved in the derepression process. *Mol. Gen. Genet.* **151:** 95.
Zipper, P. and H. Durschlag. 1978. Small angle X-ray scattering on malate synthase from baker's yeast, the native substrate free enzyme and enzyme substrate complexes. *Eur. J. Biochem.* **87:** 85.
Zitomer, R.S., D.L. Montgomery, D.L. Nichols, and B.D. Hall. 1979. Transcriptional regulation of the yeast cytochrome *c* gene. *Proc. Natl. Acad. Sci.* **76:** 3627.
Zubenko, G.S. and E.W. Jones. 1979. Catabolite inactivation of gluconeogenic enzymes in mutants of yeast deficient in proteinase B. *Proc. Natl. Acad. Sci.* **76:** 4581.

Nitrogen Metabolism in *Saccharomyces cerevisiae*

Terrance G. Cooper
Department of Biological Sciences
University of Pittsburgh
Pittsburgh, Pennsylvania 15260

1. **A Summary of the Major Catabolic Systems**
   A. Allantoin Degradation
   B. Genetics of Allantoin Degradation
   C. Induction of the Allantoin-degrading Enzymes
   D. Asparagine Degradation
   E. Regulation of Asparagine Degradation
   F. Ammonia and Glutamate Interconversion
   G. Mutants with Altered GDH Activities
   H. Proline Degradation
   I. Control of Proline Degradation
   J. Arginine Degradation
   K. Induction of the Arginine Catabolic Enzymes
   L. Other Possible Modes of Regulating Arginine Catabolism
   M. Utilization of α-aminoadipate and the Positive Selection of *lys2* and *lys5* Mutants
2. **Nitrogen Metabolism in an Adverse External Environment**
   A. Amino Acid Metabolism during Starvation
   B. Vacuolar and Cellular Homeostasis
   C. Vacuolar Constituents
   D. Vacuole Structure in Different Stages of the Cell Cycle
   E. Mobilization of Sequestered Metabolites
   F. Starvation-mediated Protein Degradation
   G. Yeast Proteinases
   H. Genetics of Intracellular Proteolysis
   I. Effects of Nitrogen Metabolism on Sporulation
3. **Integration of Nitrogen Metabolism**
   A. Short-range Modulation: Induction and Compartmentation
   B. Sequential Induction—A Means of Achieving Sensitive, Buffered Metabolic Control
   C. Long-range Modulation: Nitrogen Catabolite Repression

INTRODUCTION

In decaying fruit and vegetable matter, which form their wild habitat, yeast cells encounter a broad spectrum of compounds able to serve as nitrogen sources. These materials range from simple ammonia and amino acids to complex nucleic acids and their derivatives. In response to this heterogeneous environment, *Saccharomyces cerevisiae* has evolved equally broad degradative enzyme systems and sophisticated ways of regulating and integrating their operation.[1] In the past, enteric bacteria have served as a model for viewing the control of nitrogen metabolism in yeast. As a starting point, these past paradigms have served us well. However, they are certain to be inadequate for the future. An important area of deficiency is their failure to involve the complex internal structure of a yeast cell in the control and integration of nitrogen metabolism. Ignoring such structure and its potential role in cellular homeostasis may mortgage our understanding of a yeast cell's ability to cope with a constantly changing environment.

A SUMMARY OF THE MAJOR CATABOLIC SYSTEMS

As shown in Table 1, *S. cerevisiae* will grow fairly well on many amino acids, uracil, purine derivatives, urea, and ammonia. Sometimes the cells go through only one or two generations (this is indicated in Table 1) before growth ceases. In these cases, the conclusion that cells are able to use the compound effectively must be viewed skeptically. However, in other cases, steady-state growth can easily be maintained. Only a limited number of these compounds and their metabolic fates have been studied; they are briefly summarized in subsequent sections. For most systems, mutants that lack specific enzymes in the catabolic pathway have been isolated. In a few cases, putative control mutants have been isolated and tentatively characterized. In even fewer cases, the pertinent genes have been isolated on chimeric plasmids.

The routes of nitrogen catabolism may be conveniently divided on the basis of their end products. Several systems, such as those degrading allantoin, urea, or asparagine, generate ammonia as the final product. In these cases, nicotinamide adenine dinucleotide phosphate (NADP) glutamate dehydrogenase (GDH) is required to convert ammonia to glutamate, the predominant nitrogen donor in many biosynthetic reactions. The remaining systems, such as those participating in proline and arginine metabolism, generate glutamate directly. These considerations emphasize the central position of ammonia and glutamate at the interface between nitrogen catabolic and anabolic reactions.

Allantoin Degradation

Allantoin, a product of adenine and guanine catabolism in many organisms, can serve as a sole nitrogen source for *S. cerevisiae*. The degradation of

[1]To maintain this work at a reasonable size, the discussion has been confined, with few exceptions, to *S. cerevisiae* and covers work appearing prior to October 1982. A broad literature dealing with nitrogen metabolism in other organisms has been dealt with only peripherally due to space limitations and the availability of excellent reviews elsewhere.

Table 1 Doubling times (min) for growth on various amino acids

Amino acid	M25	M25-12b	S288C	Σ1278b	M970
Ammonia	149	140	144	148	130
Glutamine	136	155	132	144	120
Asparagine	147	162	134	173	113
Arginine	153	181	169	202	138
Glutamate	169	161	149	136	180
Serine	175	175	153	160	153
Alanine	187	194	180	163	212
Aspartate	197	170	175	171	136
Phenylalanine	193	217	224	153	164
Leucine	244	270	287	271	211
Tyrosine	278	291	326	164	265
Valine	289	205	204	212	216
Isoleucine	318	373	222	288	197
Tryptophan	319	349	260	241	322
Proline	322	284	269	268	158
Methionine	363	308	329	—	397
Histidine	531	n.d.	n.d.	423	470
Threonine	549	n.d.	312	154	176
Glycine	635[a]	n.d.	n.d.	580[a]	n.d.
Lysine	n.d.	n.d.	n.d.	n.d.	n.d.
Cysteine	n.d.	n.d.	n.d.	n.d.	n.d.
Allantoin	181	156	190	222	293
Allantoate	388	172	315	424	362
Urea	148	153	164	158	166
Citrulline	188	264	208	186	192
Ornithine	190	187	193	206	238

Wickersham's medium was used in this experiment, with 0.6% glucose as carbon source. n.d. indicates insignificant growth in which the cells failed to double one time.
[a]Cells did not continue to grow.

allantoin involves five enzymatic steps and four transport systems (Fig. 1). The hydantoin ring of the allantoin molecule is first opened in a hydrolytic reaction catalyzed by allantoinase (Lee and Roush 1964; Vogels and van der Drift 1976). Allantoate, the product of this reaction is degraded by allantoicase yielding one molecule each of urea and ureidoglycollate (Trijbels and Vogels 1967). Ureidoglycollate hydrolase then catalyzes the hydrolytic cleavage of ureidoglycollate to glyoxylate and a second molecule of urea (Gaudy et al. 1965). Although in most organisms urea is degraded to ammonia by action of urease, *S. cerevisiae* does not contain urease activity. Roon and Levenberg (1968) reported an alternative mode of urea degrada-

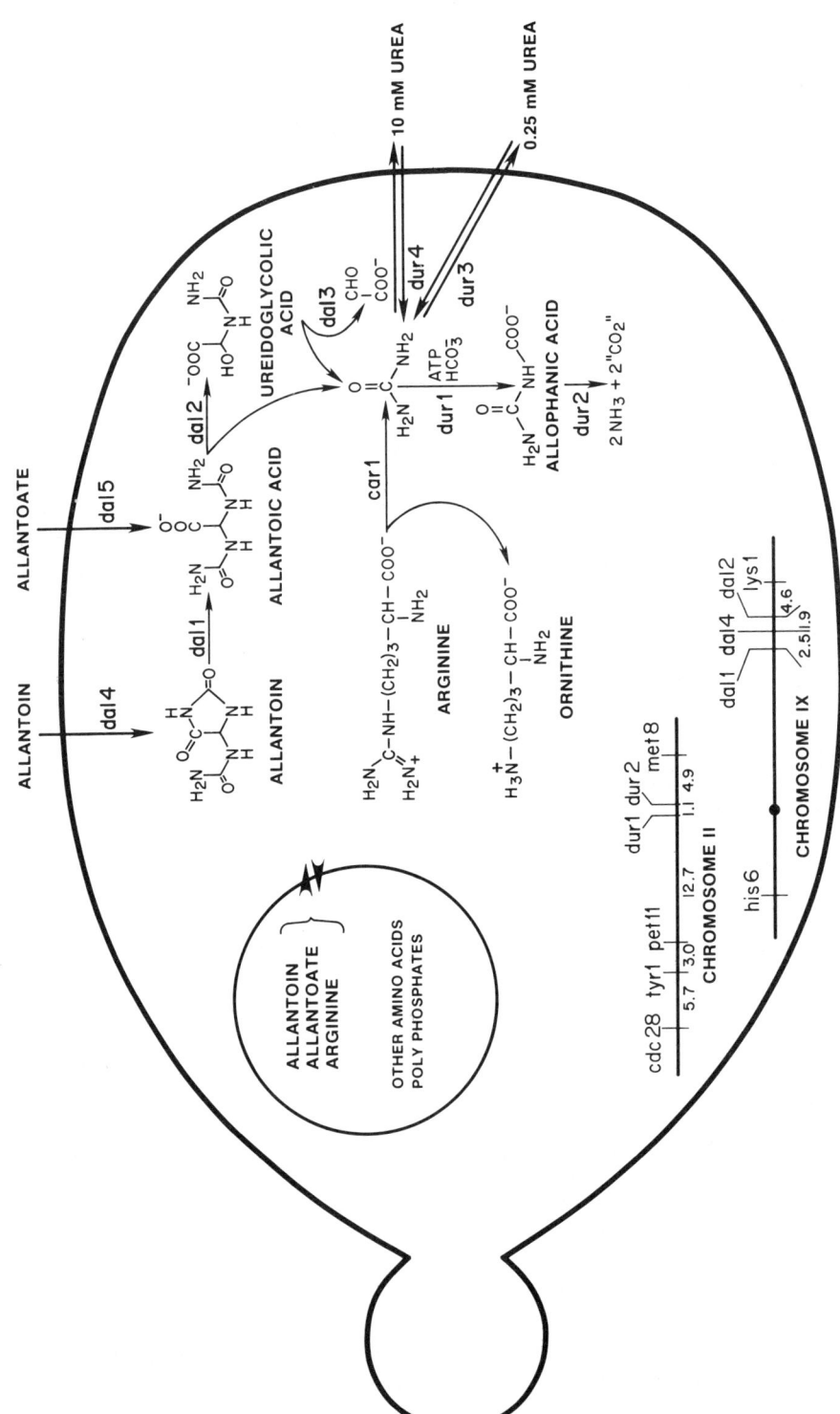

Figure 1 Reactions associated with the degradation of allantoin in *S. cerevisiae*.

tion in *Chlorella* and *Candida utilis* that was inhibited by avidin. The reaction they reported is as follows:

$$\text{urea} + \text{ATP} \xrightarrow{Mg^{++}, K^+} CO_2 + 2NH_3 + ADP + P_i$$

We were able to demonstrate a similar activity in *S. cerevisiae* (Whitney and Cooper 1970). Our difficulty in assigning a role for biotin in this reaction led us to isolate the enzyme and to discover that bicarbonate was a mandatory cofactor of the reaction (Whitney and Cooper 1970). We further demonstrated the degradative reaction to be accomplished in a multistep reaction catalyzed by two distinct enzyme activities (Whitney and Cooper 1972a,b): urea carboxylase, which catalyzes the reaction

$$\text{urea} + \text{ATP} + HCO_3^- \xrightleftharpoons{Mg^{++}, K^+} \text{allophanate} + ADP + P_i$$

and allophanate hydrolase, which catalyzes the reaction

$$\text{allophanate} \longrightarrow 2 \text{ "}CO_2\text{"} + NH_3$$

This conclusion is based on (1) differential sensitivities of the two activities to inhibitors and heating, (2) the fact that allophanate is in rapid equilibrium with allophanate in the medium, and (3) the existence of two classes of mutants, one lacking urea carboxylase and the other lacking allophanate hydrolase.

Urea carboxylase is a biotin-requiring enzyme, whose activity can be separated into two steps: (1) the activation of CO_2 by covalent linkage to the enzyme and (2) the subsequent transfer of the CO_2 to urea (Roon and Levenberg 1970; Roon et al. 1972; Whitney and Cooper 1973; Sumrada and Cooper 1982b). The requirement for biotin as a cofactor in this reaction accounts for the earlier observation that cells provided with allantoin as nitrogen source require high concentrations of biotin for maximal growth (Domnas 1962). The addition of the biotin prosthetic group to the urea carboxylase domain of the multifunctional protein also probably explains the observation that allophanate hydrolase activity appears 2-6 minutes prior to urea carboxylase, following the addition of inducer (Sumrada and Cooper 1982a).

Genetics of Allantoin Degradation

My colleagues and I have isolated several thousand mutants that are unable to degrade one or more of the allantoin pathway intermediates. Genetic mapping has indicated that most of the genes for allantoin degradation are located in one of three clusters. One cluster on the right arm of chromosome IX contains the structural genes for allantoinase (*DAL1*), the allantoin permease (*DAL4*), and allantoicase (*DAL2*), in that order (Lawther et al. 1974; Cooper et al. 1979b); a second cluster on the right arm of chromosome II contains the complementation groups for urea carboxylase (*DUR1*) and allophanate hydrolase (*DUR2*) (Cooper et al. 1980); and a third cluster, linked to the centromere of chromosome VIII, contains the genes for the active transport of urea (*DUR3*) and its facilitated diffusion (*DUR4*) (Sumrada et al. 1976; V. Turoscy et al., in prep.). We have recently found that the *DAL3* gene encoding ureidoglycollate hydrolase is situated 1 kb away from *DAL2*, adding a fourth gene to the cluster on chromosome IX (T.G. Cooper et al., in prep.). The remaining genes associated with allantoin degradation are unlinked to one another or to these clusters.

All of the allantoin pathway genes, with the exception of *DAL5* and *DUR3*, have been isolated on recombinant plasmids. The *DAL1* and *DAL4* genes were isolated together on a 5.5-kb plasmid using the transformation-complementation method (R. Buckholtz and T.G. Cooper; V. Turoscy and T.G. Cooper; both in prep.). The *DAL2* and *DAL3* genes were isolated in a similar manner (T.G. Cooper et al., in prep.). The *DUR1,2* gene possesses a coding region of 5.5–6.0 kb, which made it impossible to isolate by transformation-complementation methods using currently available libraries. Therefore, we cloned the linked *MET8* gene and "walked" to *DUR1,2*, using site-directed integration and excision procedures (F. Genbauffe and T.G. Cooper, in prep.).

For one of these loci, that containing *DUR1* and *DUR2*, the clustering appears to take the form of a single cistron encoding a multifunctional enzyme. Mutations that result in the loss of urea carboxylase are clustered at one end of the locus (Fig. 2), whereas those that result in the loss of allophanate hydrolase activity are clustered at the other end. Since *dur1* mutations complement *dur2* mutations, it was originally thought that two functional entities were involved in urea degradation (Whitney and Cooper 1972a). However, mutations have been isolated that result in the simultaneous loss of both activities and that fail to complement either *dur1* or *dur2* mutations. Furthermore, these pleiotropic mutations are scattered throughout the entire cluster (Fig. 2; Cooper et al. 1980). Finally, when this urea-degrading protein is purified to homogeneity, a single 204-kD polypeptide possessing both enzymatic activities is obtained (Sumrada and Cooper 1982b). Thus, we conclude that urea carboxylase and allophanate hydrolase are component activities of a multifunctional protein and that the *DUR1* and *DUR2* loci are two domains of a single monocistronic gene, *DUR1,2*.

Induction of the Allantoin-degrading Enzymes

All four enzymes of the allantoin pathway and the urea active transport system are inducible; allophanate, the last intermediate of the pathway

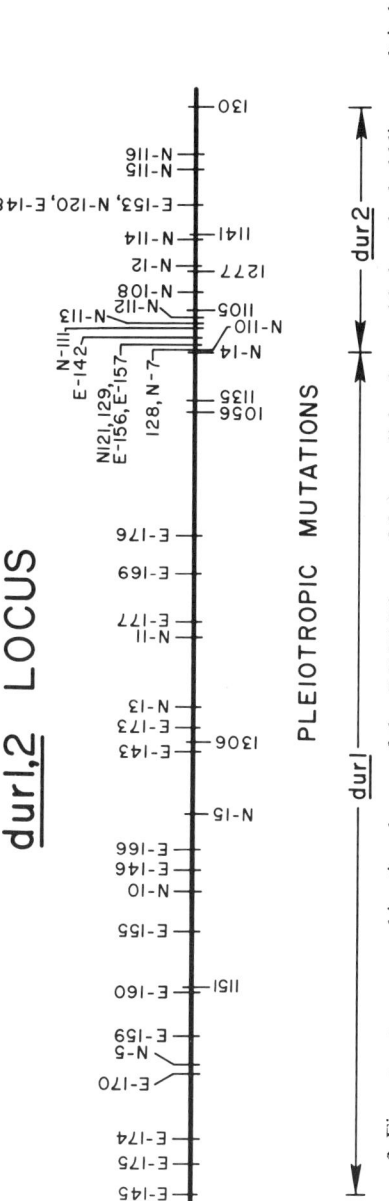

Figure 2 Fine-structure recombinational map of the *DUR1,2* locus. Mutant alleles denoted below the bold line result in loss of both enzyme activities. These alleles are also unable to complement either *dur1* or *dur2* alleles.

functions is the native inducer (Cooper and Lawther 1973a,b; Whitney et al. 1973). In addition, oxalurate can serve as a nonmetabolizable inducer (Sumrada and Cooper 1974; Sumrada et al. 1978). The allantoin transport system is also inducible, but by allantoin itself, rather than by allophanate. Hydantoin and hydantoin acetate can act as nonmetabolizable inducers (Sumrada and Cooper 1977; Sumrada et al. 1978). However, preliminary data indicate that the induction process may be more complex than originally conceived. The allantoate active transport ($DAL5$) and urea-facilitated diffusion ($DUR4$) systems both appear to be produced constitutively (Cooper and Sumrada 1975; Turoscy and Cooper 1979). Synthesis of all of the enzymes and transport systems, except urea-facilitated diffusion, is repressed when cells are grown in the presence of readily used nitrogen sources such as asparagine, serine, or glutamine (Bossinger et al. 1974; Bossinger and Cooper 1975). Repression of this and other systems will be discussed in detail later in the text.

The involvement of allophanate in the expression of five distinct genes ($DAL1$, $DAL2$, $DAL3$, $DUR1,2$, and $DUR3$) suggested that production of their cognate products was regulated by common elements. This is now supported by the isolation of three new classes of putative regulatory mutants (*dal80, dal81,* and a third group phenotypically similar to *dal81*). Strains with a lesion in the *dal80* locus (formerly *dur5*) (Cooper 1980) produce all of the allophanate-inducible gene products at high constitutive levels in the absence of inducer (Cooper 1980, 1982; Chisholm and Cooper 1982). Comparable constitutive levels were also observed in *dur1 dal80* double mutants. This and the observation that arginase activity remains at its uninduced, basal level in *dal80* mutants eliminates internal induction as the basis for constitutive enzyme synthesis. *dal80* mutations are recessive to wild-type alleles. The $DAL80$ locus is situated on the right arm of chromosome XI, unlinked to any of the structural genes whose expression is regulated by its product (V. Turoscy et al., in prep.). Synthesis of the five gene products produced constitutively in *dal80* mutants remains normally sensitive to nitrogen repression even though the *dal80* mutation is present. From these observations we concluded that production of the allantoin-degrading enzymes is subject to control by the $DAL80$-gene product and that induction and repression of enzyme synthesis involve different regulatory elements (Chisholm and Cooper 1982).

Strains containing mutations in the $DAL81$ locus, which is located on the right arm of chromosome IX, have the same basal levels of allantoin pathway enzymes whether or not inducer is present in the growth medium (Turoscy and Cooper 1982). As a consequence, such strains grow poorly when provided with any of the allantoin-related metabolites as sole nitrogen source, although they can use other nitrogen sources normally. The basal level of the allantoin pathway enzymes increases modestly when mutants are grown on poor nitrogen sources. Thus, although the specific induction system is altered by mutation of *DAL81*, the enzymes are still subject to the nitrogen repression-derepression control system. Although these data are consistent with the *DAL81*-gene product exerting positive control over synthesis of the allantoin-degrading enzymes, other interpretations are also possible. Wiame and his colleagues have isolated a mutant strain that possesses a phenotype similar to that of the *dal81* mutants. It is not known, however, whether the mutations contained in those strains are allelic with the *dal81* mutations. In addition, a class of mutations phenotypically similar but genetically unlinked to *dal81* have been identified but not further characterized (V. Turoscy et al., in prep.).

If, as we suspect, the allantoin-degradative system is regulated by multiple regulatory elements, it is important to understand which elements interact with the pathway structural genes, which interact with the inducer, and whether or not they potentially interact with one another. Mutations in the *DAL81* locus exhibit an abnormally high degree of intragenic complementation, a common characteristic seen when a gene product possesses multiple functional domains (V. Turoscy et al., in prep.). Epistasis experiments with alleles of the *DAL80* and *DAL81* genes argue that the two elements possess some degree of independence (Turoscy and Cooper 1982). Unlike most of the prokaryotic control systems where epistasis is complete, here it is not. The phenotype of one mutation does not dominate the other. Rather, the double mutant has phenotypic attributes in common with both single mutants. Our current interpretation of the data is that the *DAL80*-gene product regulates the basal level of gene expression, perhaps as a result of interactions with a regulatory target site linked to the controlled genes. The *DAL81*-gene product, on the other hand, seems to be responsible for the increase in gene expression on the addition of inducer. This may result from an interaction between the *DAL81*-gene product and allophanate. Experiments are currently in progress to determine whether or not the genes coregulated by the *DAL80*- and *DAL81*-gene products possess common sequences as targets for these regulatory elements.

Two classes of mutants that alter the level of *DUR1,2*-gene expression have been identified next to the locus (Cooper 1978; Lemoine et al. 1978). The first class of mutants express allophanate hydrolase constitutively in **a** or α cells but have essentially normal levels in **a**/α cells. Initial hopes that these mutations would identify regulatory sequences have been complicated by evidence suggesting that they are the result of Ty element insertions (Sumrada and Cooper 1982c). Mutations with similar phenotypes have been identified in several systems (Errede et al. 1980; Williamson et al. 1981).

The second class of mutants also express the *DUR1,2* gene constitutively. Other allophanate-inducible genes are completely unaffected. The mutated locus carried in these strains has been designated *DUR80* and is situated 2–3 centiMorgans (cM) distal to the *DUR1,2* gene near the *MET8* locus (G. Chisholm and T.G. Cooper, in prep.). Mutations in *DUR80* generate the constitutive *DUR1,2* phenotype in a *cis*-dominant fashion. This phenotype is not affected by the mating type of the cells as is often seen with a Ty insertion. In addition, the mutant locus is much farther away from the affected structural gene than has been previously reported. In view of these striking characteristics, we have isolated the entire region between the *MET8* and *DUR1,2* genes from wild-type cells and have used these plasmids for probing mutant genomic Southern blots. No Ty insertions were found in this region. We are presently attempting to clone the mutant sequences in an effort to identify the nature of the lesion and to determine how the alteration could exert its effects presumably over a 5–7-kb distance.

The specific biochemical events associated with induction have been extensively studied using allophanate hydrolase as a representative marker (Lawther and Cooper 1973, 1975; Lawther et al. 1975; Cooper et al. 1978). As illustrated in Figure 3, these studies suggest that induction of allophanate hydrolase includes a requirement for mRNA synthesis and de novo protein synthesis (Bossinger and Cooper 1976a,b; Cooper and Bossinger 1976). Induction of the other enzymes of this pathway are believed to occur similarly at the transcriptional level. Transcription is completed 2½ minutes after the addition of inducer at 15°C. The *RNA1*-gene product, which is presumed to participate in transport of mRNA from the nucleus to the cytoplasm, is needed for about 10 minutes; protein synthesis is initiated at 20 minutes and is terminated by 24 minutes. Two minutes later, active enzyme appears. It is anomalous that there is such a significant gap between the end of sensitivity to the mRNA synthesis inhibitors and the start of sensitivity to protein synthesis inhibition.

A second result of the induction studies was the identification of the functional half-life of allophanate-hydrolase-specific synthetic capacity (Law-

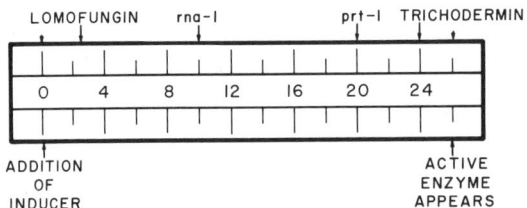

Figure 3 Sequence and timing of the macromolecular synthetic events involved in the induction of allophanate hydrolase at 15°C. The treatment to interrupt the process is indicated above the time line. The times indicated (in minutes) are the last times when the various treatments were able to inhibit the production of allophanate hydrolase. Lomofungin is an inhibitor reported to inhibit RNA synthesis, *rna1* and *prt1* indicate the use of temperature-sensitive mutants with defects in RNA metabolism and protein synthesis initiation, respectively, and trichodermin is reported to inhibit protein synthesis elongation. (Data from Bossinger and Cooper 1976a.)

ther and Cooper 1975; Lawther et al. 1975; Cooper et al. 1978). Studies of collective mRNA metabolism in *S. cerevisiae* have yielded synthetic capacity (presumed to be mRNA) half-lives of 15–20 minutes (Hutchinson et al. 1969; Kou et al. 1973; Tonneson and Friesen 1973). In contrast, we observed a value of $2\frac{1}{2}$–3 minutes for allophanate-hydrolase-specific synthetic capacity. The observed short synthetic capacity half-life was extended to several other specific systems (Bossinger and Cooper 1977; Messenguy and Cooper 1977; Cooper et al. 1978) and led to the conclusion that there is a wide spectrum of synthetic capacity stabilities ranging from those that are labile, like allophanate hydrolase, to those that are highly stable, like the glycolytic enzymes. Failure to detect the labile species in experiments measuring global turnover derived from a lack of resolution in the techniques used (Bossinger and Cooper 1977).

The possibility that these synthetic capacities reflect mRNA levels is being tested directly by measuring the amounts of mRNA capable of hybridizing to DNA fragments containing allantoin pathway genes. Little RNA derived from cultures grown in the absence of inducer is able to hybridize to *CAR1*, *DAL4*, *DAL3*, or *DUR1,2* DNA. (Sumrada and Cooper 1982c; T.G. Cooper et al.; F. Genbauffe and T.G. Cooper; both in prep.) When inducer is present in the medium, however, significant levels of hybridization are observed. Even more hybrid formation is observed when RNA, derived from *car80* or *dal80* mutants, grown in the absence of inducer, is used in place of wild-type RNA (these mutants constitutively produce the allantoin and arginine enzymes, respectively). In contrast, little if any hybridization is observed to *DAL4*, *DAL3*, or *DUR1,2* gene probes when the RNA is derived from a *dal81* mutant (T.G. Cooper et al; F. Genbauffe and T.G. Cooper; both in prep.). The level at which nitrogen repression is exerted has also been determined by ascertaining the amounts of *CAR1*, *DAL4*, *DUR1,2*, and *DAL3* mRNAs contained in *car80* and *dal80* mutants grown in the presence of asparagine, a repressing nitrogen source. In all cases, almost no *CAR*, *DUR*, or *DAL* mRNA was found. These data continue to be consistent with our earlier suggestion that induction and repression of the arginine- and allantoin-degrading systems occur at the level of transcription and suggest that the *CAR80*-, *DAL80*-, and *DAL81*-gene products exert their effects at this level as well. The fact that steady-state RNA levels were qualitatively measured (Northern blots were used) in the experiments just described requires that the results be interpreted conservatively. We are in the process of carefully measuring the rates of RNA synthesis and degradation in wild-type and mutant strains grown under various physiological conditions.

Asparagine Degradation

Asparagine catabolism in *S. cerevisiae* involves one enzyme (asparaginase I) in some strains and two (asparaginase I and II) in others. Asparaginase I is found in the cytosol. Mutants with lesions in the gene responsible for asparaginase-I production were first isolated by Jones and Mortimer (1970). These investigators began with an aspartate auxotroph resulting from the loss of glutamate-oxalacetate transaminase (*asp5*). From this strain they isolated mutants for which asparagine was unable to satisfy the auxotrophy. The locus generating this phenotype was designated *ASP1* and is 18 cM from the *TRP4* locus. Complementation studies resulted in construction of a complex map composed of six complementation groups. Fourteen alleles in the *ASP1* locus were used to construct a fine-structure recombination map (Jones 1973). Asparaginase I has not yet been purified but has been reported to possess a K_m for asparagine of approximately 0.25 mM and a pH optimum of 8.5 (Jones and Mortimer 1973). *asp1* mutants possess normal levels of L-glutaminase activity, suggesting that degradation of glutamine is independent of asparagine catabolism.

Dunlop and Roon (1975) reported a second form of L-asparaginase (form II), which has been shown to be an extracellular glycoprotein. Asparaginase II is found in some strains (i.e., X2180A, D273-10B, and XT1172) but not in others (10275, 9896, Harden and Young strain, and Σ1278b). Nitrogen repression studies have been extensively carried out by Cooper (XT-1172; Bossinger et al. 1974) and Wiame (Σ1278b; Dubois et al. 1973). Unlike the internal enzyme, asparaginase II is insensitive to *p*-hydroxymercuribenzoate. The glycoprotein has been partially purified (400-fold) and is reported to possess a K_m of approximately 0.2 mM for both D- and L-asparagine (Dunlop et al. 1976). Jones (1977a) reported the identification of two loci (*ASP2* and *ASP3*), both of which appear to be required for asparaginase-II activity. *asp2* and *asp3* mutations are recessive to wild-type alleles and complement each another. Moreover, the two loci are neither linked to each other nor to the *ASP1* locus (Jones 1977b).

Regulation of Asparagine Degradation

Asparaginase II was originally reported to be produced only during starvation for nitrogen in the presence of a carbon and energy source. However, Pauling and Jones (1980) monitored asparaginase-II activity throughout the culture cycle and observed somewhat different results. On diluting stationary-phase cells into fresh medium, a low basal activity was observed. During the course of log phase, this activity increased 21-42-fold. As cells emerged from exponential growth, asparaginase-II activity dropped precipitously to the original basal level. It was suggested that changes in nutritional state might trigger the loss of activity. This conclusion was based on the observation that the losses in asparaginase II were not observed when asparagine or glutamine was added to the culture medium as cells reached late log phase. A partial sparing of activity was also observed following the addition of aspartate. The addition of glucose, ammonium sulfate, glutamate, arginine, or proline to the medium was without effect. Subsequent kinetic studies revealed that asparagine and glutamine did not prevent asparaginase-II activity loss but rather slowed it. The loss could also be prevented if late log-phase cells were treated with cycloheximide. However, once the inactivation process began, it could proceed in the absence of protein synthesis. Pauling and Jones (1980) concluded that "the strong specific responses shown when asparagine or glutamine is added to the medium suggest that loss of activity results directly or indirectly from changes in the internal nitrogen balance of the cells as affected through the metabolism or compartmentation of the amino acids." Although this interpretation is reasonable, the data may be alternatively interpreted in another way. The data reported are consistent with the proposal that a protease is synthesized and secreted by cells as they emerge from log-phase growth. The loss of asparaginase-II activity is then a result of proteolytic degradation. Viewed in this way, the sparing effect observed on the addition of asparagine, glutamine, or aspartate might be the result of these ligands binding to asparaginase II and maintaining it in a conformation more resistant to proteolysis.

Pauling and Jones (1980) also observed an additional twofold increase in asparaginase-II activity if late log-phase cells were starved. Two lines of evidence led to the conclusion that starvation-mediated production of asparaginase II was distinct from the synthesis occurring during exponential growth. First, 3-4 hours elapse between the onset of starvation and the appearance of maximum activity, compared to 9 hours elapsing between seeding stationary cells into fresh medium and the appearance of maximum activity levels. Second, late exponential cells could be derepressed on starvation. Additional support for this hypothesis was derived from analysis of an *asp4* mutant (Pauling et al. 1980). This strain possessed only 25% of wild-type asparaginase II during exponential growth but derepressed to the same degree as the wild type. No linkage was detected between the *ASP2, ASP3,* and *ASP4* loci. Although Pauling et al. (1980) conclude that regulation of asparaginase II in rapidly growing cells can be separated genetically from regulation by nitrogen starvation, it is presently difficult to ascertain whether the *asp4* phenotype is a primary or secondary effect of the mutation.

Dunlop et al. (1980) reported that asparaginase-II synthesis is highly sensitive to nitrogen repression when wild-type cells are grown in minimal ammonia medium. However, the enzyme is produced at high levels when NADP-GDH-defective mutants (*gdh1*) are grown under similar conditions. In contrast, asparaginase-II activity is repressed 5-20-fold in both wild-type and *gdh1* mutants grown in minimal medium containing alanine, arginine, asparagine, cysteine, glutamine, glycine, histidine, isoleucine, leucine, lysine, phenylalanine, serine, or tryptophan. These observations led to the conclusion that NADP-GDH did not play a direct role in the regulation of asparaginase-II synthesis. Rather, the apparent relief of ammonia repression observed in the mutant cells was an indirect result caused by the inability of the mutant cells to assimilate ammonia. Although one can reasonably argue that the absence of asparaginase-II activity in cells grown in minimal medium containing alanine, glutamine, asparagine, or one of the other readily metabolized amino acids is the result of nitrogen repression, such an interpretation is probably not warranted in the case of glycine, cysteine, lysine, or histidine. These amino acids either are not utilized by *S. cerevisiae* or are used only very poorly. It is possible that the observed effects derive from two different physiological circumstances, depending whether or not the amino acid is a good nitrogen source.

Ammonia and Glutamate Interconversion

As mentioned earlier, the interconversion of ammonia and glutamate form the interface between biosynthesis and degradation of nitrogenous compounds. Both metabolites will serve as sole nitrogen source and both participate, along with glutamine, as the major nitrogen donors in biosynthetic reactions. The interconversion of glutamate and ammonia is catalyzed by two GDHs; one uses NADP as cofactor, and the other uses NAD.

$^+NH_4 + NADPH + \alpha$-ketoglutarate $\rightleftharpoons$ L-glutamate + NADP
L-glutamate + NAD $\rightleftharpoons$ α-ketoglutarate + NADH + $^+NH_4$

Holzer and his colleagues were among the first to study these enzymes in yeast (Holzer and Schneider 1957; Hierholzer and Holzer 1963; Bernhardt et al. 1965, 1966; Kohlhaw et al. 1965). They reported high levels of the NAD-specific enzyme when cells were grown on glutamate and low levels when ammonia was provided as nitrogen source. This prompted the conclusion that NAD-specific GDH is a catabolic enzyme. The loss of activity when cells were grown on glutamate plus ammonia was interpreted as an indication that the catabolic enzyme was subject to ammonia repression (Hierholzer and Holzer 1963).

Growth conditions that yield high expression of NADP-GDH cause low-level expression of NAD-GDH, and vice versa. This has led to the conclusion that NADP-specific GDH serves a biosynthetic function converting ammonia to glutamate (Hierholzer and Holzer 1963; Thomulka and Moat 1972; Roon and Even 1973).

The deficiencies in our understanding of these enzymes and their control is emphasized by considering the results of Roon and Even (1973). These investigators measured NADP- and NAD-specific GDH activities in cells grown on 13 different nitrogen sources. The NADP-specific enzyme was maximal on ammonia, allantoin, and urea; the latter two compounds are degraded directly to ammonia. It was twofold lower when cells were provided with a variety of amino acids, including arginine, asparagine, and glutamine; degradation of the latter three amino acids yields equimolar amounts of glutamate and ammonia. Why were the NADP enzyme levels depressed even though the cell was generating large amounts of ammonia? Does growth on glutamate repress production of the NADP enzyme? At present, these questions cannot be answered unambiguously. We may only conclude that maximal levels of NADP-GDH are obtained when ammonia is the sole nitrogen source accumulated from the medium or produced intracellularly.

NAD-specific GDH levels, on the other hand, are maximal only when glutamate, aspartate, or alanine is provided as sole nitrogen source. They are 10–15-fold lower when ammonia, urea, allantoin, asparagine, glutamine, arginine, serine, and other amino acids are provided. Here, three points seem to be important. First, the degree of modulation is much greater than that observed with biosynthetic enzymes—a typical characteristic of the nitrogen catabolic enzymes. Second, NAD-specific GDH production is probably not subject to nitrogen catabolite repression in the same way as other catabolic enzymes. The reason for this conclusion derives from the following correlations. Allantoin is one of the least repressive nitrogen sources in yeast, whereas glutamine and asparagine are the most highly repressive sources known. Yet, nearly the same levels of NAD-GDH were observed in cells provided with these three nitrogen sources; this result is not observed when other nitrogen catabolic enzymes are assayed. Third, NAD-GDH activity is low even when glutamate and ammonia are generated in equimolar amounts, as when glutamine or asparagine are provided. This indicates that the presence of ammonia may be an important direct or indirect factor for the determination of NAD-GDH levels. Consistent with this idea is the observation of Roon and Even (1973) that adding increasing amounts of ammonia (up to 1 mM) to cells growing in minimal glutamate medium results in a lowering of NAD-GDH activity. One may ask whether or not this catabolic system is repressed by ammonia in a manner similar to that exerted by amino acids or their biosynthetic pathway enzymes. What is unclear is whether the lowered activity results from a concomitant decrease in enzyme production or increased turnover.

In addition to the types of regulation discussed above, two investigators and their colleagues have reported a marked loss of NADP-specific GDH on starving cells for glucose (Mazon 1978; Satrustegui and Machado 1978). The loss is prevented by the addition of dinitrophenol (DNP), azide, iodoacetate, or cycloheximide to the starved cells. Adding glucose to the starved cells results in the reappearance of enzyme activity. Restoration may be prevented if cycloheximide is added with glucose (Mazon 1978). That these phenomena result from proteolytic loss of GDH protein and its resynthesis is supported by the observation of Mazon and Hemmings (1979) that immunoprecipitable NADP-GDH protein is also lost on glucose starvation. The universality of this phenomenon is uncertain, because Roon and Even (1978) did not observe a similar loss of NADP-GDH when they starved their cultures for glucose.

Mutants with Altered GDH Activities

Three important classes of mutants with altered levels of NAD-GDH and NADP-GDH activities have been isolated, first by Lacroute (Lacroute 1966; Drillien and Lacroute 1972; Drillien et al. 1973) and later by Grenson et al. (1970, 1974). The first class of mutants are those lacking NADP-GDH activity or possessing altered forms of the enzyme (temperature-sensitive forms or forms possessing increased K_m values for the substrates, α-ketoglutarate and ammonia). Since these are the characteristics expected to result from mutations in the NADP-GDH structural gene, we propose to designate the locus *GDH1*. Lacroute originally designated the locus *URE1*, and Grenson et al. subsequently renamed it *GDHA*.

Even though *gdh1* mutants lack NADP-GDH activity, their growth in minimal ammonia medium decreases only 2.5-fold compared to wild type (265 min vs. 110 min). Two different approaches have been made to identify the alternate pathway for ammonia assimilation that allows *gdh1* mutants to grow. Grenson et al. (1974) suggested that the residual activity derived from the operation of NAD-GDH. However, attempts to isolate strains lacking the catabolic GDH were unsuccessful. Grenson and her colleagues determined that the growth rate of a *gdh1 ama* double-mutant strain was sufficiently distinguishable from wild-type and the singly mutant strains to allow the isolation of mutations that suppress the *gdh1* defect. The new mutation isolated by this technique (*gdhCR*) proved allelic to the *URE2* locus previously described by Lacroute and his colleagues (Drillien and Lacroute 1972; Drillien et al. 1973). The recessive *ure2* mutations result in a twofold decrease in the level of NADP-GDH in cells growing in glucose-ammonia. More striking, the NAD-GDH levels in *ure2* mutants are 130–150-fold higher than in *URE2* strains when ammonia is the sole nitrogen source. This elevated level is the same in glucose-glutamate medium and represents an increase of about fourfold to fivefold in the catabolic enzyme activity. In other words, the catabolic enzyme activity is no longer depressed by growth in ammonia.

A second mutant was isolated during attempts to isolate a strain lacking catabolic GDH. This strain possessed a mutation (*gdhCS*) in the *NGL3* locus and was allelic to a mutation described earlier by Lacroute (Drillien et al. 1973). The pertinent characteristic of this strain was that it possessed the very low NAD-GDH activity seen in wild-type strains growing in ammonia medium, even though the mutant was grown in glutamate medium. Grenson et al. (1974) concluded that the strain was frozen at a repressed level of NAD-GDH. The biochemical defects of the *ure2* and *ngl3* mutants have not been identified. However, mutation of either locus results in highly pleiotropic effects, which will be discussed more fully later in the section on nitrogen repression.

Growth properties of *gdh1* mutants argue strongly in favor of multiple routes for ammonia assimilation. Although the isolation of mutants that overproduce NAD-GDH might point to that enzyme as a second means of assimilation, other possibilities are equally plausible. In this respect it is curious that so far no one has been able to isolate a mutant lacking NAD-GDH.

An alternative pathway for the interconversion of ammonia and glutamate in *S. cerevisiae* utilizes glutamine synthetase and glutamate synthase in a manner similar to that reported by Tempest et al. (1970) for *Klebsiella aerogenes*. The reactions catalyzed by these enzymes are shown below.

$$\text{L-glutamate} + \text{ATP} + {}^+\text{NH}_4 \xrightarrow{\text{glutamine synthetase}} \text{L-glutamine} + \text{ADP} + P_i$$

$$+\text{L-glutamine} + \alpha\text{-ketoglutarate} + \text{NAD(P)H} \xrightarrow{\text{glutamate synthase}} 2 \text{ glutamate} + \text{NAD(P)}$$

$$= \alpha\text{-ketoglutarate} + \text{ATP} + \text{NAD(P)H} + {}^+\text{NH}_4 \rightarrow \text{L-glutamate} + \text{ADP} + P_i + \text{NAD(P)}$$

The hallmark of this pathway is the presence of glutamate synthase. Roon et al. (1974) reported the presence of this activity in *S. cerevisiae*. It could be partially purified and was found to exhibit chromatographic characteristics different from those of the NAD- or NADP-specific GDHs. The enzyme is most active in the pH range of 7.1–7.7 and possesses K_m values of 2.6 μM, 1 mM, and 140 μM for NADH, α-ketoglutarate, and glutamine, respectively. Cells grown in minimal ammonia medium possess the greatest amount of synthase activity, but even under these conditions there was tenfold less glutamate synthase activity than NADP-GDH. This led Roon et al. to conclude that the enzyme serves as an auxiliary means of producing α-amino acids directly from the amide nitrogen of glutamine without passing through the ammonia pool. This argument, however, is based exclusively on the low activity of the enzyme in yeast.

Proline Degradation

As shown in Figure 4, proline is synthesized in four steps, catalyzed, respectively, by the enzymes glutamate kinase, glutamyl phosphate reductase, and pyrroline 5-carboxylate (P5C) reductase. Brandriss (1979) isolated

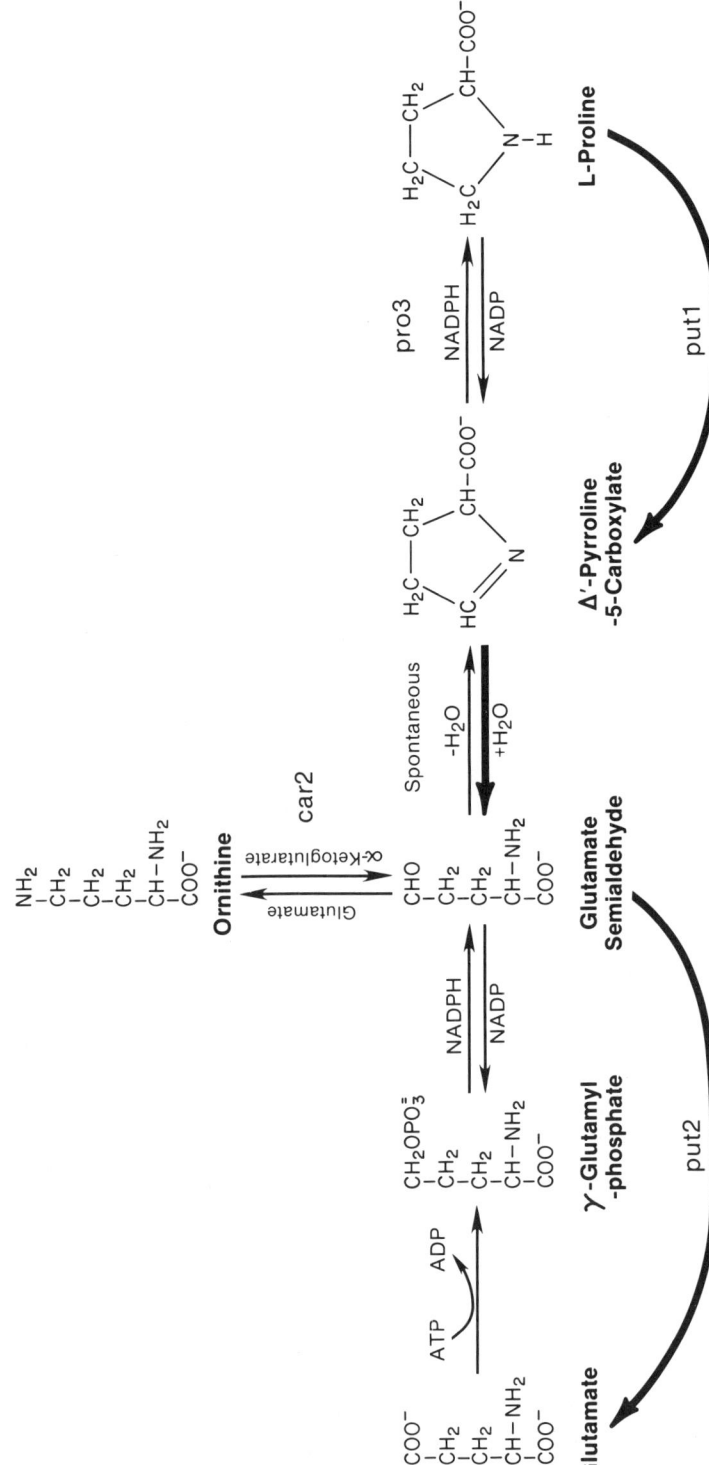

Figure 4 Reactions associated with the biosynthesis (fine arrows) and degradation (bold arrows) of proline in *S. cerevisiae*.

three unlinked classes of proline auxotrophs (*pro1, pro2, pro3*). P5C reductase activity was found to be decreased eightfold in *pro3* mutants, leading to the conclusion that the *PRO3* locus was responsible for production of that enzyme. The biochemical defects present in *pro1* and *pro2* mutants could not be assigned due to the lack of specific assays and the presence of competing enzyme activities. For some unknown reason, proline auxotrophs were unable to grow on YEPD medium, even when it was supplemented with 0.1% proline.

The degradative reactions associated with proline catabolism are just the reverse of steps involved in biosynthesis, though they are catalyzed by different enzymes (see bold arrows in Fig. 4). Proline oxidase is used in place of P5C reductase, and P5C dehydrogenase is used in place of glutamate kinase and glutamyl phosphate reductase. Mutants unable to use proline as sole nitrogen source were isolated by Brandriss and Magasanik (1979) and divided into three classes. Mutants in one class are ρ^-, a second class lack proline oxidase activity (*put1*), and the third class is deficient in P5C dehydrogenase (*put2*). Mutants of the *put2* phenotype were also isolated by Lundgren et al. (1972), but these isolates were never studied in detail. Preliminary genetic analysis revealed that the *PUT* loci are not linked to one another.

The loss of ability to use proline as sole nitrogen source observed in ρ^- strains suggest that mitochondria might play an important role in proline catabolism. This suggestion is supported by the recent observation that P5C dehydrogenase is a mitochondrial enzyme (Brandriss and Magasanik 1981). These and other data led Brandriss and Magasanik to conclude that proline biosynthesis occurs in the cytosol, and catabolism occurs in mitochondria.

Control of Proline Degradation

Using the mutants mentioned above, Brandriss and Magasanik (1981) reached the conclusion that production of proline oxidase and P5C dehydrogenase is inducible with proline acting as inducer. This conclusion was based on the observations that P5C dehydrogenase reached levels threefold higher in *put1* mutants than in the wild type and, conversely, proline oxidase attained levels threefold higher in strains lacking P5C dehydrogenase compared with the wild-type. The investigators reasoned that high proline oxidase levels in *put2* mutants derived from the accumulation of P5C, which could then be converted back to proline. Consistent with this interpretation was the observation that the addition of arginine and ornithine to the culture medium resulted in partial induction of proline oxidase. The same argument was used for induction of P5C dehydrogenase. In the absence of proline oxidase, proline was suggested to accumulate to higher levels, which resulted in increased induction.

Brandriss and Magasanik (1979b) also reported the isolation of a mutant strain designated *put3*, which possesses partially constitutive levels of the proline utilization enzymes (levels of proline oxidase ~ 10-fold higher and levels of P5C dehydrogenase 4.5-fold higher than in wild-type strains grown in minimal ammonia medium). *put3* mutants grown in proline medium, however, possess amounts of the proline-degrading enzymes twofold to threefold greater than similar cultures grown in ammonia medium. Internal induction was eliminated as the basis for constitutivity by the observation that *pro3 put3* double mutants are constitutive even though this strain is unable to synthesize proline. The effects of the mutations in the *put3* locus extend only to the proline-degrading enzymes. The levels of proline permease were not altered by *put3* mutations.

The proline degradative enzymes, though inducible, do not appear to be repressible by growth in the presence of readily used nitrogen sources (Brandriss and Magasanik 1979a). However, preliminary studies of proline transport suggest that this process, though not inducible, is markedly depressed in cells growing in minimal ammonia medium, i.e., it is sensitive to nitrogen repression. Finally, *put1* (or *put2*) mutants growing in urea-proline medium are able to transport considerably less proline than wild-type cells. This may reflect the accumulation of high levels of intracellular proline leading to "feedback inhibition" of transport. (See Cooper [Transport in *Saccharomyces cerevisiae*], this volume.)

Arginine Degradation

Middlehoven (1964) was first to study the pathway of arginine breakdown in *S. cerevisiae*. He identified the products of the arginase reaction to be ornithine and urea, as shown in Figure 5. Arginase is a trimeric enzyme of 39 kD (Penninickx et al. 1974). It has a pH optimum of 8.5–9.0 and a K_m for arginine of 5–7 mM. It is also activated by metal ions, with Mn^{++} yielding maximum activation and Co^{++}, Fe^{++}, Ni^{++}, and Mg^{++} yielding decreasing activation. Ornithine generated from arginine by arginase is converted by ornithine transaminase to glutamate semialdehyde. Initially, it was thought that glutamate semialdehyde was oxidized to glutamate by the enzyme P5C dehydrogenase. Recently, Brandriss and Magasanik (1980) reported that proline is an essential intermediate in the degradation of arginine (Fig. 5). Glutamate semialdehyde produced in the ornithine transaminase reaction is suggested to be converted to proline via P5C. Proline is in turn degraded, as discussed earlier in this paper. This conclusion was derived from two observations. First, proline-oxidase-defective mutants and strains lacking functional mitochondria (ρ^-) grew poorly compared with the wild type when ornithine was provided as sole nitrogen source. The importance of this observation is emphasized by the demonstration that P5C dehydrogenase

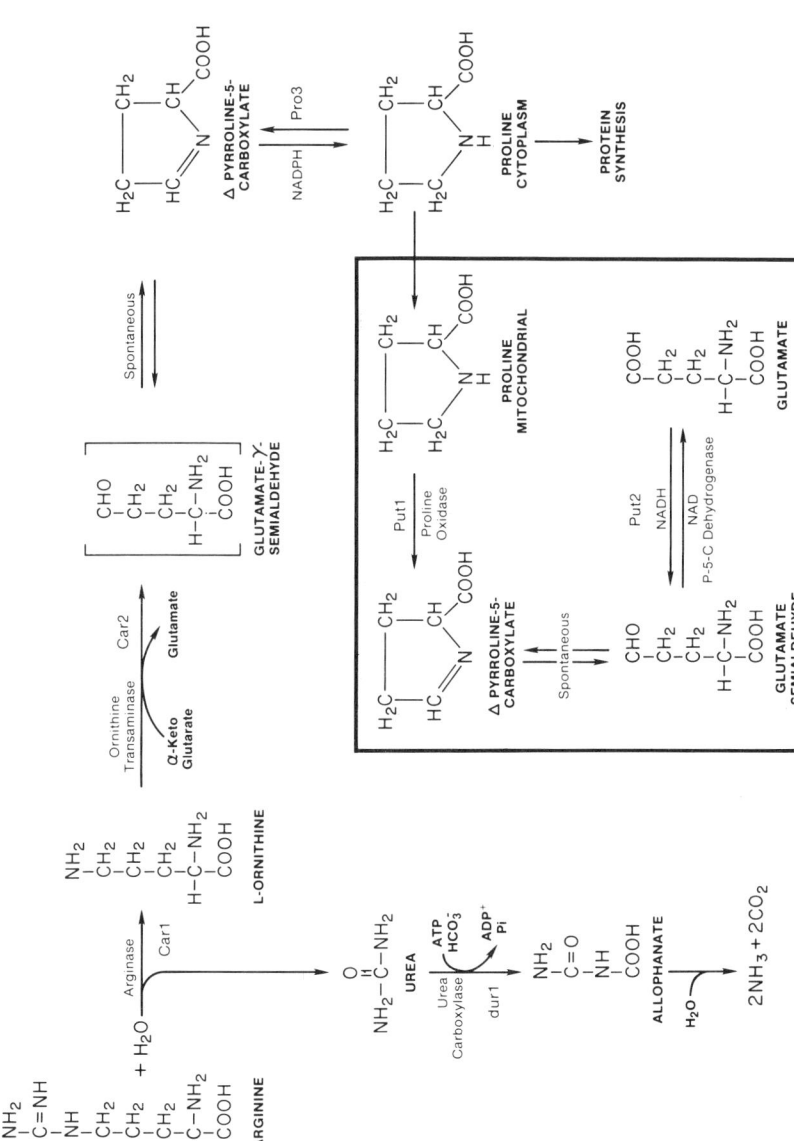

Figure 5 Proposed intracellular compartmentation of the enzymatic reactions associated with arginine degradation in *S. cerevisiae*.

was present at high concentration in *put1⁻* strains. Second, *put1*, *put2*, and ρ^- mutants all excrete proline when arginine is added to the medium.

The investigators hypothesize the presence of two P5C pools, which are structurally separated to prevent futile cycling. Arginase and ornithine transaminase have previously been reported to be cytosolic enzymes (Jauniaux et al. 1978); hence, there is probably a cytosolic pool of P5C. Brandriss and Magasanik (1981) reported that P5C reductase, the last enzyme involved in proline biosynthesis, is also cytosolic. Proline oxidase, on the other hand, is only present in ρ^+ strains and is postulated to be active only when a functional electron transport system is present. This suggests that a mitochondrial pool of P5C also exists. Additional support for the proposed scheme was provided by the demonstration that P5C dehydrogenase is a mitochondrial enzyme. It thus appears that proline is sythesized from glutamate or ornithine in the cytosol and is degraded exclusively in the mitochondria.

Induction of the Arginine Catabolic Enzymes

For the sake of simplicity, arginine catabolism will be viewed as three groups of reactions: (1) the conversion of arginine to glutamate semialdehyde and urea, (2) the degradation of urea to ammonia and "CO_2," and (3) the conversion of glutamate semialdehyde via proline to glutamate. The latter two sets of reactions have already been discussed.

The enzymes arginase and ornithine transaminase, which participate in the conversion of arginine to glutamate semialdehyde, are both inducible (Middlehoven 1964). Middlehoven observed an increase of threefold in arginase levels and sixfold in ornithine transaminase levels on the addition of arginine to minimal ammonia medium. Increases of 9- and 24-fold, respectively, were observed for these enzymes when arginine was provided as sole nitrogen source. The modest increases in enzyme activity on induction probably derived from the fact that Middlehoven pregrew his cultures in medium containing 1% casamino acids (and hence, arginine), thereby greatly increasing the basal levels of enzyme activity. Homoarginine was identified by Whitney and Magasanik (1973) as a gratuituous inducer. Both Middlehoven and Wiame (Ramos et al. 1970; Dubois et al. 1978) reported that ornithine was also an inducer of arginase and ornithine transaminase; on the addition of ornithine to cultures growing in glucose ammonia medium, arginase and ornithine transaminase activities increase to about 40% of the levels observed when arginine is added. However, ornithine-mediated induction could easily be an indirect effect resulting from intracellular conversion of ornithine to arginine. These alternatives may be distinguished by determining whether or not ornithine will serve as inducer in arginine auxotrophs lacking ornithine carbamoyltransferase (OTCase), argininosuccinate synthetase, or argininosuccinase activities. In the absence of such a determination, the assignment of ornithine as inducer of the arginine catabolic enzyme is probably unjustified.

Middlehoven (1968, 1969) also reported that the levels of arginase and ornithine transaminase increase on starvation of cells for ammonia. He concluded that these increased levels were the result of relieving repression (Middlehoven 1970). In contrast, Whitney and Magasanik (1973) argued that the increase in arginase levels observed during nitrogen starvation was not the result of derepression but, rather, derived from specific induction. They suggest that starvation promotes the liberation of the arginine pool normally sequestered in the vacuole. This internally released arginine then induces arginine catabolic enzymes. This hypothesis is substantiated by the demonstration that starvation-mediated increase of arginase activity was markedly decreased in an arginine auxotroph in which the intracellular pools of arginine were depleted. In addition, starvation also results in induction of urea amidolyase activity. This is consistent with the hypothesis, since the arginine released from the vacuole would be degraded to allophanate and, thus would induce synthesis of urea amidolyase.

Six classes of mutants that possess defects in arginase induction have been isolated. The first class of mutants (*car80*, formerly designated *cargR*) are recessive to the wild-type alleles and result in constitutive production of both arginase (*CAR1*) and ornithine transaminase (*CAR2*) (Wiame 1971). The next two classes of mutants are dominant and each is tightly linked to the gene whose expression is rendered constitutive. That is to say, *CAR10$^-$* mutants are tightly linked to *CAR1* and result in the constitutive synthesis only of arginase. Similarly, *CAR20$^-$* mutants are linked to *CAR2* and result in constitutive synthesis only of ornithine transaminase.

The remaining three mutant classes all possess the same recessive phenotype and are designated *arg80, arg81,* and *arg82* (Bechet et al. 1970; Hoet and Wiame 1974). The loci have been located on the right arm of chromosome XIII, fragment 8, and the right arm of chromosome IV, respectively (Mortimer and Schild 1980). Cooper (1982) has isolated a recombinant plasmid that complements a mutation with the same phenotype as the *arg80, arg81, arg82* series (T. Cooper and J. McKelvey, unpubl.). Mutations in the *car80* locus are epistatic to mutations in all three of the loci just described. The phenotype of *arg80, arg81,*and *arg82* mutants has two major characteristics: (1) The mutants produce arginine biosynthetic enzymes constitutively, and (2) they are unable to use arginine or ornithine as sole nitrogen source. Wiame (1971) and his colleagues have shown that mutants with defects in these three loci are unable to synthesize arginase and ornithine transaminase following the addition of inducer but appear to transport arginine into the cell normally. The effects of mutations in the *arg80, arg81,* or *arg82* loci on both arginine catabolic and anabolic enzyme synthesis prompted Wiame to conclude that these loci encode a trimeric "ambivalent repressor" made up of three nonidentical subunits (Wiame 1971; Dubois et al. 1978). A model similar to that shown in Figure 6 has been proposed to account for the observed genetic results (Wiame 1971). However, although the existence of a trimeric ambivalent repressor is certainly plausible, it is equally possible that the three gene products act independently.

A degree of formal similarity exists between the genetic characteristics of putative control mutants associated with the arginine and allantoin catabolic pathways. However, arginase is not affected by *dal80* or *dal81* mutations, nor is allophanate hydrolase activity altered in *arg80*, *arg81*, and *arg82* mutants (T.G. Cooper et al., unpubl.).

Bossinger and Cooper (1977) demonstrated in a set of indirect experiments analogous to those already described for allophanate hydrolase, that induction and repression of arginase are probably mediated at the level of transcription. The arginase-specific synthetic capacity was also shown to possess a short 4–5-minute half-life. This suggests that *CAR1*-specific mRNA is quite labile. These conclusions are derived both from measurements of arginase-specific synthetic capacity and from hybridizations of poly(A) RNA from induced and uninduced cells to plasmid DNA containing the *CAR1* gene (Sumrada and Cooper 1982c).

The inverse correlation of arginine biosynthetic and catabolic genes by the *ARG80*, *ARG81*, *ARG82* loci provides one means by which the cell avoids a futile cycle of arginine synthesis and degradation. An additional safeguard superimposed on this scheme apparently involves a direct interaction between arginase and OTCase. Several lines of evidence suggest that as arginase is synthesized upon induction, the arginase protein binds to the OTCase protein to inhibit OTCase activity (Bechet and Wiame 1965; Mes-

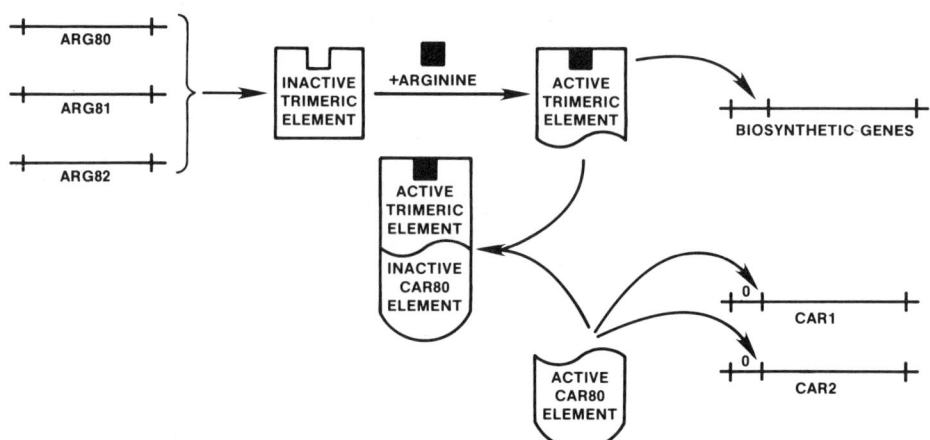

Figure 6 Molecular model proposed to account for the phenotypes of putative regulatory mutants with alterations in the levels of the arginine biosynthetic and degradative enzymes.

senguy et al. 1971). The arginase trimer and OTCase (also a trimer of 37,500 dalton subunits) form a hexamer in which arginase is active and OTCase is inactive (Penninckx 1975; Penninckx and Wiame 1976). The regulatory role of arginase on OTCase activity can be separated from its catalytic role by mutation, thermal stability, differential sensitivity to chymotrypsin, and differential response to inhibitors.

Other Possible Modes of Regulating Arginine Catabolism

In addition to specific induction of arginase production by arginine, Dubois and Wiame (1976) reported nonspecific induction by a variety of amino acids, some of which have no demonstrable relation to arginine. According to the investigators, the biological meaning of nonspecific induction is obscure, but they conclude that it is mechanistically distinct from specific induction. Specifically, if ornithine, lysine, α-aminobutyrate (but not the β derivative), or valine is added to wild-type cultures growing in minimal ammonia medium, a twofold to fivefold increase in arginase activity occurs. Similar values are also obtained with strains harboring lesions in the *arg80*, *arg81, and arg82* loci, i.e., in strains that are unable to induce arginase production upon the addition of arginine to the growth medium. Although the latter observation suggests that specific and nonspecific induction are mechanistically distinct, neither metabolite compartmentation at the vacuolar or cell membrane nor the effects of expanding a particular amino acid pool on these processes have been carefully considered as alternative explanations.

The last form of control hypothesized for the regulation of arginase activity is termed "catabolic synergism" (Dubois and Wiame 1978). This term refers to the observation that the simultaneous presence of inducer and the absence of nitrogen repression appear to bring about a strong cooperative effect on arginase production, which is greater than the summed effects of the two processes evaluated separately. However, the regulatory mutation $CAR10^h$, which defines this phenomenon, is due to insertion of a Ty element (Sumrada and Cooper 1982c, and in prep.). Since the insertion of a Ty element leads to anomolous regulatory properties of the gene next to which it is inserted, the synergism model mentioned above may have to be carefully reevaluated.

Utilization of α-aminoadipate and the Positive Selection of *lys2* and *lys5* Mutants

As shown in Figure 7, α-aminoadipate is a biosynthetic precursor of lysine in *S. cerevisiae*. Wild-type yeast cannot use α-aminoadipate as sole nitrogen source. However, Chattoo et al. (1979) observed that *lys2* and *lys5* auxo-

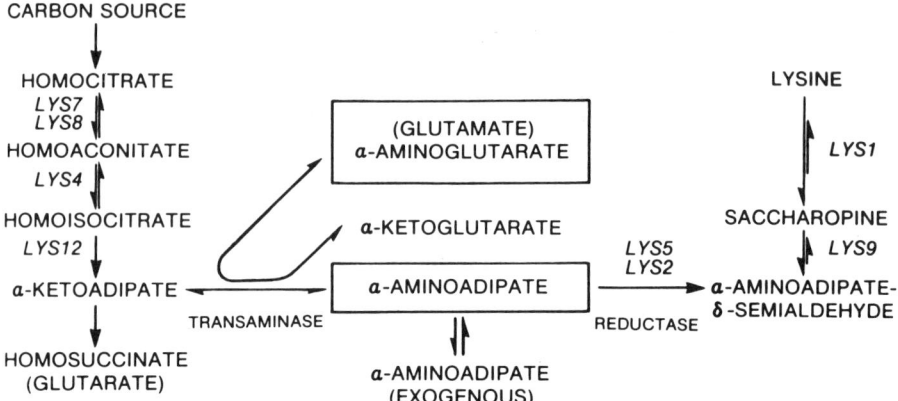

Figure 7 Reactions involved in the biosynthesis of lysine and the degradation of α-aminoadipate.

trophs were able to grow in minimal α-aminoadipate medium supplemented with a small amount of lysine. The products of both loci appear to be required for the conversion of α-aminoadipate to α-aminoadipate semialdehyde. Mutants lacking other lysine biosynthetic enzymes (*lys1*, *lys4*, *lys6–lys10*) do not behave in this manner. These observations provide a convenient means for the selection of *lys2* and *lys5* mutants.

In an effort to explain these observations, Zaret and Sherman (pers. comm.) found that wild-type yeast would not grow in minimal medium containing α-aminoadipate and a poor nitrogen source such as glutamate or proline. However, cells provided with a rich nitrogen source such as ammonia were not adversely affected by α-aminoadipate. In other words, α-aminoadipate is toxic to wild-type cells grown under conditions that derepress amino acid uptake; conversely, toxicity is relieved by the exclusion of α-aminoadipate from the cells. These findings led to the conclusion that the *lys2* and *lys5* mutants prevent accumulation of a toxic lysine precursor and point to α-aminoadipate semialdehyde as the toxic intermediate. This conclusion is consistent with the finding that *lys9* mutants, which are unable to convert α-aminoadipate semialdehyde to saccharopine, are more sensitive to α-aminoadipate than are wild-type cells.

NITROGEN METABOLISM IN AN ADVERSE EXTERNAL ENVIRONMENT

One of the most demanding tasks for living organisms is to maintain themselves in the face of adversity. *S. cerevisiae* is highly adept at this, as is shown by the following experiment. If a logarithmically growing wild-type culture of yeast is transferred from rich medium to distilled water, cell division continues, albeit more slowly, for one to two generations. Our purpose in this section is to review the operation and control of nitrogen catabolic systems in response to such metabolic stress. Our considerations are not made in the provincial perspective of an isolated metabolic system but in the context of a living cell whose existence is threatened.

Many "higher" and "lower" eukaryotic organisms control their cell-division cycle in response to the availability of organic and inorganic nutrients. *S. cerevisiae* responds to starvation for glucose, ammonia sulfate, phosphate, biotin, or potassium by arresting cell division at the unbudded G_1 state (Williamson and Scopes 1962; Johnston et al. 1977a). Johnston et al. (1977b) observed that nutrient-deprived cells situated outside of G_1 were able to complete their cell cycles. However, they neither initiated new division cycles nor completed any of the three known gene-controlled steps (*CDC28*-, *CDC4*-, or *CDC7*-gene product functions) suggested by Reid and Hartwell (1977) to be among the earliest detectable events in division. A thorough analysis of cell-cycle arrest is given in Pringle and Hartwell (1981).

Amino Acid Metabolism during Starvation

Grenson et al. (1970) showed that nitrogen starvation results in a marked increase in general amino acid permease activity. Woodward and Cirillo (1977) extended these observations by describing the fate of amino acids accumulated under these conditions. Basic amino acids were metabolized with their carbon skeletons retained in the cell. Hydrophobic amino acids, such as leucine, isoleucine, valine, methionine, phenylalanine and tyrosine, on the other hand, were taken into the cell, deaminated, and their carbon skeletons excreted. Excretion of transamination products required the presence of glucose and accounted for 50–90% of the radioactive amino acid initially accumulated. Analysis of the medium revealed the presence of α-keto acids and their fusel oil derivatives. These observations prompted Woodward and Cirillo to postulate the existence of a salvage pathway originally shown to exist in cell-free extracts of *S. cerevisiae* by Sentheshanmuganathan (1960). This pathway, as shown in Figure 8, involves transamination of an amino acid to yield its α-keto derivative. Decarboxylation

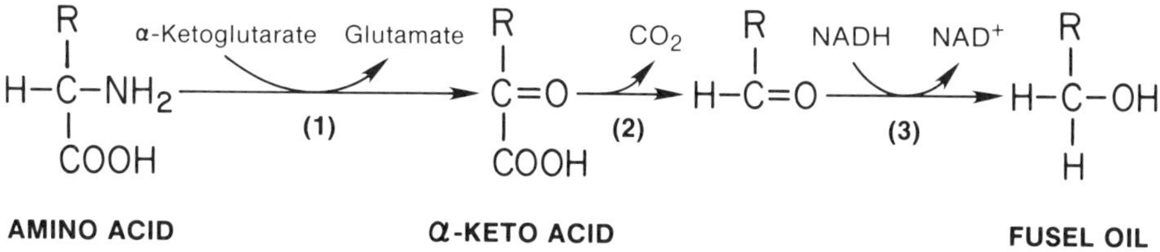

$$\underset{\substack{\text{AMINO ACID}}}{\overset{R}{\underset{\substack{|\\COOH}}{H-C-NH_2}}} \xrightarrow[(1)]{\alpha\text{-Ketoglutarate} \quad \text{Glutamate}} \underset{\substack{\alpha\text{-KETO ACID}}}{\overset{R}{\underset{\substack{|\\COOH}}{C=O}}} \xrightarrow[(2)]{CO_2} \overset{R}{\underset{\substack{|}}{H-C=O}} \xrightarrow[(3)]{NADH \quad NAD^+} \underset{\substack{\text{FUSEL OIL}}}{\overset{R}{\underset{\substack{|\\H}}{H-C-OH}}}$$

Figure 8 Proposed reactions for conversion of amino acid carbon skeletons to fusel oils.

yields an aldehyde that is reduced to a primary alcohol or fusel oil in an NADH-linked reduction. A similar mode of metabolism has been suggested for the amino acid analog, N-chloroacetyl ornithine (Larimore et al., 1980).

Vacuolar and Cellular Homeostasis

As shown above, yeast cells can employ rather drastic measures to scavenge usable nitrogen in the face of a diminished external nutrient supply. However, when deprivation becomes severe, two internal reserves must be tapped; they are the cell vacuole and the turnover of previously synthesized constituents.

Vacuolar Constituents

Biochemical characterization of yeast vacuoles derives in large part from the work of Wiemken, Matile, Schlenk, and their collaborators. Recent reviews by Matile (1978) and Schwencke (1977) are recommended for a comprehensive description of this work. Here, the discussion will focus on a few specific studies that demonstrate the composition and operation of the vacuole as the organelle most concerned with cellular homeostasis. Table 2 lists enzymes that are reported to be situated exclusively or predominantly

Table 2 Enzymes reported to be localized in the vacuole

Protease A
Protease B
Carboxypeptidase Y
Ribonuclease
α-Mannosidase
Leucyl aminopeptidase
Endopolyphosphatase
Exopolyphosphatase
Alkaline phosphatase
Adenosine-5'-monophosphoric acid
 (AMP) phosphatase
Acetylesterase
Acid phosphatase
Exo-β-1-3-gluconase
Invertase
Mg-dependent ATPase
Uricase
Polyphosphate kinase
Catalase A

in vacuoles. Almost without exception, the activities shown are associated with the degradation and salvage of cellular constituents. This correlation prompted Matile and Wiemken (1967) to hypothesize that yeast vacuoles are analogous to the lysosomes of mammalian cells.

Vacuoles also contain a variety of small metabolites. They include basic and other amino acids (Wiemken and Nurse 1973; Wiemken and Durr 1974), S-adenosylmethionine (Svihla et al. 1963; Schlenk et al. 1970, Nakamura and Schlenk 1974), polyphosphates (Indge 1968; Shabalin 1977), allantoin, and allantoate (Zacharski and Cooper 1978). The cellular distributions of amino acids found in one commonly used wild-type strain provided with either ammonia or glutamate as sole nitrogen source are shown in Table 3. Total pool sizes for all of the amino acids found in cells growing on each of 19 different amino acids have also been determined by Watson (1976). In both cases, one sees a marked increase in the pool size of the amino acid provided and those derived from it. Also, with the exception of glutamate, aspartate, and leucine, the majority of each amino acid is sequestered within the vacuole.

Table 3 Amino acid pool distribution in the wild-type strain Σ1278b

	Nitrogen source					
	ammonia			glutamate		
Amino acid	total pool (mM)	cytoplasmic pool (%)	vacuolar pool (%)	total pool (mM)	cytoplasmic pool (%)	vacuolar pool (%)
Aspartate	7.5	73	27	2.15	66	34
Threonine	4.8	25	75	3.75	23	77
Serine	5.8	27	73	8.65	28.5	71.5
Asparagine + glutamine	12.25	15.5	84.5	23.1	20	80
Glutamate	31.3	62	38	59.5	51	49
Glycine	3.8	26	74	3.58	33	67
Alanine	11.2	30	70	8.45	20	80
Valine	3.05	27.5	72.5	5.10	11	89
Isoleucine	0.88	43	57	0.6	33	67
Leucine	0.85	50	50	0.775	35	65
Tyrosine	0.65	27.5	72.5	0.30	42.5	57.5
Phenylalanine	0.4	31	69	0.225	45	55
Ornithine	6.5	9	91	9.65	7	93
Citrulline	8.25	14.5	85.5	1.5	31	69
Lysine	4.20	17	83	6.65	7	93
Histidine	1.60	10	90	1.95	10	90
Arginine	12.2	7	93	18.0	5	95

Data from Messenguy et al. (1980).

Vacuole Structure in Different Stages of the Cell Cycle

Several investigators have followed changes in vacuole morphology as *S. cerevisiae* cells traverse the cell cycle. In one study, Wiemken et al. (1970) made freeze-fracture preparations of synchronously growing cells. They observed that just prior to bud initiation, the few large vacuoles present in the mature mother cell began to shrink and fragment into numerous small vacuoles (Fig. 9). One or more of these small vacuoles migrated into the newly formed bud. There they began to expand and fuse as bud growth continued. As a result, cells ready to complete cytokinesis again contained only a few large vacuoles.

In another study, Moeller and Thomson (1979a,b) followed changes in the vacuolar membrane morphology as cells progressed from exponential to stationary phase. As shown in Figure 10A, the vacuolar membrane is uniformly covered with intramembranous particles in exponentially growing cells. As cells begin the transition to stationary phase, the particles first aggregate (Fig. 10B) and then form a regular geometric pattern (Fig. 10C). Lipid bodies were seen to enter the vacuole through invaginations of the particle-depleted regions (Fig. 11). During uptake, the lipid bodies were enclosed in a membrane derived from the particle-depleted regions of the vacuolar membrane and carried with them a clearly discernible layer of cytoplasm. The fate of these lipid globules remains obscure, but it is reasonable to inquire whether or not they are degraded in the vacuole to form a source of carbon and energy. In sum, these observations point to the vacuole as a dynamic structure in the process of almost constant morphological change.

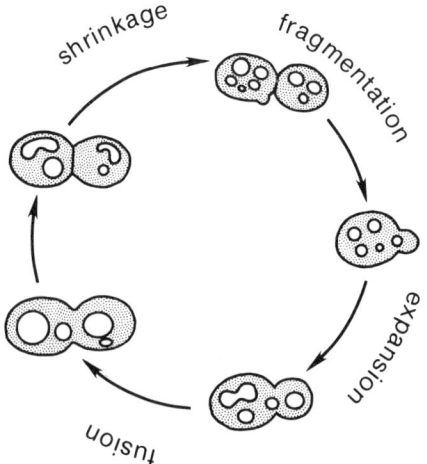

Figure 9 Summary of changes in vacuole morphology observed in yeast cells traversing the mitotic cell cycle.

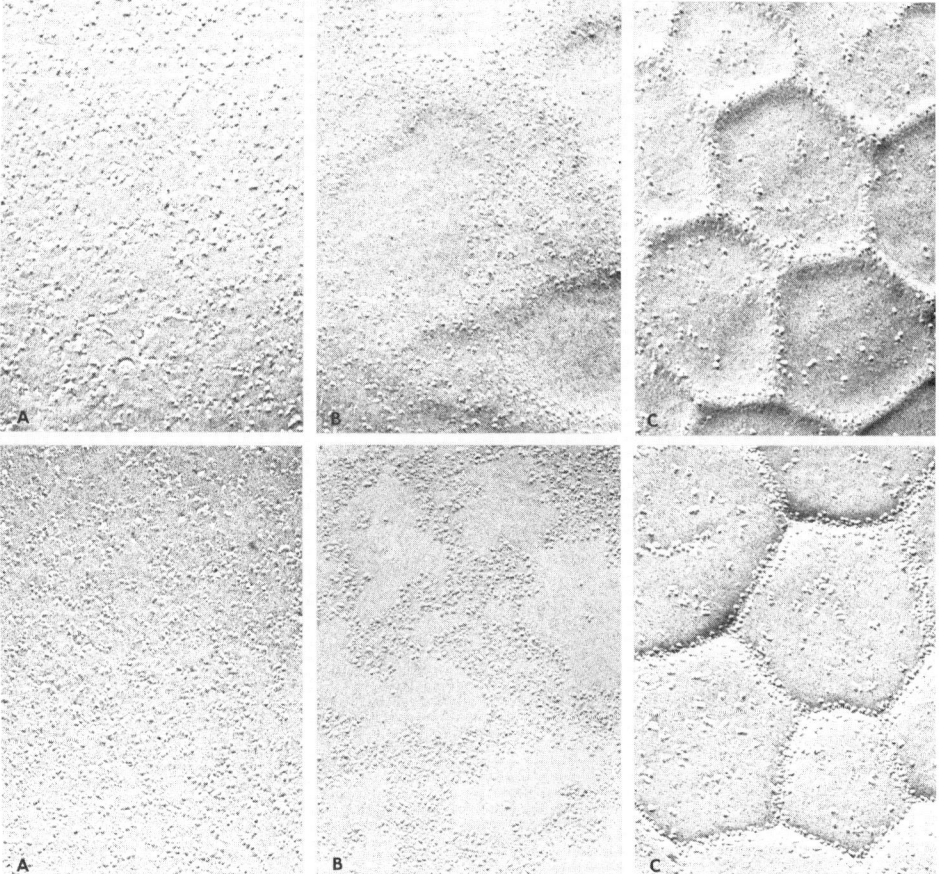

Figure 10 (Top) Representative exoplasmic fracture faces of the tonoplast showing different intramembranous particle patterns. (A) Dispersed pattern; (B) aggregated pattern; (C) geometric pattern. (Bottom) Same patterns as the top, except for the protoplasmic face of the tonoplast. (Data from Moeller and Thompson 1979a.)

Mobilization of Sequestered Metabolites

Regardless of the nitrogen source used for growth, S. cerevisiae normally contains from 1 mM to 10 mM arginine, allantoin, and allantoate. This is orders of magnitude more than necessary to fully induce the arginine- and allantoin-degradative enzymes. The low basal levels of both arginase and urea amidolyase dramatically demonstrate the efficiency of the sequestering process. When excess nitrogen is available, there is little need to catabolize these and other reserve metabolites. However, on nitrogen starvation, these

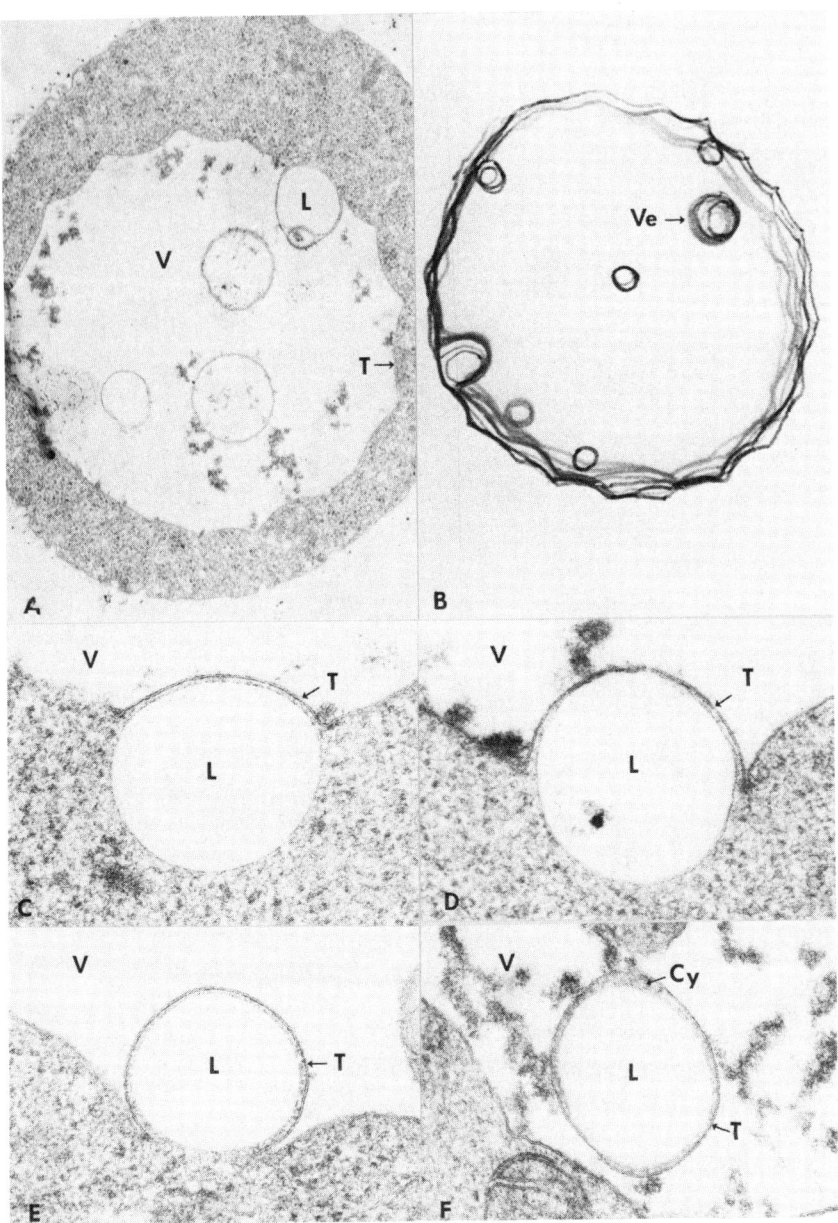

Figure 11 Thin sections of lipid body uptake by the vacuole. (*A*) The vacuole (V) is surrounded by a scalloped tonoplast (T) and has a lipid body associated with it. (*B*) Tracings obtained from nine consective serial sections of a vacuole were superimposed and photographed. A vesicle (Ve) with a lipid body is completely contained within the vacuole. (*C–F*) A probable sequence for uptake of lipid bodies by the vacuole. The lipid body does not fuse with the tonoplast but is enclosed along with some cytoplasm (Cy) in a vesicle derived from the tonoplast. (Data from Moeller and Thompson 1979b.)

large pools of arginine, allantoin, and allantoate are mobilized, resulting in the induction of the arginine- and allantoin-degrading enzymes (Whitney and Magasanik 1973; Sumrada and Cooper 1978b). The observation that blocking urea carboxylation by a *dur1,2* mutation prevents the increase in enzyme activity on starvation eliminates release from nitrogen repression as the basis for the increased enzyme synthesis. Sumrada and Cooper (1978b) also observed mobilization of sequestered metabolites on the treatment of *MATa* cells with α-factor. It will be recalled that this treatment brings about G_1 arrest. These data demonstrate the close integration existing between homeostasis throughout the cell cycle and the control of metabolite sequestration by the vacuole.

Several investigators have reported the induction of arginase following the addition of lysine to the growth medium of wild-type yeast (Whitney and Magasanik 1973; Dubois and Wiame 1976). This induction may derive from either or both of two possibilities. First, it is possible that lysine is an inducer of arginase activity. Intuitively, this suggestion seems improbable since lysine and arginine catabolism share little in common. Alternatively, the addition of lysine to the growth medium could cause an expansion of the intracellular lysine pool, which then displaces vacuolar arginine. Consistent with this suggestion is the observation that the addition of lysine to wild-type cultures growing in minimal ammonia medium results in a 9-fold increase in intracellular lysine levels and a 14-fold decrease in arginine concentrations (Sumrada and Cooper 1978a). Unfortunately, data presently available do not permit rigorous identification of the correct explanation of lysine-mediated induction of arginase.

Starvation-mediated Protein Degradation

A second possible means of coping with nitrogen starvation is the turnover of protein and other nitrogen-rich macromolecules. Sims et al. (1974) and, later, Betz (1976) reported that inhibition of protein synthesis by cycloheximide, loss of initiation and elongation factor function in temperature-sensitive mutants, and amino acid starvation resulted in a modest ($<$ twofold) increase in protein degradation. Bakalkin et al. (1976a,b) observed a similar increase in degradation as cells progressed from logarithmic to stationary-phase growth. In both cases, protein degradation was monitored as solubilization of TCA-precipitable radioactive protein. Since extensive degradation is required before TCA solubility is achieved, the values reported are most likely underestimates of the in vivo situation. The conditions triggering protein turnover have been studied in detail using more sensitive assay (loss of activity for a specific enzyme) (Sumrada and Cooper 1978b). These investigators observed a loss ($t_{1/2}$ = 300 min) of activity following starvation for nitrogen, phosphorus, potassium, glucose, sulfate, or biotin. These observations measuring enzyme activity were confirmed by Lopez and Gancedo (1979), who performed similar starvation experiments and monitored solubilization of TCA-precipitable protein.

To bring these observations together in a way that would generate new questions and suggest new experimental approaches, we proposed the working model in Figure 12. Protein synthesis provides one means of monitoring many nutrients, and those whose loss has been observed to bring about G_1 arrest are known to be associated with this process. This and the results of Unger and Hartwell (1976) led us to suggest that a product or by-product of protein synthesis may serve as the signal function for nutrient sufficiency. This signal or its absence somehow brings about the release of small molecules from the cell vacuole and initiates turnover of protein. Amino acids generated in this way would then be available for synthesis of new proteins. Concurrent onset of these processes and cell-cycle arrest likely represent a protection mechanism providing cells with an internal means of obtaining nutrients needed to complete cell division in the face of an adverse external supply of these compounds. It can be reasonably argued that signals resulting from inhibition of protein synthesis and nutrient starvation might be the same. However, similar responses were also observed after treatment with α-factor. Since protein synthesis remains functional under this circumstance, two possibilities are opened: α-factor may operate by a distinct mechanism that brings about the same result as loss of protein synthesis, or it may specifically interact with the signal at the level of protein synthesis. Whether the signal is the production of a compound or the failure to produce it is also unknown at this time. However, we now have an event (turnover of allophanate hydrolase) that is triggered by the signal and can be

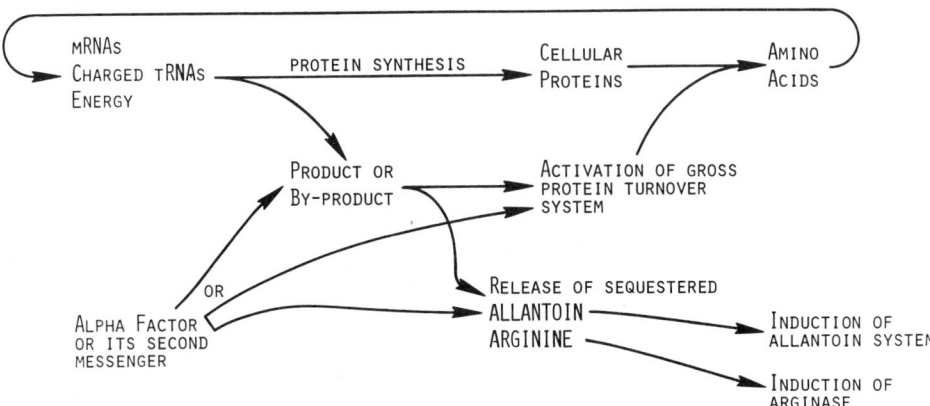

Figure 12 A model hypothesizing the control circuits that might operate in the integration of metabolic homeostasis and cell-cycle G_1 arrest mechanisms. (Data from Sumrada and Cooper 1978a.)

sensitively assayed in the absence of protein synthesis. These capabilities may provide an effective means to approach this complex problem.

Yeast Proteinases

As described in the preceding section, nitrogen starvation induces protein turnover, presumably mediated by one or more specific proteinases. The particular proteinases involved in this process have not been identified. However, as an initial step toward examining this process, the proteinases of yeast have been characterized biochemically and genetically.

Three well-characterized proteinases have been isolated from yeast: proteinases A and B (Hata et al. 1967a,b; Lenney and Dalbec 1967; Saheki and Holzer 1974; Ulane and Cabib 1976; Looze et al 1979; Sanada et al. 1979) and carboxypeptidase Y, previously called proteinase C (Doi et al. 1967; Hata et al. 1967b; Hayashi et al. 1970, 1973; Kuhn et al. 1974). Proteinase A is an endoproteinase with an acidic pH optimum. This glycoprotein is approximately 60 kD, about 10% of which is contributed by carbohydrate. Proteinase B is a serine endoproteinase with a neutral pH optimum. Estimates for the glycoprotein range from about 31 kD (Lenney and Dalbec 1967; Sanada et al. 1979) to 44 kD (Ulane and Cabib 1976; Looze et al. 1979; Zubenko 1981; G.S. Zubenko and E.W. Jones, in prep.). On the basis of the determination of 44 kD, the carbohydrate component comprises less than 0.5% of proteinase B (Ulane and Cabib 1976). Carboxypeptidase Y is a serine carboxypeptidase. This glycoprotein is approximately 60 kD, about 10 kD of which is contributed by carbohydrate. It is likely that the carbohydrate component of carboxypeptidase Y consists of four asparagine-linked oligosaccharides of the general formula (N-acetyl-glucosamine)$_2$-(mannose)$_{13}$ (Hasilik and Tanner 1978a). Although all three of these proteinases are located in the yeast vacuole (Cabib et al. 1973; Wiemken et al. 1979), peptide inhibitors of proteinase A (Lenney et al. 1974; Matern et al. 1974), proteinase B (Lenney et al. 1974; Matern et al. 1974), and carboxypeptidase Y (Matern et al. 1974) are found in the yeast cytosol.

Other proteinases exist in yeast, but they have not been as well characterized. Carboxypeptidase S is a carboxypeptidase of neutral pH optimum. Unlike carboxypeptidase Y, it is not a serine proteinase and has a requirement for a divalent cation (Wolf and Weiser 1977). Felix and Brouillet (1966) have described two yeast carboxypeptidases, which they called α and β. It is likely that carboxypeptidase α corresponds to carboxypeptidase S, and carboxypeptidase β corresponds to carboxypeptidase Y. Matile et al. (1971) have demonstrated the existence of at least four amino peptidases from yeast on the basis of differences in electrophoretic mobility and cation requirements.

Carboxypeptidase Y, proteinase A, and possibly all vacuolar hydrolases in yeast are synthesized as inactive precursors (Hasilik and Tanner 1976, 1978a,b; Hemmings et al. 1981; Zubenko et al. 1983). The biosynthesis of carboxypeptidase Y requires at least two posttranslational modification events. The first precursor is approximately 65 kD and is a glycoprotein containing approximately four endo-H-sensitive core carbohydate moieties. This precursor is converted to a larger, 67-kD intermediate by the addition of sugars to the core carbohydrates. A carbohydrate-free aminoterminal peptide is then cleaved from this intermediate to form the active 59-kD carboxypeptidase Y molecule (G.S. Zubenko et al., in prep.). A similar sequence occurs in the biosynthesis of proteinase A. A 62-kD precursor is converted to a 65-kD intermediate, which is subsequently converted to the active 58-kD proteinase-A molecule. Although carbohydrate comprises about 10% of the proteinase-A molecule, neither of its precursors contains endo-H-sensitive carbohydrate moieties. For both carboxypeptidase Y and proteinase A, conversion of the smaller to the larger percursor and conversion of the larger precursor to the mature, active enzyme occur with half-lives of about 5 minutes (G.S. Zubenko et al., in prep.).

Genetics of Intracellular Proteolysis

Zubenko et al. (1979) have described four genes *PRB1, PRB2, PRB3*, and *PRB4* whose functions are required for the production of proteinase-B activity in yeast. Of these, *PRB1* is the structural gene for proteinase B (Zubenko et al. 1980). The *PRB1* locus is situated 1.1 cM proximal to the *CAN1* locus on the left arm of chromosome V (Zubenko et al. 1980).

Mutations in any of three genes, *BIN1, BIN2*, or *BIN3*, are recessive and result in about 50% diminution in the level of proteinase-B inhibitor activity. Although mutations at any one of these loci result in an inability to use glycerol as a carbon source at 37°C, this phenotype does not derive from excess proteinase B, because double-mutant strains with mutations in both the *BIN* and *PRB1* loci exhibit the aberrant phenotype. The *BIN* loci are not closely linked to one another, to *PRB1*, or to a centromere (Zubenko 1981).

Two mutations have been isolated that result in 90-95% loss of proteinase-A activity without marked diminution of proteinase B or carboxypeptidase-Y levels. Preliminary results suggest that the two mutations are allelic and define the *PRA2* locus (Zubenko 1981; Zubenko et al. 1983).

Wolf et al. (1979) have described a mutation (unnamed) that results in a 70% decrease in the level of proteinase inhibitor activity present in cells grown at 30°C and low pH. However, the inhibitor obtained from these cells grown under restrictive conditions is not differentially sensitive to thermal denaturation or low pH. These investigators also described a mutant, *cps1*, which imparts a deficiency for carboxypeptidase-S activity.

Jones (1977) has described 16 genes whose functions are essential for the production of carboxypeptidase-Y activity. Among these were *PRC1*, the structural gene for carboxypeptidase Y (Wolf and Weiser 1977), and *PEP4*. Mutations in the *PEP4* gene are recessive and affect the expression of at least five vacuolar hydrolases. The *pep4-3* mutation results in a 90–95% reduction in the levels of activity of proteinases A and B, carboxypeptidase Y, RNase, and nonspecific repressible alkaline phosphate (Hemmings et al. 1981; Zubenko 1981; Zubenko et al., 1983). The *PEP4* and *PHO9* genes appear to be identical (Kaneko et al., in prep.). A functional *PEP4* gene is required for conversion of the larger of the two proteinase-A and carboxypeptidase-Y precursors to their respective mature, active forms (Hemmings et al. 1981; Zubenko 1981; Zubenko et al. 1983). However, conversion of both smaller precursors to the respective larger precursors does not depend on *PEP4*-gene function (G.S. Zubenko et al., in prep.).

The action of the *PEP4*-gene product is not so general, however, that it is required for expression of all *N*-acetylglucosamine-type glycoproteins or enzymes that must traverse membranes before reaching their final destinations. Secreted enzymes such as invertase or acid phosphatase, which are localized in the periplasm (Sutton and Lampen 1962; McLellan and Lampen 1963), do not seem to share the same posttranslational modification pathway because *pep4* mutations do not affect the secretion or activity of these enzymes (Zubenko et al. 1983; G.S. Zubenko and E.W. Jones, in prep.).

Effects of Nitrogen Metabolism on Sporulation

The transfer of diploid yeast cells to potassium acetate results in sporulation. This can be largely prevented, however, if ammonia is added to the medium (Miller 1963). Inhibition was also observed when a variety of amino acids were used. Surprisingly, the degree of inhibition is not correlated with either the relative usefulness of the amino acid as a nitrogen source or its ability to generate ammonia. For example, urea and glutamate are direct ammonia precursors but are reported to be less inhibitory than methionine or reduced glutathionine. Piñon (1977) reported strong inhibition by the ammonia analog methylamine and concluded that the primary inhibitory effect was exerted by ammonia itself rather than some metabolite generated from ammonia. Supporting this conclusion is Newlon's report (1979) that homozygous *gdh1* diploids sporulate to the same degree as wild-type. However, it must be kept in mind that these NADP-specific GDH-defective strains retain a good deal of their capacity for ammonia assimilation (see earlier sections of this paper).

Although one can make a reasonable case for ammonium ions acting as a primary inhibitory signal for sporulation, it may also be reasonable to consider that the triggering of sporulation by nitrogen starvation is a matter of metabolite balance. This consideration is suggested because levels of ammonia that inhibit sporulation of cells incubated in acetate by 92% have no effect when sporulation is carried out in dihydroxyacetone. It should be pointed out, however, that ammonia is a poor nitrogen source when dihydroxyacetone is provided as carbon source.

There is little rigorous mechanistic evidence concerning the inhibition of sporulation by nitrogenous compounds. Opheim (1979) reported that the addition of 10 mM ammonium sulfate to sporulation medium blocked the normal increase in proteases B and C and 1,4-amylglucosidase activities normally seen during sporulation. This raises the question, do these proteases increase in response to sporulation or to nitrogen starvation? Piñon (1977) and Durieu-Trautmann and Delavier-Klutchko (1977) reported that ammonia inhibition is exerted at two stages of sporulation, the first during DNA synthesis and the other later in sporulation.

Two sets of mutants able to sporulate in the presence of rich medium have been isolated. The first set of strains (cdc25 and cdc35) were isolated as temperature-sensitive, cell-cycle mutants that arrested growth at the unbudded G_1 state on shifting the cultures to nonpermissive temperatures (Shilo et al. 1978). A unique feature of these strains is their ability to sporulate in rich medium maintained at the nonpermissive temperature. The second strain (spd1) also sporulates in the presence of rich medium. The response of this strain to nitrogen repression has been assessed and was found to be identical to wild type (Vezihet et al. 1979). The investigators suggest that the spd1 mutant probably affects a central metabolic function concerned with the utilization of some, but not all, nonfermentable carbon sources. This induces a form of starvation that results in G_1 arrest and sporulation. If the spd1 mutant can be shown to lack a specific enzyme associated with carbon metabolism, it would again point to the balance of metabolites as an important consideration in the relationship between nitrogen metabolism and sporulation.

INTEGRATION OF NITROGEN METABOLISM

Short-range Modulation: Induction and Compartmentation

Successful integration of nitrogen metabolism in response to a constantly changing environment requires the capacity for rapid and drastic modulation of component enzyme systems. A high degree of buffering is also needed if over-response to temporary perturbation is to be avoided. All of these characteristics are found in the nitrogen catabolic systems studied thus far. Only 3–5 minutes elapse between the addition of arginine or urea to growth medium and the appearance of the arginine- and allantoin-degrading enzymes (Lawther and Cooper 1973, 1975; Bossinger and Cooper 1977).

The rapidity of induction is emphasized by the fact that the time required is the same as that needed for the induction of β-galactosidase in *Escherichia coli*, where transcription and translation are coupled. In the ensuing few minutes, both arginase and urea amidolyase levels increase 30–50-fold.

S. cerevisiae has evolved ways of selectively degrading the more useful nitrogen sources first. There are at least two means of making this selection: repression and metabolite compartmentation. In the past, most effort has focused on nitrogen catabolite repression as the preeminent selection mechanism. Increasingly complicated models attempt to explain the biochemical links between varying environmental conditions and the observed responses to them. These models will be discussed in detail later. It is possible, however, that repression, though important, is not the preeminent control mechanism. This proposal arises from our observations of the allantoin-degradative system.

The addition of asparagine to yeast cultures induced for allantoin degradation results in the rapid loss of urea amidolyase mRNA, with a half-life of about 3 minutes (Lawther and Cooper 1975). In addition, we have found that the addition of asparagine brings about a similar loss of mRNA transcribed from *DAL, DUR,* and *CAR* genes (Sumrada and Cooper 1982c; T.G. Cooper et al; F. Genbauffe and T.G. Cooper; both in prep.). Thus, repression at the mRNA level in response to asparagine addition is admittedly rapid. Nonetheless, the end result of this repression—namely, a reduction in enzyme levels—occurs much more slowly. Only after 6 hours or three generations following the addition of asparagine does the level of urea amidolyase drop to 10% of its fully induced level. In this time frame it is difficult to argue that repression provides the means for rapid, fine control of metabolic integration. Rather, such control most likely occurs at the level of intracellular/extracellular compartmentalization. For example, 5 minutes after the addition of asparagine to culture medium, cells show a 30-fold decrease in their ability to accumulate allantoin from the medium (Cooper 1982; R. Sumrada and T.G. Cooper, in prep.). Thus, it appears that control of allantoin metabolism is exerted not by loss of enzyme activity but, rather, by exclusion of its substrate from the cell.

Sequential Induction—A Means of Achieving Sensitive, Buffered Metabolic Control

It is difficult to achieve metabolic control that is at once exquisitely sensitive to changing environmental conditions yet refractile to over-response caused by minor or temporary perturbations. Nonetheless, such regulation seems to exist for control of the enzymes degrading allantoin, urea, arginine, and proline. The objective of this section is to describe the induction patterns for these enzymes and to point out the simultaneous sensitivity and buffering this regulation affords. An overview of this area of metabolism is shown in Figure 13.

Sensitivity to the external environment is maximized by transport systems for allantoin, allantoate, urea, arginine, and proline. All of these transport systems are either expressed constitutively or induced by the transported metabolite. Moreover, each system is capable of concentrating metabolites by approximately 200–25,000-fold. On the other hand, all of the transport systems are efficiently regulated by transinhibition (see Cooper [Transport in *Saccharomyces cerevisiae*], this volume) over the short term and probably by nitrogen repression over the long term.

The catabolism of arginine provides a good example of sequential induction. This term refers to a regulatory program of a metabolic pathway in which the metabolite at any particular branch point in the pathway is itself the inducer of the enzymes required for subsequent degradation. Specifically, if arginine is provided to the cell, it has two metabolic fates: It may be incorporated into protein, or it may induce synthesis of arginase and be degraded. Partitioning at this point is further controlled by the relative affinities for arginine of arginine aminoacyl tRNA synthetase and arginase. This means that the cellular arginine requirement will be satisfied prior to any degradation of the amino acid. Degradation of arginine by arginase produces urea and ornithine. Catabolism of urea is controlled by the levels of its immediate metabolite, allophanate, and is discussed below. The other product of arginine catabolism, ornithine, is converted to glutamate semialdehyde by ornithine transaminase, an enzyme also induced by arginine. Ornithine is not at a branch point of catabolic pathways and, hence, sequential induction of the enzymes degrading it offers no regulatory advantage. Glutamate semialdehyde is converted to proline, which is either used for biosynthesis or converted to glutamate, following induction by proline of the appropriate catabolic enzymes. Again, the relative K_m values of proline aminoacyl tRNA synthetase and proline oxidase most likely ensure full charging of proline tRNA before any proline is degraded. In this manner the amino acid carbon skeleton of arginine is not degraded to glutamate until the cellular requirements for both arginine and proline have been met.

As evident in Figure 13, urea is a product not only of arginine catabolism but also of allantoin metabolism and thus lies at a branch point in the metabolism of both of these compounds. However, it is allophanate, the immediate metabolic product of urea, rather than urea itself, whose levels regulate the metabolism of urea. This compound also serves to regulate the level of enzymes required to convert allantoin to urea. On the basis of the relative metabolic distance between allantoin and allophanate, one might envision difficulties in this scheme for induction of allantoin-degradative enzymes. However, the high basal levels of allantoin catabolic enzymes assure the ready conversion of allantoin to urea, the subsequent synthesis of

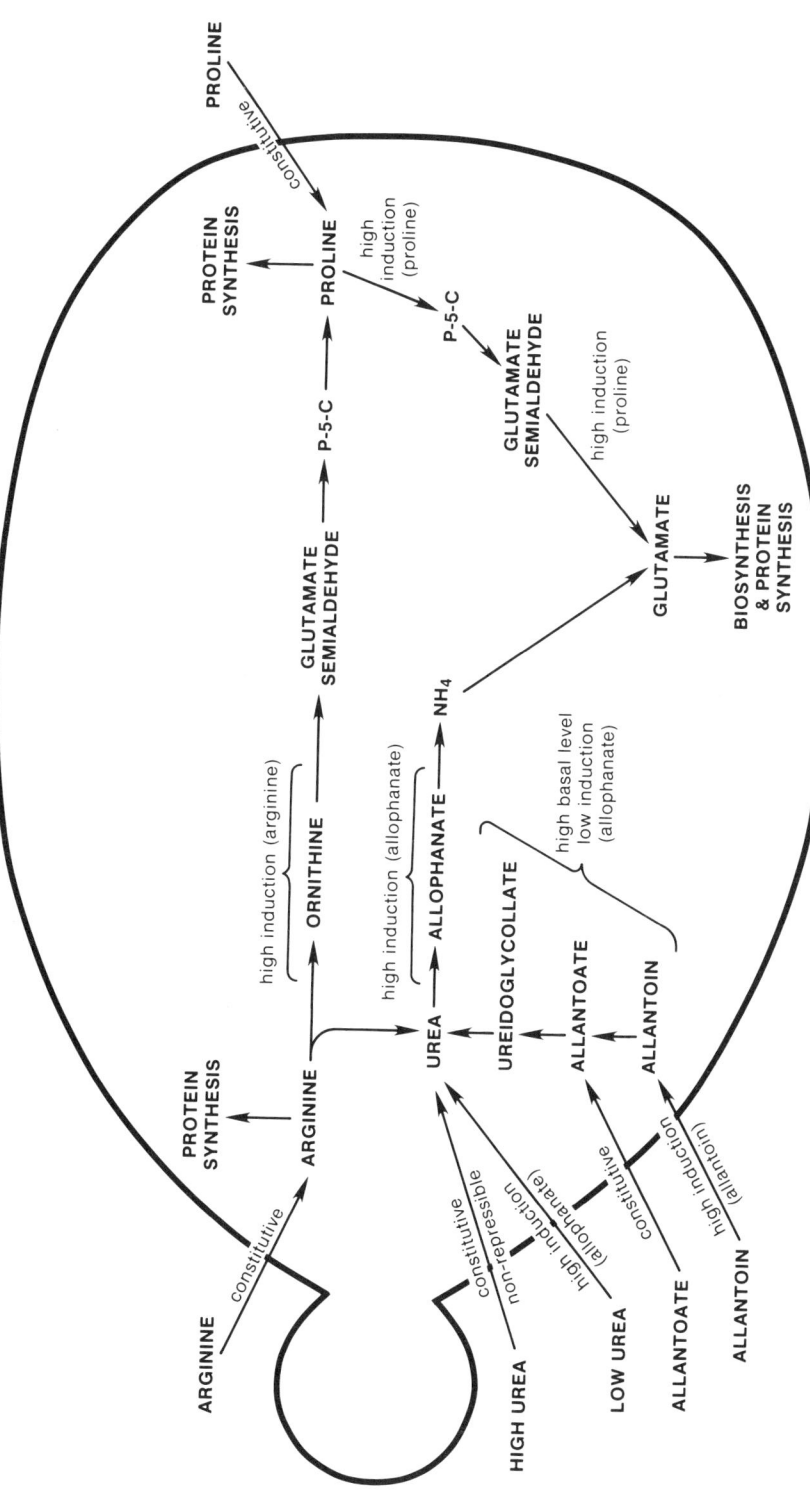

Figure 13 Summary of control patterns observed for the transport and catabolism of arginine, proline, and the allantoin pathway intermediates.

the inducer, allophanate, and the ultimate increase in the pathway to the fully induced level.

The specific nature of allophanate metabolism assures that the catabolism of allantoin and/or urea is highly buffered. Allophanate is degraded five times faster than it is synthesized. This relationship is maintained under all levels of induction, since, as discussed previously, the synthetic and degradative activities are components of the same protein. Therefore, the urea-degradative enzymes will be induced only if the concentration of a pertinent precursor increases and remains above a threshold level. Such a situation would exist only during periods of severe nitrogen limitation or following the addition of arginine or allantoin to the growth medium. In the case of severe nitrogen limitation, the large vacuolar nitrogen reserves would become available. Arginine thus released would induce arginase, thereby generating urea and, subsequently, allophanate.

In the case of allantoin addition, sequential induction operates somewhat differently with regard to generation of inducer. Allantoin reaches the cytosol either from the vacuole or by an active transport system induced by allantoin itself. Once in the cell, it is rapidly converted to urea by the high basal levels of the first three allantoin-degrading enzymes. Hence, when sufficient allantoin is present, it can be rapidly degraded to allophanate, which in turn brings the pathway to full capacity.

Long-range Modulation: Nitrogen Catabolite Repression

S. cerevisiae is able to discriminately degrade high-quality nitrogen sources over poor ones. The classification of a nitrogen source as good or poor, however, is not always trivial and is usually strain- and growth-condition-dependent. For most strains, glutamine and asparagine are high-quality nitrogen sources, whereas proline and allantoin are poor ones. Ammonia, often considered a good nitrogen source, is a special case that will be discussed later. At least three characteristics are needed for a compound to be a good nitrogen source. The compound must (1) enter the cell rapidly; (2) be converted rapidly, in as few steps as possible, to glutamate and/or ammonia; and (3) have no toxic side effects on the cell. Anything in the genetic makeup of the strain or its culture conditions that abridges these characteristics will reduce a compound's usefulness.

The allantoin pathway provides a good example of these principles in operation. As shown in Figure 1, allantoin is degraded in three steps to urea, which is then converted in two steps to ammonia. The limiting reaction in the pathway is that of urea carboxylase. Thus, one would expect that urea and allantoin would be equivalent nitrogen sources. However, urea, at a concentration of 10 mM, is considerably better than allantoin. The primary reason for this is that allantoin active transport is much slower than urea uptake by facilitated diffusion. In fact, if the urea concentration is lowered to 0.25 mM, where its entry is mediated only by a much slower active transport system, the two nitrogen sources become equivalent. A second factor can also influence the relative merits of allantoin and urea as nitrogen sources. Allantoin must traverse three reactions before yielding urea. If these enzymes are limiting, then the rate at which ammonia will be produced is decreased. Thus, in this aspect the quality of allantoin as a nitrogen source is reduced. The reason for citing this example is to emphasize that a decrease in the rate of producing biosynthetic precursors, regardless of the reason, reduces the "quality" of a potential nitrogen source.

Nitrogen catabolite repression is one manifestation of the selective use of high-quality nitrogen sources. The hallmark of nitrogen repression is the reduction of catabolic enzyme levels or transport activity for a poor nitrogen source when cells are provided with both high- and low-quality nitrogen sources together compared to that found when the poorer source is provided alone. For example, Middlehoven (1964) reported threefold less arginase and fourfold less ornithine-transaminase activity in wild-type yeast provided with arginine plus ammonia compared to the same cells growing on arginine alone. Although the degree of nitrogen repression in this study is not particularly dramatic, such ammonia repression can reach several hundredfold, depending on the growth conditions, the gene, and the strains under study. Observations parallel to those of Middlehoven have been reported for NAD-GDH (Hierholzer and Holzer 1963), the general amino acid permease (Grenson et al. 1970), ureidosuccinate transport (Drillien and Lacroute 1972), the allantoin-degrading enzymes and transport systems (Bossinger et al. 1974; Cooper and Sumrada 1975; Sumrada et al. 1978; Turoscy and Cooper 1979), proline permease (Brandriss and Magasanik 1979a), and asparaginase II (Dunlop et al. 1980). In the case of transport systems, the presence of nitrogen catabolic repression must be viewed cautiously, because it is rarely, if ever, convincingly distinguished from feedback or transinhibiton.

Biochemical models attempting to link the presence of a high quality nitrogen source in the medium to decreased enzyme production have become increasingly complex and controversial. In view of this controversy, it seems most appropriate to present both the pertinent experimental observations and the conflicting interpretations of them. The process of repression and the models proposed to explain it will be divided into a series of questions as a means of simplifying the discussion.

At What Level is Nitrogen Catabolite Repression Exerted?

Central to a discussion of repression is the question of where it occurs. Indirect measurements from the allantoin system suggest that repression occurs at the level of mRNA production. The addition of asparagine to a culture of wild-type yeast fully induced for allophanate-hydrolase synthesis results in a loss of ability for continued enzyme production (Lawther and Cooper 1975). The half-life for loss of this synthetic capacity is 3 minutes, which is the same as that found following removal of inducer from the culture medium. In addition, asparagine affects only enzyme-forming capacity; it has no effect on previously formed synthetic capacity. More recently, we have shown that there is a marked loss of mRNA capable of hybridizing to DNA containing the *DAL1, DAL2, DAL3, DAL4, DUR1,2,* and *CAR1* genes when a high-quality nitrogen source is present in the medium compared to that observed when a poor nitrogen source is provided (Cooper 1982; Sumrada and Cooper 1982c; T.G. Cooper et al.; F. Genbauffe and T.G. Cooper; both in prep.). Although quantitative mRNA half-life measurements are still needed, the present results continue to be consistent with the hypothesis that repression occurs at the level of transcription.

Do the Elements Mediating Induction and Repression Interact with One Another?

Since many of the nitrogen-repressible enzymes are also inducible, it is important to ascertain whether or not the processes of specific induction and general nitrogen repression are distinct from one another. In different terms, we must know whether the protein control elements mediating induction and repression and the target sites with which they interact are the same or different.

It is possible to address this question by asking whether or not putative regulatory mutations affecting induction of catabolic systems *in trans* also affect nitrogen repression. Two such mutant classes have been characterized in the allantoin-degradative system and four in the arginine system. Mutations in the *DAL80* locus result in constitutive production of all five allantoin-degrading enzymes. However, this constitutive enzyme production is fully repressed when glutamine, asparagine, or ammonia is provided as a nitrogen source (Cooper 1980, 1982; Chisholm and Cooper 1982). Mutations in the *DAL81* locus result in the loss of ability to induce these enzymes. But again, derepression of the basal activity by growth in proline medium remains the same as in the wild type (Cooper 1982; Turoscy and Cooper 1982). In the arginine system, growth of wild-type cells on ammonia plus arginine results in a loss of 91% of the arginase activity found in cells provided with arginine alone. In mutant strains producing arginase and ornithine transaminase constitutively (*car80*), there is a 76% loss under these conditions (Wiame 1971). Mutations in the *ARG80, ARG81,* or *ARG82* loci result in loss of ability to induce arginase. However, comparison of the arginase levels found in these strains when glutamate replaces ammonia as nitrogen source reveals a ninefold derepression. For the wild type, sixfold derepression was observed (Wiame 1973). These data argue that the *trans*-acting elements regulating induction of arginine and allantoin catabolic enzymes function independently of repression.

This conclusion is also supported by direct hybridization experiments using the *CAR1, DAL4, DAL3,* or *DUR1,2* genes as probes. mRNA synthesis from the *DAL* or *DUR* genes is fully repressed when a wild-type or *dal80* mutant strain is grown in the presence of a repressing nitrogen source. Conversely, production of these RNAs is increased when a *dal81* mutant is transferred to a derepressing nitrogen source such as proline (T.G. Cooper et al.; F. Genbauffe and T.G. Cooper; both in prep.). The same behavior is observed for *CAR1* mRNA synthesis in the wild type and *car80* mutant strains (Sumrada and Cooper 1982c).

Dubois and Wiame (1978) hypothesized that the arginase gene is regulated by an interaction between the element responsible for nitrogen repression and that responsible for specific induction. A similar but more complex hypothesis is proposed to account for repression of the enzymes of urea degradation (Lemoine et al. 1978). However, the *cis*-acting mutations (*cargA$^+$O^h* and *durO^h*) that provide sole support for this hypothesis have been shown to be insertions of Ty-like elements (Sumrada and Cooper 1982c, and in prep.). Such Ty-like elements have been shown to bestow anomolous regulatory properties to genes next to which they are inserted (Errede et al. 1980; Jones and Fink, this volume). Thus, there is not yet substantive evidence to suggest an interaction of regulatory sites at *CAR* and *DUR* loci.

Does the NADP-GDH Molecule Itself Participate as a Repression Regulatory Element?

A detailed understanding of nitrogen repression requires identification of the protein elements and effector metabolites mediating this process. Thus far, several proteins and metabolites have been nominated as regulatory elements participating in a highly complex scheme of repression. The object of this and subsequent sections is to examine the data from which these claims evolved and alternative interpretations of the data.

Wiame and his colleagues have proposed that "$^+$NH$_4$ is the signal of nitrogen catabolite repression, and the NADP-glutamate dehydrogenase being its receptor act[s] as a regulatory protein" (Dubois et al. 1977). The initial observations that eventually prompted this proposal were made by Grenson and Hou (1972), who found that ammonia inhibition of general amino acid permease activity was relieved in mutant strains lacking NADP-GDH activity (*gdh1*). The relief of "ammonia inhibition" observed in *gdh1* mutants was further extended by Dubois et al. (1973) to include the catabolic enzymes arginase, allantoinase, and urea amidolyase. In addition, these investigators found that mutations in the *GDH1* locus resulted in derepression of arginase synthesis in both *arg80–arg82* (noninducible) and *car1$^+$O$^-$* (constitutive) strains (Dubois et al. 1974; the latter mutants are phenotypically distinct—though still harboring *cis*-acting constitutive mutations linked to the structural gene—from those carrying the *car1$^+$O^h* mutations discussed earlier). From these results, Dubois et al. (1974) concluded that "the *gdh1* mutation suppressed the nitrogen catabolite repression of arginase independently of its effect on [inducer] permeability or its effect on growth rate."

The above model of nitrogen catabolite repression proposed by Wiame and his colleagues (Dubois et al. 1973) has been widely disputed. The first question was raised by Van de Poll (1973), who challenged the initial suggestion of Dubois et al. (1973) that ammonia was the important metabolic signal for nitrogen repression. Van de Poll, like the Belgian group, found that ammonia repressed allantoinase production by 11-fold in his wild-type strain, but it was an ineffective repressing agent in a corresponding *gdh1* mutant that he isolated. In contrast, he found asparagine, glutamine, and glutamate plus ammonia to repress enzyme production by up to 20-fold and to do so equally in both wild-type and *gdh1* mutant strains. From this, Van de Poll concluded that "the presence of ammonia in itself is not enough for ammonium repression.... It must be metabolized before it [ammonia] can exert a repressing effect." Bossinger et al. (1974) independently reached the same conclusion after finding no correlation whatever between the levels of intracellular ammonia generated by various nitrogen sources and their ability to bring about repression of allantoinase and allophanate-hydrolase production. The latter investigators (Bossinger and Cooper 1975) also determined the ability of various amino acids to repress allophanate-hydrolase production in the same wild-type and *gdh1* mutant strains used by Wiame and his colleagues. They found that "in strains carrying the *gdh1-6* mutation, repression was alleviated only when metabolism of the nitrogen source involved the NADP-GDH enzyme activity." Glutamate and aspartate, e.g., were more repressive nitrogen sources in the *gdh1* mutant than they were in the wild type. Similar observations (again using the strains of Wiame) were reported for asparaginase II by Dunlop et al. (1980).

The central and most controversial element of Wiame's model is his interpretation that the NADP-GDH molecule itself functions as a "regulatory element," rather than acting catalytically in the anabolism of ammonia. The strongest evidence in support of this position is the loss of ammonia repression in *gdh1* mutants or aconitase-negative mutants, which are unable to synthesize α-ketoglutarate (Ogur et al. 1964; Dubois et al. 1974). However, these observations are equally well interpreted as evidence supporting the position that relief of repression is seen in these strains because they can no longer metabolize ammonia. In fact, the conditions needed for ammonia repression are precisely the same as those required for NADP-GDH activity. As Dubois et al. (1974) pointed out, there is a highly linear relationship between the amount of NADP-GDH activity seen in leaky *gdh1* mutants and the severity of ammonia repression observed in them. These striking correlations make it difficult to ignore the catalytic activity of NADP-GDH as an explanation for the observed phenotypes of mutants lacking aconitase and NADP-GDH. The interpretation of Dubois et al. (1974) that these observations were indicative of "an identity of the regulatory and catalytic sites" seems more speculative than this alternative, especially in the absence of direct biochemical or genetic evidence concerning the hypothesized regulatory and catalytic sites.

Results obtained by Roon et al. (1975) appear to eliminate the possibility that NADP-GDH plays a direct regulatory role in ammonium repression of the general amino acid permease. They measured general amino-acid-permease activity in three strains of yeast provided with ammonia: a wild-type mutant, a *gdh1* mutant, and a *gdh1 ure2* double mutant. Mutations in the *ure2* locus were shown by Drillien et al. (1973) to result in a 30-50-fold increase in NAD-dependent GDH. Roon and his colleagues observed strong ammonia inhibition of the permease activity in the wild-type strain and relief of this inhibition in the *gdh1* mutant. However, ammonia inhibition was restored in the double mutant. The investigators reasoned that although the latter strain was still missing the anabolic (NADP) enzyme, it contained derepressed levels of NAD-GDH. The derepressed catabolic enzyme was concluded to be an adequate functional replacement for the defective anabolic enzyme, because a normal growth rate was restored to the double mutant provided with ammonia. The simultaneous improvement of growth on ammonia and restoration of ammonia sensitivity in this strain (which still possessed a defective form of NADP-GDH) prompted the conclusion that "the NADP-dependent glutamate dehydrogenase function[ed] in the inhibition of [general] amino acid transport by virtue of its catalytic capacity for incorporating ammonium ion into glutamic acid and not as a regulatory element."

As mentioned above, the GDH regulatory model proposed by J.-M. Wiame and his colleagues was initially constructed in an attempt to understand how ammonia inhibited the general amino acid permease (Grenson and Hou 1972). Throughout ensuing reports, the function of NADP-GDH in nitrogen repression was implied to be the same for the general amino acid permease and the enzymes arginase, allantoinase, and urea amidolyase. This basic implication, however, may be incorrect. Current data for nitrogen repression suggest that regulation is exerted at the level of transcription (see the two previous sections). In contrast, Grenson et al. (1970), in their initial description of how ammonia inhibited the general amino acid permease, concluded that "ammonium ions inhibit the activity of the general amino acid permease, but not its synthesis." If this conclusion is substantiated by hybridization experiments, it is difficult to conceive how regulation at the level of transport activity and gene transcription could occur by a common mechanism involving common elements. Dunlop et al. (1980) also concluded from their data that "the differences between these systems (general amino acid permease and asparaginase II) are quite significant." Hence, the extrapolation of transport data to an understanding of how enzyme production is repressed may be risky. This does not eliminate the possibility of the general amino acid permease being subject to multiple levels of regulation—repression of its production and inhibition of its activity. However, experiments needed to reach this conclusion have not yet been reported.

Finally, a set of observations by Dunlop et al. (1980) also raise doubt about the direct participation of NADP-GDH as a nitrogen-repression regulatory element. These investigators compared the ammonia-, methylamine-, and alanine-concentration dependence of asparaginase-II repression in wild-type and *gdh1* mutants. Methylamine- and alanine-concentration-dependent decreases in asparaginase-II activity were the same in wild-type and *gdh1* mutant strains. For low concentrations of ammonium sulfate, "repression" (i.e., the decrease in asparaginase-II activity) was relieved in the *gdh1* mutant but was still present in both strains at higher concentrations of ammonia. The conclusion in all three cases is that "repression" of asparaginase-II activity can be elicited even though the putative element mediating repression (NADP-GDH) has been rendered inactive by mutation.

Does the URE2-*gene Product Act as a "General Nitrogen Repressor?"*

Dubois et al. (1974, 1977) proposed that "$^+NH_4$ is the metabolic signal received by the (NADP) glutamate dehydrogenase which transmits the signal to the gene [those sensitive to nitrogen repression] with the participation of a molecule issued from the gene [*URE2*]." This proposal, which hypothesizes a direct regulatory role for the *URE2*-gene product, was based on the characteristics of strains harboring lesions in the *URE2* locus; these strains were first isolated by Drillien and Lacroute in 1972 and by Drillien et al. in 1973. The mutants were isolated as follows. Uracil auxotrophs carrying a lesion in the *URA2* locus (aspartate-transcarbamylase-deficient) are able to grow in proline minimal medium supplemented with the uracil intermediate ureidosuccinate. Growth in ammonia or in glutamate minimal medium plus ureidosuccinate is impossible, however, because this biosynthetic intermediate enters the cell by way of the nitrogen repression-sensitive allantoate transport system (Cooper 1982; V. Turoscy and T.G. Cooper, in prep.). *ure2* mutants were isolated as *ura2* strains that could grow in minimal glutamate plus ureidosuccinate medium, i.e., as mutants in which ureidosuccinate transport is resistant to nitrogen repression. Mutations allelic to those in *URE2* were also isolated 2 years later by Grenson et al. (1974), who designated the mutated locus *gdhCR*.

The phenotype of *ure2* mutants is highly pleiotropic. Drillien et al. (1973) reported that mutations in *URE2* resulted in a 3-4-fold increase in NAD-specific GDH activity in cells provided with glutamate as nitrogen source (normally a derepressing growth condition for this activity) and a striking resistance of NAD-GDH activity to repression by ammonia (30-fold more activity than that seen in the wild type grown in ammonia medium). Also NADP-GDH activity of *ure2* strains is decreased 3-4-fold more in ammonia medium (normally a derepressing medium for the anabolic enzyme) and 2-fold more in glutamate medium than that of *URE2* stains in equivalent media. Dubois and Grenson (1974) reported that *ure2* mutants possessed 10–13-fold more arginase than wild-type cells grown in ammonia medium, i.e., an apparent resistance to nitrogen repression. Similarly, high allantoinase and urea amidolyase activities were also reported for ammonia-grown *ure2* mutants (Dubois et al. 1977). In addition, the presence of the *ure2* mutation renders the production of glutamine synthetase completely resistant to repression normally observed when glutamine is added to the medium of wild-type cells (Dubois and Grenson 1974). Finally, Roon and his colleagues (Dunlop et al. 1980) found that asparaginase-II production in *ure2* mutants was partially (20–30%) resistant to nitrogen repression. Asparaginase-II production in the mutants was also partially (20–60%) resistant to high concentrations of ammonia, methylamine, and alanine; conditions previously mentioned to decrease asparaginase-II levels in *gdh1* mutants. These results led Dunlop et al. (1980) to conclude that "the [*URE2*] system seems more promising" as a putative nitrogen repression regulatory element.

Although the data cited above point out the highly pleiotropic and provocative behavior resulting from mutation of the *URE2* locus, they do not provide compelling support for the hypothesis that the "the [*ure2*] mutation might affect a gene coding for a general 'nitrogen repressor'" (Dubois and Grenson 1974). In fact, it is difficult to understand why alteration of a "general nitrogen repressor" would so drastically alter the observed levels and apparent regulation of two biosynthetic enzymes, NADP-GDH and glutamine synthetase. As an alternative view, it is possible that the *URE2*-gene product directly affects only the enzymes associated with interconversion of ammonia and glutamate by modulation of enzyme activity, enzyme synthesis, or both. Observed alterations in the pattern and extent of nitrogen repression in *ure2* strains could then be explained by the altered levels of ammonia and glutamate which result from the altered interconversions. How does one distinguish indirect, secondary effects of *ure2* mutations on nitrogen-repressive systems from a direct role at the level of gene expression as hypothesized? The data presented thus far do not address this question or provide an answer to the conceptual dilemma it raises.

Does Glutamine Synthetase Function as a Regulatory Element of Nitrogen Catabolite Repression?

In response to suggestions that the glutamine synthetase molecule participated directly in the regulation of enteric bacterial gene expression, Dubois and Grenson (1974) isolated mutants lacking glutamine synthetase activity and demonstrated that *CAR1* (arginase)-gene expression in these auxotrophs was similar to that seen in wild-type strains. In addition, the presence of a *gln1* mutation has no effect on *CAR1*-gene expression observed in *gdh1* or *ure2* mutants. This prompted the conclusion that "the escape of several enzymes from 'ammonia catabolite repression' in (*gdh1*) as well as in (*ure2*) mutations of *Saccharomyces cerevisiae* does not involve glutamine synthetase, either as a positive or as a negative control element" (Dubois and Grenson 1974).

This position was reversed when Dubois et al. (1977) isolated additional glutamine synthetase auxotrophs and found some of them to possess variably derepressed levels of allantoinase, urea amidolyase, and NAD-GDH. In some strains there was no derepression at all. An outcross of one derepressed mutant revealed the presence of a second mutation, which was designated *gnrR*. Derepression of the catabolic enzymes mentioned above was observed when the *gln1* and *gnrR* mutations were both present together. From these observations, Dubois et al. proposed the existence of a second mechanism governing nitrogen catabolite repression. This second circuit was suggested to be mediated by the *gln1* and *gnrR* elements. Glutamine or one of its derivatives was proposed to act as metabolic signal sensed by these elements.

The validity of this proposal is impossible to assess due to a lack of characterization of the mutant strains involved. It is also impossible to distinguish whether the derepression cited in this preliminary communication was a primary or secondary effect of the two mutations.

A Cytoplasmic Determinant That Affects Nitrogen-repressible Systems

During the isolation of strains in which ureidosuccinate transport is resistant to nitrogen repression, Lacroute and his collaborators discovered three mutant classes (Lacroute 1971; Drillien and Lacroute 1972). One class of mutants, *ure1*, is resistant only to repression by ammonia, and representatives of this class have been shown to possess a defective form of NADP-GDH. The remaining two classes (*ure2* and *URE3*) are phenotypically similar, i.e., both are able to transport ureidosuccinate under several conditions of nitrogen repression (e.g., growth with glutamate plus ammonia as nitrogen source). Mutations in the *URE2* locus behave in a normal Mendelian fashion with mutant alleles segregating 2+ : 2−. *URE3*-mutant alleles, on the other hand, behave as though they were cytoplasmically inherited. *URE3* mutations are also completely dominant to the wild-type *ure3* allele. Several lines of evidence have rigorously excluded the possiblility that *ure3* is a mitochondrial locus (Lacroute 1971; Aigle and Lacroute 1975). These data raise the possibility that the *URE3* determinant is plasmid borne.

Although little is known about the *URE3* determinant, Aigle and Lacroute (1975) reported an incompatibility between *ure2*- and *URE3*-mutant alleles. This incompatibility is manifested as a loss of the dominant *URE3* determinant in any spore containing the *ure2*-mutant allele. The basis of this observation remains unknown.

Also unavailable is an explanation for the nitrogen resistance observed in strains carrying the *URE3* allele. However, any detailed understanding of nitrogen catabolite repression must account for the phenotypes of these mutants and for the observed incompatibility of the *ure2-* and URE3- mutant alleles. A priori, it is difficult to construct reasonable models in which alteration of a "general nitrogen repressor, " as hypothesized for the *ure2* mutants, would result in the observed loss of a determinant whose presence in a wild-type *URE2* background generates the same phenotype as a mutation of the *URE2* locus.

Nitrogen Catabolite Repression—A Perspective

In preceding sections we described many of the important observations concerning nitrogen repression and the ways they have been interpreted. What emerges is at least seven classes of mutants, all of which possess a common phenotype: greater or lesser resistance to nitrogen repression. At least two more mutant classes have been reported by Middlehoven and his colleagues (Middlehoven 1977; Middlehoven et al. 1978; Middlehoven and Hoogkamer-te Niet 1981). It is conceptually difficult to believe that expression of nitrogen-repression-sensitive genes is directly regulated by six to ten elements with all of the elements acting at the same level. It is far more plausible to propose that at least some of the observed metabolic phenotypes derive from indirect effects, e.g., the absence or accumulation of the product normally produced or degraded by the mutant enzyme. This situation prompts the question, what is a control mutation? The following examples demonstrate the difficulty encountered when this question is addressed in a rigorous manner.

Mutants that lack allophanate-hydrolase activity (*dur2*) produce all five of the allantoin-degrading enzymes at high constitutive levels. With this information alone (the state of several putative "regulatory" mutations described above), one might feel justified in proposing that the *DUR2*-gene product functioned as a control element. However, a more detailed study reveals that there is a slow turnover of cellular arginine and allantoin that results in a low-level accumulation of allophanate. In *dur2* mutants, which are unable to degrade allophanate, this inducer of the allantoin system accumulates with attendant production for all pathway enzymes (Cooper and Lawther 1973; Whitney et al. 1973). In another case, strains carrying either $car10^h$ or $dur10^h$ mutations were the basis for constructing a detailed model of the repression-sensitive target sites linked to the *CAR1* and *DUR1,2* genes and for the elements that interact with these sites. These mutants also formed the basis for the proposed control of these genes by "synergism." On closer scrutiny the O^h-type mutations were found to be chromosomal rearrangements involving insertion of Ty sequences into sites adjacent to the genes whose regulation was altered (R. Sumrada et al., in prep.). A control mutation might be defined as any defect that alters the level of the enzyme under study. The complex integration of reactions associated with the uptake, compartmentation, catabolism, and biosynthesis of nitrogenous compounds ensures that perturbing the system at one point will cause varied, widespread responses elsewhere. It is important to know where, in this network of sensing and responding elements, the lost function operates.

A major portion of past work in the field is predicated on the premise that providing a mutant cell with the normal product of the defective enzyme completely alleviates the loss of that activity. Repair is then reflected as restoration of a "normal" growth rate. This premise has worked well in prokaryotes and is often applied to yeast. Unfortunately, it may not always be valid. The problems that can arise are demonstrated by the two following situations. Bossinger et al. (1974) attempted to correlate the doubling time of a cell supplied with a single nitrogen source with the severity of nitrogen repression elicited by that compound. Asparagine, glutamate, and ammonia repressed allophanate hydrolase synthesis by 300-, 2.7-, and 2.1-fold, respectively. Yet, the doubling times observed with these nitrogen sources were 124, 175, and 115 minutes, respectively. In other words, there was no correlation between growth rate and the severity of repression observed. If the correlation cannot be made for wild-type cells given various nitrogen sources, it is probably inappropriate to compare the degree of repression observed on supplying ammonia to a wild-type cell with that elicited by providing glutamate to a *gdh1* mutant.

A more quantitative and elegant experiment with the same message was reported by Watson (1976, 1977). This investigator grew *S. cerevisiae* in a chemostat on a growth-limiting mixture of 1 mM proline plus 1 mM glutamate. The intracellular concentration of 22 amino acids and the utilization of proline were monitored as a function of growth rate. At a growth rate of 0.30 hr^{-1}, proline and glutamate were used equally with only traces of either amino acid remaining in the culture medium. When the growth rate was increased to 0.35 hr^{-1} (an increase of only 13%), the total amino acid pool expanded a mere 10%, but 90% of the proline supplied to the culture was unused.

These examples point to the enormous flexibility and selectivity possessed by *S. cerevisiae* in choosing the nitrogen source it will use. It is remarkable that such a small increase in growth rate should alter nitrogen source utilization so dramatically. What control mechanisms are capable of this subtlety? The mutants described in this paper are the starting materials that will be used in future quests for an understanding of how such exquisite selectivity and control are accomplished. It will not be surprising if these studies lead us through layer after layer of structural and functional integration all interfaced at a central objective—the balanced provision and utilization of varied nitrogen sources from a constantly changing environment.

ACKNOWLEDGMENTS

I wish to express my appreciation to Dr. George Zubenko for providing a draft covering much of the information contained in the section on proteases. Thanks are also due to the editors and reviewers whose comments materially improved the quality of this work. Finally, I thank my colleagues at the University of Pittsburgh for their understanding. Their manuscripts were often delayed because I was working on this review.

REFERENCES

Aigle, M. and F. Lacroute. 1975. Genetical aspects of (*URE3*), a nonmitochondrial, cytoplasmically inherited mutation in yeast. *Mol. Gen. Genet.* **136**: 327.

Bakalkin, G.Y., S.L. Kalnov, A.S. Zubatov, and V.N. Luzikov. 1976a. Degradation of total cell protein at different stages of *Saccharomyces cerevisiae* yeast growth. *FEBS Lett.* **63**: 218.

———. 1976b. Degradation of intracellular proteins at different growth stages of *Saccharomyces cerevisiae* yeast. *Biokhimiya.* **41**: 1121.

Bechet, J. and J.M. Wiame. 1965. Indication of a specific regulatory binding protein for ornithinetranscarbamylase in *Saccharomyces cerevisiae. Biochem. Biophys. Res. Commun.* **21**: 226.

Bechet, J., M. Grenson, and J.M. Wiame. 1970. Mutations affecting the repressibility of arginine biosynthetic enzymes in *Saccharomyces cerevisiae. Eur. J. Biochem.* **12**: 31.

Bernhardt, W., K. Paten, and H. Holzer. 1965. Gedampfes oscillieren der synthesegeschwindigkeit von DPN-abhangiger glutamatdehydrogenase in hefezellen. *Biochim. Biophys. Acta* **99**: 531.

Bernhardt, W., M. Zink, and H. Holzer. 1966. NAD-abhangiger glutamaldehydrogenase aus reprimierter and dereprimerter backerhefe. *Biochim. Biophys. Acta* **118**: 549.

Betz, H. 1976. Inhibition of protein synthesis stimulates intracellular protein degradation in growing yeast cells. *Biochem. Biophys. Res. Commun.* **72**: 121.

Bossinger, J. and T.G. Cooper. 1975. Possible failure of NADP-glutamate dehydrogenase to participate directly in nitrogen repression of the allantoin degradative enzymes in *Saccharomyces cerevisiae. Biochem. Biophys. Res. Commun.* **66**: 889.

———. 1976a. Sequence of molecular events involved in induction of allophanate hydrolase. *J. Bacteriol.* **126**: 198.

———. 1976b. Execution times of macromolecular synthetic processes involved in induction of allophanate hydrolase at 15°C. *J. Bacteriol.* **128**: 498.

———. 1977. Molecular events associated with induction of arginase in *Saccharomyces cerevisiae. J. Bacteriol.* **131**: 163.

Bossinger, J., R.P. Lawther, and T.G. Cooper. 1974. Nitrogen repression of the allantoin degradative enzymes in *Saccharomyces cerevisiae. J. Bacteriol.* **118**: 821.

Brandriss, M.C. 1979. Isolation and preliminary characterization of *Saccharomyces cerevisiae* proline auxotrophs. *J. Bacteriol.* **138**: 816.

Brandriss, M.C. and B. Magasanik. 1979a. Genetics and physiology of proline utilization in *Saccharomyces cerevisiae:* Enzyme induction by proline. *J. Bacteriol.* **140**: 498.

———. 1979b. Genetics and physiology of proline utilization in *Saccharomyces cerevisiae:* Mutation causing constitutive enzyme expression. *J. Bacteriol.* **140**: 504.

———. 1980. Proline: An essential intermediate in arginine degradation in *Saccharomyces cerevisiae. J. Bacteriol.* **143**: 1403.

―――. 1981. Subcellular compartmentation in control of converging pathways for proline and arginine metabolism in *Saccharomyces cerevisiae*. *J. Bacteriol.* **145:** 1359.

Cabib, E., R. Ulane, and B. Bowers. 1973. Yeast chitin synthetase: Separation of the zymogen from its activating factor and recovery of the latter in the vacuole fraction. *J. Biol. Chem.* **248:** 1451.

Chattoo, B.B., F. Sherman, D.A. Azubalis, T.A. Fjellstedt, D. Mehnert, and M. Ogur. 1979. Selection of *lys2* mutants of the yeast *Saccharomyces cerevisiae* by the utilization of α-aminoadipate. *Genetics* **93:** 51.

Chisholm, G. and T.G. Cooper. 1982. Isolation and characterization of mutants that produce the allantoin degrading enzymes constitutively in *Saccharomyces cerevisiae*. *J. Molec. Cell. Biol.* **2:** 1088.

Cooper, T.G. 1978. Mutants of *Saccharomyces cerevisiae* possessing fully induced levels of urea amido-lyase in the absence of added inducers. *Biochem. Biophys. Res. Commun.* **82:** 1258.

―――. 1980. Selective gene expression and compartmentation: Two means of regulating nitrogen metabolism in yeast. *Trends Biochem. Sci.* **5:** 332.

―――. 1982. The regulation of yeast gene expression by multiple control elements. In *Genetic engineering of microorganisms for chemicals* (ed. A. Hollaender et al.), p. 143. Plenum Press, New York.

Cooper, T.G. and J. Bossinger. 1976. Selective inhibitions of protein synthesis initiation in *Saccharomyces cerevisiae* by low concentrations of cycloheximide. *J. Biol. Chem.* **251:** 7278.

Cooper, T.G. and R.P. Lawther. 1973a. Induction of the allantoin degradative enzymes by allophanic acid, the last intermediate of the pathway. *Biochem. Biophys. Res. Commun.* **52:** 137.

―――. 1973b. Induction of the allantoin degradative enzymes in *Saccharomyces cerevisiae* by the last intermediate of the pathway. *Proc. Natl. Acad. Sci.* **70:** 2340.

Cooper, T.G. and R. Sumrada. 1975. Urea transport in *Saccharomyces cerevisiae*. *J. Bacteriol.* **121:** 571.

Cooper, T.G., M. Gorski, and V. Turoscy. 1979a. A cluster of 3 genes responsible for allantoin degradation in *Saccharomyces cerevisiae*. *Genetics* **92:** 383.

Cooper, T.G., C. Lam, and V. Turoscy. 1980. Structural analysis of the *dur* loci in *S. cerevisiae*: Two domains of a single multifunctional gene. *Genetics* **94:** 555.

Cooper, T.G., G. Marcelli, and R. Sumrada. 1978. Factors influencing the observed half lives of specific synthetic capacities in *Saccharomyces cerevisiae*. *Biochem. Biophys. Acta.* **517:** 464.

Cooper, T.G. J. McKelvey, and R. Sumrada. 1979b. Oxalurate transport in *Saccharomyces cerevisiae*. *J. Bacteriol.* **139:** 917.

Cooper, T.G., C. Britton, L. Brand, and R. Sumrada. 1979c. Addition of basic amino acids prevents G1 arrest of nitrogen starved cultures of *Saccharomyces cerevisiae*. *J. Bacteriol.* **137:** 1447.

Doi, E., R. Hayashi, and T. Hata. 1967. Purification of yeast proteinases. II. Purification and some properties of yeast proteinase C. *Agric. Biol. Chem.* **31:** 160.

Domnas, A. 1962. Amide metabolism in yeast. II. The uptake of amide and amide like compounds by yeast. *J. Biochem.* **52:** 149.

Drillien, R. and F. Lacroute. 1972. Ureidosuccinic acid uptake in yeast and some aspects of its regulation. *J. Bacteriol.* **109:** 203.

Drillien, R., M. Aigle, and F. Lacroute. 1973. Yeast mutants pleiotropically impaired in the regulation of two glutamate dehydrogenases. *Biochem. Biophys. Res. Commun.* **53:** 367.

Dubois, E.L. and M. Grenson. 1974. Absence of involvement of glutamine synthetase and of NAD-linked glutamate dehydrogenase in the nitrogen catabolite repression of arginase and other enzymes in *Saccharomyces cerevisiae*. *Biochem. Biophys. Res. Commun.* **60:** 150.

Dubois, E.L. and J.M. Wiame. 1976. Nonspecific induction of arginase in *Saccharomyces cerevisiae*. *Biochimie* **58:** 207.

―――. 1978. Catabolic synergism. A cooperation between the availability of substrate and the need for nitrogen in the regulation of arginine catabolism in *Saccharomyces cerevisiae. Mol. Gen. Genet.* **164:** 275.

Dubois, E., M. Grenson, and J.M. Wiame. 1973. Release of the "ammonia effect" on three catabolic enzymes by NADP-specific glutamate dehydrogenaseless mutations in *Saccharomyces cerevisiae. Biochem. Biophys. Res. Commun.* **50:** 967.

———. 1974. The participation of the anabolic glutamate dehydrogenase in the nitrogen catabolite repression of arginase in *Saccharomyces cerevisiae. Eur. J. Biochem.* **48:** 603.

Dubois, E., D. Hiernaux, M. Grenson, and J.M. Wiame. 1978. Specific induction of catabolism and its relation to repression of biosynthesis in arginine metabolism of *Saccharomyces cerevisiae. J. Mol. Biol.* **122:** 383.

Dubois, E., S. Vissers, M. Grenson, and J.M. Wiame. 1977. Glutamine and ammonia in nitrogen catabolite repression of *Saccharomyces cerevisiae. Biochem. Biophys. Res. Commun.* **75:** 223.

Dunlop, P.C. and R.J. Roon. 1975. L-Asparaginase of *Saccharomyces cerevisiae:* An extracellular enzyme. *J. Bacteriol.* **122:** 1017.

Dunlop, P.C., G.M. Meyer, and R.J. Roon. 1980. Nitrogen catabolite repression of asparaginase II in *Saccharomyces cerevisiae. J. Bacteriol.* **143:** 422.

Dunlop, P.C., R.J. Roon, and H.L. Even. 1976. Utilization of D-asparagine by *Saccharomyces cerevisiae. J. Bacteriol.* **125:** 999.

Durieu-Trautmann, O. and C. Delavier-Klutchko. 1977. Effect of ammonia and glutamine on macromolecule synthesis and breakdown during sporulation of *Saccharomyces cerevisiae. Biochem. Biophys. Res. Commun.* **79:** 438.

Errede, B., T.S. Cardillo, F. Sherman, E. Dubois, J. Deschamps, and J.-M. Wiame. 1980. Mating signals control expression of mutations resulting from insertion of a transposable repetitive element adjacent to diverse yeast genes. *Cell* **22:** 427.

Felix, F. and N. Brouillet. 1966. Purification et proprietes de deux peptidases de levure de brasserie. *Biochim. Biophys. Acta* **122:** 127.

Gaudy, E.T., R. Bojanowski, R.C. Valentine, and R.S. Wolfe. 1965. Ureidoglycolate synthetase of *Streptococcus allantoicus. J. Bacteriol.* **90:** 1525.

Grenson, M. and C. Hou. 1972. Ammonia inhibition of the general amino acid permease and its suppression in NADPH-specific glutamate dehydrogenaseless mutants of *Saccharomyces cerevisiae. Biochem. Biophys. Res. Commun.* **48:** 749.

Grenson, M., E. Dubois, and M. Piotrowska. 1974. Ammonia assimilation in *Saccharomyces cerevisiae* as mediated by the two glutamate dehydrogenases. Evidence for the *gdhA* locus being a structural gene for the NADP-dependent glutamate dehydrogenase. *Mol. Gen. Genet.* **128:** 73.

Grenson, M., C. Hou, and M. Crabeel. 1970. Multiplicity of the amino acid permeases. *J. Bacteriol.* **103:** 770.

Hasilik, A. and W. Tanner. 1976. Biosynthesis of carboxypeptidase Y in yeast. Evidence for a precursor form of the glycoprotein. *Biochem. Biophys. Res. Commun.* **72:** 1430.

———. 1978a. Biosynthesis of the vacuolar yeast glycoprotein carboxypeptidase Y. Conversion of precursor into enzyme. *Eur. J. Biochem.* **85:** 599.

———. 1978b. Carbohydrate moiety of carboxypeptidase Y and perturbation of its biosynthesis. *Eur. J. Biochem.* **91:** 567.

Hata, T., R. Hayashi, and E. Doi. 1967a. Purification of some yeast proteinases. I. Fractionation and some properties of these proteinases. *Agric. Biol. Chem.* **31:** 150.

———. 1967b. Purification of yeast proteinases. III. Isolation and physiochemical properties of yeast proteinase A and C. *Agric. Biol. Chem.* **31:** 357.

Hayashi, R., S. Aibara, and T. Hata. 1970. A unique carboxypeptidase activity of yeast proteinase C. *Biochim. Biophys. Acta* **212:** 359.

Hayashi, R., S. Moore, and W. Stein. 1973. Carboxypeptidase from yeast. Large scale preparation and the application of COOH-terminal analysis of peptides and proteins. *J. Biol. Chem.* **248:** 2296.

Hemmings, B.A., G. S. Zubenko, A. Hasilik, and E. W. Jones. 1981. Mutant defective in processing of an enzyme located in the lysosome-like vacuole of *Saccharomyces cerevisiae*. *Proc. Natl. Acad. Sci.* **78:** 435.

Hierholzer, G. and H. Holzer. 1963. Repression der synthese von DPN-abhangiger glutaminsauredehydrogenase in *Saccharomyces cerevisiae* durch ammoniumionen. *Biochem. Z.* **339:** 175.

Hoet, P.R. and J.M. Wiame. 1974. On the nature of *argR* mutations in *Saccharomyces cerevisiae*. *Eur. J. Biochem.* **43:** 87.

Holzer, H. and S. Schneider. 1957. Anreicherung und Trennung einer DPN-spezifischen und einer TPN-spezifischen glutaminsaure-dehydrogenase aus Hefe. *Biochem. Z.* **329:** 361.

Hutchison, H.T., L.H. Hartwell, and C.S. McLaughlin. 1969. Temperature sensitive yeast mutant defective in ribonucleic acid production. *J. Bacteriol.* **99:** 807.

Indge, K.J. 1968. Polyphosphates of the yeast cell vacuole. *J. Gen. Microbiol.* **51:** 447.

Jauniaux, J.-C., L.A. Urrestarazu, and J.M. Wiame. 1978. Arginine metabolism in *Saccharomyces cerevisiae*: Subcellular localization of the enzymes. *J. Bacteriol.* **133:** 1096.

Johnston, G.C., J.R. Pringle, and L.H. Hartwell. 1977a. Coordination of growth with cell division in the yeast *Saccharomyces cerevisiae*. *Exp. Cell Res.* **105:** 79.

Johnston, G.C., R.A. Singer, and E.S. McFarlane. 1977b. Growth and cell division during nitrogen starvation of the yeast *Saccharomyces cerevisiae*. *J. Bacteriol.* **132:** 723.

Jones, E.W. 1977. Proteinase mutants of *Saccharomyces cerevisiae*. *Genetics* **85:** 23.

Jones, G.E. 1973. A fine-structure map of the yeast L-asparaginase gene. *Molec. Gen. Genet.* **121:** 9.

———. 1977a. Genetic and physiological relationships between L-asparaginase I and asparaginase II in *Saccharomyces cerevisiae*. *J. Bacteriol.* **130:** 128.

———. 1977b. Genetics of expression of asparaginase II activity in *Saccharomyces cerevisiae*. *J. Bacteriol.* **129:** 1165.

Jones, G.E. and R.K. Mortimer. 1970. L-Asparaginase-deficient mutants of yeast. *Science* **167:** 181.

———. 1973. Biochemical properties of yeast L-asparaginase. *Biochem. Gen.* **9:** 131.

Kohlhaw, G., W. Gragert, and H. Holzer. 1965. Parallel-repression der synthese von glutamin-synthetase und DPN-abhangigier glutamatedehydrogenase in hefe. *Biochem. Z.* **341:** 224.

Kuhn, R., K. Walsh, and H. Neurath. 1974. Isolation and characterization of an acid carboxypeptidase from yeast. *Biochemistry* **13:** 3871.

Kuo, S.C., F.R. Cano, and J.O. Lampen. 1973. Lomofungin, an inhibitor of ribonucleic acid synthesis in yeast protoplasts: Its effects on enzyme formation. *Antimicrob. Agents Chemother.* **3:** 716.

Lacroute, F. 1966. Ph.D. thesis, University of Paris, France.

———. 1971. Non-mendelian mutation allowing ureidosuccinic acid uptake in yeast. *J. Bacteriol.* **106:** 519.

Larimore, F.S., I. Kuist, P.M. Korkowski, and R.J. Roon. 1980. Transport and metabolism of N-δ-chloroacetyl-L-ornithine by *Saccharomyces cerevisiae*. *Arch. Biochem. Biophys.* **204:** 234.

Lawther, R.P. and T.G. Cooper. 1973. Effects of inducer addition and removal upon the level of allophanate hydrolase in *Saccharomyces cerevisiae*. *Biochem. Biophys. Res. Commun.* **55:** 1100.

———. 1975. Kinetics of induced and repressed enzyme synthesis in *Saccharomyces cerevisiae*. *J. Bacteriol.* **121:** 1064.

Lawther, R.P., S.L. Phillips, and T.G. Cooper. 1975. Lomofungin inhibition of allophanate hydrolase synthesis in *Saccharomyces cerevisiae*. *Mol. Gen. Genet.* **137:** 89.

Lawther, R.P., E. Riemer, B. Chojnacki, and T.G. Cooper. 1974. Clustering of the genes for allantoin degradation in *Saccharomyces cerevisiae*. *J. Bacteriol.* **119:** 461.

Lee, K.W. and A.H. Roush. 1964. Allantoinase assays and their application to yeast and soybean allantoinases. *Arch. Biochem. Biophys.* **108:** 460.

Lemoine, Y., E. Dubois, and J. Wiame. 1978. The regulation of urea amidolyase of *Saccharomyces cerevisiae*. *Mol. Gen. Genet.* **166**: 251.

Lenney, J.F. and J.M. Dalbec. 1967. Purification and properties of two proteinases from *Saccharomyces cerevisiae*. *Arch. Biochem. Biophys.* **120**: 42.

Lenny, J., P. Matile, A. Weimken, M. Schellenberg, and J. Meyer. 1974. Activities and cellular localization of yeast proteinases and their inhibitors. *Biochem. Biophys. Res. Commun.* **60**: 1378.

Looze, Y., L. Gillet, M. Deconinck, B. Couteaux, E. Polastro, and J. Leonis. 1979. Protease B from *Saccharomyces cerevisiae*. *Int. J. Pept. Protein Res.* **13**: 253.

Lopez, S. and J. Gancedo. 1979. Effect of metabolic condition on protein turnover in yeast. *Biochem. J.* **178**: 769.

Lundgren, D.W., M. Ogur, and S. Yuen. 1972. The isolation and characterization of a *Saccharomyces* mutant deficient in Δ^1-pyrroline-5-carboxylate dehydrogenase activity. *Biochim. Biophys. Acta* **286**: 360.

Matern, H., H. Betz, and H. Holzer. 1974. Compartmentation of inhibitors of proteinases A and B and carboxypeptidase Y in yeast. *Biochem. Biophys. Res. Commun.* **60**: 1051.

Matile, P. 1978. Biochemistry and function of vacuoles. *Ann. Res. Plant Physiol.* **29**: 193.

Matile, P. and A. Wiemken. 1967. The vacuole as the lysome of the yeast cell. *Arch. Mikrobiologie.* **56**: 148.

Matile, P., A. Wiemken, and W. Guyer. 1971. A lysosomal aminopeptidase isozyme in differentiating yeast cells and protoplasts. *Planta* **95**: 43.

Mazon, J.J. 1978. Effect of glucose starvation on the nicotinamide adenine dinucleotide phosphate-dependent glutamate dehydrogenase of yeast. *J. Bacteriol.* **133**: 780.

Mazon, M.J. and B.A. Hemmings. 1979. Regulation of *Saccharomyces cerevisiae* nicotinamide adenine dinuleotide phosphate-dependent glutamate dehydrogenase by proteolysis during carbon starvation. *J. Bacteriol.* **139**: 686.

McLellan, W.L., Jr. and J.O. Lampen. 1963. The acid phosphatase of yeast: Localization and secretion by protoplasts. *Biochim. Biophys. Acta* **67**: 324.

Messenguy, F. and T.G. Cooper. 1977. Evidence that specific and "general" control of ornithine carbamoyltransferase production occurs at the level of transcription in *Saccharomyces cerevisiae*. *J. Bacteriol.* **130**: 1253.

Messenguy, F., D. Colin, and J.-P. Ten Have. 1980. Regulation of compartmentation of amino acid pools in *Saccharomyces cerevisiae* and its effects on metabolic control. *Eur. J. Biochem.* **108**: 439.

Messenguy, F., M. Penninckx and J.-M. Wiame. 1971. Interaction between arginase and ornithine carbamoyltransferase in *Saccharomyces cerevisiae*: The regulatory site from ornithine. *Eur. J. Biochem.* **22**: 277.

Middelhoven, W.J. 1964. The pathway of arginine breakdown in *Saccharomyces cerevisiae*. *Biochim. Biophys. Acta* **93**: 650.

———. 1968. The derepression of arginase and of ornithine transaminase in nitrogen-starved baker's yeast. *Biochim. Biophys. Acta* **156**: 440.

———. 1969. Enzyme repression in the arginine pathway of *Saccharomyces cerevisiae*. *Antonie Leeuwenhoek J. Microbiol. Serol.* **35**: 215.

———. 1970. Induction and repression of arginase and ornithine transaminase in baker's yeast. *Antonie Leeuwenhoek J. Microbiol. Serol.* **36**: 1.

———. 1977. Isolation and characterization of methylammonium-resistant mutants of *Saccharomyces cerevisiae* with relieved nitrogen metabolite repression of allantoinase, arginase and ornithine transaminase synthesis. *J. Gen. Microbiol.* **100**: 257.

Middlehoven, W.J. and M.C. Hoogkamer-te Niet. 1981. Repression of catabolic NAD-specific glutamate dehydrogenase of *Saccharomyces cerevisiae* by arginine, allantoin and urea. *FEMS Microbiol. Lett.* **10**: 307.

Middlehoven, W.J., J. Van Eijk, R. Van Renesse, and J.M. Blijham. 1978. A mutant of *Saccharomyces cerevisiae* lacking catabolic NAD-specific glutamate dehydrogenase. Growth characteristics of the mutant and regulation of enzyme synthesis in the wild-type strain. *Antonie Leeuwenhoek, J. Microbiol. Serol.* **44:** 311.

Miller, J.J. 1963. The metabolism of yeast sporulation. V. Stimulation and inhibition of sporulation and growth by nitrogen compounds. *Can. J. Microbiol.* **9:** 259.

Moeller, C.H. and W.W. Thomson. 1979a. An ultrastructural study of the yeast tonoplast during the shift from exponential to stationary phase. *J. Ultrastruct. Res.* **68:** 28.

———. 1979b. Uptake of lipid bodies by the yeast vacuole involving areas of the tonoplast depleted of intramembranous particles. *J. Ultrastruct. Res.* **68:** 38.

Mortimer, R.K. and D. Schild. 1980. Genetic map of *Saccharomyces cerevisiae*. *Microbiol. Rev.* **44:** 519.

Nakamura, K.D. and F. Schlenk. 1974. Active transport of exogenous S-adenosylmethionine and related compounds into cells and vacuoles of *Saccharomyces cerevisiae*. *J. Bacteriol.* **120:** 482.

Newlon, M.C. 1979. NADP-specific glutamate dehydrogenase is not involved in repression of yeast sporulation by ammonia. *Mol. Gen. Genet.* **176:** 297.

Ogur, M., L. Coker, and S. Ogur. 1964. Glutamate auxotrophs in *Saccharomyces*. I. The biochemical lesion in the glt_1 mutants. *Biochem. Biophys. Res. Commun.* **14:** 193.

Opheim, D.J. 1979. Effects of ammonium ions on activity of hydrolytic enzymes during sporulation of yeast. *J. Bacteriol.* **138:** 1022.

Pauling, K.D. and G.E. Jones. 1980. Asparaginase II of *Saccharomyces cerevisiae*: Dynamics of accumulation and loss in rapidly growing cells. *J. Gen. Microbiol.* **117:** 423.

Pauling, K.D., J.E. Hann, and G.E. Jones. 1980. Asparaginase II of *Saccharomyces cerevisiae*: Characterization of a mutation that affects expression in rapidly growing cells. *J. Gen. Microbiol.* **119:** 539.

Penninckx, M. 1975. Interaction between arginase and L-ornithine carbamoyltransferase in *Saccharomyces cerevisiae*. *Eur. J. Biochem.* **58:** 533.

Penninckx, M. and J.M. Wiame. 1976. Affinity of arginase for ornithine carbamoyltransferase in *Saccharomyces cerevisiae*. *J. Mol. Biol.* **104:** 819.

Penninckx, M., J.-P. Simon, and J.M. Wiame. 1974. Interaction between arginase and L-ornithine carbamoyltransferase in *Saccharomyces cerevisiae*. *Eur. J. Biochem.* **49:** 429.

Piñon, R. 1977. Effects of ammonium ions on sporulation of *Saccharomyces cerevisae*. *Exp. Cell Res.* **105:** 367.

Pringle, J.R. and L.H. Hartwell. 1981. The *Saccharomyces cerevisiae* cell cycle. In *The molecular biology of the yeast* Saccharomyces. I. *Life cycle and inheritance* (ed. J. Strathern et al.), p. 97. Cold Spring Harbor Laboratory, Cold Spring Harbor, New York.

Ramos, F., P. Thuriaux, J.M. Wiame, and J. Bechet. 1970. The participation of ornithine and citrulline in the regulation of arginine metabolism in *Saccharomyces cerevisiae*. *Eur. J. Biochem.* **12:** 40.

Reid, B.J. and L.H. Hartwell. 1977. Regulation of mating in the cell cycle of *Saccharomyces cerevisiae*. *J. Cell Biol.* **75:** 355.

Roon, R.J. and H.L. Even. 1978. Regulation of the nicotinamide adenine dinucleotide- and nicotinamide adenine dinucleotide phospate-dependent glutamate dehydrogenases of *Saccharomyces cerevisiae*. *J. Bacteriol.* **116:** 36.

Roon, R.J. and B. Levenberg. 1968. An adenosine triphosphate-dependent, avidin-sensitive enzymatic cleavage of urea in yeast and green algae. *J. Biol. Chem.* **243:** 5213.

Roon, R.J. and B. Levenberg. 1970. CO_2 fixation and the involvement of allophanate in the biotin-enzyme-catalyzed cleavage of urea. *J. Biol. Chem.* **245:** 4593.

Roon, R.J., H.L. Even, and F. Larimore. 1974. Glutamate synthase: Properties of the reduced nicotinamide adenine dinucleotide-dependent enzyme from *Saccharomyces cerevisiae*. *J. Bacteriol.* **118:** 89.

Roon, R.J., J. Hampshire, and B. Levenberg. 1972. Urea amidolyase: The involvement of biotin in urea cleavage. *J. Biol. Chem.* **247:** 7539.

Roon, R.J., F. Larimore, and J.S. Levy. 1975. Inhibition of amino acid transport by ammonium ion in *Saccharomyces cerevisiae. J. Bacteriol.* **124:** 325.
Saheki, T. and H. Holzer. 1969. Comparisons of the tryptophan synthase inactivating enzymes with proteinases from yeast. *Eur. J. Biochem.* **42:** 621.
Sanada, Y., K. Fujishiro, H. Tanaka, and N. Katunuma. 1979. Isolation and characterization of yeast protease B. *Biochem. Biophys. Res. Commun.* **86:** 815.
Satrustegui, J. and A. Machado. 1978. Specific inactivation of NADP-dependent glutamate dehydrogenase from *Saccharomyces cerevisiae. FEMS Microbiol. Lett.* **4:** 171.
Schlenk, F., J.L. Dainko, and G. Svihla. 1970. The accumulation and intracellular distribution of biological sulfonium compounds in yeast. *Arch. Biochem. Biophys.* **140:** 228.
Schwencke, J. 1977. Characteristics and integration of the yeast vacuole with cellular functions. *Physiol. Veg.* **15:** 491.
Sentheshanmuganathan, S. 1960. The mechanism of the formation of higher alcohols from amino acids by *Saccharomyces cerevisiae. J. Gen. Microbiol.* **74:** 568.
Shabalin, Yu.A., V.I. Vagabov, A.B. Tsiomenko, O.A. Zemlyanukhina, and I.S. Kulaev. 1977. Polyphosphate kinase activity in vacuoles of yeasts. *Biokhimiya* **42:** 1642.
Shilo, V., G. Simchen, and B. Shilo. 1978. Initiation of meiosis in cell cycle initiation mutants of *Saccharomyces cerevisiae. Exp. Cell Res.* **112:** 241.
Sims, A.P., J. Toone, and V. Box. 1974. The regulation of glutamine metabolism in *Candida utilis:* Mechanisms of control of glutamine synthetase. *J. Gen. Microbiol.* **84:** 149.
Sumrada, R. and T.G. Cooper. 1974. Oxaluric acid: A non-metabolizable inducer of the allantoin degradative enzymes in *Saccharomyces cerevisiae. J. Bacteriol.* **117:** 1240.
———. 1976. Basic amino acid inhibition of growth in *Saccharomyces cerevisiae. Biochem. Biophys. Res. Commun.* **68:** 598.
———. 1977. Allantoin transport in *Saccharomyces cerevisiae. J. Bacteriol.* **131:** 839.
———. 1978a. Basic amino acid inhibition of cell division and macromolecular synthesis in *Saccharomyces cerevisiae. J. Gen. Microbiol.* **108:** 45.
———. 1978b. Control of vacuole permeability and protein degradation by the cell cylce arrest signal in *Saccharomyces cerevisiae. J. Bacteriol.* **136:** 234.
———. 1982a. Post-translational processing of urea amidolyase in *Saccharomyces cerevisiae. J. Mol. Cell. Biol.* **2:** 800.
———. 1982b. Urea carboxylase and allophanate hydrolyase are components of a multifunctional protein in yeast. *J. Biol. Chem.* **257:** 9119.
———. 1982c. Isolation of the *CAR1* gene from *Saccharomyces cerevisiae* and analysis of its expression. *J. Mol. Cell. Biol.* (in press.).
Sumrada, R., M. Gorski, and T.G. Cooper. 1976. Urea transport defective strains of *Saccharomyces cerevisiae. J. Bacteriol.* **125:** 1048.
Sumrada, R., C.A. Zacharski, V. Turoscy, and T.G. Cooper. 1978. Induction and inhibition of the allantoin permease in *Saccharomyces cerevisiae. J. Bacteriol.* **135:** 498.
Sutton, D.D. and J.O. Lampen. 1962. Localization of sucrose and maltose fermenting systems in *Saccharomyces cerevisiae. Biochim. Biophys. Acta* **56:** 303.
Svihla, G., J.L. Dainko, and F. Schlenk. 1963. Ultraviolet microscopy of purine compounds in the yeast vacuole. *J. Bacteriol.* **85:** 399.
Tempest, D.W., J.L. Meers, and C.M. Brown. 1970. Glutamine (amide): 2-oxoglutarate aminotransferase oxido-reductase (NADP), an enzyme involved in the synthesis of glutamate by some bacteria. *J. Gen. Microbiol.* **64:** 187.
Thomulka, K.W. and A.G. Moat. 1972. Inorganic nitrogen assimilation in yeasts: Alteration in enzyme activities associated with changes in cultural conditions and growth phase *J. Bacteriol.* **109:** 25.
Tonnesen, T. and J.D. Friesen. 1973. Inhibition of ribonucleic acid synthesis in *Saccharomyces cerevisiae*: Decay rate of messenger ribonucleic acid. *J. Bacteriol.* **115:** 889.
Trijbels, F. and G.D. Vogels. 1967. Allantoate and ureidoglycolate degradation by *Pseudomonas aeruginosa. Biochim. Biophys. Acta* **132:** 115.

Turoscy, V. and T.G. Cooper. 1979. Allantoate transport in *Saccharomyces cerevisiae*. *J. Bacteriol.* **140:** 971.

———. 1982. Pleiotropic control of five eucaryotic genes by multiple regulatory elements. *J. Bacteriol.* **151:** 1237.

Ulane, R.E. and E. Cabib. 1976. The activating system of chitin synthetase from *Saccharomyces cerevisiae*. *J. Biol. Chem.* **251:** 3367.

Unger, M.W. and L.H. Hartwell. 1976. Control of cell division in *Saccharomyces cerevisiae* by methionyl-tRNA. *Proc. Natl. Acad. Sci.* **73:** 1664.

Van de Poll, K.W. 1973. Ammonium repression in a mutant of *Saccharomyces carlsbergenesis* lacking NADP dependent glutamate dehydrogenase activity. *FEBS Lett.* **32:** 265.

Vezihet, F., J.H. Kinnaird, and I.W. Dawes. 1979. The physiology of mutants derepressed for sporulation in *Saccharomyces cerevisiae*. *J. Gen. Microbiol.* **115:** 391.

Vogels, G.D. and C. van der Drift. 1976. Degradation of purines and pyrimidines by microorganisms. *Bacteriol. Rev.* **40:** 403.

Watson, T.G. 1976. Amino-acid pool composition of *Saccharomyces cerevisiae* as a function of growth rate and amino-acid nitrogen source. *J. Gen. Microbiol.* **96:** 263.

———. 1977. Inhibition of proline utilization by glutamate during steady-state growth of *Saccharomyces cerevisiae*. *J. Gen. Microbiol.* **103:** 123.

Whitney, P.A. and T.G. Cooper. 1970. Requirement for HCO_3^- by ATP: Urea amidolyase in yeast. *Biochem. Biophys. Res. Commun.* **40:** 814.

———. 1972a. Urea carboxylase and allophanate hydrolase: Two components of ATP:urea amido-lyase in *Saccharomyces cerevisiae*. *J. Biol. Chem.* **247:** 1349.

———. 1972b. Urea carboxylase and allophanate hydrolase: Two components of a multi-enzyme complex in *Saccaromyces cerevisiae*. *Biochem. Biophys. Res. Commun.* **49:** 45.

———. 1973. Urea carboxylase from *Saccharomyces cerevisiae*. Evidence for a minimal two-step reaction sequence. *J. Biol. Chem.* **248:** 325.

Whitney, P.A. and B. Magasanik. 1973. The induction of arginase in *Saccharomyces cerevisiae*. *J. Biol. Chem.* **248:** 6197.

Whitney, P.A., T.G. Cooper, and B. Magasanik. 1973. Allophanate, the inducer of urea carboxylase and allophanate hydrolase in *Saccharomyces cerevisiae*. *J. Biol. Chem.* **248:** 6203.

Wiame, J.M. 1971. The regulation of arginine metabolism in *Saccharomyces cerevisiae*: Exclusion mechanisms. *Curr. Top. Cell. Regul.* **4:** 1.

———. 1973. The regulation of the biosynthesis and the degradation of arginine in *Saccharomyces cerevisiae*. In *Proceedings of the Third International Specialized Symposium on Yeasts*, Helsinki (ed. H. Soumalainen and C. Waller), part II, p. 307.

Wiemken, A. and M. Durr. 1974. Characterization of amino acid pools in the vacuolar compartment of *S. cerevisiae*. *Arch. Micorbiol.* **101:** 45.

Wiemken, A. and P. Nurse. 1973. The vacuole as a compartment of amino acid pools in yeast. In *Proceedings of the Third International Specialized Symposium on Yeast*, Helsinki (ed. H. Soumalainen and C. Waller), part II, p. 331.

Wiemken, A., P. Matile, and H. Moor. 1970. Vacuolar dynamics in synchronously budding yeast. *Arch. Mikrobiol.* **70:** 89.

Wiemken, A., M. Schellenberg, and K. Urech. 1979. Vacuoles: The sole compartments of digestive enzymes in yeast (*Saccharomyces cerevisiae*)? *Arch. Microbiol.* **123:** 23.

Williamson, D.H. and A.W. Scopes. 1962. A rapid method for synchronizing division in the yeast *Saccharomyces cerevisiae*. *Nature* **193:** 256.

Williamson, V.M., E.T. Young, and M. Ciriacy. 1981. Transposable elements associated with constitutive expression of yeast alcohol dehydrogenase II. *Cell* **23:** 605.

Wolf, D. and U. Weiser. 1977. Proteinase C (carboxypeptidase Y) mutant of yeast and evidence for a second carboxypeptidase activity. *Eur. J. Biochem.* **73:** 553.

Wolf, D., I. Beck, and C. Ehman. 1979. The intracellular proteinase and their inhibitors in yeast: A genetic and biochemical approach to their function. In *Limited proteolysis in microorganisms* (ed. H. Holzer and G. Cohen), Department of Health, Education and Welfare Publication no. 79-1591, p. 61.

Woodward, J.R. and V.P. Cirillo. 1977. Amino acid transport and metabolism in nitrogen-starved cells of *Saccharomyces cerevisiae*. *J. Bacteriol.* **130:** 714.

Zacharski, C.A. and T.G. Cooper. 1978. Metabolite compartmentation in *Saccharomyces cerevisiae*. *J. Bacteriol.* **135:** 490.

Zubenko, G.S. 1981. "A genetic approach to the study of intracellular proteolysis in *Saccharomyces cerevisiae*." Ph.D. thesis, Carnegie-Mellon University, Pittsburgh.

Zubenko, G.S., A.P. Mitchell, and E.W. Jones. 1979. Septum formation, cell division and sporulation in mutants of yeast deficient in proteinases B. *Proc. Natl. Acad. Sci.* **76:** 2395.

———. 1980. Mapping of the proteinase B structural gene, *PRB1*, in *Saccharomyces cerevisiae* and identification of nonsense alleles within the locus. *Genetics* **96:** 137.

Zubenko, G.S., F.J. Park, and E.W. Jones. 1983. Mutations in the *PEP4* locus of *Saccharomyces cerevisiae* block the final step in the maturation of two vacuolar hydrolases. *Proc. Natl. Acad. Sci.* (in press).

Membrane Lipids of Yeast: Biochemical and Genetic Studies

Susan Armstrong Henry
Departments of Genetics and Molecular Biology
Albert Einstein College of Medicine of Yeshiva University
Bronx, New York 10461

1. Distribution of Lipids and Lipid Biosynthetic Activities into Various Membrane Fractions
   A. Distribution of Lipid Biosynthetic Activities
   B. Lipid Composition of Yeast Membrane Fractions
2. Sterol Biosynthesis and Requirements
   A. Experiments with Anaerobic Cells
   B. Sterol Mutant Studies
   C. Experiments Using Inhibitors of Sterol Biosynthesis
3. Fatty Acid Biosynthesis and Requirements
   A. Mutants Defective in Fatty Acid Biosynthesis
   B. Regulation of Fatty Acid Biosynthesis
   C. Experimental Manipulation of Fatty Acid Mutants
4. Phospholipid Biosynthesis
   A. Regulation of Phospholipid Biosynthesis
   B. Mutants Defective in Phospholipid Synthesis
5. Summary and Conclusions

INTRODUCTION

Saccharomyces cerevisiae is a fairly typical eukaryotic organism with respect to the membranous organelles it contains, as well as the lipids that comprise its membranes. Like other eukaryotes, it synthesizes and incorporates sterol into its membranes. In addition, its membranes contain a typically eukaryotic mixture of phospholipids, including sphingolipids, cardiolipin, phosphatidylserine, phosphatidylethanolamine, phosphatidylcholine, and phosphatidylinositol (see Fig. 5). The pathways for the synthesis of yeast membrane lipids are increasingly well characterized. In addition, a number of mutations affecting these pathways have been characterized, and reliable methodologies for isolation of some subcellular membrane fractions are being developed. As a result, *S. cerevisiae* is an attractive organism in which to conduct studies on the roles of specific lipids in membrane biogenesis and membrane-mediated processes.

In bacteria (for reviews, see Silbert et al. 1974; Silbert 1975; Raetz 1978), lipid mutants have been used to study the role of lipid synthesis on membrane biogenesis and to manipulate membrane lipid composition. Such experiments permit assessment of the role(s) of various lipids in membrane function and allow the physical properties of the membrane lipid bilayer to be experimentally manipulated in vivo. Since *S. cerevisiae* is a eukaryotic organism, yeast lipid mutants can be used for probing the role of lipids in the biogenesis and functioning of a variety of membranous organelles. In this paper studies of this kind are reviewed together with genetic and biochemical studies on the pathways involved in lipid synthesis in yeast.

DISTRIBUTION OF LIPIDS AND LIPID BIOSYNTHETIC ACTIVITIES INTO VARIOUS MEMBRANE FRACTIONS

The lipid composition of yeast cells and membranes varies considerably, depending upon cultural conditions. Relevant variables include temperature (Hunter and Rose 1972; Okuyama et al. 1979), carbon source (Jakovcic et al. 1971), and lipid precursor supplement (Waechter and Lester 1971; Ratcliffe et al. 1973). Anaerobic growth, which prevents the synthesis of ergosterol and unsaturated fatty acid (Andreason and Stier 1953, 1954), and mutants with specific defects in lipid metabolism can also be employed to produce substantial alterations in lipid composition and metabolism. The studies involving mutants and special growth conditions are discussed in subsequent sections. The purpose of this introductory section is to point out differences in lipid composition and metabolism that have been detected in different membrane fractions isolated from "wild-type" yeast strains under "normal" growth conditions.

The yeast cell contains the usual array of eukaryotic membranes and membranous organelles (Hunter and Rose 1971), including nucleus, vacuole, mitochondria, vesicles, endoplasmic reticulum, and plasma membrane. At present, not all of these different organelles and membranes have been successfully isolated from yeast in a highly purified form. However, mitochondria can be routinely isolated (Guerin et al. 1979; Trembath and Tzagoloff 1979) and have been studied a great deal, with respect to distribution of lipids and lipid biosynthetic activities. In addition, considerable progress has been made in developing techniques for the isolation and purification of membrane fractions enriched in plasma membrane. Cabib and his coworkers have shown that chitin synthetase can be used as a marker for the plasma membrane (Duran et al. 1975; Duran and Cabib 1978). Using this marker enzyme, Duran et al. (1975) achieved substantial enrichment of plasma membrane derived from glucose-grown wild-type yeast by stabilizing the membrane of spheroplasts with concanavalin A. As yet, no study has been made of the phospholipid biosynthetic activities associated with plasma membrane isolated by this method. However, since many phospholipid biosynthetic activities are presumed to be associated with the endoplasmic reticulum, several studies in yeast have attempted to distinguish "microsomal" fractions (Cobon et al. 1974; Schlossman and Bell 1978). It must be recognized that these microsomal fractions really only exclude mitochondrial activity and can be presumed to be composed of a large array of vesiculated cellular membrane components. These and similar studies on the relative distribution of lipids and lipid biosynthetic activities in the yeast cell are the topic of the following discussion and are summarized in Tables 1 and 2.

Distribution of Lipid Biosynthetic Activities

The biosynthesis of certain lipid precursors, such as long-chain saturated fatty acids (Lynen 1967) and inositol (Culbertson et al. 1976a), occurs in the cytoplasm of $S.\ cerevisiae$. The synthesis of the sterol precursor mevalonic acid from acetyl-CoA involves enzymes located both in the cytoplasm and in the mitochondrion. For example, one enzyme involved in mevalonate synthesis (β-hydroxy-β-methyl-glutaryl synthase) appears to be located exclusively in the cytoplasm (Trocha and Sprinson 1976). On the other hand, acetoacetyl-CoA thiolase exists in both a mitochondrial and a cytoplasmic form in yeast (Kornblatt and Rudney 1971). The enzyme β-hydroxy-β-methyl-glutaryl reductase, is found exclusively in mitochondria (Boll et al. 1975; Trocha and Sprinson 1976). It is interesting that this enzyme was found associated with promitochondria in anaerobically grown cells, as well as with mitochondria in aerobic cells (Boll et al. 1975). This and other mitochondrially located steps in sterol biosynthesis may be among the essential functions performed by the promitochondrial structure retained in ρ^0 cells (see Stevens 1981). Other steps in sterol biosynthesis have been reported to be associated with a microsomal fraction. For example, the synthesis of presqualene and squalene from farnesyl pyrophosphate (for a review of the steps from mevalonate to squalene, see Karst and Lacroute [1977] and Parks [1978]) have been reported to be associated with a microsomal fraction in yeast (Shechter and Bloch 1971; Agnew and Popjak 1978). Likewise, the enzyme Δ^5-desaturase responsible for the introduction of the Δ^5-double bond (reaction 9, Figs. 1 and 2), has been reported to be associated with a microsomal fraction in yeast that was shown to be free of mitochondrial contamination (Osumi et al. 1979). On the other hand, the Δ^{24}-sterol methyltransferase, an enzyme catalyzing one of the reactions that distinguishes ergosterol from cholesterol (reaction 7, Figs. 1 and 2) is located in the mitochondrion (Thompson et al. 1974). The subcellular locations of many of the reactions leading to ergosterol (Fig. 2) have not been exhaustively studied, but it is assumed that they are located in membrane fractions. Ergosterol can also be esterified and deesterified by yeast. Taketani et al. (1976) demonstrated that the sterol ester synthetase is located in a microsomal fraction, whereas the sterol ester hydrolase is located in the mitochondria of aerobic cells (Taketani et al. 1978). The subcellular locations of these enzymes of sterol and sterol ester synthesis are summarized in Table 1.

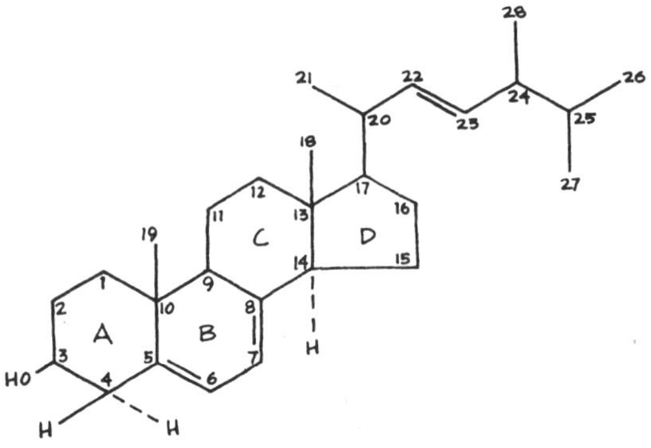

Figure 1 Structures of ergosterol, the predominant yeast sterol, and cholesterol. Numbering of carbons and lettering of rings is standard nomenclature (Parks 1978). Note structural differences between the two sterols in the B ring and in the side chain.

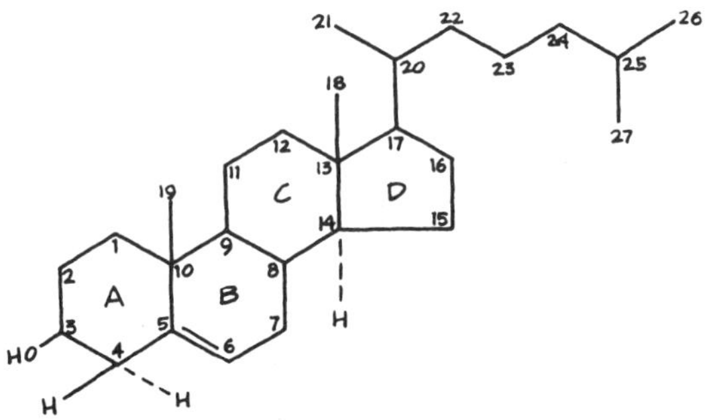

Unfractionated total yeast membranes have been shown to carry out most of the phospholipid biosynthetic reactions depicted in Figure 4 (Waechter and Lester 1971; Steiner and Lester 1972c). Cobon et al. (1974) studied the phospholipid synthesizing capabilities of membrane fractions purified from yeast and provided evidence that the mitochondrial fraction is approximately ten times more active in the synthesis of phosphatidylglycerol and cardiolipin (reactions 7 and 8, Fig. 4) than the microsomal fractions. This general conclusion is consistent with the distribution of the product lipids, which are discussed in the following text. In addition, Cobon et al. (1974) reported that synthesis of phosphatidylcholine either via methylation of phosphatidylethanolamine (reaction 6, Fig. 4) or via incorporation of CDP-choline is very reduced in the mitochondrial fraction as compared with the microsomal fractions. They concluded that the amount of synthesis of phosphatidylcholine observed in the mitochondrial fraction may be due solely to cross-contamination by microsomal material. However, they could

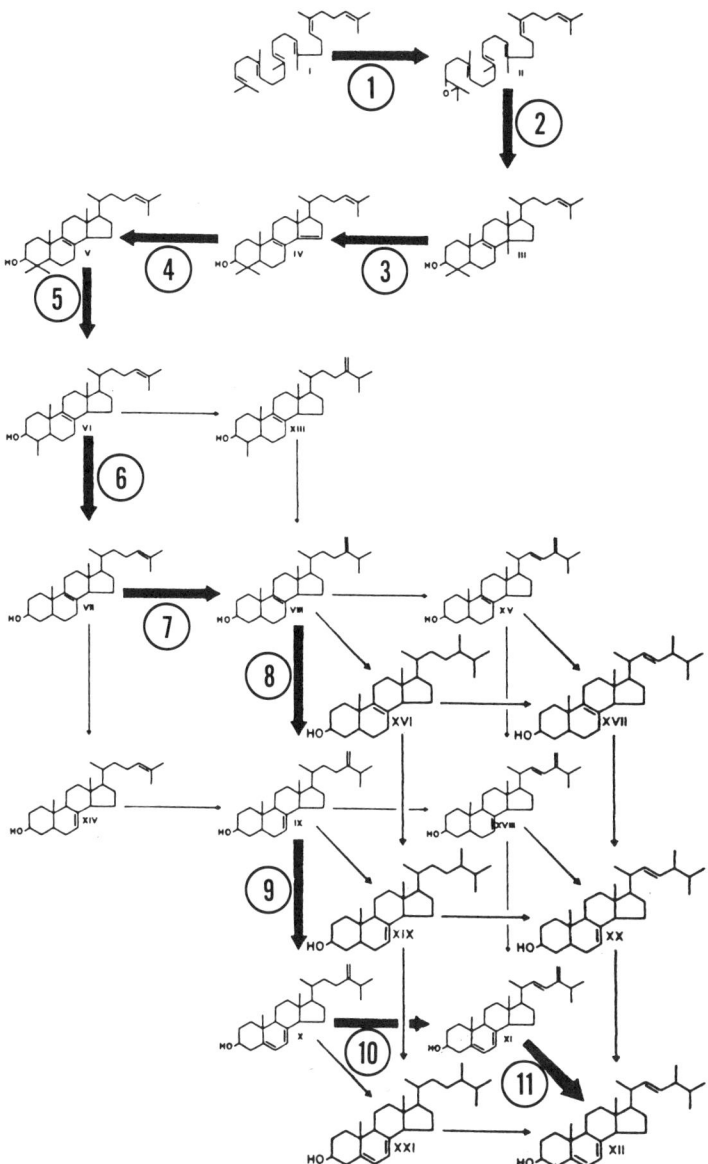

Figure 2 Transformation sequences from squalene to ergosterol. The compounds described with Roman numerals are: (I) squalene; (II) squalene expoxide; (III) lanosterol; (IV) 4,4-dimethylcholesta-8,14,24-trienol; (V) 14-dimethyllanosterol; (VI) 4α-methylzymosterol; (VII) zymosterol; (VIII) fecosterol, ergosta-8,24(28)-dienol; (IX) episterol, ergosta-7,24(28)-dienol; (X) ergosta-5,7,24(28)-trienol; (XI) ergosta-5,7,22,24(28)-tetraenol; (XII) ergosterol ergosta-5,7,22-trienol; (XIII) 4α-methylfecosterol, 4α-methylergosta-8,24(28)-dienol; (XIV) cholesta-7,24-dienol; (XV) ergosta-8,22-dienol; (XVI) ergosta-8-enol; (XVII) ergosta-7,22,24(28)-trienol; (XIX) ergosta-7-enol; (XX) ergosta-7,22-dienol; and (XXI) ergosta-5,7-dienol. (Reprinted, with permission, from Parks 1978.) Roman numerals are used to identify the individual sterol structures. Circled arabic numbers refer to the adjacent reactions.

Table 1 Cellular localization of lipid biosynthetic activities in yeast

	Activity	Cellular location	Reference
Sterol and sterol ester biosynthesis	acetoacetyl-CoA thiolase	cytoplasm; mitochondrion	Kornblatt and Rudney (1971); Trocha and Sprinson (1976); Parks (1978)
	β-hydroxy-β methylglutaryl-CoA synthase	cytoplasm	Trocha and Sprinson (1976); Parks (1978)
	β-hydroxy-β methylglutaryl-CoA reductase	mitochondrion	Boll et al. (1975); Trocha and Sprinson (1976); Parks (1978)
	squalene synthetase	microsomal membrane	Shechter and Bloch (1971); Agnew and Popjak (1978)
	Δ^5-desaturase (reaction 9, Fig. 2)	microsomal membrane	Osumi et al. (1979)
	S-adenosylmethionine Δ^{24}-methyltransferase (reaction 7, Fig. 2)	mitochondrion	Thompson et al. (1974); Parks (1978)
Fatty acid biosynthesis	fatty acid synthetase	cytoplasm	Lynen (1967); Schweizer et al. (1978); Stoops et al. (1978)
	fatty acid desaturase	microsomal membrane	Bloomfield and Bloch (1960)
Phospholipid biosynthesis	glycerol-3-phosphate acyl-transferase (reaction 1, Fig. 4)	microsomal membrane	Schlossman and Bell (1978)
	CTP-phosphatidic acid cytidyl-transferase (reaction 2, Fig. 4)	mitochondrion	Belenduik et al. (1978)
		unfractionated membrane from glucose-grown cells	Steiner and Lester (1972c)
	synthesis of phosphatidyl-inositol (reaction 3, Fig. 4)	unfractionated membrane from glucose-grown cells	Steiner and Lester (1972c)

Reaction	Location	References
synthesis of phosphatidylserine (reaction 4, Fig. 4)	microsomal membrane	Cobon et al. (1974); Carman and Matas (1981)
	mitochondrion	Cobon et al. (1974)
	unfractionated membrane from glucose-grown cells	Steiner and Lester (1972c); Atkinson et al. (1980b); Carson and Waechter (pers. comm.)
decarboxylation of phosphatidylserine (reaction 5, Fig. 4)	microsomal membrane	Cobon et al. (1974); Carman and Matas (1981)
	mitochondrion	Cobon et al. (1974)
	unfractionated membrane from glucose-grown cells	Steiner and Lester (1972c); Atkinson et al. (1980b)
	microsomal membrane	Cobon et al. (1974)
	mitochondrion (rate of synthesis of phosphatidylethanolamine reported at least two times greater in mitochondrion compared with microsomal fraction)	Cobon et al. (1974)
methylation of phosphatidylethanolamine, PMME, and PDME (reaction 6, Fig. 4)[a]	unfractionated total membrane from glucose-grown yeast cells	Waechter and Lester (1973)
	microsomal membrane (reported these activities absent from the mitochondrial fraction)	Cobon et al. (1974)
	mitochondrion (reported these activities absent from the microsomal fraction)	Cobon et al. (1974)
synthesis of phosphatidylglycerol and cardiolipin (reactions 7 and 8, Fig. 4)		
I-1-P synthase (Fig. 6)	cytoplasm	Culbertson et al. (1976a); Donahue and Henry (1981a)

[a] PMME, phosphatidylmonomethylethanolamine; PDME, phosphatidyldimethylethanolamine.

not exlude the possibility of some synthesis of phosphatidylcholine by the mitochondrial fraction. The available data on the distribution of lipid biosynthetic activities in subcellular fractions are summarized in Table 1.

Lipid Composition of Yeast Membrane Fractions

Nurminen et al. (1976) examined yeast membrane fractions derived from glucose-repressed yeast separated on density gradients. They assessed a number of membrane-bound enzyme activities as well as total sterol and phospholipid. Over 80% of the total cellular phospholipid and sterol from such cells was found at a density corresponding to the main peak of oligomycin-insensitive Mg-ATPase. Presumably, this fraction contained primarily plasma membrane. However, the study did not investigate the relative proportions of the different phospholipids present in the preparation. Atkinson (1978) investigated the phospholipid composition of plasma membranes from glucose-grown cells prepared by a modification of the method of Duran et al. (1975). She found no significant difference between total cellular phospholipid composition and the composition of the plasma membrane. Rank et al. (1978) examined the lipid compositions of plasma membrane prepared by several different methods from several strains, all grown in glucose-containing medium. Although they did not compare the lipid compositions of their plasma-membrane fraction to any other membrane fraction, the proportions of the various lipids are quite similar to those obtained by Atkinson (1978), and both sets of data are displayed in Table 2.

The sterol content of the mitochondrion of *S. cerevisiae* has been examined by Bottema and Parks (1980), with respect to the content of the inner and outer mitochondrial membrane. In other organisms it has been reported that sterols are exclusively located in the outer mitochondrial membrane (Colbeau et al. 1971; Hallermayer and Neupert 1974; Comte et al. 1976) and are not detected in the inner membrane. However, Bottema and Parks (1980) did not find any difference in the phospholipid-to-sterol ratio in the inner and outer mitochondrial membrane of wild-type *S. cerevisiae* grown with ethanol as a carbon source. Jakovcic et al. (1971) compared the phospholipid compositions of mitochondrial membranes derived from respiratory-sufficient and -deficient yeast strains grown in several carbon sources. Cardiolipin (diphosphatidylglycerol; Figs. 4 and 5) was found to be enriched in the mitochondrial membrane. In mitochondria derived from galactose-grown, respiratory-sufficient cells, this lipid represents 15.6% of all phospholipid, whereas it represents only 6.5% of the phospholipid derived from unfractionated membrane. In general, the proportion of cardiolipin in yeast cells was found to be related to the degree of mitochondrial activity. Thus, the proportion of cardiolipin is higher in respiratory-sufficient (ρ^+) cells

Table 2 Phospholipid composition of yeast cells and membrane fractions

Genotype of strain	Cellular fraction	Growth condition	Phospholipids (%)[a]										Reference
			PA	CDP-DG	PI	PS	PE	PMME	PDME	PC	PG	CL	
Wild type (ρ^+)	total cell homogenate	stationary galactose-grown culture	1.5	*	12.7	4.6	14.1	*	1.7	28.6	2.7	6.5	Jakovcic et al. (1971)
Wild type (ρ^+)	mitochondria	stationary galactose-grown culture	1.0	*	10.7	3.3	22.0	*	0.9	32.5	2.0	15.6	Jakovcic et al. (1971)
Wild type (ρ^+)	microsomes	stationary galactose-grown culture	0.7	*	27.8	7.3	12.4	*	1.4	40.5	*	1.6	Jakovcic et al. (1971)
Wild type (ρ^+)	total cell lipid extract	late log-phase glucose-grown culture	0.8	*	20.3	7.5	21.6	*	1.4	41.6	*	3.3	Jakovcic et al. (1971)
ρ^-	total cell homogenate	galactose-grown cells	2.2	*	23.9	9.7	14.7	*	1.1	37.6	trace	4.9	Jakovcic et al. (1971)
Wild type (ρ^+)	total cell extract	glucose-grown log-phase cells	2.5	*	28.7	8.0	20.5	*	*	29.2	*	*	Atkinson (1978)
Wild type (ρ^+)	plasma membrane	glucose-grown log-phase cells	1.6	*	35.4	7.5	21.9	*	*	26.0	*	*	Atkinson (1978)
Wild type (ρ^+)	"intracellular" membrane	glucose-grown log-phase cells	3.6	*	33.0	5.8	23.2	*	*	26.0	*	*	Atkinson (1978)
Wild type (ρ^+)	plasma membrane	glucose-grown cells	2.0	*	27.5	*	24.3	*	*	34.3	*	4.2	Rank et al. (1978)

Table 2 (Continued)

Genotype of strain	Cellular fraction	Growth condition	Phospholipids (%)[a]										Reference
			PA	CDP-DG	PI	PS	PE	PMME	PDME	PC	PG	CL	
Wild type (ρ^+)	total cell lipid extract	glucose-grown cells, defined medium (no lipid precursor supplement added)	*	*	*	13.7	21.0	0.5	2.5	36.2	*	*	Waechter and Lester (1973)
Wild type (ρ^+)	total cell lipid extract	(1 mM MME)	*	*	*	14.6	15.7	3.8	12.1	32.1	*	*	Waechter and Lester (1973)
Wild type (ρ^+)	total cell lipid extract	(1 mM DME)	*	*	*	14.8	13.9	trace	22.2	31.7	*	*	Waechter and Lester (1973)
Wild type	total cell lipid extract	(1 mM choline)	*	*	*	14.7	16.3	trace	0.5	51.3	*	*	Waechter and Lester (1973)
ino1-13	total cell lipid extract	glucose-grown cells (75 μM inositol)	*	*	28.5	4.8	12.9	0.8	1.9	37.6	*	*	Henry et al. (1981)
	total cell lipid extract	(starved for inositol for 4 hr prior to lipid extraction, cells still viable)	*	*	7.7	10.6	19.5	0.9	1.3	47.6	*	*	Henry et al. (1981)

Strain	Sample	Condition										Reference
cho1 ino1-13	total cell lipid extract	(75 μM inositol, 1 mM choline)	*	29.0	n.d.	4.8	n.d.	*	n.d.	57.8	*	*	Atkinson et al. (1980a)
cho1 ino1-13	total cell lipid extract	(75 μM inositol, 1 mM ethanolamine)	*	32.8	n.d.	15.0	1.0	*	1.1	41.7	*	*	Atkinson et al. (1980a)
ino4-BSI	total cell lipid extract	(75 μM inositol)	*	34.4	4.1	19.8	5.5	*	13.6	13.6	*	*	Henry et al. (1981)
ino4-26	total cell lipid extract	(75 μM inositol)	*	37.4	4.5	18.3	1.3	*	7.1	17.6	*	*	Henry et al. (1981)
ino4-40	total cell lipid extract	(75 μM inositol)	*	32.2	5.9	21.0	1.9	*	10.7	15.3	*	*	Henry et al. (1981)
opi3	total cell lipid extract	(75 μM inositol)	*	26.8	5.9	5.8	29.7	*	15.2	5.7	*	*	Henry et al. (1981)
ino2-2	total cell lipid extract	(75 μM inositol)	3.8	38.3	4.9	27.9	1.3	*	8.2	11.3	*	*	S. Henry (unpubl.)
ino2-21	total cell lipid extract	(75 μM inositol)	4.9	36.5	4.6	28.3	1.3	*	9.5	11.9	*	*	S. Henry (unpubl.)

Abbreviations and symbols: PA, phosphatidic acid; CDP-DG, cytidine diphosphate diglyceride; PI, phosphatidylinositol; PS, phosphatidylserine; PE, phosphatidylethanolamine; PMME, phosphatidylmonomethylethanolamine; PDME, phosphatidyldimethylethanolamine; PC, phosphatidylcholine; PG, phosphatidylglycerine; CL, cardiolipin; MME, N-monomethylethanolamine; DME, N,N-dimethylethanolamine; n.d., not detected; lipid reported absent; *, no data reported.

[a] Percentage of total lipid phosphorous associated with the individual lipid fraction. The data are either taken directly or calculated from data in the stated reference. The percentages do not add up to 100% in most cases. This is because not all detectable phospholipids were surveyed in all cases. Also, in some cases, notably those obtained from Jakovcic et al. (1971), variable proportions of lysophosphatides (i.e., partially deacylated phospholipids) were reported in the various preparations and are not listed.

compared with respiratory-deficient (ρ^-) cells, and in galatose-grown cells as opposed to glucose-grown cells. Another feature of the mitochondrial membrane is its relatively high content of phosphatidylethanolamine. In the mitochondria of galactose-grown ρ^+ cells, phosphatidylethanolamine represented approximately 22% of all phospholipid, as compared with 14% in the total unfractionated membrane (Jakovcic et al. 1971). Some of these proportions are displayed in Table 2, along with total cellular phospholipid compositions reported for yeast wild-type and mutant strains grown under a variety of conditions.

STEROL BIOSYNTHESIS AND REQUIREMENTS

The principal yeast sterol is ergosterol (Fieser and Fieser 1959). Ergosterol shares many structural similarities with cholesterol (Fig. 1), but it differs in several regards. Ergosterol has a conjugated diene system in the B ring, an unsaturated side chain, and a methyl group attached to C-24 on the side chain. The basic pathways for sterol biosynthesis in yeast have recently been reviewed in detail (Nes and McKean 1977; Parks 1978) and, for this reason, will be discussed only briefly here. The pathways for ergosterol synthesis have much in common with the pathways for cholesterol biosynthesis, which have been well characterized (Bloch 1965). Ergosterol, like cholesterol, derives most of its carbon from acetate (Ottke et al. 1951) via acetyl-CoA. Acetyl-CoA serves as the precursor of mevalonic acid, which, in a series of reactions similar to those in cholesterol biosynthesis, is converted to squalene. The synthesis of squalene can occur anaerobically in yeast (Klein et al. 1954). However, the conversion of squalene to ergosterol (depicted in Fig. 2) requires molecular oxygen for the formation of squalene epoxide (structure II, Fig. 2) and the demethylation of lanosterol (structure III, Fig. 2) to form zymosterol (structure VII, Fig. 2) (Moore and Gaylor 1968; Miller and Gaylor 1970; Schechter et al. 1970; Miller et al. 1971; Alexander et al. 1974; Barton et al. 1974c, 1975b; Parks et al. 1974; Mitropoulos et al. 1976; Parks 1978). Beyond zymosterol, the pathway diverges, providing several routes to the end product ergosterol (structure XII, Figs. 1 and 2). This divergence of the pathway is supported by studies of mutants defective in sterol biosynthesis (Parks et al. 1972; Parks 1978).

The remainder of this section focuses upon experiments using yeast cells to probe the structural specificity of the sterol requirements of various membranes. The experiments that have been performed to probe membrane sterol requirements in yeast have been carried out on anaerobically grown cells, sterol mutants, or cells treated with inhibitors of sterol biosynthesis. The rationale for each type of experiment is somewhat different, and each is discussed in turn.

Experiments with Anaerobic Cells

The synthesis of unsaturated fatty acids and ergosterol in yeast occurs only in the presence of oxygen. Strictly anaerobically grown cells, therefore, require both unsaturated fatty acid and sterol (Andreason and Stier 1953, 1954). A similar requirement for both sterol and unsaturated fatty acid is observed in yeast mutants with defects in porphyrin biosynthesis (*ole* mutants; see Table 3) due to the involvement of hematin in the synthesis of both classes of compounds (Bard et al. 1974; Parks 1978; Haslam and Astin 1979). Anaerobically grown wild-type cells accumulate squalene and synthesize no sterol (Klein 1955). The ability of a particular sterol to support growth of anaerobic yeast provides a test of the structural specificity of the membrane sterol requirement.

Proudlock et al. (1968) tested the overall ability of various sterols to support growth of anaerobic yeast and found that cholesterol could be substituted for the naturally occurring ergosterol. However, Nes and his colleagues have presented evidence that cholesterol is somewhat less effective in stimulating anaerobic growth of yeast than ergosterol (Nes et al. 1976). In keeping with this observation, Hossack and Rose (1976) reported that spheroplasts made from anaerobically grown cells enriched with a sterol such as ergosterol were more resistant to osmotic stretching than cells grown with a sterol such as cholesterol, which has a saturated side chain. A. Rose and his colleagues also found that the unsaturated alkyl side chain in the sterol was important in conferring ethanol tolerance, thought to be a property of the plasma membrane (Thomas et al. 1978). Since cholesterol and ergosterol differ in a number of structural features (Fig. 1), Nes et al. (1978) compared the anaerobic growth stimulation of pairs of sterols that differed only by a single structural feature. They found that the structure of the alkyl side chain is of major importance, including the methyl group attached to C-24 and the double bond between C-22 and C-23 (Δ^{22}). Of much less importance is the existence of a $\Delta^{5,7}$-diene system in the B ring. In addition, it was found that the introduction of a methyl group at C-4 or C-14 rendered the sterol unable to support growth. With the exception of one specific analog (Δ^7 analog of cholesterol), all of the sterol growth supplements were recovered unmodified from the lipids extracted from the anaerobically grown cells, which indicates that the growth of the cells reflects the properties of the supplied sterols.

Furthermore, Nes et al. (1978) reported that removal of the hydroxyl group of C-3 obliterated the biological activity of the sterol and rendered it unable to support detectable anaerobic growth. However, Lala et al. (1979) found that sterol methyl ethers blocked at the hydroxyl group were effective in supporting growth of a mutant (GL7; Table 3) that synthesizes no sterol but accumulates oxidosqualene (Gollub et al. 1977). It was shown that, under strictly anaerobic conditions, the yeast mutant cells were unable to

Table 3 Yeast lipid mutants

	Genotype	Phenotype and identified biochemical defects	Reference
Sterol mutants	erg1	requires ergosterol, accumulates squalene, defective in squalene epoxidase. One allele is temperature-sensitive (defect: reaction 1, Fig. 2).	Karst and Lacroute (1974, 1977)
	erg2 (pol2)	nystatin-resistant, no auxotrophic requirement, defective in $\Delta^8 \rightarrow \Delta^7$-isomerase, accumulates a mixture of sterol intermediates but makes no ergosterol (defect: reaction 8, Fig. 2).	Molzahn and Woods (1972); Bard et al. (1977); Parks (1978)
	erg3 (pol3, nys3)	nystatin-resistant, no auxotrophic requirement for ergosterol, defective in 5-dehydrogenase; accumulates a mixture of sterol intermediates but makes no ergosterol (defect: reaction 9, Fig. 2).	Molzahn and Woods (1972); Bard et al. (1977); Parks (1978)
	erg5 (pol5)	nystatin-resistant, no auxotrophic requirement for ergosterol, defective in 22-dehydrogenase; accumulates a mixture of sterol intermediates but makes no ergosterol (defect: reaction 10, Fig. 2).	Molzahn and Woods (1972); Bard et al. (1977); Parks (1978)
	erg6 (pol1, nys1)	nystatin-resistant, no auxotrophic requirement for ergosterol, defective in 24-methyltransferase; accumulates mixture of sterol intermediates but makes no ergosterol (defect: reaction 7, Fig. 2).	Molzahn and Woods (1972); Bard et al. (1977); Parks (1978)
	erg7	temperature-sensitive growth (36°C); not restored by ergosterol at 36°C. Requires ergosterol for growth at intermediate temperature; growth-independent of ergosterol at 22°C.	Karst and Lacroute (1977)

	erg8–erg12	Defect in 2,3-oxidosqualene cyclase; accumulates oxidosqualene. (Mutant described as erg12 by Gollub et al. [1977] has a similar, but nonconditional, biochemical lesion.) (defect: reaction 2, Fig. 2). temperature-sensitive growth similar to erg7. Accumulates no squalene. Mutants erg9–erg11 are defective in conversion of farnesyl pyrophosphate to squalene. Mutants erg8 and erg12 have lesions in conversion of mevalonic acid to mevalonic acid pyrophosphate (defects: prior to reaction sequence, Fig. 2).	Karst and Lacroute (1977)
	SG1 (has been referred to as nys1 and erg11)	nystatin-resistant; accumulates mixtures of sterols, blocked in demethylation of lanosterol at C-14 (reaction 3, Fig. 2).	Gollub et al. (1974); Trocha et al. (1977)
	GL7 (has been referred to as erg12)	this mutant has two genetic lesions: hem3 plus an additional lesion in ergosterol biosynthesis that leads to accumulation of oxidosqualene (possibly similar to erg7). The mutant's aerobic and anaerobic growth requirements have been studied extensively (Buttke and Bloch 1980; Lala et al. 1980). The genetic lesion in ergosterol biosynthesis appears to be lethal outside the hem background (defect: reaction 2, Fig. 2).	Gollub et al. (1977)
Heme mutants	cyc4	cytochrome deficiency; reversed by δ-aminolevulinate.	Sanders et al. (1973)

Table 3 (Continued)

	Genotype	Phenotype and identified biochemical defects	Reference
Heme mutants (continued)	ole3, hem1 (cyd1)	defective in δ-aminolevulinate synthetase; cytochrome-deficient; requires δ-aminolevulinate for restoration of cytochrome function; requires oleic acid and methionine for growth; growth enhanced by ergosterol; nystatin-resistant.	Resnick and Mortimer (1966); Bard (1972); Sanders et al. (1973); Bard et al. (1974); Woods et al. (1975); Gollub et al. (1977); Haslam and Astin (1979)
	ole2, hem3, olerg2	requires oleic acid and methionine for growth; nystatin-resistant; growth enhanced by ergosterol; cytochrome deficiency reversed by protoporphyrin IX, hematoporphyrin. (hem3 mutant reported allelic ole2 and deficient in uroporphyrin-I synthase.)	Resnick and Mortimer (1966); Bard (1972); Karst and Lacroute (1973); Bard et al. (1974); Gollub et al. (1977); Haslam and Astin (1979)
	ole4, hem2 (olerg4)	requires oleic acid and methionine; nystatin-resistant; growth enhanced by ergosterol; cytochrome deficiency not reversed by prophyrin intermediates. (hem2 mutant reported allelic to ole4; deficient in δ-aminolevulinate dehydratase and reversible by heme and protoporphyrin.)	Resnick and Mortimer (1966); Bard (1972); Karst and Lacroute (1973); Bard et al. (1974); Gollub et al. (1977); Haslam and Astin (1979)
	hem4, hem5	requires oleic acid and ergosterol but not methionine, growth restored by heme. hem4 responds to protoporphyrin, but hem5 does not.	Gollub et al. (1977)
	(olerg1–olerg6)	growth requirements and phenotype similar to ole2, ole3, and ole4. olerg2 and olerg4 allelic to ole2 and ole4, respectively.	Karst and Lacroute (1973); Haslam and Astin (1979)

Category	Gene	Description	References
Fatty acid mutants	ole1	requires oleic acid or other unsaturated fatty acid (see Table 4) (defective in fatty acid desaturase).	Resnick and Mortimer (1966); Keith et al. (1969)
	fas1, fas2	requires saturated fatty acid for growth (see Table 4); structural genes for the two subunits of fatty acid synthetase.	Schweizer and Bolling (1970); Henry and Keith (1971); Schweizer et al. (1974, 1977, 1978); Dietlein and Schweizer (1975)
	acc1	requires saturated fatty acid for growth; lacks acetyl-CoA carboxylase activity.	Mishina et al. (1980)
	acc2	requires saturated fatty acid for growth; lacks biotin; apoacetyl-CoA carboxylase ligase.	Mishina et al. (1980)
	B-201, B-53	fails to repress acetyl-CoA carboxylase in the presence of exogenous fatty acid; defective in acyl-CoA synthase.	Kamiryo et al. (1976)
Inositol-requiring mutants	ino1	requires inositol; defective in the structural gene for I-1-P synthase (defective in reaction depicted in Fig. 6).	Culbertson and Henry (1975); Donahue (1979); Donahue and Henry (1981a).
	ino2, ino4	requires inositol, lacks I-1-P synthase activity in crude extracts, has pleiotropic defects in phospholipid methylation; accumulates less phosphatidylcholine than wild type (see Table 2) (defective in reaction 6, Fig. 4; also in reaction depicted in Fig. 6).	Culbertson and Henry (1975); Culbertson et al. (1976b); Henry et al. (1981)
	ino3, ino5–ino10	requires inositol, lacks I-1-P synthase activity in crude extracts; some are respiratory-deficient; not otherwise characterized.	Culbertson and Henry (1975); Culbertson et al. (1976b); Donahue et al. (1978)

Table 3 (Continued)

	Genotype	Phenotype and identified biochemical defects	Reference
Inositol-secreting mutants	opi1, opi2, opi4	secretes inositol, constitutive for I-1-P synthase (defect: regulation of reaction depicted in Fig. 6).	Greenberg (1980)
	opi3	secretes inositol, linked to ino4; not constitutive for I-1-P synthase; has defects in phosphatidylcholine biosynthesis (Table 2) (defect: reaction 6, Fig. 4).	Greenberg (1980); Henry et al. (1981); Greenberg et al. (1982)
Choline/ethanolamine-requiring mutants	cho1	requires choline or ethanolamine for growth; defective in phosphatidylserine synthesis; secretes inositol (defect: reaction 4, Fig. 4).	Atkinson et al. (1980a,b); Kovac et al. (1980); Letts (1980)
	172α	requires choline when grown in the presence of inositol. Phosphatidyl methyltransferase activity deficient when cells are grown in the presence of inositol (defect: reaction 6, Fig. 4).	Yamashita and Oshima (1980)
Mutants temperature-sensitive for phospholipid synthesis	ts-dam303	temperature-sensitive for total phospholipid synthesis; believed to be defective in glycerol-3-phosphate acylation. Secretes inositol at the permissive temperature (defect: reaction 1, Fig. 4).	Letts (1980)

demethylate the sterol methyl ethers to form free sterol. Since no free sterol was available for esterification, these results suggest that esterification of the 3-hydroxyl group is not necessary for the biological function of sterols in yeast. Sterol esters are primarily located in lipid particles rather than in the membrane in yeast and may serve strictly as a storage function (Parks 1978; Parks et al. 1978; Taylor and Parks 1978).

Nes et al. (1978) also reported that lanosterol (structure III, Fig. 2) will not support growth of anaerobic wild-type yeast cells. Likewise, cycloartenol, the 9,19-cyclopropane isomer of lanosterol, was not growth-supporting in their studies. In constrast, Buttke and Bloch (1980) report that both of these sterols support growth of anaerobically grown cells of a yeast mutant strain (GL7) that synthesizes no sterol but accumulates oxidosqualene (Gollub et al. 1977). The C-14 methyl derivative of cycloartenol, cyclolaudenol, was reported to be a more effective anaerobic growth supplement for the mutant than the other two sterols. However, under aerobic conditions, lanosterol was not able to support growth of the mutant, whereas the other two sterols were growth-supporting. All three sterols were recovered unmodified from the lipids of the anaerobically grown cells, whereas under aerobic conditions, the cells did carry out some modification of the supplemented sterols. These results suggest that anaerobic cells have a limited capacity to modify existing sterols and that the sterol requirements for aerobic and anaerobic growth may be quite different.

Under anaerobic growth conditions, the yeast cell contains only promitochondria (Jollow et al. 1968) and necessarily relies upon fermentative growth. Therefore, experiments testing the ability of particular sterols to promote anaerobic growth of yeast do not reflect the requirements of the mitochondrion. However, the promitochondria of anaerobically grown cells supplemented with sterol and unsaturated fatty acid have better developed inner membranes and cristae than do the promitochondria from lipid-depleted cells (Morpurgo et al. 1964; Lukins et al. 1966; Jollow et al. 1968; Wallace et al. 1968; Criddle and Schatz 1969; Paltauf and Schatz 1979; Plattner and Schatz 1979). Furthermore, lipid synthesis is required for the conversion of anaerobic promitochondria to mitochondria, and the required lipid synthesis is dependent upon protein synthesis both in the mitochondria and in the cytoplasm (Clark-Walker and Linnane 1967; Vary et al. 1970; Gordon and Stewart 1971, 1972; Gordon et al. 1972a,b). However, if the cells are adequately lipid-supplemented during anaerobic growth immediately prior to the shift, ongoing lipid synthesis during the adaptation to aerobic conditions has been reported to be unnecessary (Rouslin 1979).

Sterol Mutant Studies

Yeast sterol mutants have been isolated as auxotrophs requiring ergosterol and/or as polyene-antibiotic-resistant mutants (Table 3). Originally, the auxotrophic mutants were designated *erg*, and the polyene-antibiotic-resistant mutants were designated *pol* or *nys* (for nystatin-resistant). However, all yeast sterol mutants are now being referred to as *erg*. For a discussion of the yeast sterol mutant nomenclature, see Bard et al. (1977) or Parks (1978); also see Table 3.

Polyene antibiotics including nystatin are known to alter the permeability of natural and artificial membranes (Cirillo et al. 1964; Lampen 1966; Kinsky 1970; Holz 1974). The polyene antibiotics are thought to interact with specific membrane sterols (Cass et al. 1970; Norman et al. 1972) and are not known to specifically inhibit lipid synthesis. For this reason, the polyene antibiotics have been extremely useful in the selection of yeast sterol mutants. Yeast mutants with altered sterol compositions are notably resistant to the effects of the polyene antibiotics that include nystatin (Woods 1971; Bard 1972; Molzahn and Woods 1972; Fryberg et al. 1974). (Similar resistance to polyene antibiotics has been reported in sterol mutants of other fungi as well [Grindle 1973].) Many polyene-resistant mutants (*erg*) synthesize no ergosterol but are not auxotrophic for ergosterol or other sterols. They have altered sterol compositions and may accumulate several intermediates of ergosterol biosynthesis (Thompson et al. 1971; Woods 1971; Molzahn and Woods 1972; Parks et al. 1972; Barton et al. 1974a,b, 1975a; Trocha et al. 1974, 1977; Bailey et al. 1976b; Parks 1978; Parks et al. 1978). The fact that these mutants synthesize no ergosterol and yet grow with no sterol supplement suggests that the intermediates accumulated by them satisfy the basic membrane sterol requirement. These mixtures of intermediates can be quite complex (Bard et al. 1977; Parks 1978). Because these mutants accumulate intermediates, they cannot be used in experiments designed to systematically manipulate sterol composition. However, they can, under some circumstances, provide insight into the effects of altered sterol composition on various cellular membranes. For example, Kleinhans et al. (1978) reported altered membrane permeability to nickel ion in a yeast sterol double mutant (*erg2, erg6*) that makes no ergosterol and accumulates, primarily, the intermediate zymosterol (structure VII, Fig. 2). In another mutant (*erg3*) that makes no ergosterol, but that accumulated several intermediates of sterol biosynthesis (notably $\Delta^{7,22}$-ergostadienol; structure XX, Fig. 2), Parks et al. (1978) found that transition temperatures of a number of mitochondrial enzymes were shifted compared with wild type. This same mutant shows temperature-sensitive growth on nonfermentable carbon sources, whereas fermentative growth was relatively unaffected (Thompson and Parks 1974). These studies indicate that sterol composition is critical for mitochondrial function and that the mitochondrion is more sensitive to sterol composition changes than other cellular membranes. However, a variety of yeast sterol mutants accumulating different mixtures of intermediates were found by Parks et al. (1981) to be respiratory-competent despite alterations in the activities of a number of mitochondrial enzymes.

The sterol mutants described above, designated *erg*, are not auxotrophic for ergosterol but have defects in the actual pathways for ergosterol biosynthesis. Paradoxically, the mutants that demonstrate a requirement for ergosterol (or whose growth is enhanced by ergosterol) are for the most part not sterol mutants but heme mutants (Bard et al. 1974). The heme mutants of yeast (*ole2, ole3, ole4; hem1-hem5; cyc4; olerg1-olerg6*) display a complex phenotype, which is summarized in Table 3. The mutants are auxotrophic for unsaturated fatty acid and methionine. Although not all heme mutants are ergosterol auxotrophs, their growth is enhanced by sterol supplement. They are also respiratory-deficient and nystatin-resistant (Bard et al. 1974). The first mutants of this type to be isolated (Resnick and Mortimer 1966) were designated *ole2, ole3,* and *ole4* because they were auxotrophic for oleic acid. Other series of similar mutants were isolated on the basis of a simultaneous requirement for ergosterol and unsaturated fatty acid (Karst and Lacroute 1973; Gollub et al. 1974, 1977). A similar phenotype was also reported for mutants defective in δ-aminolevulinate synthetase (Sanders et al. 1973). The auxotrophic requirements of these mutants all reflect the involvement of cytochrome or hematin coenzymes in the various metabolic pathways (Bard et al. 1974; Gollub et al. 1977). The heme mutants are blocked in the demethylation of lanosterol (Bard et al. 1974) due to the involvement of cytochrome in this reaction. The basis of the oleic acid requirement will be discussed in the section on fatty acid biosynthesis. If δ-aminolevulinic acid or other porphorin intermediates are supplied, the requirements are all alleviated (Woods et al. 1975; Gollub et al. 1977). The heme-deficient phenotype and the various loci that confer it are discussed in detail by Gollub et al. (1977) and Haslam and Astin (1979). The nomenclature associated with these mutants is unfortunately extremely confused, due to the complex phenotype and the many allelic sets of mutants that have been isolated. Haslam and Astin (1979) suggest the use of *ole*, which has historical precedence, since it was first used by Resnick and Mortimer (1966). However, Gollub et al. (1977) suggest the use of *hem* (for heme). This nomenclature is more descriptive of the biochemical lesions involved. It would be helpful if a single usage could be universally adopted. (The loci and their designations are listed in Table 3.) Virtually all of the mutants isolated in yeast as ergosterol-requiring are heme mutants. The apparent exceptions are mutants isolated by Karst and Lacroute (1974, 1977), which require only ergosterol and accumulate the intermediate squalene (*erg1*; Table 3). Karst and Lacroute (1977) have also described a series of conditional-lethal temperature-sensitive mutants, *erg7-erg12* (Table 3), defective in metabolic steps prior to squalene (structure I, Fig. 2). In addition, Gollub et al. (1977) described a genetic lesion (GL7) that results in accumulation of oxidosqualene (structure II, Fig. 2). However, the growth requirements of this mutant have only been analyzed in a *hem3* (*ole2*) background (Table 2) since, for reasons that are not understood, this mutation appears to be lethal in a wild-type (HEM^+) background.

The pleiotropic heme mutants can be manipulated to study mitochondrial functions despite the respiratory deficiency that exists when the mutants are not supplemented with an intermediate in porphyrin biosynthesis. When the *ole3* mutant is fed low levels of δ-aminolevulinate, respiratory activity is restored and sterol biosynthesis is not (Astin et al. 1977; Haslam and Astin 1979). Thus, in appropriately manipulated *ole3* cells, the sterol requirements for mitochondrial functions can be assessed. In these experiments, fatty acid supplementation is held constant, and sterol composition is systematically manipulated. Basically, it is observed that sterol deprivation raises the transition temperatures of mitochondrial-bound enzymes (Haslam and Astin 1979) and uncouples oxidative phosphorylation due to loss of the proton gradient (Astin and Haslam 1977). Thus, it is clear that many mitochondrial functions are highly sensitive to the sterol content of the mitochondrial membrane.

Experiments Using Inhibitors of Sterol Biosynthesis

Azasterols are sterol analogs that inhibit several steps of sterol biosynthesis. Yeast cells generating their energy by respiration are much more sensitive to the effects of azasterol than are cells growing fermentatively (Bailey et al. 1976a). The inhibition of sterol biosynthesis is concentration-dependent, some steps are more sensitive than others (Hays et al. 1977), and the concentrations of azasterol can be manipulated in order to change the proportions and types of accumulated intermediates (Parks et al. 1978). Mitochondria were isolated from a wild-type yeast culture treated with a concentration of azasterol that caused the intermediate ignosterol (8,14-ergostadien-3β-ol) to accumulate to approximately 50% of the total cellular sterol (Parks et al. 1978). The accumulation of ignosterol is due to the inhibition of the Δ^{14}-reductase (reaction 4, Fig. 2) (Hays et al. 1977). The intermediate ignosterol, which is not normally seen in yeast, is presumably produced by subsequent alterations of the sterol intermediate retaining the double bond at Δ^{14}. Despite the fact that approximately half of the cellular sterol was ignosterol, the purified mitochondria from the azasterol-treated cells contained virtually none of this intermediate and instead had accumulated ergosterol almost exclusively (Parks et al. 1978). On the basis of this result and the observation that cholesterol is replaced by ergosterol in the mitochondria of cells making the transition from anaerobic to aerobic growth (Gordon and Stewart 1971), Parks and his collaborators suggest that there is selective use of ergosterol in the mitochondria and perhaps in other membranes as well. They suggest that the selectivity may reside in the formation and conversion of sterol esters (Bailey and Parks 1975; Parks 1978; Taylor and Parks 1978). In cells blocked with azasterol at concentrations sufficient to totally prevent the conversion of the intermediate ignosterol to ergosterol, prelabeled esterified ergosterol was rapidly converted to the free form, presumably for preferential use in the cellular membranes (Parks et al. 1978). Recently, McCammon and Parks (1981) also reported the use of S-adenosylhomocysteine analogs to inhibit the enzyme S-adenosylmethionine: Δ^{24}-sterol-C-methyltransferase (reaction 7, Fig. 2). One such analog, sinefungin, inhibited growth of *S. cerevisiae* and led to the accumulation of zymosterol (structure VII, Fig. 2) in vivo. However, it is clear that many S-adenosylmethionine-dependent methylation reactions, in addition to those involved in sterol biosynthesis, are affected, since cell growth was found to be inhibited at concentrations of the analog that did not affect sterol biosynthesis (McCammon and Parks 1981). (S-adenosylmethionine-dependent methylation is also involved in synthesis of phosphatidylcholine in yeast, a subject to be discussed later in the text.)

FATTY ACID BIOSYNTHESIS AND REQUIREMENTS

Fatty acids, together with glycerol-3-phosphate, are the general precursors of the phospholipids. Fatty acid chain elongation and desaturation in yeast occur in a manner similar to other eukaryotes rather than bacteria. In eukaryotes, the de novo formation of long-chain fatty acids is catalyzed by fatty acid synthetase, a large multifunctional enzyme complex found in the cytoplasm (Lynen 1969, 1980; Oesterhelt et al. 1969; Stoops et al. 1978). In yeast, both genetic and biochemical evidence has shown that the fatty acid synthetase complex is composed of two nonidentical subunits (Schweizer et al. 1978; Stoops and Wakil 1978; Stoops et al. 1978). This enzyme complex, together with the biotin-dependent enzyme acetyl-CoA carboxylase, carries out the formation of long-chain saturated fatty acids. The carboxylase is responsible for the formation of malonyl-CoA, the precursor used by fatty acid synthetase in each successive two-carbon addition to the growing fatty acid chain. The mechanism of this reaction sequence and the structural characteristics of fatty acid synthetase are among the best studied in all of yeast biochemistry due to the classic work of Lynen and his co-workers (Lynen 1967, 1969, 1980). The long-chain saturated fatty acids released from the fatty acid synthetase complex as CoA derivatives are then desaturated by an oxygen-requiring desaturase that introduces a double bond in the preformed hydrocarbon chain (Bloomfield and Bloch 1960). In contrast, fatty acid synthesis in *Escherichia coli* is carried out by readily separable enzymes not contained in a multifunctional complex (Vagelos et al. 1969; Prescott and Vagelos 1972). Also, in *E. coli*, the desaturation step is integral to the process of fatty acid chain elongation (Helmkamp and Bloch 1969). Silbert (1975) provides discussion of the differences in the *E. coli* and *S. cerevisiae* pathways for fatty acid synthesis and desaturation.

Yeast synthesizes primarily 16- and 18-carbon fatty acids (Keith et al. 1972; Okuyama et al. 1979; Lynen 1980), which are depicted in Figure 3. In addition, some longer-chain fatty acids have been detected (Nurminen and

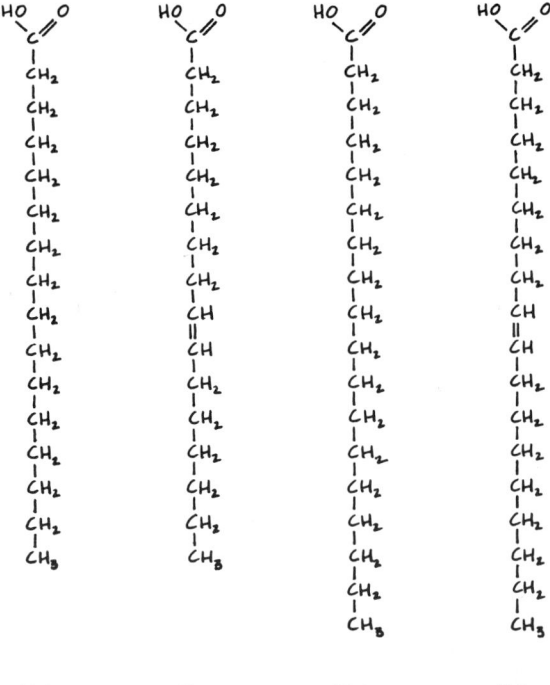

16:0   **16:1**   **18:0**   **18:1**

Figure 3 Basic structures of the commonly occurring yeast fatty acids. (16:0) Palmitic acid; (16:1) palmitoleic acid (16:1Δ^9 *cis*); (18:0) stearic acid; (18:1) oleic acid (18:1Δ^9 *cis*).

Suomalainen 1971; Welch and Burlingame 1973). The synthesis mechanism of fatty acids longer than 18 carbons is not clear since yeast fatty acid synthetase produces a mixture of 16- and 18-carbon fatty acids (Okuyama et al. 1979; Lynen 1980). However, it is known that yeast is capable of fatty-acid-synthetase-independent elongation of preformed fatty acids (Henry and Keith 1971; Orme et al. 1972). This elongation pathway is malonyl-CoA-independent (Schweizer et al. 1974) and takes place outside of the mitochondria. Although yeast apparently does not carry out total degradation of fatty acids (Bloomfield and Bloch 1960), it is capable of limited shortening of exogenously supplied fatty acids under some circumstances (Schweizer et al. 1974).

Mutants Defective in Fatty Acid Biosynthesis

Both saturated- and unsaturated-fatty-acid-requiring yeast mutants have been isolated. Complementation and linkage analysis of yeast-saturated fatty acid auxotrophs originally identified two loci, designated *fas1* and *fas2* (Table 3) (Schweizer and Bolling 1970; Henry and Fogel 1971; Burkl et al. 1972; Kühn et al. 1972; Schweizer et al. 1977, 1978). In addition, Mishina et al. (1980) have described a series of saturated-fatty-acid-requiring yeast mutants (*acc1, acc2*) that represent two additional loci, unlinked to each other or to *fas1* and *fas2*. The *acc1* mutants have been shown to be defective in acetyl-CoA carboxylase, and the *acc2* mutants are defective in biotin: apocarboxylase ligase (see Table 3).

Schweizer and his colleagues have shown that the *fas1* and *fas2* loci are the structural genes for the two fatty acid synthase subunits (Kühn et al. 1972). Many, but not all, *fas1* and *fas2* mutants have material that cross-reacts with antibody prepared against purified wild-type fatty acid synthetase (Dietlien and Schweizer 1975). The defective enzyme complexes purified from such mutants retain some of the partial reactions of fatty acid chain elongation (Schweizer et al. 1978). Many of the mutants totally lose only one of the eight identifiable partial reactions, and members of each such class of mutants show a characteristic genetic complementation pattern. In such combined genetic and biochemical analyses, it has been possible for Schweizer and his co-workers to assign all of the partial activities of the complex to specific genetic complementation groups. In addition, many classes of mutants have pleiotropic changes in the partial activities, involving increases or decreases in activities other than the one totally obliterated by the mutation. These additional changes no doubt reflect the interactions between the different domains in the multifunctional enzyme complex. A detailed interpretation of the complexities of the genetic complementation and the partial activity pattern of *fas1* and *fas2* mutants is given by Schweizer et al. (1974, 1978).

Some *fas1* and *fas2* mutants fail to demonstrate any interallelic complementation. Such mutants may contain nonsense mutations (Henry and Fogel 1971; Tauro et al. 1974; Schweizer et al. 1977, 1978). However, clearly not all such mutants are of the nonsense type despite the fact that virtually all of the noncomplementing mutants fail to produce immunologically cross-reacting material or else produce cross-reacting material of molecular weight much lower than the native complex (Dietlein and Schweizer 1975; Schweizer et al. 1978). The antibody used in these studies was prepared against the holoenzyme (Dietlein and Schweizer 1975). Yet, the noncomplementing mutants, carrying a genetic lesion of only one of the two subunits of the enzyme complex, fail to produce any cross-reacting material. The subunit that is unaffected by mutation is not detected immunologically. It has been proposed that, in such mutants, the mutationally altered subunit, as well as the unaltered subunit, may be susceptible to proteolytic degradation because it cannot associate into an enzyme complex (Dietlein and Schweizer 1975; Schweizer et al. 1978). However, Schweizer et al. (1978) point out that similar results could be obtained if the synthesis of both subunits were under mutual positive control. It is known that in wild-type yeast, the enzyme complex is constitutively synthesized. However, the mechanism whereby the wild-type cell coordinates the synthesis of equal amounts of the products of two unlinked genes remains to be elucidated.

The yeast mutants that require unsaturated fatty acid (see *ole* mutants, Table 3) (Resnick and Mortimer 1966) are of two types: those that result in a simple defect in fatty acid desaturation and those that result in a defect in heme biosynthesis. The *ole1* mutants have a primary defect in fatty acid desaturation (Keith et al. 1969) and are unable to introduce a double bond at the Δ^9 position of 16- and 18-carbon fatty acids (see Fig. 4 for the fatty acid structures and double bond positions). These mutants are presumably defective in a structural gene for a component of the fatty acid desaturase (Keith et al. 1969), a membrane-bound enzyme (Bloomfield and Bloch 1969) that is presumably microsomal in location. Since *ole1* mutants are equally

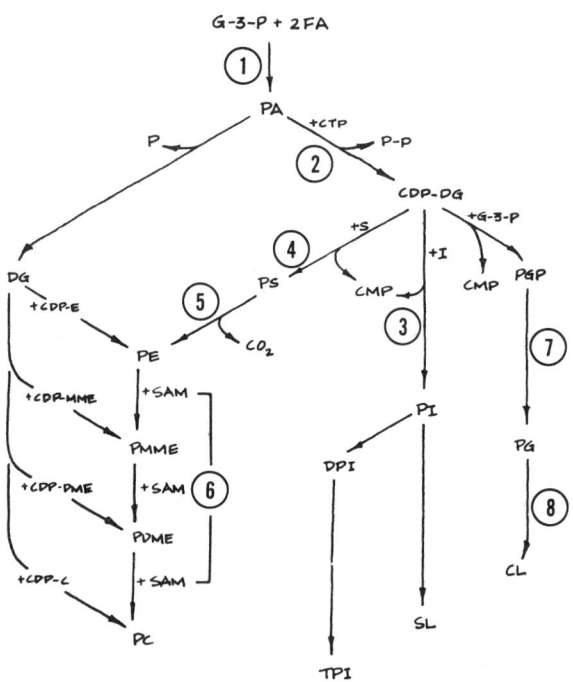

Figure 4 Principal pathways for phospholipid biosynthesis in yeast. (G-3-P) Glycerol-3-phosphate; (FA) fatty acyl-CoA; (PA) phosphatidic acid; (CTP) cytidine triphosphate; (P-P) pyrophosphate; (CDP-DG) cytidine diphosphate diglyceride; (MP) cytidine monophosphate; (PGP) phosphatidylglycerol phosphate; (PG) phosphatidylglycerol; (CL) cardiolipin; (S) serine; (I) inositol; (PI) phosphatidylinositol; (DPI) diphosphoinositide; (TPI) triphosphoinositide; (SL) sphingolipid; (PS) phosphatidylserine; (PE) phosphatidylethanolamine; (SAM) S-adenosylmethionine; (PMME) phosphatidylmonomethylethanolamine; (PDME) phosphatidyldimethylethanolamine; (PC) phosphatidylcholine; (DG) diglyceride; (CDP-E) cytidine diphosphate-ethanolamine; (CDP-MME) cytidine diphosphate-monomethylethanolamine; (CDP-DME) cytidine-diphosphate-dimethylethanolamine; (CDP-C) cytidine diphosphate choline. Circled numbers refer to the adjacent reactions.

defective in introducing a double bond in 16- and 18-carbon fatty acids, it is presumed that a single enzyme carries out these two desaturations. Strains carrying the *ole2*, *ole3*, or *ole4* mutations are defective in heme biosynthesis and have been discussed in the section on sterol metabolism. Their fatty acid requirement stems from the fact that the yeast fatty acid desaturase is a mixed-function oxygenase having a cytochrome-*b* component (Oshino et al. 1966; Tamura et al. 1976). For a discussion of the phenotype and defects present in these mutants, see Haslam and Astin (1979). The locus designations of all of the fatty acid mutants are given in Table 3.

Regulation of Fatty Acid Biosynthesis

There are many variables involved in the synthesis of the fatty acid mixture of yeast cellular membranes, including control of chain length, degree of unsaturation, and the response of these factors to temperature and physiological changes. Furthermore, the different classes of fatty acids must be partitioned into the various classes of phospholipids and, finally, the phospholipids must be distributed into the cellular organelles. It is known that lipid composition in yeast varies with temperature (Hunter and Rose 1972) and growth conditions (Getz et al. 1970; Keith et al. 1972). It is widely assumed that regulation of the lipid composition is a critical factor in maintaining the physiological state of the cellular membranes. There are many complex factors to be considered in such regulation. For example, how is membrane fluidity regulated in response to temperature? It may be regulated partly by the products of the fatty acid synthetase reaction. Okuyama et al. (1979) reported that the in vitro products of the fatty acid synthetase reaction are a varying proportion of 16- and 18-carbon saturated fatty acids. The lower the reaction temperature, the greater the proportion of palmitic acid (16:0) there is compared with stearic acid (18:0). The synthesis of a higher proportion of shorter fatty acids was also observed in vivo at lower temperatures. However, the organism may have many other mechanisms for coping with such physiological requirements. As discussed, yeast possesses many enzymatic activities for the modification of preexisting fatty acids (Orme et al. 1972; Schweizer et al. 1974), which suggests that there is considerable potential for modification and regulation independent of the fatty acid synthetase reaction. Furthermore, Waechter and Lester (1973) showed that the proportions of the various fatty acids in the total cellular phospholipids varied depending upon which of the two alternative pathways (Fig. 4) was employed in the biosynthesis of phosphatidylcholine. The pathways of phospholipid synthesis are discussed in a subsequent section.

Carboxylation of acetyl-CoA is probably the key biosynthetic step in regulating the total level of fatty acid synthesis (as opposed to relative proportions of different chain lengths discussed above). Kamiryo et al. (1976) demonstrated that the addition of long-chain fatty acids to the growth medium of yeast cells leads to reduction of acetyl-CoA carboxylase activity. Furthermore, these investigators isolated mutants defective in acyl-CoA synthetase (i.e., the mutants are defective in the formation of CoA derivatives of exogenously supplied fatty acids; Table 3). In these mutants, repression of the acetyl-CoA carboxylase did not occur in the presence of exogenous fatty acids, which suggests that derivatives of the fatty acids, and not the fatty acids themselves, are responsible for the repression.

Experimental Manipulation of Fatty Acid Mutants

Both unsaturated- and saturated-fatty-acid-requiring mutants have been used in experimental manipulation of membrane fatty acid composition. These experiments take two main forms: (1) fatty acid replacement, in which fatty acids not normally present in wild-type yeast are tested for their ability to support growth and/or their effect upon membrane properties of fatty-acid-requiring mutants, and (2) fatty-acid-deprivation studies, in which fatty-acid-requiring mutants are starved for fatty acids and the effect of fatty acid deprivation on membrane functions is assessed. Some results of such feeding experiments are summarized in Table 4. In most but not all of these experiments, the incorporation and subsequent fate of the supplemented fatty acid was followed by extraction of the cellular lipids and gas liquid chromatographic analysis of the incorporated fatty acids. In the studies on *ole1* mutants, the data in the column headed *ole1*[a] are compiled from a previous review (Keith et al. 1972) and, in most but not all cases, the fatty acids of the supplemented cells were extracted and analyzed. In the columns headed *ole1*[b] and *ole1*[c], the fate of the incorporated fatty acids was exhaustively examined. In the case of unsaturated fatty acid supplements, the supplied fatty acids were recovered, for the most part, in an unaltered form from the supplemented cells (Proudlock et al. 1971; Walenga and Lands 1975). However, in the saturated fatty acid auxotrophs, as discussed below, considerable alteration of the supplied fatty acid did occur in many cases. In both sets of experiments involving saturated fatty acid mutants (columns headed *fas*[d] and *fas*[e]), extensive analysis of fatty acids incorporated into the mutant cells was carried out.

Saturated fatty acids of 13, 14, 15, 17, and, in some cases, 12 carbons in length have all been reported to support growth of saturated fatty acid mutants (Table 4) (Schweizer and Bolling 1970; Henry and Keith 1971; Orme et al. 1972; Schweizer et al. 1974). It is interesting that *fas* mutants were able to grow when supplemented with odd-chain fatty acids 13:0, 15:0, and 17:0, even though such fatty acids are not normally found in yeast. Furthermore, such odd-chain fatty acids were readily desaturated (Henry and Keith 1971; Orme et al. 1972). The feeding experiments with odd-chain fatty acids, as well as isotopic labeling experiments in fatty acid synthetase mutants, have clearly shown that preformed long-chain fatty acids can be elongated by the addition of 2 carbon units (Henry and Keith 1971; Orme et al. 1972). These experiments establish the existence of a fatty acid chain elongation pathway that is independent of fatty acid synthetase. Furthermore, *fas* mutants can apparently shorten preformed fatty acids to some degree (Schweizer et al. 1974). Thus, there is a limit to the degree of experimental control over membrane fatty acid composition that can be achieved by feeding different fatty acids to these yeast mutants. Nonetheless, the fatty acid composition of *fas* mutants does differ depending upon the fatty acid supplied in the growth medium. Membrane properties, as measured by spin-labeling techniques, differed depending upon the saturated fatty acid supplements supplied to a *fas* strain (Henry and Keith 1971). It has been reported that some (Schweizer et al. 1974) though not all (Henry and Keith 1971) *fas* strains can grow when supplied with oleic acid, an unsaturated fatty acid. If such strains were unable to modify this fatty acid, this observation would require the organism to construct membranes consisting only of unsaturated fatty acid. This is apparently not the case, however, since it is reported that *fas* mutants that grow under these conditions convert oleic acid (18:1Δ^9 *cis*) to palmitic (16:0) or myristic (14:0) acid. The conversion apparently does not involve direct reduction of the double bond but most likely is the result of partial degradation and reelongation (Schweizer et al. 1974). Overall, feeding and fatty acid composition experiments with *fas* mutants suggest that some saturated fatty acid is required in the total fatty acid composition of the cell for maximal growth.

Much literature exists on the subject of fatty acid deprivation and replacement studies in yeast fatty acid desaturase (*ole1*) mutants (Wisnieski et al. 1970; Haslam et al. 1971; Proudlock et al. 1971; Keith et al. 1972; Linnane et al. 1972; Wisnieski and Kiyomoto 1972; Barber and Lands 1973; Walenga and Lands 1975; Haslam and Fellows 1977; Marzuki and Linnane 1979). The results of some of the fatty acid replacement studies are displayed in Table 4. One impression gained from inspection of the data in Table 4 is that a wide range of unsaturated fatty acids will support growth of the *ole1* mutant. The growth-supporting fatty acids include *cis* unsaturated fatty acids from 14 to 20 carbons in chain length, whereas the naturally occurring yeast fatty acids are predominantly 16 and 18 carbons long. Furthermore, the position of the double bond in naturally occurring yeast lipids is at the Δ^9 position, whereas unsaturated fatty acids with double bonds at other positions are capable of supporting growth. In comparing a large series of positional isomers of *cis* octadecanoic acid (18:1), Walenga and Lands (1975) found that isomers with double bonds at positions 2–4 and 13–15 did not support growth of the *ole1* mutant. Other positional isomers varied in their ability to support growth of the *ole1* mutant, with a peak effectiveness at the naturally occurring Δ^9-positional isomer, oleic acid. Furthermore, they found that increasing desaturation of the fatty acid was correlated with

Table 4 Growth of *ole1* and *fas* mutants of yeast in response to exogenous fatty acids

Supplemented fatty acid	$ole1^a$	$ole1^b$	$ole1^c$	fas^d	fas^e
12:0	0			+*	+^{n}
14:0	0			+	+
15:0				+	
16:0	0		0	+	+
17:0				+	
18:0	0		0	0	+*
20:0					
14:1Δ^5 cis	+				
14:1Δ^9 cis	+			0	
16:1Δ^9 cis	+	+	+	0	
16:1Δ^9 trans	+				
18:1Δ^{2-4}			0		
18:1Δ^5			+*		
18:1Δ^6 cis	+	+	+*		
18:1Δ^9 cis	+	+	+	0	+^{n}
18:1Δ^9 trans	0	0			
18:1Δ^{10} cis			+*		
18:1Δ^{11} cis	+		+*		
18:1Δ^{11} trans	0				
18:1Δ^{12} cis			+*		
18:1Δ^{13-15}			0		
18:2$\Delta^{9,12}$ cis, cis	+	+	++	0	
18:2$\Delta^{9,12}$ trans, trans	+				

Symbols represent reported growth of the mutants *ole1* and *fas1* or *fas2* in defined glucose medium in response to supplemented fatty acids. ++, growth response greater than standard response in each case (+) given below. +*, growth response poor compared to standard (+) given for each case below. +^{n}, some, but not all, strains tested responded. 0, no detectable growth. Effectiveness of growth supplements was based upon observed doubling time and/or total cell yield in response to the supplement.
^{a}+, growth response reported equivalent to wild type (Wisnieski and Kiyomoto 1972; Keith et al. 1972).

more effective growth support, despite the fact that naturally occurring yeast fatty acids are monoenoic. Walenga and Lands (1975) also analyzed the fate of the supplemented fatty acid, demonstrating in each case the incorporation of the fatty acid in its original form into phospholipid.

The *trans* analogs of the fatty acids have not been tested as extensively as the *cis* analogs. Although 16:1Δ^9 *trans* has been reported capable of promoting growth, the 18:1Δ^9 and 18:1Δ^{11} *trans* analogs were reported not to stimulate growth of the *ole1* mutants (see Table 4). A number of fatty acid analogs, such as hydroxystearic acids (Proudlock et al. 1971; Wisnieski and Kiyomoto 1972) and cyclopropane fatty acid (Lands et al. 1978) will also support growth of the yeast *ole1* mutant. Yeast cells therefore exhibit con-

Table 4 (Continued)

Supplemented fatty acid	$ole1^a$	$ole1^b$	$ole1^c$	fas^d	fas^e
$18:3\Delta^{6,9,12}$ cis, cis, cis	+				
$18:3\Delta^{9,12,15}$ cis, cis, cis	+	+	++	0	
$20:1\Delta^{11}$ cis	+*	+	+*		
$22:1\Delta^{13}$ cis	0		+*		
$24:1\Delta^{15}$ cis	0				
$20:2\Delta^{11,14}$ cis, cis	+*		+*		
$20:3\Delta^{11,14,17}$ cis, cis, cis	+*		+*		
$20:4\Delta^{5,8,11,14}$ cis, cis, cis, cis	+	+	++		
$20:5\Delta^{5,8,11,14,17}$			++		
$22:6\Delta^{4,7,10,13,16,19}$			+*		
$18:1\Delta^9$ cis-ol	0				
$18:1\Delta^9$P cis-ol-PO$_4$	0				
$18:1\Delta^{9\equiv}$	+				
$18:1\Delta^9$ cis, 12OH	+	+			
$18:1\Delta^9$ trans, 12OH	+*				
$18:1\Delta^9$ cis, 12-acetoxy	0				
Cyclopropane fatty acids: positional isomers of cis methylene octadecanoic acid					
position 1–4			0		
5–7			+*		
8			+		
9			++		
10–15			+*		
16–17			0		

^{b}+, growth response reported equivalent to oleic acid $18:1\Delta^9$ cis (Proudlock et al. 1971).

^{c}+, growth response reported equivalent to oleic acid $18:1\Delta^9$ cis (Barber and Lands 1973; Walenga and Lands 1975; Lands et al. 1978).

^{d}+, growth response equivalent to palmitic acid (Henry and Keith 1971).

^{e}+, growth response equivalent to palmitic acid (Schweizer and Bolling 1970; Schweizer et al. 1974).

siderable flexibility in the structure of fatty acids, which are capable of fulfilling the requirements for unsaturated fatty acids.

However, mitochondrial functions appear to be more sensitive to fatty acid compositional changes than other cellular processes. When *ole1* cells are grown on a nonfermentable carbon source in the absence of fatty acid supplement, the proportion of unsaturated fatty acids will not drop below 20%, because the cells stop growing (Marzuki and Linnane 1979) due to loss of oxidative phosphorylation. However, if the cells are shifted to a fermentable carbon source, they will continue to grow until the total level of unsaturated fatty acid reaches 5%. Furthermore, the frequency of respiratory-deficient (petite) mutants rises in unsaturated-fatty-acid-depleted cells (Marzuki et al. 1974). Haslam et al. (1971) and Proudlock et al. (1971) reported that oxidative phosphorylation is absent in mitochondria depleted of unsaturated fatty acids but that both cytochrome content and respiration are normal in such cells. They suggest that the lesion lies in the coupling of respiration and phosphorylation. However, Walenga and Lands (1975) reported that *ole1* cells supplemented with 20:1Δ^{11}, although capable of fermentative growth, had decreased cytochrome content of the mitochondrial membrane, as well as impaired respiration. Furthermore, using the *ole1* mutant, Tustanoff and his colleagues (Ainsworth et al. 1972, 1974; Janki et al. 1974, 1975; Aithal and Tustanoff 1975) have shown that the transition temperatures of several mitochondrial enzymes are affected by the fatty acid composition of the mitochondrial membrane. In general, these experiments, like those involving changes in sterol composition (see previous section), led to the conclusion that mitochondrial functions in yeast are particularly sensitive to alterations in lipid composition.

Finally, a double-mutant strain *ole1 fas1* (Henry 1973) was constructed that requires both saturated and unsaturated fatty acids for growth on glucose-containing media. This double mutant has been employed in both fatty acid deprivation studies (Henry 1973) and fatty acid replacement experiments (Esfahani et al. 1981). In the *ole1 fas1* strain, starvation for fatty acids leads to precipitous death similar to "inositolless death" (discussed in a subsequent section). The use of the double-mutant strain allows greater experimental control over membrane fatty acid composition than the use of either single mutant. Esfahani et al. (1981) employed the *ole1 fas1* strain to show that the effectiveness of growth support depended upon both the saturated and the unsaturated fatty acids employed and the combination in which they were used. Although the effectiveness of different fatty acid combinations in supporting growth varied considerably, there was little change in membrane fluidity of cells grown under different supplemental conditions. These studies suggest that the difference in effectiveness of growth support of the different fatty acid mixtures may not lie either in membrane fluidity requirements or in the efficiency with which certain fatty acids can be incorporated into phospholipids. Rather, it was proposed that the optimal width of the membrane bilayer requires phospholipids containing saturated fatty acids of 14:0–16:0 carbons.

PHOSPHOLIPID BIOSYNTHESIS

The phospholipids, together with the free sterols, are the major lipid components of the eukaryotic membrane. The pathways for phospholipid synthesis in yeast are similar to those in other eukaryotic organisms. However, the individual reactions of phospholipid synthesis have not been as completely characterized in yeast as they have in mammalian tissues. (For a review of the mammalian pathways of phospholipid synthesis, see Snyder [1977], especially the introductory chapter by L. Van Golde and S.G. Van den Bergh.) The major reactions of phospholipid biosynthesis in yeast are illustrated in Figure 4. Phospholipid synthesis starts with the acylation of glycerol-3-phosphate with long-chain fatty acids to form phosphatidic acid (reaction 1, Fig. 4) (Kühn and Lynen 1965). In mammalian systems, dihydroxyacetone-phosphate (DHAP) can also serve as a substrate for acylation (Hajra 1968; La Belle and Hajra 1974). Using labeled glycerol-3-phosphate as a precursor, Steiner and Lester (1972c) detected the formation of the precursor's phosphatidic acid and CDP-diglyceride in vitro using an unfractionated membrane preparation from glucose-repressed cells. Schlossman and Bell (1978) investigated the acylation of glycerol-3-phosphate and DHAP in a yeast membrane fraction derived from anaerobically grown cells, in which the contribution of mitochondrial enzymes was minimal. They concluded that a single membrane-bound enzyme catalyzes the acylation of both glycerol-3-phosphate and DHAP. However, on the basis of physiological arguments and the absence of detectable acyl-DHAP oxidoreductase activity in their yeast membrane preparations, they suggested that glycerol-3-phosphate is the major phospholipid precursor in yeast. Belendiuk et al. (1978) characterized the formation of CDP-diglyceride from CTP and phosphatidic acid (reaction 2, Fig. 4) in yeast mitochondrial membranes and partially purified the membrane-bound mitochondrial CTP-phosphatidic acid cytidyltransferase. Presumably, a similar reaction occurs in microsomal membranes as well, although enzymes derived from microsomal and mitochondrial sources were not compared.

In yeast, CDP-diglyceride serves as the precursor for the synthesis of several phospholipids including phosphatidylinositol (Figs. 4 and 5; Steiner and Lester 1972c). Carman and Matas (1981) recently reported the solubilization and characterization of phosphatidylinositol synthase from "microsomes" of glucose-grown wild-type yeast cells. Inositol-containing phosphatides are widely found in eukaryotic organisms, including yeast (Steiner and Lester 1972b; Lester et al. 1978; Wells and Eisenberg 1978), but are not generally present in bacteria. In yeast, phosphatidylinositol serves as the precursor to the synthesis of other inositol-containing lipids, including diphosphoinositides and triphosphoinositides (Lester and Steiner 1968; Steiner and Lester 1972a) and several sphingolipids (Steiner et al. 1969; Angus and Lester 1972; Lester and Steiner 1972b; Lester et al. 1978). Steiner and Lester (1972c) also provided evidence that CDP-diglyceride reacts with serine in yeast membranes to form phosphatidylserine (reaction 4, Fig. 4). Their experiments with labeled CDP-diglyceride and ^{14}C-, ^{32}P-labeled glycerol-3-phosphate support the claim that this is the principal route for phosphatidylserine biosynthesis in this organism. Carson et al. (1982) have studied this enzymatic activity in greater detail in isolated yeast membranes

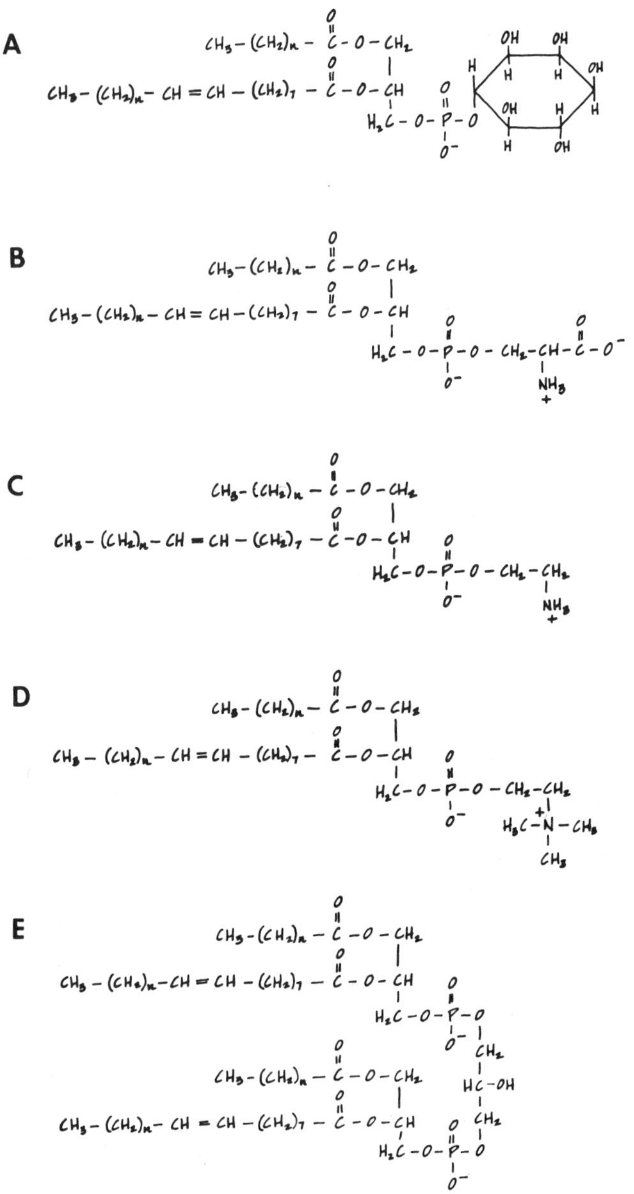

Figure 5 Structures of yeast phospholipids. *(A)* Phosphatidylinositol, *(B)* phosphatidylserine *(C)* phosphatidylethanolamine, *(D)* phosphatidylcholine, *(E)* cardiolipin.

treated with hydroxylamine (to inhibit the subsequent decarboxylation of phosphatidylserine) and have shown synthesis of phosphatidylserine to be dependent upon CDP-diglyceride. Furthermore, phosphatidylserine synthase from microsomes of wild-type glucose-grown yeast has now been solubilized and characterized (Carman and Matas 1981; Carson et al. 1982). In contrast, in mammals, phosphatidylserine is synthesized via an exchange reaction between serine and phosphatidylethanolamine (Van Golde and Van den Bergh 1977), and the reaction of the serine with CDP-diglyceride has not been detected. However, the reaction of CDP-diglyceride and serine appears to be the principal route by which phosphatidylserine is made in bacteria (Kaufer and Kennedy 1964). Other phospholipids to which CDP-diglyceride serves as precursor in yeast (Steiner and Lester 1972c), as well as in mammals (Snyder 1977) and bacteria (Silbert 1975), are phosphatidylglycerol phosphate, phosphatidylglycerol and diphosphatidylglycerol (cardiolipin). As mentioned previously, the synthesis of cardiolipin occurs predominantly in the mitochondrial membrane in yeast (Cobon et al. 1974).

In yeast (Steiner and Lester 1972c), as in mammals (Snyder 1977) and bacteria (Kaufer and Kennedy 1964), phosphatidylserine is decarboxylated to form phosphatidylethanolamine (reaction 5, Fig. 4), whereas the decarboxylation of free serine has not been described. Phosphatidylethanolamine in yeast is sequentially methylated, using S-adenosylmethionine as a methyl donor (reaction 6, Fig. 4), to form phosphatidylcholine (Waechter and Lester 1971, 1973); the methylation of free ethanolamine has not been described. In mammals, ethanolamine, choline, and the two methylated intermediates, N-methylethanolamine and N,N-dimethylethanolamine (DME), are also directly incorporated in phospholipids via the pathway described by Kennedy and Weiss (1956). In this pathway, α,β-diglyceride reacts with a CDP derivative of ethanolamine or one of its methylated bases (Fig. 4). The existence of this pathway in yeast can be inferred from results of various labeling experiments in vivo and in vitro (Waechter et al. 1969; Waechter and Lester 1971, 1973; Steiner and Lester 1972c; Cobon et al. 1974; Atkinson et al. 1980a). These experiments show that yeast cells incorporate exogenous ethanolamine and its N-methylated bases into phospholipids with high efficiency in vivo. Isotopic labeling experiments in vitro are consistent with a reaction mechanism involving α,β-diglyceride as the immediate precursor (Steiner and Lester 1972c).

Regulation of Phospholipid Biosynthesis

Little is known about the regulation of the membrane-bound enzymes of phospholipid biosynthesis, yet it is clear that regulation must exist. During cell division, e.g., various membranous organelles such as the nucleus and plasma membrane must increase in surface area at discrete times during the cell-division cycle, and the formation of membrane-bound vesicles derived from endoplasmic reticulum at the time of bud initiation has been reported (Moor 1967). Any differential synthesis of membrane in such fashion must necessarily involve regulation of availability of the phospholipids of the membrane matrix. Perhaps the best-studied example of regulated membrane biogenesis in yeast is mitochondrial derepression during the transition from anaerobic to aerobic growth (discussed in the section on sterol biosynthesis), a process known to involve and require the synthesis of specific sterols, fatty acids, and phospholipids. The increased synthesis of the phospholipid cardiolipin (Jakovcic et al. 1971) during mitochondrial depression has already been mentioned. In general, however, little is known about the regulation of the activities of the enzymes directly involved in phospholipid synthesis, let alone the relationship of such regulation to membrane biogenesis. The existing information concerns the effect of precursor availability on the pattern of phospholipid synthesis.

The presence of the phospholipid precursors choline, DME, or N-methylethanolamine in the growth medium has been shown to have an effect upon the activity of the membrane-bound S-adenosylmethionine-dependent N-methyltransferases involved in the successive methylations of phosphatidylethanolamine (reaction 6, Fig. 4) (Waechter and Lester 1971, 1973). There are at least two separate enzymes involved in the conversion of phosphatidylethanolamine to phosphatidylcholine, and the enzymes are differentially regulated. Waechter and Lester (1971, 1973) showed that inclusion of DME in the growth medium causes a reduction in specific activity of the N-methyltransferase(s) involved in methylation of phosphatidylethanolamine and phosphatidylmonomethylethanolamine (PMME). However, the final methylation step, phoshatidyldimethylethanolamine (PDME) to phosphatidylcholine is unaffected by the presence of exogenous DME. If choline, on the other hand, is included in the growth medium, all three methylation reactions are repressed. Carson et al. (1982) have demonstrated that the addition of choline to the growth medium also reduces the activity of phosphatidylserine synthetase to about 25% of its activity in cells grown in the absence of choline.

In addition, phosphatidylethanolamine methylation (reaction 6, Fig. 4) is affected by the presence of exogenous inositol. Yamashita and Oshima (1980) have reported reduction in phosphatidylethanolamine methyltransferase activity when inositol is added to the medium of wild-type cells. The regulation of phosphatidylethanolamine methylation in response to choline is also altered by the presence or absence of inositol. B. Loewy and V. Letts (unpubl.) found that when exogenous choline is added to inositol-supplemented cells, methylation of phosphatidylethanolamine is reduced in keeping with the data of Waechter and Lester (1971, 1973). However, they have shown that if wild-type cells are grown in the absence of inositol, phosphatidylethanolamine methylation is no longer reduced by the addition of exogenous choline. Both genetic and biochemical studies (summarized in Table 5) suggest an intricate coordination of phosphatidylcholine synthesis with the synthesis of phosphatidylinositol and its precursor inositol. This coordination will be discussed further in the section on phospholipid mutants.

The availability of precursors of phospholipid synthesis also affects the relative proportions of the phospholipids accumulated in the membranes.

The inclusion of choline, DME, or *N*-monomethylethanolamine (MME) in the growth medium leads to the elevated accumulation in the membranes of the lipids phosphatidylcholine, PDME, and PMME, respectively (Waechter and Lester 1973). Likewise, the presence of exogenous ethanolamine leads to an increased content of phosphatidylethanolamine in yeast (Ratcliffe et al. 1973), and the addition of exogenous inositol to a wild-type culture that has been growing in the absence of inositol leads to an immediate increase in the rate of synthesis of phosphatidylinositol and to an increase in the proportion of this lipid in the total cellular phospholipid (S. Henry, unpubl.). The synthesis of the precursor inositol is highly regulated in wild-type cells, as is discussed in the following text.

Mutants Defective in Phospholipid Synthesis

In bacteria, much use has been made of temperature-sensitive mutants defective in various reactions of phospholipid biosynthesis (Raetz 1978). A temperature-sensitive mutant of yeast, defective in acylation of glycerol-3-phosphate has been identified (tsDAM, Table 3) (Letts 1980), and work on other temperature-sensitive mutants defective in phospholipid synthesis is in progress (R. Lester and G. Getz, pers. comm.). However, none of these mutants is sufficiently characterized to permit its use in manipulation of membrane lipid synthesis. At present, two major classes of mutants, inositol-requiring (*ino*) (Culbertson and Henry 1975) and ethanolamine/choline-requiring (*cho*) (Atkinson et al. 1980b; Kováč et al. 1980; Letts 1980), permit experimental manipulation of the polar components of the membrane phospholipids in yeast and analysis of genetic regulation involved in membrane lipid synthesis.

Inositol-requiring Mutants

Inositol auxotrophs of *S. cerevisiae* were described by Culbertson and Henry (1975). Inositol-1-phosphate, the immediate precursor of inositol, is synthesized from glucose-6-phosphate in a reaction catalyzed by the cytoplasmic enzyme, inositol-1-phosphate synthase (I-1-P synthase). The reaction has been detected in a variety of eukaryotes, including fungi, higher plants, and animals (Wells and Eisenberg 1978). The precise reaction mechanism has not been elucidated, but the existence of three partial reactions and two intermediates, 5-ketoglucose-6-phosphate and inosose-1-phosphate, has been postulated (Barnett et al. 1973; Eisenberg 1978; Fig. 6). The enzyme that carries out this reaction sequence has been purified from a variety of organisms and, in all cases, it has been found to be a large multimer consisting of either identical subunits or two types of subunits (Hofmann-Ostenhoff et al. 1978; Maeda and Eisenberg 1980). The yeast enzyme has been purified (Donahue 1979; Donahue and Henry 1981a) and was found to be a tetramer of approximately 240,000 daltons, consisting of identical sub-

Table 5 Phosphatidylcholine synthesis and inositol metabolism: Summary of genetic and biochemical evidence of coordinate regulation

Observation	Reference
A. Starvation of *chol* mutants for ethanolamine/choline leads to a coupled decline in the rate of synthesis of phosphatidylcholine and phosphatidylinositol. The decline in phosphatidylinositol synthesis cannot be reversed by exogenous inositol.	Atkinson et al. (1980b); Letts (1980); V. Letts (in prep.)
B. Starvation of *ino1* mutants for inositol leads to an immediate decline in the rate of synthesis of phosphatidylinositol, but there is no coupled decrease in phosphatidylcholine synthesis. Phosphatidylcholine synthesis is (if anything) actually increased.	Becker and Lester (1977); Henry et al. (1977)
C. In wild-type cells grown in inositol-containing medium; addition of choline to the growth medium leads to repression of the three methylation reactions involved in the synthesis of phosphatidylcholine from phosphatidylethanolamine. However, if the cells are grown in inositol-free medium, the addition of choline has little or no effect upon the methylation of phosphatidylethanolamine.	Waechter and Lester (1971, 1973); V. Letts and B. Loewy (unpubl.)
D. Addition of inositol to the growth medium of wild-type cells leads to a decrease in the methylation of phosphatidylethanolamine. A mutant has been described in which this effect is enhanced compared with wild type, leading to a choline requirement when the mutant cells are grown in the presence of inositol.	Yamashita and Oshima (1980)
E. Mutants of the *ino4* and *ino2* loci have a pleiotropic phenotype. Such mutants have a substantial deficiency in the formation of phosphatidylcholine via methylation of phosphatidylethanolamine. In addition, these mutants fail to produce the cytoplasmic enzyme I-1-P synthase, gene product of the *ino1* locus.	Henry et al. (1981); B. Loewy and S. Henry (in prep.)

Table 5 (Continued)

Observation	Reference
F. Mutants have been isolated that allow the synthesis of I-1-P synthase in an *ino2* or *ino4* genetic background. These "regulatory" mutants map to a third locus. They have no effect upon the primary methylation defect in *ino2* and *ino4* strains but lead to constitutive synthesis of I-1-P synthase.	B. Loewy (unpubl.)
G. The *opi3* mutant excretes inositol but is not constitutive for the cytoplasmic enzyme I-1-P synthase. The mutant has an almost total deficiency in the final methylation leading to phosphatidylcholine synthesis, and the second methylation is impaired as well. The *opi3* mutant is tightly linked to locus *ino4* (see E, above).	Greenberg (1980, and unpubl.); Henry et al. (1981)

units of approximately 62,000 daltons. Antibody directed specifically against the wild-type yeast holoenzyme has been prepared as has antibody produced in response to the denatured purified subunit (Donahue 1979; Donahue and Henry 1981a). The amount of enzyme, assayed either by activity or immunoprecipitation, is reduced at least 60-fold in crude extracts

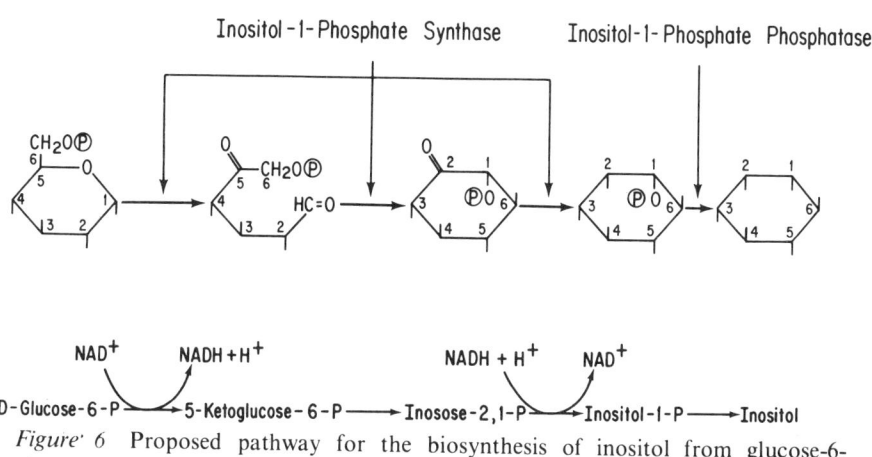

Figure 6 Proposed pathway for the biosynthesis of inositol from glucose-6-phosphate in yeast and other eukaryotes. (Reprinted, with permission, from Donahue 1979; based upon the proposed pathway discussed by Eisenberg 1978.)

prepared from wild-type yeast grown in the presence of inositol concentrations of 50 µM or greater, compared with cells grown in the absence of inositol (Culbertson et al. 1976a; Donahue 1979; Donahue and Henry 1981a). Greenberg (1980) studied the kinetics of repression of the wild-type enzyme following addition of inositol to the culture by pulse-labeling total protein with [^{35}S]methionine followed by immunoprecipitation of the enzyme complex. She found that synthesis of the enzyme decayed with a half-life of approximately 15 minutes following the addition of inositol. An estimate of the half-life of the mRNA for this protein was obtained by carrying out similar pulse-labeling of I-1-P synthase in the RNA processing mutant ts136 (Hutchinson et al. 1969). The messenger half-life for the subunit of I-1-P synthase estimated in this fashion was approximately 15 minutes, identical to the half-life of decay of enzyme synthesis during repression, which suggests that regulation occurs at the transcriptional or messenger processing level (Greenberg 1980).

The *INO1* locus is the structural locus for the single subunit of I-1-P synthase (Donahue 1979; Donahue and Henry 1981a). Mutations at the *INO1* locus eliminate I-1-P synthase activity (Culbertson et al. 1976b); and approximately half of all *ino1* mutants produce cross-reacting material (CRM) that cross-reacts with antibody prepared against the purified wild-type I-1-P synthase (Donahue 1979; Majumder et al. 1981). Over 70 alleles of this locus have been screened for suppressibility (Donahue 1979; Donahue and Henry 1981b), yet not one UAA, UAG, or UGA mutation has been detected. A very complex pattern of interallelic complementation was observed at the *INO1* locus (Culbertson and Henry 1975; Majumder et al. 1981), which could arise from the interaction of subunits in hybrid multimers (Crick and Orgel 1964) or from the interaction of different domains in the multifunctional protein as observed for fatty acid synthetase (Schweizer et al. 1978). However, these explanations do not account for all of the data, because many of the CRM⁻ mutants, known not to carry nonsense mutations, show extensive interallelic complementation (Majumder et al. 1981). It has been proposed that the *ino1*-gene products in these mutants are very susceptible to proteolysis, and the interallelic complementation may result from stabilization of the mutant gene product in a hybrid multimer (Majumder et al. 1981). A similar phenomenon is thought to account for some complementation among β-galactosidase mutants of *E. coli* (Lin and Zaben 1972; Villarejo et al. 1972).

Inositol-requiring mutants, representing at least nine loci in addition to *INO1*, have been isolated (Culbertson and Henry 1975; Donahue et al. 1978; Donahue and Henry 1981b). Only the *INO1, INO2,* and *INO4* loci are represented by more than one allele (Donahue 1979). All of the Ino⁻ mutants lack I-1-P synthase (Culbertson et al. 1976b). The loci, other than *INO1*, are presumed to be involved in the regulation of inositol biosynthesis or in the coordination of inositol biosynthesis with basic metabolic pathways of the cell. Some of the other mutations cause pleiotropic phenotypes such as respiratory deficiency and the inability to utilize certain carbon sources (Donahue et al. 1978), and the *ino2* and *ino4* mutants cause pleiotropic defects in phospholipid metabolism (Henry et al. 1981); these mutants are discussed later in the text.

Mutants constitutive for I-1-P synthase (*opi*) have been isolated as colonies that excrete inositol (Greenberg 1980; Greenberg et al. 1982b; Table 3). One such recessive constitutive mutant (*opi1-1*) produces approximately twice as much enzyme under both repressed and derepressed growth conditions compared with the wild-type strain grown under the derepressed condition (Greenberg 1980; Greenberg et al. 1982a). The mutation conferring the constitutive phenotype is unlinked to *INO1*, the structural gene for I-1-P synthase. Mutations in two other loci (*OPI2* and *OPI4*) confer a similar recessive constitutive phenotype and are not linked to *INO1*, *OPI1*, or each other (Greenberg 1980; Greenberg et al. 1982b). However, the *opi3* mutant was found to be closely linked to *INO4*. The *opi3* mutant secretes inositol, is not constitutive for I-1-P synthase, but does exhibit pleiotropic defects of phospholipid metabolism (Greenberg 1980; Henry et al. 1981; Greenberg et al. 1982b). This mutant is discussed, together with the *ino4* and *ino2* mutants, in a subsequent section.

Inositol-deprivation Experiments

Virtually all of the inositol in yeast cells is lipid-bound, and the pool of free inositol is very small (Angus and Lester 1972). Consequently, in cells deprived of inositol, synthesis and accumulation of phosphatidylinositol stops very rapidly, but other phospholipids continue to be made (Becker and Lester 1977; Henry et al. 1977). Thus, the inositol-requiring mutants provide an opportunity for experimentally manipulating the relative quantity of inositol-containing phospholipids in the cellular membranes. In inositol-starved *ino1* cells, the proportion of phosphatidylinositol can be reduced to about 8% of total cellular phospholipid compared with 28% in inositol-supplemented cells (Table 2). However, the proportion cannot be further reduced because the inositol-starved cells rapidly die. Inositolless death has been described in *S. cerevisiae* (Culbertson and Henry 1975), in other fungi (Ridgeway and Douglas 1958; Strauss 1958; Thomas 1972), and in cultured animal cells (Jackson and Shin 1980). The fact that the phenomenon is so widely observed suggests that it reflects some aspect of the basic role of inositol in eukaryotic cells. Since the bulk of the inositol in the cell is incorporated into lipids that are components of the cellular membranes, it has been widely assumed that the cause of inositolless death must lie in some defect of membrane composition, structure, function, or assembly (Shatkin and Tatum 1961; Matile 1966; Sullivan and Debusk 1973; Henry et al. 1977; Ulaszewski et al. 1978).

The precise timing of the events during inositol starvation in yeast varies, depending upon cultural conditions, but the relative order of events is quite reproducible (Henry et al. 1977; Hanson and Lester 1980). First, within minutes, in growing cells deprived of inositol, the rate of phosphatidylinositol synthesis drops (Henry et al. 1977), and there is no further accumulation of this lipid (Becker and Lester 1977). Immediately and concomitantly, the rate of synthesis of the major cell-wall carbohydrates mannan and glucan drops (Hanson and Lester 1980). The synthesis of phospholipids that do not contain inositol, including phosphatidylserine, phosphatidylethanolamine, and phosphatidylcholine, continues at normal or even slightly elevated rates (Becker and Lester 1977; Henry et al. 1977). Termination of cell division occurs within one generation but is not preceded by any slowing of macromolecular synthesis. Even after cell division has stopped, electron microscopy of thin sections of inositol-starved cells reveals no apparent change in any of the membraneous organelles. As in dividing cells (Moor 1967), vesicles are observed in the bud tip of the inositol-starved cells even after they have stopped dividing (S. Henry, unpubl.). The vesicles neither disappear nor do they accumulate, as in the case of the *sec* mutants isolated by Novick and Schekman (1979); (also see Novick et al. 1980; Schekman and Novick, this volume). Furthermore, inositol-starved cells continue to secrete invertase normally until they begin to lose viability (R. Schekman et al., unpubl.). Spheroplasts made from inositol-starved cells maintain their integrity in medium isotonic to normal cells and metabolize normally until the cells stop dividing. At that precise time (which precedes cell death by several hours), the inositol-starved spheroplasts begin to lyse. The lysis can be prevented by increasing the osmotic concentration of the supporting medium (Atkinson et al. 1977). Both the cell growth and spheroplast studies suggest that inositol-starved cells are capable of only limited cell-volume expansion. Once the volume has been reached, continuing metabolism causes the osmolarity of the cytoplasm to exceed that of the isotonic supporting medium.

Within a period of time equivalent to one generation after the cells stop dividing, they begin to die (Henry et al. 1977). At the time when viability is lost, many metabolic functions fail almost simultaneously. A variety of transport functions fail (Ulaszweski et al. 1978; S. A. Henry and T. Cooper, in prep.), the cells become permeable to small molecules (Ulaszewski et al. 1978), potassium leaks out of the cells (Henry et al. 1977), ATP levels drop, and macromolecular synthesis stops (Henry et al. 1977). During the period when the cells continue to divide, they may be protected against future viability loss by any treatment that blocks macromolecular synthesis or metabolism in general. These observations have led to the development of potent mutant enrichment procedures based on inositolless death (Henry et al. 1976). Several explanations of inositolless death have been proposed. The hypotheses take two main forms: (1) Cell death is caused by loss of selective permeability of the plasma membrane/and or the vacuole, presumably brought about by changes in membrane composition (Matile 1966; Sullivan and Debusk 1973; Ulaszewski et al. 1978); and (2) cell death results from an imbalance in the rate of growth of the cell surface compared with the accumulation of cytoplasmic components—so-called "unbalanced growth" (Shatkin and Tatum 1961; Henry et al. 1977). The mechanism by which inositol starvation would lead to either a block in membrane and general surface expansion and/or gross failure of membrane functions is obscure at present.

Choline/Ethanolamine-requiring Mutants

Several independent series of mutants auxotrophic for choline or ethanolamine have been isolated (Lindegren et al. 1962; Letts and Dawes 1979; Atkinson et al. 1980b; Letts 1980). All of the mutants isolated on the basis of a simple choline or ethanolamine auxotrophy have proved to be alleles of locus *cho1* (Lindegren et al. 1962; Atkinson, et al. 1980b; Kovač et al. 1980; Letts 1980). The *cho1* mutants confer an auxotrophic requirement that is satisfied by ethanolamine, choline, MME, or DME (Atkinson et al. 1980b; Letts 1980). All *cho1* mutants have a reduced capacity to synthesize phosphatidylserine, and some *cho1* mutants completely lack phosphatidylserine (Atkinson et al. 1980a,b; Kováč et al. 1980; Letts 1980). The *cho1* defect is presumed to reside in phosphatidylserine synthetase, the enzyme that catalyzes the reaction of CDP-diglyceride and free serine to yield phosphatidylserine in bacteria (Kaufer and Kennedy 1964). The results obtained in in vitro assays of membrane preparations derived from the *cho1* strains and a wild-type control are entirely consistent with a defect in a phosphatidylserine synthetase (Atkinson et al. 1980b; Kováč et al. 1980; Letts 1980). Since phosphatidylserine is not available as a precursor in *cho1* cells, phosphatidylethanolamine must be synthesized via the reaction of CDP-ethanolamine and diglyceride (Fig. 4), the pathway described by Kennedy and Weiss (1956). When choline is used as a supplement, phosphatidylethanolamine is also bypassed as an intermediate, and the percentage of this lipid drops to 5% of the total phospholipid (Table 2).

The phospholipid composition and growth properties of the *cho1* mutant allow startling conclusions to be drawn about the phospholipid requirements of yeast cells and membranes. First, phosphatidylserine is essentially expendable for vegetative growth on glucose. This situation is quite different from the absolute requirement for ongoing synthesis of phosphatidylinositol, a phospholipid that carries the same net negative charge as phosphatidylserine. The total amount of phosphatidylinositol observed in the *cho1* lipid composition is approximately equal to the combined total of phosphatidylinositol plus phosphatidylserine in the wild-type strain under a variety of growth conditions (Table 2) (Atkinson et al. 1980a,b). The increase in phosphatidylinositol could be the result of increased synthesis of this lipid from the CDP-diglyceride precursor due to lack of competition from phosphatidylserine synthesis. Alternatively, or in addition, it may reflect maintenance of net lipid charge. Maintenance of net lipid charge has been observed in yeast during inositol starvation as well (Becker and Lester 1977). The absence of phosphatidylserine in the phospholipid composition of the *cho1* mutant is but one of its aberrant characteristics (Table 2). With choline as a growth supplement, the *cho1* mutants also grow with very little phosphatidylethanolamine in their membranes (Table 2). Although *cho1* mutants grow vegetatively on glucose with a choline or ethanolamine supplement, one *cho1* mutant exhibited defective growth under conditions requiring mitochondrial function (Atkinson et al. 1980b).

When starved for ethanolamine/choline, the *cho1* mutants do not die like the inositol-starved mutants (Atkinson et al. 1980a, Kováč et al. 1980). During choline starvation, the slowing of cell division appears to be coupled to a slowing of total phospholipid synthesis. Furthermore, phosphatidylinositol synthesis stops as rapidly as phosphatidylcholine synthesis (Atkinson et al. 1980b; Letts 1980; Henry et al. 1981). The tight coupling of phosphatidylinositol synthesis to phosphatidylcholine synthesis during choline/ethanolamine starvation in *cho1* mutants is another indication of the coordination of synthesis of the two lipids (Table 5). It is clear that the slowing of phosphatidylinositol synthesis in choline/ethanolamine-starved cells is not due to lack of the inositol precursor. Reduction in phosphatidylinositol synthesis occurs even in the presence of high concentrations of exogenous inositol (Atkinson et al. 1980b; V. Letts, unpubl.) and, in *cho1* strains, inositol is actually secreted into the growth medium during choline/ethanolamine starvation (S. Henry and V. Letts, unpubl.). The coupling of phosphatidylinositol and phosphatidylcholine synthesis during choline/ethanolamine starvation is to be contrasted to inositol starvation, where slowing of phosphatidylinositol synthesis is not coupled to slowing of synthesis of other lipids or any other metabolic function (Becker and Lester 1977; Henry et al. 1977).

Mutants Defective in Synthesis of Phosphatidylcholine

Choline/ethanolamine (*cho1*) auxotrophs of yeast are defective in the synthesis of phosphatidylserine but retain normal methylation of phosphatidylethanolamine (Atkinson et al. 1980a,b), whereas choline auxotrophs of *Neurospora* are deficient in phospholipid methylation (Scarborough and Nyc 1967). Recently, however, a number of yeast mutants defective in methylation of phosphatidylethanolamine have been identified. These mutants have been obtained in several different studies, and each has a complex phenotype involving abnormalities in inositol metabolism or regulation in addition to a deficiency in the formation of phosphatidylcholine via methylation of phosphatidylethanolamine (reaction 6, Fig. 4). The pleiotropic phenotypes of these mutants no doubt reflect the complex coordinate regulation of phosphatidylcholine synthesis and inositol metabolism (Table 5).

The various mutants that will be discussed in this context are *ino4, ino2,* and *opi3* and the mutants isolated by Yamashita and Oshima (1980) (see Tables 3 and 5 for a summary of phenotypes.)

The *ino4* and *ino2* mutants were first isolated by Culbertson and Henry (1975) as inositol auxotrophs and, as previously discussed, lack the cytoplasmic enzyme, I-1-P synthase (Culbertson et al. 1976b; Donahue and Henry 1981a). No other defect in these mutants was suspected until a new allele of the *ino4* locus (*ino4-BS1*) was isolated by Brenda Loewy using a *chol* strain as the parental background in a mutant search designed to detect mutants with a specific requirement for choline (as opposed to the *chol* requirement, which is satisfied by ethanolamine as well as choline). A mutant (BS1) that had the appropriate phenotype was identified after mutagenesis of the *chol* parental strain. This mutant was found to retain only about 15% of wildtype capacity to incorporate [^{14}C]methyl-labeled methionine into phosphatidylcholine (B. Loewy and S. Henry, in prep.). The BS1 mutant, when crossed out of the *chol* genetic background, retained the phospholipid methylation deficiency but had no choline requirement. (In other words, a *CHO1*, BS1 strain requires neither ethanolamine nor choline, despite a substantial reduction in its capacity to synthesize phosphatidylcholine de novo.) However, genetic analysis of the BS1 mutant revealed an inositol auxotrophy that cosegregated in crosses with the phospholipid methylation deficiency. Genetic complementation analysis showed that the new inositol mutant was an allele of the *ino4* locus (designated *ino4-BS1*). Subsequently, a number of independent allelic representatives of the *ino4* locus (isolated on the basis of an inositol requirement [Culbertson and Henry 1975]) were tested for deficiency in phosphatidylcholine synthesis. All were found to be similar to *ino4-BS1* and to have decreased levels of phosphatidylcholine and elevated levels of phosphatidylethanolamine and monomethylated and dimethylated phosphatidylethanolamine. Phospholipid compositions of the *ino4-BS1* mutant and two other allelic representatives of locus *ino4* are shown in Table 2. In vitro analysis of the phospholipid biosynthetic capacity of membranes derived from *ino4* strains has confirmed that these mutants have greatly impaired capacity to carry out the methylation of phosphatidylethanolamine (V. Letts, unpubl.). Analysis of *ino2* mutants revealed a defect in phosphatidylcholine synthesis similar to that observed in the *ino4* mutants (V. Letts and B. Loewy, unpubl.). The phospholipid composition of two *ino2* mutants is given in Table 2.

If, as seems likely, the primary defect(s) in the *ino2* and *ino4* mutants lies in phospholipid methylation, why do these mutants exhibit abnormal inositol metabolism? In the case of the *ino4* and *ino2* mutants, it has been demonstrated that I-1-P synthase is not produced (Culbertson et al. 1976b) even though the structural gene for the enzyme is present and unaffected (Donahue 1979; Donahue and Henry 1981a). The explanation for the complex phenotype may lie in the regulation that controls the relative production of the various phospholipids. As discussed previously, experiments performed using *cho1* mutants demonstrated tight coupling of phosphatidylinositol synthesis to phosphatidylcholine synthesis (Atkinson et al. 1980b; Letts 1980; V. Letts, unpubl.). The phenotypes of the *ino2* and *ino4* mutants suggest that synthesis of the precursor inositol is also coordinated with phosphatidylcholine synthesis (Table 5) and that, in the phospholipid-methylation-deficient *ino2* and *ino4* genetic backgrounds, the biosynthetic enzyme I-1-P synthase cannot be synthesized. Genetic evidence in support of this hypothesis has been obtained by B. Loewy (unpubl.). By selecting for inositol prototrophy in an *ino2, ino4* haploid strain, she has isolated mutants that permit synthesis of I-1-P synthase in an *ino2* or *ino4* genetic background. Some of these mutants are unlinked to *ino2* or *ino4* and result in an inositol-excretion phenotype. These "regulatory" mutants synthesize I-1-P synthase constitutively but do not effect the *ino2* or *ino4* phospholipid methylation deficiency (Table 5). The relationship of these constitutive mutants to the *opi* mutants isolated by Greenberg et al. (1982) and their linkage to *INO1*, the structural gene for I-1-P synthase, is under investigation.

Other mutants with complex phenotypes involving simultaneous abnormalities in the metabolism of phosphatidylcholine and inositol have been described. Yamashita and Oshima (1980) have reported the isolation of an auxotroph that requires choline, but only in the presence of exogenous inositol. They demonstrated that addition of inositol to the growth medium of this strain causes a reduction in phosphatidylethanolamine methyltransferase activity. However, as discussed previously, a similar, though not as extreme, depression of this activity is also observed in wild-type cells exposed to inositol (Table 5).

Removal of inositol leads to an increase in phosphatidylethanolamine methylation in the mutant. The relationship between the mutant of Yamashita and Oshima and other mutants described in this paper is not known. Another mutant, *opi3*, described by Greenberg (1980), has an almost total deficiency in phosphatidylcholine and a reduced phosphatidylethanolamine content (Table 2). Instead, the *opi3* mutant accumulates PMME and PDME (Greenberg 1980; Henry et al. 1981; M. Greenberg and L. Klig, unpubl.). In vitro assay of *opi3* membranes confirmed that this mutant is defective in the second and, possibly, the third methylation in the synthesis of phosphatidylcholine. The conversion of phosphatidylethanolamine to PMME is unaffected in the *opi3* mutant and may, in fact, be somewhat elevated (V. Letts, unpubl.). The *opi3* mutant is tightly linked to locus *ino4* (Greenberg 1980; B. Loewy, unpubl.). Unlike the *ino4* mutant, the *opi3* mutant is not an inositol auxotroph, but an inositol excretor. Paradoxically, the growth of the *opi3* mutant is enhanced by the addition of inositol to the growth medium (L. Klig and M. Greenberg, unpubl.). The complex phenotypes of these mutants, although not yet completely understood, no doubt reflect some aspect of the intricate coordination between inositol metabolism and phosphatidylcholine synthesis (Table 5).

The *opi3* mutant, like the *ino4* and *ino2* mutants, has no auxotrophic requirement for choline or ethanolamine. The phospholipid composition data for the *ino2*, *ino4*, and *opi3* mutants (Table 2) indicate that the yeast cell tolerates, with no substantial inhibition of growth, major changes in the relative proportion of the four phospholipids: phosphatidylethanolamine, phosphatidylcholine, PMME, and PDME. The ability of the *ino2*, *ino4*, and *opi3* mutants to grow without exogenous choline supplementation despite substantial defects in phosphatidylcholine synthesis probably explains why direct selection for choline auxotrophs, in several independent attempts, has resulted in the isolation of *cho1* mutants but not in the isolation of phospholipid methylation mutants (Lindegren et al. 1962; Atkinson et al. 1980b; Letts 1980).

SUMMARY AND CONCLUSIONS

Experiments employing yeast lipid mutants, as well as special metabolic conditions and inhibitors of lipid biosynthesis, suggest that the mitochondrion is the organelle in yeast most sensitive to shifts in lipid composition. The mitochondrion is also the only yeast membrane that has been shown to have a distinctive phospholipid composition, consisting of higher proportions of cardiolipin and phosphatidylethanolamine than other cellular fractions. It is interesting that these lipids are generally found in prokaryotic as well as eukaryotic systems (Silbert 1975). Yeast cells grown fermentatively or anaerobically do not require mitochondrial function and will accept a wide range of sterols and fatty acids as substitutes for the naturally occurring ergosterol and 16- and 18-carbon fatty acids (Table 4). Furthermore, a wide range of phospholipid compositions are compatible with growth on glucose (Table 2). Thus, it would appear that yeast membranes other than the mitochondrion are relatively flexible in the lipid substitutions they can tolerate without loss of function.

The availability of well-characterized lipid mutants and the ability of the yeast cell to tolerate alterations in lipid composition and mitochondrial function provide opportunities for in vivo manipulation of the lipid component of the cellular membranes which, at present, exceed that of any other eukaryotic organism. The yeast lipid mutants may provide unique opportunities for analysis of the mechanisms by which the eukaryotic cell regulates the synthesis of the complex array of membrane lipid components.

ACKNOWLEDGMENTS

I am indebted to Miss Lucy Romano for her expert professional assistance in preparing the manuscript. I am especially grateful to Dr. Leo Parks for his generous permission to use the figure from his article, for many informative discussions on sterol metabolism and, also, for making available manuscripts in press and in preparation. I am grateful to Drs. Fred Taylor and Martin Bard for helpful discussions on the heme mutants. Dr. Robert Lester made available unpublished manuscripts and has provided many useful discussions and ideas about yeast phospholipid biochemistry. I thank Dr. Charles Waechter and his colleague, Mary Anne Carson for permission to include discussion of their unpublished data. Most especially, I am indebted to my colleagues, Katharine Atkinson, Brian Cooperman, Michael Culbertson, Thomas Donahue, Paul Goldwasser, Miriam Greenberg, Margaret Johnson, Lisa Klig, Anita Kolat, Verity Letts, Arun Majumder, Leonard Moss, and Brenda Shicker Loewy, who have made available their data and invaluable ideas.

S. A. H. is supported by grants (GM-19629 and GM-11301) from the National Institutes of Health and is the recipient of an Irma T. Hirschl Faculty Award.

REFERENCES

Agnew, W.S. and G. Popjak. 1978. Squalene synthetase. *J. Biol. Chem.* **253**:4566.

Ainsworth, P.J., E.R. Tustanoff, and A.J.S. Ball. 1972. Membrane phase transitions as a diagnostic tool for studying mitochondriogenesis. *Biochem. Biophys. Research Commun.* **47**:1299.

Ainsworth, P.J., R.M. Janki, E.R. Tustanoff, and A.J.S. Ball. 1974. The incorporation of cytochrome oxidase into newly forming yeast. *J. Bioenerg.* **6**:135.

Aithal, H.N. and E.R. Tustanoff. 1975. Assembly of complex III into newly developing mitochondrial membranes. *Can. J. Biochem.* **53**:1278.

Alexander, K.T.W., K.A. Mitropoulos, and G.F. Gibbons. 1974. A possible role for cytochrome P-450 during the biosynthesis of zymosterol from lanosterol by *Saccharomyces cerevisiae*. *Biochem. Biophys. Res. Commun.* **60**:460.

Andreasen, A. and T. Stier. 1953. Anaerobic nutrition of *Saccharomyces cerevisiae*. I. Ergosterol requirements for growth in a defined medium. *J. Cell Comp. Physiol.* **41**:23.

———. 1954. Anaerobic nutrition of *Saccharomyces cerevisiae*. II. Unsaturated fatty acid requirement for growth in defined medium. *J. Cell. Comp. Physiol.* **43**:271.

Angus, W. and R. Lester. 1972. Turnover of inositol and phosphorus containing lipids in *Saccharomyces cerevisiae:* Extracellular accumulation of glycerophosphorylinositol derived from phosphatidylinositol. *Arch. Biochem. Biophys.* **151**:483.

Astin, A. and J. Haslam. 1977. The effects of altered membrane sterol composition on oxidative phosphorylation in a haem mutant of *Saccharomyces cerevisiae*. *Biochem. J.* **166**:287.

Astin, A., J. Haslam, and R.A. Woods. 1977. The manipulation of cellular cytochrome and lipid composition in a haem mutant of *Saccharomyces cerevisiae*. *Biochem. J.* **166**:275.

Atkinson, K. 1978. "Regulation of membrane biogenesis by inositol containing phospholipids in *Saccharomyces cerevisiae*." Ph.D. thesis, Albert Einstein College of Medicine, Bronx, New York.

Atkinson, K., S. Fogel, and S. Henry. 1980a. Yeast mutants defective in phosphatidylserine synthesis. *J. Biol. Chem.* **255**:6653.

Atkinson, K., A. Kolat, and S. Henry. 1977. Osmotic imbalance in inositol-starved spheroplasts of *Saccharomyces cerevisiae*. *J. Bacteriol.* **132**:806.

Atkinson, K., B. Jensen, E. Storm, A. Kolat, S. Henry, and S. Fogel. 1980b. Yeast mutants auxotrophic for ethanolamine or choline. *J. Bacteriol.* **141**:558.

Bailey, R. and L. Parks. 1975. Yeast sterol esters and their relationship to the growth of yeast. *J. Bacteriol.* **124:** 606.
Bailey, R., P. Hays, and L. Parks. 1976a. Homoazasterol-mediated inhibition of yeast sterol biosynthesis. *J. Bacteriol.* **128:** 730.
Bailey, R., L. Miller, and L. Parks. 1976b. Enzymatic analysis of C_{27} sterol-accumulating yeast strains. *J. Bacteriol.* **126:** 1012.
Barber, E. and W. Lands. 1973. Quantitative measurement of the effectiveness of unsaturated fatty acids required for growth of *Saccharomyces cerevisiae. J. Bacteriol.* **115:** 543.
Bard, M. 1972. Biochemical and genetic aspects of nystatin resistance in *Saccharomyces cerevisiae. J. Bacteriol.* **111:** 649.
Bard, M., R. Woods, and J. Haslam. 1974. Porphyrin mutants of *Saccharomyces cerevisiae* correlated lesions in sterol and fatty acid biosynthesis. *Biochem. Biophys. Res. Commun.* **56:** 324.
Bard, M., R. Woods, D. Barton, J. Corrie, and D. Widdowson. 1977. Sterol mutants of *Saccharomyces cerevisiae:* Chromatographic analyses. *Lipids* **12:** 645.
Barnett, J., A. Rasheed, and D. Corina. 1973. Partial reactions of D-glucose-6-phosphate-1L-myoinositol-1-phosphate cyclase. *Biochem. J.* **131:** 21.
Barton, D.H.R., J.E.T. Corrie, D.A. Widdowson, M. Bard, and R.A. Woods. 1974a. Biosynthesis of terpenes and steroids. IX. The sterols of some mutant yeast and their relationship to the biosynthesis of ergosterol. *J. Chem. Soc. Perkin Trans.* **1:** 1326.
―――. 1974b. Biosynthetic implications of the sterol content of ergosterol-deficient mutants of yeasts. *J. Chem. Soc. Chem. Commun.* **1:** 30.
Barton, D.H.R., A.A.L. Gunatilaka, T.R. Jarman, D.A. Widdowson, M. Bard, and R.A. Woods. 1975a. Biosynthesis of terpenes and steroids. X. The sterols of some yeast mutants double defective in ergosterol biosynthesis. *J. Chem. Soc. Perkin Trans.* **1:** 88.
Barton, D.H.R., T.R. Jarman, K.C. Watson, D.A. Widdowson, R.B. Boar, and K. Damps. 1974c. Assimilation of the antipodal forms of squalene 2,3-oxide by mammalian, yeast and plant systems. *J. Chem. Soc. Chem. Commun.* **21:** 861.
―――. 1975b. Investigation on the biosynthesis of steroids and terpenoids. XII. Biosynthesis of 3β-hydroxy triterpenoids and 3-hydroxy steroids from 3S-2,3 expo-2,3-dihydrosqualene. *J. Chem. Soc. Perkin Trans.* **1:** 1134.
Becker, G.W. and R.L. Lester. 1977. Changes in phospholipids of *Saccharomyces cerevisiae* associated with inositolless death. *J. Biol. Chem.* **252:** 8684.
Belendiuk, G., D. Mangnall, B. Tung, J. Westley, and G. Getz. 1978. CTP-phosphatidic acid cytidyltransferase from *Saccharomyces cerevisiae:* Partial purification, characterization and kinetic behavior. *J. Biol. Chem.* **253:** 4555.
Bloch, K. 1965. The biological synthesis of cholesterol. *Science* **150:** 19.
Bloomfield, D.K. and K. Bloch. 1960. The formation of Δ^9-unsaturaturated fatty acids. *J. Biol. Chem.* **235:** 337.
Boll, M., M. Lowel, J. Still, and J. Berndt. 1975. Sterol biosynthesis in yeast, 3-hydroxy-3-methylglutaryl coenzyme A reductase as a regulatory enzyme. *Eur. J. Biochem.* **54:** 435.
Bottema, C.K. and L.W. Parks. 1980. Sterol analysis of the inner and outer mitochondrial membranes in yeast. *Lipids* **15:** 987.
Burkl, G., H. Castorph, and E. Schweizer. 1972. Mapping of a complex gene locus coding for part of the *Saccharomyces cerevisiae* fatty acid synthetase multienzyme complex. *Mol. Gen. Genet.* **119:** 315.
Buttke, T.M. and K. Bloch. 1980. Response of yeast mutant strain GL-7 to lanosterol cycloartenol and cyclolaudenol. *Biochem. Biophys. Res. Commun.* **92:** 229.
Carman, G.M. and J. Matas. 1981. Solubilization of microsomal-associated phosphatidylserine synthase and phosphatidylinositol synthase from *Saccharomyces cerevisiae. Can. J. Microbiol.* **27:** 1140.
Carson, M., K.D. Atkinson, and C.J. Waechter. 1982. Properties of particulate and solubilized

phosphatidylserine synthase activity from *Saccharomyces cerevisiae:* Inhibitory effect of choline in the growth medium. *J. Biol. Chem.* (in press).

Cass, A., A. Finkelstein, and V. Krespi. 1970. The ion permeability induced in thin lipid membranes by the polyene antibiotics nystatin and amphotericin B. *J. Gen. Physiol.* **56:** 100.

Charalompous, F. and I. Chen. 1966. Inositol-1-phosphate synthetase and inositol-1-phosphate phosphatase from yeast. *Methods Enzymol.* **9:**698.

Clark-Walker, G. and A. Linnane. 1967. The biogenesis of mitochondria in *Saccharomyces cerevisiae:* A comparison between cytoplasmic respiratory-deficient mutant yeast and chloramphenicol inhibited wild type yeast. *J. Cell Biol.* **34:**1.

Cirillo, V., M. Harsch, and J. Lampen. 1964. Action of the polyene antibiotics filipin, nystatin and *N*-acetylcandidin on the yeast cell membrane. *J. Gen. Microbiol.* **35:**249.

Cobon, G., P. Crowfoot, and A. Linnane. 1974. Biogenesis of mitochondria. Phospholipid synthesis *in vitro* by yeast mitochondrial and microsomal fractions. *Biochem. J.* **144:**265.

Colbeau, A., J. Nachbaur, and P.M. Vignais. 1971. Enzymatic characterization and lipid composition of rat liver subcellular membranes. *Biochim. Biophys. Acta.* **249:**462.

Comte, J.M., B. Maisterrena, and D.C. Gautheron. 1976. Lipid composition and protein profiles of outer and inner membranes from pig heart mitochondria: Comparison with microsomes. *Biochim. Biophys. Acta* **419:**271.

Crick, F. and L. Orgel. 1964. The theory of interallelic complementation. *J. Mol. Biol.* **8:**161.

Criddle, R. and G. Schatz. 1969. Promitochondria of anaerobically grown yeast. I. Isolation and biochemistry. *Biochemistry* **8:**322.

Culbertson, M. 1975. "Regulation of inositol biosynthesis in *Saccharomyces cerevisiae*." Ph.D. thesis, Albert Einstein College of Medicine, Bronx, New York.

Culbertson, M. and S. Henry. 1975. Inositol requiring mutants of *Saccharomyces cerevisiae. Genetics* **80:**23.

Culbertson, M., T. Donahue, and S. Henry. 1976a. Control of inositol biosynthesis in *Saccharomyces cerevisiae*. I. Properties of a repressible enzyme system in extracts of wild-type (Ino$^+$) cells. *J. Bacteriol.* **126:**232.

———. 1976b. Control of inositol biosynthesis in *Saccharomyces cerevisiae*. I. Inositol-1-phosphate synthetase mutants. *J. Bacteriol.* **126:**243.

Dietlein, G. and E. Schweizer. 1975. Control of fatty acid biosynthesis in *Saccharomyces cerevisiae. Eur. J. Biochem.* **58:**177.

Donahue, T. 1979. "Biochemical and genetic analysis of inositol-1-phosphate synthase mutants of yeast." Ph.D. thesis, Albert Einstein College of Medicine, Bronx, New York.

Donahue, T. and S. Henry. 1981a. Myoinositol-1-phosphate synthase: Characteristics of the enzyme and identification of its structural gene in yeast. *J. Biol. Chem.* **256:**7077.

———. 1981b. Inositol requiring mutants of yeast: Mapping of the *ino1* locus and characterizing alleles of loci *ino1, ino2* and *ino4. Genetics* **98:**491.

Donahue, T., K. Atkinson, A. Kolat, and S. Henry. 1978. Inositol-1-phosphate synthase mutants of yeast *Saccharomyces cerevisiae.* In *Cyclitols and phosphoinositides* (ed. F. Eisenberg and W. Wells), p. 311. Academic Press, New York.

Duran, A. and E. Cabib. 1978. Solubilization and partial purification of yeast chitin synthetase. *J. Biol. Chem.* **253:**4419.

Duran, A., B. Bowers, and E. Cabib. 1975. Chitin synthetase zymogen is attached to the yeast plasma membrane. *Proc. Natl. Acad. Sci.* **72:**3952.

Eisenberg, F. 1978. Intermediates in the myo-inositol-1-phosphate synthase reaction. In *Cyclitols and phosphoinositides* (ed. W. Wells and F. Eisenberg), p. 269. Academic Press, New York.

Esfahani, M., E. Kucirka, F. Timmons, S. Tyagi, A. Lord, and S. Henry. 1981. Effect of exogenous fatty acids on growth, membrane fluidity, and phospholipid fatty acid composition in yeast. *J. Supramol. Struct. Cell. Biochem.* **15:**119.

Fieser, L. and M. Fieser. 1959. *Steroids.* Reinhold, New York.
Fryberg, M., A. Oehlschlager, and A. Unrau. 1974. Sterol biosynthesis in antibiotic resistant yeast: Nystatin. *Arch. Biochem. Biophys.* **160**: 83.
Getz, G. 1970. Lipids in membrane development. *Adv. Lipid Res.* **8**: 175.
Getz, G.S., S. Jakovcic, J. Heywood, J. Frank, and M. Rabinowitz. 1970. A two dimensional thin layer chromatographic system for phospholipid separation. The analysis of yeast phospholipids. *Biochim. Biophys. Acta* **218**: 441.
Gollub, E., P. Trocha, P. Liu, and D. Sprinson. 1974. Yeast mutants requiring ergosterol as only lipid supplement. *Biochem. Biophys. Res. Commun.* **56**: 471.
Gollub, E., K. Lin, J. Doyan, M. Adlersberg, and D. Sprinson. 1977. Yeast mutants deficient in heme biosynthesis and a heme mutant additionally blocked in cyclization of 2,3 oxido squalene. *J. Biol. Chem.* **252**: 2846.
Gordon, P. and P. Stewart. 1971. The effect of antibiotics on lipid synthesis during respiratory development in *Saccharomyces cerevisiae. Microbios* **4**: 115.
———. 1972. The effect of lipid status on cytoplasmic and mitochondrial protein synthesis in anaerobic cultures of *Saccharomyces cerevisiae. J. Gen. Microbiol.* **72**: 231.
Gordon, P., M. Lowdon, and P. Stewart. 1972a. Effects of chloramphenicol isomers and erythromycin on enzyme and lipid synthesis induced by oxygen in wild-type and petite yeast. *J. Bacteriol.* **110**: 504.
———. 1972b. Effect of unsaturated fatty acids on the development of respiration and on protein synthesis in an unsaturated fatty acid mutant of *Saccharomyces cerevisiae. J. Bacteriol.* **110**: 511.
Greenberg, M. 1980. "Genetic regulation of the biosynthesis and utilization of inositol in yeast." Ph.D. thesis, Albert Einstein College of Medicine, Bronx, New York.
Greenberg, M., P. Goldwasser, and S. Henry. 1982a. Characterization of a yeast regulatory mutant constitutive for inositol-1-phosphate synthase. *Mol. Gen. Genet.* (in press).
Greenberg, M., B. Reiner, and S.A. Henry. 1982b. Regulatory mutations of inositol biosynthesis in yeast: Isolation of inositol excreting mutants. *Genetics* **100**: 19.
Grindle, M. 1973. Sterol mutants of *Neurospora crassa:* Their isolation, growth characteristics and resistance to polyene antibiotics. *Mol. Gen. Genet.* **120**: 283.
Guerin, B., P. Lable, and M. Somlo. 1979. Preparation of yeast mitochondria (*Saccharomyces cerevisiae*) with good P/O respiratory control ratios. *Methods Enzymol.* **60**: 149.
Hajra, A. 1968. Biosynthesis of phosphatidic acid from dihydroxyacetone phosphate. *Biochem. Biophys. Res. Commun.* **33**: 929.
Hallermayer, G. and W. Neupert. 1974. Lipid composition of mitochondrial outer and inner membranes of *Neurospora crassa. Hoppe-Seyler's Z. Physiol. Chem.* **355**: 279.
Hanson, B. and R. Lester. 1980. The effects of inositol starvation on phospholipid and glycan synthesis in *Saccharomyces cerevisiae. J. Bacteriol.* **142**: 79.
Haslam, J. and A. Astin. 1979. The use of heme-deficient mutants to investigate mitochondrial function and biogenesis in yeast. *Methods Enzymol.* **61**: 558.
Haslam, J. and N. Fellows. 1977. The effects of unsaturated fatty acid depletion on the proton permeability and energetic functions of yeast mitochondria. *Biochem. J.* **166**: 565.
Haslam, J., J. Proudlock, and A. Linnane. 1971. Biogenesis of mitochondria. 20. The effects of altered membrane lipid composition on mitochondrial oxidative phosphorylation in *Saccharomyces cerevisiae. Bioenergetics* **2**: 351.
Hays, P., L. Parks, H. Pierce, and A. Oehlschlager. 1977. Accumulation of Ergosta-1, 14-dien-3-β-01 by *Saccharomyces cerevisiae* cultured with an azasterol antimycotic agent. *Lipids* **12**: 666.
Helmkamp, G.M. and K. Bloch. 1969. β-hydroxyldecanoyl thioester dehydratase: Studies on molecular structure and active site. *J. Biol. Chem.* **244**: 6014.
Henry, S. 1973. Death resulting from fatty acid starvation in yeast. *J. Bacteriol.* **116**: 1293.
Henry, S. and S. Fogel. 1971. Saturated fatty acid mutants in yeast. *Mol. Gen. Genet.* **113**: 1.

Henry, S. and A. Keith. 1971. Membrane properties of saturated fatty acid mutants of yeast revealed by spin labels. *Chem. Phys. Lipids* **7:** 245.
Henry, S., T. Donahue, and M. Culbertson. 1975. Selection of spontaneous mutants by inositol starvation in *Saccharomyces cerevisiae. Mol. Gen. Genet.* **143:** 5.
Henry, S., A. Keith, and W. Snipes. 1976. Changes in the restriction of molecular rotational diffusion of water soluble spin labels during fatty acid starvation in yeast. *Biophys. J.* **16:** 641.
Henry, S., K. Atkinson, A. Kolat, and M. Culbertson. 1977. Growth and metabolism of inositol-starved cells of *Saccharomyces cerevisiae. J. Bacteriol.* **140:** 472.
Henry, S.A., M. Greenberg, V. Letts, B. Shicker, L. Klig, and K.D. Atkinson. 1981. Genetic regulation of phospholipid synthesis in yeast. In *Current developments in yeast research: Advances in biotechnology. Proceedings of the 5th International Yeast Symposium* (1980) (ed. G. Stewart and I. Russell), p. 311. Pergamon, Toronto.
Hoffmann-Ostenhof, O., F. Pittner, and F. Koller. 1978. Some enzymes of inositol metabolism, their purification and mechanism of action. In *Cyclitols and phosphoinositides* (ed. W. Wells and F. Eisenberg), p. 233. Academic Press, New York.
Holz, R.W. 1974. The effects of the polyene antibiotics nystatin and amphotericin B on thin lipid membranes. *Ann. N.Y. Acad. Sci.* **235:** 469.
Hossack, J. and A. Rose. 1976. Fragility of plasma membranes in *Saccharomyces* enriched with different sterols. *J. Bacteriol.* **127:** 67.
Hunter, K. and A. Rose. 1971. Yeast lipids and membranes. In *The yeasts* (ed. A.H. Rose and J.S. Harrison), vol. 2, p. 211. Academic Press, New York.
―――. 1972. Lipid composition of *Saccharomyces cerevisiae* as influenced by growth temperature. *Biochim. Biophys. Acta* **260:** 639.
Hutchinson, H.T., L. Hartwell, and C.S. McLaughlin. 1969. Temperature-sensitive yeast mutant defective in ribonucleic acid production. *J. Bacteriol.* **99:** 807.
Jackson, M. and S. Shin. 1980. Analysis of inositol starvation in an inositol requiring mutant of CHO-K1. *J. Cell Biol.* **87:** 293a.
Jakovcic, S., G. Getz, M. Rabinowitz, H. Jakob, and H. Swift. 1971. Cardiolipin contents of wild type and mutant yeasts in relation to mitochondrial function and development. *J. Cell. Biol.* **48:** 490.
Janki, R.M., H.N. Aithal, W.C. McMurray, and E.R. Tustanoff. 1974. The effect of altered membrane lipid composition on enzyme activities of outer and inner mitochondrial membranes of *Saccharomyces cerevisiae. Biochem. Biophys. Res. Commun.* **56:** 1078.
Janki, R.M., H.N. Aithal, E.R. Tustanoff, and A.J.S. Ball. 1975. The biogenesis of mitochondrial membranes in the yeast *Saccharomyces cerevisiae. Biochim. Biophys. Acta* **375:** 446.
Jollow, D., G. Kellerman, and A. Linnane. 1968. The biogenesis of mitochondria. III. The lipid composition of aerobically and anaerobically grown *Saccharomyces cerevisiae* as related to the membrane systems of the cells. *J. Cell Biol.* **37:** 221.
Kamiryo, T., S. Parthasarathy, and S. Numa. 1976. Evidence that acyl coenzyme A synthetase is required for repression of yeast acetyl coenzyme A carboxylase by exogenous fatty acids. *Proc. Natl. Acad. Sci.* **73:** 386.
Karst, F. and F. Lacroute. 1973. Isolation of pleiotropic yeast mutants requiring ergosterol for growth. *Biochem. Biophys Res. Commun.* **52:** 741.
―――. 1974. Yeast mutant requiring only a sterol as growth supplement. *Biochem. Biophys. Res. Commun.* **59:** 370.
―――. 1977. Ergosterol biosynthesis in *Saccharomyces cerevisiae:* Mutants deficient in the early steps of the pathway. *Mol. Gen. Genet.* **154:** 269.
Kaufer, J. and E. Kennedy. 1964. Metabolism and function of bacterial lipids. II. Biosynthesis of phospholipids in *Escherichia coli. J. Biol. Chem.* **239:** 1720.
Keith, A., M. Resnick. and A. Haley. 1969. Fatty acid desaturase mutants of *Saccharomyces cerevisiae. J. Bacteriol.* **98:** 415.

Keith, A., B. Wisnieski, S. Henry, and J. Williams. 1972. Membranes of yeast and *Neurospora*: Lipid mutants and physical studies. In *Lipids and biomembranes of eukaryotic microorganisms* (ed. J.A. Erwin), p. 259. Academic Press, New York.

Kennedy, E. and S. Weiss. 1956. The function of cytidine coenzymes in the biosynthesis of phospholipids. *J. Biol. Chem.* **222:** 193.

Kinsky, S.C. 1970. Antibiotic interaction with model membranes. *Annu. Rev. Pharmacol.* **10:** 119.

Klein, H. 1955. Synthesis of lipids in resting cell of *Saccharomyces cerevisiae*. *J. Bacteriol.* **69:** 620.

Klein, H., N. Eaton, and J. Murphy. 1954. Net synthesis of sterols in resting cells of *Saccharomyces cerevisiae*. *Biochim. Biophys. Acta* **13:** 591.

Kleinhans, F., N. Lees, M. Bard, R. Haak, and R. Woods. 1978. ESR determinants of membrane permeability in a yeast sterol mutant. *Chem. Phys. Lipids* **23:** 143.

Kornblatt, J.A. and H. Rudney. 1971. Two forms of acetoacetyl coenzyme A thiolase in yeast. II. Intracellular location and relationship to growth. *J. Biol. Chem.* **246:** 4424.

Ková č, L., I. Gbelská, V. Poliachová, J. Subik, and V. Kováčová. 1980. Membrane mutants: A yeast mutant with a lesion in phosphatidyl serine biosynthesis. *Eur. J. Biochem.* **111:** 491.

Kühn, N. and F. Lynen. 1965. Phosphatidic acid synthesis in yeast. *Biochem. J.* **94:** 240.

Kühn, L., E. Schweizer, and H. Castorph. 1972. A new gene cluster in yeast: The fatty acid synthetase system. *Hoppe-Seylers Z. Physiol. Chem.* **352:** 3277.

LaBelle, E., Jr. and A. Hajra. 1974. Purification and kinetic properties of acyl and alkyl dihydroxyacetone phosphate oxidoreductase. *J. Biol. Chem.* **249:** 6939.

Lala, A., T. Buttke, and K. Block. 1979. On the role of the sterol hydroxyl group in membranes. *J. Biol. Chem.* **254:** 10582.

Lampen, J. 1966. Interference by polyenic antifungal antibiotics (especially nystatin and filipin with specific membrane functions. *Symp. Soc. Gen. Microbiol.* **16:** 111.

Lands, W., R. Sacks, J. Sauter, and F. Gunstone. 1978. Selective effects of fatty acids upon cell growth and metabolic regulation. *Lipids* **13:** 878.

Lester, R. and M. Steiner. 1968. The occurrence of diphosphoinositide and triphosphoinositides in *Saccharomyces cerevisiae*. *J. Biol. Chem.* **243:** 4889.

Lester, R., G. Becker, and K. Kaul. 1978. Phosphoinositides of fungi and plants. In *Cyclitols and phosphoinositides* (ed. W. Wells and F. Eisenberg), p. 83. Academic Press, New York.

Letts, V. 1980. "Mutants of *Saccharomyces cerevisiae* defective in lipid biosynthesis." Ph.D. thesis, University of Edinburgh, Scotland.

Letts, V. and I. Dawes. 1979. Mutations affecting lipid biosynthesis of *Saccharomyces cerevisiae:* Isolation of ethanolamine auxotrophs. *Biochem. Soc. Trans.* **7:** 976.

Lin, S. and I. Zabin. 1972. Beta-galactosidase rates of synthesis and degradation of incomplete chains. *J. Biol. Chem.* **247:** 2205.

Lindegren, C., E. Shult, Y.L. Hwang. 1962. Centromeres, sites of affinity and gene loci on the chromosomes of *Saccharomyces*. *Nature* **194:** 260.

Linnane, A., J. Haslam, H. Lukins, and P. Nagley. 1972. The biogenesis of mitochondria in microorganisms. *Annu. Rev. Microbiol.* **26:** 163.

Lukins, H., S. Thain, P. Wallace, and A. Linnane. 1966. Correlation of membrane bound succinate dehydrogenase with the occurrence of mitochondrial profiles in *Saccharomyces cerevisiae*. *Biochem. Biophys. Res. Commun.* **23:** 363.

Lynen, F. 1967. The role of biotin dependent carboxylations in biosynthetic reactions. *Biochem. J.* **102:** 381.

———. 1969. Fatty acid synthetase. *Methods Enzymol.* **14:** 17.

———. 1980. On the structure of fatty acid synthetase of yeast. *Eur. J. Biochem.* **112:** 431.

Maeda, T. and F. Eisenberg. 1980. Purification, structure and catalytic properties of L-myoinositol-1-phosphate synthase from rat testis. *J. Biol. Chem.* **255:** 8458.

Majumder, A., S. Duttagupta, P. Goldwasser, T. Donahue, and S. Henry. 1981. The mechanism of interallelic complementation at the *INO1* locus in yeast: Immunological analysis of mutants. *Mol. Gen. Genet.* **184:** 347.

Marzuki, S. and A. Linnane. 1979. Modification of yeast mitochondria by diet in specific mutants. *Methods Enzymol.* **61:** 568.

Marzuki, S., R. Hall, and A. Linnane. 1974. Induction of respiratory incompetent mutants by unsaturated fatty acid depletion in *Saccharomyces cerevisiae. Biochem. Biophys. Res. Commun.* **57:** 372.

Matile, P. 1966. Inositol deficiency resulting in death: An explanation of its occurrence in *Neurospora crassa. Science* **151:** 86.

McCammon, M.T. and L.W. Parks. 1981. Inhibition of sterol transmethylation by analogs of S-adenosylhomo-cysteine. *J. Bacteriol.* **145:** 106.

Miller, W.L. and J.L. Gaylor. 1970. Investigation of the component reactions of oxidative sterol demethylation. *J. Biol. Chem.* **245:** 5369.

Miller, W.L., D.R. Brady, and J.L. Gaylor. 1971. Investigation of the component reactions of oxidative demethylation of sterols. Metabolism of 4α hydroxy methyl steroids. *J. Biol. Chem.* **246:** 5147.

Mishina, M., R. Roggenkamp, and E. Schweizer. 1980. Yeast mutants defective in acetyl-coenzyme A carboxylase and biotin: Apocarboxylase ligase. *Eur. J. Biochem.* **111:** 79.

Mitropoulos, K.A., G.F. Gibbons, and B.E.A. Reeves. 1976. Lanosterol 14α demethylase: Similarity of the enzyme system from yeast and rat liver. *Steroids* **27:** 821.

Molzahn, S. and R. Woods. 1972. Polyene resistance and the isolation of sterol mutants in *Saccharomyces cerevisiae. J. Gen. Microbiol.* **72:** 339.

Moor, H. 1967. Endoplasmic reticulum as the initiator of bud formation in yeast. *Arch. Microbiol.* **57:** 135

Moore, J.T. and J.L. Gaylor. 1968. Investigation of the component reactions of sterol demethylation: Preparation and properties of yeast demethylase activities. *Arch. Biochem. Biophys.* **124:** 167.

Morpurgo, G., G. Serlupi-Crescenzi, G. Tecce, F. Valente, and D. Venetacci. 1964. The influence of ergosterol on the physiology and ultrastructure of *Saccharomyces cerevisiae. Nature* **201:** 897.

Nes, W. and M. McKean. 1977. *Biochemistry of steroids and other isopentenoids.* University Park Press, Baltimore.

Nes, W., J. Adler, B. Sekula, and K. Krevitz. 1976. Discrimination between cholesterol and ergosterol by yeast membranes. *Biochem. Biophys. Res. Commun.* **71:** 1296.

Nes, W., B. Sekula, W.D. Nes, and J. Adler. 1978. The functional importance of structural features of ergosterol in yeast. *J. Biol. Chem.* **253:** 6218.

Norman, A.W., R.A. Demel, B. DeKruyff, W.S.M. Geurts Van Kessel, and L.L.M. Van Deenen. 1972. Studies on the polyene antibiotics comparison of the other polyenes with filipin in their ability to interact specifically with sterol. *Biochim. Biophys. Acta* **290:** 1.

Novick, P. and R. Schekman. 1979. Secretion and cell surface growth are blocked in a temperature sensitive mutant of *Saccharomyces cerevisiae. Proc. Natl. Acad. Sci.* **76:** 1858.

Novick, P., C. Field, and R. Schekman. 1980. The identification of 23 complementation groups required for post-translational events in the yeast secretory pathway. *Cell* **21:** 205.

Nurminen, T. and H. Suomalainen. 1971. Occurrence of long-chain fatty acids and glycolipids in the cell envelope fractions of baker's yeast. *Biochem. J.* **125:** 963.

Nurminen, T., L. Taskinen, and H. Suomalainen. 1976. Distribution of membranes, especially of plasma membrane fragments, during zonal centrifugation of homogenates from glucose-repressed *Saccharomyces cerevisiae. Biochem. J.* **154:** 751.

Oesterhelt, D., H. Bauer, and F. Lynen. 1969. Crystallization of a multienzyme complex, fatty acid synthetase from yeast. *Proc. Natl. Acad. Sci.* **63:** 1377.

Okuyama, H., M. Saito, V. Joski, S. Gunsberg, and S. Wakil. 1979. Regulation by temperature of chain length of fatty acids in yeast. *J. Biol. Chem.* **254:** 12281.

Orme, T., J. McIntyre, F. Lynen, L. Kühn, and E. Schweizer. 1972. Fatty acid elongation in a mutant of *Saccharomyces cerevisiae* deficient in fatty-acid synthetase. *Eur. J. Biochem.* **24:** 407.

Oshino, M., Y. Imai, and R. Sato. 1966. Electron-transfer mechanism associated with fatty acid desaturation catalyzed by liver microsomes. *Biochim. Biophys. Acta* **128:** 13.

Osumi, T., T. Nishino, and H. Katsuki. 1979. Studies on the Δ^5 desaturation in ergosterol biosynthesis in yeast. *J. Biochem.* **85:** 819.

Ottke, R.C., E.L. Tatum, I. Zabin, and K. Block. 1951. Isotopic acetate and isovalerate in the synthesis of ergosterol in *Neurospora*. *J. Biol. Chem.* **189.:** 429.

Paltauf, F. and G. Schatz. 1969. Promitochondria of anaerobically grown yeast. II. Lipid composition. *Biochemistry* **8:** 335.

Parks, L. 1978. Metabolism of sterols in yeast. *CRC Crit. Rev. Microbiol.* **6:** 301.

Parks, L.W., C. Anding, and G. Ourisson. 1974. Sterol transmethylation during aerobic adaptation of yeast. *Eur. J. Biochem.* **43:** 451.

Parks, L., C.A. McLean-Bowen, and M.T. McCammon. 1981. Correlation of sterol structural changes with membrane fluidity and enzymatic activities. In *Current developments in yeast research: Advances in biotechnology. Proceedings of the 5th International Yeast Symposium* (1980) (ed. G. Stewart and I. Russell), p. 325. Pergamon, Toronto.

Parks, L.W., F.T. Bond, E.D. Thompson, and P.R. Starr. 1972. $\Delta^{8,22}$ ergostadiene 3β-ol, an ergosterol precursor accumulated in wild type and mutants of yeast. *J. Lipid. Res.* **13:** 311.

Parks, L., C. McLean-Bowen, F. Taylor, and S. Hough. 1978. Sterols in yeast subcellular fractions. *Lipids* **13:** 730.

Plattner, H. and G. Schatz. 1969. Promitochondria of anaerobically grown yeast. III. Morphology. *Biochemistry* **8:** 339.

Prescott, D.J. and P.R. Vagelos. 1972. Acyl carrier protein. *Adv. Enzymol.* **36:** 269.

Proudlock, J., J. Haslam, and A. Linnane. 1971. Biogenesis of mitochondria. The effects of unsaturated fatty acid depletion on lipid composition and energy metabolism of a fatty acid desaturase mutant of *Saccharomyces cerevisiae*. *Bioenergetics* **2:** 327.

Proudlock, J., L. Wheeldon, D. Jollow, and A. Linnane. 1968. Role of sterols in *Saccharomyces cerevisiae*. *Biochim. Biophys. Acta* **152:** 434.

Raetz, C. 1978. Enzymology, genetics, and regulation of membrane phospholipid synthesis in *Escherichia coli*. *Microbiol. Rev.* **42:** 614.

Rank, G., A. Robertson, and H. Bussey. 1978. The viscosity and lipid composition of the plasma membrane of multiple drug resistant and sensitive yeast strains. *Can. J. Biochem.* **56:** 1036.

Ratcliffe, J., G. Hossack, G. Wheeler, and A. Rose. 1973. Modifications to the phospholipid composition of *Saccharomyces cerevisiae* induced by exogenous ethanolamine. *J. Gen. Microbiol.* **76:** 445.

Resnick, M. and R. Mortimer. 1966. Unsaturated fatty acid mutants of *Saccharomyces cerevisiae*. *J. Bacteriol.* **92:** 597.

Ridgeway, C. and H. Douglas. 1958. Unbalanced growth of yeast due to inositol deficiency. *J. Bacteriol.* **76:** 163.

Rouslin, W. 1979. Effects of cerulenin upon the synthesis of lipids and protein and upon the formation of respiratory enzymes in adapting lipid limited *Saccharomyces cerevisiae*. *J. Bacteriol.* **139:** 502.

Sanders, H.K., P. A. Mied, M. Briquet, J. Hernandez-Rodriguez, R.F. Gottal, and J.B. Mattoon. 1973. Regulation of mitochondrial biogenesis: Yeast mutants deficient in synthesis of δ aminolevulinic acid. *J. Mol. Biol.* **80:** 17.

Scarborough, G. and J. Nyc. 1967. Methylation of ethanolamine phosphatides by microsomes from normal and mutant strains of *Neurospora crassa*. *J. Biol. Chem.* **242:** 238.

Schlossman, D. and R. Bell. 1978. Glycerolipid biosynthesis in *Saccharomyces cerevisiae*: sn-glycerol-3-phosphate and dihydroxyacetone phosphate acyltransferase activities. *J. Bacteriol.* **133:** 1368.

Schweizer, E. and H. Bolling. 1970. A *Saccharomyces cerevisiae* mutant defective in saturated fatty acid biosynthesis. *Proc. Natl. Acad. Sci.* **67:** 660.

Schweizer, E., K. Werkmeister, and M. Jain. 1978. Fatty acid biosynthesis in yeast. *Mol. Cell. Biochem.* **21:** 95.

Schweizer, E., V. Holzner, K. Meyer, P. Tauro, and M. Schweizer. 1974. The genetics of fatty acid biosynthesis in yeast. In *Comparative biochemistry and physiology of transport; Proceedings of the 1973 International Conference on Biological Membranes* (ed. L. Bolis et al.), p. 219. North-Holland, Amsterdam.

Schweizer, E., K. Meyer, M. Schweizer, K. Werkmeister and W. Fischer. 1977. Regulation of fatty acid synthesis in yeast. *FEBS Fed. Eur. Biochem. Soc. Symp.* **46:** 11.

Shatkin, A. and E. Tatum. 1961. The relationship of M-inositol to morphology of *Neurospora crassa*. *Am. J. Bot.* **48:** 760.

Shechter, I. and K. Bloch. 1971. Solubilization and purification of trans-farnesyl pyrophosphate squalene synthetase. *J. Biol. Chem.* **246:** 7690.

Shechter, I., F.W. Sweat, and K. Bloch. 1970. Comparative properties of 2,3-oxido-squalene-lanosterol cyclase from yeast and liver. *Biochim. Biophys. Acta* **220:** 463.

Silbert, D. 1975. Genetic modification of membrane lipid. *Annu. Rev. Biochem.* **44:** 315.

Silbert, D., J. Cronan, I. Beacham, and M. Larder. 1974. Genetic engineering of membrane lipid. *Fed. Proc.* **33:** 1725.

Synder, F., ed. 1977. *Lipid synthesis in mammals*. Plenum Press, New York.

Steiner, S. and R. Lester. 1972a. Metabolism of diphosphoinositide and triphosphoinositide in *Saccharomyces cerevisiae*. *Biochim. Biophys. Acta* **260:** 82.

———. 1972b. Studies on the diversity of inositol-containing yeast phospholipids: Incorporation of 2-deoxyglucose into lipids. *J. Bacteriol.* **109:** 81.

———. 1972c. *In vitro* studies of phospholipid biosynthesis in *Saccharomyces cerevisiae*. *Biochim. Biophys. Acta* **260:** 222.

Steiner, S., S. Smith, C. Wachter, and R. Lester. 1969. Isolation and partial purification of a major inositol containing lipid in baker's yeast, mannosylodi-inositol, diphosphoryl ceremide. *Proc. Natl. Acad. Sci.* **64:** 1042.

Stevens, B. 1981. Mitochondrial structure. In *The molecular biology of the yeast* Saccharomyces. *Life cycle and inheritance* (ed. J. Strathern et al.), p. 471. Cold Spring Harbor Laboratory, Cold Spring Harbor, New York.

Stoops, J. and S. Wakil. 1978. The isolation of the two subunits of yeast fatty acid synthetase. *Biochim. Biophys. Res. Commun.* **84:** 225.

Stoops, J., E. Awad, M. Arslanian, S. Gunsberg, S. Wakil, and R. Oliver. 1978. Studies on the yeast fatty acid synthetase: Subunit composition and structural organization of a large multifunctional enzyme complex. *J. Biol. Chem.* **253:** 4464.

Strauss, B.S. 1958. Cell death and unbalanced growth in *Neurospora*. *J. Gen. Microbiol.* **18:** 658.

Sullivan, J. and A. Debusk. 1973. Inositolless death in *Neurospora* and cellular ageing. *Nat. New Biol.* **243:** 72.

Taketani, S., T. Nishino, and H. Katsuki. 1979. Characterization of sterol-ester synthetase in *Saccharomyces cerevisiae*. *Biochim. Biophys. Acta* **575:** 148.

Taketani, S., T. Osumi, and H. Katsuki. 1978. Characterization of sterol ester hydrolase in *Saccharomyces cerevisiae*. *Biochim. Biophys. Acta* **525:** 87.

Tamura, Y., Y. Yoshida, R. Sato, and H. Kumaoka. 1976. Fatty acid desaturate system of yeast microsomes: Involvement of Cyt b5 containing electron transport chain. *Arch. Biochem. Biophys.* **175:** 284.

Tauro, P., U. Holzner, H. Castorph, F. Hill, and E. Schweizer. 1974. Genetic analysis of noncomplementing fatty acid synthetase mutants in *Saccharomyces cerevisiae*. *Mol. Gen. Genet.* **129:** 131.

Taylor, F. and L. Parks. 1978. Metabolic interconversion of free sterols and steryl esters in *Saccharomyces cerevisiae*. *J. Bacteriol.* **136:** 531.

Thomas, D., J. Hossack, and A. Rose. 1978. Plasma membrane lipid composition and ethanol tolerance in *Saccharomyces cerevisiae*. *Arch. Microbiol.* **117**: 239.
Thomas, P. 1972. Increased frequency of auxotrophic mutants of *Ustilago hordei* after combined UV irradiation and inositol starvation. *Can. J. Genet. Cytol.* **14**: 785.
Thompson, E. and L. Parks. 1974. Effect of altered sterol composition on growth characteristics of *Saccharomyces cerevisiae*. *J. Bacteriol.* **120**: 779.
Thompson, E., R.B. Bailey, and L.W. Parks. 1974. The subcellular location of S-adenosyl methionine Δ^{24}-sterol methyltransferase in *Saccharomyces cerevisiae*. *Biochim. Biophys. Acta* **334**: 116.
Thompson, E., P. Starr, and L. Parks. 1971. Sterol accumulation in a mutant of *Saccharomyces cerevisiae* defective in ergosterol production. *Biochem. Biophys. Res. Commun.* **43**: 1304.
Trembath, M. and A. Tzagoloff. 1979. Large and small scale preparation of yeast mitochondria. *Methods Enzymol.* **60**: 160.
Trocha, P.J. and D.B. Sprinson. 1976. Location and regulation of early enzymes of sterol biosynthesis in yeast. *Arch. Biochem. Biophys.* **174**: 45.
Trocha, P., S. Jasne, and D. Sprinson. 1974. Novel sterols in ergosterol deficient yeast mutants. *Biochem. Biophys. Res. Commun.* **59**: 666.
―――. 1977. Yeast mutants blocked in removal of lanosterol at C-14. Separation of sterols by high pressure liquid chromatography. *Biochemistry* **16**: 4721.
Ulaszewski, S., J. Woodward, and V. Cirillo. 1978. Membrane damage associated with inositolless death in *Saccharomyces cerevisiae*. *J. Bacteriol.* **136**: 49.
Vagelos, R., A. Alberts, and P. Majerus. 1969. Mechanism of saturated fatty acid biosynthesis in *Escherichia coli*. *Methods Enzymol.* **14**: 39.
Van Golde, L. and S.G. Van den Bergh. 1977. Introduction: General pathways in the metabolism of lipids in mammalian tissues. In *Lipid synthesis in mammals* (ed. F. Snyder), vol. 1, p. 1. Plenum Press, New York.
Vary, M., P. Stewart, and A. Linnane. 1970. Biogenesis of mitochondria: Role of mitochondrial and cytoplasmic ribosomal protein synthesis in the oxygen-induced formation of yeast mitochondrial enzyme. *Arch. Biochem. Biophys.* **141**: 430.
Villarejo, M., P.J. Zamenhof, and I. Zabin. 1972. Beta-galactosidase: In vivo complementation. *J. Biol. Chem.* **247**: 2212.
Waechter, C. and R. Lester. 1971. Regulation of phosphatidylcholine biosynthesis in *Saccharomyces cerevisiae*. *J. Bacteriol.* **105**: 837.
―――. 1973. Differential regulation of the N-methyl-transferases responsible for phosphatidylcholine synthesis in *Saccharomyces cerevisiae*. *Arch. Biochem. Biophys.* **158**: 401.
Waechter, C., M. Steiner, and R. Lester. 1969. Regulation of phosphatidylcholine biosynthesis by the methylation pathway in *Saccharomyces cerevisiae*. *J. Biol. Chem.* **244**: 3419.
Walenga, R. and W. Lands. 1975. Selectiveness of various unsaturated fatty acids in supporting growth and respiration in *Saccharomyces cerevisiae*. *J. Biol. Chem.* **250**: 9121.
Wallace, P., M. Huang, and A. Linnane. 1968. Biogenesis of mitochondria. II. The influence of medium composition on the cytology of anaerobically grown *Saccharomyces cerevisiae*. *J. Cell Biol.* **37**: 207.
Welch, J. and A. Burlingame. 1973. Very long chain fatty acids in yeast. *J. Bacteriol.* **115**: 464.
Wells, W. and F. Eisenberg, eds. 1978. *Cyclitols and phosphoinositides*. Academic Press, New York.
Wisnieski, B. and R. Kiyomoto. 1972. Fatty acid desaturase mutants of yeast: Growth requirements and electron spin resonance distribution. *J. Bacteriol.* **109**: 186.
Wisnieski, B., A. Keith, and M. Resnick. 1970. Double bond requirement in a fatty acid desaturase mutant of *Saccharomyces cerevisiae*. *J. Bacteriol.* **101**: 160.
Woods, R. 1971. Nystatin resistant mutants of yeast: Alterations in sterol content. *J. Bacteriol.* **108**: 69.
Woods, R., H. Sanders, M. Briquet, F. Foury, B. Drysdale, and J. Mattoon. 1975. Regulation

of mitochondrial biogenesis: Enzymatic changes in cytochrome-deficient yeast mutants requiring δ-aminolevulinic acid. *J. Biol. Chem.* **250**: 9090.

Yamashita, S. and A. Oshima. 1980. Regulation of phosphatidylethanolamine methyltransferase level by myo inositol in *Saccharomyces cerevisiae*. *Eur. J. Biochem.* **104**: 611.

Regulatory Circuits for Gene Expression: The Metabolism of Galactose and Phosphate

Yasuji Oshima
Department of Fermentation Technology
Faculty of Engineering
Osaka University
Yamadaoka, Suita-shi, Osaka 565, Japan

1. **Genes Involved in Galactose Utilization**
   A. Metabolism of Galactose
   B. Galactose Genes
   C. Douglas-Hawthorne Model
2. **Characterization of the Galatose Regulatory Genes**
   A. Function of the *GAL80* Gene
   B. Characterization of the *gal81-GAL4* Locus
3. **Regulatory Circuit for the Galactose Pathway Enzymes**
4. **Regulation of Phosphatases**
   A. Phosphatases in Yeast
   B. Phosphatase Genes
   C. Regulatory Circuit for Phosphatase
5. **Discussion**

INTRODUCTION

The enzyme system responsible for the initial steps of galactose utilization offers considerable advantages over other systems in the study of genetic regulatory mechanisms involved in enzyme synthesis in *Saccharomyces cerevisiae*. One advantage is that the synthesis of these enzymes is regulated in response to a simple external effector, galactose as inducer, whereas the activities are affected by carbon catabolite repression and inhibition. In practice, there are neither inherently programmed control mechanisms, like the master timing mechanism suggested in the cell-division cycle (Hartwell et al. 1974), nor complex interactions between various effectors or between different enzyme proteins, like those in the regulation of some biosynthetic pathway enzymes in microorganisms (Calvo and Fink 1971; Metzenberg 1979). Likewise, the galactose system does not involve duplicated genes for a particular gene function, which is common in maltose and sucrose utilization (Mortimer and Hawthorne 1969), nor housekeeping enzymes like those observed in acid and alkaline phosphatases in *Neurospora crassa* (Metzenberg 1979) and *S. cerevisiae* (Toh-e et al. 1973, 1976). Having these advantages, the galactose system has been subjected to intensive genetic analysis, and a model of genetic regulation has been proposed by Douglas and Hawthorne (1966). This provided an initial clue to the understanding of one genetic regulatory mechanism in yeast.

Also under investigation in our laboratory is the system regulating the synthesis of phosphatases in *S. cerevisiae*. Although this system involves housekeeping enzymes, most of the activities are due to the repressible enzymes; their activities are controlled in response to a simple effector: inorganic phosphate (Suomalainen et al. 1960; Schmidt et al. 1963; Schurr and Yagil 1971; Toh-e et al. 1973). This system also has an advantage in genetic study, in that the acid phosphatase is located in the periplasmic space of the cell envelope, thus facilitating phenotype determination (Toh-e et al. 1973; Toh-e and Oshima 1974). Although the alkaline phosphatase is located wholly inside the cell envelope, this enzyme activity is also easily detectable by permeabilization of cells to substrate.

This paper summarizes the experimental evidence upon which genetic regulatory models have been constructed for the galactose and phosphatase systems in *S. cerevisiae*. Since several recent findings, mostly concerning the galactose system, contradict some of the predictions that arise from the Douglas-Hawthorne model, a revised model is proposed. Lastly, the striking similarities in the regulatory mechanism of synthesis between the galactose pathway enzymes in *S. cerevisiae* and the acid and alkaline phosphatases in *S. cerevisiae* and *N. crassa* are discussed.

GENES INVOLVED IN GALACTOSE UTILIZATION

Metabolism of Galactose

In *S. cerevisiae*, the utilization of exogenous galactose requires the induction of an enzyme system consisting of a specific galactose transport activity (Douglas and Condie 1954; Cirillo 1968) and the Leloir pathway enzymes, namely, galactokinase (Enzyme Commission [EC] number 2.7.1.6), α-D-galactose-1-phosphate uridyltransferase (transferase; EC number 2.7.7.12), and uridine diphosphoglucose 4-epimerase (epimerase; EC number 5.1.3.2) (Kosterlitz 1943; Leloir 1951; Kalckar et al. 1953; Douglas and Hawthorne 1964). These inducible enzymes convert galactose to glucose 1-phosphate, which is then converted to glucose 6-phosphate by a constitutively produced isozyme of phosphoglucomutase (EC number 2.7.5.1) (Douglas 1961; Bevan and Douglas 1969). Glucose 6-phosphate then enters the glycolytic pathway.

These galactose pathway enzymes are inducible by galactose and susceptible to glucose repression. Their repression by carbon catabolites involves a characteristic hysteresis (Adams 1972). The enzymes, especially the galactose transport activity, are also sensitive to carbon catabolite inhibition (Matern and Holzer 1977).

Galactose Genes

The Leloir pathway enzymes galactokinase, transferase, and epimerase are coded for by three genes designated, respectively, *GAL1, GAL7,* and *GAL10* (Douglas and Hawthorne 1964). These three genes are tightly linked in the order centromere *GAL7-GAL10-GAL1* on chromosome II (Bassell and Mortimer 1971). This gene order deduced from recombination analyses is in accord with the DNA sequence data of the gene cluster cloned on phage λ DNA (T.P. St. John and R.W. Davis, pers. comm.). The specific galactose transport protein is coded for by the *GAL2* gene (Douglas and Condie 1954; Cirillo 1968) on chromosome XII (Mortimer and Hawthorne 1966), and the *GAL5* gene controlling the phosphoglucomutase activity (Douglas 1961) is mapped on fragment 6 (Mortimer and Hawthorne 1973), a chromosomal segment whose connection with one of the established chromosomes has not been identified.

These structural genes are regulated by four regulatory genes: *GAL4, GAL80, gal81*[1], and *GAL3*. Either a recessive *gal80* or a dominant *GAL81* mutation (previous designations were *i* [Douglas and Pelroy 1963] and *C* [Douglas and Hawthorne 1966], respectively) gives rise to constitutive expression of the three Leloir pathway enzymes (Douglas and Hawthorne 1966) and possibly the galactose transport activity (V.P. Cirillo [unpubl.] cited in Kew and Douglas [1976]), whereas a recessive *gal4* mutation simultaneously blocks the synthesis of galactokinase, transferase, and epimerase (Douglas and Hawthorne 1964). Since there are suppressible (nonsense) *gal80* and *gal4* mutations (Douglas and Hawthorne 1972; Matsumoto et al. 1980), these regulatory genes each encode a corresponding polypeptide. *GAL81* and *gal4* mutations are known to be closely linked. The constitutive phenotype produced by the *GAL81* mutation is expressed only if the mutant allele lies in *cis* position with respect to a functional allele of *GAL4* (Douglas and Hawthorne 1966). These regulatory genes also control α-galactosidase (EC number 3.2.1.22) production (Kew and Douglas 1976). The *gal81-GAL4* cluster and the *GAL80* locus have been mapped separately on chromosome XVI remote from its centromere and on fragment 8 (Mortimer and Schild 1980).

The recessive *gal3* mutation confers a long-term adaptation phenotype; i.e., although *GAL3* strains ferment galactose rapidly (within 1-2 days), *gal3* strains are able to ferment galactose only following a period of adaptation of approximately 1 week (Winge and Roberts 1948; Spiegelman et al. 1950; Mundkar 1952). This phenotype is suppressed by mutations in the *GAL80* locus (Douglas and Pelroy 1963). The *gal3* mutation is also associated with pleiotropic impairment in the utilization of melibiose and maltose, as well as

---

[1]As described later in this paper, *GAL81* mutations are within the *GAL4* gene and generate an activator protein insensitive to inhibition by *GAL80*; hence, they are generally dominant.

galactose, but not sucrose (Kew and Douglas 1976). The *gal3* locus has been mapped on chromosome IV close to the centromere (Hawthorne and Mortimer 1960).

Douglas-Hawthorne Model

Douglas and Hawthorne (1966) have proposed a model for the function of regulatory genes in the galactose system that conforms with the Jacob-Monod operon concept of the β-galactosidase system of *Escherichia coli*. The Douglas-Hawthorne model posits that the *GAL80* gene produces a repressor that represses the expression of the *GAL4* gene by interacting at the *gal81* site in the absence of galactose. In the presence of galactose, the repressor is inactivated and the *GAL4* gene is expressed to produce a diffusible positive intermediary. This in turn exerts positive control over the expression of the complex of the three linked genes that specify the structure of Leloir pathway enzymes and, possibly, *GAL2*, which specifies the galactose transport system. The model implies that the *GAL81* site plays essentially the same role as the *lac* operator, as the site of repressor recognition, and controls the transcription of the contiguous structural gene. In the bacterial *lac* system, the contiguous structural genes code directly for the enzymatic functions, whereas in the yeast system the contiguous structural gene *GAL4* acts as an intermediary between the *GAL81* site and the structural genes for the enzymes. Whether the product of *GAL4* controls the synthesis of the enzymes at the level of mRNA synthesis or at some subsequent step in protein synthesis is not specified in the model.

The Douglas-Hawthorne model is consistent with the genetic data so far presented. However, equally consistent with these data is a model in which the *GAL4* protein is synthesized constitutively, but its activity is modulated through interaction with the *GAL80* protein. In this model *GAL81* mutations do not delineate an operator region contiguous to *GAL4* but rather lie within the *GAL4* gene and define the affinity site in the *GAL4* protein at which the *GAL80* protein acts. In the next section, more recent analyses of the *GAL80* and *gal81-GAL4* loci are discussed in terms of the Douglas-Hawthorne model, and the results from several experiments are presented that suggest that this second model, rather than the Douglas-Hawthorne model, more accurately describes the mechanism of regulation of galactose enzymes in yeast.

CHARACTERIZATION OF THE GALACTOSE REGULATORY GENES

Function of the *GAL80* Gene

The Douglas-Hawthorne model posits that *GAL80* produces a repressor that acts at the site of *gal81*. Thus, the relationship between *GAL80* and *gal81* is analogous to the repressor-operator relationship in bacterial systems. Hence, one might expect to find a mutant in the yeast system that is equivalent to the i^s (super-repressible) mutants described in the *E. coli* lactose regulatory gene (Bourgeois and Jobe 1970). Such dominant galactose-negative mutants of *S. cerevisiae* arising from mutations at the *GAL80* locus indeed have been isolated (Douglas and Hawthorne 1972; Nogi et al. 1977). These mutations, designated *GAL80^s* (previous designation was i^s [Douglas and Hawthorne 1972]), are *trans* dominant over the wild-type allele *GAL80$^+$* and mutant *gal80* alleles.

Several mutants showing the galactose-positive phenotype have been isolated as revertants from the *GAL80^s* mutants. The secondary mutations were identified as occurring either at the *GAL80* or *gal81* loci: a mutation tightly linked to the original *GAL80^s* allele to give a recessive *gal80* phenotype, or a mutation at the *gal81* locus insensitive to the *GAL80^s* super-repressor (Douglas and Hawthorne 1972; Nogi et al. 1977). This resembles the bacterial case, in which the two types of Lac$^+$ revertants were obtained from the i^s mutant: i^- and O^c (Willson et al. 1964). Although the dominant constitutive *GAL81* mutation is epistatic to the dominant super-repressible *GAL80^s* allele (Douglas and Hawthorne 1972), it was shown later that double *GAL80^s GAL81* mutants exhibit a variety of phenotypes—inducible, uninducible, or constitutive—depending on the combination of mutant alleles in the two loci (Nogi et al. 1977).

Kinetic study of galactokinase induction with a temperature-sensitive *gal80* mutant, whose product is labile at 35°C and, because it was derived from a dominant super-repressible *GAL80^s* mutant, is super-repressible at 25°C, showed an abrupt increase in the enzyme activity immediately after the temperature shift from 25°C to 35°C under both growing and nongrowing conditions (Matsumoto et al. 1978). This observation is compatible with the proposed role of the *GAL80*-gene product as repressor.

Characterization of the *gal81-GAL4* Locus

Although Douglas and Hawthorne (1966) proposed that *GAL4* exerts positive control in coordinate expression of the genes, it is not clear whether the *GAL4* gene functions at the transcriptional, translational, or postranslational level. From an analysis of deadaptation kinetics in galactose-adapted cells, Tsuyumu and Adams (1974) proposed that the Leloir pathway enzymes are aggregated in vivo, and the *GAL4*-gene product joins to the aggregate upon exposure to inducer and facilitates the function of the enzymes in the aggregate. Klar and Halvorson (1974) found, however, contrary to the aggregation mechanism, that a temperature-sensitive *gal4* mutation did not affect the in vivo thermolability or temperature optimum of the epimerase activity. Their subsequent finding that *GAL4* had no dosage effects on the level of epimerase activity led them to favor a mechanism in which the *GAL4*-gene product is required for the synthesis but not the activity of the Leloir enzymes (Klar and Halvorson 1976). These arguments, however, have been resolved by the recent biochemical evidences advanced by Hopper and his colleagues.

These investigators observed that inducible transferase activity specified by the *GAL7* gene (Hopper et al. 1978) and galactokinase activity specified by the *GAL1* gene (Hopper and Rowe 1978) appeared following the de novo synthesis of their respective polypeptides, based on the de novo appearance of mRNA species detectable by a wheat germ in vitro translation system, and that the appearance of those mRNA species was dependent on the *GAL4*-gene function. In a cloning experiment to isolate galactose-inducible DNA sequences of *S. cerevisiae* as viable molecular hybrids of bacteriophage λ DNA by differential plaque filter hybridization, St. John and Davis (1979) obtained evidence supporting a similar conclusion: The function of the *GAL4* gene is necessary for the appearance of at least four specifically galactose-induced RNA species in yeast cells.

The central feature of the Douglas-Hawthorne model—the concept that *GAL80* protein controls expression of the *GAL4* gene—has been called into question by recent evidence indicating that *GAL4* is expressed constitutively. Matsumoto et al. (1978) isolated a *gal4* mutant whose product is thermolabile. The mutant cells pregrown at restrictive temperature (35° C) in glycerol nutrient medium showed a prolonged lag period (35 min) in the induction of galactokinase activity at the permissive temperature (25° C), in comparison with that (15 min) of the same mutant cells pregrown at the permissive temperature. This strongly suggests that de novo synthesis of the $gal4^{ts}$-gene product is required for the expression of the *GAL1* gene in the mutant cells pregrown at 35° C, whereas the same cells pregrown at 25° C might have a sufficient amount of the $gal4^{ts}$-gene product to induce the enzyme.

Further evidence of the constitutive expression of the *GAL4* gene is the semidominancy of *GAL81* mutations (Matsumoto et al. 1980). If the *GAL4* gene contiguous to $gal81^+$ is not expressed in the uninduced condition, as predicted by the Douglas-Hawthorne model, then diploids having heterozygous genotypes of the forms *GAL81 GAL4*/$gal81^+$ *GAL4* and *GAL81 GAL4*/$gal81^+$ *gal4* should always show the same level of galactokinase activity, irrespective of the *gal4* allele contiguous to $gal81^+$. On testing this hypothesis, significantly diverse levels of galactokinase activity were found in the diploids, depending on the *gal4* allele contiguous to $gal81^+$, thus, indicating that the *GAL4* gene is expressed constitutively. This argument, however, requires that active *GAL4* protein be composed of several identical subunits.

Constitutive synthesis of the *GAL4* protein was demonstrated more elegantly by Perlman and Hopper (1979). The mRNAs encoding the galactokinase and transferase were synthesized in wild-type cells by the addition of galactose in the presence of cycloheximide. The prompt induction of the mRNAs and the sustained accumulation activity (detected in the wheat germ in vitro translation system), following the cessation of all detectable protein synthesis in vivo, indicate that the *GAL4* regulatory protein and any other positive factors of transcription of the galactose genes exist prior to induction. The same conclusion was reached by synchronous mating of galactose-negative cells (Perlman and Hopper 1979), whereby haploid a *gal1 GAL4 GAL80* cells were mated with α *GAL1 gal4 GAL80* or α *GAL1 gal4 GAL80^s* cells. The galactokinase activity was uninducible in the haploid strains when they were cultivated alone. All of the cells were cultured in the presence of galactose to allow for the possibility of inducible synthesis of the *GAL4*-specified protein. According to the Douglas-Hawthorne model, both the *GAL80/GAL80* and *GAL80/GAL80^s* zygotes should produce galactokinase activity, because the *GAL4* protein necessary for expression of the *GAL1* allele is available from the mating partner. In the experiments, however, galactokinase activity appeared only in the *GAL80/GAL80* zygote, not in the *GAL80/GAL80^s* zygote. This indicates that the *GAL80^s*-gene product, a super-repressor, can nullify any existing *GAL4* protein activity brought into the zygote from the other partner. In other words, the *GAL80*-gene product (and inducer) does not control the expression of the *GAL4* gene, but it does control the function of the *GAL4*-gene product. This argument obviously implies that the expression of the *GAL4* gene is constitutive.

The fact that the *GAL4* gene is expressed constitutively suggests that the *gal81* site is not an operator but a part of the *GAL4* gene. This possibility was recently studied by Matsumoto et al. (1980) by meiotic fine-structure mapping of the *gal4* and *GAL81* mutations and characterization of mutations at the *gal81* site susceptible to nonsense suppressors. Twenty-eight independently isolated *gal4* mutant alleles were located in a single linear map (Fig. 1) by scoring the frequencies of supposed meiotic recombination in randomly isolated ascospores from various diploids constructed by pairwise combination of the *gal4* mutant alleles. The map covers a region expressible by a recombination frequency of 0.44%. Four of the mutations (*gal4-2, gal4-76, gal4-77,* and *gal4-78*) were found to be suppressible by a tyrosine-inserting UAA suppressor, (*SUP3*), and two closely linked mutations, *gal4-62* and *gal4-69*, were suppressed by a newly isolated dominant, presumed UGA suppressor, a recessive omnipotent suppressor, and two uncharacterized suppressors. Since these nonsense mutations were located at various sites on the map, it is possible to conclude that the *GAL4* gene codes for a protein molecule. The *gal4-62* and *gal4-69* alleles, which might occupy the same site on the map, could revert to give the constitutive galactose-positive phenotype, presumably by a single secondary mutation (spontaneous reversion frequencies were ~4 × 10^{-7}) at or close to the *gal4-62* or *gal4-69* allele, or through the action of extragenic suppressors as described

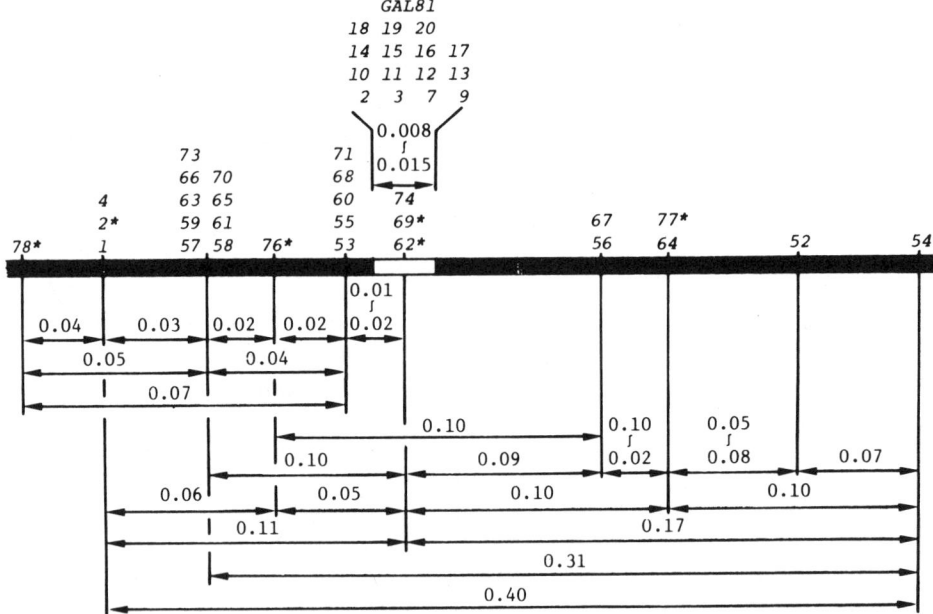

Figure 1 Meiotic fine-structure map of the *GAL4* locus. (*) Mutant alleles susceptible to nonsense suppressors. Map distances are expressed as appearance of supposed recombinant spores (%). (Reprinted, with permission, from Matsumoto et al. 1980.)

above. Of 19 *GAL81* mutations, 15 were located in a narrow region with a recombination frequency of at most 0.015%. (The remaining four could not be analyzed due to their ambiguous phenotypes.) The *gal4-62* and *gal4-69* sites appeared to lie in the *GAL81* region, as *gal4-62* was mapped at the same region as *GAL81* by three-point crosses. Thus, the *GAL81* region is located inside the *GAL4* locus and occupies not more than 4% of its total length. The fact that the *gal4-62* and *gal4-69* alleles lie in the *GAL81* region and these alleles are probably UGA nonsense mutations, supports the view that the *GAL81* region is not an operator but codes for a portion of the *GAL4* protein. It is noteworthy that the *gal4-62* and *gal4-69* alleles give rise to the constitutive phenotype for galactokinase synthesis when suppressed with two uncharacterized suppressors, but result in the inducible phenotype when combined with a dominant UGA suppressor or a recessive omnipotent suppressor.

REGULATORY CIRCUIT FOR THE GALACTOSE PATHWAY ENZYMES

The above evidence indicates that the revision of the Douglas-Hawthorne model is necessary. In the revised model, the expression of the *GAL4* gene would be constitutive and not under the control of *GAL80* and inducer (galactose). The simplest mechanism that fits the observations is one in which the *GAL81* site codes for an affinity site in the *GAL4* protein—the positive factor—through which it interacts in the cytoplasm (or in the nucleoplasm) with the *GAL80*-gene product—the negative factor, as illustrated in Figure 2. The presence of inducer inhibits the function of the negative factor or dissociates it from the positive factor, allowing the positive factor to activate expression of the structural genes of the galactose pathway enzymes at the transcriptional level. Although no conclusive evidence is available at present, the negative factor may be dissociated from the positive factor in the presence of inducer, since synchronous mating experiments have shown that preexisting *GAL4* activity is immediately blocked by the introduction of the *GAL80^s* product (Perlman and Hopper 1979).

Several other mechanisms for galactose regulation have been postulated and dismissed (Perlman and Hopper 1979; Matsumoto et al. 1980). In one model, a presumptive promoter contiguous to the Leloir enzyme structural genes has two controlling sites, one of which has affinity for the repressor (*GAL80* protein) and the other, for the positive factor (*GAL4* protein). In a second model, the *GAL4*-gene product functions as an inducer-activated antirepressor that modulates the affinity of *GAL80* protein for a site contiguous to the Leloir enzyme genes. A third model postulates that the *GAL4* protein metabolizes galactose to a hypothetical true inducer of the pathway to alleviate *GAL80*-mediated repression. However, all of these models are

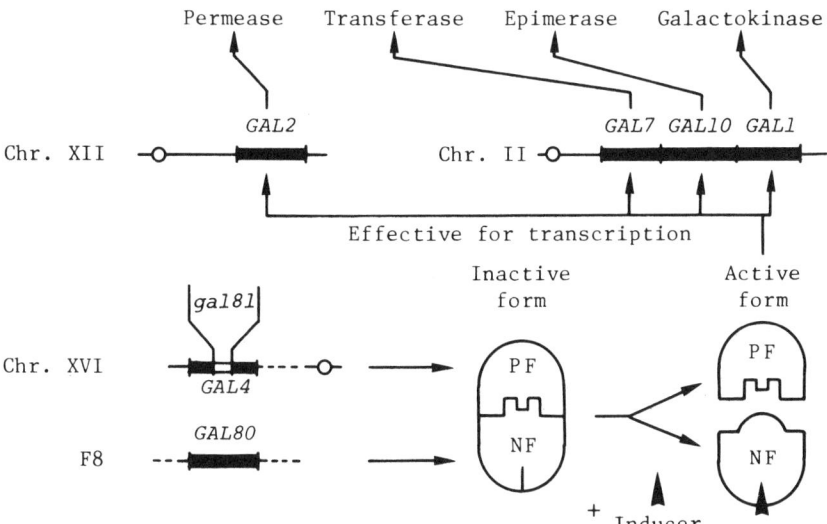

Figure 2 Model for the role of the *GAL4*- (positive factor [PF]) and *GAL80*- (negative factor [NF]) gene products in regulation of the Leloir pathway enzymes.

improbable since none of them is consistent both with the epistasis-hypostasis relationship of *GAL4* and *GAL80* and with the existence of *GAL81* mutations.

Does the *GAL7-GAL10-GAL1* cluster encode a single polycistronic mRNA, as observed in the lactose operon? Hopper and Rowe (1978) found that the polysomal mRNAs that encode the transferase and galactokinase polypeptides differ distinctly in their migration behavior on polyacrylamide gel electrophoresis. There are three distinct mRNA species encoding the three Leloir pathway enzymes. That these arise by separate transcription of the three structural genes, and not by cleavage of a multicistronic precursor transcript, has been established by analysis of in vivo transcription of the *GAL7-GAL10-GAL1* cluster, using as probe DNA containing the three genes carried on a hybrid λ phage (St. John and Davis 1979, and pers. comm.). *GAL1* is transcribed in a divergent direction from a strand opposite that from which *GAL7* and *GAL10* are transcribed. In addition, although under certain conditions a transcript spanning both *GAL7* and *GAL10* can be observed in vivo, the kinetics of RNA appearance upon induction suggest that the two genes are transcribed from separate promoters and do not constitute a single transcription unit. Thus, each of the structural genes appears to have its own promoter.

The essential difference between the Douglas-Hawthorne model and the revised mechanism is that the former implies a protein-to-DNA interaction between *GAL80* and *GAL4*, and the latter implies a protein-to-protein interaction. The hierarchical relationship between the *GAL80* and *GAL4* genes deduced from the epistasis-hypostasis relationship seen in the genetic data is explained by the stepwise expression of the *GAL4* and structural genes under the control of *GAL80* and inducer in the Douglas-Hawthorne model, but by the special consecutive interaction or steric relationships in the cytoplasmic aggregate of the regulatory factors in the revised model.

REGULATION OF PHOSPHATASES

Phosphatases in Yeast

Acid phosphatase (orthophosphoric monoester phosphohydrolase; EC number 3.1.3.2) in yeast is a highly glycosylated enzyme (Boer and Steyn-Parvé 1966) located in the periplasmic space (Arnold 1972), whereas alkaline phosphatase (EC number 3.1.3.1) is located intracellularly (Suomalainen et al. 1960). Genetic and biochemical analyses of a wild-type strain of *S. cerevisiae* have revealed the existence of two species of acid phosphatase: one, constitutive and the other, repressible by inorganic phosphate (Toh-e et al. 1973). On the other hand, cellular extracts of *S. cerevisiae* have been found to contain two species of phosphatase that hydrolyze *p*-nitrophenylphosphate at alkaline pH: One is a nonspecific alkaline phosphatase, and the other is a specific *p*-

nitrophenylphosphatase (Gorman and Hu 1969; Attias and Bonnet 1972; Toh-e et al. 1976). The nonspecific alkaline phosphatase is repressible by inorganic phosphate (Toh-e et al. 1976) and appears to be a glycoprotein of approximately 90,000 daltons (W. A. Fonzi and D. J. Opheim, pers. comm.).

Phosphatase Genes

The constitutive acid phosphatase is coded for by the *PHO3* gene[2], and requires two additional complementary genes, *PHO6* and *PHO7*, for its synthesis (Toh-e et al. 1975a). In the case of the repressible acid phosphatase, a genetic system consisting of at least four genes, *PHO2, PHO80, PHO81,* and *PHO85* and a gene cluster, *pho82-PHO4* (Toh-e et al. 1973; Toh-e and Oshima 1974; Ueda et al. 1975), participates in the expression of the structural gene for the enzyme *PHO5* (Toh- et al. 1975b) in response to the presence or absence of exogenous inorganic phosphate.

Although Schweingruber and Schweingruber (1979) observed no significant difference in the amount of translatable mRNA for repressible acid phosphatase between repressed and derepressed cells, Bostian et al. (1980) have provided strong evidence from in vitro translation and immunoprecipitation assays that derepression of acid phosphatase is primarily the result of de novo synthesis of at least three polypeptides, due to the de novo appearance of three corresponding translatable acid phosphatase mRNAs. Therefore, regulation appears to be at the level of transcription of *PHO5*.

In addition to *pho5* mutants, strains containing *pho2, pho4,* or *pho81* mutations lack repressible acid phosphatase activity (Toh-e et al. 1973). Recessive *pho80* and *pho85* mutants produce the enzyme in high phosphate medium as well as in low phosphate medium (Toh-e et al. 1973; Ueda et al. 1975). A mutation at the *pho82* site, which is contiguous and closely linked to *PHO4*, is semidominant over the wild-type allele and results in the constitutive synthesis of the enzyme when it is in *cis* position to a functional allele of *PHO4* (Toh-e and Oshima 1974). It is noteworthy that dominant, constitutive mutations also occur at or close to the *PHO81* locus (Toh-e and Oshima 1974).

The two enzyme structural genes, *PHO5* (*ACP1* in the designation of Hansche et al. 1978) and *PHO3*, are clustered but separable by recombination (Toh-e et al. 1975b). They are located on chromosome II (Toh-e et al. 1973) and map between *tsm134* and *lys2* (Hansche et al. 1978). The *PHO80* locus has been mapped on chromosome XV close to the centromere (V. Beres, pers. comm.); *PHO2* has been mapped on the left arm of chromosome IV, *PHO4* has been mapped on the right arm of chromosome VI close

---

[2]The numerical gene designations used in this paper are consistent with current yeast genetic nomenclature. The correspondence of the numerical designation with previous letter designation, as well as function of each gene, is given in Table 1.

Table 1 Genes and their possible functions in the phosphatase system

Gene	Former designation	Probable function of gene or its product
PHO1	PHOA	required for synthesis of both acid and alkaline phosphatases (Schurr and Yagil 1971)
PHO2	PHOB	specific factor for expression of PHO5 and inorganic phosphate permease gene
PHO3	PHOC	constitutive acid phosphatase; structural gene
PHO4	PHOD	positive factor
PHO5	PHOE	repressible acid phosphatase; structural gene
PHO6	PHOF	required for PHO3 expression
PHO7	PHOG	required for PHO3 expression
PHO8	PHOH	repressible, nonspecific alkaline phosphatase; structural gene
PHO9	PHOI	specific factor for PHO8 expression
PHO80	PHOR	negative factor
PHO81	PHOS	mediator
pho82[a]	phoO	region with positive factor defining negative-factor interaction site
pho83[a]	phoP	regulatory site contiguous to PHO5
PHO84	PHOT	component of inorganic phosphate transport system
PHO85	PHOU	negative factor

Gene symbols are described in wild-type form.
[a] Mutations in the locus are, in general, dominant or semidominant over the wild-type counterpart.

to *met10*, and *PHO85* has been mapped on the left arm of chromosome XVI (Toh-e 1980).

In addition to these regulatory genes, a controlling site for *PHO5* expression has been suggested by the occurrence of a *cis*-dominant constitutive mutation, *PHO83*, at a site contiguous to *PHO5* (Toh-e and Oshima 1975). The *PHO83* mutation is epistatic to mutations in any of the other regulatory genes. Preliminary mapping of the *PHO83* mutation with several *pho5* and *pho3* mutant alleles by random spore analysis indicates that *pho83* is located at a site contiguous to the *PHO5* locus on the side opposite *PHO3*.

The nonspecific alkaline phosphatase, which is coded for by *PHO8* on chromosome IV (9 centiMorgans [cM] distal to *rna3* [Kaneko et al. 1979]) is repressible by inorganic phosphate, though a significant basal level of the enzyme activity is retained. The specific *p*-nitrophenylphosphatase, on the other hand, is produced constitutively (Toh-e et al. 1976). All of the genes required for regulation of *PHO5* (acid phosphatase)-gene expression, except for *PHO2* and *pho83*, are required for regulation of *PHO8* (alkaline phosphatase) expression (Toh-e et al. 1976). Recently, another regulatory gene, *PHO9*, indispensible for the alkaline phosphatase derepression but not for the repressible acid phosphatase, has been identified (Kaneko et al. 1979).

Another gene, *PHO84*, originally identified as a recessive constitutive mutation for repressible acid phosphatase synthesis (Ueda et al. 1975), is believed to be concerned with a system for the active transport of inorganic phosphate (Ueda and Oshima 1975).

To study the epistasis-hypostasis relationship between mutations showing the constitutive phenotype and others lacking the enzyme activity, various double mutants were constructed by recombination from strains marked with a *pho3* allele to eliminate the acid phosphatase activity due to the constitutive enzyme, and their phenotypes were compared with that of the parent. The *pho80 pho81*, *pho85 pho81*, and *PHO82 pho81* strains showed constitutive production of the repressible acid phosphatase, whereas the *pho80 pho2*, *pho80 pho4*, *pho80 pho5*, *pho85 pho2*, *pho85 pho4*, *pho85 pho5*, *PHO82 pho2*, *PHO82 pho4*, and *PHO82 pho5* strains all showed the same phenotype as the *pho2*, *pho4*, and *pho5* mutants, i.e., they lacked the repressible acid phosphatase activity (Toh-e et al. 1973; Toh-e and Oshima 1974; Ueda et al. 1975). Since these regulatory genes are hypostatic to the *PHO83* mutation and *pho83$^+$* is supposed to be a controlling site of the structural gene of the enzyme, all of the regulatory genes might exert their function at the level of transcription of the structural genes.

Regulatory Circuit for Phosphatase

The results of epistasis and hypostasis tests of the regulatory genes clearly indicated a cascade of molecular events in the transmission of signals caused by the presence or absence of inorganic phosphate to the structural genes, as summarized in Figure 3. To explain these observations a model was first proposed involving the sequential function of the regulatory genes. The *PHO81* gene controls either the expression of the *PHO80* and/or *PHO85* gene (Toh-e et al. 1973; Ueda et al. 1975) or the function of the *PHO80* and/or *PHO85* product(s)—the repressor (Ueda et al. 1975). The repressor prevents the expression of the *PHO4* gene by binding at *pho82$^+$*, the operator site of *PHO4* (Toh-e and Oshima 1974). The *PHO4*-gene product positively controls the expression of *PHO5* (Toh-e and Oshima 1974) by interacting at *pho83$^+$*, the controlling site for the *PHO5* locus (Toh-e and Oshima 1975). Active *PHO2*-gene product or its metabolite exerts its function at the same step as the *PHO4*-gene product or at a later step (Toh-e and Oshima 1974). Inorganic phosphate acts presumably via the *PHO81*-gene product either to permit expression of the *PHO80* and/or *PHO85* genes or

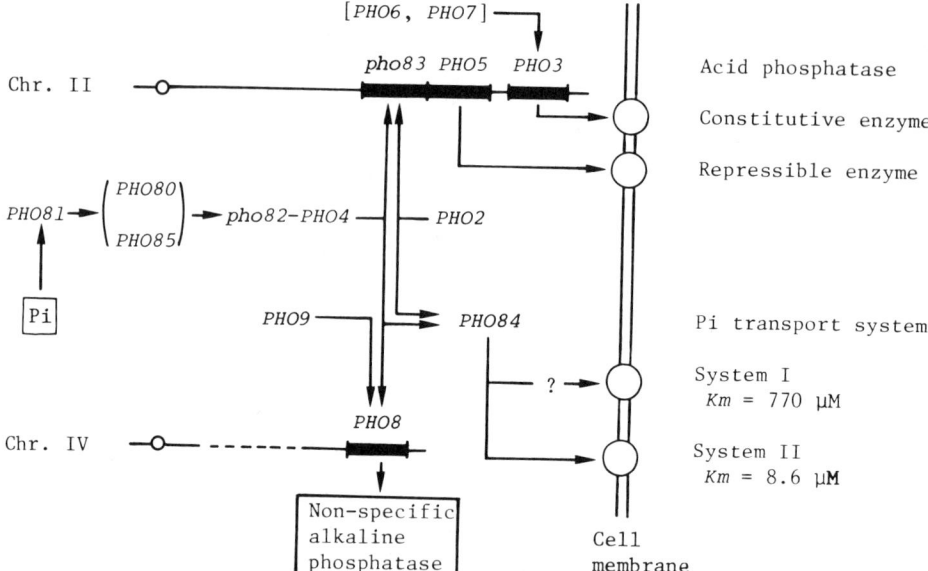

Figure 3 Genetic hierarchy of the regulatory genes in the phosphatase system. The regulatory scheme for the inorganic phosphate (Pi) transport system is based on the results of Y. Tamai et al. (unpubl.).

to activate a preexistent inactive repressor complex formed from the *PHO80* and *PHO85* proteins.

This model is essentially the same as the Douglas-Hawthorne model, though the galactose system does not have genes corresponding to *PHO81* and *PHO2* or *PHO9*. Certain subsequent findings, however, are difficult to reconcile with predictions arising from the model. The regulatory circuit for *PHO5* expression is found to be disturbed at acidic pH (below 3.0) but still effective for *PHO8* expression. Thus, cells grown at pH 3.0 in low phosphate medium lack acid phosphatase activity but possess alkaline phosphatase activity. These cells produce acid phosphatase activity immediately upon shifting the pH of the medium to 4.0 or above. The appearance of the enzyme activity probably represents de novo synthesis of the enzyme, since the appearance of enzyme activity was completely abolished if the pH shift experiment was conducted in the presence of cycloheximide. An experiment involving coupled temperature and pH shifts with a temperature-sensitive *pho81* mutant has suggested that the *PHO81*-gene function is indispensible for the function of the *PHO4*- and/or *PHO2*-gene product(s) but not for the expression of the *PHO4* and/or *PHO2* gene(s) (Toh-e et al. 1978). Random spore and tetrad analyses have shown that the *PHO82* alleles map inside the *PHO4* locus, and doubly heterozygous diploids *PHO82 PHO4/pho82$^+$ pho4* have been found to possess widely different activities in the repressible condition depending on the combination of the *PHO82* and *pho4* alleles (A. Toh-e et al., unpubl.). These findings strongly suggest close parallels between the *pho82-PHO4* and *gal81-GAL4* loci and between the *PHO80* and *PHO85* genes and the *GAL80* gene.

Thus, phosphatase synthesis may well be controlled by a similar genetic regulatory mechanism to the revised model in the galactose system; the simplest interpretation is summarized in Figure 4. A few molecules of a positive factor specified by *PHO4* are produced constitutively, as is the case for *GAL4*. Under repressed conditions, the *PHO4*-gene product is aggregated with a complex of the *PHO80*- and PHO85-gene products (negative factor) and hence is unable to activate *PHO5* and *PHO8* transcription. Under derepressed conditions (absence of inorganic phosphate) the *PHO81*-gene product aggregates with the *PHO80*- and/or *PHO85*-gene products, thereby releasing the positive factor. The released positive factor can then combine with the *PHO2* protein (specific factor) to turn on *PHO5* (acid phosphatase) and with *PHO9* (the other specific factor) to turn on *PHO8* (alkaline phosphatase). This repression or derepression of phosphatase is ultimately controlled by the interaction of the *PHO81*-gene product with the negative factor. That is, when the *PHO81*-gene product is bound to the negative factor, the positive factor is free to activate transcription, and when the *PHO81*-gene product is not bound, the positive factor is bound with the negative factor and hence is unable to activate transcription. The binding of the *PHO81*-gene product to the negative factor is regulated by the effector, inorganic phosphate, or a metabolite thereof. In the absence of the effector, the *PHO81*-gene product is bound with the negative factor and, in its presence, the *PHO81*-gene product is unbound. In this model, the dominant, constitutive mutations at or close to the *PHO81* locus (Toh-e and Oshima 1974) are explained as $PHO81^s$ mutations which, like the $GAL80^s$ mutations, render the gene product unresponsive to the effector and cause it to remain bound with the negative factor.

DISCUSSION

The evidence summarized in this paper indicates significant similarities between the regulatory circuits in the two systems, with respect to the mode of function of the regulatory genes and the cascade of signal transfers. On the basis of these similarities, the regulatory genes have been grouped into four classes, as listed in Table 2. The similarities also extend to the regulatory circuit for synthesis of acid and alkaline phosphatases in *N. crassa*, elucidated by Metzenberg and his colleagues. These enzymes are derepressed in low phosphate medium and repressed in high phosphate medium, as in the same enzyme system of *S. cerevisiae*. The synthesis of these enzymes is mediated by three regulatory genes: *nuc-2, preg,* and *nuc-1* (Littlewood et al. 1975). Their functions are exerted in a definite sequence, in

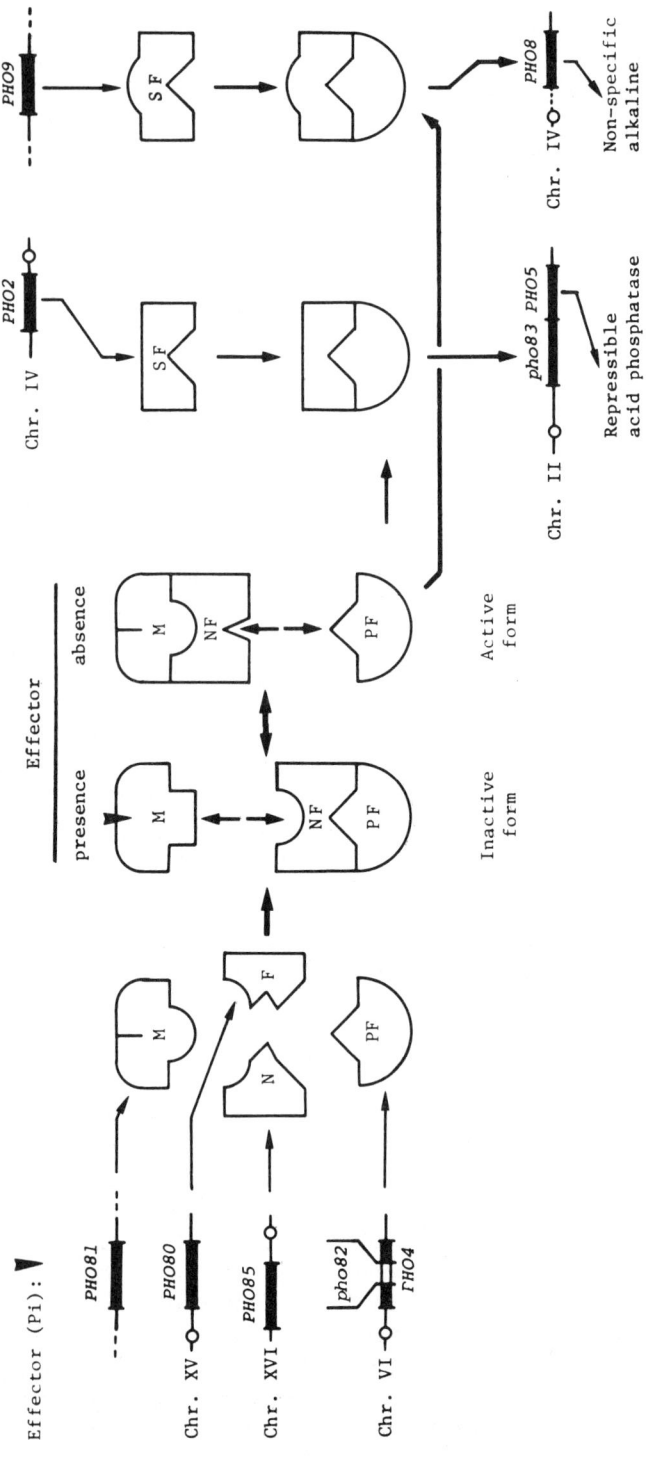

Figure 4 Model for the function of regulatory factors in the phosphatase system. Regulatory genes are expressed constitutively and produce respective cytoplasmic factors, i.e., positive factor (PF; specified by *PHO4*), negative factor (NF; *PHO80* and *PHO85*), mediator (M; *PHO81*), and specific factor (SF; *PHO2* and *PHO9*). Their interaction is ultimately modulated by the effector (inorganic phosphate or a metabolite thereof), and the specific PF-SF complex activates the transcription of the respective *PHO* structural gene.

Table 2 Regulatory genes in various enzyme systems

Organism	Enzyme	I (mediator)	II (negative)	III (positive)	IV (specific)
S. cerevisiae	Leloir pathway enzymes and galactose permease	—	*GAL80*	*gal81-GAL4*	—
S. cerevisiae	acid phosphatase and active transport[a] of inorganic phosphate	*PHO81*	*PHO80* *PHO85*	*pho82-PHO4*	*PHO2*
S. cerevisiae	nonspecific alkaline phosphatase	*PHO81*	*PHO80* *PHO85*	*pho82-PHO4*	*PHO9*
N. crassa	acid and alkaline phosphatases	*nuc-2*	*preg*	*nuc-1*	—

[a] Evidenced by Y. Tamai et al. (unpubl.).

which the action of the *nuc-1*-gene product is the final and indispensible step in turning on *PHO-2* and *PHO-3* (Metzenberg et al. 1974), the structural genes for alkaline and acid phosphatases, respectively (Lehman and Metzenberg 1976; Nelson et al. 1976). A constitutive mutation at *preg* is epistatic to *nuc-2*, and a *nuc-2* mutation can block the enzyme synthesis if the wild-type allele of *preg* is present. Although no super-repressible mutations were found at the *preg* locus (Metzenberg and Chia 1979), two mutant alleles have been described: $pcon^c$, which is allelic with *nuc-2* (Littlewood et al. 1975), and $nuc-1^c$ (Metzenberg and Nelson 1977; Metzenberg and Chia 1979); both alleles resemble, respectively, the dominant constitutive mutations at or close to *PHO81* (Toh-e and Oshima 1974) and *PHO82* or *GAL81* in the yeast systems.

Thus, the regulatory circuit for the phosphatase system of *N. crassa* seems to consist of essentially the same genetic factors as the corresponding system of *S. cerevisiae* (Table 2). To explain their observations, Metzenberg and his associates (Metzenberg and Nelson 1977; Metzenberg 1979; Metzenberg and Chia 1979) proposed a dosage titration model, by which cytoplasmic products specified by the regulatory genes are consecutively titrated. The structural genes for acid and alkaline phosphatases, respectively, are under positive control by the product of the $nuc-1^+$ gene; this product is inactivated by the $preg^+$-gene product; and this in turn is inactivated by the $nuc-2^+$-gene product, which in turn is inactivated or repressed by inorganic phosphate or a metabolite thereof.

This model appears to accord, in general, with the models proposed for the galactose (Fig. 2) and phosphatase (Fig. 4) systems in *S. cerevisiae*. In their interpretation of different levels of enzyme activities in response to different gene dosages at the *nuc-2, preg* (Metzenberg and Nelson 1977), and *nuc-1* (Metzenberg and Chia 1979) loci with the respective mutant alleles, these investigators suggested that the mutations would arise from a quantitative, but not a qualitative, modification of a regulatory factor by the mutation. Hence, the stoichiometric titration of one gene product with the product of another gene would not be possible in the mutant cells. The above explanation, however, might not hold in the case of *S. cerevisiae*, at least in the galactose system, since all of the *gal4* mutations, including the *GAL81* mutations mapped on the locus (Fig. 1), occurred in the region that (as evidenced by the occurrence of nonsense mutations) encodes the protein structure but not at a promoter region. We interpreted the semidominance of the *GAL81* mutation over the wild-type and *gal4* mutant alleles—an analogous phenomenon to the variable level of the enzyme activity with gene dosage in *N. crassa*—by supposing a subunit structure for the positive factor coded for by *GAL4* (Matsumoto et al. 1980). The semidominance of the *PHO82* mutation was interpreted similarly (A. Toh-e et al., unpubl.).

How do the regulatory gene products exert their function? It is possible to conclude that there may be two cytoplasmic factors for regulation of enzyme synthesis: one positive and one negative. Access of the effector to the negative factor may cause its separation from the positive factor, allowing it to activate the structural gene for expression, as suggested in Figures 2 and 4. Alternatively, it may be speculated that the positive and negative factors, and possibly other factors, e.g., the *PHO81*- and *PHO2*- or *PHO9*-gene products in the phosphatase system, are always aggregated to form a cytoplasmic regulatory complex. Access of the effector to a domain originated from the negative factor (or from the *PHO81*-gene product) may cause alteration of the molecular configuration of the complex and may ultimately affect the function of another domain of the complex—the positive factor. Since preexisting *GAL4* activity was immediately blocked by the introduction of the $GAL80^s$-gene product in the mating experiment (Perlman and Hopper 1979) and all of the nonsense *gal80* mutants so far isolated have shown constitutive synthesis of the Leloir pathway enzymes, the *GAL80*-gene product (negative factor) might be separated from and unnecessary for the function of the *GAL4*-gene product (positive factor) in the presence of inducer. These facts suggest that the latter model is unlikely.

In systems in which the effector acts as corepressor rather than inducer, an additional genetic factor, the *PHO81*- or $nuc-2^+$-gene product, mediates between the effector and the negative factor. This factor (mediator) functions apparently to reverse the signals, due to the presence or absence of the effector in the regulatory circuit, and to switch the enzyme synthesis off or on, respectively.

Since a single signal, due to the change of a certain metabolic status in the cell, results in the alteration of various enzyme activities in the cell, the regulatory system is presumed to mediate the signal to many unlinked structural genes that it controls. The genetic evidence strongly suggests that the

PHO2-gene product makes the regulatory system specific for *PHO5*, the structural gene of the repressible acid phosphatase, and possibly for an unidentified permease gene, as these two activities are coordinately lost with the *pho2* mutation (Y. Tamai et al., unpubl.). By the same mechanism, the *PHO9*-gene product is presumed to make the regulatory system specific for expression of *PHO8*, the structural gene for repressible alkaline phosphatase. Whether similar factors are involved in the galactose system of *S. cerevisiae* and in the phosphatase system of *N. crassa* is obscure.

ACKNOWLEDGMENTS

I thank Akio Toh-e and Kunihiro Matsumoto for discussions. Appreciation is extended to Ronald W. Davis and Thomas P. St. John for communication of unpublished results. Our original works cited in this paper were supported by grants 048164 and 238032 from the Ministry of Education, Science, and Culture of Japan.

REFERENCES

Adams, B.G. 1972. Induction of galactokinase in *Saccharomyces cerevisiae*: Kinetics of induction and glucose effects. *J. Bacteriol.* **111**: 308.

Arnold, W.L. 1972. Location of acid phosphatase and β-fructofuranosidase within yeast cell envelopes. *J. Bacteriol.* **112**: 1346.

Attias, J. and J.L. Bonnet. 1972. A specific alkaline *p*-nitrophenylphosphatase activity from baker's yeast. *Biochim. Biophys. Acta* **268**: 422.

Bassel, J. and R. Mortimer. 1971. Genetic order of the galactose structural genes in *Saccharomyces cerevisiae*. *J. Bacteriol.* **108**: 179.

Bevan, P. and H.C. Douglas. 1969. Genetic control of phosphoglucomutase variants in *Saccharomyces cerevisiae*. *J. Bacteriol.* **98**: 532.

Boer, P. and E.P. Steyn-Parvé. 1966. Isolation and purification of an acid phosphatase from baker's yeast (*Saccharomyces cerevisiae*). *Biochim. Biophys. Acta* **128**: 400.

Bostian, K.A., J.M. Lemire, L.E. Cannon, and H.O. Halvorson. 1980. In vitro synthesis of repressible yeast acid phosphatase: Identification of multiple mRNAs and products. *Proc. Natl. Acad. Sci.* **77**: 4504.

Bourgeios, S. and A. Jobe. 1970. Superrepressors of the *lac* operon. In *The lactose operon* (ed. J.R. Beckwith and D. Zipser), p. 325. Cold Spring Harbor Laboratory, Cold Spring Harbor, New York.

Calvo, J.M. and G.R. Fink. 1971. Regulation of biosynthetic pathways in bacteria and fungi. *Annu. Rev. Biochem.* **40**: 943.

Cirillo, V.P. 1968. Galactose transport in *Saccharomyces cerevisiae*. I. Nonmetabolized sugars as substrates and inducers of the galactose transport system. *J. Bacteriol.* **95**: 1727.

Douglas, H.C. 1961. A mutation in *Saccharomyces* that affects phosphoglucomutase activity and galactose utilization. *Biochim. Biophys. Acta* **52**: 209.

Douglas, H.C. and F. Condie. 1954. The genetic control of galactose utilization in *Saccharomyces*. *J. Bacteriol.* **68**: 662.

Douglas, H.C. and D.C. Hawthorne. 1964. Enzymatic expression and genetic linkage of genes controlling galactose utilization in Saccharomyces. *Genetics* **49**: 837.

———. 1966. Regulation of genes controlling synthesis of the galactose pathway enzymes in yeast. *Genetics* **54:** 911.

———. 1972. Uninducible mutants in the *gal i* locus of *Saccharomyces cerevisiae*. *J. Bacteriol.* **109:** 1139.

Douglas, H.C. and G. Pelroy. 1963. A gene controlling inducibility of the galactose pathway enzymes in *Saccharomyces*. *Biochim. Biophys. Acta* **68:** 155.

Gorman, J.A. and A.S.L. Hu. 1969. The separation and partial characterization of L-histidinol phosphatase and an alkaline phosphatase of *Saccharomyces cerevisiae J. Biol. Chem.* **244:** 1645.

Hansche, P.E., V. Beres, and P. Lange. 1978. Gene duplication in *Saccharomyces cerevisiae*. *Genetics* **88:** 673.

Hartwell, L.H., J. Culotti, J.R. Pringle, and B.J. Reid. 1974. Genetic control of the cell division cycle in yeast: A model to account for the order of cell cycle events is deduced from the phenotypes of yeast mutants. *Science* **183:** 46.

Hawthorne, D.C. and R.K. Mortimer. 1960. Chromosome mapping in *Saccharomyces*: Centromere-linked genes. *Genetics* **45:** 1085.

Hopper, J.E. and L.B. Rowe. 1978. Molecular expression and regulation of the galactose pathway genes in *Saccharomyces cerevisiae*. *J. Biol. Chem.* **253:** 7566.

Hopper, J.E., J.R. Broach, and L.B. Rowe. 1978. Regulation of the galactose pathway in *Saccharomyces cerevisiae*: Induction of uridyl transferase mRNA and dependency of *GAL4* gene function. *Proc. Natl. Acad. Sci.* **75:** 2878.

Kalckar, H., B. Braganca, and A. Munch-Petersen. 1953. Uridyl transferases and the formation of uridine diphosphogalactose. *Nature* **172:** 1038.

Kaneko, Y., A. Toh-e, and Y. Oshima. 1979. Characterization of yeast mutants lacking nonspecific alkaline phosphatase. *Jpn. J. Genet.* **54:** 441.

Kew, O.M. and H.C. Douglas. 1976. Genetic co-regulation of galactose and melibiose utilization in *Saccharomyces*. *J. Bacteriol.* **125:** 33.

Klar, A.J.S. and H.O. Halvorson. 1974. Studies on the positive regulatory gene, *GAL4*, in regulation of galactose catabolic enzymes in *Saccharomyces cerevisiae*. *Mol. Gen. Genet.* **135:** 203.

———. 1976. Effect of *GAL4* gene dosage on the level of galactose catabolic enzymes in *Saccharomyces cerevisiae*. *J. Bacteriol.* **125:** 379.

Kosterlitz, H.W. 1943. The fermentation of galactose and galactose-1-PO_4. *Biochem. J.* **37:** 322.

Lehman, J.F. and R.L. Metzenberg. 1976. Regulation of phosphate metabolism in *Neurospora crassa*: Identification of the structural gene for repressible alkaline phosphatase. *Genetics* **84:** 175.

Leloir, L.F. 1951. The enzymatic transformation of uridine diphosphate glucose into a galactose derivative. *Arch. Biochem.* **33:** 186.

Littlewood, B.S., W. Chia, and R.L. Metzenberg. 1975. Genetic control of phosphate-metabolizing enzymes in *Neurospora crassa*: Relationships among regulatory mutations. *Genetics* **79:** 419.

Matern, H. and H. Holzer. 1977. Catabolite inactivation of the galactose uptake system in yeast. *J. Biol. Chem.* **252:** 6399.

Matsumoto, K., A. Toh-e, and Y. Oshima. 1978. Genetic control of galactokinase synthesis in *Saccharomyces cerevisiae*: Evidence for constitutive expression of the positive regulatory gene *gal4*. *J. Bacteriol.* **134:** 446.

Matsumoto, K., Y. Adachi, A. Toh-e, and Y. Oshima. 1980. Function of positive regulatory gene *gal4* in the synthesis of galactose pathway enzymes in *Saccharomyces cerevisiae*: Evidence that the *GAL81* region codes for part of the *gal4* protein. *J. Bacteriol.* **141:** 508.

Metzenberg, R.L. 1979. Implication of some genetic control mechanisms in *Neurospora*. *Microbiol. Rev.* **43:** 361.

Metzenberg, R.L. and W. Chia. 1979. Genetic control of phosphorus assimilation in *Neuros-*

pora crassa: Dose-dependent dominance and recessiveness in constitutive mutants. *Genetics* **93**: 625.

Metzenberg, R.L. and R.E. Nelson. 1977. Genetic control of phosphorus metabolism in Neurospora. *ICN-UCLA Symp. Mol. Cell. Biol.* **8**: 253.

Metzenberg, R.L., M.K. Gleason, and B.S. Littlewood. 1974. Genetic control of alkaline phosphatase synthesis in Neurospora: The use of partial diploids in dominance studies. *Genetics* **77**: 25.

Mortimer, R.K. and D.C. Hawthorne. 1966. Genetic mapping in Saccharomyces. *Genetics* **53**: 165.

———. 1969. Yeast genetics. In *The yeasts* (ed. A.H. Rose and J.S. Harrison), vol. 1, p. 385. Academic Press, New York.

———. 1973. Genetic mapping in Saccharomyces. IV. Mapping of temperature sensitive genes and use of disomic strains in localizing genes. *Genetics* **74**: 33.

Mortimer, R.K. and D. Schild. 1980. Genetic maps of *Saccharomyces cerevisiae*. *Microbiol. Rev.* **44**: 519.

Mundkur, B.D. 1952. Long term adaptation to galactose by yeast. *Genetics* **37**: 484.

Nelson, R.E., J.F. Lehman, and R.L. Metzenberg. 1976. Regulation of phosphate metabolism in *Neurospora crassa*: Identification of the structural gene for repressible acid phosphatase. *Genetics* **84**: 183.

Nogi, Y., K. Matsumoto, A. Toh-e, and Y. Oshima. 1977. Interaction of super-repressible and dominant constitutive mutations for the synthesis of galactose pathway enzymes in *Saccharomyces cerevisiae*. *Mol. Gen. Genet.* **152**: 137.

Perlman, D. and J.E. Hopper. 1979. Constitutive synthesis of the *GAL4* protein, a galactose pathway regulator in *Saccharomyces cerevisiae*. *Cell* **16**: 89.

St. John, T.P. and R.W. Davis. 1979. Isolation of galactose-inducible DNA sequences from *Saccharomyces cerevisiae* by differential plaque filter hybridization. *Cell* **16**: 443.

Schmidt, G., G. Bartsch, M.-C. Laumont, T. Herman, and M. Liss. 1963. Acid phosphatase of baker's yeast: An enzyme of the external cell surface. *Biochemistry* **2**: 126.

Schurr, A. and E. Yagil. 1971. Regulation and characterization of acid and alkaline phosphatase in yeast. *J. Gen. Microbiol.* **65**: 291.

Schweingruber, M.E. and A.-M. Schweingruber. 1979. Posttranslational regulation of repressible acid phosphatase in yeast. *Mol. Gen. Genet.* **173**: 349.

Spiegelman, S., R.R. Sussman, and E. Pinska. 1950. On the cytoplasmic nature of long-term adaptation in yeast. *Proc. Natl. Acad. Sci.* **36**: 591.

Suomalainen, H., M. Linko, and E. Oura. 1960. Changes in the phosphatase activity of baker's yeast during the growth phase and location of the phosphatases in the yeast cell. *Biochim. Biophys. Acta* **37**: 428.

Toh-e, A. 1980. Genetic mapping of the *pho2*, *PHO82-pho4* and *pho85* loci of yeast. *Genetics* **94**: 929.

Toh-e, A. and Y. Oshima. 1974. Characterization of a dominant, constitutive mutation, *PHOO*, for the repressible acid phosphatase synthesis in *Saccharomyces cerevisiae*. *J. Bacteriol.* **120**: 608.

———. 1975. Regulation of acid phosphatase synthesis in *Saccharomyces cerevisiae*. In *Proceedings of the 1st Intersectional Congress of International Association of Microbiological Societies* (ed. T. Hasegawa), vol. 1, p. 396. Science Council of Japan, Tokyo.

Toh-e, A., S. Kakimoto, and Y. Oshima. 1975a. Two new genes controlling the constitutive acid phosphatase synthesis in *Saccharomyces cerevisiae*. *Mol. Gen. Genet.* **141**: 81.

———. 1975b. Genes coding for the structure of the acid phosphatases in *Saccharomyces cerevisiae*. *Mol. Gen. Genet.* **143**: 65.

Toh-e, A., S. Kobayashi, and Y. Oshima. 1978. Disturbance of the machinery for the gene expression by acidic pH in the repressible acid phosphatase system of *Saccharomyces cerevisiae*. *Mol. Gen. Genet.* **162**: 139.

Toh-e, A., H. Nakamura, and Y. Oshima. 1976. A gene controlling the synthesis of non-specific alkaline phosphatase in *Saccharomyces cerevisiae*. *Biochim. Biophys. Acta* **428**: 182.

Toh-e, A., Y. Ueda, S. Kakimoto, and Y. Oshima. 1973. Isolation and characterization of acid phosphatase mutants in *Saccharomyces cerevisiae*. *J. Bacteriol.* **113**: 727.

Tsuyumu, S. and B.G. Adams. 1974. Dilution kinetic studies of yeast populations: In vivo aggregation of galactose utilizing enzymes and positive regulator molecules. *Genetics* **77**: 491.

Ueda, Y. and Y. Oshima. 1975. A constitutive mutation, *phoT*, of the repressible acid phosphatase synthesis with inability to transport inorganic phosphate in *Saccharomyces cerevisiae*. *Mol. Gen. Genet.* **136**: 255.

Ueda, Y., A. Toh-e, and Y. Oshima. 1975. Isolation and characterization of recessive, constitutive mutations for repressible acid phosphatase synthesis in *Saccharomyces cerevisiae*. *J. Bacteriol.* **122**: 911.

Willson, C., D. Perrin, M. Cohn, F. Jacob, and J. Monod. 1964. Non-inducible mutants of the regulator gene in the lactose system of *Escherichia coli*. *J. Mol. Biol.* **8**: 582.

Winge, Ö. and C. Roberts. 1948. Inheritance of enzymatic characters in yeast, and the phenomenon of long-term adaptation. *C. R. Trav. Lab.* **24**: 263.

Regulation of Amino Acid and Nucleotide Biosynthesis in Yeast

Elizabeth W. Jones
Department of Biological Sciences
Carnegie-Mellon University
Pittsburgh, Pennsylvania 15213

Gerald R. Fink
Department of Biology
Massachusetts Institute of Technology
Cambridge, Massachusetts 02139

1. **Amino Acid Biosynthesis**
   A. Glutamate Family
   B. Aromatic Family
   C. Serine Family
   D. Aspartate Family
   E. Pyruvate Family
   F. Histidine Biosynthesis
2. **Specific and General Control Systems for Regulating Synthesis of Enzymes of Amino Acid Biosynthesis**
   A. Specific Control Systems and Their Mutations
   B. General Control Systems
3. **Magnitude of Repression-Derepression Responses for Enzymes of Amino Acid Biosynthesis**
4. **Nucleotide Biosynthesis**
   A. Biosynthesis of Pyrimidine Nucleotides
   B. Biosynthesis of Purine Nucleotides
5. **Single Carbon Metabolism**
6. **No Polycistronic Operons in Yeast**
7. **Conclusions**

INTRODUCTION

In this paper we attempt to summarize the current status of our knowledge of regulation of anabolic pathways. The discussion includes pathways of biosynthesis of all 20 amino acids, of purine and pyrimidine nucleotides, and of single carbon metabolism.

Some discussion is devoted to the pathways themselves. We consider this essential, for, in our opinion, doubt exists as to what the actual pathway of synthesis is for some metabolites (e.g., methionine and cysteine). We have also devoted some attention to assignment of genetic blocks. For many genes, assignment is based on enzyme assay, but for a substantial number, assignments are based on indirect evidence such as accumulations and feeding tests.

Regulation of biosynthesis occurs at two levels: the regulation of enzyme formation by control of gene expression and the regulation of enzyme activity that controls flow of metabolites. It seems likely that an essential aspect of the latter control is the regulation of the flow of metabolites between compartments of the cell. Substantial portions of the intracellular amino acids are compartmentalized within the cell, largely in the vacuole (for review, see Wiemken 1980; also see Messenguy et al. 1980). Redistribution of amino acids between compartments in response to metabolic signals has been demonstrated (Messenguy et al. 1980; Wiemken 1980). To what extent compartmentalization of metabolites may play a role in regulation for pathways other than amino acid biosynthesis is unknown.

Compartmentalization within the mitochondrion has also been demonstrated for some anabolic enzymes. Only for the arginine and branched-chain amino acid pathways is the location of essentially all enzymes known. Until data are available for enzymes of other pathways, we will be unable to assess the significance of enzyme location for anabolism.

Conventional wisdom has it that yeast shows limited derepression responses for anabolic enzymes. The extent to which this is true has been examined, and a compilation of available data is presented.

An interesting feature of regulation in yeast is the existence of suprapathway controls. The best known example is cross-pathway control of synthesis of the enzymes of amino acid biosynthesis. Regulation of metabolic flow into selected pathways of amino acid biosynthesis, effected by control of enzyme activity by coenzyme A (CoA), and glucose repression of certain enzymes of amino acid biosynthesis, each of which provides a linkage to carbon metabolism, may be other examples of suprapathway controls.

This paper is our view of the status of anabolic regulation in yeast. We have speculated fairly extensively, in order to suggest plausible interpretations where the data are unclear or missing. Our intent is to stimulate, not to enrage.

AMINO ACID BIOSYNTHESIS

Families of amino acids that derive from a common molecule are discussed together, following the pattern used in reviews of prokaryotic amino acid biosynthesis (Umbarger 1978). In the discussions on repression and/or induction, no attempt will be made to relate enzyme levels to levels of amino acid pools. As was inferred by Cowie and McClure (1959) and de Robichon-Szulmajster (1967), and later shown to be the case (Wiemken and Durr 1974), amino acids are compartmentalized in yeast, largely in the vacuoles (Table 1). With the exception of the acidic amino acids, more than half of each amino acid is in the vacuole (Table 1) (Messenguy et al. 1980). Moreover, the distribution within compartments varies with nutritional condi-

Table 1 Concentration and distribution of amino acid pools in yeast

Amino acid	Σ1278b[a]			S288C[a]		
	total pool (mM)	location		total pool (mM)	location	
		cytosol (%)	vacuole[b] (%)		cytosol (%)	vacuole[b] (%)
Alanine	11.2	30	70	18.0	35	65
Arginine	12.2	7	93	16.2	4	96
Asparagine + glutamine	12.3	16	84	15.0	27	73
Aspartate	7.5	73	27	11.7	70	30
Citrulline	8.3	15	85	2.4	35	65
Glutamate	31.3	62	38	42.7	57	43
Glycine	3.8	26	74	7.1	40	60
Histidine	1.6	10	90	3.1	7	93
Isoleucine	0.9	43	57	1.2	43	57
Leucine	0.9	50	50	0.8	58	42
Lysine	4.2	17	83	10.1	5	95
Ornithine	6.5	9	91	0.9	14	86
Phenylalanine	0.4	31	69	0.6	41	59
Serine	5.8	27	73	5.7	46	54
Threonine	4.8	25	75	5.2	23	27
Tyrosine	0.7	28	72	0.5	55	45
Valine	3.1	28	72	5.5	30	70
Tryptophan				0.02[c]		
Methionine	0.4[d]					

Yeast was grown in minimal medium with glucose as carbon source and NH_4^+ as nitrogen source. (Data from Messenguy et al. 1980.)
[a]Σ1278b is J.-M. Wiame's wild type; S288C is R.K. Mortimer's wild type.
[b]Not in the cytosol; could include compartments other than vacuoles, such as mitochondria.
[c]Calculated from Fantes et al. (1976) for X2180-1A (S288C)
[d]Calculated from Cherest et al. (1973a) for 4094B.

tions (Messenguy et al. 1980; Wiemken 1980). Addition of histidine or lysine to the medium can chase amino acids from the vacuole into the cytosol. Similarly, either reduction in ammonia assimilation or in protein synthesis or growth on nitrogen sources other than ammonia drives the amino acids from the cytosol into compartments. In evaluating metabolic flow, these distributions will eventually have to be taken into account, especially for pathways (such as those of arginine, isoleucine-valine, leucine, and lysine biosynthesis) where the feedback-sensitive enzymes are located in the mitochondrion (Ryan et al. 1973; Tracy and Kohlhaw 1975; Wipf and Leisinger 1977).

Glutamate Family

The glutamate family of amino acids (Fig. 1; Table 2) includes the amino acids that derive all or most of their carbon skeleton from glutamate or its immediate precursor, α-ketoglutarate: glutamate, glutamine, proline, argi-

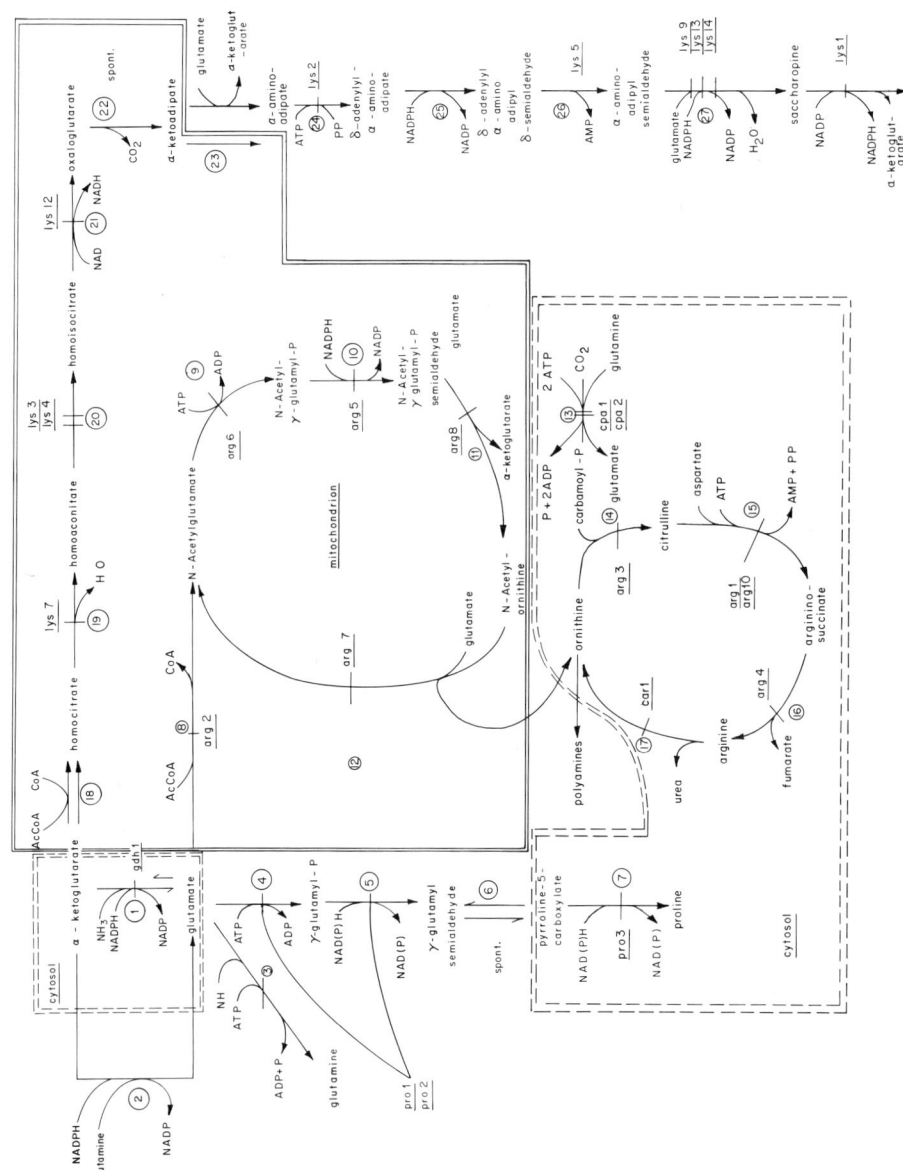

nine, and lysine. α-Ketoglutarate is derived from the tricarboxylic acid cycle and, indeed, mutants that are unable to produce α-ketoglutarate, such as those deficient in citrate synthase (*glu3*; Burand et al. 1975) or aconitate hydratase (*glu1*; Ogur et al. 1964a), require glutamate.

Glutamate and Glutamine Biosynthesis

Pathway. Glutamate dehydrogenase catalyzes an ammonia-dependent amination of α-ketoglutarate to yield glutamate (reaction 1). Two species of enzyme exist in yeast: one dependent on nicotinamide adenine dinucleotide phosphate (NADP), the other on NAD. The NADP-dependent glutamate dehydrogenase (GDH) is responsible for assimilation of ammonia. Mutants defective in the structural gene for this enzyme, *GDH1*, grow slowly unless a source of amino nitrogen is supplied or the level of the NAD-dependent GDH is raised by mutation (Grenson et al. 1974) (for the role of the NAD-linked enzyme, see Cooper [Nitrogen Metabolism in *Saccharomyces cerevisiae*], this volume). Both GDHs are in the cytosol (Hollenberg et al. 1970; Perlman and Mahler 1970). Glutamate synthase, which catalyzes the NADPH-linked reaction of α-ketoglutarate and glutamine to yield two molecules of glutamate (reaction 2), is present at low levels in yeast (Roon et al. 1974), but its significance is unclear. The amino group of glutamate is transferred to other amino acids by transamination.

Glutamine is synthesized from glutamate and ammonia in a reaction that requires ATP, catalyzed by glutamine synthetase (reaction 3). Mutants deficient in the activity require glutamine (Dubois and Grenson 1974), and the structural gene *GLN1* has been identified (A. Mitchell, pers. comm.). As *gln1* mutants require glutamine, glutamine synthetase must provide the only route for synthesis of glutamine.

Regulation of Enzyme Levels. The NADP-GDH appears constitutive (Dubois and Grenson 1974), although the levels may be somewhat higher when ammonia rather than glutamate is the nitrogen source (Roon et al. 1974). NADP-GDH is not under cross-pathway control (Wolfner et al. 1975). Glutamate synthase levels are somewhat higher when cells are grown on ammonia rather than glutamate (Roon et al. 1974).

Glutamine synthetase is important in catabolism as well as anabolism, and its level appears to be regulated by the nature and quality of the nitrogen source provided to the cells. The levels are very low when glutamine is the nitrogen source, intermediate when ammonia is supplied, and high when poor nitrogen sources such as glutamate, proline, or urea are provided (Dubois and Grenson 1974; A. Mitchell, pers. comm.). The levels can change by a factor of at least 150-fold by varying nutritional conditions.

Figure 1 (See facing page) Biosynthesis of the glutamate family of amino acids. Genetic blocks are indicated. Enclosure by solid line indicates location in the mitochondrion; enclosure by dashed line indicates location in the cytosol; unenclosed, location unknown. (See Table 2.)

Table 2 Genes and enzymes for biosynthesis of the glutamate family of amino acids

Reaction	Enzyme	Locus	References for gene assignment
1	glutamate dehydrogenase (L-glutamate: NADP oxidoreductase [deaminating], 1.4.1.4[a])	gdh1 (gdhA)	Grenson et al. (1974)
2	glutamate synthase (L-glutamate: NADP oxidoreductase [transaminating], 1.4.1.13)		
3	glutamine synthetase (L-glutamate: ammonia ligase [ADP-forming], 6.3.1.2)	gln1	Dubois and Grenson (1974); A. Mitchell (pers. comm.)
4	glutamate kinase (ATP: L-glutamate γ-phosphotransferase, 2.7.2.11)	pro1 or pro2	Brandriss (1979)
5	glutamyl-P reductase[a] (L-glutamate-γ-semialdehyde: NAD(P) oxidoreductase [phosphorylating], 1.2.1.41)	pro1 or pro2	Brandriss (1979)
7	P5C reductase (L-proline: NAD(P) 5-oxidoreductase, 1.5.1.2)	pro3	Brandriss (1979)
8	acetylglutamate synthase[b] (acetyl-CoA:L-glutamate N-acetyltransferase, 2.3.1.1)	arg2 (argA)	see Messenguy (1976)
9	acetylglutamate kinase (ATP: N-acetyl-L-glutamate 5-phosphotransferase, 2.7.2.8)	arg6 (argB)	Hilger et al. (1973); Minet et al. (1979)
10	N-Acetyl-γ-glutamyl-P reductase (N-Acetyl-L-glutamate-5 semialdehyde: NADP oxidoreductase [phosphorylating], 1.2.1.38)	arg5 (argC)	Minet et al. (1979)
11	acetylornithine aminotransferase (N²-acetyl-L-ornithine: 2-oxoglutarate aminotransferase, 2.6.1.11)	arg8 (argD)	Wiame (1971b)
12	acetylornithine-glutamate acetyltransferase[b] (N²-acetyl-L-ornithine: L-glutamate-N-acetyltransferase, 2.3.1.35)	arg7 (argE)	DeDeken (1963)
13	carbamoyl-P synthetase (Carbon dioxide: L-glutamine amidoligase [ADP-forming, carbamate phosphorylating], 6.3.5.5)	cpa1 (cpaI) cpa2 (cpaII)	Lacroute et al. (1965) Lacroute et al. (1965)

#	Enzyme	Gene	Reference
14	OTCase[b] (carbamoyl-P: L-ornithine carbamoyltransferase, 2.1.3.3)	arg3 (argF)	Lacroute et al. (1965); Bechet et al. (1970)
15	argininosuccinate synthetase (L-citrulline: L-aspartate ligase [AMP-forming], 6.3.4.5)	arg1 (argGI) arg10 (argGII)	Wiame (1971b); Messenguy (1976)
16	argininosuccinate lyase (L-argininosuccinate arginine lysase, 4.3.2.1)	arg4 (argH)	Wiame (1971b); Hilger et al. (1973)
17	arginase (L-arginine amidohydrolase, 3.5.3.1)	car1 (carA)	Bechet et al. (1970)
18	homocitrate synthase (3-hydroxy-3-carboxyadipate 2-oxoglutarate-lyase [CoA-acetylating], 4.1.3.21) two isozymes		
19	homoaconitate hydratase (2-hydroxy-3-carboxyadipate hydro-lyase, 4.2.1.36)	lys7	Bhattacharjee et al. (1968)
20	homoaconitase[b] (2-hydroxy-3-carboxyadipate hydro-lyase, 4.2.1.36)	lys3	Maragoudakis and Strassman (1966)
21	homoisocitrate dehydrogenase[b] (2-hydroxy-3-carboxyadipate: NAD oxidoreductase [decarboxylating], 1.1.1.87)	lys4	Maragoudakis and Strassman (1966)
23	2-aminoadipate aminotransferase (L-2-aminoadipate: 2-oxoglutarate aminotransferase, 2.6.1.39) two isozymes	lys12	Bhattacharjee and Strassman (1967)
24–26	2-aminoadipate reductase[b] (L-2-amino adipate-6-semialdehyde: NAD(P) oxidoreductase, 1.2.1.31)	lys2	Chatoo et al. (1979)
		lys5	Sinha and Bhattacharjee (1971)
27	saccharopine reductase[a] (N^5-[1,3-dicarboxypropyl]-L-lysine: NADP oxidoreductase [1-glutamate-forming], 1.5.1.10)	lys9	Jones and Broquist (1965)
		lys13	Bhattacharjee and Sinha (1972)
28	saccharopine dehydrogenase[b] (N^5-[1,3-dicarboxypropyl]-L-lysine: NADP oxidoreductase (L-lysine-forming], 1.5.1.8)	lys14	Bhattacharjee and Sinha (1972)
		lys1	Jones and Broquist (1965)

[a]Enzyme Commission number.
[b]Not the International Union of Pure and Applied Chemistry-International Union of Biochemistry (IUPAC-IUB) recommended name.

Mutants unable to derepress glutamine synthetase upon medium shift have been identified (*gln3* mutants; A. Mitchell, pers. comm.). Glutamine synthetase is not under cross-pathway control (Wolfner et al. 1975).

Proline Biosynthesis

Proline is synthesized via activation and reduction of the γ-carboxyl group of glutamate to yield glutamyl-γ-semialdehyde, which spontaneously cyclizes to give pyrroline-5-carboxylate (P5C) (reactions 4–6). The P5C is reduced to yield proline (reaction 7) (Brandriss 1979). *pro3* mutants lack P5C reductase (reaction 7) and grow only if supplied with proline. P5C reductase (reaction 7), in addition to being required for proline biosynthesis, is also required for degradation of arginine (Brandriss and Magasanik 1980; also see Cooper [Nitrogen Metabolism in *Saccharomyces cerevisiae*], this volume). The *pro1* and *pro2* mutants cannot make proline from glutamate but can be fed by proline, ornithine, or arginine. All three compounds are degraded to yield P5C. Hence, *pro1* and *pro2* mutants are blocked between glutamate and P5C. Blocks for *pro1* and *pro2* have not been assigned, since unique assays for these reactions do not exist. The Pro⁻ mutants are unusual in that they cannot grow on rich media such as YEPD for reasons that are unknown.

P5C reductase (reaction 7) is located in the cytosol, whereas P5C dehydrogenase, which catalyzes oxidation of P5C to glutamate, is located in the mitochondrion (Brandriss and Magasanik 1981). The two P5C pools, presumably located in the cytosol and the mitochondrion, apparently do not mix (Brandriss and Magasanik 1980). Little is known about regulation of the enzymes of proline biosynthesis except that growth on proline-containing medium does not lead to repression of P5C reductase (Brandriss 1979).

Arginine Biosynthesis

Due principally to the efforts of J.-M. Wiame, M. Grenson, F. Messenguy, and their collaborators, we know more about regulation of arginine biosynthesis than we do about regulation of any other pathway of amino acid biosynthesis. They have identified the structural genes for the enzymes and determined their genetic organization, identified regulatory mutations, determined that regulation involves arginine-specific as well as cross-pathway responses, and have reported the properties of many and the intracellular locations of all of the enzymes. From these data and a knowledge of the regulation of carbon flow, a picture of regulation of this pathway is beginning to emerge.

Pathway. The pathway from glutamate to arginine, like that for synthesis of proline, involves activation and reduction of the γ-carboxyl of glutamate (reactions 9 and 10). Prior to activation, the α-amino group of glutamate is blocked with an acetyl group from acetyl-CoA (reaction 8) so that cyclization of the product γ-semialdehyde will not occur. After transamination onto the aldehyde group of acetyl- γ-glutamyl semialdehyde to yield acetylornithine (reaction 11), the acetyl group is transferred, in an energy-efficient cycle, to glutamate (reaction 12), yielding ornithine and acetylglutamate to begin the cycle again. All of the enzymes of this cycle are located in the mitochondrion (Wipf and Leisinger 1977; Jauniaux et al. 1978). The ornithine produced in the above cycle must exit the mitochondrion, for the enzymes of the latter half of the pathway are located in the cytosol (Jauniaux et al. 1978).

Carbamoylphosphate is synthesized from CO_2, ATP, and the amide group of glutamine (reaction 13). It reacts with ornithine to yield citrulline, in a reaction catalyzed by ornithine carbamoyltransferase (OTCase) (reaction 14). Aspartate is transferred in toto to citrulline, yielding argininosuccinate, followed by cleavage with release of fumarate and arginine (reactions 15 and 16). Arginase catalyzes cleavage of arginine to ornithine and urea (reaction 17). This reaction is part of the pathway of arginine degradation (see Cooper [Nitrogen Metabolism in *Saccharomyces cerevisiae*], this volume). The assignment of genes to enzyme deficiencies is based on enzyme assays.

Metabolic Flow. Acetylglutamate synthase (reaction 8) fills an anaplerotic function, i.e., it serves to replenish the supply of acetylglutamate, as this is diluted by cell division, and to expand pools when the cells derepress. Acetylglutamate synthase is subject to end-product inhibition by arginine. This inhibition is synergistically enhanced by acetylglutamate which, in a sense, can also be viewed as an end product of the "pathway." The synthase is inhibited by CoA, and this inhibition is additive with that of arginine, thus providing for a coupling of arginine biosynthesis to carbon metabolism, as does the pattern of derepression of synthase (see below). Acetylglutamate synthase is not inactivated by CoA (Wipf and Leisinger 1979). (The distinction between inhibition and inactivation is that inhibition and recovery from inhibition occur immediately upon addition or removal of the effector, whereas inactivation and recovery from inactivation are time-dependent after addition or removal of the effector.)

Acetylglutamate kinase is also end-product-inhibited by the end product arginine both in vitro and in vivo (De Deken 1962; Hilger et al. 1973). Dosage studies have shown that the level of acetylglutamate kinase is rate-limiting for production of arginine in vivo (Hilger et al. 1973). To effect feedback inhibition of the synthase and the kinase, arginine, which is made in the cytosol, must flow back into the mitochondrion where these enzymes are located.

Two species of carbamoylphosphate synthetase are found in yeast cells. One species is repressed by growth in arginine-containing media, the other by growth in uracil-containing media. The latter activity is inhibited by UTP. Mutants lacking one or both of these species can be isolated, allowing studies of each species in isolation (Lacroute et al. 1965). Both enzymes are located in the cytosol (Jauniaux et al. 1978), and the carbamoylphosphate produced from these two isozymes forms a single pool (Lacroute et al. 1965). The level of the arginine-specific carbamoylphosphate synthetase has been shown to be rate-limiting for production of arginine in vivo when the pyrimidine-specific enzyme is absent (Hilger et al. 1973; Piérard et al. 1979).

In the presence of moderate concentrations of arginine, the enzymes of arginine degradation, including arginase (reaction 17), are induced (Middelhoven 1964). Under this condition, flow from glutamate into the pathway would cease. Were the cycle that includes reactions 13 and 14–17 to run freely in arginine excess, the cell would lose six ATPs per cycle. This wasteful cycle is prevented by a phenomenon called epiarginasic regulation (Wiame 1971a). Arginase molecules bind stoichiometrically to molecules of ornithine carbamoyltransferase in the presence of arginine and ornithine and/or citrulline (Messenguy et al. 1971; Wiame 1971a; Penninckx 1975; Simon and Stalon 1978). The complex possesses arginase catalytic activity but lacks catalytic activity for OTCase. The catabolic pathway is intact; the anabolic pathway is interrupted. Epiarginasic regulation seems to occur only when arginase and OTCase are in the same cellular compartment. In other yeasts and in *Neurospora*, prevention of the cycle is achieved by physical separation of intermediates in different compartments (for discussion, see Urrestarazu et al. 1977; Jauniaux et al. 1978). So far, no mutations that affect the association between arginase and OTCase or epiarginasic regulation have been uncovered.

Regulation of Arginine Biosynthetic Enzyme Levels

Arginine-specific Repression. If arginine (or ornithine) is added to cells growing in minimal medium, the levels of the enzymes catalyzing reactions 8–11 and 13–15, the products of the genes *ARG2, ARG6, ARG5, ARG8, CPA1, ARG3*, and *ARG1,10* are repressed severalfold (Lacroute et al. 1965; Bechet et al. 1970; Wiame 1971b; Messenguy 1976; Minet et al. 1979; Piérard et al. 1979; Wipf and Leisinger 1979). Levels of acetylornithine acetyltransferase (reaction 12), argininosuccinate lyase (reaction 16), and one subunit of carbamoylphosphate synthetase, the products of genes *ARG7, ARG4*, and *CPA2*, do not respond to such medium additions (Minet 1971; Wiame 1971b; Delforge et al. 1975; Piérard et al. 1979; Wipf and Leisinger 1979).

The repressive effects of arginine on synthesis of all of the above enzymes, except carbamoylphosphate synthetase (*CPA1*-encoded subunit), depend upon the integrity of three genes: *ARG80* (ARGRI), *ARG81* (ARGRII), and *ARG82* (ARGRIII) (Bechet et al. 1970; Minet 1971; Wiame 1971b; Jacobs et al. 1980). Mutations in these three loci are recessive and are unlinked to one another and to the genes that they control (Bechet et al. 1970). In mutants bearing tight mutations in these regulatory genes, the enzyme levels are elevated about twofold on minimal medium compared with wild type and are completely insensitive to the addition of arginine.

Multiply mutant strains, bearing mutations at two or even all three of the loci, have phenotypes indistinguishable from single mutants and are unimpaired for growth. Mutations in *ARG80, ARG81,* and *ARG82* not only lead to derepressed synthesis of six enzymes of arginine biosynthesis but also result in the inability to induce the arginine-degradative enzymes. Because both pathways are affected, but in opposite directions, the *arg80, arg81,* and *arg82* mutations lead to canavanine resistance and inability to use arginine or ornithine as a nitrogen source (Bechet et al. 1970). Wiame and his collaborators have interpreted these results to mean that an aporepressor is formed from the products of the *ARG80, ARG81,* and *ARG82* genes; it functions negatively, in combination with its corepressor, to reduce synthesis of the biosynthetic enzymes under its control and, through two successive negative controls, to enhance the synthesis of the catabolic enzymes (e.g., see Dubois et al. 1978). An alternative explanation is that the *arg80, arg81,* and *arg82* mutations interrupt synthesis of some small molecule(s) that is an effector for repression of anabolic enzymes and induction of catabolic enzymes. Obvious candidates for the small molecule effector(s) are polyamines.

Mutations that are *cis* dominant and lead to constitutive synthesis of OTCase (Messenguy 1976) or of acetylglutamate kinase and acetylglutamylphosphate reductase, encoded by the two-gene cluster *ARG5-ARG6* (Jacobs et al. 1980), have been obtained. The putative operator mutation (O^-) for OTCase is tightly linked to mutations in *ARG3* (the OTCase structural gene), is *cis* dominant, does not affect levels of two other arginine biosynthetic enzymes, and does not affect the response of *ARG3* to cross-pathway control (Messenguy 1976). The dominance of this O^- mutation is expressed in a/α diploids, in contrast to what has been seen for some, but not all, other putative operator mutations (for a/α effects and their relation to Ty insertions, see below, *Cis*-dominant Regulatory Elements; Dubois et al. 1978; Errede et al. 1980).

Jacobs et al. (1980) have described a *cis*-dominant mutation that elevates the levels of acetylglutamate kinase and acetylglutamylphosphate reductase. These two enzymatic activities are physically separable in extracts but are encoded by *ARG6* and *ARG5*, respectively, in adjacent regions of the chromosome (Minet et al. 1979). Nonsense mutations in *ARG6* result in deficiency for both activities, whereas nonsense mutations in *ARG5* eliminate only the reductase activity. The two enzyme activities are expressed perfectly coordinately through an 85-fold range of activity (Minet et al. 1979). These findings led to the alternative suggestions that the two genes comprise an operon or that a single polypeptide is synthesized and subsequently cleaved to give two activities. (For further consideration of this question, see below, No Polycistronic Operons in Yeast.) A putative operator mutation (O^c) has been identified that elevates the levels of both activities comparably. The mutation is tightly linked to *arg5* and *arg6* mutations and, indeed, maps proximal to a mutation in *ARG6* (the part of the locus that gives rise to polar nonsense mutations). The O^c mutation is *cis* dominant, does not affect levels of OTCase activity, does not affect the response of *ARG6* and *ARG5* to cross-pathway control, and is dominant in a/α diploids (Jacobs et al. 1980).

The structure of the arginine-specific carbamoylphosphate synthetase is determined by two unlinked genes, *CPA1* and *CPA2*. Mutations in either of these genes leads to a requirement for arginine if uracil is present in the growth medium (Lacroute et al. 1965). The enzyme is composed of two subunits: a large one encoded by the *CPA2* gene and a small one encoded by the *CPA1* gene (Piérard and Schröter 1978). The mixing of extracts of *cpa1* and *cpa2* mutants results in active enzyme. This in vitro complementation assay allows measurement of levels of the two gene products independently of each other (Lacroute et al. 1965).

Arginine represses the synthesis of the *activity* of the arginine-specific carbamoyl-phosphate synthetase. This repression is not mediated by the *ARG80*, *ARG81*, and *ARG82* genes (Thuriaux et al. 1972). Mutants with altered levels of the arginine-specific carbamoylphosphate synthetase were isolated as derivatives of *ura2C* mutants resistant to growth inhibition by arginine (Thuriaux et al. 1972). (*ura2C* mutants lack the pyrimidine-specific carbamoylphosphate synthetase [Lacroute et al. 1965].) Two classes of mutations were isolated: one in which the mutations were recessive (*cpa81* or *cpaR-2* mutants), and a second in which they were dominant (*CPA80* or *cpa1O* mutants) (Thuriaux et al. 1972).

Piérard et al. (1979) have studied regulation of the *CPA1*- and *CPA2*-gene products. Synthesis of both gene products is subject to cross-pathway control. Synthesis of the *CPA1*-gene product is repressed by arginine, but that of the *CPA2*-gene product is not. When repressing amounts of arginine are present in wild-type cells, 80% of the large subunit is apparently free in the cytoplasm, uncombined with small subunit. The specific repression of the *CPA1*-gene product by arginine depends upon the integrity of the *CPA81* gene. Mutations in this gene are recessive and reduce substantially the repression of *CPA1*-gene expression by arginine. Gene expression from *CPA1* and *CPA2* is somewhat elevated in the *cpa81* mutants. *cpa81* mutations do not alter the levels of OTCase, an enzyme regulated by the *ARG80*, *ARG81*, and *ARG82* genes (Thuriaux et al. 1972).

CPA80 mutations are *cis* dominant, tightly linked to the *CPA1* gene, are expressed in a/α diploids (Thuriaux et al. 1972), and render synthesis of the *CPA1*-gene product virtually resistant to repression by arginine (Piérard et al. 1979). The mutations have no effect on synthesis of the *CPA2*-gene product and do not cause elevation of the nonrepressed levels of either polypeptide. Cross-pathway control is intact in the *CPA80* mutant (Piérard et al. 1979). Because both gene products are subject to cross-pathway control, but only the *CPA1*-gene product is subject to arginine-specific control, derepression of the two gene products can be uncoupled. If the cells are starved for an amino acid other than arginine, the levels of the arginine pools will rise (Delforge et al. 1975). During such a starvation, synthesis of the *CPA2*-gene product will be derepressed due to relief of cross-pathway repression, but synthesis of the *CPA1*-gene product will be repressed because of the high arginine levels. The net result is that synthesis of carbamoylphosphate synthetase *activity* will not derepress profoundly. Since carbamoylphosphate synthetase levels are rate-limiting for synthesis of arginine, at least in the absence of the pyrimidine enzyme (Hilger et al. 1973; Piérard et al. 1979), this interactive regulation may well regulate flow of metabolites into the pathway. (Entry of metabolites into the other cycle would be prevented by feedback inhibition.) Whether such bottlenecks are generated in other pathways by interactions of this type is unknown.

The pattern observed for arginine-specific regulation in yeast has all the appearances of a negatively controlled system with repressors and operators (Jacob and Monod 1961). Whether this similarity will be as great at the molecular level is unknown at this time. Indirect evidence suggests that arginine-specific regulation is achieved at the level of transcription (Messenguy and Cooper 1977).

Cross-pathway Regulation. All ten of the polypeptides involved in arginine biosynthesis are subject to cross-pathway control (Schürch et al. 1974; Delforge et al. 1975; Wolfner et al. 1975; Messenguy 1979; Piérard et al. 1979; Wipf and Leisinger 1979; Jacobs et al. 1980). Whether the high degree of repression of arginine biosynthetic enzymes effected by growth on casamino acids or YEPD is a reflection of cross-pathway control is as yet unknown but would seem unlikely, since an *arg81* mutant, which has lost arginine-specific control but retains cross-pathway control, is not repressed by YEPD for synthesis of OTCase (Messenguy 1979). Possibly, the YEPD effect on arginine biosynthetic enzymes is a variation of the arginine-specific control in situations where arginine is a dominating amino acid in a mixture.

Glucose Repression. Acetylglutamate synthase appears to be subject to glucose repression (Wipf and Leisinger 1979). When cells grow on glucose, they exhibit diauxie, growing fermentatively until the glucose is exhausted, and then, after a lag, they grow respiratively at a slower rate on the ethanol produced from glucose. During the lag phase, synthesis of acetylglutamate synthase (reaction 7) is derepressed threefold. The derepression is prevented by the addition of arginine. Synthesis of acetylornithine acetyltransferase (reaction 11) (subject only to cross-pathway control) is unaffected by this regimen (Wipf and Leisinger 1979). It is tempting to speculate that this mode of derepression of synthase, as well as the inhibition of the synthase by

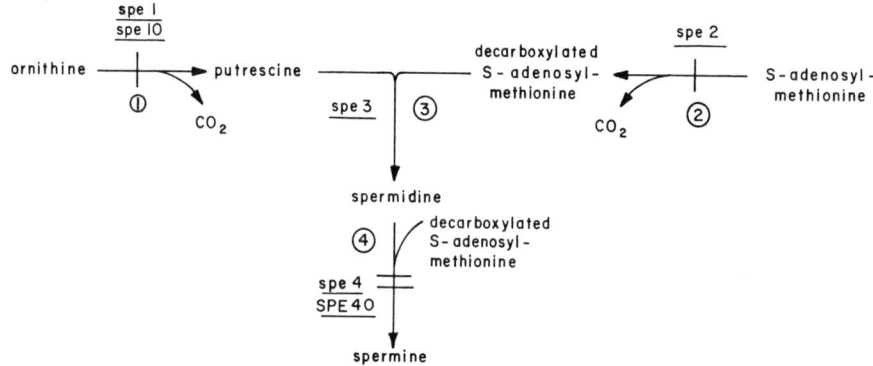

Figure 2 Biosynthesis of polyamines. Genetic blocks are indicated. (See Table 3.)

CoA, provides for linkage of flux through the arginine biosynthetic pathway to carbon and energy metabolism.

Polyamine Biosynthesis

Polyamines are synthesized from ornithine and *S*-adenosylmethionine (SAM) (Figs. 1 and 2; Table 3). Each of the molecules is decarboxylated by its decarboxylase. An aminopropyl group is transferred from the decarboxylated SAM to putrescine to yield spermidine. A second aminopropyl transfer yields spermine. This is apparently the only route for synthesis in yeast (Whitney and Morris 1978). *spe1* and *spe10* mutants have reduced levels of ornithine decarboxylase activity (Whitney and Morris 1978; Cohn et al. 1980) and can grow if putrescine, spermine, or spermidine is supplied. *spe2* and *SPE40* mutants lack spermine synthase but do not require polyamines. They were detected as suppressors of *spe10* mutations, and *spe10* and *SPE40* are tightly linked (Cohn et al. 1980).

The two research groups have used the mutants to try to determine cellular roles for polyamines. The picture that obtains from these experiments is first, that the polyamine pools are enormous relative to the cell's absolute needs. However, starvation for many generations can deplete the pools of the mutants. Experiments with cells having depleted pools have shown that some level of spermine and/or spermidine is essential for growth, sporulation, and maintenance of the killer plasmid (Cohn et al. 1978, 1980; Whitney and Morris 1978).

The levels of ornithine decarboxylase are related to the phase of growth. The level rises as the cells enter growth phase and, after the cells cease to divide, the enzyme activity disappears more rapidly than dilution would achieve (Kay et al. 1980). Variants have been obtained, in a two-step selection, that have 1000-fold higher levels of ornithine decarboxylase (Cohn et al. 1980).

Table 3 Genes and enzymes of polyamine biosynthesis

Reaction	Enzyme	Locus	References for gene assignment
1	ornithine decarboxylase (L-ornithine carboxy-lyase, 4.1.1.17[a])	spe1	Whitney and Morris (1978)
		spe10	Cohn et al. (1980)
2	adenosylmethionine decarboxylase (S-adenosyl-L-methionine carboxy-lyase, 4.1.1.50)	spe2	Whitney and Morris (1978); Cohn et al. (1978)
3	spermidine synthase (putrescine aminopropyltransferase)	spe3	Whitney and Morris (1978)
4	spermine synthase (spermidine aminopropyltransferase)	spe4	Cohn et al. (1980)
		SPE40	Cohn et al. (1980)

[a] Enzyme Commission number.

Lysine Biosynthesis

Pathway. Yeast, like other fungi, synthesizes lysine via the homocitric acid α-aminoadipic acid pathway (Fig. 1; Table 2), first proposed by Strassman and Weinhouse (1953), rather than from aspartate as do the *Enterobacteriaciae.* α-Ketoglutarate is condensed with acetyl-CoA to yield homocitrate (reaction 18), a reaction catalyzed by homocitrate synthase. By a series of reactions homologous to the corresponding tricarboxylic acid cycle reactions, homocitrate is converted to α-ketoadipate (reactions 19–22); α-ketoadipate is then transaminated to yield α-aminoadipate (reaction 23), which is converted to α-aminoadipyl semialdehyde in a three-step reaction (reactions 24–26) involving activation of the δ-carboxyl by formation of an adenylyl derivative, followed by reduction and eventual elimination of AMP with concomitant formation of the semialdehyde (Sagisaka and Shimura 1962; Fjellstedt and Ogur 1970). The three steps together (reactions 24–26) are referred to as α-aminoadipate reductase. Condensation of glutamate with the semialdehyde and reduction yields saccharopine (reaction 27), catalyzed by saccharopine reductase. Saccharopine is oxidized, with release of α-ketoglutarate, to yield lysine (reaction 28; saccharopine dehydrogenase).

The enzymes for the first portion of the pathway, from homocitrate synthase through one of two species of α-aminoadipate aminotransferase, are located in the mitochondrion (Betterton et al. 1968; Matsuda and Ogur 1969a; Tracy and Kohlhaw 1975). The intracellular location of the enzymes catalyzing steps between α-aminoadipate and lysine has not been reported.

lys3, lys4, lys7, and *lys12* mutants are blocked early in the pathway, since their lysine requirements can also be met by α-aminoadipate. Assignment of these blocks is based primarily on accumulation studies (Maragoudakis and Strassman 1966; Bhattacharjee and Strassman 1967; Bhattacharjee et al. 1968; Bhattacharjee and Sinha 1972). It seems possible that, unlike the bifunctional aconitase in the tricarboxylic acid cycle, homoconitate hydratase and homoaconitase are separate enzymes or polypeptides, since the *LYS4* and *LYS7* genes are on separate chromosomes.

Both *lys2* and *lys5* mutants are defective in α-aminoadipate reductase, for extracts of neither strain catalyze formation of α-aminoadipyl semialdehyde from α-aminoadipate. *lys2* mutants are presumably defective in the activation step (reaction 24), for *lys2* mutants do not catalyze α-aminoadipate-dependent oxidation of NADPH (reaction 25), and one *aau* (*lys2*) mutant has an enzyme with an altered K_m for α-aminoadipate (Chattoo et al. 1979). *lys5* mutants are probably defective in the hydrolysis step (reaction 26), for extracts catalyze α-aminoadipate-dependent oxidation of NADPH but do not catalyze formation of the semialdehyde (Sinha and Bhattacharjee 1971). Mutants defective in the actual reduction step presumably exist, for a Lys⁻ mutant of unknown genotype secreted what was presumed to be δ-adenylyl-α-aminoadipate when starved for lysine (Mattoon et al. 1961). Strains carrying mutations in the *LYS2* gene are able to grow on α-aminoadipate as sole nitrogen source, whereas Lys$^+$ strains are not. This difference in growth provides the basis for a positive selection for *lys2* mutants (Chattoo et al. 1979). *lys9*, *lys13*, and *lys14* mutants are all defective in saccharopine reductase (reaction 27) (which may be a two-step reaction) (Jones and Broquist 1965; Bhattacharjee and Sinha 1972), and *lys1* mutants are deficient in saccharopine dehydrogenase (reaction 28) (Jones and Broquist 1965). *LYS9* and *LYS1* are structural genes, for suppression of nonsense mutations can lead to altered kinetic constants for the enzymes (Fjellstedt and Ogur 1970). *lys6* and *lys8* mutants require both glutamate and lysine and lack enzyme activities for both the citric and homocitric acid cycles (Ogur et al. 1964b; Scheifinger et al. 1966). Conspicuous by their absence are mutants deficient in homocitrate synthase (reaction 18) and α-aminoadipate aminotransferase (reaction 23), possibly because there are two enzyme species capable of catalyzing each of these reactions (Matsuda and Ogur 1969a; Tucci and Ceci 1972a). It is possible that mutants deficient in the mitochondrial species of α-aminoadipate aminotransferase might be detected among Lys$^-$ mutants if mutant extracts were examined at the level of enzyme species rather than at the level of enzyme activity.

Metabolic Flow. Both species of homocitrate synthase (reaction 18) are feedback-inhibited by lysine (Maragoudakis et al. 1967; Tucci and Ceci 1972a). Considerable variation in the concentration of lysine needed to inhibit was reported, possibly reflecting the high lability of allosteric sites of mitochondrially located enzymes (Magee and de Robichon-Szulmajster 1968b; Ryan and Kohlhaw 1974). Mutants in which homocitrate synthase is at least partially resistant to feedback inhibition by lysine have been isolated as strains resistant to the lysine analogs *S*-(3-aminoethyl)-L-cysteine and hydroxylysine (Gray and Bhattacharjee 1976). Mutants lacking one or both species of homocitrate synthase could obviously be generated from these mutants, if desired, by isolating analog-sensitive derivatives of the analog-resistant strain followed by isolation of new lysine auxotrophs that complement existing ones. The principle here is that the resistance to the analog is a mutation in the structural gene for one of the homocitrate synthases and, therefore, a subsequent mutation that destroys the activity of that synthase will render the cells analog-sensitive. A similar strategy is being used for the first enzyme of leucine biosynthesis which, possibly coincidentally, catalyzes a homologous reaction (see below, Pyruvate Family; Isoleucine, Valine, and Leucine Biosynthesis).

Homocitrate synthase is reversibly inactivated by CoA by a time-dependent process. Protection against inactivation is afforded by the substrate α-ketoglutarate and the pathway end product, L-lysine (Tracy and Kohlhaw 1975). Inactivation by CoA may provide the cell with a means of coordinating its energy needs with its needs for other metabolites whose production results in consumption of acetyl-CoA (see below, General Control Systems, Glucose Repression and CoA; Tracy and Kohlhaw 1975).

Tucci and Ceci (1972b) reported that there is a step between α-aminoadipate and lysine that is feedback-inhibited by lysine. The results can as easily be explained by an interference with uptake of α-aminoadipate by lysine.

Regulation of Enzyme Levels. Growth in lysine-containing medium results in repression of saccharopine reductase (reaction 26) (Sinha et al. 1971) and one of the two isozymes of homocitrate synthase (Maragoudakis et al. 1967; Tucci and Ceci 1972a). Levels of saccharopine dehydrogenase and α-aminoadipate reductase (Sinha et al. 1971) and both species of α-aminoadipate aminotransferase (Matsuda and Ogur 1969b) are indifferent to lysine addition. Whether the repression of saccharopine reductase and homocitrate synthase reflects specific control by lysine is unknown but likely if regulation in the arginine pathway is a valid precedent. Saccharopine dehydrogenase and α-aminoadipate reductase are under cross-pathway control (Delforge et al. 1975; Wolfner et al. 1975). The cytosolic aminotransferase is induced by growth on α-aminoadipate and repressed by increased glucose concentration (Matsuda and Ogur 1969b). The information available on regulation of the enzymes of lysine biosynthesis is fragmentary. Few of the enzymes have been examined for repression by lysine or glucose or for cross-pathway control. Without such information, a clear picture of regulation of this pathway cannot emerge.

Aromatic Family

The aromatic family includes phenylalanine, tyrosine, and tryptophan (Figs. 3 and 4; Table 4). Part of the pathway for synthesis of the vitamins *p*-aminobenzoate and *p*-hydroxybenzoate is shared with the aromatic amino acids. Synthesis of these vitamins will not be discussed.

Pathway. Synthesis of phenylalanine, tyrosine, and tryptophan proceeds via a common pathway to chorismate, at which point the pathway branches. One branch leads to tryptophan; the other to phenylalanine and tyrosine. The common pathway to chorismate begins with condensation of erythrose-4-phosphate and phospho*enol*pyruvate to yield deoxy-D-*arabino*-D-heptulosonate-phosphate (DAHP) (reaction 1). In the next step, there is removal of a phosphate and an internal oxidation reduction, which results in cyclization to yield 5-dehydroquinate (reaction 2). Dehydration, followed by reduction and phosphorylation, yields shikimate phosphate (reactions 3–5). Shikimate phosphate condenses with a second molecule of phospho*enol*pyruvate (reaction 6). Chorismate is generated by removal of a phosphate with introduction of a second double bond (reaction 7).

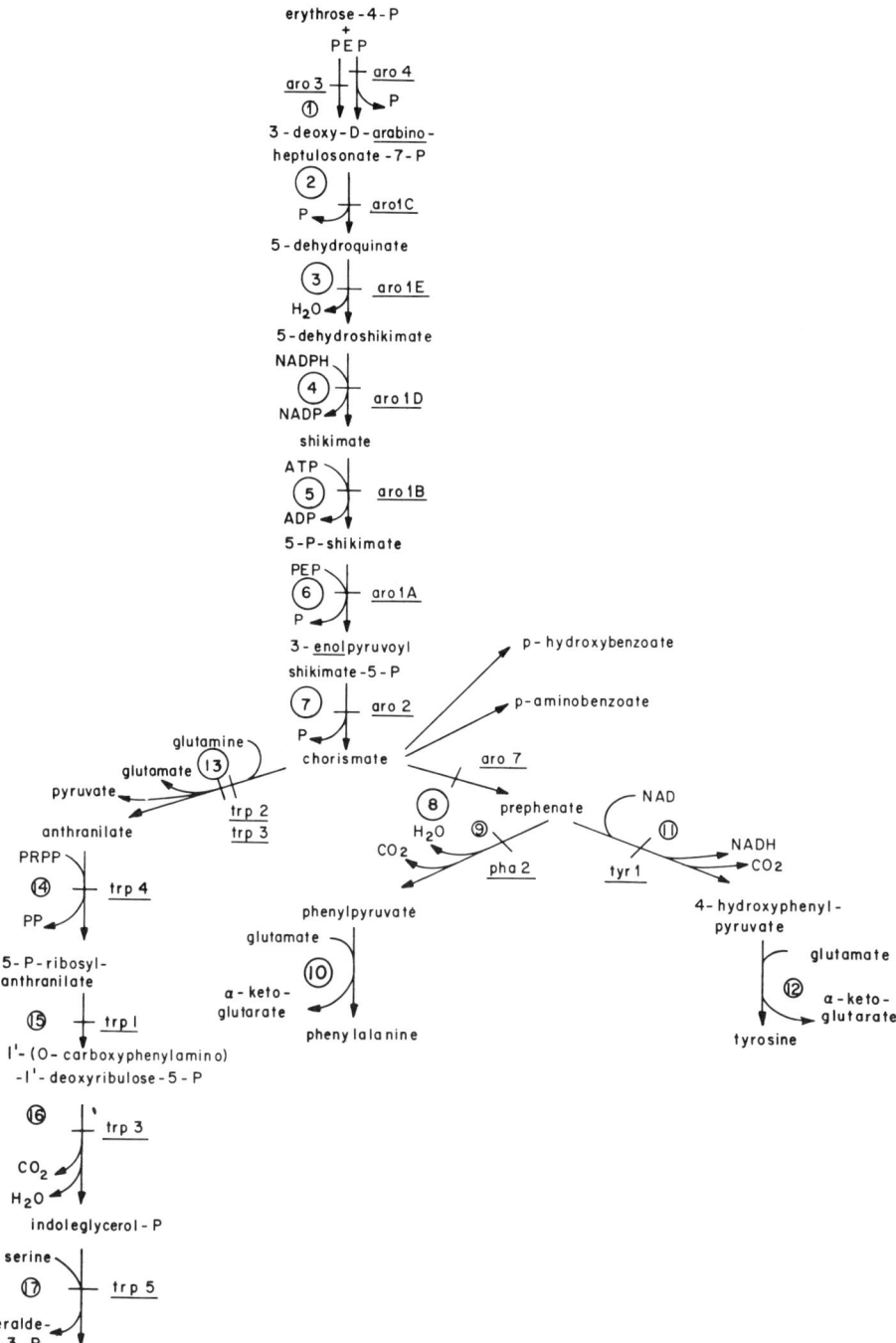

Figure 3 Biosynthesis of the aromatic amino acids. Genetic blocks are indicated. (See Table 4.)

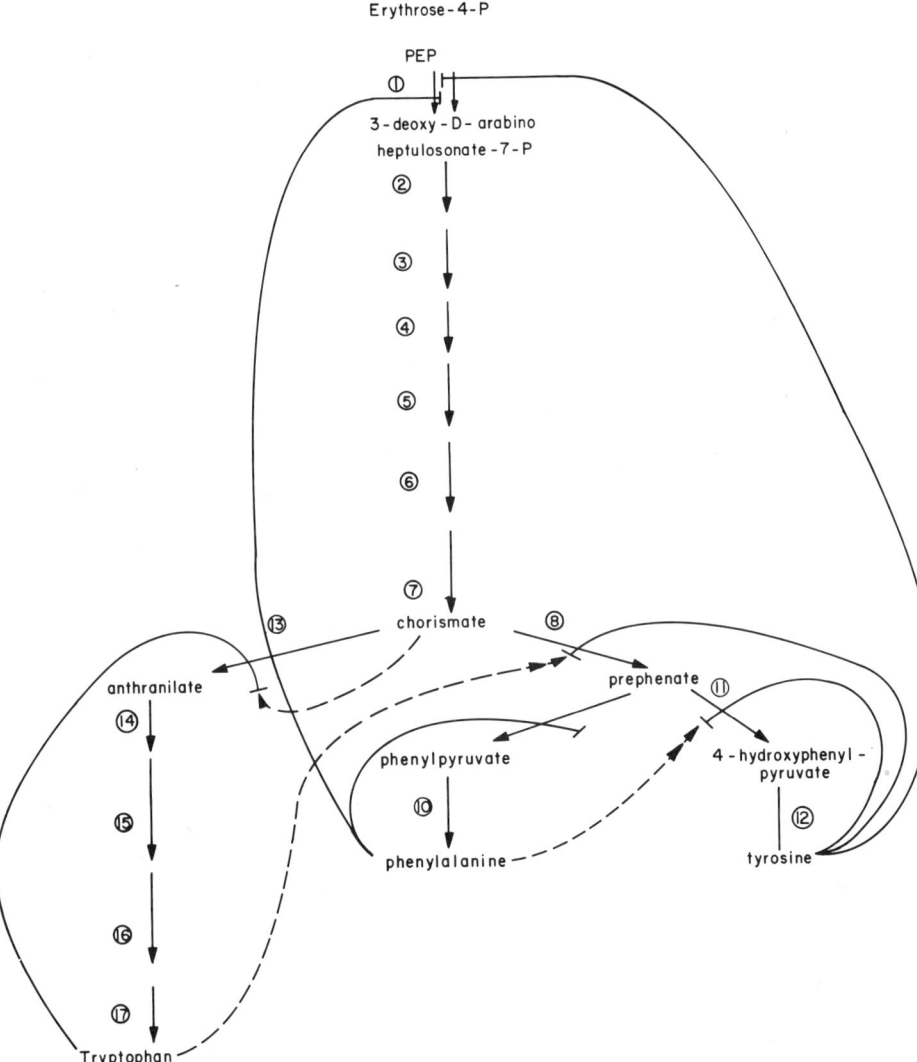

Figure 4 Regulation of carbon flow in the aromatic amino acid pathways. (|——) Feedback inhibition; (→ |——) counteracts feedback inhibition; (↠ |——) overcomes feedback inhibition and activates the enzyme by normalizing the kinetics.

The branch to phenylalanine and tyrosine proceeds by rearrangement of chorismate to yield prephenate (reaction 8). Simultaneous dehydration and decarboxylation yields phenylpyruvate, the α-keto acid precursor of phenylalanine (reaction 9). Transamination yields phenylalanine. Simultaneous dehydrogenation and decarboxylation of prephenate yields p-hydroxyphenylpyruvate, the α-keto acid precursor of tyrosine (reaction 11). Transamination yields tyrosine.

The branch to tryptophan begins with the glutamine-dependent conversion of chorismate to anthranilate (reaction 13). A phosphoribosyl group is transferred from phosphoribosylpyrophosphate (PRPP) onto the amino group of anthranilate (reaction 14). Isomerization followed by cyclization yields indoleglycerolphosphate (reactions 15-16). The final reaction consists of cleavage of indoleglycerolphosphate followed by condensation of indole with serine to yield tryptophan (reaction 17).

Two DAHP synthase isozymes are present in yeast. One is inhibited by tyrosine; the other by phenylalanine (Lingens et al. 1966). *aro3* mutants are prototrophs, lack the phenylalanine-sensitive isozyme, and are inhibited for growth by tyrosine. *aro4* mutants are prototrophs, lack the tyrosine-sensitive isozyme, and are inhibited for growth by phenylalanine (Meuris 1967). The double mutant has an absolute requirement for all three aromatic amino acids as well as p-aminobenzoate and p-hydroxybenzoate. Enzymatic activities for the next five steps (reactions 2-6) are encoded by the *ARO1* locus (de Leeuw 1967). *ARO1* is thought to encode a multifunctional polypeptide by analogy with *Neurospora* (Lumsden and Coggins 1977). All gene assignments for the aromatic pathway are based on enzyme assays.

Lingens et al. (1966) reported that a mutant requiring only tyrosine lacked prephenate dehydrogenase, that a mutant requiring only phenylalanine lacked prephenate dehydratase, and that two mutants requiring only phenylalanine and tyrosine lacked chorismate mutase. Although allelism tests have not been done, we have placed *tyr1, pha2,* and *aro7,* respectively, at positions on the pathway reflecting these blocks, for mutations in these genes cause the comparable requirements. Mutations that render chorismate mutase resistant to feedback inhibition by tyrosine are tightly linked to *aro7* mutations, confirming the assignment of the *aro7* block (Kradolfer et al. 1977). Additional mutants that required phenylalanine and tyrosine but were not defective in chorismate mutase were also reported (Lingens et al. 1966). These could arise by partial blocks in enzymes of the common pathway to chorismate (see Davis 1952).

The number of aminotransferase species involved in transamination of phenylalanine and tyrosine is unknown. Auxotrophies assignable to transamination deficiencies for these amino acids have not been reported.

trp2 mutants lack anthranilate synthase activity (reaction 13), and *trp3* mutants lack activity for anthranilate synthase and/or indoleglycerolphosphate synthase (reaction 16) (De Moss 1965; Schürch-Rathgeb 1972). The gene products of *TRP2* and *TRP3* aggregate to form an enzyme that possesses both enzyme activities (De Moss 1965). Tryptophan synthase (reaction 17), which catalyzes a two-step reaction with the intermediate, indole, remaining enzyme bound, is apparently a single polypeptide (Manney 1968;

Table 4 Genes and enzymes for biosynthesis of the aromatic family of amino acids

Reaction	Enzyme	Locus	References for gene assignment
1	deoxy-D-arabino-D-heptulosonatephosphate synthase[a] (7-P-2-keto-3-deoxy-D-arabinoheptonate D-erythrose-4-P-lyase [pyruvate phosphorylating], 4.1.2.15[d])	aro3[b] aro4[c]	Meuris (1967) Meuris (1967)
2	dehydroquinate synthase (7-phospho-3-deoxy-D-arabinoheptulosonate P-lyase [cyclizing], 4.6.1.3)	aro1C	de Leeuw (1967)
3	dehydroquinate dehydratase (3-dehydroquinate hydro-lyase, 4.2.1.10)	aro1E	de Leeuw (1967)
4	shikimate dehydrogenase (shikimate: NADP oxidoreductase, 1.1.1.25)	aro1D	de Leeuw (1967)
5	shikimate kinase (ATP: shikimate 5-phosphotransferase, 2.7.1.71)	aro1B	de Leeuw (1967)
6	3-enolpyruvoylshikimate-5-P synthase (P-enolpyruvate: shikimate-5-P enolpyruvoyltransferase 2.5.1.19)	aro1A	de Leeuw (1967)
7	chorismate synthase (3-P-5-enolpyruvoylshikimate P-lyase, 4.6.1.4)	aro2	de Leeuw (1967)
8	chorismate mutase (chorismate pyruvate mutase, 5.4.99.5)	aro7	Lingens et al. (1966); Kradolfer et al. (1977)

9	prephenate dehydratase (prephenate hydro-lyase [decarboxylating], 4.2.1.51)	pha2	Lingens et al. (1966)
10	phenylalanine aminotransferase		
11	prephenate dehydrogenase (prephenate: NADP oxidoreductase [decarboxylating], 1.3.1.13)	tyr1	Lingens et al. (1966)
12	tyrosine aminotransferase		
13	anthranilate synthase (chorismate pyruvate-lyase [aminoaccepting] 4.1.3.27)	trp2 trp3	De Moss (1965); Doy and Cooper (1966) De Moss (1965); Doy and Cooper (1966); Schürch-Rathgeb (1972)
14	anthranilate PR-transferase (N-[5′-PR]-anthranilate: pyrophosphate PR-transferase, 2.4.2.18)	trp4	De Moss (1965); Doy and Cooper (1966)
15	PR-anthranilate isomerase (N-[5′-PR]-anthranilate ketolisomerase)	trp1	De Moss (1965); Doy and Cooper (1966)
16	indole-3-glycerol-P synthase (1-[2-carboxyphenylamino]-1-deoxyribulose-5-P carboxy-lyase [cyclizing], 4.1.1.48)	trp3	De Moss (1965); Doy and Cooper (1966); Schürch-Rathgeb (1972)
17	tryptophan synthase (L-serine hydro-lyase [adding indoleglycerolphosphate], 4.2.1.20)	trp5	Manney (1964); De Moss (1965); Doy and Cooper (1966)

[a]Not the IUPAC-IUB recommended name.
[b]Phenylalanine-sensitive isozyme.
[c]Tyrosine-sensitive isozyme.
[d]Enzyme Commission number.

Manney et al. 1969; Dettwiler and Kirschner 1979; Zalkin and Yanofsky 1982).

Metabolic flow. Flow of carbon through the aromatic pathway is regulated by a combination of feedback inhibition and activation of enzyme activity effected by normalization of sigmoid kinetics and by reversal of inhibitions. Two species of DAHP synthase (reaction 1) exist. One species is inhibited by tyrosine (Lingens et al. 1966; Meuris 1967) and by tyrosyl-tRNA^{Tyr} (Meuris 1973). The second is inhibited by phenylalanine (Lingens et al. 1966; Meuris 1967) and by charged and uncharged tRNA^{Phe} (Meuris 1973). Inhibition by uncharged tRNA^{Phe} would appear to be counterproductive for the cell, and the in vivo relevance is unknown. This observation has not been confirmed. Mutants resistant to feedback inhibition by phenylalanine (Meuris 1973) and by tyrosine (Meuris 1974) have been isolated. The mutations map to the respective structural genes, *ARO3* and *ARO4*. So far as is known, tryptophan plays no part in this regulation but rather exerts control in an indirect fashion (see below).

In the absence of aromatic amino acids, chorismate will preferentially flow into the tryptophan branch of the pathway. This occurs because of the kinetics of the enzymes, anthranilate synthase and chorismate mutase, and the relative affinities of each enzyme for their common substrate chorismate (Lingens et al. 1967). Anthranilate synthase is feedback-inhibited by tryptophan (Doy and Cooper 1966; Lingens et al. 1966). This inhibition is counteracted by chorismate (Kradolfer et al. 1977). Mutations that render anthranilate synthase resistant to feedback inhibition have been described and mapped to *TRP2*. Such mutations lead to tryptophan excretion (Fantes et al. 1976).

Flow into the phenylalanine-tyrosine branch of the pathway proceeds via chorismate mutase (reaction 8). As mentioned above, chorismate will preferentially pass into the tryptophan branch at low substrate concentrations because of the kinetics and kinetic constants of the enzymes. Chorismat mutase is feedback-inhibited by tyrosine but not by phenylalanine (Lingens et al. 1966). Tryptophan cancels the inhibition by tyrosine and normalizes the sigmoid kinetics of chorismate mutase in the absence of tyrosine (i.e., tryptophan activates the enzyme) (Lingens et al. 1967). Thus, flow of chorismate into the tyrosine-phenylalanine branch is enhanced when tryptophan is in excess, because tryptophan inhibits flow of chorismate into the tryptophan branch by inhibiting anthranilate synthase (reaction 13) and activates chorismate mutase (reacion 8).

Prephenate is at the branch point to phenylalanine and tyrosine. Prephenate dehydratase (reaction 9) displays normal Michaelis-Menten kinetics, whereas prephenate dehydrogenase (reaction 11) shows positive cooperativity (Lingens et al. 1967). Because of the kinetics and the relative affinities of each enzyme for prephenate, prephenate will preferentially pass into the phenylalanine arm at low substrate concentrations. Prephenate dehydratase (reaction 9) is feedback-inhibited by phenylalanine but not by tyrosine, whereas prephenate dehydrogenase (reaction 11) is feedback-inhibited by tyrosine only (Lingens et al. 1966). Phenylalanine eliminates the inhibition by tyrosine and normalizes the sigmoid kinetics of prephenate dehydrogenase in the absence of tyrosine (Lingens et al. 1967). Thus, when phenylalanine is in excess, flow of prephenate is diverted into the tyrosine arm, both because the dehydratase is inhibited and because activity of the dehydrogenase is enhanced. When phenylalanine is present at low levels, prephenate will pass into the phenylalanine arm, whether or not tyrosine is present, because of the low affinity of the dehydrogenase for prephenate when phenylalanine is absent.

Because of the combined effects of the inhibitions and activations, the pools for the aromatic amino acids fill in the order of tryptophan first, followed by phenylalanine, and then tyrosine. Because the tryptophan pool always fills first, control by tryptophan at the level of DAHP synthase (reaction 1) is unnecessary and is not found.

That regulation of flow may work in vivo in the pattern derived from the in vitro studies is suggested by analyses of mutants that possess chorismate mutase activities resistant to feedback inhibition by tyrosine (Kradolfer et al. 1977). All three mutations analyzed result in high levels of chorismate mutase activity, which is feedback-resistant. This elevation in activity levels is apparently the result of a conformational change that renders the mutase nearly fully active in the absence of tryptophan. In the mutants, there is enhanced flux of chorismate into the phenylalanine-tyrosine branch, causing reduced chorismate pools and enhanced sensitivity of the tryptophan pathway to feedback inhibition by the false feedback inhibitor 5-methyltryptophan (5MT). The latter effect presumably occurs because chorismate is not present to counter the feedback inhibition of anthranilate synthase (the mutants were isolated as sensitive to 5MT).

Regulation of Enzyme Levels. The DAHP synthase isozymes (reaction 1) are not repressed by any of the aromatic amino acids singly or in combination (Doy and Cooper 1966; Lingens et al. 1967), nor is dehydroquinate dehydratase (reaction 3) (Doy and Cooper 1966). Of the enzymes of the phenylalanine-tyrosine branch, chorismate mutase (reaction 8) is induced by tryptophan, whether or not phenylalanine and tyrosine are present, prephenate dehydrogenase (reaction 11) is induced by phenylalanine, and prephenate dehydratase (reaction 9) is repressed by phenylalanine (Lingens et al. 1967). All of these effects are on the order of twofold to threefold. These effects are difficult to understand and the biological significance is unclear.

Contradictory data have been reported for anthranilate synthase (reaction 13). Lingens et al. (1967) report a 3.6-fold repression of this enzyme activity by 1 mM tryptophan for strain S288C, whereas Fantes et al. (1976) report no repression of this enzyme for strain X2180-1A at the same tryptophan concentration. S288C and X2180-1A are isogenic. The discernible differences in protocol are somewhat different media (vitamin content and pH) and a use of crude extracts (Fantes et al. 1976) versus an ammonium sulfate cut (Lingens et al. 1967). The two sets of data cannot be reconciled. Neither indoleglycerolphosphate synthase (reaction 16) nor tryptophan synthase (reaction 17) is repressed by tryptophan (Manney 1968; Miozzari et al. 1978).

Analyses of the responses of anthranilate synthase and indoleglycerolphosphate synthase to regulatory signals is not straightforward because the products of the *TRP2* and *TRP3* genes aggregate to form a complex that possesses both enzyme activities (De Moss 1965; Schürch-Rathgeb 1972). The results reported by Schürch-Rathgeb (1972) appear to indicate that free *TRP2*-gene product, uncomplexed to *TRP3*-gene product, is inactive but that free *TRP3* subunit possesses indoleglycerolphosphate synthase activity. Examination of sucrose gradient profiles (Schürch-Rathgeb 1972, Fig. 3) reveals that, in wild-type extracts, a substantial portion of indoleglycerolphosphate synthase activity may be uncomplexed with *TRP2*-gene product.

Dosage analyses of tetraploids bearing different numbers of wild-type alleles for each of the *TRP* genes were reported (Miozzari et al. 1978). Dosage analysis of *TRP2* reveals that the output per wild-type allele rises steadily as the dose of the wild-type allele decreases. The outputs per wild-type allele were 0.35, 0.47, 0.55, and 0.60 for tetraploids with 4, 3, 2, and 1 dose of the wild-type allele, respectively. That is, anthranilate synthase activity is being derepressed. Clearly anthranilate synthase activity is the rate-limiting step. Yet the indoleglycerolphosphate synthase levels, involving the same polypeptides, are constant through this dosage series. Both anthranilate synthase and indoleglycerolphosphate synthase are responsive to cross-pathway control (Schürch et al. 1975). This observation of a response of one activity in the absence of response of the other raises the possibility that this derepression of anthranilate synthase is a tryptophan-specific response rather than a response to cross-pathway control.

There are two possible explanations for the observation that anthranilate synthase activity can derepress without derepression of indoleglycerolphosphate synthase activity. The first is that the *TRP3*-gene product is normally limiting, that the *TRP2*-gene product is produced in excess, and that the *TRP2*-gene product must aggregate for production of anthranilate synthase activity. At reduced dosage levels of the *TRP2* allele, the gene product of wild-type *TRP2* allele might be assembled into the aggregate more efficiently than would be the product of the mutant allele. This explanation is unlikely because there is no effect of dosage of *TRP3* alleles on levels of anthranilate synthase, and the output per *TRP3* allele with respect to anthranilate synthase and indoleglycerolphosphate synthase is constant through the *TRP3* dosage series.

A more likely possibility is that the *TRP3*-gene product has activity as indoleglycerolphosphate synthase in the absence of the *TRP2* subunit and that the *TRP3*-gene product is produced in excess, as the data of Schürch-Rathgeb (1972) suggest. Derepression of anthranilate synthase could then occur without derepression of indoleglycerolphosphate synthase. The additional *TRP2*-gene product would drive more *TRP3*-gene product into an aggregated form, leaving the levels of indoleglycerolphosphate synthase unaltered but levels of anthranilate synthase elevated. It is of interest that these phenomena are observed for the crucial first activity of the pathway. Precedent for different levels of two subunits of an enzyme and for differential control of the two subunits is found for the arginine-specific carbamoylphosphate synthetase, which possibly coincidentally, is also an enzyme that controls flow into a pathway (Piérard et al. 1979).

Both anthranilate synthase and indoleglycerolphosphate synthase activities are derepressed about twofold in tetraploids bearing one wild-type allele of *TRP4* (which encodes phosphoribosyltransferase; reaction 14) (Miozzari et al. 1978). This effect must reflect derepression of both the *TRP2* and *TRP3* genes, if *TRP3*-gene product is normally produced in excess as suggested above.

All of the enzymes of tryptophan biosynthesis save the isomerase (reaction 15) respond to cross-pathway control (Schürch et al. 1974; Miozzari et al. 1978) and can be derepressed by starvation for histidine, leucine, arginine, or tryptophan. Mutations that lead to derepressed synthesis for these four enzymes have been reported (Miozzari et al. 1978). Also reported were mutations that result in derepressed levels of anthranilate synthase (the only tryptophan enzyme assayed), as well as derepressed levels for enzymes of other pathways (*tra3* mutations) (Wolfner et al. 1975). Schürch et al. (1974) and Wolfner et al. (1975) have described mutations that eliminate derepression of enzymes of several pathways in response to tryptophan or histidine starvation.

Serine Family

The serine family of amino acids consists of serine, glycine, and cysteine (Fig. 5; Table 5). Serine is also a major source of the single carbon pool needed for synthesis of purine nucleotides, thymidylate, and methionine and contributes its carbon chain to phospholipids and tryptophan. Two carbons of its chain are incorporated into purines and heme via glycine.

Serine and Glycine Biosynthesis

Biosynthesis of serine proceeds via a phosphorylated pathway (reactions 1–3). Mutant blocks have been assigned by enzyme assay (Ulane and Ogur 1972). All *ser2* mutants are leaky, presumably because nonspecific phosphatases can act upon the substrate (cf. *his2* mutants). No mutant lacking

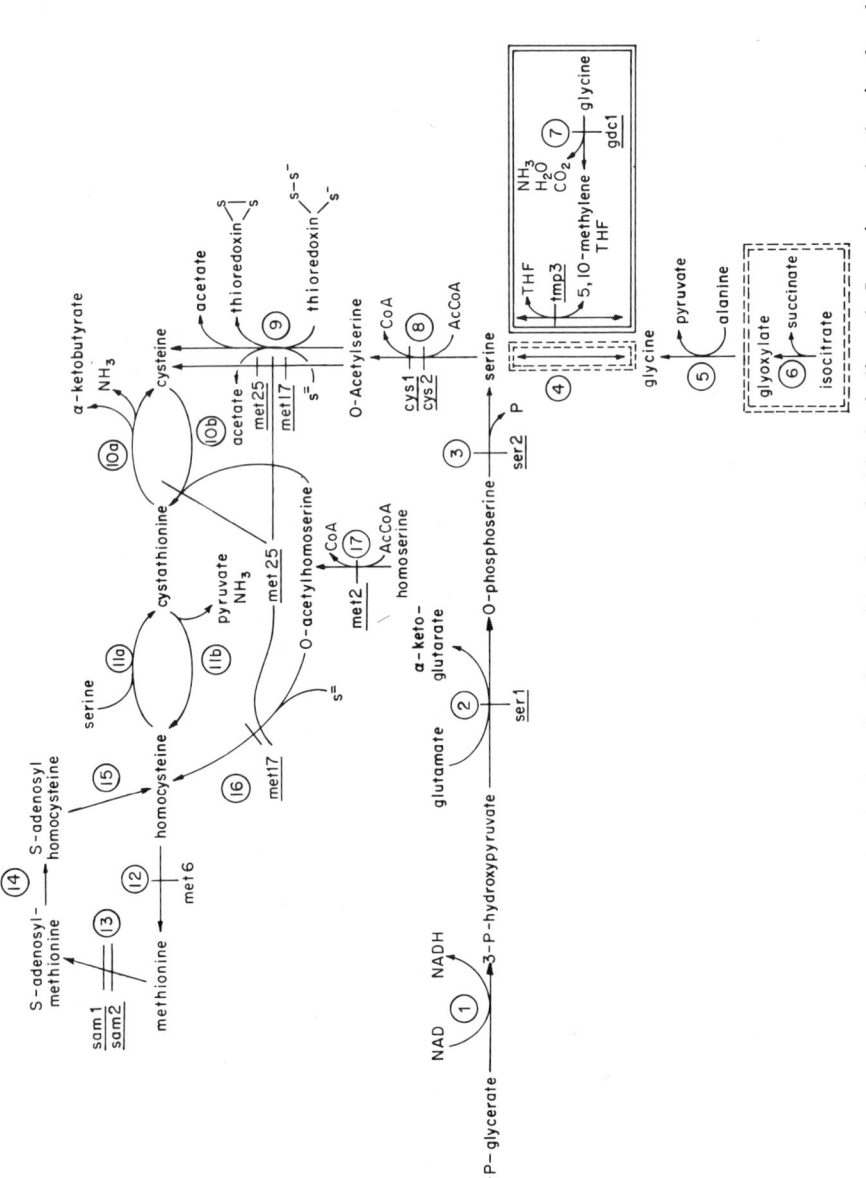

Figure 5 Biosynthesis of the serine family of amino acids. Genetic blocks indicated. Locations: (——) mitochondrion; (----) cytosol; no enclosure, unknown. (See Table 5.)

phosphoglycerate dehydrogenase (reaction 1) has been reported. The dehydrogenase (reaction 1) shows end-product inhibition by serine (Ulane and Ogur 1972). Control of enzyme levels has not been investigated.

Glycine can be generated from serine by serine hydroxymethyltransferase (reaction 4). Two species of the enzyme exist: one located in the mitochondrion, the other in the cytosol. Activity for the mitochondrial species (and other related enzymes) cannot be detected in the *tmp3* mutant (Zelikson and Luzzatti 1976, 1977). The *tmp3* mutant requires thymidylate, methionine, adenine, and histidine but does not require glycine. Presumably, glycine-requiring mutants could be derived from the *tmp3* mutant, since only one species of hydroxymethyltransferase remains. Methionine and SAM each inhibit part of the total serine hydroxymethyltransferase activity (Botsford and Parks 1969). The effects of each on the individual species have not been examined. The level of hydroxymethyltransferase activity may rise in response to glycine supplementation (Botsford and Parks 1969) but is not repressed by methionine.

When cells grow on acetate, an alternative pathway to glycine and serine is present. Isocitrate lyase (reaction 6) and glycine decarboxylase (reaction 7) are derepressed in this condition. Glyoxylate is produced from isocitrate (reaction 6) and transaminated to yield glycine (reaction 5). A portion of the glycine is decarboxylated (reaction 7), with release of CO_2 and formation of 5,10-methylene tetrahydrofolate (THF) from the α carbon of glycine. Reversal of the hydroxymethyltransferase reaction results in production of serine. The glycine generated by transamination serves as both a donor of single carbons via glycine decarboxylase (reaction 7) and a recipient for the single carbon for production of serine via the hydroxymethyltransferase (reaction 4). *ser1 gdc1* double mutants require formate as a source of single carbons when growing on glycine, as a source of serine, since glycine decarboxylase is absent (Ogur et al. 1977).

Cysteine Biosynthesis

Potentially, two pathways exist for synthesis of cysteine. The first (reactions 8 and 9) proceeds by acetylation of serine to yield *O*-acetylserine, followed by sulfhydrylation (either by free sulfide or via a transsulfurylation from a thioredoxin carrier; see below, Methionine Biosynthesis) to yield cysteine. Two enzymes exist that can catalyze formation of cysteine from *O*-acetylserine. One catalyzes only cysteine synthesis (reaction 9) (Yamagata 1980), and the second catalyzes synthesis of both cysteine and homocysteine (reactions 9 and 16) (Yamagata and Takeshima 1976). *met25* mutants appear to lack the latter bifunctional enzyme (Cherest et al. 1969; Masselot and de Robichon-Szulmajster 1975), as well as activity for γ-cystathionine synthase (reaction 10b) (Savin and Flavin 1972). *met17* mutants have very low levels of the bifunctional homocysteine-cysteine synthase but retain the specific synthase (Yamagata 1980) and have very low levels of phosphoadenylylsul-

Table 5 Genes and enzymes for biosynthesis of the serine family of amino acids

Reaction	Enzyme	Locus	References for gene assignment
1	P-glycerate dehydrogenase (3-P-glycerate: NAD 2-oxidoreductase, 1.1.1.95[a])		
2	P-serine aminotransferase (O-P-L-serine: 2-oxoglutarate aminotransferase, 2.6.1.52)	ser1	Ulane and Ogur (1972)
3	P-serine phosphatase (O-P-serine P-hydrolase, 3.1.3.3)	ser2	Ulane and Ogur (1972)
4	serine hydroxymethyltransferase (5,10-methyleneTHF:glycine hydroxymethyltransferase, 2.1.2.1)	tmp3[b]	Zelikson and Luzzatti (1976)
5	alanine-glyoxylate aminotransferase (L-alanine:glyoxylate aminotransferase, 2.6.1.44)		
6	isocitrate lyase (threo-D_5-isocitrate glyoxylate-lyase, 4.1.3.1)		
7	glycine decarboxylase[c] (5,10-methyleneTHF:ammonia hydroxymethyltransferase [carboxylating, reducing], 2.1.2.10)	gdc1	Ogur et al. (1977)
8	serine acetyltransferase (acetyl-CoA:L-serine O-acetyltransferase, 2.3.1.30)	cys1[d] cys2[d]	S. Halos and S. Fogel (pers. comm.) S. Halos and S. Fogel (pers. comm.)
9	cysteine synthase[c] (O-acetyl-L-serine acetate-lyase [adding sulfide], 4.2.99.8) two isozymes	met25[e]	Cherest et al. (1969) (corrected from met8)
10a	γ-cystathionase[c] (L-cystathionine cysteine-lyase [deaminating], 4.4.1.1)	met17[e]	Yamagata et al. (1974); see Yamagata (1980)

10b	γ-cystathionine synthase[c] (O-acetyl-L-homoserine acetate-lyase [adding cysteine], 4.2.99.9)	met25[e]	Savin and Flavin (1972) (corrected from met8); see Masselot and de Robichon-Szulmajster (1975)
11a	β-cystathionine synthase[c] (L-serine hydro-lyase [adding homocysteine], 4.2.1.22)		
11b	β-cystathionase[c] (cystathionine L-homocysteine-lyase [deaminating], 4.4.1.8)		
12	homocysteine methyltransferase[c] (5-methyltetrahydropteroyl-tri-L-glutamate: L-homocysteine S-methyltransferase, 2.1.1.14)	met6	Selhub (1972) (cited in Masselot and de Robichon-Szulmajster [1975])
13	SAM synthetase[c] (ATP: L-methionine S-adenosyltransferase, 2.5.1.6) two isozymes	eth2 (sam2) eth10 (sam1)	Cherest et al. (1978) Cherest et al. (1978)
14	SAM demethylase		
15	adenosylhomocysteinase (S-adenosyl-L-homocysteine hydrolase, 3.3.1.1)		
16	homocysteine synthase[c] (O-acetyl-L-homoserine acetate-lyase [adding H₂S], 4.2.99.10)	met25[e]	Cherest et al. (1969) (corrected from met8; see Masselot and de Robichon-Szulmajster [1975])
17	homoserine acetyltransferase (acetyl-CoA: L-homoserine O-acetyltransferase, 2.3.1.31)	met17[e] met2 (me˙a)	Yamagata et al. (1974) de Robichon-Szulmajster and Cherest (1967) (for locus assignment, see Masselot and de Robichon-Szulmajster [1975])

[a]Enzyme Commission number.
[b]Mitochondrial species.
[c]Not the IUPAC-IUB recommended name.
[d]See text.
[e]Pleiotropic; see text.

fate (PAPS) reductase (Masselot and Surdin-Kerjan 1977; see below, Methionine Biosynthesis).

The second pathway proceeds from homocysteine by condensation with serine to yield cystathionine (reaction 11a), a reaction catalyzed by β-cystathionine synthase, followed by cleavage, catalyzed by γ-cystathionase, to yield cysteine (reaction 10a). Existence of this complete transsulfuration pathway from methionine to cysteine (reactions 13-15, 11a, 10a) is inferred from the fact that mutants unable to reduce sulfate can grow if supplied with methionine, cysteine, or homocysteine (Masselot and de Robichon-Szulmajster 1975; Masselot and Surdin-Kerjan 1977). That transsulfuration from cysteine to methionine can also occur (reactions 10b and 11b) is inferred from the finding that all Met$^-$ mutants able to use homocysteine to satisfy a methionine requirement (save *met2;* reaction 17) can also use cysteine in its place (Masselot and de Robichon-Szulmajster 1975).

According to the pathways in Figure 5, no single defect for any reaction of the group 8-17 should result in a requirement that can only be satisfied by cysteine. This is because the transsulfuration pathway exists. Yet it is undeniable that *cys1* and *cys2* mutants exist, grow only if supplied with cysteine or *O*-acetylserine, and excrete H_2S (S. Halos and S. Fogel, pers. comm.). There are at least two explanations, however, for *cys1* and *cys2* mutants, which take into account the transsulfuration pathway. The mutations may, like the *met25* mutants described above, be pleiotropic and eliminate activity for serine acetyltransferase (reaction 8) and for an enzyme of the transsulfuration pathway (reactions 11a and 10a). Alternatively, the true substrate for β-cystathionine synthase (reaction 11a) may be homocysteine and *O*-acetylserine, rather than homocysteine and serine. If the latter hypothesis were true and if *cys1* and *cys2* mutants were simply deficient for the acetyltransferase, they would produce no *O*-acetylserine and would lack substrate for the transsulfuration pathway as well as for cysteine synthase (reaction 9). Addition of *O*-acetylserine would restore both pathways.

Little is known about regulation of cysteine biosynthesis. The bifunctional homocysteine-cysteine synthase is inhibited by methionine whether *O*-acetylhomoserine (Cherest et al. 1969) or *O*-acetylserine (Yamagata et al. 1975) is the substrate. Cysteine was not tested. This enzyme is repressed by growth in media containing methionine or SAM and will be discussed below in Methionine Biosynthesis.

Aspartate Family

The aspartate family of amino acids includes aspartate itself, asparagine, threonine, methionine, and isoleucine (Figs. 6 and 7; Table 6). Lysine is part of this family in prokaryotes but is part of the glutamate family in fungi. It is synthesized by different pathways in the two groups (for prokaryotes, see

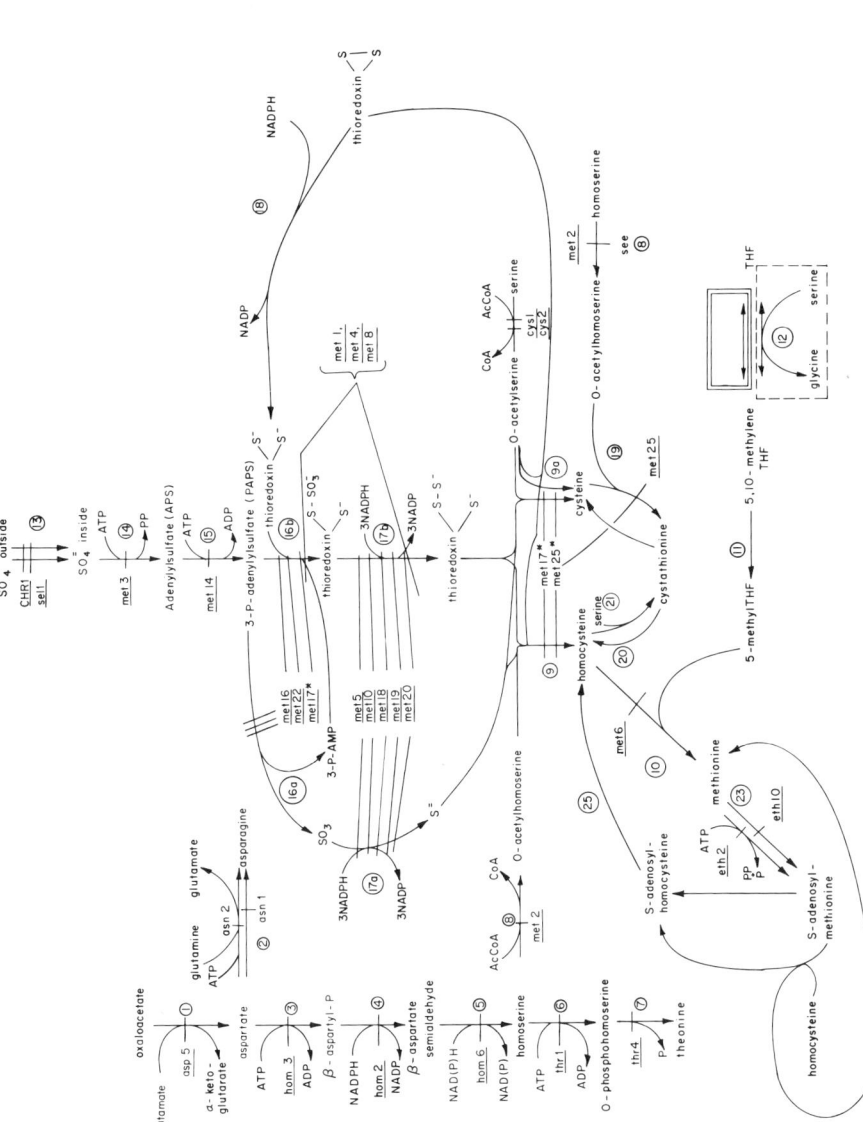

Figure 6 Biosynthesis of the aspartate family of amino acids. Genetic blocks indicated. (∗) Note pleiotropy. Location: (——) mitochondrion; (---) cytosol; unenclosed, unknown. (See Table 6; see text for differentiation between known and postulated reactions.)

Umbarger 1978). Isoleucine biosynthesis is considered with the pyruvate family since part of its chain is derived from pyruvate and four of its biosynthetic enzymes are shared with the valine pathway.

Aspartate and Asparagine Biosynthesis

Synthesis of aspartate occurs by transamination from glutamate onto oxalacetate (reaction 1). *asp5* mutants are deficient for this aspartate aminotransferase (de Robichon-Szulmajster et al. 1966). *asp5* mutants, including those bearing nonsense mutations, can be fed by aspartate or asparagine or, somewhat surprisingly, by a combination of threonine and methionine. The last observation is unexpected since aspartate is required for synthesis of arginine, pyrimidines, purines, and so on. All *asp5* mutants are somewhat leaky, implying that other aminotransferases can catalyze this reaction to at least a limited extent. *asp5* mutants do not grow well unless adenine is added to the medium, and colonies turn pink if adenine is omitted from the medium. This is not unexpected for cells deficient in aspartate, for this deficiency could result in reduced conversion of phosphoribosylaminoimidazolecarboxylate (CAIR) to phosphoribosylsuccinocarboxamide aminoimidazole (SAICAR) (see Fig. 13, purine biosynthesis), resulting in an *ade1* phenocopy.

The aminotransferase levels are somewhat higher when cells are grown with glutamate as nitrogen source rather than ammonia (de Robichon-Szulmajster et al. 1966). The enzyme is neither repressed nor inhibited by methionine, threonine, or aspartate (de Robichon-Szulmajster 1967).

Asparagine is synthesized in a glutamine-dependent reaction (reaction 2) (Ramos and Wiame 1979). Two equivalents of high-energy phosphate are required, as compared to one for glutamine synthesis. A requirement for asparagine occurs only if two genes bear mutations, *ASN1* and *ASN2* (Jones 1978) or *ASNA* and *ASNB* (Ramos and Wiame 1979). The equivalences of these genes are unknown. Two isozymes of asparagine synthetase exist. *asnA* mutants lack one of the species; *asnB* mutants lack the other (Ramos and Wiame 1980), but deficiency for asparagine synthetase activity can be detected in crude extracts only of the *asnA* mutants (Ramos and Wiame 1979). Both isozymes are inhibited by asparagine, and both are derepressed when asparagine synthesis is limiting for growth (Ramos and Wiame 1980).

Common Pathway for Synthesis of Threonine and Methionine via Homoserine

Pathway. A common pathway for synthesis of threonine and methionine (and eventually isoleucine) consists of three reactions leading from aspartate to homoserine. The β-carboxyl of aspartate is activated by a reaction catalyzed by aspartokinase (reaction 3), followed by reduction of β-aspartylphosphate to aspartate semialdehyde (reaction 4), catalyzed by aspartate

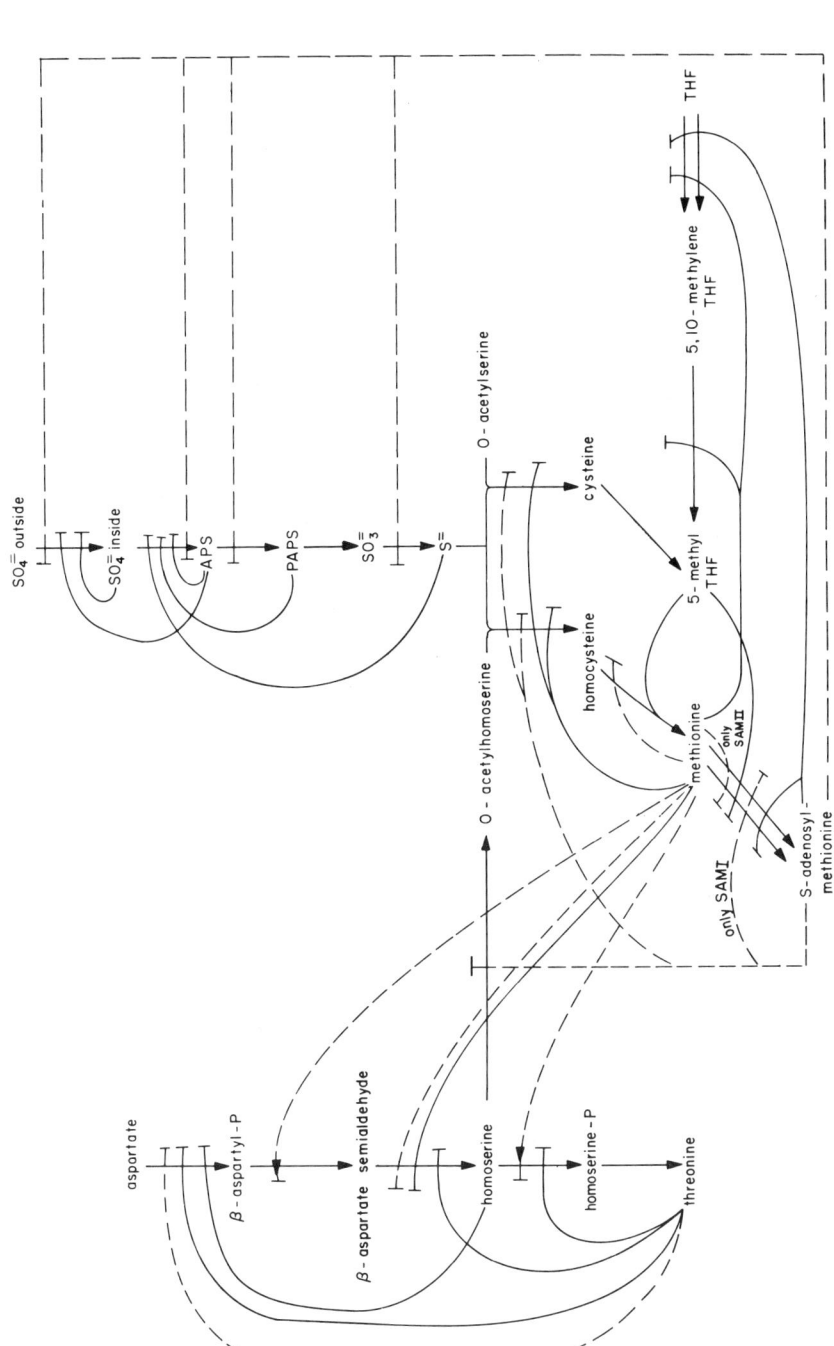

Figure 7 Regulation of carbon flow and enzyme synthesis for the aspartate family of amino acids. (——|) Feedback inhibition; (— — —|) repression of enzyme synthesis; (— — —→) induction of enzyme synthesis.

Table 6 Genes and enzymes for biosynthesis of the aspartate family of amino acids

Reaction	Enzyme	Locus	References for gene assignment
1	aspartate aminotransferase (L-aspartate: 2-oxoglutarate aminotransferase, 2.6.1.1[a])	asp5	de Robichon-Szulmajster et al. (1966)
2	asparagine synthetase (L-aspartate: L-glutamine amidoligase [AMP-forming], 6.3.5.4) two isozymes	asn1 (asnA) asn2 (asnB[b])	Jones (1978); Ramos and Wiame (1979)
3	aspartate kinase (ATP: L-aspartate 4-P-transferase, 2.7.2.4)	hom3	de Robichon-Szulmajster et al. (1966)
4	aspartate-semialdehyde dehydrogenase (L-aspartate-β-semialdehyde:NADP oxidoreductase [phosphorylating], 1.2.1.11)	hom2	de Robichon-Szulmajster et al. (1966)
5	homoserine dehydrogenase (L-homoserine:NAD(P) oxidoreductase, 1.1.1.3)	hom6	de Robichon-Szulmajster et al. (1966)
6	homoserine kinase (ATP:L-homoserine-O-P-transferase, 2.7.1.39)	thr1	de Robichon-Szulmajster (1967); de Robichon-Szulmajster et al. (1973)
7	threonine synthase (O-P-homoserine P-lyase [adding water], 4.2.99.2)	thr4	assumed
8	homoserine acetyltransferase (acetyl-CoA:L-homoserine O-acetyltransferase, 2.3.1.31)	met2 (meta)	de Robichon-Szulmajster and Cherest (1967); see Masselot and de Robichon-Szulmajster (1975)
9	homocysteine synthase[c] (O-acetyl-L homoserine acetate-lyase [adding H$_2$S], 4.2.99.10)	met17[d] met25[d]	Yamagata et al. (1974) Cherest et al. (1969) (corrected from met8; see Masselot and de Robichon-Szulmajster [1975])
9a	cysteine synthase[c] (O-acetyl-L-serine acetate-lyase [adding H$_2$S], 4.2.99.8) two isozymes	met17[d] met25[d]	Yamagata et al. (1974) (see Yamagata [1980]) Cherest et al. (1969)

#	Enzyme	Gene	Reference
10	homocysteine methyltransferase[c] (5-methyltetrahydropteroyl-tri-L-glutamate:L-homocysteine S-methyltransferase, 2.1.1.14)	met6	Selhub (1972) (cited in Masselot and de Robichon-Szulmajster [1975])
11	5,10-methyleneTHF reductase (5-methylTHF: [acceptor] oxidoreductase, 1.1.99.15)		
12	serine hydroxymethyltransferase (5,10-methyleneTHF: glycine hydroxymethyltransferase, 2.1.2.1.)	tmp3[e]	Zelikson and Luzzatti (1976)
13	sulfate permease (two)	CHR1	Breton and Surdin-Kerjan (1977)
		sel1	Breton and Surdin-Kerjan (1977)
14	ATP sulfurylase[c] (ATP:sulfate adenylyltransferase, 2.7.7.4)	met3	Naiki (1964); Masselot and Surdin-Kerjan (1977)
15	APS kinase[c] (ATP:adenylylsulfate-3′-P-transferase, 2.7.1.25)	met14	Naiki (1964); Masselot and Surdin-Kerjan (1977)
16	PAPS reductase	met16	Masselot and Surdin-Kerjan (1977)
		met17[d]	Masselot and Surdin-Kerjan (1977)
		met22	Masselot and Surdin-Kerjan (1977)
		met1[d]	Masselot and Surdin-Kerjan (1977)
		met4[d]	Masselot and Surdin-Kerjan (1977)
		met8[d]	Masselot and Surdin-Kerjan (1977)
17	sulfite reductase (hydrogen sulfide:NADP oxidoreductase, 1.8.1.2)	met5	Naiki (1965); Masselot and Surdin-Kerjan (1977)
		met10	Naiki (1965); Yoshimoto and Sato (1968b); Masselot and Surdin-Kerjan (1977)
		met18	Masselot and Surdin-Kerjan (1977)
		met19	Masselot and Surdin-Kerjan (1977)
		met20	Masselot and Surdin-Kerjan (1977)
		met1[d]	Masselot and Surdin-Kerjan (1977)
		met4[d]	Masselot and Surdin-Kerjan (1977)
		met8[d]	Masselot and Surdin-Kerjan (1977)

Table 6 (Continued)

Reaction	Enzyme	Locus	References for gene assignment
18	thioredoxin reductase (NADP:oxidized thioredoxin oxidoreductase, 1.6.4.5)		
19	γ-cystathionine synthase[c] (O-acetyl-L-homoserine acetate-lyase [adding cysteine], 4.2.99.9)	met25[d]	Savin and Flavin (1972) (for correction from met8, see Masselot and de Robichon-Szulmajster [1975])
20	β-cystathionase[c] (cystathionine L-homocysteine-lyase [deaminating], 4.4.1.8)		
21	β-cystathionine synthase[c] (L-serine hydro-lyase [adding homocysteine], 4.2.1.22)		
22	γ-cystathionase[c] (L-cystathionine cysteine-lyase [deaminating], 4.4.1.1)		
23	SAM synthetase[c] (ATP:L-methionine S-adenosyltransferase, 2.5.1.6) two isozymes, MATI and MATII	sam1 (eth10)- MATII sam2 (eth2)- MATI	Cherest et al. (1978) Cherest et al. (1978)
24	SAM demethylase		
25	adenosylhomocysteinase (S-adenosyl-L-homocysteine hydrolase, 3.3.1.1)		
26	homocysteine methyltransferase (S-adenosyl-L-methionine: L-homocysteine S-methyltransferase, 2.1.1.10)		

[a]Enzyme Commission number.
[b]See text.
[c]Not the IUPAC-IUB recommended name.
[d]Pleiotropic; see text.
[e]Mitochondrial species.

semialdehyde dehydrogenase. Reduction of the semialdehyde, catalyzed by homoserine dehydrogenase (reaction 5), yields homoserine. From this point, the pathways for methionine and threonine biosynthesis diverge.

Metabolic Flow. Aspartokinase (reaction 3) is end-product-inhibited by threonine and homoserine but not by methionine, unless the cells have grown in the presence of methionine or lysine (de Robichon-Szulmajster and Corrivaux 1963, 1964). Dominant mutations, designated *BOR1* (borrelidin-resistant), have been isolated, which cause aspartokinase to be resistant to feedback inhibition by threonine (borrelidin is a threonine analog). These map very near mutations in *HOM3*, a likely candidate for the structural gene for aspartokinase (Nass and Poralla 1976). The effects of *BOR1* mutations on carbon flow have not been reported. Homoserine dehydrogenase is competitively inhibited by threonine and methionine (Karassevitch and de Robichon-Szulmajster 1963). Thus, threonine seems to be the primary compound for regulating flow of carbon into and through the common pathway.

Regulation of Enzyme Levels. For this section, the observations will be presented first, followed by discussion. Aspartokinase is repressed fairly strongly by addition of threonine or homoserine to the medium (4–7-fold). If the results observed for the arginine pathway are any guide, this response must indicate a threonine-specific regulation. Aspartokinase is also, peculiarly, repressed threefold by lysine (de Robichon-Szulmajster and Corrivaux 1963, 1964). Starvation for threonine, effected by starvation of a threonine auxotroph (de Robichon-Szulmajster et al. 1973) or by addition of borrelidin, an inhibitor of threonyl-tRNA synthetase (Nass and Hasenbank 1970), causes derepressed synthesis of apartokinase. Thus, repression (from levels on minimal medium) seems to involve a threonine-specific signal, presumably threonyl-tRNA^{Thr}. The derepression responses could reflect either a specific or general control signal.

Methionine causes only a weak repression of aspartokinase (30%). Starvation of a methionine auxotroph for methionine leads to derepression (Cherest et al. 1969, 1971). Growth near the restrictive temperature of a methionyl-tRNA synthetase conditional mutant, *mes1*, results in derepression of aspartokinase to the same extent as that seen for the NAD-dependent GDH (Cherest et al. 1971). This could mean that the effect on levels of these enzymes is nonspecific or, alternatively, that this is a derepression response effected through the cross-pathway control system, which is known to cause the NAD-dependent GDH to derepress (Delforge et al. 1975). The second alternative seems more likely, in view of the possible involvement of tRNA^{Thr} in repression.

Aspartate semialdehyde dehydrogenase (reaction 4) levels are unaffected by addition of threonine to the medium. Addition of methionine leads to an apparent 1.5–2-fold induction of the enzyme, and starvation for methionine leads to repression (de Robichon-Szulmajster et al. 1973). As starvation of a threonine auxotroph leads to derepression of the enzyme (Surdin 1967; de Robichon-Szulmajster et al. 1973), the inductive effects of methionine are thought to arise as a consequence of threonine starvation brought about by repression and/or feedback inhibition of homoserine dehydrogenase by methionine (de Robichon-Szulmajster 1967; Surdin 1967; de Robichon-Szulmajster et al. 1973). Again, one cannot distinguish between a threonine-specific response and a response mediated by the cross-pathway control system. The latter might seem more likely since threonine addition does not repress. On the other hand, if cross-pathway control is the mechanism, one would expect methionine starvation to effect derepression. However, no such effect is seen.

Homoserine dehydrogenase (reaction 5) levels are repressed twofold by addition of methionine but not threonine (de Robichon-Szulmajster and Corrivaux 1964). Starvation of a methionine auxotroph leads to derepression (Cherest et al. 1971; de Robichon-Szulmajster et al. 1973), but starvation of a threonine auxotroph does not alter enzyme levels (de Robichon-Szulmajster et al. 1973). The available evidence suggests that this enzyme is controlled solely by means of a methionine-specific signal and is not subject to cross-pathway control.

Growth on rich medium (YPGA = YEPD + adenine) may be somewhat more repressing for homoserine dehydrogenase than is methionine alone, but little effect of this medium is seen on aspartokinase levels beyond the repression effected by threonine (Cherest et al. 1971; Cherest and de Robichon-Szulmajster 1973).

A mutant, B6, was isolated as resistant to methionine (we have been unable to trace a genetic designation). In the mutant, aspartokinase is no longer repressed by threonine, and aspartate semialdehyde dehydrogenase is no longer induced by methionine. Homoserine dehydrogenase levels are unaffected in the mutant. The regulatory alteration is not expressed in heterozygotes (de Robichon-Szulmajster et al. 1965; de Robichon-Szulmajster 1967).

For this common part of the pathway, the circuitry of regulation is complex. It would appear that aspartokinase responds to a threonine-specific signal (threonine or, more likely, charged threonyl-tRNA) or possibly to multivalent repression (threonine and methionine). If the signal is threonine-specific, rather than multivalent, derepression in response to methionine starvation might reflect the cross-pathway control system.

Control of aspartate semialdehyde dehydrogenase would appear not to respond to a specific signal since neither end product represses. The response to threonine starvation might then be a manifestation of cross-pathway control. If true, the repressive effect of methionine starvation remains unexplained. The B6 mutant could be altered in the cross-pathway control system or in a threonine-specific element. If the latter is true, the response to threonine starvation would have to be reevaluated.

Finally, homoserine dehydrogenase would appear to be under methionine-specific control since it is repressed by methionine. If this inference is correct, it may not be in the same control circuit as the other methionine biosynthetic enzymes (see below), for it may still be repressible by methionine in *eth2* mutants (Cherest et al. 1969; but see Masselot and de Robichon-Szulmajster 1972).

In the discussion above and to come, possible tie-ins to the cross-pathway control system are necessarily inferential or conjectural. This work was carried out before the cross-pathway control system had been detected. Because of the death of Huguette de Robichon-Szulmajster, experiments that might clarify the role of the cross-pathway system in regulating synthesis of these enzymes have not been carried out. Her presence as the guiding force in studies of regulation of these pathways is sorely missed, for her contributions were outstanding. Reading the papers that emerged from her research group during preparation of this review was a pleasure, for the data are hard and the insights plentiful.

Threonine Biosynthesis

Homoserine is phosphorylated by homoserine kinase (reaction 6). *O*-phosphohomoserine is hydrolyzed in a reaction catalyzed by threonine synthase. Assignment of *THR1* to homoserine kinase is by enzyme assay; assignment of *THR4* to threonine synthase is by inference (it is the only other auxotrophy satisfied only by threonine). Flow into the threonine-specific arm is controlled by end-product inhibition of homoserine kinase by threonine. Like aspartate semialdehyde dehydrogenase, levels of homoserine kinase are unaffected by threonine addition but are induced by addition of methionine to the medium (de Robichon-Szulmajster et al. 1973). Levels of these enzymes of the threonine-specific arm in the B6 mutant and in cells starved of threonine or methionine have not been reported.

Methionine Biosynthesis

Pathway. Biosynthesis of methionine begins with an acetyl-CoA-dependent acetylation of homoserine to yield *O*-acetylhomoserine. *met2* mutants lack activity for homoserine *O*-acetyltransferase (de Robichon-Szulmajster and Cherest 1967) and require *O*-acetylhomoserine, homocysteine, or methionine (Naiki and Yamagata 1973; Masselot and de Robichon-Szulmajster 1975). Two possible routes exist for conversion of *O*-acetylhomoserine to homocysteine. The first is a direct sulfhydrylation of *O*-acetylhomoserine to yield homocysteine (reaction 9) with the sulfhydryl donor being free sulfide or, possibly, a carrier such as thioredoxin (see below). The second route involves condensation of *O*-acetylhomoserine with cysteine to yield cystathionine (reaction 9), followed by cleavage by β-cystathionase (reaction 20) to yield homocysteine and pyruvate. Homocysteine is the acceptor for a methyl group donated by 5-tetrahydropteroyl-triglutamate (reaction 10). The methyltransferase requires the diglutamyl or triglutamyl derivative of 5-methyl-

THF (Burton et al. 1969) and is missing in *met6* mutants (cited in Masselot and de Robichon-Szulmajster 1975).

Sulfide (free or bound) is generated by reduction of sulfate (reactions 14–17). Sulfate is activated in two steps (reactions 14 and 15) by reaction with ATP, yielding PAPS. PAPS is reduced to sulfite by PAPS reductase (reaction 16a). Sulfite is then reduced to sulfide by sulfite reductase (reaction 17a). In vitro the reduction of PAPS can be catalyzed by a mixture of three proteins (Asahi et al. 1961; Wilson et al. 1961). The sulfite produced is a bound intermediate (Torii and Bandurski 1967). Porqué et al. (1970b) have shown that the three proteins required for in vitro reduction of PAPS are thioredoxin, thioredoxin reductase, and PAPS reductase (PAPS sulfotransferase), corresponding to fractions C, A, and B, respectively, from Wilson et al. (1961).

Tsang and Schiff (1976) have suggested that assimilatory sulfate reduction involves bound intermediates throughout the pathway in *Escherichia coli*, yeast, and *Chlorella*. They suggest that the sulfate group of PAPS is transferred directly to thioredoxin (reaction 16b) to yield a thiosulfonate (thioredoxin-S-SO_3^-), which is then reduced by an NADPH-dependent thiosulfonate reductase (reaction 17b) to yield a persulfide (thioredoxin-S-S^-). Thiosulfonate reductase (reaction 17b) in this sequence corresponds to sulfite reductase in the usual sequence (reaction 17a). The persulfide would then donate sulfide to the acetylated amino acid acceptors (reactions 9 and 9a) releasing oxidized thioredoxin, which would be reduced by thioredoxin reductase (reaction 18) to complete the cycle. Whether reactions 16b and 17b, involving thioredoxin, exist and function in yeast as drawn is conjectural.

met3 mutants lack activity for ATP sulfurylase (reaction 14). *met14* mutants are inferred to lack adenylylsulfate (APS) kinase activity (reactions 15) because they possess activity for ATP sulfurylase but do not catalyze synthesis of PAPS from sulfate (Naiki 1964; Masselot and Surdin-Kerjan 1977). *met16*, *met22*, and the pleiotropic *met17* mutants are deficient for PAPS reductase activity (reactions 16a and 16b) (Masselot and Surdin-Kerjan 1977).

Sulfite reductase has been purified from yeast (Yoshimoto and Sato 1968a,b, 1970). It catalyzes reduction of free sulfite and does not catalyze reduction of PAPS by NADH even if supplemented with one or two of the A, B, and C proteins of Wilson et al. (1961; Yoshimoto and Sato 1968a). The significance of this negative observation is unclear. *met5* and *met10* mutants (Naiki 1965; Masselot and Surdin-Kerjan 1977) and *met18*, *met19*, and *met20* mutants (Masselot and Surdin-Kerjan 1977) all lack sulfite reductase activity, and two *met10* mutants (mutants 6 and 20 of N. Naiki, which fail to complement other *met10* alleles) have sulfite reductase proteins that lack flavin mononucleotide (FMN) and are smaller than usual (Yoshimoto and Sato 1968b). Two other mutants (mutants 11 and 21 of N. Naiki) also have sulfite reductase with altered structure, with that of mutant 21 lacking flavin adenine dinucleotide (FAD) as well as FMN (Yoshimoto and Sato 1968b). The complementation groups of mutants 11 and 21 have not been reported. These observations imply that there are at least two different subunits in sulfite reductase. Certainly, *MET10* and, probably, *MET5, MET18, MET19,* and *MET20* are involved in determining the structure of sulfite reductase. Whether these genes encode polypeptides that end up in the enzyme or that are required for attachment of cofactors to the enzyme has not been determined.

Free sulfide can certainly be incorporated into cysteine and homocysteine in vitro and in vivo. Thus, there is no compelling evidence to suggest that the pathway of bound intermediates exists and functions in yeast. However, a pathway involving bound intermediates would allow one to account for the properties of *met1, met4,* and *met8* mutants. These mutations are pleiotropic, and the mutants bearing them are unable to reduce PAPS to sulfite (reaction 16) and are also unable to reduce sulfite to sulfide (reaction 17) (Masselot and Surdin-Kerjan 1977). Possibly, one or more of these mutants is defective in thioredoxin or thioredoxin reductase and, hence, defective for reduction of both PAPS and sulfite. Mutants of *E. coli* that lack thioredoxin or thioredoxin reductase have been obtained and are viable (Chamberlin 1974; Fuchs 1977; Mark et al. 1977). Other explanations for the properties of the *met1, met4,* and *met8* mutants are, of course, possible. They could be regulatory mutants, or their pleiotropy could derive from participation of the corresponding normal gene products in a larger complex. A tantalizing hint of such a complex is provided by the properties of one *met14* mutant. This mutant is inferred to lack APS kinase, but it also possesses an unstable sulfite reductase (Masselot and Surdin-Kerjan 1977).

Cherest et al. (1969) concluded that synthesis of methionine proceeds solely by condensation of *O*-acetylhomoserine with sulfide to yield homocysteine, because *met25* mutants (Masselot and de Robichon-Szulmajster 1975) lack homocysteine synthase (reaction 9) and have a requirement for homocysteine or methionine (or cysteine) (Masselot and de Robichon-Szulmajster 1975). Cherest et al. (1969) argued that *met25* mutants should not have had a requirement if there were a route from cysteine to homocysteine via cystathionine, for this route would provide a bypass for the lesion. It has since been shown, however, that the homocysteine synthase is bifunctional and catalyzes sulfhydrylation of acetylserine as well as acetylhomoserine (Yamagata et al. 1974). Formation of cysteine would presumably rest with the specific cysteine synthase (Yamagata 1980) if it is still present in the *met25* mutant. Moreover, the *met25* mutant was later shown to be deficient for γ-cystathionine synthase (reaction 19), as well as the bifunctional homocysteine-cysteine synthase (Savin and Flavin 1972). Because of the pleiotropy, both routes to homocysteine are blocked. (How the mutant grows on cysteine is mysterious.) Because of the pleiotropic effects of the *met25* mutation, these experiments do not provide evidence on whether the cystathionine pathway exists. *met17* mutants also have very low levels of the bifunctional cysteine-homocysteine synthase (Yamagata et al. 1974). They are also deficient in PAPS reductase (Masselot and Surdin-Kerjan 1977). Hence, they too have both routes to homocysteine blocked. One other mutant lacking homocysteine synthase has been described, mutant 16, of Yamagata et al. (1975). Its requirement can be satisfied by homocysteine or methionine or by the combination of O-acetylhomoserine and cysteine, substrates of cystathionine γ-synthase (reaction 19). Mutant 16 not only lacks homocysteine synthase activity (reaction 9) but also bears a *met2* mutation, which eliminates homoserine O-acetyltransferase activity (Masselot and de Robichon-Szulmajster 1975). It is inferred to be a double mutant. Again, both routes to homocysteine are blocked. The direct sulfhydrylation pathway is blocked because homocysteine synthase and homoserine O-acetyltransferase are absent; the cystathionine pathway is blocked because one of the substrates of γ-cystathionine synthase, namely O-acetylhomoserine, is not made.

These results provide circumstantial evidence that two pathways—via sulfhydrylation and via γ-cystathionine—exist for the synthesis of homocysteine. That is, Met$^-$ mutants that lack homocysteine synthase are only recovered if, as a consequence of a second lesion or pleiotropy, other routes to homocysteine are also blocked (*met25*, the route via cystathionine; *met17*, the route to cysteine via sulfate reduction; mutant 16, the route to O-acetylhomoserine). This implies to us that more than one route to homocysteine exists and that loss of homocysteine synthase activity does not result in a requirement for methionine. It would be interesting to know whether the mutation in mutant 16 that causes deficiency for homocysteine synthase would result in a methionine requirement once separated from the *met2* mutation present in the strain. If two routes to homocysteine exist, mutations in the cystathionine and/or transsulfuration pathways could be readily obtained by isolating derivatives of mutants defective in sulfate reduction whose requirement could now be satisfied solely by methionine or solely by cysteine, where the parent strain utilized either. Those able to grow only if supplied with methionine should have an intact route from homocysteine to cysteine (reactions 21 and 22) but a defect in the route from cysteine to homocysteine (reactions 19 and 20). Those able to grow only if supplied with cysteine should have defects in the route from methionine to cysteine (reactions 23–25, 21 and 22) but be intact for the path from cysteine to methionine (reactions 19 and 20). Such an analysis would allow resolution of the question of the number of different routes to homocysteine as well as providing information on cysteine biosynthesis. Precedent for synthesis of methionine via two pathways exists in *Aspergillus* (see Paszewski and Grabski 1974) and in *S. lipolytica* (Morzycka and Paszewski 1979).

Cystathionine will not satisfy the methionine requirement of methionine auxotrophs, despite the fact that it is taken up and that extracts can be shown to have β-cystathionase activity (Sorsoli et al. 1975). These observa-

Amino Acid and Nucleotide Biosynthesis 225

tions are difficult to assess until we know whether any of the methionine enzymes are compartmentalized and, if so, whether cystathionine can penetrate to a compartment that would allow its utilization.

Unaccounted for in Figure 6 or the text are mutations in the *MET7, MET13, MET21, MET23,* and *MET24* genes. Mutations in these genes cause an absolute requirement for methionine as do mutations in *MET6.* However, *met7* mutants also require adenine (Lowenstein 1973). They will be discussed more fully in the section on single carbon metabolism. *met6* mutants lack activity for homocysteine methyltransferase (cited in Masselot and de Robichon-Szulmajster 1975). Possibly these five loci are concerned with synthesis or reduction of 5,10-methyleneTHF to yield 5-methylTHF or in synthesis of the polyglutamyl sidechain needed if the cofactor is to function (Burton et al. 1969). Of the six loci which, when mutant, cause an absolute methionine requirement, only *met6, met7,* and *met13* mutations have been mapped (see Appendix I, Genetic Map of *Saccharomyces cerevisiae*). *met21, met23,* and *met24* mutations need to be mapped relative to these and to one another, since *met7* and *met21* are relatively rare alleles (Singh and Sherman 1974b; Masselot and de Robichon-Szulmajster 1975), and there seem to be too few functions for so many loci. The growth and the enzymatic properties of *met15* mutants have not been studied, although extensive genetic studies on this locus have been carried out (Singh and Sherman 1975). *met9, met11,* and *met12* mutations may not exist (Singh and Sherman 1974a). Positive selections exist for mutations in *MET2* and *MET15* (Singh and Sherman 1974b) and in *MET3, MET14,* and *MET16* and the sulfate permease (Breton and Surdin-Kerjan 1977).

There are two isozymes for conversion of methionine to SAM (reaction 23) (Cherest et al. 1978). *eth2* (now *sam2*) mutants lack the species of activity designated MATI (*m*ethionine *a*denosyl*t*ransferase *I*); *eth10* (now *sam1*) mutants lack the species designated MATII. Doubly mutant *sam1 sam2* strains bearing tight mutations at each locus require SAM.

The transsulfuration pathway from methionine to cysteine proceeds via SAM. The sequence involves SAM synthetase, SAM demethylase, S-adenosylhomocysteine lyase (reactions 23–25), cystathionine β-synthase, and γ-cystathionase (reactions 21 and 22).

Metabolic Flow. Initial reports indicated that SAM inhibits flow of carbon from homoserine into the methionine-specific pathway at the level of homoserine acetyltransferase (de Robichon-Szulmajster and Cherest 1967), but more recent work indicates that neither SAM nor methionine inhibits the acetyltransferase (de Robichon-Szulmajster and Cherest 1967; de Robichon-Szulmajster et al. 1973). (For a summary of regulation of the aspartate family, see Fig. 7.) Thus, flow of carbon into the pathway appears unregulated. One could account for this surprising finding if homoserine acetyltransferase were located in the mitochondrion, as are a number of acetyltransferases involved in amino acid biosynthesis (Ryan et al. 1973; Tracy and Kohlhaw 1975; Wipf and Leisinger 1979), and if the binding site for the effector on the acetyltransferase did not survive disruption of the organelle. The binding site for valine on acetohydroxy acid synthase, a mitochondrially located enzyme, shows substantial but not complete lability of this type (Ryan and Kohlhaw 1974). Methionine does inhibit flow from O-acetylserine to homocysteine (Cherest et al. 1969).

Both methionine and SAM are involved in regulating the flow of single carbons from serine into methionine. Methionine inhibits methyleneTHF reductase (reaction 11) somewhat. SAM was not tested (Lor and Cossins 1972). Both methionine and SAM inhibit serine hydroxymethyltransferase (reaction 12) (Botsford and Parks 1969; Lor and Cossins 1972). Whether the inhibitions are effected on the same or different isozymes is unknown, but the pattern of inhibition is suggestive of inhibition of only one species by methionine and the other or both by SAM (Botsford and Parks 1969).

The feedback inhibitions on the path of sulfate reduction appear mostly to be competitive inhibitions and would not appear to involve what are normally considered end products (Fig. 7) (de Vito and Dreyfuss 1964). Whether they are relevant to regulation is unknown.

Mutants selected as resistant to methyl mercury arise by mutations in *MET2* and *MET15* (Singh and Sherman 1974a,b). Moreover, methyl mercury alleviates the methionine requirement of both *MET2* and *MET15* mutants. The data are consistent with a model that proposes that a compound, X, directly or indirectly involved in the biosynthesis of methionine (reaction 8), can cause detoxification of methyl mercury. The mutations in *MET2* and *MET15* lead to the accumulation of X and, therefore, to resistance to mercury compounds. The promotion of the growth of *MET2* and *MET15* mutants by methyl mercury is more difficult to explain. One proposal is that the *met2* mutation does not affect homoserine-O-acetyltransferase itself, but causes the accumulation of X, which inhibits the enzyme's activity or biosynthesis (Singh and Sherman 1974a).

Regulation of Enzyme Levels. Three separate biosynthetic sequences, in addition to the common pathway to homoserine, must be considered when analyzing regulation of methionine biosynthesis. These are synthesis and addition of the methyl group, reduction of sulfate to sulfide (bound or free), and the acetylation of homoserine followed by sulfhydrylation. All of the evidence indicates that the sulfate reduction pathway and the acetylation and sulfhydrylation sequences are controlled similarly, whereas the single carbon pathway responds to different controls. Enzymes of the transsulfuration (cystathionine) pathways have not been examined.

Regulation of the enzymes of sulfate reduction and of the acetylation and sulfhydrylation steps involves SAM and methionyl tRNA, which will be discussed in turn. SAM is added to wild-type cells growing in minimal

Amino Acid and Nucleotide Biosynthesis 227

medium, the levels of sulfate permease (reaction 13) (Breton and Surdin-Kerjan 1977), ATP sulfurylase (reaction 14), sulfite reductase (reaction 17), and homocysteine synthase (reaction 9) (Cherest et al. 1973a; Ferro and Spence 1973) are repressed severalfold just as they are if methionine is added to the medium (Cherest et al. 1969, 1971; Ferro and Spence 1973; Breton and Surdin-Kerjan 1977). APS kinase (reaction 15) and homoserine acetyltransferase (reaction 8) are repressed by methionine but have not been tested for repression by SAM (de Robichon-Szulmajster and Cherest 1967; Cherest et al. 1971; Antoniewski and de Robichon-Szulmajster 1973). However, these two enzymes are thought to be regulated by a circuit that involves SAM. Enzymes in this circuit are referred to as group-I enzymes. In mutants resistant to ethionine as a result of mutations at the *ETH2* (*SAM2,* see below), or *ETH10* (*SAM1*) loci, the levels of ATP sulfurylase, sulfite reductase, homoserine acetyltransferase, and homocysteine synthase are not repressible by methionine (Cherest et al. 1969, 1971, 1973b; Ferro and Spence 1973) although, for all enzymes tested, the levels remain normally repressible by SAM (Cherest et al. 1973b).

Two isozymes are known that catalyze synthesis of SAM (reaction 23). In the *eth2-2* (*sam2-12*) mutant, which bears an ochre mutation, the *SAMI* isozyme of *S-a*denosyl*m*ethionine synthetase (called SAM synthetase, rather than *m*ethionine *a*denosyl*t*ransferase [MAT] in this section to avoid confusion) is absent (Cherest et al. 1978). In the *eth10-2* (*sam1-12*) mutant, the SAMII isozyme is present at greatly reduced levels. Analyses of SAM and methionine pools have been carried out for the wild type and the mutants and indicate that repression of the levels of group-I enzymes by methionine is effected by conversion of methionine to SAM, which is the effective molecule (Cherest et al. 1973b). Addition of methionine to wild-type cells growing in minimal medium causes the pools of methionine and SAM to rise; addition of SAM results in increased SAM, but not methionine, pools. In the *eth2-2* or *eth10-2* mutants, the SAM pools are lower than normal and increase very little if methionine is added to the medium. Methionine pools are tenfold higher in the mutants on minimal medium and rise when methionine is added. Yet no repression of group-I enzymes occurs. Addition of SAM to the mutants causes increased SAM pools but results in restoration of methionine pools to normal levels and repression of group-I enzymes. This lowering of methionine pools upon SAM addition presumably occurs because SAM represses homoserine acetyltransferase and homocysteine synthase. It is not at all clear why loss of activity for one of two SAM synthetase isozymes should lead to such marked effects on levels of SAM in the pools and on the ability of methionine to repress group-I enzymes via conversion to SAM. One would naively think that the presence of one particular species of SAM synthetase or the presence of either would allow maintenance and/or expansion of the SAM pool. Possibly, there are two SAM pools, each derived from one isozyme and each (or a derivative of it) required for SAM-mediated repression. Obvious candidates for metabolic derivatives are polyamines. Evidence for at least two pools of SAM in yeast has been obtained (Cherest et al. 1973a).

Repression of SAM synthetase shares some features with repression of group-I enzymes. In the wild type, repression can be effected by addition of SAM or methionine. However, what evidence there is suggests that the SAMI isozyme is repressed by methionine, not SAM, for the SAMI isozyme retained in the *eth10-2* mutant is completely repressible by methionine under conditions where the SAM pool is not elevated (Cherest et al. 1973b). The levels of the SAMII isozyme may be controlled by levels of SAM, for the enzyme retained in the *eth2-2* mutant is not repressed by methionine (Mertz and Spence 1972; Cherest et al. 1973b).

In the *eth3-1* mutant, neither methionine nor SAM effectively represses group-I enzymes (Cherest et al. 1973b). Possibly the *ETH3*-gene product intermediates repression by SAM. Alternatively, the *eth3* mutant may be defective in conversion of SAM to other metabolites.

Repression by SAM is thought to occur posttranscriptionally, for repression and derepression occur very rapidly; and addition of lomofungin, an inhibitor of RNA synthesis, coincident with initiation of derepression, is not immediately effective in preventing derepression, whereas addition of cycloheximide is (Surdin-Kerjan and de Robichon-Szulmajster 1975). The *eth2-2* mutation, which causes deficiency in SAM production, also causes a pronounced undermethylation of tRNAs including $tRNA^{Met}$. However, the undermethylation is a result of the abnormal regulation of methionine biosynthesis rather than a cause (Fesneau et al. 1975).

In the *eth2* and *eth10* mutants, in which methionine does not effect repression, the levels of group-I enzymes for cells growing in minimal medium are normal. They are not derepressed. Derepression of group-I enzymes can be brought about by starving methionine auxotrophs for methionine (Cherest et al. 1971) or by growing a *mes1* mutant, bearing a thermolabile methionyl-tRNA synthetase, at permissive or semipermissive temperatures (Cherest et al. 1971). In the *mes1* mutant, group-I enzymes can be partially repressed if the methionine level is raised sufficiently. In the wild type growing on minimal medium, the $tRNA^{Met}$ is nearly completely acylated. In the *mes1* mutant, a substantial portion of the $tRNA^{Met}$ is uncharged, even at permissive temperatures. Supplementation with high levels of methionine results in an increase in the fraction of charged $tRNA^{Met}$ in the mutant. These results clearly indicate that methionyl $tRNA^{Met}$ participates in repression of group-I enzymes and is probably involved in maintaining the enzymes at the intermediate level characteristic of cells growing in minimal medium (Cherest et al. 1971).

Cross-pathway regulation seems to involve charged tRNAs (Messenguy and Delforge 1976), at least for isoleucyl $tRNA^{Ileu}$. Derepression of methionine group-I enzymes in response to low levels of methionyl $tRNA^{Met}$ might be reflective of methionine-specific regulation, or of cross-pathway control, or both. Sulfite reductase, one of the group-I enzymes that is derepressed in the *mes1* mutant, appears not to be under cross-pathway control (Wolfner et al. 1975). This suggests that there is a methionine-specific regulatory circuit that involves methionyl tRNA^{Met}.

To summarize, regulation of group-I enzymes, i.e., enzymes involved in sulfate reduction, acetylation, and sulfhydrylation, appears to involve methionyl tRNA^{Met} and SAM, at least. The levels of enzymes found in cells growing on minimal medium (taken as a neutral point) correspond to the fully repressed level effected by methionyl tRNA^{Met}. SAM is involved in modulating levels below this neutral point.

de Vito and Dreyfuss (1964) have reported that cysteine induces synthesis of ATP sulfurylase and represses synthesis of sulfite reductase when cysteine is the only sulfur source. This pattern is clearly different from that reported above for these enzymes. Possibly, regulation is effected by different circuitry under these conditions.

Only fragmentary information is available about the regulation of the pathway by which the single carbon unit is produced and transferred to homocysteine. MethyleneTHF reductase (reaction 11) is not repressed by methionine. SAM was not tested. 5-MethylTHF-dependent homocysteine methyltransferase is repressed by methionine. SAM was not tested (Lor and Cossins 1972).

There appears to be a minimum of three regulatory circuits for enzymes involved in methionine biosynthesis. The first involves aspartokinase and aspartate semialdehyde dehydrogenase and seems primarily to be regulated by levels of threonine or threonyl tRNA^{Thr}. Levels of both enzymes are altered in the B6 mutant. The second circuit involves homoserine dehydrogenase. This enzyme is repressed by methionine and may be unaffected by mutations in *ETH2,* although the data are somewhat inconclusive on this point (Cherest et al. 1969; Masselot and de Robichon-Szulmajster 1972). The third circuit involves group-I enzymes and has SAM and methionyl tRNA^{Met} as effectors. Whether or where the other enzymes fit into this circuitry is unclear.

Of the methionine biosynthesis enzymes, only sulfite reductase has been examined for cross-pathway regulation. It appears not to be susceptible to this regulatory system (Wolfner et al. 1975). Because *tra* mutants are resistant and *aas* mutants are sensitive to analogs of methionine, it has been argued that enzyme(s) of methionine biosynthesis are susceptible to crosspathway control (Wolfner et al. 1975). Candidates for susceptible enzymes include aspartokinase which, although only repressible by threonine, is derepressed by methionine starvation and aspartate semialdehyde dehydrogenase which, although not repressible by threonine, is derepressed upon threonine starvation. Of obvious interest is whether there will prove to be a dichotomy, with enzymes of sulfate reduction immune to cross-pathway control and enzymes involved in generation and transformation of the carbon chain susceptible to this regulation. Only sulfite reductase seems to be repressed by growth in rich medium (Cherest et al. 1971). As sulfite reductase is not regulated by the cross-pathway control system, its repression by growth in YEPD implies that the YEPD response is separate from cross-pathway control.

Nothing is known about the intracellular location of the enzymes of methionine biosynthesis. Of obvious interest is whether the homoserine and serine acetyltransferases are located in the mitochondrion and whether they are susceptible to glucose repression and regulation by CoA as are all other acetyltransferases involved in amino acid biosynthesis (Ryan and Kohlhaw 1973; Tracy and Kohlhaw 1975; Wipf and Leisinger 1979).

Pyruvate Family

The pyruvate family of amino acids includes alanine, valine, and leucine, for they derive a major portion of their carbon chains from pyruvate (Fig. 8; Table 7). Isoleucine will be discussed with this family, for although it derives four of its carbons from aspartate and only two from pyruvate, its biosynthetic pathway shares at least three enzymes with that of valine.

Isoleucine, Valine, and Leucine Biosynthesis

Pathway. Isoleucine biosynthesis starts with deamination of threonine to yield α-ketobutyrate (reaction 1). α-Ketobutyrate and pyruvate serve as acceptors of an "active aldehyde" derivative of thiamine pyrophosphate (Holzer and Kohlhaw 1961) to yield the acetohydroxy acids acetolactate and α-acetohydroxybutyrate, which are precursors of valine and isoleucine, respectively (reactions 2 and 2a). The two acetohydroxy acids undergo an isomerization and reduction to yield dihydroxy acid precursors for valine and isoleucine (reactions 3 and 3a). The α,β-dihydroxy acids undergo dehydration to give α-ketoisovalerate and α-keto-β-methylvalerate, α-keto acid precursors of valine and isoleucine, respectively (reactions 4 and 4a). The keto acids are transaminated to yield valine and isoleucine (reactions 5 and 5a). The actual amino donor for transamination has not be identified. In crude extracts, leucine, valine, and isoleucine will serve as donors for one another's aminotransferases, as will glutamate (Kakar and Wagner 1964).

Gene assignments were made by Kakar and Wagner (1964) and Magee and de Robichon-Szulmajster (1968a). Mutations in *ILV1*, which encodes threonine deaminase (reaction 1), can lead to an isoleucine requirement or to feedback resistance to the isoleucine analog thiaisoleucine (Kakar and Wagner 1964; Betz et al. 1971). The enzyme acetohydroxy acid synthase (AHAS), missing in *ilv2* mutants (Magee and de Robichon-Szulmajster 1968a), catalyzes reactions 2 and 2a. Mutation in *ILV2* can lead to an isoleucine-valine requirement or to growth inhibition by valine, which is

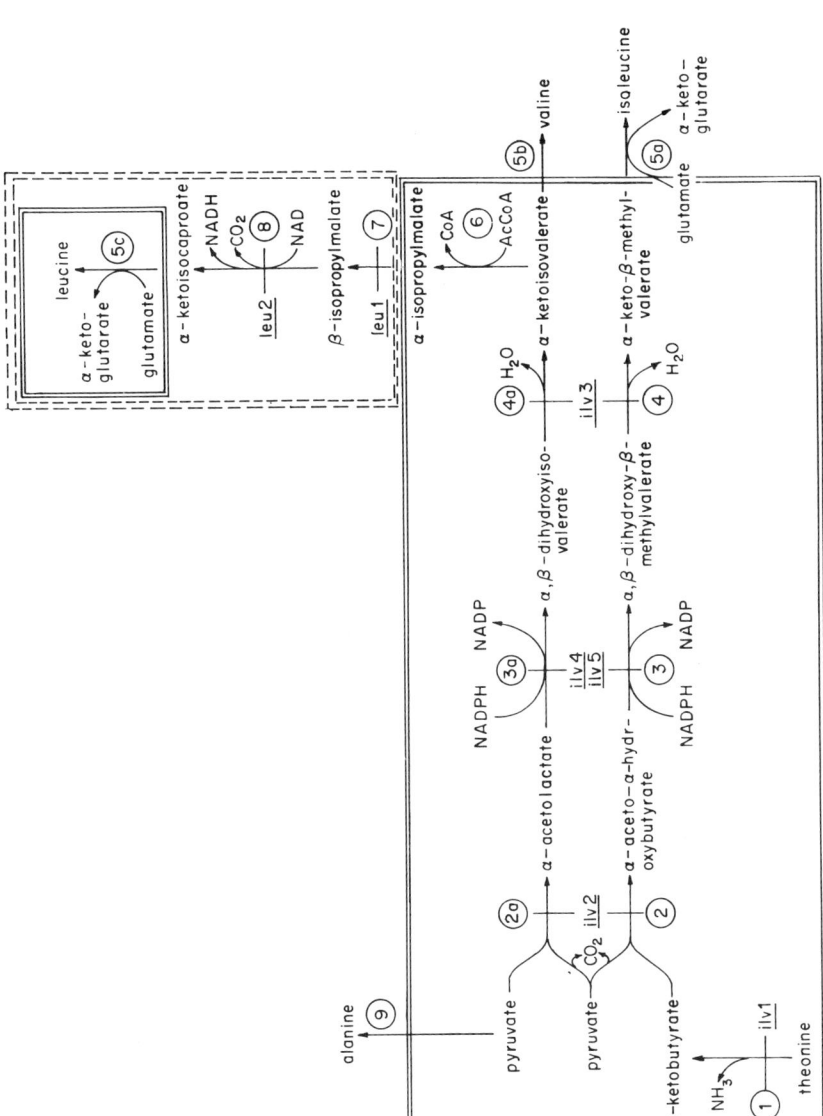

Figure 8 Biosynthesis of the pyruvate family of amino acids. Genetic blocks indicated. Locations: (———) mitochondrion; (- - -) cytosol; unenclosed, unknown. (See Table 7.)

Table 7 Genes and enzymes for biosynthesis of the pyruvate family of amino acids

Reaction	Enzyme	Locus	References for gene assignment
1	threonine deaminase[a] (L-threonine hydro-lyase [deaminating], 4.2.1.16[b])	ilv1	Kakar and Wagner (1964)
2	AHAS[a] (acetolactate pyruvate-lyase [carboxylating], 4.1.3.18)	ilv2	Magee and de Robichon-Szulmajster (1968a)
3	acetohydroxy acid isomeroreductase[a] (2-acetolactate methylmutase, 5.4.99.3)	ilv4 ilv5	Kakar and Wagner (1964) Kakar and Wagner (1964)
4	dihydroxyacid dehydratase (2,3-dihydroxyacid hydro-lyase, 4.2.1.9)	ilv3	Kakar and Wagner (1964)
5	branched-chain amino acid aminotransferase		
6	α-IPM synthase[a] (3-hydroxy 4-methyl-3-carboxyvalerate:2-oxo-3-methylbutyrate-lyase (CoA-acetylating], 4.1.3.12)	leu4? leu5?	G. Kohlhaw (pers. comm.)
7	α-IPM isomerase[a] (2-hydroxy-4-methyl-3-carboxyvalerate hydro-lyase, 4.2.1.33)	leu1	Satyanarayana et al. (1968a); G. Kohlhaw (pers. comm.)
8	β-IPM dehydrogenase[a] (2-hydroxy-4-methyl-3-carboxyvalerate NAD oxidoreductase, 1.1.1.85)	leu2	Ratzkin and Carbon (1977); Hinnen et al. (1978)
9	alanine aminotransferase		

[a]Not the IUPAC-IUB recommended name.
[b]Enzyme Commission number.

reversed by isoleucine (Kakar and Wagner 1964; Meuris et al. 1967; Meuris 1969). A single isomeroreductase catalyzes reactions 3 and 3a. *ilv4* and *ilv5* mutants are each incapable of catalyzing acetohydroxy acid-dependent oxidation of NADPH, and mutations in the two genes are tightly linked (Kakar and Wagner 1964). As only a single allele has been reported for each locus, it is impossible to discern whether the complementation observed reflects intragenic complementation for subunits of a bifunctional enzyme or intergenic complementation of separate polypeptides catalyzing two separate reactions. The dihydroxy acid dehydratase catalyzes reactions 4 and 4a and is missing in *ilv3* mutants. Whether there is a single enzyme responsible for transamination of the branched chain amino acids is unknown. No mutants with a requirement for isoleucine, valine, or leucine or a combination of these as a consequence of aminotransferase deficiency have been reported. All of the enzymes required for synthesis of the α-keto acid precursors of isoleucine and valine from pyruvate and threonine are located in the mitochondrion (Ryan and Kohlhaw 1974). Location of the aminotransferase(s) was not examined.

The α-keto acid precursor of valine, α-ketoisovalerate, is the source of most of the carbon chain of leucine, for it is a substrate for a condensation reaction with acetyl-CoA, catalyzed by α-isopropylmalate (IPM) synthase (reaction 6). α-IPM undergoes an isomerization to β-IPM, followed by oxidative decarboxylation to yield the α-keto acid precursor of leucine, α-ketoisocaproate (reactions 7 and 8). The condensation, isomerization, and oxidative decarboxylation reactions are analogous to similar reactions found in the tricarboxylic acid cycle and in lysine biosynthesis in fungi. The entire isomerization reaction in leucine biosynthesis appears to be catalyzed by a single polypeptide chain (Bigelis and Umbarger 1976), in possible contrast to the findings for lysine biosynthesis. The final step in leucine biosynthesis involves transamination of the α-keto acid as described above.

Confusion exists for assignment of genes to enzymes for the leucine pathway. Enzyme levels were reported for mutants in ten genes that cause leucine requirement (Satyanarayana et al. 1968a). The data indicating ten different genes were not presented. The more customary number found is three. It is certain that *LEU2* encodes β-IPM dehydrogenase (reaction 8) (Ratzkin and Carbon 1977; Hinnen et al. 1978; Hsu and Kohlhaw 1982). *LEU1* is thought to encode isomerase (reaction 7) (Satyanarayana et al. 1968a; G. Kohlhaw, pers. comm.). *leu3* mutants have low levels of isomerase and dehydrogenase (G. Kohlhaw, pers. comm.) and may be defective in the induction process (see below).

The genetic evidence available (G. Kohlhaw, pers. comm.) suggests that there may be two genes that encode the first enzyme of the pathway, α-IPM synthase. (This enzyme contains only one type of subunit.) Starting from a strain (SK103) bearing a trifluoroleucine (feedback)-resistant IPM synthase, a derivative (SK413) was isolated, which was sensitive to trifluoroleucine, had low levels of all enzymes of the leucine pathway, and had a leaky requirement for leucine. From this bradytroph, a leucine-requiring strain was isolated (SK605), which had no detectable IPM synthase and possessed low but inducible levels of isomerase and dehydrogenase. Crosses of SK605 to *leu1* or *leu2* mutants yield prototrophic diploids. SK103 is inferred to contain a mutation in one of two structural genes for IPM synthase, which renders that synthase feedback-resistant. SK413 is inferred to carry the same mutation and a second mutation in the same gene that results in no or inactive gene product. SK605 would then carry a third mutation in a second, IPM-synthase-encoding gene as well (G. Kohlhaw, pers. comm.). We (and Dr. Kohlhaw) have designated the gene mutated in SK103 to be *LEU4*. Should all the above inferences prove correct, the gene mutated only in SK605 will be designated *LEU5*. The low levels of isomerase and dehydrogenase observed in SK413 and SK605 are presumed to reflect lack of induction (see below).

α-IPM synthase (reaction 6) is located in the mitochondrion, whereas IPM isomerase and β-IPM dehydrogenase are in the cytosol (Ryan et al. 1973). Glutamate-α-ketoisocaproate aminotransferase activity was found in soluble and particulate (mitochondrial) fractions (Ryan and Kohlhaw 1974).

Metabolic Flow. Regulation of flow of carbon into the isoleucine and valine pathways involves feedback inhibition and enzyme activation. The pattern somewhat resembles that seen for regulation of flow in aromatic amino acid biosynthesis.

Threonine deaminase (reaction 1) shows positive cooperativity with respect to threonine concentration. Isoleucine activates the enzyme at low concentrations but is inhibitory at higher concentrations. Valine reverses the inhibition by isoleucine and normalizes the kinetics for binding of threonine (de Robichon-Szulmajster and Magee 1968). When both isoleucine and valine are present at high concentrations, the enzymes become inhibited. Thus, the balance between isoleucine and valine levels is important for regulating flow from threonine into the isoleucine pathway.

Flow into the pathways is also regulated at the next step (AHAS) (reaction 2). This is the sole control point for flow into the valine pathway. AHAS displays typical Michaelis-Menten kinetics for synthesis of acetolactate from pyruvate (synthesis from α-ketobutyrate, for the isoleucine pathway, was not examined). Valine is a noncompetitive inhibitor of AHAS; isoleucine and leucine are without effect (Magee and de Robichon-Szulmajster 1968b). Presumably, the mutations in *ILV2,* the AHAS gene, which lead to growth inhibition by valine, which is, in turn, reversible by isoleucine, reflect changes in sensitivity of AHAS to feedback inhibition by valine (Meuris et al. 1967; Meuris 1969). The activity of AHAS and the binding site for valine are very labile in cell extracts (Magee and de Robichon-Szulmajster 1968b; Ryan and Kohlhaw 1974). When assayed in permeabilized cells, the activity detected is greater than that found in extracts and is more sensitive to inhibition by valine. These differences presumably reflect the fact that, in permeabilized cells, the enzyme is in its normal mitochondrial environment.

Only an incomplete, rather speculative picture of regulation of flow can be constructed since information on α-ketobutyrate in relation to AHAS is lacking. Our guess is the following. When the effective pools of isoleucine, valine, and threonine are low, little α-ketobutyrate will be synthesized because threonine deaminase will be relatively inactive because of its sigmoid kinetics. Hence, flow through AHAS will predominantly be along the valine pathway since ample pyruvate should be available. High threonine pools would result in synthesis of α-ketobutyrate because of positive cooperativity of threonine deaminase. Alternatively, as the levels of valine rise, threonine deaminase will be activated, α-ketobutyrate will be synthesized, and flow through AHAS along the isoleucine pathway will increase. In the absence of information on the interactions of valine, α-ketobutyrate, pyruvate, and AHAS, one cannot speculate on what the relative proportions of flow into the two pathways might be. When both pools have filled, both threonine deaminase and AHAS will be inhibited and synthesis through both pathways will be reduced. The effective pools of amino acids, both as substrate (threonine) and inhibitors (isoleucine and valine), must be in the mitochondrion where the enzymes are located (Ryan and Kohlhaw 1974).

Flow of carbon into the leucine pathway seems to be regulated by feedback inhibition of α-IPM synthase by leucine (Satyanarayana et al. 1968b; Ulm et al. 1972). Bussey (1970) reported isolation of dominant trifluoroleucine-resistant mutations, which lead to leucine-insensitive α-IPM synthase activity. These mutants secrete leucine (Kohlhaw 1982). CoA causes reversible inactivation of synthase activity. Inactivation is antagonized by leucine or the substrate of the enzyme, α-ketoisovalerate. ATP reverses the CoA inactivation (Hampsey and Kohlhaw 1981). Two binding sites exist for CoA on the enzyme: one at the product site, the second at a "regulatory site" (Tracy and Kohlhaw 1975, 1977). The inactivation of the enzyme by CoA and reversal of the inactivation by ATP is consonant with the mitochondrial location of the enzyme and presumably serves to link carbon flow into biosynthetic pathways with the energy needs of the cell.

Regulation of Isoleucine-valine Enzyme Levels. Conflicting data exist as to the regulation of enzymes of isoleucine biosynthesis. Possibly, these differences reflect strain differences. Taking as basal levels those levels found in cells growing in minimal medium, Bussey and Umbarger (1969) have concluded that repression of the five enzymes of isoleucine-valine biosynthesis is effected by addition of isoleucine, valine, and leucine to the minimal medium and, hence, resembles the multivalent repression seen in bacteria (Freundlich et al. 1962). Threonine can replace isoleucine for repression, possibly, as suggested by Bussey and Umbarger, because it can be converted to isoleucine and thus elevate the isoleucine pools. An alternative explanation is that both threonine and isoleucine are required for repression. Feeding threonine could cause elevation of pools of both amino acids—directly for the threonine pool and by conversion for the isoleucine pool. Feeding isoleucine could also raise both pools—directly for the isoleucine pool and indirectly for the threonine pool by decreasing flux of threonine to isoleucine by feedback inhibition of threonine deaminase. Starvation for any of the three amino acids in auxotrophs did not lead to derepression in these studies, in contrast to the results of Magee and Hereford (1969). Magee and Hereford (1969) also concluded that multivalent repression exists for these enzymes. Their conclusions derive from the observation that starvation for any of the three amino acids in appropriate auxotrophs leads to derepression. Bollon (1975) reached somewhat different conclusions. His results indicate that threonine deaminase can be repressed solely by isoleucine, whereas AHAS, isomeroreductase, and dehydratase (reactions 2-4) are subject to multivalent repression by isoleucine, valine, and leucine. The effects of threonine were not examined. Because of pool considerations mentioned above, threonine cannot be excluded from participation in these repression phenomena. Multivalent repression of the isoleucine-valine enzymes requires the activity of isoleucyl-tRNA synthetase (McLaughlin et al. 1969), although the effects seen could also be due to the effects of isoleucyl-tRNA synthetase on cross-pathway control (see below, General Control Systems, Cross-pathway Control).

Some of the derepressions seen upon starvation could be a consequence of cross-pathway control. Derepression of AHAS, at least, may reflect a control specific to isoleucine-valine biosynthesis, for AHAS does not respond to cross-pathway control (Wolfner et al. 1975).

Alterations of the regulatory pattern for isoleucine-valine enzymes have been observed in mutants bearing lesions in *ILV1*, the structural gene for threonine deaminase, and have led to the suggestion that threonine deaminase, in addition to its role in catalysis, plays a positive role in regulation of the isoleucine-valine enzymes, possibly by binding tRNA (Bollon and Magee 1971, 1973; Bollon 1974, 1975, 1980), although there is no direct evidence for such binding in yeast. Briefly, the observations are as follows. Strains bearing a mutation that renders threonine deaminase 100-fold less sensitive to feedback inhibition by isoleucine (Betz et al. 1971) have elevated levels of AHAS, isomeroreductase, dehydratase, and α-IPM synthase (the first leucine enzyme) when cells are grown on minimal medium (Bollon and Magee 1971, 1973). Threonine deaminase itself shows normal regulation in this strain (Bollon 1975). Normal enzyme levels are restored if 5 mM isoleucine is added to the medium. Diploids bearing a feedback-insensitive threonine deaminase activity as a result of intragenic complementation show a similarly altered repression pattern (Bollon 1975). Strains bearing nonsense mutations, which map to the middle but not to the presumed carboxyterminal end of the *ILV1* gene (Thuriaux et al. 1971), have repressed levels of AHAS, isomeroreductase, and dehydratase when growing on minimal medium. These levels are not further repressed by addition of amino acids to the medium, and derepression can only be effected by starvation for leucine (but not isoleucine or valine) (Bollon and Magee 1971, 1973). These results have been interpreted to mean that threonine deaminase has a catalytic and a regulatory role; that the carboxyterminal region of the threonine deaminase polypeptide is essential if derepression is to occur, possibly by binding a regulatory species of charged leucyl tRNA. This would explain the failure of strains bearing *ilv1* nonsense mutations to derepress, since the leucyl tRNA would not be bound and would be available for repression. Derepression in the feedback-resistant strains would be explained because threonine deaminase would bind up charged leucyl tRNA, a binding unchallenged by isoleucine because of the reduced affinity of the enzyme for isoleucine (Bollon and Magee 1973).

As we point out in the following paragraphs, though, alternative explanations of these observations are plausible. The first observation is that strains with feedback-resistant threonine deaminase activity contain elevated levels of enzymes of isoleucine-valine biosynthesis. However, one expected consequence of a feedback-resistant threonine deaminase is that the threonine pools would be low since flux of threonine through the isoleucine pathway would be unimpeded. If threonine were required for repression of the isoleucine-valine enzymes, as it is for two leucine enzymes (Satyanarayana et al. 1968b; Bollon and Magee 1973; Brown et al. 1975; Kohlhaw et al. 1980; G. Kohlhaw, pers. comm.), derepression of the isoleucine enzymes would be expected in a mutant with depleted threonine pools. Of relevance to this argument may be the observation that threonine can substitute for isoleucine in repression of isoleucine-valine enzymes (Bussey and Umbarger 1969). Compatible with this hypothesis is the finding that 5 mM isoleucine restores normal enzyme levels (Bollon and Magee 1971, 1973), for high levels of isoleucine might shut off threonine deaminase, allowing restoration of normal threonine pools. It should be borne in mind that the location of the pools may be important. Unregulated threonine deaminase in the mitochondrion could drain the *cytosolic* threonine pools, which might be of paramount importance in repression.

That centrally mapping nonsense mutations in *ILV1* result in repressed synthesis of isoleucine-valine enzymes and a failure to respond to normal repression signals may be related to the fact that all of the isoleucine-valine enzymes so far examined are located in the mitochondrion (Ryan and Kohlhaw 1974). It seems possible that derepression of the relevant enzyme activities (as opposed to polypeptides) may be coupled to transport of the threonine deaminase polypeptide into the mitochondrion. One obvious possibility for coupling is that the isoleucine-valine enzymes form a complex in the mitochondrion. If transport of the ochre fragments of threonine deaminase into the mitochondrion were abnormally slow, or complex formation inefficient, derepression of activities might not be seen in response to normal signals. If this explanation has any validity, nonsense mutations in the other *ILV* structural genes might have similar effects on regulation.

We would raise one final point with respect to a role for threonine deaminase in regulation of gene expression. It is in the wrong compartment to play the role envisioned for it, for it is located in the mitochondrion.

Mutations that are *cis* dominant, tightly linked to *ILV1*, and cause derepressed synthesis of threonine deaminase have been isolated (Bollon 1980). The levels of threonine deaminase found in these mutants are substantially higher than those that can be achieved by isoleucine starvation. Elevated levels of AHAS are found in these mutants, a result expected by either of the hypotheses presented above. The dominant constitutivity is expressed in a/α diploids, reducing, but not excluding, the possibility that the mutation is a Ty insertion (see below, Specific Control Systems and Their Mutations, *Cis*-dominant Regulatory Elements).

Regulation of Leucine Enzyme Levels. Synthesis of α-IPM synthase (reaction 6) is slightly repressed below basal levels if both threonine and leucine are added to minimal medium (Satyanarayana et al. 1968b; Bollon and Magee 1973; Brown et al. 1975; G. Kohlhaw, pers. comm.). Single addition of leucine or threonine results in a slight elevation of synthase activity (Satyanarayana et al. 1968b; G. Kohlhaw, pers. comm.).

Threonine or leucine represses synthesis of IPM isomerase, but maximum repression is achieved only when both amino acids are present (Brown et al. 1975; G. Kohlhaw, pers. comm.). Maximal repression of IPM dehydrogenase activity requires both threonine and leucine. Threonine alone appears to induce, but leucine alone slightly represses synthesis of dehydrogenase (Brown et al. 1975; G. Kohlhaw, pers. comm.). If exogenous leucine levels are raised to 2 mM, leucine alone will cause repression of isomerase and dehydrogenase (G. Kohlhaw, pers. comm.).

However, synthesis of IPM isomerase and IPM dehydrogenase seems to be primarily controlled by induction, with α-IPM being the inducer (G. Kohlhaw, pers. comm.). Repression by leucine or leucine plus threonine would appear to be effected by reducing the endogenous level of the inducer α-IPM. If α-IPM is supplied exogenously, the combination of leucine plus threonine no longer effects repression of isomerase and dehydrogenase.

Regulation of the leucine biosynthetic enzymes is altered in a number of different mutants. Mutants that possess feedback-resistant forms of IPM synthase (isolated as resistant to trifluoroleucine) show elevated levels of all three enzymes of leucine biosynthesis, with isomerase and dehydrogenase showing the greatest effects. Addition of leucine plus threonine is quite ineffective in repressing enzyme synthesis in these mutants (G. Kohlhaw, pers. comm.). The simplest explanation of the phenotypes of these mutants is that the cells have elevated levels of α-IPM, which induces isomerase and dehydrogenase.

α-IPM synthase levels are derepressed in mutants that possess feedback-resistant threonine deaminase activity, as was seen for the isoleucine-valine enzymes. In this mutant, repression requires isoleucine as well as leucine and threonine (Bollon and Magee 1973). Possible mechanisms for this altered regulation are discussed in the section on isoleucine-valine regulation. It is also possible that this effect is a cross-pathway-caused derepression brought about by threonine starvation, for IPM synthase is subject to cross-pathway control (Hsu et al. 1982). Synthase cannot be repressed in certain mutants altered in uptake of leucine and other amino acids (Bussey and Umbarger 1970).

Levels of the leucine biosynthetic enzymes are responsive to the carbon source on which cells are grown. Synthase levels are threefold higher in acetate-grown cells compared to glucose-grown cells, whereas levels of isomerase and dehydrogenase are about twofold lower in acetate, compared to glucose-grown cells. Studies on enzyme levels during diauxic growth on glucose have not been reported but might be interesting in view of the results reported for acetylglutamate synthase of the arginine pathway, where derepression of the enzyme occurs during adaptation to the respiratory mode of growth (Wipf and Leisinger 1979). If glucose is added to cultures growing on lactate, about two thirds of the synthase activity is rapidly lost; the rest is stable. The suggestion was made that there are two forms of the enzyme in acetate-grown cells with one form susceptible to proteolysis and the other not susceptible (Brown et al. 1975). These effects on synthase are of interest because of the mitochondrial location of this enzyme (Ryan et al. 1973) and the possible relevance to integration of carbon metabolism and anabolism.

Cross-pathway Control. Of the isoleucine-valine enzymes tested, transaminase B (reaction 5a) is subject to cross-pathway control (Delforge et al. 1975), whereas AHAS is not (Wolfner et al. 1975). Of the three leucine biosynthetic enzymes tested, IPM synthase (reaction 7) is subject to cross-pathway control; IPM isomerase and IPM dehydrogenase are not (Hsu et al. 1982).

Alanine Biosynthesis

Biosynthesis of alanine is presumed to take place by transamination of pyruvate (reaction 9). To our knowledge, nothing is known about the enzyme(s) or regulation of alanine biosynthesis.

Histidine Biosynthesis

Pathway. Histidine biosynthesis (Fig. 9; Table 8) begins with condensation of the ribosylphosphate moiety of PRPP onto ring nitrogen 1 of ATP (reaction 1) to give phosphoribosyl-ATP (PR-ATP). Pyrophosphorolysis of PR-ATP yields phosphoribosyl-AMP (PR-AMP) (reaction 2), which is followed by opening of the purine ring between N1 and C6 (reaction 3), giving a compound referred to as BBMII (*B*ound *B*ratton-*M*arshall *II:* acid hydrolysis releases phosphoribosylaminoimidazolecarboxamide [AICAR], which is a diazotizable amine). The ribose moiety in BBMII that originated from PRPP undergoes isomerization to the ribulose group of BBMIII (reaction 4). Amido transfer from glutamine and closure of the new imidazole ring (reactions 5 and 6) results in formation of imidazoleglycerolphosphate and the purine intermediate AICAR. Dehydration of imidazoleglycerolphosphate yields imidazoleacetolphosphate (reaction 7), which undergoes transamination to histidinolphosphate (reaction 8). Histidinol is produced by dephosphorylation (reaction 9). Histidinol is oxidized in two steps (via histidinal) to yield histidine.

Gene assignments were made by Fink (1964, 1966) and E. Jones (unpubl.). The *HIS4* locus encodes a trifunctional polypeptide that possesses enzymatic activities for PR-AMP cyclohydrolase (*HIS4A;* reaction 3), PR-ATP pyrophosphohydrolase (*HIS4B;* reaction 2), and histidinol dehydrogenase (*HIS4C;* reaction 10) (Keesey et al. 1979). All known *his2* mutants are leaky. Their leakiness is presumed to result from the action of nonspecific phosphatases on the substrate histidinol-phosphate. Indeed, these nonspecific phosphatases lead to cleavage of the substrate in vitro. However, their action is inhibited by beryllium (G. Fink, unpubl.). *ade3* mutants also require histidine (in addition to adenine) (Roman 1956). They are defective in single carbon metabolism, however (Jones and Magasanik 1967; Lazowska and Luzzatti 1970a). *ade3* mutants accumulate 5,10-methyleneTHF (Luzzatti 1975), which can be shown to inhibit the fourth enzyme of histidine biosynthesis in vitro and may do so in vivo, for the cells accumulate the substrate of this enzyme, BBMII (E. Jones, unpubl.).

Metabolic Flow. Flow of carbon into the pathway is regulated by feedback inhibition by histidine of the phosphoribosyltransferase (reaction 1) (Rasse-Messenguy and Fink 1973). Mutations to triazolealanine resistance, *TRA1*, map to *HIS1,* the structural gene for the first enzyme. *TRA1* mutations may be dominant or recessive (Rasse-Messenguy and Fink 1973; Lax et al. 1979). Some, but not all, of the *TRA1* mutants excrete histidine. The phosphoribosyltransferase activity in *TRA1* mutants is resistant to feedback inhibition by histidine (Rasse-Messenguy and Fink 1973). That no nonsense mutations of *HIS1* were found in a sample of 52 (Korch and Snow 1973) continues to be a puzzling observation, since viable deletions of the locus have been constructed using recombinant DNA techniques (A. Hinnebusch and G. Fink, unpubl.).

Regulation of Enzyme Levels. Addition of histidine to the medium does not result in repression of the enzymes catalyzing the first, third, ninth, and tenth steps of the pathway (Wolfner et al. 1975), the only steps for which data have been reported. If 3-amino-1,2,4-triazole (3AT), an inhibitor of dehydratase, is added to wild-type cells, starvation for histidine ensues and derepression occurs for at least the enzymes catalyzing the first, third, eighth, ninth, and tenth steps (Rasse-Messenguy and Fink 1973; Schürch et al. 1974; Wolfner et al. 1975). The derepressions are small but may reach nearly tenfold for the cyclohydrolase (Rasse-Messenguy and Fink 1973). Prolonged starvation of histidine auxotrophs also leads to derepression, with cyclohydrolase (reaction 3) derepressing as much as 18-fold (Rasse-Messenguy and Fink 1973; Delforge et al. 1975). These derepression responses can be eliminated by mutations that alter cross-pathway control and therefore seem to be entirely controlled by this system (Schürch et al. 1974; Wolfner et al. 1975). There is no evidence for histidine-specific control.

Cross-pathway Control. Six of the enzymes of histidine biosynthesis have been examined for regulation by the cross-pathway control system (those catalyzing reactions 1,3, 7-10), and all are controlled by this system as evidenced by their showing one or more of the following characteristics. (1) They are derepressed by starvation for tryptophan, lysine, arginine, or valine (Schürch et al. 1974; Delforge et al. 1975; Wolfner et al. 1975; Struhl 1979). (2) They, and enzymes of other pathways controlled by this system, cannot be derepressed in *aas* mutants or certain classes of 5MT-sensitive mutants (Schürch et al. 1974; Wolfner et al. 1975.) (3) They, and enzymes of other pathways controlled by this system, are derepressed in *tra3* mutants (Wolfner et al. 1975).

SPECIFIC AND GENERAL CONTROL SYSTEMS FOR REGULATING SYNTHESIS OF ENZYMES OF AMINO ACID BIOSYNTHESIS

Specific Control Systems and Their Mutations

Trans-*acting "Regulatory" Genes*

Mutations in *trans*-acting regulatory genes have been described that alter regulation of enzymes of single pathways. The affected pathways are those of arginine, threonine, and methionine biosynthesis. In no case has it been shown that the "regulatory" gene encodes a classical repressor.

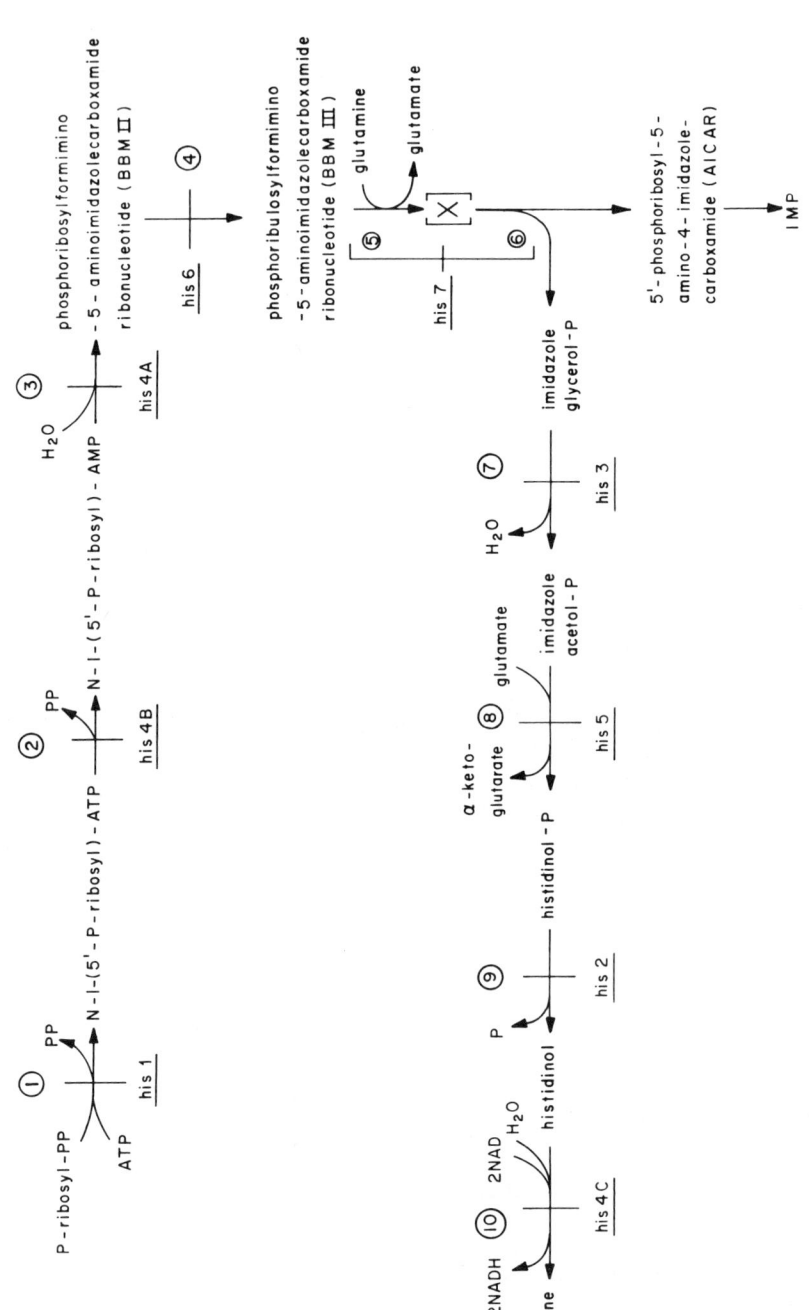

Figure 9 Biosynthesis of histidine. Genetic blocks indicated. (See Table 8.)

Table 8 Genes and enzymes for biosynthesis of histidine

Reaction	Enzyme	Locus	References for gene assignment
1	ATP PR-transferase (1-[5'-PR]-ATP:pyro-P PR-transferase, 2.4.2.17[a])	his1	Fink (1964)
2	PR-ATP pyrophosphohydrolase	his4B	Fink (1966)
3	PR-AMP cyclohydrolase (1-N-[5'-phospho-D-ribosyl]-AMP 1,6-hydrolase, 3.5.4.19)	his4A	Fink (1966)
4	BBMII isomerase[b] (N-[5'-phospho-D-ribosylformimino]-5-amino-1-[5''-PR]-4-imidazolecarboxamide ketolisomerase, 5.3.1.16)	his6	E. Jones (unpubl.)
5–6	glutamine amidotransferase catalyzing conversion of BBMIII to imidazoleglycerol-P	his7	E. Jones (unpubl.)
7	imidazoleglycerol-P dehydratase (D-erythro-imidazoleglycerol-P hydro-lyase, 4.2.1.19)	his3	Fink (1964)
8	histidinol-P aminotransferase (L-histidinol-P:2-oxoglutarate aminotransferase, 2.6.1.9)	his5	Fink (1964)
9	histidinol-P phosphatase (L-histidinol-P P-hydrolase, 3.1.3.15)	his2	Fink (1964)
10	histidinol dehydrogenase (L-histidinol: NAD oxidoreductase, 1.1.1.23)	his4C	Fink (1966)

[a]Enzyme Commission number.
[b]Not the IUPAC-IUB recommended name.

Repression by arginine of synthesis of the gene products of the *ARG2, ARG3, ARG5, ARG6, ARG8,* and *ARG1,10* loci depends on the integrity of the *ARG80 (ARGRI), ARG81 (ARGRII),* and *ARG82 (ARGRIII)* genes. Mutations in *ARG80, ARG81,* and *ARG82* are recessive and render synthesis of these six gene products nonrepressible by arginine. In addition, enzyme levels on minimal medium are elevated twofold compared with wild type. Multiply mutant strains, bearing mutations at two or even all three of these regulatory loci have enzyme levels indistinguishable from those of single mutants. The *arg80, arg81,* and *arg82* mutations eliminate repression for only these six genes. They also lead to noninducibility of the arginine-degradative enzymes arginase and ornithine aminotransferase (Bechet et al. 1970; Minet 1971; Wiame 1971b; Messenguy 1976; Jacobs et al. 1980).

Repression by arginine of the synthesis of the subunit of the arginine-specific carbamoylphosphate synthetase encoded by *CPA1* is dependent on the integrity of *CPA81*. Mutations in this gene are recessive and reduce substantially the repression of *CPA1*-gene product by arginine. Gene expression from *CPA1* and *CPA2* is somewhat elevated in the mutants (Piérard et al. 1979).

Limited information is available on a mutant, B6, isolated as resistant to ethionine, which appears to be affected in regulation of two enzymes common to threonine and methionine biosynthesis, namely aspartokinase and aspartate semialdehyde dehydrogenase, which are apparently regulated by threonine levels (de Robichon-Szulmajster 1967; Surdin 1967). In the B6 mutant, aspartokinase is no longer repressible by threonine and the dehydrogenase is no longer inducible by methionine. The phenotype is recessive. Levels of the third enzyme of the common pathway are unaffected by the mutation (de Robichon-Szulmajster et al. 1965; de Robichon-Szulmajster 1967).

Repression of enzymes of sulfate reduction and acetylation and sulfhydrylation of homoserine depends on the integrity of the *ETH2 (SAM2), ETH3,* and *ETH10 (SAM1)* genes. Mutations in the *ETH2* and *ETH10* genes are recessive and result in enzyme synthesis that is nonrepressible by methionine but normally repressible by SAM (Cherest et al. 1969, 1971, 1973b; Ferro and Spence 1973). It seems clear, however, that *ETH2 (SAM2)* and *ETH10 (SAM1)* are not regulatory genes per se. They encode isozymes for SAM synthetase (Cherest et al. 1978). Nonrepressibility of methionine enzymes in *eth2* and *eth10* mutants seems to be a metabolic artifact, resulting from inability of the cells to produce levels of SAM sufficient for repression. In the *eth3* mutants, however, neither methionine nor SAM effectively represses synthesis (Cherest et al. 1973b). The *ETH3* gene remains a candidate for a *trans*-acting regulatory gene.

Cis-dominant Regulatory Elements

Cis-dominant mutations that lead to elevated enzyme levels for their contiguous structural genes have been reported for genes of arginine, phenylalanine-tyrosine, and isoleucine biosynthesis. The structural genes whose expression is affected are *CPA1, ARG3, ARG6-ARG5, ARO7,* and *ILV1*. In none of these cases have mapping studies or DNA sequence analysis progressed to the point where mutation in a structural gene can be excluded.

Several *cis*-acting mutations affecting synthesis of enzymes of arginine biosynthesis have been described. Each is *cis*-dominant, is closely linked to the affected structural gene(s), is operative in **a**/α diploids (see the following text on Ty insertions), and leaves the response to cross-pathway control intact.

CPA80 mutations map close to the *CPA1* gene and render synthesis of the small subunit of the arginine-specific carbamoylphosphate synthetase virtually nonrepressible by arginine. Levels of gene product produced under nonrepressing conditions are about the same in the mutant and the wild type (Piérard et al. 1979). A mutation, designated O^-, which maps to *ARG3*, renders synthesis of OTCase virtually nonrepressible by arginine. Levels of the enzyme produced under nonrepressing conditions are about twofold greater than wild-type levels in the same medium and quite comparable to the levels found in the *arg80, arg81,* and *arg82* mutants. A mutation, called O^c, which maps to the *ARG6-ARG5* locus, renders synthesis of acetylglutamate kinase and acetylglutamylphosphate reductase nonrepressible by arginine. Levels of both enzymatic activities are elevated twofold under nonrepressing conditions, or about comparable to levels observed in *arg80, arg81,* and *arg82* mutants (Jacobs et al. 1980). Kinetic parameters of the kinase and reductase appear normal in the O^c mutant.

Several *cis*-dominant mutations that lead to elevated levels of chorismate mutase have been described (Kradolfer et al. 1977). They map to the *ARO7* locus. However, these mutations are known to alter the structure of the enzyme such that it is fully active in the absence of its activator, tryptophan, and no longer sensitive to feedback inhibition by tyrosine. In regulation phenotype, these mutations mimic operator constitutive mutations. Only the kinetic analyses revealed how misleading this phenotype was.

Several *cis*-dominant mutations have been reported to result in elevated levels of threonine deaminase, which are not repressible by isoleucine. The mutations, designated $ilv1\text{-}OP^c$, map to the structural gene *ILV1*. The levels of threonine deaminase in these mutants are quite high and, indeed, severalfold higher than those found in auxotrophs starving for isoleucine (Bollon 1980). Threonine deaminase shows complicated kinetic properties (de Robichon-Szulmajster and Magee 1968). The possibility must be entertained that these $ilv1\text{-}OP^c$ mutations are analogous to the *aro7* mutations described by Kradolfer et al. (1977) and discussed above. The kinetic data reported by Bollon (1980) do not permit exclusion of this possibility. An alternative possibility is that the $ilv1\text{-}OP^c$ mutations are Ty insertions (see below).

Cis-dominant regulatory mutations that lower transcription are rare in yeast. This scarcity could be related to the difficulty in the identification of mutations of this type by classical genetic and biochemical techniques. A promoter mutation that destroys the expression of a gene is not phenotypically different from nonsense or frameshift mutations located early within that gene. Moreover, genetic fine-structure analysis does not have the resolution to delineate the coding and noncoding regions. Identification of regulatory mutations defective in transcription therefore requires a cloned segment of the gene to measure the mRNA levels by hybridization and to localize the mutation by DNA sequence analysis. In addition, the availability of the cloned gene permits the creation of mutations by in vitro mutagenesis procedures. These recombinant DNA procedures have been used to identify *cis*-dominant regulatory mutations at *HIS4* and *HIS3*.

Although all of 1400 mutagen-induced *his4* mutations are structural, 2 of the 20 spontaneous *his4* mutations are regulatory. These mutations, *his4-912* and *his4-917*, map at the 5' end of the *HIS4* gene outside all mutations of the structural gene. DNA from the *his4-912* and *his4-917* mutants and their wild-type parent was analyzed by Southern hybridization using the wild-type *HIS4* gene as probe (Fink et al. 1980; Roeder and Fink 1980; Roeder et al. 1980). This analysis indicates that both mutations result from insertion of 6000 bp of foreign DNA into the *HIS4* region. Studies of the cloned *his4-912* insertion element demonstrate that this element shares extensive homology with Ty1, a repeated yeast DNA sequence characterized by Cameron et al. (1979). The *his4-917* insertion element is homologous to Ty1 for only one third of its length. The remaining two thirds, which fails to hybridize to Ty1, represents another class of yeast repetitive DNA (Roeder et al. 1980).

Proof that the *his4-912* and *his4-917* mutations are in fact regulatory mutations comes from both DNA sequence analysis and transcription studies. The wild-type *HIS4* region has been cloned and sequenced. The position of the *HIS4* coding sequence within this region could be determined, since the amino acid sequence at the amino terminus of the *HIS4* protein was known. That the assigned reading frame was correct was confirmed by DNA sequence analysis of several *HIS4* structural gene mutations (T.F. Donahue et al., in prep.). Sequencing of DNA from the *his4-912* and *his4-917* mutations demonstrated that these mutations lie outside the *HIS4* coding sequence at the 5' end of the gene (Farabaugh and Fink 1980; Roeder et al. 1980). Recent studies have shown that no *HIS4* mRNA is produced in the *his4-912* and *his4-917* mutant strains, indicating that these mutations block gene expression at the transcriptional level (S. Silverman and G. Fink, unpubl.).

Ty insertions are found frequently when regulatory mutations are sought. *Cis*-acting regulatory mutations resulting from the insertion of Ty elements have also been found at the *ADR2* and *CYC7* loci. Seven out of nine *cis*-acting *ADR2* mutations leading to constitutive derepression of alcohol dehydrogenase II result from the insertion of Ty elements into the *ADR2* 5'-noncoding region (Williamson et al. 1981). Of three *CYC7* mutations leading to constitutive overproduction of iso-2-cytochrome *c*, one mutation, *CYC7H2*, results from insertion of a Ty element into the *CYC7* regulation region (Errede et al. 1980). In all of the *ADR2* constitutive mutations and the *CYC7H2* mutation, the Ty element is inserted in the same orientation with respect to the structural gene coding sequence. This orientation is opposite to that of the *his4-912* insertion element. The properties of the *HIS4*, *ADR2*, and *CYC7* insertion mutations are summarized in Table 9. These observations could be interpreted to mean that Ty elements enhance gene expression when inserted in one direction and abolish gene expression when inserted in the other direction. However, the orientation of a Ty element is clearly not the only factor affecting the transcription of the adjacent gene. The structure of the particular Ty, the precise location of the Ty with respect to the start of the adjacent coding sequence, and the nature of the normal regulatory mechanism (positive or negative) could be factors in determining the effect of Ty on transcription. For example, the Ty element in *his4-917*, which is inserted in the same orientation as the *ADR2* and *CYC7* elements, abolishes *HIS4* expression. The *his4-917* insertion element differs from most other Ty elements by a substitution of 4000 bp.

A striking feature of the *CYC7H2* mutation is the dependence of iso-2-cytochrome *c* overproduction upon homozygosity at the mating-type locus (Errede et al. 1980; Rothstein and Sherman 1980). Overproduction occurs in haploid strains and in diploid strains that are homozygous at the mating-type locus. However, this overproduction does not occur in diploid strains that carry both *MATa* and *MATα* alleles. This result suggests that the inserted Ty element is sensitive to mating-type control. *Cis*-acting constitutive mutations at the *DUR1-DUR2* (Lemoine et al. 1978), *CARGA*, and *CARGB* (Deschamps and Wiame 1979) loci are also subject to mating-type control and are thus thought to result from mutational alterations similar to the *CYC7H2* insertion mutation. In the case of the $DURO^h$ mutations, Southern hybridization analysis using the wild-type gene as probe indicates that the constitutive mutations are in fact insertion mutations (M. Crabeel et al., pers. comm.). One striking feature of MAT control over these genes is that transcription of the Ty element in all cases is in the direction opposite to that of the adjacent gene. However, not all Ty elements in this orientation place the adjacent gene under MAT control (Roeder and Fink 1982).

Most of the *cis*-acting regulatory mutations obtained by classical means are either Ty insertions or gross chromosomal aberrations. *Cis*-dominant regulatory mutations that lower transcription of *HIS3* (Struhl 1979, 1981; Sherer and Davis 1980; Struhl and Davis 1980) and *HIS4* (T.F. Donahue et al., in prep.) have been made in vitro. These mutations result from extensive deletion of sequences in the 5'-noncoding region adjacent to the structural gene. Presumably, mutations of this type have not been uncovered in standard mutant hunts because in vivo deletions are rare in yeast. Taken together, these observations suggest that simple point mutations do not

Table 9 Cis-acting regulatory mutations resulting from the insertion of yeast transposable elements

Mutation	Type of mutation	Distance from start of coding sequence (bp)	Orientation of element	Reference
CYC7-H2	constitutive overproduction	140–270	Ty transcription proceeds away from CYC7 gene	Errede et al. (1980); Rothstein and Sherman (1979)
ADR2-1ᶜ, ADR2-2ᶜ, ADR2-3ᶜ, ADR2-6ᶜ, ADR2-7ᶜ, ADR2-8ᶜ, ADR2-9ᶜ, his4-912	constitutive overproduction	50–150	Ty transcription proceeds away from ADR2 gene	Williamson et al. (1981)
	promoter negative	161	Ty transcription proceeds toward HIS4 gene	Farabaugh and Fink (1980); Roeder and Fink (1980); Roeder et al. (1980)
his4-917	promoter negative	63	Ty transcription proceeds away from HIS4 gene	Roeder et al. (1980)

result in regulatory defects of sufficient magnitude to yield a mutant phenotype. One possible explanation is that the regulatory signals are repeated in front of these genes, making it unlikely that the change of a single base pair will have a large effect.

General Control Systems
Cross-pathway Control

Characteristics. Many of the enzymes of amino acid biosynthesis are under a complex general control (Schürch et al. 1974; Delforge et al. 1975; Wolfner et al. 1975). Starvation for histidine leads not only to elevated levels of histidine biosynthetic enzymes but also to elevated levels of enzymes of tryptophan, arginine, lysine, and isoleucine-valine biosynthesis. Starvation for a number of amino acids other than histidine will lead to derepression of the same spectrum of enzymes of several pathways. On the basis of the properties of mutants, Wolfner et al. (1975) have inferred that enzymes of methionine, cysteine, tyrosine, phenylalanine, and proline biosynthesis are likely to be under the general control system. As it is difficult to be sure that the resistances or sensitivities of mutants reflect changes in enzyme levels, rather than changes in pool concentrations or intracellular distributions of pools brought about by changes in the fluxes of compounds of connected pathways, this inference must be treated cautiously. (An obvious example is that elevated levels of tryptophan in derepressed mutants [Fantes et al. 1976] could increase flux of chorismate into the phenylalanine-tyrosine arm by activating chorismate mutase. Increased pools of phenylalanine and tyrosine and consequent analog resistance might result. Similiar arguments could be made for the interconnected methionine and cysteine pools or arginine and proline pools.) A compilation of pathways and enzymes that have been examined for this cross-pathway response and the results found are in Table 10; references are in the table.

To date, enzymes of the pathways for synthesis of tryptophan, arginine, histidine, lysine, glutamine, glutamate, isoleucine-valine, leucine, and methionine have been examined. All of the enzymes of arginine biosynthesis are under cross-pathway control as are all of the enzymes of histidine and lysine biosynthesis so far examined. For the leucine pathway, only the first enzyme of the pathway, α-IPM synthase, is subject to cross-pathway control (Hsu et al. 1982). For the tryptophan pathway, all enzymes except phosphoribosylanthranilate isomerase are under the general control. Conflicting data are available for the isomerase, but dosage analysis of tetraploids and the properties of constitutive mutants tend to exclude the isomerase from participation (Schürch et al. 1974; Miozzari et al. 1978). Within the isoleucine-valine pathway, transaminase B is under general control, but AHAS is not. As AHAS shows the largest derepression response of any *ilv* enzyme, an

Table 10 Pathways and enzymes examined for cross-pathway regulation of amino acid biosynthesis

Pathway	Enzyme	genetic locus	Cross-pathway control	references[a]
Trp	Anthranilate synthase	trp2	+	1,2,3,4
		trp3		
	Anthranilate PR-transferase	trp4	+	4
	PR-anthranilate isomerase	trp1	−	4 (but see 1)
	indoleglycerol-P synthase	trp3	+	1,4
	tryptophan synthase	trp5	+	4
Arg	acetylglutamate synthase	arg2	+	5,6
	acetylglutamate kinase	arg6	+	5,7
	acetylglutamyl-P reductase	arg5	+	5,7
	acetylornithine aminotransferase	arg8	+	1,5
	acetylornithine acetyltransferase	arg7	+	5,6
	OTCase	arg3	+	1,2,3,5
	argininosuccinate synthetase	arg1	+	2,5
	argininosuccinate lyase	arg4	+	2,5
	carbamoyl-P synthetase	cpa1	+	8
		cpa2	+	8
His	ATP PR-transferase	his1	+	2,3
	PR-AMP cyclohydrolase	his4A	+	3
	imidazoleglycerol-P dehydratase	his3	+	9
	histidinol-P aminotransferase	his5	+	1
	histidinol-P phosphatase	his2	+	1,3
	histidinol dehydrogenase	his4C	+	1,3

Lys	saccharopine dehydrogenase	*lys1*	+	2
	α-aminoadipate reductase	*lys2*	+	3
		lys5		
Gln	glutamine synthetase	*gln1*	–	3
Ilv	transaminase B	*ilv2*	+	2
	AHAS		–	3
Glt	NADP-GDH (anabolic)	*gdh1*	–	2,3
Leu	α-IPM synthase	*leu4*	+	10
		leu5		
	α-IPM isomerase	*leu1*	–	10
	β-IPM dehydrogenase	*leu2*	–	10
Met	sulfite reductase	*met10*	–	3
		met5		
		met18		
		met19		
		met20		
Miscellaneous				
	fumarose		+	2
	NAD-GDH (catabolic)		+	2

[a] (1) Schürch et al. (1974); (2) Delforge et al. (1975); (3) Wolfner et al. (1975); (4) Miozzari et al. (1978); (5) Messenguy (1979); (6) Wipf and Leisinger (1979); (7) Jacobs et al. (1980); (8) Piérard et al. (1979); (9) Struhl (1979); (10) Hsu et al. (1982).

isoleucine-valine-specific control is implied (Magee and Hereford 1969). The only enzyme of methionine biosynthesis examined for this response, namely sulfite reductase, is not regulated by this general control. However, aspartokinase, which catalyzes the first step in the common threonine-methionine pathway, is a good candidate for inclusion within this system. It is repressed by threonine but not by methionine, but is derepressed upon methionine starvation (de Robichon-Szulmajster and Corrivaux 1963, 1964; Cherest et al. 1969). Neither of the main enzymes for ammonia assimilation, namely NADP-GDH and glutamine synthetase, are subject to cross-pathway control. Peculiarly enough, the catabolic NAD-GDH and fumarase respond to the cross-pathway derepression signals.

That cross-pathway control may be mediated by levels of charged tRNAs is suggested by experiments involving mutants with thermosensitive aminoacyl-tRNA synthetases. If an *ils1-1* mutant, bearing a thermosensitive isoleucyl-tRNA synthetase, is incubated at the restrictive temperature, enzymes of arginine, histidine, and isoleucine-valine biosynthesis are derepressed. At this temperature, the amino acid pools are not lowered, but the isoleucyl tRNA is poorly charged (Messenguy and Delforge 1976). Similarly, incubation of a methionyl-tRNA synthetase mutant, *mes1-1*, at temperatures at which the methionyl tRNA is poorly charged results in derepression of aspartokinase and NAD-GDH (Cherest et al. 1969).

The increased specific activities of biosynthetic enzymes upon amino acid starvation appear to result from increased amounts of enzyme protein (Piérard et al. 1979; R. Bigelis and G. Fink, unpubl.). Derepression of enzymes is noncoordinate, even when cross-pathway control seems to be the only control regulating synthesis of that polypeptide. In the tryptophan pathway, the gene products of *TRP2* and *TRP3* derepress 2.5-fold, whereas those of *TRP4* and *TRP5* derepress 1.5-fold. In the histidine pathway, the products of *HIS1* and *HIS4A* derepress 4-5-fold; that of *HIS2* about 2-fold. In the arginine pathway, where the full range of derepression has been examined, the *CPA2*-gene product is capable of a 64-fold derepression (Piérard et al. 1979), whereas the *ARG4*-gene product can derepress 26-fold (Messenguy 1979). With the possible exception of *TRP2,* all of the above genes are controlled solely by the cross-pathway system, so far as is known.

Recent experiments using cloned genes to measure levels of specific mRNAs show that amino acid starvation increases mRNA levels 5-10-fold for the *HIS3, HIS4,* and *ARG3* genes (Struhl 1979; Silverman et al. 1982; F. Messenguy, pers. comm.). Thus, cross-pathway control seems to regulate the expression of unlinked genes of different pathways at the level of transcription.

In certain cases, cross-pathway and pathway-specific controls may interact to temper the response of enzymes for a specific pathway to signals originating external to that pathway. The best example involves the *CPA1* and *CPA2* genes, whose products aggregate to form the arginine-specific carbamoylphosphate synthetase. Starvation for an amino acid other than arginine leads to derepression of the *CPA2*-gene product, but the rise in the arginine pools attendant to this starvation works to repress synthesis of the *CPA1*-gene product. The specific repression tends to override the cross-pathway-triggered derepression. The extent of derepression of carbamoylphosphate synthetase *activity* is the outcome of these competing responses (Piérard et al. 1979).

Mutations that Affect the Cross-pathway Control System(s). A compilation of mutations known to affect enzymes subject to cross-pathway regulation is in Table 11. Mutations that affect cross-pathway regulation are generally of two classes: those that result in constitutive levels of the enzymes, and those that prevent derepression. Schürch et al. (1974) described 5MT-sensitive mutants, which are unable to derepress enzymes of tryptophan, histidine, and arginine biosynthesis upon starvation for histidine or tryptophan (induced by growth in 3AT or 5MT). These mutants are designated Ndr (for *nonderepressible*) and include members of at least two complementation groups (RH 487 and RH 376 are examples). The *aas1*, *aas2*, and *aas3* mutants of Wolfner et al. (1975) show comparable phenotypes, are sensitive to a number of amino acid analogs, and affect enzymes of lysine biosynthesis as well. Recently, the mutation in RH 487, *ndr2-1*, has

Table 11 Mutations that affect cross-pathway regulation

Designation	Phenotype[a]	Pathways of affected enzymes[a]					Comments	References[b]
		Trp	His	Arg	Lys	Leu		
RH 487 (*ndr2-1*)	Ndr	Ndr[c]	Ndr[d]	Ndr[d]	−	−	allelic to *aas1*	1,6
RH 376	Ndr	Ndr[c]	Ndr[d]	−	−	−		1
RH 375 (*ndr1-1*)	Ndr	Ndr[c]	Dr[d]	−	−	−		1,6
RH 558	Cdr	Cdr	−	−	−	−	no Cdc⁻ phenotype	2
aas1	Ndr	Ndr[d]	Ndr[d]	Ndr[d]	Ndr[d]	−	*aas2* and *aas3* similar	3
tra3	Cdr	Cdr	Cdr	Cdr	Cdr	−	Cdc⁻ (ts)[e] phenotype; *tra5* similar	3
tfl	Cdr	Cdr	−	−	−	Cdr		4
gen[c]	Ind	−	−	Ind	−	−		5

[a]Ndr, nonderepressible; Cdr, constitutively derepressed; Dr, derepressible; Ind, enzymes induced by growth in rich media; −, not tested.
[b](1) Schürch et al. (1974); (2) Miozzari et al. (1978); (3) Wolfner et al. (1975); (4) Roeder (1980); (5) Messenguy (1979); (6) Niederberger et al. (1981).
[c]Both 5MT and 3AT used to derepress.
[d]3AT used to derepress.
[e]ts, temperature sensitive.

been reported to be allelic to an *aas1* mutation (Niederberger et al. 1981). Other allelism tests have not been reported. One of the mutants reported by Schürch et al. (1974), RH 375, has lost the capacity to derepress the tryptophan enzymes in response to tryptophan or histidine starvation but retains the ability to derepress the histidine enzymes in response to histidine starvation (the effect of tryptophan starvation was not reported). Whether the mutation inactivates a gene product that intermediates response of the tryptophan enzymes to the cross-pathway control system is an interesting question.

The *tra3* and *tra5* mutants, designated Cdr (for *c*onstitutively *d*erepressed), selected as triazolealanine-resistant, seem to be the complementary class to the *aas* mutants and RH 376, for they lead to derepressed levels of the same enzymes that are nonderepressible in *aas* mutants and resistance to the same analogs to which *aas* mutants are sensitive. The *tra3* mutants are temperature-sensitive for growth and arrest in G_1 of the cell cycle at restrictive temperatures (Wolfner et al. 1975). Possibly, the *TRA3* gene encodes a product involved in monitoring the level of required precursors for protein synthesis so that cells will not embark upon a cell cycle unless they have sufficient precursors to complete the cycle (Unger and Hartwell 1976). It seems likely that the mutations that lead to resistance to trifluoroleucine and other analogs and result in elevated levels of all three leucine biosynthetic enzymes (the last two probably as a result of internal induction by α-IPM), as well as two tryptophan enzymes, are also of this class, although temperature sensitivity was not reported as part of the phenotype (Roeder 1980). Whether the mutation reported by Miozzari et al. (1978) is also of this group is uncertain, for, although it leads to elevated levels of tryptophan enzymes, enzymes of other pathways were not examined. This mutant, RH 558, differs from *tra3* mutants in that it does not result in a *cdc* phenotype at elevated temperatures.

The *gen*^c mutation described by Messenguy (1979) also appears to affect a general control system. In the *gen*^c mutant, levels of enzymes appear to be induced when cells are grown in rich medium.

Caution should be used in interpreting any mutation as regulatory, simply because it gives elevated levels of amino acid biosynthetic enzymes. A leaky mutation affecting any biosynthetic pathway or an aminoacyl tRNA synthetase mutation could result in this phenotype.

In all cases involving cross-pathway control where analysis has been made, the mutations leading to constitutive synthesis are epistatic to mutations leading to nonderepressible synthesis (Wolfner et al. 1975; Miozzari et al. 1978; Messenguy 1979). This is true whether the mutations leading to derepressed synthesis are part of the general control system (*tra3*, RH 558, *gen*^c) or a pathway-specific regulatory system (*arg81*). Thus, *tra3 aas1* double mutants show derepressed enzyme levels, are resistant to analogs, and are temperature sensitive for growth (Wolfner et al. 1975). *gen*^c *aas* double mutants show high enzyme levels on rich media, and *arg81 aas* double mutants show derepressed enzyme levels (Messenguy 1979). Apparently, in the absence of gene products that act negatively (*ARG81, TRA2, GEN^c*), the products of genes that act positively (*AAS1, AAS2, AAS3*) are not required.

The coincident control of a large number of genes involved in amino acid biosynthesis suggests a signal common to all of the genes involved in this control. Recent experiments using the cloned genes coding for *HIS3* (Struhl 1979), *ARG3* (F. Messenguy, unpubl.), and *HIS4* (Silverman et al. 1982) have shown that the general control acts at the level of transcription. In each case, the steady-state levels of mRNA were measured by hybridization of the structural gene to mRNA immobilized on filters. The beginning of the *HIS4* gene has been fused to the structural gene for *E. coli* β-galactosidase (Silverman et al. 1982). This fusion, which contains *HIS4* DNA from −732 to +30 relative to the AUG that initiates *HIS4* translation, places the β-galactosidase under the general control. This result suggests that all the sequences required for the general control reside in the 5′-noncoding region adjacent to *HIS4*. The simplest assumption is that there is a similar nucleotide sequence in the 5′-noncoding regions of all of the genes under the general control and that this sequence is the recognition site for the molecules that mediate the control.

One can test the common regulatory sequence hypothesis directly by comparing the DNA sequence of genes under the general control. The 5′-noncoding regions of the *HIS3* and *HIS4* genes are available for such comparison. These genes may be expected to have two regions of homologous sequence, a region recognized by the *AAS-TRA* general control system and a region recognized by RNA polymerase. To identify these regions, we have compared the 5′-noncoding regions of *HIS3* and *HIS4* with each other and with the 5′-noncoding regions of the *CYC1* (Smith et al. 1979) and *GAP491* (Holland and Holland 1979) genes. These latter two genes, which encode iso-1-cytochrome *c* and glyceraldehyde-phosphate dehydrogenase, respectively, are expected to have an RNA polymerase interaction site but no *AAS-TRA* recognition site. To look for biologically relevant homologies, we aligned the four DNA sequences with respect to the initiation site of RNA synthesis. The sites of RNA initiation of the *HIS3* (Struhl 1979), *HIS4* (Donahue et al. 1982), *CYC1* (Szostak et al. 1977), and *GAP491* (M.J. Holland, pers. comm.) genes have been mapped. We have identified the most probable site for RNA initiation in each case using two additional criteria. First, two oligonucleotide sequences are commonly found in the region of the 5′ side of the initiation site. An oligonucleotide similar to the sequence 5′-TATAAATA-3′ (the "Hogness box") lies approximately 40 bp from the initiation site (Gannon et al. 1979), and a sequence similar to 5′-CATTA-3′ (the "cap box") lies within several base pairs of the 5′ end of the transcript (Sures et al. 1978). Second, since eukaryotic protein synthesis initiates at the first AUG in a transcript (Kozak 1981), the site of initiation was chosen such that the first AUG in the putative transcript is the initial AUG of the structural gene.

The regions on the 5' side of these sites of RNA initiation show slight sequence homology (Fig. 10). Both *HIS3* and *HIS4* have sequences similar to the Hogness box over 30 bp from the RNA initiation site (in *HIS3*-5'-TATATAA-3' and in *HIS4*-5'-TATATTCA-3'). These genes also have a potential cap box sequence 5–6 bp from the initiation site (5'-CTAAA-3' in *HIS3* and 5'-CATAA-3' in *HIS4*). Two genes that are not under the general control also include such sequences—*CYC1* with the sequences 5'-TATA-TAAA-3' and 5'-CTAAA-3', and *GAP491* with the sequences 5'-TATA-AAGA-3' and 5'-GTAAA-3'. Other than these two oligonucleotides, the regions between the Hogness box and the initial nucleotides of these genes are similar only in that they have a high A + T content, averaging approximately 70%, or slightly higher than the average for yeast (65%).

Since the sequences common to *HIS3* and *HIS4* near the site of initiation of RNA synthesis are also found in this region adjacent to genes not under the general control, we have searched for homologies to the 5' side of the Hogness box. When the sequences in this segment are compared, only small regions of homology are found. The largest regions are a run of seven Ts and the nearby 6-base sequence 5-TGACTC-3' (Fig. 11). The regions of homology are so short that it is difficult to ascribe any significance to them by simple inspection.

Deletion analysis of the 5'-noncoding region of the *HIS4* gene has revealed that the 5'-TGACTC-3' sequence is important for regulation (T.F. Donahue et al., in prep.). Deletions were made in vitro and inserted back into the chromosome by yeast transformation. The deletions all start in the 5'-noncoding region, 588 bp upstream from the site of transcription initiation and delete sequences between this site and the start of *HIS4* transcription. All deletions show normal regulation of *HIS4* except for those that delete the sequence 5'-TGACTC-3' identified in Figure 11. One deletion that

HIS4   TATATTCACCTCCGATGTGTGTTGTACATACATAAAAATA

HIS3   TATATAAAGTAATGTGATTTCTTCGAAGAATATACTAAAAATGA

CYC1   TATATAAAACTCTTGTTTTCTTCTTTTCTCTAAATA

GAP491 TATAAAGAACGGTAGGTATTGATTGTAATTCTGTAAATCTA

Figure 10 The putative promoter regions of the *HIS4, HIS3, CYC1,* and *GAP491* genes. The sequences have been aligned with respect to the site of transcriptional initiation, which is marked by an asterisk. The putative RNA initiation site lies to the 5' side of the initial ATG of the gene, a distance of 63 nucleotides in *HIS4*, 41 nucleotides in *HIS3*, 88 nucleotides in *CYC1*, and 101 nucleotides in *GAP491*.

```
              ^
HIS4  TTAATTAATTGCTAAACCCAT-GCACAGTGACTCACGTTTTTTTATCAGTC-ATT ACTG-TGT-ATATAATAGAGAGGGAACGTTATATTCA
                            ******
HIS3  TTCATTTTTTTTTTTCCCC-TAGCGGA-TGACTC-TTTTTTTTTCTTAG-CGATTGGC-ATTATCACATAAT-GA-ATTATACATTATATAAA
```

Figure 11 The regions to the 5' side of the *HIS4* and *HIS3* Hogness box. The side-by-side format is presented so that a visual comparison can be made. To obtain even this comparison, we had to loop out a 44-bp region of *HIS4*. The position of this region is indicated by the carat. Base-pair gaps are indicated by dashes. The Hogness box in each sequence is underlined at the extreme right. The asterisks represent the sequence at *HIS4* and *HIS3* that appears to be required for regulation.

removes the 5'-T of this sequence is transcribed but is no longer regulated. Revertants that restore the 5'-TGACTC-3' sequence restore regulation.

This short sequence is a likely candidate for the common regulatory sequence involved in general control. This sequence is not only present in front of *HIS4* but is also present in front of *TRP5* (Zalkin and Yanofsky 1982), *HIS3* (Struhl 1981), and *HIS1* (A. Hinnebusch and G.W. Fink, in prep.)—all genes under the general control. The sequence is not present in front of *CYC1* or *GAP491*—genes not under general control. The role of this sequence in the regulation of genes other than *HIS4* awaits a similar mutational analysis.

A good start has been made in delineating the outlines of the general control system. A number of questions remain. Is it restricted to amino acid biosynthesis? Which pathways are under cross-pathway control? Where some enzymes are under cross-pathway control and some are not, is there a pattern to participation? Are enzymes catalyzing reactions involved in transformation of the carbon skeleton for methionine biosynthesis subject to general control but those involved in sulfate reduction not subject to this control, for example? Do the molecules that mediate the general control exert this control by interaction at specific sites on the DNA adjacent to the genes they control? Are there other instances where interactions between cross-pathway control and pathway-specific control temper the response of a pathway to signals external to that pathway? Does the intracellular location of an enzyme have any bearing on what control an enzyme will be subject to? For arginine and leucine biosynthesis, the answer would appear to be negative. Is flux of amino acids into and out of the vacuole related to this control? What is the relation between the cell cycle and cross-pathway control? Is the repressive effect of YEPD exerted through the cross-pathway control system or, as the evidence suggests, is this exerted through a different circuit?

Glucose Repression and CoA

There are five amino acid biosynthetic pathways that begin with condensation of acetyl-CoA with a receptor for the acetyl group, resulting in production of an acetylated compound and CoA. The five pathways are arginine, leucine, lysine, cysteine, and methionine (the methionine-specific branch begins with the condensation). The reactions catalyzed can be subdivided into those that consume acetyl-CoA (catalyzed by α-IPM synthase, leucine; homocitrate synthase, lysine; serine acetyltransferase, cysteine; homoserine acetyltransferase, methionine) and one that does not consume acetyl-CoA but rather fulfills an anaplerotic function, since the product acetylglutamate is part of a cycle and need be replenished only upon cell division or derepression (acetylglutamate synthase, arginine). Three of the enzymes catalyzing these reactions, namely homocitrate synthase, α-IPM synthase, and acetylglutamate synthase, have been investigated in some detail. All three are located in the mitochondrion (Ryan et al. 1973; Tracy and Kohlhaw 1975; Wipf and Leisinger 1977) as are the enzymes of the tricarboxylic acid cycle, through which acetyl-CoA is consumed during aerobic energy production (Linnane and Still 1955; Vitols and Linnane 1961). Acetylglutamate synthase, the anaplerotic enzyme, is inhibited but not inactivated by CoA, and the inhibition is additive with that of the end product arginine (Wipf and Leisinger 1979). Homocitrate synthase and α-IPM synthase, which consume acetyl-CoA, are reversibly inactivated by CoA in a time-dependent fashion. The inactivations are antagonized by substrate or by the pathway end product (lysine or leucine, respectively). Citrate synthase, which consumes acetyl-CoA but is necessary for energy production, is not subject to inactivation by CoA (Tracy and Kohlhaw 1975). As suggested by Tracy and Kohlhaw (1975), CoA, through its effects on these enzymes, could reduce or prevent utilization of acetyl-CoA for biosynthesis when the cell needs the acetyl-CoA for energy production. ATP activates the CoA-inactivated α-IPM synthase (Hampsey and Kohlhaw 1981), providing a means for increasing the flux to the biosynthetic pathway when energy needs have been met. One might also speculate that the dichotomy between anaplerotic and consuming reactions is significant. Since acetylglutamate synthase may rarely draw heavily on the acetyl-CoA pool because of its anaplerotic function, inhibition by CoA may provide sufficient control. For the enzymes that draw heavily on the pool, a more extended (in time) control might be needed and could be provided by the inactivations. Inactivation would only be necessary if three conditions are met: (1) levels of pathway end product are insufficient to inhibit the enzyme, (2) levels of enzyme substrates are low, and (3) ATP levels are low. The first would indicate a need for the end product; the last two a general deficit in carbon metabolism. If these three conditions were met, long-term shutdown for acetyl-CoA-consuming biosynthetic pathways would be in order and allow the cell to use acetyl-CoA for carbon metabolism and energy production.

The levels of α-IPM synthase and acetylglutamate synthase are controlled, at least in part, by the carbon source provided to cells and by whether the cells are in a respiratory mode. The levels are increased when

Amino Acid and Nucleotide Biosynthesis 259

cells are in a respiratory mode (Brown et al. 1975; Wipf and Leisinger 1979). Homocitrate synthase has not been examined for this response. Glucose repression of enzyme synthesis will also serve to keep flux through biosynthetic pathways in line with the cell's ability to maintain acetyl-CoA levels.

Two other biosynthetic enzymes, namely homoserine acetyltransferase and serine acetyltransferase of the methionine and cysteine pathways, respectively, catalyze the first steps in their respective pathways and consume acetyl-CoA. Of obvious interest is whether they, too, will prove to be located in the mitochondrion and be subject to similar regulations of their activities and biosyntheses. One final question raised by these observations is whether there is a *common* control system responsible for causing glucose repression of all target biosynthetic enzymes.

MAGNITUDE OF REPRESSION-DEREPRESSION RESPONSES FOR ENZYMES OF AMINO ACID BIOSYNTHESIS

In Table 12 a compilation of the repression-derepression responses for enzymes of amino acid biosynthesis is presented. If there is hard (+) or suggestive (+?) evidence that an enzyme is regulated by a pathway-specific or cross-pathway control system, the appropriate entry is included. The conditions in which the maximum and minimum enzyme levels are found are presented, as are the ratios generated from them. The data presented are not necessarily comparable between pathways or even between enzymes in the same pathway, for few enzymes have been examined for their levels at what one might expect to be extreme limits of their ranges, namely growth in YEPD and growth where cells are starved for the pathway end product(s). For some pathways, growth in YEPD might not result in fully repressed enzyme synthesis, for YEPD contains low levels of some amino acids such as tryptophan. The parent sections contain additional information.

The overall impression that one gains is that the levels of enzymes of certain pathways or portions of pathways can vary a great deal, whereas the levels of enzymes of other pathways may show only limited variations. In general, the enzymes of the glutamine-, arginine-, and methionine-specific pathways show large derepressions. Those of the aromatic and branched chain amino acid pathways show much smaller ranges.

A problem in studying regulation of a biosynthetic pathway in yeast is that the structural genes are unlinked and each may be separately regulated by a different set of signals as well as by different receptors for the same signals. In bacteria, where the genes for a given pathway are linked in an operon, all respond to the same metabolic signals mediated through the same receptors (operators and promoters). For this reason the analysis of one gene's regulation in *E. coli* is sufficient to provide information on the regulation of an entire pathway. In yeast, each gene of a biosynthetic pathway may be regulated differently.

Table 12 Maximum repression-derepression responses for enzymes of amino acid biosynthesis

Reaction		Locus	Subject to pathway-specific control[b]	Subject to cross-pathway control[c]	Ratio of highest level to lowest level at steady state	Growth conditions for low and high levels and comments[a]	
						low	high
3	glutamine synthetase	*gln1*	+?[1,2]	−[3]	150[1]	Gln	urea or proline
							(N source)
			Arginine pathway (Fig. 1)				
8	acetylglutamate synthase	*arg2*	+[4]	+[4]	52[4]	glucose + Arg	ethanol (diauxie)
9	acetylglutamate kinase	*arg6*	+[5-7]	+[7]	85[6]	+ Arg	− Arg
10	acetylglutamyl-P reductase	*arg5*	+[5-7]	+[7]	79[6]	+ Arg	− Arg
11	acetylornithine aminotransferase	*arg8*	+[5]	+[8]	2[8]	MV	− His (3AT)
12	acetylornithine acetyltransferase	*arg7*	−[5]	+[4]	4[4]	+ Leu + Arg	− Leu + Arg (auxo)
13	carbamoyl-P synthetase	*cpa1*	+[9,10]	+[10]	103[10]	YEPD	− Arg
13	carbamoyl-P synthetase	*cpa2*	−[9,10]	+[10]	64[10]	YEPD	− Arg
14	OTCase	*arg3*	+[5,11]	+[12]	186[13]	YEPD	− Arg
15	argininosuccinate synthetase	*arg1,10*	+[5]	+[13]	4[14]	+ Arg	MV
16	argininosuccinate lyase	*arg4*	−[5]	+[12]	26[13]	YEPD	− Arg
			Lysine pathway (Fig. 1)				
18	homocitrate synthase	*lys2*	+?[15]		~5[15]	+ Lys	MV
						one isozyme only is repressed	
24–26	α-aminoadipate reductase	*lys5*	−?[16]	+[3]	7[3]	MV	− His (3AT)

#	Enzyme	Gene					
27	saccharopine reductase	lys9, lys13, lys14	+?[16]		~5[16]	+Lys	MV
28	saccharopine dehydrogenase	lys1	−?[16]	+[12]	5[12]	MV	−Arg (auxo)

Aromatic pathway (Fig. 3)

#	Enzyme	Gene					
1	DAHP synthase	aro3, aro4	−?[17,18]		1[17,18]	+Phe,Trp,Tyr	MV
3	dehydroquinate dehydrogenase	aro1E	−?[19]		1[19]	+Phe,Trp,Tyr	MV
8	chorismate mutase	aro7			3[17]	MV	+Trp
9	prephenate dehydratase	pha2	+?		3[17]	+Phe	MV
11	prephenate dehydrogenase	tyr1			2[17]	MV	+Phe
13	anthranilate synthase	trp2	+?[20]	+[3,8,12,20]	9[19](31)[17,19]	MV	−Trp (9)
		trp3	−[20]	+[3,8,12,20]		+Trp	−Trp (31)
						(the repression in ref. 17 is not seen by others)
14	anthranilate PR-transferase	trp4	−[20]	+[20]	3[20]	MV (WT)[d]	MV (Cdr mutant)
15	PR-anthranilate isomerase	trp1	−[20]	−[20]	1	MV (WT)	MV (Cdr mutant)
16	indoleglycerol-P synthase	trp3	−[20,21]	+[3,8,12,20]	5[8]	MV	−Trp (5MT)
17	tryptophan synthase	trp5	−[20,21]	+[20]	3[20]	MV (WT)	MV (Cdr mutant)

Threonine-Methionine pathways (Fig. 6)

Common pathway

#	Enzyme	Gene					
3	aspartate kinase	hom3	+?Thr[22,23] −?Met[22,23]		21[22,23]	+Thr	−Met
4	aspartate semialdehyde dehydrogenase	hom2	unclear		5[24]	−Met	+Met
5	homoserine dehydrogenase	hom6	+?Met[23] −?Thr		5[23]	+Met	−Met

Table 12 (Continued)

Reaction		Locus	Subject to pathway-specific control[b]	Subject to cross-pathway control[c]	Ratio of highest level to lowest level at steady state	Growth conditions for low and high levels and comments[a]	
						low	high
Threonine-specific pathway (Fig. 6)							
6	homoserine kinase	thr1	unclear		~2[24]	MV	+Met
Methionine-specific pathway (Fig. 6)							
8	homoserine acetyltransferase	met2	+[25]		24[23]	+Met	−Met
9	homocysteine synthase	met17	+[26]		55[23]	+Met	−Met
		met25					
14	ATP sulfurylase	met3	+[26]		140[23]	+Met	−Met
17	sulfite reductase	met5	+[26]	−[3]	>250[23,27]	YEPD	−Met (250)
		met10				SAM	−Met∞
		met16					
		met19					
		met20					
23	SAM synthetase	sam1			}28[26,27]	+SAM	−Met
		sam2					
10	homocysteine methyltransferase	met6	+?[28]		3[28]	+Met	MV
Branched-chain amino acid pathway (Fig. 8)							
1	threonine deaminase	ilv1	+[29-31]		5[29]	+Ile	−Ile
					15		ilvOP[c]
2	AHAS	ilv2	+?[29,31,32]	−[3]	29[33]	+Ile,Val,Leu	−Ile+Val,Leu
3	acetohydroxy acid isomeroreductase	ilv4	+?[31,32]		4[33]	+Ile,Val,Leu	−Ile+Val,Leu
		ilv5					

4	dihydroxy acid dehydratase	ilv3	+,?[32]	—Ile+Val,Leu
5a	transaminase B		+,?[34]	—His (auxo)
6	α-IPM synthase	leu4	+,?[32]	acetate+Leu
		leu5		$MV(leu4^{fbr})^d$
7	α-IPM isomerase	leu1	—[35]	73[37]	MV+Leu, Thr(WT)	—His (auxo)
8	β-IPM dehydrogenase	leu2	—[35]	39[37]	MV+Leu, Thr(WT)	low Leu $(leu4^{fbr}leu1)^d$

Histidine pathway (Fig. 9)

1	ATP PR-transferase	his1	—?[3]	+[3,12]	5[38]	MV	—His (auxo)
3	PR-AMP cyclohydrolase	his4A	—?[3]	+[3]	18[37]	MV	—His (auxo)
7	imidazoleglycerol-P dehydratase	his3		+[39]	5–10[39]	MV	—His (3AT)
8	histidinol-P aminotransferase	his5		+[8]	2[8]	MV	—His (3AT)
9	histidinol-P phosphatase	his2	—?[3]	+[3,8]	2[8]	MV	—His (3AT)
10	histidinol dehydrogenase	his4C	—?[3]	+[3,8]	4[3]	MV	—His (auxo)

Citations are indicated by superscript numbers. [1]A. Mitchell (pers. comm.); [2]Dubois and Grenson (1974); [3]Wolfner et al. (1975); [4]Wipf and Leisinger (1979); [5]Wiame (1971b); [6]Minet et al. (1979); [7]Jacobs et al. (1980); [8]Schürch et al. (1974); [9]Thuriaux et al. (1972); [10]Piérard et al. (1979); [11]Bechet et al. (1970); [12]Delforge et al. (1975); [13]Messenguy (1979); [14]Messenguy (1976); [15]Tucci and Ceci (1972a); [16]Sinha et al. (1971); [17]Lingens et al. (1967); [18]Doy (1968); [19]Doy and Cooper (1966); [20]Miozzari et al. (1978); [21]Manney (1968); [22]de Robichon-Szulmajster et al. (1965); [23]Cherest et al. (1971); [24]de Robichon-Szulmajster et al. (1973); [25]Cherest et al. (1969); [26]Cherest (1973b); [27]Cherest et al. (1973a); [28]Lor and Cossins (1972); [29]Bollon (1980); [30]Bollon and Magee (1971); [31]Bollon (1975) [32]Bollon and Magee (1973); [33]Magee and Hereford (1969); [34]Bussey and Umbarger (1969); [35]Roeder (1980); [36]Brown et al. (1975); [37]G. Kohlhaw (pers. comm.); [38]Rasse-Messenguy and Fink (1973); [39]Struhl (1979).

[a] +Amino acid signifies addition to minimal medium; — amino acid signifies starvation of an auxotroph in a chemostat or by removal of the amino acid; —amino acid (analog) signifies analog-induced starvation for that amino acid; MV, minimal medium.

[b] +, repression by pathway end product(s), no repression in one or more pathway-specific "regulatory" mutants; +?, repression by pathway end product(s), no known pathway-specific "regulatory" mutants; —, no repression by pathway end product(s), no effect of pathway-specific "regulatory" mutants; —?, no repression by pathway-specific "regulatory" mutants.

[c] +, derepression in response to starvation (auxotroph- or analog-induced) for amino acid other than the pathway end product and/or affected by mutations that affect the cross-pathway control system; —, do not derepress in above conditions and/or unaffected by above mutations.

[d] WT, wild type; fbr, feedback resistant.

Four polypeptides of the arginine pathway have been studied through the full range, from YEPD to arginine starvation. The maximum derepression responses for those subject to both arginine-specific and cross-pathway control, namely OTCase and the small subunit of carbamoylphosphate synthetase (*CPA1*), show large responses of 186-fold and 103-fold, respectively. The two subject only to cross-pathway control, namely argininosuccinate lyase and the large subunit of carbamoylphosphate synthetase (*CPA2*), show somewhat smaller but still large ratios of 26-fold and 64-fold, respectively.

Enzymes of the methionine-specific branches of the threonine-methionine pathway have also been analyzed through nearly the complete range. (The ratio for the acetyltransferase may be an underestimate, for neither SAM nor YEPD repression has been examined.) Again, the ratios are very large. For sulfite reductase, the ratio becomes essentially infinite, for the enzyme is virtually undetectable in SAM-grown cells. This enzyme is not subject to cross-pathway control. Of interest is the finding that enzymes of the common threonine-methionine part of the methionine pathway, with the possible exception of aspartokinase, are capable of only limited changes in level in contrast to the large responses of the methionine-specific branches.

As only one of the tryptophan biosynthetic enzymes has been assayed after the cells have been subjected to a rigorous tryptophan starvation, it is difficult to get a feel for the extent to which the enzymes might derepress if such a starvation were imposed. Certainly anthranilate synthase levels rise to a greater extent in response to starvation of an auxotroph as compared to a 5MT-induced tryptophan starvation. Possibly, analog-induced starvation is less effective in general than starvation of auxotrophs. A possible basis for such a difference is that pool distributions might be different in the two situations.

The derepression ratios presented for the enzymes of the branched-chain amino acid pathway should be representative of nearly the full range. Yet the ratios, with the possible exception of AHAS, are modest. This is in clear contrast to enzymes of the arginine and methionine pathways.

Although the information available is not nearly complete, it looks as though enzymes of some pathways show large responses, whereas those of other pathways show limited responses. What significance is to be attached to such findings is quite unclear.

NUCLEOTIDE BIOSYNTHESIS

Biosynthesis of Pyrimidine Nucleotides

The family of pyrimidine nucleotides includes uridylate and cytidylate and their deoxy derivatives, and thymidylate (Fig. 12; Tables 13 and 14). Cytidylate and thymidylate, or their nucleoside triphosphate derivatives, are derived ultimately from uridylate.

Amino Acid and Nucleotide Biosynthesis   265

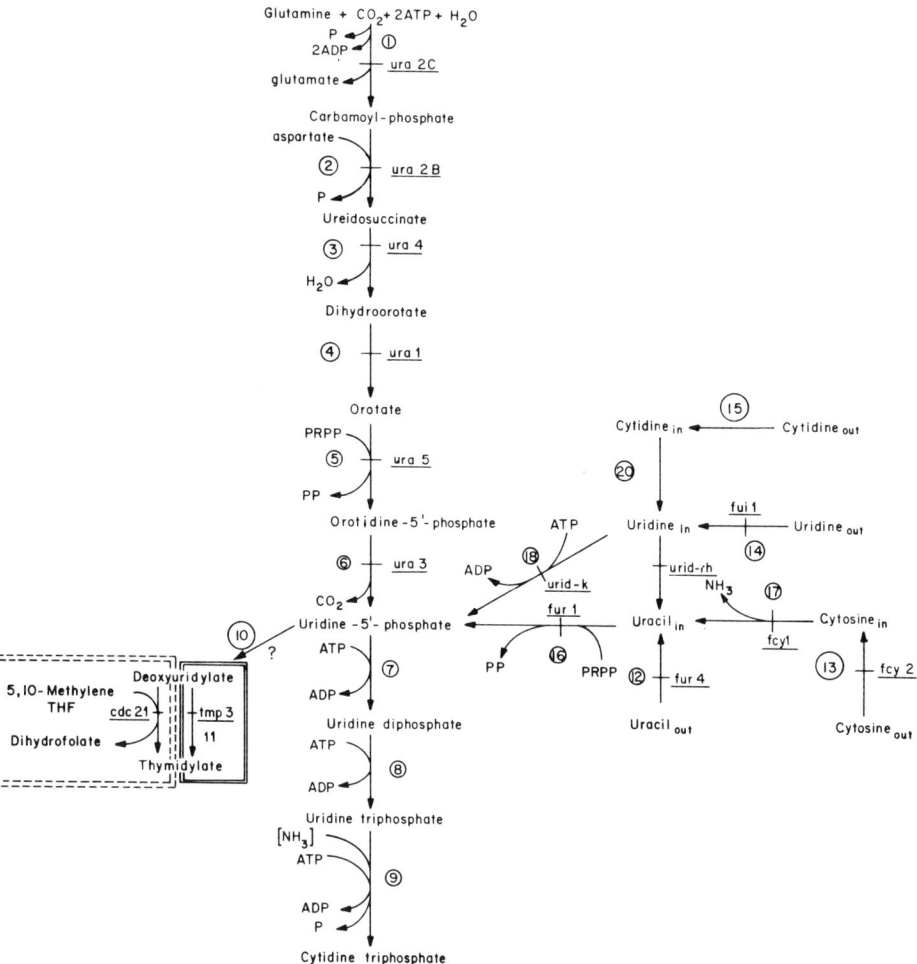

Figure 12 Biosynthesis of pyrimidine nucleotides. Genetic blocks indicated. Locations: (——) mitochondrion; (----) cytosol; unenclosed, unknown. (See Table 13.)

Pathway. Synthesis of uridylate begins with synthesis of carbamoylphosphate from glutamine, CO_2, and ATP (reaction 1). The pools of carbamoylphosphate synthesized by pyrimidine-specific and arginine-specific enzymes apparently mix, for loss of either carbamoylphosphate synthetase activity alone does not result in auxotrophy (Lacroute et al. 1965). Carbamoylphosphate condenses with aspartate to yield ureidosuccinate (reaction 2), catalyzed by aspartate carbamoyltransferase. Aspartate carbamoyltransferase (and presumably the pyrimidine-specific carbamoylphosphate synthetase, since the two activities copurify; see below) is located in the nucleus (Nagy et

Table 13 Genes and enzymes for biosynthesis of pyrimidine nucleotides

Reaction	Enzyme	Locus	References for gene assignment
1	carbamoyl-P synthetase (carbon dioxide:L-glutamine amido-ligase [ADP-forming, carbamate phosphorylating], 6.3.5.5^a)	ura2C (cpu)	Lacroute et al. (1965); Lacroute (1968)
2	aspartate carbamoyltransferase (carbamoyl-P:L-aspartate carbamoyltransferase, 2.1.3.2)	ura2B	Lacroute (1968); Denis-Duphil and Lacroute (1971)
3	dihydroorotase (L-5,6-dihydroorotate amidohydrolase, 3.5.2.3)	ura4	Lacroute (1968)
4	dihydroorotate dehydrogenase^b(L-5,6-dihydroorotate: oxidoreductase, 1.3.1.14)	ura1	Lacroute (1968)
5	orotate PR-transferase (orotidine-5′-P:pyro-P PR-transferase, 2.4.2.10) (also OMP pyrophosphorylase)	ura5	Jund and Lacroute (1972)
6	orotidine-5′-P decarboxylase (orotidine-5′-P carboxy-lyase, 4.1.1.23)	ura3	Lacroute (1968)
7	nucleosidemono-P kinase (ATP:nucleosidemono-P P-transferase, 2.7.4.4.)		
8	nucleosidedi-P kinase (ATP:nucleosidedi-P P-transferase, 2.7.4.6)		

9	cytidylate synthetase (UTP:ammonia ligase [ADP-forming], 6.3.4.2)		Bisson and Thorner (1977)
11	thymidylate synthase (5,10-methyleneTHF:dUMP C-methyltransferase, 2.1.1.45)	$cdc21$ $(tmp1^c)$ $tmp3^d$	Zelikson and Luzzatti (1977)
12	uracil permease	$fur4^e$ $(ura-p)$	Grenson (1969); Jund and Lacroute (1970)
13	cytosine permease	$fcy2^e$ $(cyt-p)$	Grenson (1969); Jund and Lacroute (1970)
14	uridine permease	$fui1^e$ $(urid-p)$	Grenson (1969); Jund and Lacroute (1970)
15	cytidine permease		
16	uracil PR-transferase (UMP:pyro-P PR-transferase, 2.4.2.9) (also UMP pyrophosphorylase)	$fur1^e$ (ups)	Grenson (1969); Jund and Lacroute (1970)
17	cytosine deaminase (cytosine aminohydrolase, 3.5.4.1)	$fcy1^e$ $(cyt-da)$	Grenson (1969); Jund and Lacroute (1970)
18	uridine kinase (ATP:uridine 5′-P-transferase, 2.7.1.48)	$urid-k$	Grenson (1969)
19	uridine nucleosidase (uridine ribohydrolase, 3.2.2.3)	$urid-rh$	Grenson (1969)
20	cytidine deaminase (cytidine aminohydrolase, 3.5.4.5)		

[a] Enzyme Commission number.
[b] Not the IUPAC-IUB recommended name.
[c] Cytosol.
[d] Mitochondrion.
[e] Lacroute's designations are retained. He first described some of the mutants (for references, see Jund and Lacroute [1970]) and uses standard nomenclature.

Table 14 Maximum repression, derepression, or induction responses for enzymes of pyrimidine biosynthesis

Reaction	Enzyme	Locus	Highest/lowest level	Nature of response	Reference
1	carbamoyl-P synthetase	ura2C	3	repression only	Lacroute et al. (1965)
2	aspartate carbamoyltransferase	ura2B	6	repression-derepression	Jund and Lacroute (1972); Lacroute (1968)
3	dihydroorotase	ura4	14	induction by ureidosuccinate	Lacroute (1968)
4	dihydroorotate dehydrogenase	ura1	14	induction by dihydroorotate	Lacroute (1968)
5	OMP pyrophosphorylase	ura5	1	no response	Jund and Lacroute (1972)
6	OMP decarboxylase	ura3	5	induction by dihydroorotate	Lacroute (1968)

al. 1982). Ring closure, accompanied by loss of water, yields dihydroorotate (reaction 3) and is catalyzed by dihydroorotase. Dihydroorotate is oxidized to orotate by dihydroorotate dehydrogenase (reaction 4). The nucleotide, orotidine-5'-phosphate, is formed by condensation of orotate with PRPP (reaction 5), catalyzed by orotate phosphoribosylhydrolase (also called OMP pyrophosphorylase). Decarboxylation of OMP (reaction 6), catalyzed by OMP decarboxylase, yields uridylate.

Synthesis of cytidine triphosphate is presumed to occur via reactions 7–9, although no evidence is available. The amino donor for CTP synthetase is unknown but is ammonia in bacteria and glutamine in animal cells.

Presumably, synthesis of deoxyribonucleotides involves reduction of nucleotides or nucleoside diphosphates or triphosphates. In other systems, reduction requires thioredoxin and thioredoxin reductase. Yeast possesses thioredoxin and thioredoxin reductase (Porqué et al. 1970a), but involvement of these proteins in generation of deoxyribonucleotides in yeast has not been examined.

Thymidilate is synthesized from dUMP by thymidylate synthase (reaction 11). The activity is present in the cytoplasm and the mitochondrion (Zelikson and Luzzatti 1977). These are probably different species of enzyme since mutations can inactivate single species. Mutations in the *CDC21* (*TMP1*) and *TMP3* loci lead to requirement for TMP (Brendel and Fäth 1974; Luzzatti 1975; Game 1976). *CDC21* is apparently the structural gene for thymidylate synthase (probably the cytosolic species) (Bisson and Thorner 1977). The *tmp3* mutant has reduced activity for the mitochondrial activity, as well as for other activities of THF metabolism (Zelikson and Luzzatti 1977). Thymidylate can satisfy these auxotrophies only if the cells have been made permeable to the nucleotide genetically (Wickner 1975).

Some exogenous pyrimidines and pyrimidine nucleosides can be utilized after uptake by specific permeases (see Cooper [Transport in *Saccharomyces cerevisiae*], this volume). Exogenous uracil is converted to UMP by UMP pyrophosphorylase (reaction 16). Cytosine is utilized after conversion to uracil by cytosine deaminase (reaction 17). Uridine can be utilized by direct conversion to UMP, catalyzed by uridine kinase (reaction 18), or by conversion to uracil (reaction 19), catalyzed by uridine ribohydrolase (see Grenson 1969; Jund and Lacroute 1970).

Assignment of genetic blocks is due primarily to Lacroute and collaborators (Lacroute et al. 1965; Lacroute 1968; Grenson 1969; Jund and Lacroute 1970, 1972). Direct selections exist for mutations in the *URA1, URA2, URA3,* and *URA5* loci, although the efficiency of the selections depends upon genetic background (Bach and Lacroute 1972). *URA2* is a complex locus that encodes carbamoylphosphate synthetase and aspartate carbamoyltransferase (reactions 1 and 2) (Denis-Duphil and Lacroute 1971; Denis-Duphil and Kaplan 1976). Mutation in the *URA2* locus can lead to several different phenotypes, including resistance to fluorouracil due to loss of feedback inhibition (Lacroute 1968). The two enzyme activities encoded by *URA2* usually copurify (Lue and Kaplan 1969, 1971), but it is unclear whether the product of the locus is a multifunctional polypeptide or two polypeptides that aggregate. The finding that the molecular size and susceptibility of the enzyme to feedback inhibition are altered in vitro by proteinase B activity present in extracts (Denis-Duphil et al. 1981) suggests that the unambiguous determination of the molecular substructure of the enzyme may only be obtained by DNA sequence analysis. Residence of the two activities in the same protein apparently provides for channeling into the pyrimidine pathway of carbamoylphosphate synthesized by this isozyme (Lue and Kaplan 1970).

All mutations in the *URA5* locus (OMP pyrophosphorylase, reaction 5) are leaky. This leakiness comes about because UMP pyrophosphorylase (reaction 16) can catalyze conversion of orotate to OMP with low efficiency. The double mutant *ura5 fur1*, lacking both OMP and UMP pyrophosphorylase, does not survive (Jund and Lacroute 1972).

Metabolic Flow. Both the pyrimidine-specific carbamoylphosphate synthetase and aspartate carbamoyltransferase are feedback-inhibited by UTP (Lacroute et al. 1965; Kaplan et al. 1967; Lacroute 1968). The latter inhibition presumably is required because the carbamoylphosphate pools produced for the pyrimidine and arginine pathways are not segregated (Lacroute et al. 1965). Burns (1966) reported that high purine nucleotide pools appeared to reverse feedback inhibition by UTP in vivo, at least for synthesis of ureidosuccinate. The finding that ATP can counteract feedback inhibition by UTP of aspartate carbamoyltransferase in vitro (Kaplan et al. 1967) accounts for Burns' observations.

Regulation of Enzyme Levels. Carbamoylphosphate synthetase and aspartate carbamoyltransferase, product(s) of the *URA2* locus, are repressed up to twofold by addition of uracil to the medium and derepressed about threefold by starvation of uracil auxotrophs (Lacroute et al. 1965; Lacroute 1968). Dihydroorotase (reaction 3) is not repressible by pyrimidines but is induced by its substrate ureidosuccinate. Dihydroorotate dehydrogenase (reaction 4) and OMP decarboxylase (reaction 6) are likewise nonrepressible by pyrimidines but are induced by dihydroorotate, substrate of the dehydrogenase. Whether orotate can induce is unclear (Lacroute 1968). The OMP pyrophosphorylase enzyme (reaction 5) is only slightly repressible by growth in pyrimidines, if at all, and is not induced by any of the pyrimidine pathway intermediates (Jund and Lacroute 1972). The *for1* mutant, isolated as resistant to fluoroorotate, has slightly elevated levels of OMP pyrophosphorylase and OMP decarboxylase (reactions 5 and 6).

The induction of OMP decarboxylase brought about by dihydroorotate is effected at the level of transcription, for levels of *URA3* mRNA, measured by hybridization to the cloned *URA3* gene, vary coordinately with levels of the enzyme (Bach et al. 1979). Mutations in the *PPR1* gene lead to induced levels (5×) of OMP decarboxylase and dihydroorotate dehydrogenase in the absence of the inducer dihydroorotate (Loison et al. 1980). The *ppr1* mutation is semidominant and unlinked to either *URA3* (decarboxylase) or *URA1* (dehydrogenase), which genes, themselves, are unlinked. The mutant *ppr1*-gene product can effect induced levels of synthesis for all 50 copies of the *URA3* gene in strains bearing the *URA3*-containing plasmid. This result implies that the *PPR1*-gene product is present in great excess, or acts catalytically. The relative levels of OMP decarboxylase in strains of the constitution given are wild type, 1; *ppr1*, 5; wild-type bearing the *URA3*-containing plasmid, 44; *ppr1* bearing the *URA3* plasmid, 171 (Loison et al. 1980).

The maximum repression, derepression, and induction responses for those genes that have been examined range from about 5- to 15-fold (Table 14). The finding that the cells are capable of very high levels of expression of the *URA3* gene when the gene is on a plasmid (Loison et al. 1980; Rose et al. 1981) clearly indicates that the wild-type cells show limited derepression responses from choice, not from necessity.

Biosynthesis of Purine Nucleotides

Purine nucleotides include adenylate and guanylate; the two nucleotides are derived from inosinate, which is synthesized de novo from PRPP (Fig. 13; Table 15). This pathway contrasts sharply with pyrimidine nucleotide biosynthesis in which addition of the ribosylphosphate moiety occurs late in the pathway.

Pathway. PRPP reacts with glutamine and water such that the amino group from glutamine displaces the pyrophosphate group of PRPP yielding phosphoribosylamine, a reaction catalyzed by amidophosphoribosyltransferase (reaction 1). The carboxyl group of glycine reacts with this amino group to give an amide linkage in phosphoribosylglycinamide (reaction 2). A single carbon at the formate level of oxidation is added from 5,10-methenylTHF to the free α-amino group to yield phosphoribosyl-*N*-formylglycinamide (reaction 3). An amino group, derived from glutamine, is added to the amide carbon to yield phosphoribosylglycinamidine (reaction 4) in a reaction that requires ATP. Closure of the imidazole ring to yield phosphoribosylaminoimidazole (also called AIR) follows and requires ATP (reaction 5). The red pigment that accumulates in *ade1* and *ade2* mutants apparently derives from this intermediate. Carboxylation occurs to form CAIR (reaction 6) to which aspartate is added in toto (reaction 7). Cleavage, with release of fumarate

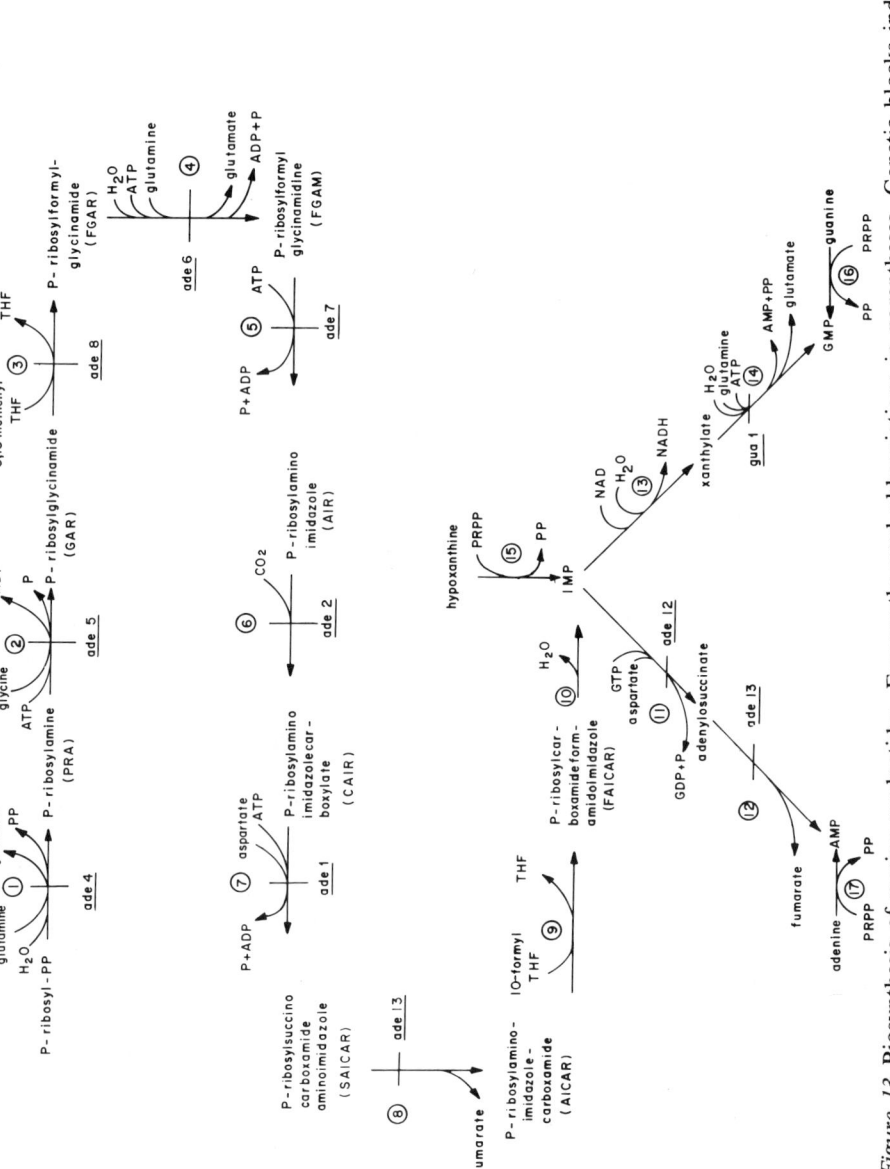

Figure 13 Biosynthesis of purine nucleotides. Frequently used abbreviations in parentheses. Genetic blocks indicated. (See Table 15.)

(reaction 8), produces phosphoribosylaminoimidazolecarboxamide, also called aminoimidazolecarboxamide ribonucleotide (AICAR). AICAR is a by-product of histidine biosynthesis (see Fig. 9). Adenylosuccinate lyase, which catalyzes the cleavage step (reaction 8), catalyzes a similar reaction in a later step specific to synthesis of adenylate (reaction 12). The remaining ring atom is introduced from 10-formylTHF onto the free amino group (reaction 9). Ring closure completes synthesis of the purine ring with generation of IMP (inosinic acid) (reaction 10).

Synthesis of AMP from IMP proceeds in two steps similar to reactions 7 and 8. Aspartate is incorporated in toto, followed by cleavage and release of fumarate (reactions 11 and 12). Incorporation of aspartate is catalyzed by adenylosuccinate synthetase; cleavage by adenylosuccinate lyase (also used in reaction 8). In most organisms, GTP is required for synthesis of adenylosuccinate.

Synthesis of GMP from IMP proceeds by oxidation of IMP to yield XMP, catalyzed by IMP dehydrogenase, followed by amination of XMP, with glutamine as donor, to yield GMP. The GMP synthetase reaction requires ATP, which is cleaved to yield AMP and pyrophosphate.

Exogenously supplied hypoxanthine, adenine, and guanine are presumably converted to their corresponding nucleotides through action of the respective phosphoribosyltransferases (reactions 15-17). In most organisms formation of IMP and GMP is catalyzed by a single phosphoribosyltransferase.

Guanine cannot satisfy the purine requirement of mutants blocked in steps leading to synthesis of IMP (Demain 1964). This has been interpreted to mean that yeast lacks a guanylate reductase, which catalyzes conversion of GMP to IMP. Since yeast possesses at least a limited ability to convert added guanine to adenine nucleotides, the explanation of this failure of guanine to supplement these auxotrophs may lie elsewhere (Burridge et al. 1977). Adenine can satisfy the requirement of auxotrophs with early blocks and, hence, must serve as a source of guanine. Two routes are probably involved: one through the histidine pathway through AICAR and IMP (AICAR, derived from ATP, is a by-product of reaction 6 of histidine biosynthesis; see Fig. 9); the second through AMP, adenosine, inosine, hypoxanthine, and IMP (Burridge et al. 1977). The second postulated route is based on analyses of intracellular pools. The enzymes required for these conversions, presumably a phosphatase, deaminase, and ribohydrolase (in addition to the phosphoribosyltransferase) have not been studied. Normally, nucleosides cannot satisfy purine requirements (Demain 1964), although selection for utilization of adenosine, which results in sensitivity to cordycepin, has been successful (Anderson and Roth 1974).

Assignment of genetic blocks is based on enzyme assay or intermediates accumulated. As the latter tests have sometimes given results at variance with enzyme analyses, the assignments based on accumulations must be

Table 15 Genes and enzymes for biosynthesis of purine nucleotides

Reaction	Enzyme	Locus	References for gene assignment
1	amido-PR-transferase (5-PR-amine:pyro-P PR-transferase [glutamate-amidating], 2.4.2.14)	ade4 (pur6) (dap), (su-pur)	Gross and Woods (1971); Lomax and Woods (1973)
2	PR-glycinamide synthetase (5-PR-amine:glycine ligase [ADP-forming], 6.3.4.13)	ade5	Gross and Woods (1971)
3	PR-glycinamide formyltransferase (5,10-methenylTHF:5′-PR-glycinamide formyltransferase, 2.1.2.2)	ade8	Woods and Jackson (1973)
4	PR-formylglycinamidine synthetase (5′-PR-formylglycinamide:L-glutamine amidoligase [ADP-forming], 6.3.5.3)	ade6	Silver and Eaton (1968)
5	PR-aminoimidazole synthetase (5′-PR-formylglycinamidine cycloligase [ADP-forming], 6.3.3.1)	ade7	Silver and Eaton (1968)
6	PR-aminoimidazole carboxylase (5′-PR-5-amino-4-imidazolecarboxylate carboxy-lyase, 4.1.1.21)	ade2	Silver and Eaton (1969); Woods (1969)
7	PR-aminoimidazolesuccinocarboxamide synthetase (5′-PR-4-carboxy-5-aminoimidazole:L-aspartate ligase [ADP-forming], 6.3.2.6)	ade1	Silver and Eaton (1969); Fisher (1969)

8	adenylosuccinate lyase (adenylosuccinate AMP-lyase, 4.3.2.2) (see reaction 12)	*ade13*	Dorfman (1969)
9	PR-aminoimidazolecarboxamide formyltransferase (10-formylTHF:5′-PR-5-amino-4-imidazolecarboxamide formyltransferase, 2.1.2.3)		
10	IMP cyclohydrolase (IMP 1,2-hydrolase [decyclizing], 3.5.4.10)		
11	adenylosuccinate synthetase (IMP:L-aspartate ligase [GDP-forming], 6.3.4.4)	*ade12 (pur1)*	Dorfman (1969); Lomax and Woods (1971)
12	adenylosuccinate lyase (adenylosuccinate AMP-lyase,, 4.3.2.2) (see reaction 8)	*ade13*	Dorfman (1969)
13	IMP dehydrogenase (IMP:NAD oxidoreductase 1.2.1.14)	*pur5?*	Gardner and Woods (cited in Burridge et al. [1978])
14	GMP synthetase (xanthosine-5′-P:L-glutamine amidoligase [AMP-forming], 6.3.5.2)	*gua1*	Reichert and Winter (1975)
15	hypoxanthine PR-transferase (IMP:pyrophosphate PR-transferase 2.4.2.8)		
16	adenine PR-transferase (AMP:pyrophosphate PR-transferase, 2.4.2.7)		
17	guanine PR-transferase (IMP:pyrophosphate PR-transferase, 2.4.2.8)		

[a]Enzyme Commisssion number.

regarded as provisional. Enzyme analyses provide the basis for assignments of the mutations *ade1* (reaction 1; Fisher 1969), *ade4* (reaction 1), *ade5* (reaction 2) (Gross and Woods 1971), *ade13* (reactions 8 and 12; Dorfman 1969), and *gua1* (reaction 14; Reichert and Winter 1975). Accumulations provide the basis for assignment of mutations *ade6* (reaction 4), *ade7* (reaction 5) (Silver and Eaton 1968), and *ade8* (reaction 3; Woods and Jackson 1973). The *ade2* assignment is based on accumulations (Silver and Eaton 1969) and the fact that growth of certain *ade2* mutants is stimulated by CO_2 (Woods 1969). *ade12* mutants are inferred to lack adenylosuccinate synthetase because they grow only if supplied with adenine, accumulate inosine, and possess adenylosuccinate lyase activity (Dorfman 1969). *ade3* mutants require adenine or hypoxanthine (and histidine) but are defective in synthesis of derivatives of THF at the formate level of oxidation (Jones and Magasanik 1967; Lazowska and Luzzatti 1970a; Lomax et al. 1971). The absence of 10-formyl- and 5,10-methenylTHF in *ade3* mutants results in blockage of the third and ninth steps of purine nucleotide biosynthesis.

The enzyme defects in *gua2* and *gua3* mutants are unknown. They are *not* defective in IMP dehydrogenase (reaction 13) (Gardner and Woods 1979). Indeed, these investigators suggest that absence of IMP dehydrogenase is lethal. Growth of guanine auxotrophs is inhibited by adenine, presumably because of competition for uptake (Reichert and Winter 1975; Gardner and Woods 1979).

The *ADE5* and *ADE7* loci are tightly linked (Roman 1956). Mutants bearing mutations that fail to complement either *ade5* or *ade7* mutations can be isolated. These *ade5,7* mutations map within the *ADE5* portion of the complex *ADE5-ADE7* locus (Dorfman 1964). The number of gene products encoded by this complex locus is unknown. The finding that *ade7* mutants are partially deficient for the enzyme activity missing in *ade5* mutants (reaction 2) (Gross and Woods 1971) suggests that the enzyme activities for the second and fifth steps might reside in a single polypeptide or an enzyme complex.

Mutations at the *ADE4* locus can lead to a number of different phenotypes. These include requirement for adenine or hypoxanthine (*ade4*), excretion of purines and resistance to 8-azaadenine and 8-azaguanine (*pur6*), sensitivity to 2,6-diaminopurine (*dap*), suppression of purine excretion (*supur*) (Lomax and Woods 1973), and resistance to ureidosuccinate (Korch et al. 1974). As *ADE4* appears to encode the first enzyme of purine nucleotide biosynthesis, which is subject to feedback inhibition (see below), such a diversity of phenotypes as a consequence of mutation is not unexpected.

Mutations in the *ADE12* locus, thought to encode adenylosuccinate synthetase, can likewise lead to different phenotypes: a purine requirement that can only be satisfied by adenine accompanied by excretion of hypoxanthine and intense red pigmentation of *ade1 ade12* or *ade2 ade12* double mutants with all three effects assignable to the *ade12* mutation (Dorfman 1969), and excretion of purines unaccompanied by a purine requirement (*pur1* mutations; Lomax and Woods 1971). Interestingly enough, *ade12* mutations not only fail to complement *pur1* mutations with respect to purine excretion but also fail to complement *pur6* mutations, which lie in the *ADE4* locus, for purine excretion (Lomax and Woods 1971). *ADE4*, as mentioned, encodes the first feedback-sensitive enzyme of purine nucleotide biosynthesis. Heteroallelic *ade12* diploids may show allelic complementation for growth and/or intensification of red pigmentation. All 96 alleles examined showed complementation for at least one of the two characteristics (Dorfman 1971). Few of the strains carrying an *ade12* allele can survive if the strain does not also carry an *ade1* or *ade2* allele (Dorfman 1969). This observation, coupled with the finding that all 96 *ade12* alleles show intragenic complementation, suggests that total absence of the activity (or is it polypeptide?) would be lethal. It is of some interest that the evidence (admittedly negative) suggests that absence of activity for the first enzymes of the specific branches to GMP (Gardner and Woods 1979) and AMP (see above) may be lethal, for these pathways are intertwined metabolically.

Metabolic Flow. The first enzyme of the common pathway, amidophosphoribosyltransferase, is subject to feedback inhibition by AMP and to a lesser extent by ATP and GTP (Satyanarayana and Kaplan 1971). These same investigators have reported that GMP does not inhibit (Satyanarayana and Kaplan 1971), but contrary evidence has been presented (Smolina et al. 1978). Feedback inhibition of purine nucleotide synthesis can be demonstrated in vivo (Burns 1964; Smolina et al. 1978). These in vivo studies suggest that an adenine nucleotide must be the prime feedback inhibitor for the common pathway (Smolina et al. [1978], and see Burridge et al. [1977] for nucleotide concentrations after feeding various purines). The stimulatory effect of ATP on pyrimidine nucleotide biosynthesis previously described is not a reciprocal relationship, for pyrimidine nucleotides are without effect on the first enzyme of purine nucleotide biosynthesis (Satyanarayana and Kaplan 1971). However, the finding that mutations in the *ADE4* locus can lead to resistance to ureidosuccinate, the first specific intermediate in pyrimidine biosynthesis, raises the possibility that purine nucleotide biosynthesis may be linked to pyrimidine biosynthesis via ureidosuccinate (Korch et al. 1974). Whether ureidosuccinate levels ever reach a concentration sufficient to exert such control in normal cells remains conjectural.

In many organisms, flow into the specific GMP and AMP arms is controlled by feedback inhibition, with AMP inhibiting adenylosuccinate synthetase (reaction 11) and GMP inhibiting IMP dehydrogenase (reaction 13). Direct studies of the enzymes have not been reported for yeast, but the labeling patterns reported by Smolina et al. (1978) imply that a guanine nucleotide inhibits flow from IMP into the guanylate arm. *pur5* mutants, which secrete purines (Armitt and Woods 1970), may be altered in IMP dehydrogenase function and have an alteration in regulation of flow into this branch (Gardner and Woods, cited in Burridge et al. 1978).

It seems likely that the *ADE12*-gene product, thought to be adenylosuccinate synthetase, is involved somehow in regulating flow into the common part of the purine nucleotide biosynthetic pathway, for mutations in this locus lead to excretion of hypoxanthine (Dorfman 1969; Lomax and Woods 1971) and increased red pigmentation in *ade1 ade12* double mutants (Dorfman 1969). Although this effect has been ascribed to alteration in repression (Dorfman 1969, 1971), it seems likely that it reflects alterations in feedback inhibition comparable to similar effects seen for mutations at the *ADE4* locus.

Regulation of Enzyme Levels. Little has been reported on regulation of levels of enzymes of purine biosynthesis. Dorfman and colleagues (Dorfman 1969; Dorfman et al. 1970) have proposed that the *ADE12* locus encodes a bifunctional protein that possesses catalytic activity for adenylosuccinate synthetase and repressor activity for the purine loci. This postulate derives from the observation of enhanced production of pathway intermediates in *ade12* mutants. To our knowledge, there are no reports of levels of purine biosynthetic enzymes in such mutants and, hence, no justification for concluding that the levels of purine enzymes are derepressed in such mutants. It seems more likely that the *ADE12*-gene product may interact with the *ADE4*-gene product (carrying catalytic activity for the first enzyme). Alterations in the structure of the *ADE12* polypeptide could then interfere with feedback inhibition of the first enzyme, leading to overproduction of purines. Failure of *pur1* mutations (in the *ADE12* locus) to complement *pur6* mutations (in the *ADE4* locus) for purine excretion would be understandable on this model.

Gross and Woods (1972) reported that adenine represses the first enzyme of purine biosynthesis, amidophosphoribosyltransferase, sixfold. The only other report on regulation of enzyme levels for purine biosynthesis concerns enzymes of the guanylate arm, IMP dehydrogenase and GMP synthetase. GMP synthetase appears to be unregulated, for addition of guanine to the medium does not repress levels of GMP synthetase, and dosage studies employing a *gua1* mutation (deficient for synthetase) reveal a constant output per wild-type allele despite slowing of growth due to GMP insufficiency. Synthesis of IMP dehydrogenase seems to be repressed by a guanine nucleotide, for addition of guanine to the medium results in 2.4-fold repression of IMP dehydrogenase, and dosage studies employing the *gua1* mutation reveal a 2.8-fold derepression of IMP dehydrogenase when the dose of wild-type alleles for GMP synthetase is changed from four to one in tetraploids. Thus, the total derepression range for IMP dehydrogenase is about 7-fold (Reichert and Winter 1975).

SINGLE CARBON METABOLISM

Single carbon derivatives of THF are required for synthesis of purine nucleotides, thymidylate, and methionine and, in the mitochondrion, for synthesis of N-formylmethionine as well (Figs. 14 and 15; Table 16). A preponderance of each of the single carbon adducts of THF exist as the polyglutamyl derivative (Lor and Cossins 1972).

Pathway. Synthesis begins with reduction of dihydrofolate (DHF), catalyzed by DHF reductase (reaction 1). The β carbon of serine is transferred onto THF in a reaction catalyzed by serine hydroxymethyltransferase (reaction 2), yielding glycine and 5,10-methyleneTHF at the formaldehyde level of oxidation. 5,10-MethyleneTHF serves both as a single carbon donor and reductant for synthesis of TMP from dUMP, catalyzed by thymidylate synthase (reaction 8). The DHF generated is recycled following reduction. 5,10-MethyleneTHF, after reduction to 5-methylTHF (reaction 3), serves as a single carbon donor for generation of methionine from homocysteine (reaction 10). As the diglutamyl or triglutamyl derivative of 5-methylTHF is required as cofactor for this reaction (Burton et al. 1969), at some point in the sequence glutamyl moieties must be added. The same must be true for the other derivatives, since they exist largely as polyglutamyl derivatives.

The remaining two derivatives, 5,10-methenylTHF and 10-formylTHF, both at the formate level of oxidation, can be produced from 5,10-methyleneTHF through oxidation, catalyzed by methyleneTHF dehydrogenase (reaction 4), followed by ring opening, catalyzed by methenylTHF cyclohydrolase (reaction 5). Alternatively, 10-formylTHF can be synthesized directly from formate and THF in an ATP-requiring reaction catalyzed by formylTHF synthetase (reaction 6). 5,10-MethenylTHF is required for the third step of purine nucleotide biosynthesis; 10-formylTHF for the ninth step (Fig. 13). When cells are growing aerobically, glycine can also serve as a source of single carbons through action of glycine decarboxylase (reaction 9), which is located in the mitochondrion (Ogur et al. 1977). Since the dehydrogenase and cyclohydrolase reactions are reversible, serine, formate, or glycine can serve as source of all single carbon derivatives of THF. Yeast is inferred to possess 5,10-methenylTHF synthetase (reaction 7), since 5-formylTHF can supply needed single carbons in certain mutants (see below) and because mitochondrial extracts can use 5-formylTHF as a source of 10-formylTHF for synthesis of N-formylmethionine (Halbreich and Rabinowitz 1971).

The methyleneTHF dehydrogenase, methenylTHF cyclohydrolase, and formylTHF synthetase (reactions 4-6) activities are contained within a single polypeptide chain (Paukert et al. 1977) encoded by the *ADE3* locus (Jones and Magasanik 1967; Lazowska and Luzzatti 1970a; McKenzie and

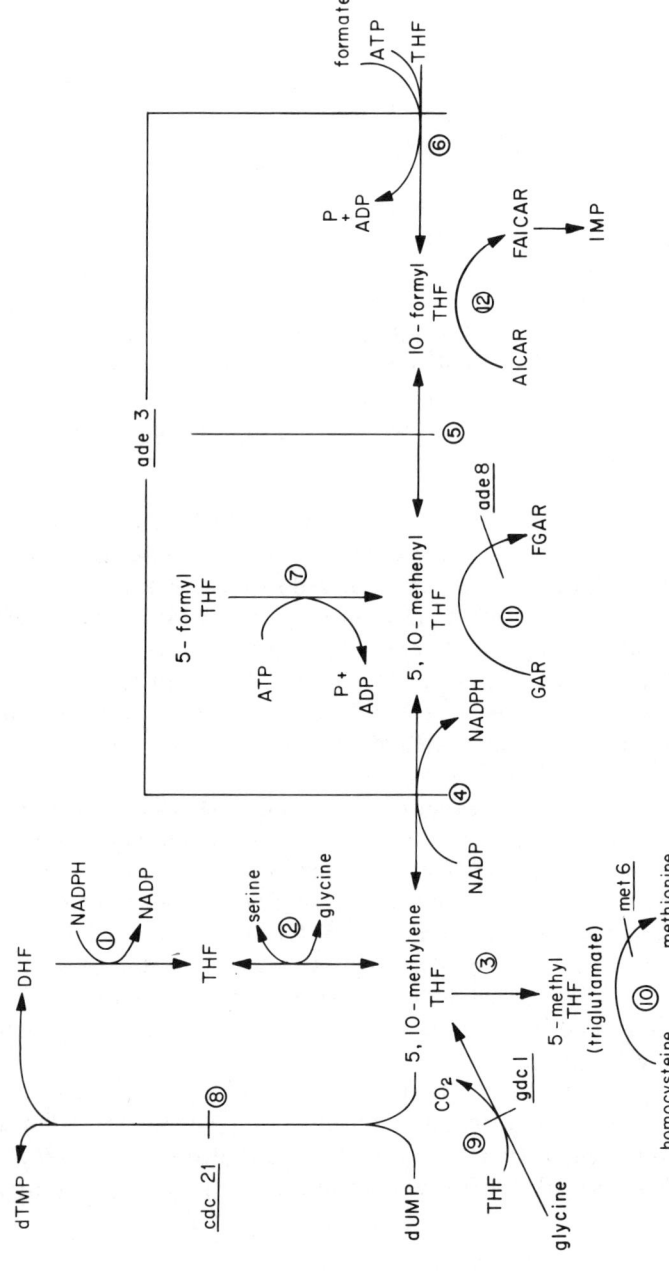

Figure 14 Synthesis and interconversion of single carbon derivatives of THF. The reactions where they are used are indicated. Except for reaction 9 and, possibly, 7, all enzymes and genetic blocks indicated are for cytosolic enzymes. (See Table 16.)

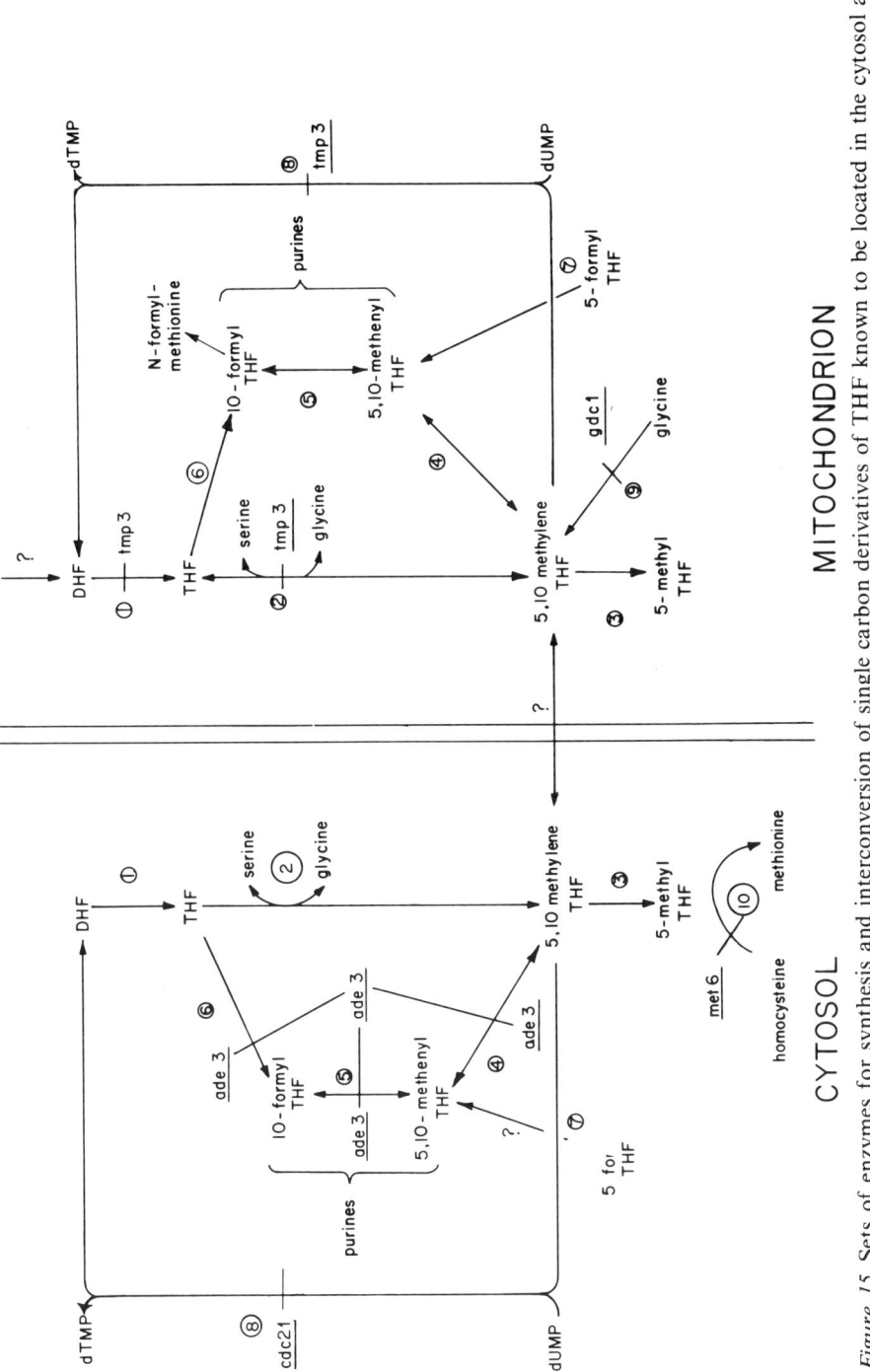

Figure 15 Sets of enzymes for synthesis and interconversion of single carbon derivatives of THF known to be located in the cytosol and mitochondrion. Genetic blocks for each set are indicated. (See Table 16.)

Table 16 Genes and enzymes for biosynthesis of single carbon derivatives of THF

Reaction	Enzyme	Locus	References for gene assignment
1	DHF reductase[a] (5,6,7,8-THF:NADP oxidoreductase, 1.5.1.3[b])		
2	serine hydroxymethyltransferase (5,10-methyleneTHF:glycine hydroxymethyltransferase, 2.1.2.1)	$tmp3^c$	Zelikson and Luzzatti (1976, 1977)
3	5,10-methyleneTHF reductase (5-methylTHF:(acceptor) oxidoreductase, 1.1.99.15)		
4	methyleneTHF dehydrogenase (5,10-methyleneTHF:NADP oxidoreductase, 1.5.1.5)	$ade3^d$	Jones and Magasanik (1967); Lazowska and Luzzatti (1970a)
5	methenylTHF cyclohydrolase (5,10-methenylTHF 5-hydrolase [decyclizing], 3.5.4.9)	$ade3^d$	Jones and Magasanik (1967); Lazowska and Luzzatti (1970a)
6	formylTHF synthetase (formate: THF ligase [ADP-forming], 6.3.4.3)	$ade3^d$	Jones and Magasanik (1967); Luzowska and Luzzatti (1970a)
7	5,10-methenylTHF synthetase (5-formylTHF cycloligase [ADP-forming], 6.3.3.2)		
8	thymidylate synthase (5,10-methyleneTHF:dUMP C-methyltransferase, 2.1.1.45)	$cdc21$ ($tmp1^d$) $tmp3^c$	Bisson and Thorner (1977) Zelikson and Luzzatti (1977)
9	glycine decarboxylase[a] (5,10-methyleneTHF ammonia hydroxymethyltransferase [carboxylating, reducing], 2.1.2.10)	$gdc1$	Ogur et al. (1977)
10	homocysteine methyltransferase (5-methyltetrahydropteroyl-tri-L-glutamate:L-homocysteine S-methyltransferase, 2.1.1.14)	$met6$	Selhub (1972) (cited in Masselot and de Robichon-Szulmajster [1975])
11	PR-glycinamide formyltransferase (5,10-methenylTHF:5′-PR-glycinamide formyltransferase, 2.1.2.2)	$ade8$	Woods and Jackson (1973)
12	PR-aminoimidazolecarboxamide formyltransferase (10-formylTHF:5′-PR-5-amino-4-imidazolecarboxamide formyltransferase, 2.1.2.3)		

[a] Not IUPAC-IUB recommended name.
[b] Enzyme Commission number.
[c] Mitochondrial species.
[d] Cytosolic species.

Jones 1977; de Mata and Rabinowitz 1980). *ade3* mutants require adenine and histidine and usually lack all three enzymatic activities. The adenine requirement stems from failure of the *ade3* mutants to make derivatives of THF at the formate level of oxidation (Fig. 13). The histidine requirement occurs because the mutants accumulate 5,10-methyleneTHF (Luzzatti 1975), which inhibits the fourth enzyme of histidine biosynthesis in vitro and causes accumulation of BBMII, the substrate of this enzyme, in vivo (E. Jones, unpubl.).

S. cerevisiae appears to possess two complete sets of folate interconversion enzymes: one in the cytosol, the other in the mitochondrion (Lazowska and Luzzatti 1970b; Zelikson and Luzzatti 1977). The three activities encoded by the *ADE3* locus are located in the cytosol (Zelikson and Luzzatti 1977).

Luzzatti (1975) reported isolation of the *tmp3* mutant, which requires TMP, methionine, histidine, and adenine and is a nuclear and ρ^- petite mutant. The first three of these requirements (and sometimes the fourth) can be met by 5-formylTHF. The *tmp3* mutant has very low mitochondrial activities for DHF reductase, serine hydroxymethyltransferase, and thymidylate synthase (reactions 1, 2, and 8) (Zelikson and Luzzatti 1977) and lacks the mitochondrial species of serine hydroxymethyltransferase (Zelikson and Luzzatti 1976). To account for the properties of the mutant, Zelikson and Luzzatti (1977) proposed that all DHF is synthesized in the mitochondrion and must pass to the cytosol, possibly in the form of 5,10-methyleneTHF. Because the mitochondrial path from DHF is blocked in *tmp3* cells, the cells are starved for THF derivatives in both cytoplasm and mitochondrion. If this explanation is correct, the histidine requirement of the *tmp3* mutant cannot have the same origin as that of the *ade3* mutant, for no accumulated 5,10-methyleneTHF should be present in the *tmp3* mutant to cause inhibition of the fourth enzyme of histidine biosynthesis.

met7 mutants share some characteristics with the *tmp3* mutant. They require adenine as well as methionine, and the *met7 ser1* double mutant cannot grow on glycine plus formate as a source of serine. The *met7* mutation segregates as a nuclear petite as well. Whether *met7* is allelic to *tmp3* has not been reported (Lowenstein 1973).

Metabolic flow. Methionine and SAM inhibit serine hydroxymethyltransferase activity (Botsford and Parks 1969; Lor and Cossins 1972). As the experiments were carried out on crude extracts, the effects on individual enzyme species are unknown. The pattern observed suggests that one species is inhibitable by methionine, the other or both by SAM (Botsford and Parks 1969). Methionine inhibits methyleneTHF reductase somewhat (Lor and Cossins 1972). No other information on regulation of flow in these pathways or on flow of intermediates between the cytosol and the mitochondrion is available.

Regulation of Enzyme Levels. Little information is available. That levels of at least the two serine hydroxymethyltransferases may be regulated is indicated by the finding that levels of both species are elevated in *ade3* mutants (Luzzatti 1975; Zelikson and Luzzatti 1976, 1977).

NO POLYCISTRONIC OPERONS IN YEAST

In yeast the genes encoding enzymes of a biosynthetic pathway are not usually found linked to one another. The absence of linkage is particularly striking by comparison with the situation in enteric bacteria, where genes of related function are often closely linked and organized into an operon, a group of genes that are cotranscribed and coregulated. There are, however, a few instances in yeast where the information for several enzymes in a biosynthetic pathway is clustered in a single region.

The biochemical evidence for several of these clusters shows clearly that they are monocistronic and therefore not similar to polycistronic bacterial operons. Purification of the *TRP5-*, *HIS4-*, and *ADE3*-gene products reveals in each case a single protein with multiple enzymatic activities (Dettwiler and Kirschner 1979; Keesey et al. 1979; de Mata and Rabinowitz 1980). These proteins are organized into separate domains of function as shown by the fact that individual activities can be physically separated from the others either by proteolysis or by mutation (Manney et al. 1969; Paukert et al. 1977; Bigelis and Fink 1978, 1981; Keesey et al. 1979). DNA sequence analysis of *HIS4* and *TRP5* supports the protein studies (Zalkin and Yanofsky 1982; Donahue et al. 1982). *HIS4* is an uninterrupted coding sequence of 2397 nucleotides, which specifies a single polypeptide with three enzymatic activities (Donahue et al. 1982). Sequence analysis of deletions that retain one of the three activities has identified three separate regions of the gene, each responsible for one of the three functions. The linear array of the three domains within a single gene explains the genetic and complementation results that earlier had led to comparisons of *HIS4* with bacterial operons. *TRP5* encodes a single polypeptide carrying two activities borne by two polypeptides, α and β, in *E. coli* (Zalkin and Yanofsky 1982). Sequence comparisons reveal that *TRP5* probably arose by fusion of two genes homologous to α and β, in the order α-β, an order opposite to that found in bacterial operons. These findings lend credence to the idea that fungal gene clusters arose during evolution by gene fusions. Table 17 contains a list of the known biosynthetic gene clusters that encode multifunctional proteins.

Table 17 Loci encoding multifunctional proteins

Locus	Enzyme activities encoded by the locus[a]	Method of analysis[b]
ADE3	reactions 4–6 (Fig. 15; Table 16)	1,2
ADE5,7	reactions 2 and 5 (Fig. 13; Table 15)	1
ARO1	reactions 2–6 (Fig. 3; Table 4)	1
HIS4	reactions 2,3, and 10 (Fig. 9; Table 8)	1,2,3
TRP5	reaction 17 (Fig. 3; Table 4)	1,2,3

[a]Enzyme activities encoded by the locus and their roles in the pathway can be found in the figures and tables given.
[b]The methods of determining the association between the enzyme activities are (1) enzyme assay of defective mutants; (2) purification of enzyme complex to homogeneity; (3) DNA sequence of the gene.

The evidence available for the *ARG5,6* gene cluster does not permit a definitive picture of its structure. Two enzymatic activities are encoded by *ARG5,6* (Minet et al. 1979). Nonsense mutations in *ARG6* result in a deficiency for both activities, whereas nonsense mutations in *ARG5* eliminate only one of these. The two enzymatic activities encoded by this region are easily separable in extracts where care has been taken to avoid in vitro proteolysis (Minet et al. 1979; Jacobs et al. 1980). The distribution of mutations on the genetic map, together with the absence of any evidence for a physical association between the two enzymatic activities, is compatible with the independent translation of two proteins from a single polycistronic mRNA. Several experimental facts argue against polycistrony for *ARG5,6*. *ARG6* nonsense mutations are absolutely polar, lacking all *ARG5* enzyme activity. If the message were polycistronic, many nonpolar *ARG6* nonsense mutations would be expected since ribosomes could reinitiate at the beginning of *ARG5*. The absolute polarity of nonsense mutations is typical of the other yeast gene clusters (*ADE3, HIS4, TRP5, URA2*). Two general observations on protein synthesis in yeast also argue against the operon interpretation for *ARG5,6*. First, kinetic studies have shown that most yeast messages are monocistronic (Peterson and McLaughlin 1973). Second, yeast ribosomes are apparently able to initiate protein synthesis only at the first 5′ AUG in a message (Sherman et al. 1980). This constraint on initiation would prevent the independent translation of *ARG5* from a single polycistronic message encoding both *ARG5* and *ARG6*. In view of these considerations we think that *ARG5* and *ARG6* are not organized into a bacterial-type operon.

There are several differences between the gene product(s) of *ARG5,6* and those of the other yeast gene clusters. There is no evidence for any physical association between the mature *ARG5*- and *ARG6*-gene products. Moreover, the *ARG5* and *ARG6* enzymes are found in the mitochondrion, whereas the enzymes specified by the other gene clusters are found in the cytosol.

Two interpretations of the *ARG5,6* cluster fit all of the data. (1) A single mRNA carryng information for both *ARG5* and *ARG6* is synthesized and then subsequently cleaved. Additional constraints would have to be added to the RNA cleavage model in order to explain the absolute polarity of *ARG6* nonsense mutations. (2) A single polypeptide containing both activities is synthesized, and this polypeptide is cleaved in vivo posttranscriptionally. We favor the second model because it could provide a physiological role for the cotranslation of *ARG5,6*. Proteolytic cleavage could be required for insertion of the proteins into the mitochondrion.

The *URA2* locus encodes aspartate carbamoyltransferase and the pyrimidine-specific carbamoylphosphate synthetase (Denis-Duphil and Lacroute 1971; Denis-Duphil and Kaplan 1976). Although the two activities usually copurify, they are separable under certain conditions (Lue and Kaplan 1969, 1971). The subunit structure of the enzyme has not been reported. As the enzyme is exquisitely sensitive to the proteinase B present in extracts (Denis-Duphil et al. 1981), possible in vitro cleavage during purification presents a considerate technical problem to this determination. All of the *ura2* mutants that lack both enzyme activities bear nonsense mutations or deletions to one end of the locus (Denis-Duphil and Kaplan 1976). Since the mutations that map to this same region but cause deficiency only for the synthetase activity have not been examined for suppressibility by nonsense suppressors, it is not known whether all nonsense mutations mapping in this region are completely polar. Whether the product of the *URA2* locus is one or more than one polypeptide remains of considerable interest since the activities encoded by *URA2*, like those encoded by the *ARG5,6* cluster, are compartmentalized, albeit in the nucleus rather than the mitochondrion.

Evidence that gene clusters do not always encode multifunctional proteins has been presented for the coregulated *GAL1, GAL7,* and *GAL10* genes. However, these three genes are clearly transcribed into three distinct mRNAs (Oshima, this volume). Moreover, transcription of *GAL1* occurs in a direction opposite to that of *GAL7* and *GAL10*.

CONCLUSIONS

Even a hasty reading of this paper will reveal that our knowledge of regulation for various pathways ranges from little information to quite a lot. In no case do we know what the molecular basis of the regulation is. Enzymes of an amino acid pathway may be controlled by cross-pathway regulation or by a combination of cross-pathway and pathway-specific regulation. Entire amino acid pathways may be controlled solely by cross-pathway regulation. The DNA sequences in the putative control regions of coregulated genes have not yet provided a clue to the basis for coregulation. No control comparable to cross-pathway regulation is known to exist for nucleotide biosynthesis, but none has been sought. And we have no insight as to whether metabolic systems, such as single carbon metabolism, which supply both amino acid and nucleotide synthetic pathways, will have controls that interact with those of the respective receptor pathways. We are only beginning to appreciate that compartmentalization of enzymes and/or metabolites may play a significant role in regulation of metabolism. It is hoped that this review will help in understanding and appreciating what the current status of regulation of anabolism is in yeast and will provide a stimulus to investigation.

ACKNOWLEDGMENTS

We thank Aaron Mitchell, Carol Newlon, George Zubenko, Tom Donahue, and Phil Farabaugh for their helpful discussions and suggestions about the contents and organization of this paper. The research in our laboratories has been supported by research grants from the National Institutes of Health.

REFERENCES

Anderson, J.M. and R.M. Roth. 1974. Cordycepin sensitivity in adenosine utilizing mutants of *Saccharomyces cerevisiae. Biochim. Biophys. Acta* **335**: 285.

Antoniewski, J. and H. de Robichon-Szulmajster. 1973. Biosynthesis of methionine and its control in wild type and regulatory mutants of *Saccharomyces cerevisiae. Biochimie* **55**: 529.

Armitt, S. and R.A. Woods. 1970. Purine-excreting mutants of *Saccharomyces cerevisiae*. I. Isolation and genetic analysis. *Genet. Res.* **15**: 7.

Asahi, T., R. Bandurski, and L. Wilson. 1961. Yeast sulfate-reducing system. II. Enzymatic reduction of protein disulfide. *J. Biol. Chem.* **263**: 1830.

Bach, M. and F. Lacroute. 1972. Direct selective techniques for the isolation of pyrimidine auxotrophs in yeast. *Mol. Gen. Genet.* **115**: 126.

Bach, M., F. Lacroute, and D. Botstein. 1979. Evidence for transcriptional regulation of orotidine-5′-phosphate decarboxylase in yeast by hybridization of mRNA to the yeast structural gene cloned in *Escherichia coli. Proc. Natl. Acad. Sci.* **76**: 386.

Bechet, J., M. Grenson, and J.-M. Wiame. 1970. Mutations affecting the repressibility of arginine biosynthetic enzymes in *Saccharomyces cerevisiae. Eur. J. Biochem.* **12**: 31.

Betterton, H., T. Fjellstedt, M. Matsuda, M. Ogur, and R. Tate. 1968. Localization of the homocitrate pathway. *Biochim. Biophys. Acta* **170**: 459.

Betz, J.L., L.M. Hereford, and P.T. Magee. 1971. Threonine deaminases from *Saccharomyces cerevisiae* mutationally altered in regulatory properties. *Biochemistry* **10**: 1818.

Bhattacharjee, J.K. and A.K. Sinha. 1972. Relationship among the genes, enzymes, and intermediates of the biosynthetic pathway of lysine in *Saccharomyces. Mol. Gen. Genet.* **115**: 26.

Bhattacharjee, J.K. and M. Strassman. 1967. Accumulation of tricarboxylic acids related to lysine biosynthesis in a yeast mutant. *J. Biol. Chem.* **242**: 2542.

Bhattacharjee, J.K., A.F. Tucci, and M. Strassman. 1968. Accumulation of α-ketoglutaric acid in yeast mutants requiring lysine. *Arch. Biochem. Biophys.* **123**: 235.

Bigelis, R. and G.R. Fink. 1978. The use of iodinated antibody in immunodiffusion analysis to detect nonsense termination fragments. *J. Immunol. Methods* **22**: 393.

———. 1981. The *HIS4* multifunctional protein: Immunochemistry of the wild type protein and altered forms. *J. Biol. Chem.* **256**: 5144.

Bigelis, R. and H.E. Umbarger. 1976. Yeast α-isopropylmalate isomerase: Factors affecting stability and enzyme activity. *J. Biol. Chem.* **251**: 3545.

Bisson, L. and J. Thorner. 1977. Thymidine 5′-monophosphate-requiring mutants of *Saccharomyces cerevisiae* are deficient in thymidylate synthetase. *J. Bacteriol.* **132**:44.

Bollon, A.P. 1974. Fine structure analysis of a eukaryotic multifunctional gene. *Nature* **250**:630.

———. 1975. Regulation of the *ilv1* multifunctional gene in *Saccharomyces cerevisiae*. *Mol. Gen. Genet.* **142**:1.

———. 1980 Analysis of yeast *ilv1* CIS control and domain mutants. *Mol. Gen. Genet.* **177**:283.

Bollon, A.P. and P.T. Magee. 1971. Involvement of threonine deaminase in multivalent repression of the isoleucine-valine pathway in *Saccharomyces cerevisiae*. *Proc. Natl. Acad. Sci.* **68**:2169.

———. 1973. Involvement of threonine deaminase in repression of the isoleucine-valine and leucine pathways in *Saccharomyces cerevisiae*. *J. Bacteriol.* **113**:1333.

Botsford, J.L. and L.W. Parks. 1969. Serine transhydroxymethylase in methionine biosynthesis in *Saccharomyces cerevisiae*. *J. Bacteriol.* **97**:1176.

Brandriss, M. 1979. Isolation and preliminary characterization of *Saccharomyces cerevisiae* proline auxotrophs. *J. Bacteriol.* **138**:816.

Brandriss, M. and B. Magasanik. 1980. Proline: An essential intermediate in arginine degradation in *Saccharomyces cerevisiae*. *J. Bacteriol.* **143**:1403.

———. 1981. Subcellular compartmentation in control of converging pathways for proline and arginine metabolism in *Saccharomyces cerevisiae*. *J. Bacteriol.* **145**:1359.

Brendel, M. and W.W. Fäth. 1974. Isolation and characterization of mutants of *Saccharomyces cerevisiae* auxotropic and conditionally auxotrophic for 5′-dTMP. *Z. Naturforsch.* **29**:733.

Breton, A. and Y. Surdin-Kerjan. 1977. Sulfate uptake in *Saccharomyces cerevisiae*: Biochemical and genetic study. *J. Bacteriol.* **132**:224.

Brown, H., T. Satyanarayana, and H. Umbarger. 1975. Biosynthesis of branched-chain amino acids in yeast: Effect of carbon sources on leucine biosynthetic enzymes. *J. Bacteriol.* **121**:959.

Burand, J.P., R. Drillien, and J.K. Bhattacharjee. 1975. Citrate synthaseless glutamic acid auxotroph of *Saccharomyces cerevisiae*. *Mol. Gen. Genet.* **139**:303.

Burns, V.W. 1964. Regulation and coordination of purine and pyrimidine biosynthesis in yeast; I. Regulation of purine biosynthesis and its relation to transient changes in intracellular nucleotide levels. *Biophys. J.* **4**:151.

———. 1966. Regulation of pyrimidine biosynthesis and its strong coupling to the purine system. *Biophys. J.* **6**:787.

Burridge, P.W., R.A. Woods, and J.F. Henderson. 1977. Purine metabolism in *Saccharomyces cerevisiae*. *Can. J. Biochem.* **55**:935.

———. 1978. Altered purine metabolism in regulatory mutations of *Saccharomyces cerevisiae*. *J. Gen. Microbiol.* **107**:403.

Burton, E., J. Selhub, and W. Sakami. 1969. The substrate specificity of 5-methyltetrahydropteroyltriglutamate-homocysteine methyltransferase. *Biochem. J.* **111**:793.

Bussey, H. 1970. Simple selection for end-product inhibitor insensitive mutants in yeast. *J. Bacteriol.* **101**:1081.

Bussey, H. and H. Umbarger. 1969. Biosynthesis of branched-chain amino acids in yeast: Regulation of synthesis of the enzymes of isoleucine and valine biosynthesis. *J. Bacteriol.* **98**:623.

———. 1970. Biosynthesis of the branched-chain amino acids in yeast: A trifluoroleucine-resistant mutant with altered regulation of leucine uptake. *J. Bacteriol.* **103**:286.

Cameron, J.R., E.Y. Loh, and R.W. Davis. 1979. Evidence of transposition of dispersed repetitive DNA families in yeast. *Cell* **16**:739.

Chamberlin, M. 1974. Isolation and characterization of prototrophic mutants of *Escherichia coli* unable to support the intracellular growth of T7. *J. Virol.* **14**:509.

Chattoo, B.B., F. Sherman, D.A. Azubalis, T.A. Fjellstedt, D. Mehnert, and M. Ogur. 1979. Selection of *lys2* mutants of the yeast *Saccharomyces cerevisiae* by the utilization of α-aminoadipate. *Genetics* **93**:51.

Cherest, H. and H. de Robichon-Szulmajster. 1973. The role of two independent genes in one of the regulatory systems involved in methionine biosynthesis in *Saccharomyces cerevisiae*. In *Genetics of industrial microorganisms: Actinomycetes and fungi* (ed. Z. Vaněk et al.), vol. 2, p. 165. Elsevier, Amsterdam.

Cherest, H., F. Eichler, and H. de Robichon-Szulmajster. 1969. Genetic and regulatory aspects of methionine biosynthesis in *Saccharomyces cerevisiae*. *J. Bacteriol.* **97**:328.

Cherest, H., Y. Surdin-Kerjan, and H. de Robichon-Szulmajster. 1971. Methionine-mediated repression in *Saccharomyces cerevisiae:* A pleiotropic regulatory system involving methionyltransfer ribonucleic acid and the product of gene *eth2*. *J. Bacteriol.* **106**:758.

Cherest, H., Y. Surdin-Kerjan, J. Antoniewski, and H. de Robichon-Szulmajster. 1973a. S-adenosylmethionine-mediated repression of methionine biosynthetic enzymes in *Saccharomyces cerevisiae*. *J. Bacteriol.* **114**:928.

———.1973b. Effects of regulatory mutations upon methionione biosynthesis in *Saccharomyces cerevisiae:* Loci eth2-eth3-eth10. *J. Bacteriol.* **115**:1084.

Cherest, H., Y. Surdin-Kerjan, F. Exinger, and F. Lacroute. 1978. S-adenosylmethionine requiring mutants in *Saccharomyces cerevisiae:* Evidences for the existence of two methionine adenosyltransferases. *Mol. Gen. Genet.* **163**:153.

Cohn, M.S., C.W. Tabor, and H. Tabor. 1978. Isolation and characterization of *Saccharomyces cerevisiae* mutants deficient in S-adenosylmethionine decarboxylase, spermidine and spermine. *J. Bacteriol.* **134**:208.

———. 1980. Regulatory mutations affecting ornithine decarboxylase activity in *Saccharomyces cerevisiae*. *J. Bacteriol.* **142**:791.

Cowie, D.R. and F.T. McClure. 1959. Metabolic pools and the synthesis of macromolecules. *Biochim. Biophys. Acta* **31**:236.

Davis, B. 1952. Aromatic biosynthesis. IV. Preferential conversion, in incompletely blocked mutants, of a common precursor of several metabolites. *J. Bacteriol.* **64**:729.

De Deken, R.H. 1962. Pathway of arginine biosynthesis in yeast. *Biochem. Biophys. Res. Commun.* **8**:462.

———. 1963. Biosynthèse de l'arginine chez la levure. I. Le sort de la N-α-acetylornithine. *Biochim. Biophys. Acta* **78**:606.

de Leeuw, A. 1967. Gene enzyme relationships in polyaromatic auxotrophic mutants in *Saccharomyces cerevisiae*. *Genetics* **56**:554.

Delforge, J., F. Messenguy, and J. Wiame. 1975. The regulation of arginine biosynthesis in *Saccharomyces cerevisiae:* The specificity of *argR* mutations and the general control of amino acid biosynthesis. *Eur. J. Biochem.* **57**:231.

Demain, A.L. 1964. Nutrition of "adenineless" auxotrophs of yeast. *J. Bacteriol.* **88**:339.

de Mata, Z.S. and J.C. Rabinowitz. 1980. Formyl-methenyl-methylenetetrahydrofolate synthetase (combined) from yeast. *J. Biol. Chem.* **255**:2569.

De Moss, J. 1965. Biochemical diversity in the tryptophan pathway. *Biochem. Biophys. Res. Commun.* **18**:850.

Denis-Duphil, M. and J.G. Kaplan. 1976. Fine structure of the *URA2* locus in *Saccharomyces cerevisiae*. II. Meiotic and mitotic mapping studies. *Mol. Gen. Genet.* **145**:259.

Denis-Duphill, M. and F. Lacroute. 1971. Fine structure of the *ura2* locus in *Saccharomyces cerevisiae*. I. In vivo complementation studies. *Mol. Gen. Genet.* **112**:354.

Denis-Duphil, M., Y. Mathien-Shire, and G. Hervé. 1981. Proteolytically induced changes in the molecular form of the carbamylphosphate synthetase-uracil-transcarbamylase complex coded for by the *URA2* locus in *Saccharomyces cerevisiae*. *J. Bacteriol.* **148**:659.

de Robichon-Szulmajster, H. 1967. Régulation du fonctionnement de deux chaines de biosynthèse chez *Saccharomyes cerevisiae:* Thréonine-méthionine et isoleucine-valine. *Bull. Soc. Chim. Biol.* **49**:1431.

de Robichon-Szulmajster, H. and H. Cherest. 1967. Regulation of homoserine O-transacetylase, first step in methionine biosynthesis in *Saccharomyces cerevisiae*. *Biochem. Biophys. Res. Commun.* **28**: 256.

de Robichon-Szulmajster, H. and D. Corrivaux. 1963. Régulations métaboliques de la biosynthèse de la méthionine et de la thréonine chez *Saccharomyces cerevisiae*. I. Répression et rétro-inhibition de l'aspartokinase. *Biochim. Biophys. Acta* **73**: 248.

―――. 1964. Régulations métaboliques de la biosynthèse de la méthionine et de la thréonine chez *Saccharomyces cerevisiae*. III. Etude cinétique de la répression et de la dérepression des trois premiers enzymes de la chaine. *Biochim. Biophys. Acta* **92**: 1.

de Robichon-Szulmajster, H. and P.T. Magee. 1968. The regulation of isoleucine-valine biosynthesis in *Saccharomyces cerevisiae*. I. Threonine deaminase. *Eur. J. Biochem.* **3**: 492.

de Robichon-Szulmajster, H.Y. Surdin, and R.K. Mortimer. 1966. Genetic and biochemical studies of genes controlling the synthesis of threonine and methionine in *Saccharomyces*. *Genetics* **53**: 609.

de Robichon-Szulmajster, H., Y. Surdin-Kerjan, and H. Cherest. 1973. Regulatory aspects of threonine and methionine biosynthesis in *Saccharomyces cerevisiae*. In *Genetics of industrial microorganisms: Actinomycetes and fungi* (ed. Z. Vaněk et al.), vol. 2, p. 149. Elsevier, Amsterdam.

de Robichon-Szulmajster, H., Y. Surdin, Y. Karassevitch, and D. Corrivaux. 1965. Régulations métaboliques chez *Saccharomyces cerevisiae*. Biosynthèse de la méthionine et de la thréonine. *Colloq. Int. Cen. Natl. Rech. Sci.* **124**: 255.

Deschamps, J. and J.M. Wiame. 1979. Mating-type effect on cis mutations leading to constitutivity of ornithine transaminase in diploid cells of *Saccharomyces cerevisiae*. *Genetics* **92**: 749.

Dettwiler, M. and K. Kirschner. 1979. Tryptophan synthase from *Saccharomyces cerevisiae* is a dimer of two polypeptide chains of M_r 76,000 each. *Eur. J. Biochem.* **102**: 159.

de Vito, P. and J. Dreyfuss. 1964. Metabolic regulation of adenosine triphosphate sulfurylase in yeast. *J. Bacteriol.* **88**: 1341.

Donahue, T.F., P.J. Farabaugh, and G.R. Fink. 1982. The nucleotide sequence of the *HIS4* region of yeast. *Gene* **18**: 47.

Dorfman, B. 1964. Allelic complementation at the *ade5/7* locus in yeast. *Genetics* **50**: 1231.

―――. 1969. The isolation of adenylosuccinate synthetase mutants in yeast by selection for constitutive behavior in pigmented strains. *Genetics* **61**: 377.

―――. 1971. Allelic variability in comparative complementation confirming that the *ade12*-specified protein of yeast is bifunctional. *J. Bacteriol.* **107**: 646.

Dorfman, B., B.A. Goldfinger, and M. Berger. 1970. Partial reversion in yeast: Genetic evidence for a new type of bifunctional protein. *Science* **168**: 1482.

Doy, C. 1968. Aromatic biosynthesis in yeast. II. Feedback inhibition and repression of 3-deoxy-D-*arabino*heptulosonic acid 7-phosphate synthase. *Biochim. Biophys. Acta* **151**: 293.

Doy, C. and J. Cooper. 1966. Aromatic biosynthesis in yeast. I. The synthesis of tryptophan and the regulation of this pathway. *Biochim. Biophys. Acta* **127**: 302.

Dubois, E.L. and M. Grenson. 1974. Absence of involvement of glutamine synthetase and of NAD-linked glutamate dehydrogenase in the nitrogen catabolite repression of arginase and other enzymes in *Saccharomyces cerevisiae*. *Biochem. Biophys. Res. Commun.* **60**: 150.

Dubois, E., D. Hiernaux, M. Grenson, and J.-M. Wiame. 1978. Specific induction of catabolism and its relation to repression of biosynthesis in arginine metabolism of *Saccharomyces cerevisiae*. *J. Mol. Biol.* **122**: 383.

Errede, B., T.S. Cardillo, F. Sherman, E. Dubois, T. Deschamps, and J.-M. Wiame. 1980. Mating signals control expression of mutations resulting from insertion of a transposable repetitive element adjacent to diverse yeast genes. *Cell* **22**: 427.

Fantes, P.A., L.M. Roberts, and R. Huetter. 1976. Free tryptophan pool and tryptophan biosynthetic enzymes in *Saccharomyces cerevisiae*. *Arch. Microbiol.* **107**: 207.

Farabaugh, P.J. and G.R. Fink. 1980. Insertion of the eukaryotic transposable element Ty1 creates a 5 base pair duplication. *Nature* **286**: 352.

Ferro, A.J. and K.D. Spence. 1973. Induction and repression in the *S*-adenosylmethionine and methionine biosynthetic systems of *Saccharomyces cerevisiae*. *J. Bacteriol.* **116:** 812.

Fesneau, C., H. de Robichon-Szulmajster, A. Fradin, and H. Feldmann. 1975. tRNAs undermethylation in a met-regulatory mutant of *Saccharomyces cerevisiae*. *Biochimie.* **57:** 49.

Fink, G. 1964. Gene-enzyme relations in histidine biosynthesis in yeast. *Science* **146:** 525.

———. 1966. A cluster of genes controlling three enzymes in histidine biosynthesis in *Saccharomyces cerevisiae*. *Genetics* **53:** 445.

Fink, G., P. Farabaugh, G. Roeder, and D. Chaleff. 1980. Transposable elements (Ty) in yeast. *Cold Spring Harbor Symp. Quant. Biol.* **45:** 575.

Fisher, C.R. 1969. Enzymology of the pigmented adenine-requiring mutants of *Saccharomyces* and *Schizosaccharomyces*. *Biochem. Biophys. Res. Commun.* **34:** 306.

Fjellstedt, T.A. and M. Ogur. 1970. Effects of supersuppressor genes on enzymes controlling lysine biosynthesis in *Saccharomyces*. *J. Bacteriol.* **101:** 108.

Freundlich, M., R. Burns, and H. Umbarger. 1962. Control of isoleucine, valine, and leucine biosynthesis. I. Multivalent repression. *Proc. Natl. Acad. Sci.* **48:** 1804.

Fuchs, J. 1977. Isolation of an *Escherichia coli* mutant deficient in thioredoxin reductase. *J. Bacteriol.* **129:** 967.

Game, J.C. 1976. Yeast cell-cycle mutant *cdc21* is a temperature-sensitive thymidylate auxotroph. *Mol. Gen. Genet.* **146:** 313.

Gannon, F., K. O'Hare, F. Perrin, J.P. LePennec, C. Benoist, M. Cochet, R. Breathnach, A. Royal, A. Garapin, B. Cami, and P. Chambon. 1979. Organization and sequences at the 5' end of a cloned complete ovalbumin gene. *Nature* **278:** 428.

Gardner, W.J.E. and R.A. Woods. 1979. Isolation and characterization of guanine auxotrophs in *Saccharomyces cerevisiae*. *Can. J. Microbiol.* **25:** 380.

Gray, G. and J.K. Bhattacharjee. 1976. Biosynthesis of lysine in *Saccharomyces cerevisiae*: Regulation of homocitrate synthase in analogue-resistant mutants. *J. Gen. Microbiol.* **97:** 117.

Grenson, M. 1969. The utilization of exogenous pyrimidines and the recycling of uridine-5'-phosphate derivatives in *Saccharomyces cerevisiae*, as studied by means of mutants affected in pyrimidine uptake and metabolism. *Eur. J. Biochem.* **11:** 249.

Grenson, M., E. Dubois, M. Piotrowska, R. Drillien, and M. Aigle. 1974. Ammonia assimilation in *Saccharomyces cerevisiae* as mediated by the two glutamate dehydrogenases; evidence for the *gdhA* locus being a structural gene for the NADP-dependent glutamate dehydrogenase. *Mol. Gen. Genet.* **128:** 73.

Gross, T.S. and R.A. Woods. 1971. Identification of mutants defective in the first and second steps of *de novo* purine synthesis in *Saccharomyces cerevisiae*. *Biochim. Biophys. Acta* **274:** 13.

———. 1972. Regulation of *de novo* purine nucleotide synthesis by enzyme repression in *Saccharomyces cerevisiae*. *Heredity* **28:** 275.

Halbreich, A. and M. Rabinowitz. 1971. Isolation of *Saccharomyces cerevisiae* mitochondrial formyltetrahydrofolic acid: Methionyl-tRNA transformylase and the hybridization of mitochondrial fMet-tRNA with mitochondrial DNA. *Proc. Natl. Acad. Sci.* **68:** 295.

Hampsey, D.M. and G.B. Kohlhaw. 1981. Inactivation of yeast α-isopropylmalate synthase by CoA. Antagonism between CoA and adenylates and the mechanism of CoA inactivation. *J. Biol. Chem.* **256:** 3791.

Hilger, F., M. Culot, M. Minet, A. Piérard, M. Grenson, and J.-M. Wiame. 1973. Studies on the kinetics of the enzyme sequence mediating arginine synthesis in *Saccharomyces cerevisiae*. *J. Gen. Microbiol.* **75:** 33.

Hinnen, A., J.B. Hicks, and G.R. Fink. 1978. Transformation of yeast. *Proc. Natl. Acad. Sci.* **75:** 1929.

Holland, J.P. and M.J. Holland. 1979. The primary structure of a glyceraldehyde-3-phosphate dehydrogenase gene from *Saccharomyces cerevisiae*. *J. Biol. Chem.* **254:** 9839.

Hollenberg, C.P., W.F. Riks, and P. Borst. 1970. The glutamate dehydrogenases of yeast: Extra-mitochondrial enzymes. *Biochim. Biophys. Acta* **201:** 13.

Holzer, H. and G. Kohlhaw. 1961. Enzymatic formation of α-acetolactate from α-hydroxyethyl-2-thiamine pyrophosphate ("active aldehyde") and pyruvate. *Biochem. Biophys. Res. Commun.* **5:** 452.

Hsu, Y.-P., G. Kohlhaw, and P. Niederberger. 1982. Evidence that α-isopropylmalate synthase of *Saccharomyces cerevisiae* is under the "general" control of amino acid biosynthesis. *J. Bacteriol.* **150:** 969.

Jacob, F. and J. Monod. 1961. Genetic regulatory mechanisms in the synthesis of proteins. *J. Mol. Biol.* **3:** 318.

Jacobs, P., J. Jauniaux, and M. Grenson. 1980 A *cis*-dominant regulatory mutation linked to the *arg*B-*arg*C gene cluster in *Saccharomyces cerevisiae*. *J. Mol. Biol.* **139:** 691.

Jauniaux, J., L. Urrestarazu, and J. Wiame. 1978. Arginine metabolism in *Saccharomyces cerevisiae*. Subcellular localization of the enzymes. *J. Bacteriol.* **133:** 1096.

Jones, E.E. and H.P. Broquist. 1965. Saccharopine, an intermediate of the aminoadipic acid pathway of lysine biosynthesis. II. Studies in *Saccharomyces cerevisiae*. *J. Biol. Chem.* **240:** 2531.

Jones, E.W. and B. Magasanik. 1967. Genetic block in the interconversion of folic acid coenzymes in *Saccharomyces cerevisiae*. *Bacteriol. Proc.*, p. 127.

Jones, G.E. 1978. L-asparagine auxotrophs of *Saccharomyces cerevisiae*: Genetic and phenotypic characterization. *J. Bacteriol.* **134:** 200.

Jund, R. and F. Lacroute. 1970. Genetic and physiological aspects of resistance to 5-fluoropyrimidines in *Saccharomyces cerevisiae*. *J. Bacteriol.* **102:** 607.

―――. 1972. Regulation of orotidylic acid pyrophosphorylase in *Saccharomyces cerevisiae*. *J. Bacteriol.* **109:** 196.

Kakar, S.N. and R.P. Wagner. 1964. Genetic and biochemical analysis of isoleucine-valine mutants of yeast. *Genetics* **49:** 213.

Kaplan, J.G., M. Duphil, and F. Lacroute. 1967. A study of the aspartate transcarbamylase activity of yeast. *Arch. Biochem. Biophys.* **119:** 541.

Karassevitch, Y. and H. de Robichon-Szulmajster. 1963. Régulation métabolique de la biosynthèse de la méthionine et de la thréonine chez *Saccharomyces cerevisiae*. *Biochim. Biophys. Acta* **73:** 414.

Kay, D.G., R.A. Singer, and G.C. Johnston. 1980. Ornithine decarboxylase activity and cell cycle regulation in *Saccharomyces cerevisiae*. *J. Bacteriol.* **141:** 1041.

Keesey, J., Jr., R. Bigelis, and G. Fink. 1979. The product of the *his4* gene cluster in *Saccharomyces cerevisiae*: A trifunctional polypeptide. *J. Biol. Chem.* **254:** 7427.

Kohlhaw, G. 1982. Regulation of leucine biosynthesis in lower eukaryotes. In *Biotechnology: Amino acid biosynthesis and regulation* (ed. K.H. Herrman and R.S. Somerville), vol. 1. Addison-Wesley. (In press).

Kohlhaw, G.B., Y. Hsu, R.D. Lemmon, and T.D. Petes. 1980. Transposed *LEU2* gene of *Saccharomyces cerevisiae* is regulated normally. *J. Bacteriol.* **144:** 852.

Korch, C. and R. Snow. 1973. Allelic complementation in the first gene for histidine biosynthesis in *Saccharomyces cerevisiae*. I. Characteristics of mutants and genetic mapping of alleles. *Genetics* **74:** 287.

Korch, C.T., F. Lacroute, and F. Exinger. 1974. A regulatory interaction between pyrimidine and purine biosynthesis via ureidosuccinic acid. *Mol. Gen. Genet.* **133:** 63.

Kozak, M. 1981. Mechanism of mRNA recognition by eukaryotic ribosomes during initiation of protein synthesis. *Curr. Top. Microbiol. Immunol.* **93:** 81.

Kradolfer, P., J. Zeyer, G. Miozzari, and R. Huetter. 1977. Dominant regulatory mutants in chorismate mutase of *Saccharomyces cerevisiae*. *FEMS Microbiol. Lett.* **2:** 211.

Lacroute, F. 1968. Regulation of pyrimidine biosynthesis in *Saccharomyces cerevisiae*. *J. Bacteriol.* **95:** 824.

Lacroute, F., A. Piérard, M. Grenson, and J.-M. Wiame. 1965. The biosynthesis of carbamoyl phosphate in *Saccharomyces cerevisiae*. *J. Gen. Microbiol.* **40:** 127.

Lax, C., S. Fogel, and C. Cramer. 1979. Regulatory mutants at the *his1* locus of yeast. *Genetics* **92:** 363.

Lazowska, J. and M. Luzzatti. 1970a. Biochemical deficiency associated with *ad3* mutations in *Saccharomyces cerevisiae*. I. Levels of three enzymes of tetrahydrofolate metabolism. *Biochem. Biophys. Res. Commun.* **39:**34.

———. 1970b. Biochemical deficiency associated with *ad3* mutations in *Saccharomyces cerevisiae*. II. Separation of two forms of methylene tetrahydrofolate dehydrogenase. *Biochem. Biophys. Res. Commun.* **39:**40.

Lemoine, Y., E. Dubois, and J.-M. Wiame. 1978. The regulation of urea amidolyase of *Saccharomyces cerevisiae*. *Mol. Gen. Genet.* **166:**251.

Lingens, F., W. Goebel, and H. Uesseler. 1966. Regulation der biosynthese der aromatischen aminosauren in *Saccharomyces cerevisiae*. *Biochem. Z.* **346:**357.

———. 1967. Regulation der biosynthese der aromatischen aminosauren in *Saccharomyces cerevisiae*. 2. Repression, induktion, und aktivierung. *Eur. J. Biochem.* **1:**363.

Linnane, A.W. and J.L. Still. 1955. The isolation of respiring mitochondria from baker's yeast. *Arch. Biochem. Biophys.* **59:**383.

Loison, G., R. Losson, and F. Lacroute. 1980. Constitutive mutants for orotidine 5 phosphate decarboxylase and dihydroorotic acid dehydrogenase in *Saccharomyces cerevisiae*. *Curr. Genet.* **2:**39.

Lomax, C.A. and R.A. Woods. 1971. Prototrophic regulatory mutants of adenylosuccinate synthetase in yeast. *Nat. New Biol.* **229:**116.

———. 1973. A complex genetic locus controlling purine nucleotide biosynthesis in yeast. *Mol. Gen. Genet.* **120:**139.

Lomax, C.A., T.S. Gross, and R.A. Woods. 1971. New mutant types at the *ade3* locus of *Saccharomyces cerevisiae*. *J. Bacteriol.* **107:**1.

Lor, K.L. and E.A. Cossins. 1972. Regulation of C_1 metabolism by L-methionine in *Saccharomyces cerevisiae*. *Biochem. J.* **130:**773.

Lowenstein, R.E. 1973. "Genetic studies in *Saccharomyces cerevisiae*." Ph.D. thesis, Cornell University, Ithaca, New York.

Lue, P.F. and J.G. Kaplan. 1969. The aspartate transcarbamylase and carbamoyl phosphate synthetase of yeast: A multifunctional enzyme complex. *Biochem. Biophys. Res. Commun.* **34:**426.

———. 1970. Metabolic compartmentation at the molecular level: The function of a multienzyme aggregate in the pyrimidine pathway of yeast. *Biochim. Biophys. Acta* **220:**365.

———. 1971. Aggregation states of a regulating enzyme complex catalyzing the early steps of pyrimidine biosynthesis in baker's yeast. *Can. J. Biochem.* **49:**403.

Lumsden, J. and J. Coggins. 1977. The subunit structure of the *arom* multienzyme complex of *Neurospora crassa*: A possible pentafunctional polypeptide chain. *Biochem. J.* **161:**599.

Luzzatti, M. 1975. Isolation and properties of a thymidylateless mutant in *Saccharomyces cerevisiae*. *Eur. J. Biochem.* **56:**533.

Magee, P. and L. Hereford. 1969. Multivalent repression of isoleucine-valine biosynthesis in *Saccharomyces cerevisiae*. *J. Bacteriol.* **98:**857.

Magee, P.T. and H. de Robichon-Szulmajster. 1968a. The regulation of isoleucine-valine biosynthesis in *Saccharomyces cerevisiae*. 2. Identification and characterization of mutants lacking the acetohydroxyacid synthetase. *Eur. J. Biochem.* **3:**502.

———. 1968b. The regulation of isoleucine-valine biosynthesis in *Saccharomyces cerevisiae*. 3. Properties and regulation of the activity of acetohydroxyacid synthetase. *Eur. J. Biochem.* **3:**507.

Manney, T. 1964. Action of a super-suppressor in yeast in relation to allelic mapping and complementation. *Genetics* **50:**109.

———. 1968. Evidence for chain termination by super-suppressible mutants in yeast. *Genetics* **60:**719.

Manney, T., W. Duntze, N. Janosko, and J. Salazar. 1969. Genetic and biochemical studies of partially active tryptophan synthetase mutants of *Saccharomyces cerevisiae*. *J. Bacteriol.* **99:**590.

Maragoudakis, M.E. and M. Strassman. 1966. Homocitric acid accumulation by a lysine-requiring yeast mutant. *J. Biol. Chem.* **241:** 695.

Maragoudakis, M.E., H. Holmes, and M. Strassman. 1967. Control of lysine biosynthesis in yeast by a feedback mechanism. *J. Bacteriol.* **93:** 1677.

Mark, D., J. Chase, and C. Richardson. 1977. Genetic mapping of *trxA*, a gene affecting thioredoxin in *Escherichia coli* K12. *Mol. Gen. Genet.* **155:** 145.

Masselot, M. and H. de Robichon-Szulmajster. 1972. Nonsense mutation in the regulatory gene *ETH2* involved in methionine biosynthesis in *Saccharomyces cerevisiae*. *Genetics* **71:** 535.

———. 1975. Methionine biosynthesis in *Saccharomyces cerevisiae*. I. Genetic analysis of auxotrophic mutants. *Mol. Gen. Genet.* **139:** 121.

Masselot, M. and Y. Surdin-Kerjan. 1977. Methionine biosynthesis in *Saccharomyces cerevisiae*. II. Gene-enzyme relationships in the sulfate assimilation pathway. *Mol. Gen. Genet.* **154:** 23.

Matsuda, M. and M. Ogur. 1969a. Separation and specificity of the yeast glutamate-α-ketoadipate transaminase. *J. Biol. Chem.* **244:** 3352.

———. 1969b. Enzymatic and physiological properties of the yeast glutamate-α-ketoadipate transaminase. *J. Biol. Chem.* **244:** 5153.

Mattoon, J.R., T.A. Moshier, and T.H. Kreiser. 1961. Separation and partial characterization of a lysine precursor produced by a yeast mutant. *Biochim. Biophys. Acta* **51:** 615.

McKenzie, K.Q. and E.W. Jones. 1977. Mutants of the formyltetrahydrofolate interconversion pathway of *Saccharomyces cerevisiae*. *Genetics* **86:** 85.

McLaughlin, C.S., P.T. Magee, and L.H. Hartwell. 1969. Role of isoleucyltransfer ribonucleic acid synthetase in ribonucleic acid synthesis and enzyme repression in yeast. *J. Bacteriol.* **100:** 579.

Mertz, J.E. and K.D. Spence. 1972. Methionine adenosyltransferase and ethionine resistance in *Saccharomyces cerevisiae*. *J. Bacteriol.* **111:** 778.

Messenguy, F. 1976. Regulation of arginine biosynthesis in *Saccharomyces cerevisiae:* Isolation of a *cis*-dominant, constitutive mutant for ornithine carbamoyltransferase synthesis. *J. Bacteriol.* **128:** 49.

——— 1979. Concerted repression of the synthesis of the arginine biosynthetic enzymes by amino acids: A comparison between the regulatory mechanisms controlling amino acid biosyntheses in bacteria and in yeast. *Mol. Gen. Genet.* **169:** 85.

Messenguy, F. and T.G. Cooper. 1977. Evidence that specific and "general" control of ornithine carbamoyltransferase production occurs at the level of transcription in *Saccharomyces cerevisiae*. *J. Bacteriol.* **130:** 1253.

Messenguy, F. and J. Delforge. 1976. Role of transfer ribonucleic acids in the regulation of several biosyntheses in *Saccharomyces cerevisiae*. *Eur. J. Biochem.* **67:** 335.

Messenguy, F., D. Colin, and J. Ten Have. 1980. Regulation of compartmentation of amino acid pools in *Saccharomyces cerevisiae* and its effects on metabolic control. *Eur. J. Biochem.* **108:** 439.

Messenguy, F., M. Penninckx, and J.-M. Wiame. 1971. Interaction between arginase and ornithine carbamoyltransferase in *Saccharomyces cerevisiae*: The regulatory site for ornithine. *Eur. J. Biochem.* **22:** 277.

Meuris, P. 1967. Regulation de la première étape de la chaine de biosynthése des composés aromatiques chez *Saccharomyces cerevisiae*. *Bull. Soc. Chim. Biol.* **49:** 1573.

———. 1969. Studies of mutants inhibited by their own metabolites in *Saccharomyces cerevisiae*. II. Genetic and enzymatic analysis of three classes of mutants. *Genetics* **63:** 569.

———. 1973. Feedback inhibition of the DAHP synthetase by tRNA in *Saccharomyces cerevisiae*. *Mol. Gen. Genet.* **121:** 207.

———. 1974. Feedback-insensitive mutants of the gene for the tyrosine-inhibited DAHP synthetase in yeast. *Genetics* **76:** 735.

Meuris, P., F. Lacroute, and P. Slonimski. 1967. Étude systematique de mutants inhibes par leurs propres metabolites chez la levure *Saccharomyces cerevisiae*. 1. Obtention et characterisation des differences classes de mutants. *Genetics.* **56:** 140.

Middelhoven, W.J. 1964. The pathway of arginine breakdown in *Saccharomyces cerevisiae*. *Biochim. Biophys. Acta* **93:** 650.

Minet, M. 1971. "La biosynthèse de l'arginine chez *S. cerevisiae:* Contribution a l'étude de son determinisme génétique et de sa regulation." Ph.D. thesis, University of Brussels, Belgium.

Minet, M., J.C. Jauniaux, P. Thuriaux, M. Grenson, and J.-M. Wiame. 1979. Organization and expression of a two-gene cluster in the arginine biosynthesis of *Saccharomyces cerevisiae*. *Mol. Gen. Genet.* **168:** 299.

Miozzari, G., P. Niederberger, and R. Hütter. 1978. Tryptophan biosynthesis in *Saccharomyces cerevisiae:* Control of the flux through the pathway. *J. Bacteriol.* **134:** 48.

Morzycka, E. and A. Paszewski. 1979. Two pathways of cysteine biosynthesis in *Saccharomyces lipolytica*. *FEBS Lett.* **101:** 97.

Nagy, M., J. Laporte, B. Penverne, and G. Hervé. 1982. Nuclear localization of aspartate transcarbamylase in *Saccharomyces cerevisiae*. *J. Cell Biol.* **92:** 790.

Naiki, N. 1964. Enzymatic defects in sulfate reducing system of sulfite-less yeast mutants. *Plant Cell Physiol.* **5:** 71.

———. 1965. Some properties of sulfite reductase from yeast. *Plant Cell Physiol.* **6:** 179.

Naiki, N. and S. Yamagata. 1973. Isolation of O-acetylhomoserine from culture broth of a methionine auxotroph of *Saccharomyces cerevisiae*. *Plant Cell Physiol.* **14:** 1193.

Nass, G. and R. Hasenbank. 1970. Effect of borrelidin on the threonyl-tRNA-synthetase activity and the regulation of threonine-biosynthetic enzymes in *Saccharomyces cerevisiae*. *Mol. Gen. Genet.* **108:** 28.

Nass, G. and K. Poralla. 1976. Genetics of borrelidin resistant mutants of *Saccharomyces cerevisiae* and properties of their threonyl-tRNA synthetase. *Mol. Gen. Genet.* **147:** 39.

Niederberger, R., G. Miozzari, and R. Hütter. 1981. Biological role of the general control of amino acid biosynthesis in *Saccharomyces cerevisiae*. *Mol. Cell. Biol.* **1:** 584.

Ogur, M., L. Coker, and S. Ogur. 1964a. Glutamate auxotrophs in *Saccharomyces*. I. The biochemical lesion in the *glt* mutants. *Biochem. Biophys. Res. Commun.* **14:** 193.

Ogur, M., A. Roshanmanesh, and S. Ogur. 1964b. Tricarboxylic acid cycle mutants in *Saccharomyces:* Comparison of independently derived mutants. *Science* **147:** 1590.

Ogur, M., T.N. Liu, I. Cheung, I. Paulavicius, W. Wales, D. Mehnert, and D. Blaise. 1977. "Active" one-carbon generation in *Saccharomyces cerevisiae*. *J. Bacteriol.* **129:** 926.

Paszewski, A. and J. Grabski. 1974. Regulation of S-amino acid biosynthesis in *Aspergillus nidulans*; role of cysteine and/or homocysteine as regulatory effectors. *Mol. Gen. Genet.* **132:** 307.

Paukert, J.L., G.R. Williams, and J.C. Rabinowitz. 1977. Formyl-methenyl-methylenetetrahydrofolate synthetase (combined): Correlation of enzymic activities with limited proteolytic degradation of the protein from yeast. *Biochem. Biophys. Res. Commun.* **77:** 147.

Peterson, N.S. and C. McLaughlin. 1973. Monocistronic mRNA in yeast. *J. Mol. Biol.* **81:** 33.

Penninckx, M. 1975. Interaction between arginase and L-ornithine carbamoyltransferase in *Saccharomyces cerevisiae*. *Eur. J. Biochem.* **58:** 533.

Perlman, P.S. and H.R. Mahler. 1970. Intracellular localization of enzymes in yeast. *Arch. Biochem. Biophys.* **136:** 245.

Piérard, A. and B. Schröter. 1978. Structure-function relationships in the arginine pathway carbamoyl phosphate synthase of *Saccharomyces cerevisiae*. *J. Bacteriol.* **134:** 167.

Piérard A., F. Messenguy, A. Feller, and F. Hilger. 1979. Dual regulation of the synthesis of the arginine pathway carbamoyl phosphate synthase of *Saccharomyces cerevisiae* by specific and general controls of amino acid biosynthesis. *Mol. Gen. Genet.* **174:** 163.

Porqué, P., A. Baldesten, and P. Reichard. 1970a. Purification of a thioredoxin system from yeast. *J. Biol. Chem.* **245:** 2363.

———. 1970b. The involvement of the thioredoxin system in the reduction of methionine sulfoxide and sulfate. *J. Biol. Chem.* **245:** 2371.

Ramos, F. and J.-M. Wiame. 1970. Synthesis and activation of asparagine in asparagine auxotrophs of *Saccharomyces cerevisiae*. *Eur. J. Biochem.* **94:**409.

———. 1980. Two asparagine synthetases in *Saccharomyces cerevisiae*. *Arch. Int. Physiol. Biochim.* **88:**B96.

Rasse-Messenguy, F. and G. Fink. 1973. Feedback resistant mutants of histidine biosynthesis in yeast. In *Genes, enzymes, and populations* (ed. A.M. Srb), p. 85. Plenum Press, New York.

Ratzkin, B. and J. Carbon. 1977. Functional expression of cloned yeast DNA in *Escherichia coli*. *Proc. Natl. Acad. Sci.* **74:**487.

Reichert, U. and M. Winter. 1975. Gene dosage effects in polyploid strains of *Saccharomyces cerevisiae* containing *gua-1* wild-type and mutant alleles. *J. Bacteriol.* **124:**1041.

Roeder, G.S. and G.R. Fink. 1980. DNA rearrangements associated with a transposable element in yeast. *Cell* **21:**239.

———. 1982. Movement of yeast transposable elements by gene conversion. *Proc. Natl. Acad. Sci.* **79:**5621.

Roeder, G.S., P.J. Farabaugh, D.T. Chaleff, and G.R. Fink. 1980. The origins of gene instability in yeast. *Science* **209:**1375.

Roeder, P. 1980. Studies on α-isopropylmalate synthase and regulation of leucine biosynthetic enzymes in yeast. *Dis. Abstr. Int.* **41:**176-B.

Roman, H. 1956. A system selective for mutations affecting the synthesis of adenine in yeast. *C.R. Trav. Carlsberg. Ser. Physiol.* **26:**299.

Roon, R.J., H.L. Even, and F. Larimore. 1974. Glutamate synthase: Properties of the reduced nicotinamide adenine dinucleotide-dependent enzyme from *Saccharomyces cerevisiae*. *J. Bacteriol.* **118:**89.

Rose, M., M.J. Casadaban, and D. Botstein. 1981. Yeast genes fused to β-galactosidase in *Escherichia coli* can be expressed normally in yeast. *Proc. Natl. Acad. Sci.* **78:**2460.

Rothstein, R.J. and F. Sherman. 1979. Dependence on mating type for the overproduction of iso-2-cytochrome *c* in the yeast mutant *CYC7-H2*. *Genetics* **94:**891.

Ryan, E.D. and G.B. Kohlhaw. 1974. Subcellular localization of isoleucine-valine biosynthetic enzymes in yeast. *J. Bacteriol.* **120:**631.

Ryan, E.D., J.W. Tracy, and G.B. Kohlhaw. 1973. Subcellular localization of leucine biosynthetic enzymes in yeast. *J. Bacteriol.* **116:**222.

Sagisaka, S. and K. Shimura. 1962. Studies in lysine biosynthesis. III. Enzymatic reduction of α-amino adipic acid: Isolation and some properties of the enzyme. *J. Biochem.* **51:**398.

Satyanarayana, T. and J.G. Kaplan. 1971. Regulation of the purine pathway in baker's yeast: Activity and feedback inhibition of phosphoribosylpyrophosphate amidotransferase. *Arch. Biochem. Biophys.* **142:**40.

Satayanarayana, T., H.E. Umbarger, and G. Lindegren. 1968a. Biosynthesis of branched-chain amino acids in yeast: Correlation of biochemical blocks and genetic lesions in leucine auxotrophs. *J. Bacteriol.* **96:**2012.

———. 1968b. Biosynthesis of branched-chain amino acids in yeast: Regulation of leucine biosynthesis in prototrophic and leucine auxotrophic strains. *J. Bacteriol.* **96:**2018.

Savin, M. and M. Flavin. 1972. Cystathionine synthesis in yeast: An alternative pathway for homocysteine biosynthesis. *J. Bacteriol.* **112:**299.

Scheifinger, C., S. Ogur, and M. Ogur. 1966. Genetic and metabolic control of homoaconitase biosynthesis. *Fed. Proc.* **25:**710.

Schürch, A., J. Miozzari, and R. Hütter. 1974. Regulation of tryptophan biosynthesis in *Saccharomyces cerevisiae*: Mode of action of 5-methyl-tryptophan and 5-methyl-tryptophan-sensitive mutants. *J. Bacteriol.* **117:**1131.

Schürch-Rathgeb, Y. 1972. Der *trp3*-locus von *Saccharomyces cerevisiae*. *Arch. Genet.* **45:**129.

Sherer, S. and R.W. Davis. 1980. Recombination of dispersed repeated DNA sequences in yeast. *Science* **209:**1380.

Sherman, F., J.W. Stewart, and A.M. Schweingruber. 1980. Mutants of yeast initiating translation of cytochrome *c* within a region spanning 37 nucleotides. *Cell* **20**:215.
Silver, J.M. and N.R. Eaton. 1968. Biosynthetic blocks of eight non-allelic adenine-requiring mutants of *Saccharomyces cerevisiae*. *Genetics* **60**:225.
———. 1969. Functional blocks of the *ad1* and *ad2* mutants of *Saccharomyces cerevisiae*. *Biochem. Biophys. Res. Commun.* **34**:301.
Silverman, S.J., M. Rose, D. Botstein, and G.R. Fink. 1982. Regulation of *HIS4-LACz* fusions in yeast. *Gene* (in press).
Simon, J.-P. and V. Stalon. 1978. Epiarginasic regulation in *Saccharomyces cerevisiae*. Citrulline, the third effector, acts at a specific binding site on the ornithine carbamoyl transferase. *Eur. J. Biochem.* **88**:287.
Singh, A. and F. Sherman. 1974a. Association of methionine requirement with methyl mercury resistant mutants of yeast. *Nature* **247**:227.
———. 1974b. Characteristics and relationships of mercury-resistant mutants and methionine auxotrophs of yeast. *J. Bacteriol.* **118**:911.
———. 1975. Genetic and physiological characteristics of *met15* mutants of *Saccharomyces cerevisiae*. A selective system for forward and reverse mutations. *Genetics* **81**:75.
Sinha, A.K. and J.K. Bhattacharjee. 1971. Lysine biosynthesis in *Saccharomyces;* conversion of α-aminoadipate into α-aminoadipic δ-semialdehyde. *Biochem. J.* **125**:743.
Sinha, A.K., M. Kurtz, and J.K. Bhattacharjee. 1971. Effects of hydroxylysine on the biosynthesis of lysine in *Saccharomyces*. *J. Bacteriol.* **108**:715.
Smith, M., D.W. Leung, S. Gillam, and C.R. Astell. 1979. Sequence of the gene for iso-1-cytochrome *c* in *Saccharomyces cerevisiae*. *Cell* **16**:753.
Smolina, V.S., V.M. Andrianova, and M.L. Bekker. 1978. Regulation of the biosynthesis of purine nucleotides in mutants of the yeast *Saccharomyces cerevisiae* with increased sensitivity of the de novo pathway of synthesis to inhibition by exogenous guanine. *Genetika* **14**:1495.
Sorsoli, W.A., M. Buettner, and L.W. Parks. 1968. Cystathionine metabolism in methionine auxotrophic and wild-type strains of *Saccharomyces cerevisiae*. *J. Bacteriol.* **95**:1024.
Strassman, M. and S. Weinhouse. 1963. Biosynthetic pathway. III. The biosynthesis of lysine by *Torulopsis utilis*. *J. Am. Chem. Soc.* **75**:1680.
Struhl, K. 1979. "The yeast *his3* gene." Ph.D. thesis, Stanford University, California.
———. 1981. Deletion mapping a eukaryotic promoter. *Proc Natl. Acad. Sci.* **78**:4461.
Struhl, K. and R. Davis. 1980. A physical and transcriptional map of the cloned *his3* region of *Saccharomyces cerevisiae*. *J. Mol. Biol.* **136**:309.
Surdin, Y. 1967. La semi-aldéhyde aspartique déshydrogénase chez *Saccharomyces cerevisiae*: Propriétés et regulation. *Eur. J. Biochem.* **2**:341.
Surdin-Kerjan, Y. and H. de Robichon-Szulmajster. 1975. Existence of two levels of repression in the biosynthesis of methionine in *Saccharomyces cerevisiae*: Effect of lomofungin on enzyme synthesis. *J. Bacteriol.* **122**:367.
Sures, I., J. Lowry, and L.H. Kedes. 1978. The DNA sequence of sea urchin (*S. purpuratus*) H2A, H2B and H3 histone coding and spacer regions. *Cell* **15**:1033.
Szostak, J.W., J.I. Stiles, C.P. Bahl, and R. Wu. 1977. Specific binding of a synthetic oligodeoxyribonucleotide to yeast cytochrome *c* mRNA. *Nature* **265**:61.
Thuriaux, P., M. Minet, A.M.A. ten Berge, and F.F. Zimmermann. 1971. Genetic fine structure and function of mutants at the *ilv1*-gene locus of *Saccharomyces cerevisiae*. *Mol. Gen. Genet.* **112**:60.
Thuriaux, P., F. Ramos, A. Piérard, M. Grenson, and J.-M. Wiame. 1972. Regulation of the carbamoyl phosphate synthetase belonging to the arginine pathway of *Saccharomyces cerevisiae*. *J. Mol. Biol.* **67**:277.
Torii, K. and R. Bandurski. 1967. Yeast sulfate-reducing system. III. An intermediate in the reduction of 3'-phosphoryl-5'-adenosinephosphosulfate to sulfite. *Biochim. Biophys. Acta* **136**:286.

Tracy, J.W. and G.B. Kohlhaw. 1975. Reversible coenzyme-A-mediated inactivation of biosynthetic condensing enzymes in yeast: A possible regulatory mechanism. *Proc. Natl. Acad. Sci.* **72**: 1802.

———. 1977. Evidence for two distinct CoA binding sites on yeast α-isopropylmalate synthase. *J. Biol. Chem.* **252**: 4085.

Tsang, M. and J. Schiff. 1976. Sulfate reducing pathway in *Escherichia coli* involving bound intermediates. *J. Bacteriol.* **125**: 923.

Tucci, A.F. and L.N. Ceci. 1972a. Homocitrate synthase from yeast. *Arch. Biochem. Biophys.* **153**: 742.

———. 1972b. Control of lysine biosynthesis in yeast. *Arch. Biochem. Biophys.* **153**: 751.

Ulane, R. and M. Ogur. 1972. Genetic and physiological control of serine and glycine biosynthesis in *Saccharomyces*. *J. Bacteriol.* **109**: 34.

Ulm, E., R. Bohme, and G. Kohlhaw. 1972. α-isopropylmalate synthase from yeast: Purification, kinetic studies and effect of ligands on stability. *J. Bacteriol.* **110**: 1118.

Umbarger, H.E. 1978. Amino acid biosynthesis and its regulation. *Annu. Rev. Biochem.* **47**: 533.

Unger, M. and L. Hartwell. 1976. Control of cell division in *Saccharomyces cerevisiae* by methionyl-tRNA. *Proc. Natl. Acad. Sci.* **73**: 1664.

Urrestarazu, L.A., S. Vissers, and J.-M. Wiame. 1977. Change in location of ornithine carbamoyltransferase and carbamoyl phosphate synthetase among yeast in relation to the arginase/ornithine carbamoyltransferase regulatory complex and the energy status of the cells. *Eur. J. Biochem.* **79**: 473.

Vitols, E. and A.W. Linnane. 1961. Studies on the oxidative metabolism of *Saccharomyces cerevisiae*: The morphology and oxidative phosphorylation capacity of mitochondria and derived particles from baker's yeast. *J. Biophys. Biochem. Cytol.* **9**: 701.

Whitney, P.A. and D.R. Morris. 1978. Polyamine auxotrophs of *Saccharomyces cerevisiae*. *J. Bacteriol.* **134**: 214.

Wiame, J.-M. 1971a. The regulation of arginine metabolism in *Saccharomyces cerevisiae*: Exclusion mechanisms. *Curr. Top. Cell. Regul.* **4**: 1.

———. 1971b. Mechanism of the interaction between anabolism and catabolism of arginine in *Saccharomyces cerevisiae*. In *Recent advances in microbiology* (ed. A. P'erez-Miravete and D. Pel'aez), p. 243. Association Mexicana Microbiologia, Mexico City.

Wickner, R.B. 1975. Mutants of *Saccharomyces cerevisiae* that incorporate deoxythymidine 5'-monophosphate into DNA *in vivo*. *Methods Cell Biol.* **11**: 295.

Wiemken, A. 1980. Compartmentation and control of amino acid utilization in yeast. In *Cell compartmentation and metabolic channeling* (ed. L. Nover et al.), p. 225. Elsevier/North Holland Biomedical Press, Amsterdam.

Wiemken, A. and M. Durr. 1974. Characterization of amino acid pools in the vacuolar compartment of *Saccharomyces cerevisiae*. *Arch. Microbiol.* **101**: 45.

Williamson, V.M., E.T. Young, and M. Ciriacy. 1981. Transposable elements associated with constitutive expression of yeast alcohol dehydrogenase II. *Cell* **23**: 605.

Wilson, L., T. Asahi, and R. Bandurski. 1961. Yeast sulfate-reducing system. I. Reduction of sulfate to sulfite. *J. Biol. Chem.* **236**: 1822.

Wipf, B. and T. Leisinger. 1977. Compartmentation of arginine biosynthesis in *Saccharomyces cerevisiae*. *FEMS Microbiol. Lett.* **2**: 239.

———. 1979. Regulation of activity and synthesis of N-acetylglutamate synthase from *Saccharomyces cerevisiae*. *J. Bacteriol.* **140**: 874.

Wolfner, M., D. Yep, F. Messenguy, and G. Fink. 1975. Integration of amino acid biosynthesis into the cell cycle of *Saccharomyces cerevisiae*. *J. Mol. Biol.* **96**: 273.

Woods, R.A. 1969. Response of *ad-2* mutants of *Saccharomyces cerevisiae* to carbon dioxide. *Mol. Gen. Genet.* **105**: 314.

Woods, R.A. and I.E. Jackson. 1973. The accumulation of glycinamide ribotide by *ade3* and *ade8* mutants of *Saccharomyces cerevisiae*. *Biochem. Biophys. Res. Commun.* **53**: 787.

Yamagata. S. 1980. Occurrence of low molecular weight O-acetylserine sulfhydrylase in the yeast *Saccharomyces cerevisiae*. *J. Biochem.* **88**: 1419.

Yamagata, S. and K. Takeshima. 1976. O-acetylserine and O-acetylhomoserine sulfhydrylase of yeast; further purification and characterization as a pyridoxal enzyme. *J. Biochem.* **80**: 777.

Yamagata, S., K. Takeshima, and N. Naiki. 1974. Evidence for the identity of O-acetylserine sulfhydrylase with O-acetylhomoserine sulfhydrylase in yeast. *J. Biochem.* **75**: 1221.

———. 1975. O-acetylserine and O-acetylhomoserine sulfhydrylase of yeast: Studies with methionine auxotrophs. *J. Biochem.* **77**: 1029.

Yoshimoto, A. and R. Sato. 1968a. Studies on yeast sulfite reductase. I. Purification and characterization. *Biochim. Biophys. Acta* **153**: 555.

———. 1968b. Studies on yeast sulfite reductase. II. Partial purification and properties of genetically incomplete sulfite reductases. *Biochim. Biophys. Acta* **153**: 576.

———. 1970. Studies on yeast sulfite reductase. III. Further characterization. *Biochim. Biophys. Acta* **220**: 190.

Zalkin, H. and C. Yanofsky. 1982. Yeast gene *TRP5:* Structure, function, regulation. *J. Biol. Chem.* **257**: 1491.

Zelikson, R. and M. Luzzatti. 1976. Two forms of serine transhydroxymethylase, one absent in a thymidylate-less mutant in *Saccharomyces cerevisiae*. *Eur. J. Biochem.* **64**: 7.

———. 1977. Mitochondrial and cytoplasmic distribution in *Saccharomyces cerevisiae* of enzymes involved in folate-coenzyme-mediated one-carbon-group transfer; a genetic and biochemical study of the enzyme deficiencies in mutants *tmp3* and *ade3*. *Eur. J. Biochem.* **79**: 285.

Mutations Altering Initiation of Translation of Yeast Iso-1-cytochrome c; Contrasts between the Eukaryotic and Prokaryotic Initiation Process

Fred Sherman and John W. Stewart
Department of Radiation Biology and Biophysics
University of Rochester School of Medicine and Dentistry
Rochester, New York 14642

1. Mutations of the AUG Initiation Codon
2. Methionine Aminopeptidase
3. Construction of Sequences by Mutation and Recombination
4. Translation Initiating within a 37-nucleotide Region
5. AUG Is the Only Initiation Codon
6. Translation Initiates Only at the Most 5'-AUG Codon
7. Mutations Influencing Efficiency of Translation
8. *cyc1-362* Mutation in the Leader Region
9. Contrasts between Eukaryotes and Prokaryotes
Appendix

INTRODUCTION

Although the yeast *Saccharomyces cerevisiae* is a eukaryotic organism that has a greater genetic complexity than bacteria, it shares many of the technical advantages that permit rapid progress in determining fundamental mechanisms of biological processes. One of the major impacts of yeast in molecular biology is its use for investigating processes that are particularly amenable to genetic analysis and that are different in prokaryotes and eukaryotes. A systematic analysis of mutational alterations at the end of the *CYC1* gene in the yeast *S. cerevisiae* has revealed the nucleotide sequences that effect initiation of translation. The analysis of these findings led us to suggest that certain essential features of the translational process may be fundamentally different from the analogous features in *Escherichia coli*. Furthermore, we have suggested that these differences between yeast and *E. coli* reflect differences between eukaryotes and prokaryotes. These findings and conclusions concerning the initiation process are briefly reviewed in this paper.

MUTATIONS OF THE AUG INITIATION CODON

Protein analysis of mutationally altered forms of iso-1-cytochrome *c*, along with genetic analysis of the corresponding mutant strains, established that the primary structure of this mitochondrial protein is determined by the *CYC1* gene (Sherman et al. 1966), which is located on the right arm of chromosome X (Lawrence et al. 1975). Over 400 independently derived *cyc1* mutants that lack iso-1-cytochrome *c* or that contain nonfunctional iso-1-cytochrome *c* have been isolated by one of three following procedures: (1) a spectroscopic scanning procedure (Sherman 1964), (2) a benzidine staining procedure (Sherman et al. 1968), and (3) a procedure using chlorolactate medium (Sherman et al. 1974). The vast majority of the *cyc1* mutants contain single-site mutations that could be assigned to at least 47 different sites within the translated portion of the *CYC1* locus (Sherman et al. 1975; F. Sherman and S. Consaul, unpubl.). The positions of the mutational lesions were determined from the rates of X-ray-induced recombination of crosses of these *cyc1* mutants with a series of *cyc1* deletions (Sherman et al. 1975). In addition, the sites could be precisely located from the results of two-point crosses of the unknown *cyc1* mutants with tester *cyc1* strains (Sherman et al. 1974). Fine-structure mapping with *cyc1* deletions and *cyc1* point mutations resulted in the assignment of 17 out of 490 *cyc1* mutations to the site corresponding to the AUG initiator codon. Furthermore, the analysis of altered iso-1-cytochromes *c* from intragenic revertants (summarized in Table 1) indicated that the deficiency of iso-1-cytochrome *c* was due to the alteration of the AUG codon that is required for initiation of translation (Stewart et al. 1971; F. Sherman and J.W. Stewart, unpubl.). Structural analysis of 64 revertant proteins from these 17 *cyc1* mutants indicated that some of the reverse mutations introduced abnormal initiator codons at new sites. Each of the *cyc1* mutants gave rise to revertant iso-1-cytochrome *c*, which had one of the following five additions at the amino terminus of the otherwise normal protein: Met-Ile-, Met-Leu-, Met-Arg-, Met-Lys-, and a mixture of approximately 85% Val- and 15% Met-Val. A given *cyc1* mutant always gave rise to revertants having the same longer form. In addition, other revertants contained either the normal protein or a short form that lacked the four normal aminoterminal residues (Table 1).

These results are adequately explained by the mutational pathways that are presented in Figure 1, illustrating the formation and reversion of the *cyc1-131* mutant. The mRNA of normal iso-1-cytochrome *c* (Fig. 1) contains an AUG codon that is required for initiation of translation and encodes a methionine residue that is excised in vivo by an aminopeptidase (see the next section). The *cyc1-131* mutant lacks iso-1-cytochrome *c* because the AUG initiator codon is mutated to the valine codon GUG. A normal protein in the revertant *CYC1-131-A* is the result of an exact reverse mutation. The long form of iso-1-cytochrome *c* in the *CYC1-131-C* revertant is generated by an AT → GC transition in the AUA triplet preceding the normal initiator codon to produce an abnormal AUG initiator codon, as shown in FIgure 1. With this protein, the methionine aminopeptidase is only partially active, and the long form contains approximately 85% Val- and

Table 1 cyc1 mutations of the AUG initiation codon

Nucleotide position	Mutational change in mRNA	Allele number	Inducing mutagen[a]	Revertants[b] normal	Revertants[b] short form	Revertants[b] long form
1	AUG → GUG	cyc1-131	UV	8	1	1
		cyc1-497	None	1	0	1
1	AUG → CUG[c]	cyc1-51	UV	1	1	2
		cyc1-100	None	0	0	1
		cyc1-181	UV	1	0	1
2	AUG → AGG	cyc1-133	UV	1	0	2
2	AUG → AAG	cyc1-493	None	0	0	3
3	AUG → AUA[d]	cyc1-13	NIL	10	2	10
		cyc1-667	NQO	0	0	3
		cyc1-687	NQO	0	0	2
3	AUG → AUA	cyc1-74	UV	0	0	1
	or AUU	cyc1-85	UV	0	0	1
	or AUC	cyc1-163	None	0	0	1
		cyc1-490	None	0	0	3
		cyc1-513	None	0	0	2
		cyc1-515	None	0	0	2
		cyc1-610[e]	DEB	1	0	2

(Data from Stewart et al. 1971; F. Sherman and J.W. Stewart, unpubl.)

[a] The cyc1 mutants were isolated from the strain D311-3A either after no treatment (None) or after treatments with ultraviolet light (UV), 1-nitrosoimidazolidone-2 (NIL), 4-nitroquinoline-1-oxide (NQO), or DL-diepoxybutane (DEB).

[b] The short form lacks the four aminoterminal residues of the normal protein. The long form carries the following appendices at the amino terminus of the otherwise normal protein: Met-Val- and Val- (cyc1-131, etc.); Met-Leu- (cyc1-51, etc.); Met-Arg- (cyc1-133); Met-Lys- (cyc1-493); Met-Ile- (cyc1-13, cyc1-74, etc.).

[c] The mutation is assumed to be CUG and not UUG because the cyc1-51, cyc1-100, and cyc1-181 mutants are reverted with NQO and because NQO is believed to preferentially act on GC base pairs (Prakash et al. 1974).

[d] The cyc1-13, cyc1-667, and cyc1-687 mutations are assumed to be AUA and not AUU or AUC, because NIL and NQO preferentially induce GC → AT transitions (Prakash and Sherman 1973; Prakash et al. 1974).

[e] The cyc1-610 mutation also contains a substitution that results in a CAG codon instead of the normal AAG codon at amino acid position 4.

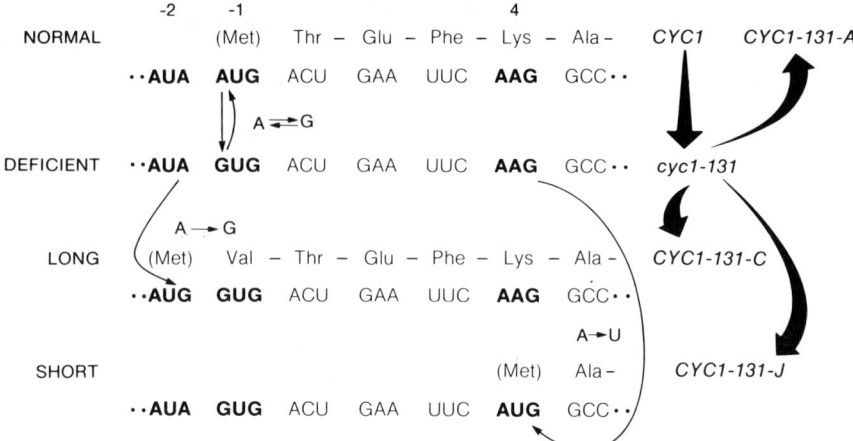

Figure 1 The mutational pathway leading to the formation of the *cyc1-131* mutation and to the revertants containing the normal (*CYC1-131-A*), long (*CYC1-131-C*), and short (*CYC1-131-J*) forms of iso-1-cytochrome *c*. The codons at the normal site, −1, and the abnormal sites, −2 and 4, of initiation of translation are shown in boldface type. Arrows indicate single base-pair substitutions in the DNA that encode the mRNA sequences shown. The methionine residues (in parentheses) are excised from ~85% of the long form of iso-1-cytochrome *c* in the *CYC1-131-C* strain and from all of the normal and short forms of iso-1-cytochrome *c* in the *CYC1* and *CYC1-131-J* strains, respectively. (Adapted from Stewart et al. 1971.)

15% Met-Val- at the amino terminus. Still another initiating AUG codon is provided by AT → TA transversion in the AAG codon for lysine 4, resulting in the short protein in the *CYC1-131-J* revertant after excision of the aminoterminal methionine.

Except for threonine, all amino acids whose codons differ from AUG by single nucleotides were observed in the different long forms of iso-1-cytochrome *c*. The lack of long proteins with threonine at the amino terminus is reasonably explained by the improbability of isolating representatives of all nine possible single-base changes in a group of 17 *cyc1* mutants. Thus, the formation of all 17 *cyc1* mutants and all of the revertants is simply explained by single base-pair changes and by the ability to form effective AUG initiation codons at the sites corresponding to positions −2, −1, and 4.

METHIONINE AMINOPEPTIDASE

Methionine animopeptidase activities that remove aminoterminal residues of methionine from certain proteins have been observed in both prokaryotes (Adams 1968; Takeda and Webster 1968; Matheson et al. 1970; Vogt 1970) and eukaryotes (Kerwar et al. 1971; Kolehmainen and Mikola 1971; Yoshida and Lin 1972). The amino termini of iso-1-cytochromes c from revertants of the initiation mutations described above suggest that the yeast aminopeptidase removes aminoterminal residues of methionine when they precede certain amino acids (Stewart et al. 1971). Revertants of initiation mutants and of other $cyc1$ mutants having iso-1-cytochromes c with altered amino termini are listed in Table 2. The aminoterminal sequences of the iso-1-cytochromes c indicate that the hypothetical aminopeptidase excises methionine from residues of threonine, alanine, and glycine but not from isoleucine, leucine, methionine, asparagine, lysine, arginine, or glutamine; the aminopeptidase appears to be only partially active in excising methionine from residues of valine. The residue at the position penultimate to the aminoterminal methionine residue does not appear to influence the specificity of the aminopeptidase. This is most clearly demonstrated with the series of iso-1-cytochrome c having threonine at the amino terminus but having one of ten different adjacent residues (Table 2).

Table 2 Aminoterminal sequences of iso-1-cytochromes c

Thr-Glu-Phe-	*CYC1* (normal)	Met-Ile-Thr-Glu-	*CYC1-13-A* etc.
Thr-Gln-Phe-	*CYC1-9-A* etc.	Met-Ile-Glu-Phe-	*CYC1-9-AU*
Thr-Leu-Phe-	*CYC1-9-Q* etc.	X-Met-Ile-Arg-Ile-	*CYC1-31-Y*
Thr-Tyr-Phe-	*CYC1-9-S* etc.	Met-Lys-Thr-Glu-	*CYC1-493-A*
Thr-Lys-Phe-	*CYC1-9-P* etc.	Met-Lys-Gly-Ala-	*CYC1-345-F*
Thr-Ser-Phe-	*CYC1-9-AA* etc.	Met-Met-Ile-Met-	*CYC1-242-O*
Thr-Pro-Phe-	*CYC1-9-CA*	X-Met-Met-Asn-Ser-	*CYC1-183-T*
Thr-Asp-Ser-	*CYC1-183-U*	Met-Asn-Glu-Lys-	*CYC1-31-N*
Thr-Gly-Ile-	*CYC1-239-Y*	Met-Asn-Gln-Phe-	*CYC1-9-CB*
Thr-Ala-Lys-	*CYC1-31-K*	X-Met-Asn-Asn-Asn-	*CYC1-345-H*
Ala-Gly-Ser-	*CYC1-13-S* etc.	Met-Leu-Thr-Glu-	*CYC1-51-C* etc.
Gly-Ser-Ala-	*CYC1-345-J*	Met-Arg-Thr-Glu-	*CYC1-133-A*
		Met-Gln-Ala-Gly-	*CYC1-345-C*
		Met-Val-Thr-Glu- 15%	*CYC1-131-C* etc.
		Val-Thr-Glu- 85%	

X denotes an unknown blocking group that may possibly be acetyl. (Data from Stewart et al. 1971; 1972; J.W. Stewart and F. Sherman, unpubl.)

The methionine aminopeptidase specificity that is deduced from the results of the *cyc1* mutants is consistent with the results that are observed for amino termini of other proteins from a wide range of prokaryotes and eukaryotes. Vogt (1970) suggested a specificity that was based on the amino termini of proteins from *E. coli* and its bacteriophages. However, the deduced specificity of methionine aminopeptidase from only amino termini of mature proteins is ambiguous because of the possibility of additional posttranslational processing. A more reliable means of deducing the specificity of methionine aminopeptidases is based on the aminoterminal sequences formed at initiation sites. Such data have become available by sequencing both proteins and the corresponding RNAs or by sequencing proteins that were translated in cell-free systems lacking processing activities or deformylase activity. Recently, we uncovered 40 pertinent cases from *E. coli,* several mammalian species, several higher plants, chickens, and sea urchins after a systematic search of the literature (J.W. Stewart and F. Sherman, in prep.). The data from this search, in conjunction with the data on *CYC1* described above, led us to conclude that aminoterminal methionine is cleaved when it precedes residues of alanine, glycine, valine, proline, serine, or threonine but not when it precedes residues of leucine, isoleucine, methionine, asparagine, lysine, arginine, glutamic acid, or aspartic acid.

CONSTRUCTION OF SEQUENCES BY MUTATION AND RECOMBINATION

The revertant iso-1-cytochromes *c* from the *cyc1* mutants discussed above demonstrated that translation can initiate at the sites corresponding to amino acid positions -2, -1, and 4, as illustrated in Figure 1. To extend the investigation of the requirements for initiation of translation, we have systematically altered the 5' region of the *CYC1* locus and examined the corresponding amounts of iso-1-cytochrome *c* in the mutant strains. Sequences were constructed for investigating numerous questions, including the extent of the region in which translation can initiate, the utilization of codons other than AUG for initiation, sites of initiation in sequences containing more than one AUG codon, and the influence of UAA, frameshift mutations, and deletion mutations on initiation. Mutational and recombinational steps used in constructing some of the mutant sequences are briefly outlined in the Appendix to this chapter.

TRANSLATION INITIATING WITHIN A 37-NUCLEOTIDE REGION

By the analysis of revertants of initiator mutations (see, e.g., *cyc1-131* shown in Fig. 1), we have shown that translation of iso-1-cytochrome *c* can initiate at amino acid positions -2, -1, and 4. We have extended this analysis with the use of the mutant *cyc1-345,* which is described below. Relevant to this analysis is the finding that the aminoterminal region encompassing residues 1 through 10 is not essential for function (Sherman and Stewart 1973; Stewart and Sherman 1974). Every one of the residues at positions 1 through 10 were altered, and deletions of up to 6 residues within the aminoterminal region were observed among at least partially functional iso-1-cytochromes c from approximately 50 revertants. Although some of the mutants contained iso-1-cytochrome c that was slightly diminished in amount or activity, all of the mutants were selected on lactate medium and had sufficient activities to allow significant utilization of lactate as a sole carbon source. Thus, it appeared as if at least partially functional iso-1-cytochrome c should be formed if initiation of translation occurred at the site of any AUG codon preceding amino acid position 11.

As shown in Table 3, the *cyc1-345* mutant contains an AUA codon instead of the normal AUG initiating codon at position -1, a UAA ochre codon instead of the GAA codon at position 2, and a CAA codon instead of the AAG codon at position 4. Intragenic revertants having normal amounts of iso-1-cytochrome c could not be derived from the *cyc1-345* mutants by single base-pair changes. Mutation of the AUA codons at position -1 or -2 to AUG does not allow formation of iso-1-cytochrome c because of the UAA codon at position 2 (see below). In contrast to simple initiation mutants of the type illustrated in Figure 1, AUG initiation codons cannot be formed at position 4 by single base-pair changes because the codon at this position in the *cyc1-345* mutant is CAA.

Revertants of *cyc1-345* with normal amounts of iso-1-cytochrome c are rare and arise only after mutagenic treatments. The primary structures of the iso-1-cytochromes c from 12 *cyc1-345* revertants were determined by Sherman et al. (1980b); four of these are listed in Table 3. The *CYC1-345-C* revertant appeared to have arisen by a two base-pair substitution that changed the UUC codon at amino acid position 3 to AUG. The sequence of the *CYC1-345-H* revertant appears to have arisen by a deletion of nucleotide 12 or 13 and a U → G substitution at nucleotide positon -5, forming an AUG initiation codon at amino acid position -3. The *CYC1-345-J* revertant cannot be explained by a two base-pair change and requires three base-pair changes at amino acid position 5 or a complex rearrangement.

In addition to uncovering rare revertants that arose by multiple base-pair changes and that have normal amounts of iso-1-cytochrome c, the *cyc1-345* mutant gave rise to a revertant, *CYC1-345-F,* that arose by a single base-pair substitution but that contained only approximately 20% of the normal amount of iso-1-cytochrome c (Sherman et al. 1980b). The amino acid sequence of iso-1-cytochrome c from the *CYC1-345-F* revertant (presented in Table 3) suggests that the reversion occurred by a single AT → TA transversion of the codon at position 9. Although it is not known why the *CYC1-345-F* revertant contains a subnormal amount of iso-1-cytochrome c, other evidence discussed below indicates that partial deficiencies cannot be

Table 3 Aminoterminal sequences of iso-1-cytochrome c from intragenic revertants of the cyc1-13 and cyc1-345 mutants

Allele	Sequence
	-3 -2 -1 3 4 5 9
CYC1+	(Met)Thr-Glu-Phe-Lys-Ala-Gly-Ser-Ala-Lys-Gly-Ala-Thr-Leu-Phe-Lys-Thr-
	A UUA AUA AUG ACU GAA UUC AAG GCC GGU UCU GCU AAG AAA GGU GCU ACA CUU UUC AAG ACU
	-5 1 12 30
cyc1-13	A UUA AUA _AUA_ ACU GAA UUC _AAG_ GCC GGU UCU GCU _AAG_ AAA GGU CCU ACA CUU UUL AAG ACU
cyc1-345	A UUA AUA _AUA_ ACU _UAA_ UUC _CAA_ GCC GGU UCU GCU _AAG_ AAA GGU GCU ACA CUU UUC AAG ACU
CYC1-345-H	X-*Met-Asn-Asn-Asn-Leu-Ile-Gln*-Ala-Gly-Ser-Ala-Lys-Lys-Gly-
CYC1-13-A	*Met-Ile*-Thr-Glu-Phe-Lys-Ala-Gly-Ser-Ala-Lys-Lys-Gly-
CYC1+	Thr-Glu-Phe-Lys-Ala-Gly-Ser-Ala-Lys-Lys-Gly-
CYC1-345-C	*Met-Gln*-Ala-Gly-Ser-Ala-Lys-Lys-Gly-
CYC1-13-S	Ala-Gly-Ser-Ala-Lys-Lys-Gly-
CYC1-345-J	Gly-Ser-Ala-Lys-Lys-Gly-
CYC1-345-F	*Met*-Lys-Gly-Ala-Thr-Leu-Phe-Lys-Thr-

Portions of the protein and mRNA sequences of the normal iso-1-cytochrome c (CYC1+) and the inferred mRNA sequence of the cyc1-13 and cyc1-345 mutants are shown at the top. Sites of initiation of translation appear to have been derived by single nucleotide changes in the triplets underlined in the cyc1-13 mRNA sequence and by single or multiple nucleotide changes in the triplets underlined in the cyc1-345 mRNA sequence. X in the CYC1-345-H sequence refers to an unknown blocking group. Altered residues and codons are denoted in italics. All revertants contained normal or near normal amounts of iso-1-cytochrome c, except the CYC1-13-S and CYC1-345-F revertants, which contained, respectively, ~ 50% and ~ 20% of the normal amount. (Adapted from Stewart et al. 1971; Sherman et al. 1980b.)

simply related to the distance between normal and abnormal initiation sites. We have suggested that the unusually low level of iso-1-cytochrome c in the CYC1-345-F revertant could reflect an additional requirement for the aminoterminal region of the protein, such as for the proper attachment of the heme group (Sherman et al. 1980b).

The amino acid sequences of the cyc1-345 revertants, along with the amino acid sequences of the cyc1-13 revertants (presented in Table 3), indicate that translation can initiate at sites corresponding to amino acid positions -3, -2, -1, 3, 4, 5, and 9, a region that encompasses 37 nucleotides. All of the iso-1-cytochrome c occurred at normal levels, except for the iso-1-cytochromes c initiating at positions 4 and 9, which contained, respectively, 50% and 20% of the normal level. Because the reduced amount of iso-1-cytochrome c observed in the mutant that initiates at position 9 may not necessarily be due to an inefficiency of translation, translation appears to be able to initiate with normal or almost normal efficiency at any site within the region spanning 37 nucleotides and presumably at any site in an extended region preceding and following the site of the normal initiation codon. The relocation of initiation sites establishes that there is no absolute requirement for a unique nucleotide sequence immediately adjacent to the AUG initiation codon.

AUG IS THE ONLY INITIATION CODON

The single base-pair changes that inactivated and formed AUG initiation codons and the pathways of reversion of certain cyc1 mutants led Sherman et al. (1980a) to conclude that codons other than AUG were unable to initiate translation of iso-1-cytochrome c. Iso-1-cytochrome c is absent when the AUG initiation codon is mutated to GUG, AUA, CUG, AGG, or AAG (Table 1), thus establishing that none of these codons can initiate translation. Because translation can be initiated at least between amino acid positions -3 through 9 if an initiator codon is present and because, e.g., the cyc1-13 mutant completely lacks iso-1-cytochrome c, the following additional codons are unable to initiate translation: UUA, ACU, GAA, UUC, GCC, GGU, UCU, and GCU (see Table 3). Furthermore, numerous other codons were deduced to be incapable of initiating translation by considering the pathway of reversion of the two frameshift mutants cyc1-183 (Table 8) and cyc1-239 (Table 9) and the cyc1-345 mutant (Table 3), and by considering that reversion of initiator mutants is restricted to amino acid positions -2, -1, and 4 (Table 1; Sherman et al. 1980a). A detailed analysis of the revertants of these cyc1 mutants suggested that no codons other than AUG could initiate translation with normal or near normal efficiencies; only the UGG codon was not excluded as being a potential initiator codon (Sherman et al. 1980a).

The observation that only the AUG codon is capable of initiating translation of iso-1-cytochrome c is consistent with the observation that only AUG codons initiate translation in over 60 other eukaryotic mRNAs (Kozak 1981). Although AUG is the most commonly found initiation codon in prokaryotes (Steitz 1979), GUG normally initiates translation of the products of the *A* gene in bacteriophage MS2 (Fiers et al. 1975) and of the *lacI* gene in *E. coli* (Steege 1977); in addition, UUG may reinitiate translation when the *lacI* gene contains a certain mutation (Files et al. 1975; Weber and Geisler 1978). The sole use of AUG for initiation in eukaryotes and the use of both AUG and GUG for initiation in prokaryotes have been attributed to the presence of the hypermodified base t^6A in eukaryotic initiator tRNA and its absence in prokaryotic initiator tRNA (Sherman et al. 1980a).

TRANSLATION INITIATES ONLY AT THE MOST 5'-AUG CODON

Although translation can initiate within the region corresponding to amino acid positions −3 through 9 (Table 3), no abnormal initiation sites were uncovered after examining proteins from over 130 intragenic revertants of the *cyc1-9* mutant, which contain a UAA codon at amino acid position 2 (Table 8). Because the various *cyc1-9* revertants contained every single base-pair change of the UAA codon, as well as several instances of multiple base-pair changes (Stewart et al. 1972; Lawrence et al. 1974; Sherman and Stewart 1974), the lack of initiation at the lysine 4 codon did not appear to be due to the lack of mutation of this AAG codon to AUG. It appeared more likely that reinitiation could not occur after a nonsense codon, although such types of reinitiation occur in mutants of the *lacI* gene of *E. coli* (Platt et al. 1972; Ganem et al. 1973; Files et al. 1975; Weber and Geisler 1978). The lack of *cyc1-9* revertants initiating at position 4 could be due solely to the UAA codon at position 2 or to the presence of the normal AUG codon in conjunction with the UAA codon.

The lack of *cyc1-9* revertants with reinitiation sites prompted us to systematically examine sequences containing two AUG codons with or without an intervening UAA mutation. These sequences (listed in Table 4) include the *CYC1-340a* and *cyc1-341* recombinants (Table 9) and the *CYC1-341-D* revertant (Table 10). Examination of iso-1-cytochrome c from the *CYC1-340a* mutant indicates that translation initiates predominantly, if not exclusively, at the normal positon −1 even when an AUG codon is at position 4. The lack of iso-1-cytochrome c in the *cyc1-341* mutant and the presence of iso-1-cytochrome c in the *CYC1-341-D* revertant indicates that initiation does not occur at the AUG at position 4 if the normal initiation codon at position −1 is still present. The results show that translation will initiate only at the most 5'-AUG codon even when two AUG codons are separated by UAA codons. These conclusions were corroborated with still other sequences,

Table 4 Sequences containing various combinations of AUG codons at positions -1, 4, and 9 and UAA codons at position 2

	-1 2 4 9	
CYC1	AUG ACU GAA UUC AAG GCC GGU UCU GCU AAG AAA GGU (Met)Thr-Glu-Phe-Lys-Ala-Gly-Ser-Ala-Lys-Lys-Gly- ├──────────────────────────────────→	100%
CYC1-340a	AUG ACU GAA UUC AUG GCC GGU UCU GCU AAG AAA GGU (Met)Thr-Glu-Phe-Met-Ala-Gly-Ser-Ala-Lys-Lys-Gly- ├──────────────────────────────────→	100%
CYC1-599-A	AUG ACU CAA UUC CAA GCC GGU UCU GCU AUG AAA GGU (Met)Thr-Gln-Phe-Gln-Ala-Gly-Ser-Ala-Met-Lys-Gly- ├──────────────────────────────────→	100%
cyc1-341	AUG ACU UAA UUC AUG GCC GGU UCU GCU AAG AAA GGU Met-Thr ├────→	0%
cyc1-599	AUG ACU UAA UUC CAA GCC GGU UCU GCU AUG AAA GGU Met-Thr ├────→	0%
CYC1-341-D	AUA ACU UAA UUC AUG GCC GGU UCU GCU AAG AAA GGU (Met)Ala-Gly-Ser-Ala-Lys-Lys-Gly- ├──────────────────────────→	50%
CYC1-599-B	AUA ACU UAA UUC CAA GCC GGU UCU GCU AUG AAA GGU Met-Lys-Gly- ├──────→	20%

All amino acid sequences were directly determined experimentally except for the Met-Thr peptide, which is suggested to occur in the *cyc1-341* and *cyc1-599* mutants. The percentages of the normal amounts of iso-1-cytochrome *c* are indicated at the right of the arrows. (See footnote to Table 3 for other designations.) (Adapted from Sherman and Stewart 1973; F. Sherman and J.W. Stewart, in prep.)

cyc1-599, CYC1-599-A, and *CYC1-599-B,* which were constructed by the procedures outlined in Table 11 (see Appendix to this chapter). Thus, translation can initiate at positions 4 and 9 only when other AUG codons are not 5' to these sites.

The lack of reinitiation at position 4 was further considered with the frameshift mutations listed in Table 5. The frameshift mutant *cyc1-331* contains no recognizable iso-1-cytochrome *c* because of a deletion of a single adenylate residue at nucleotide position 8 or 9. This single base-pair deletion also occurs in the *cyc1-333* mutant, which also has a base-pair substitution that creates an in-phase AUG triplet near the site corresponding to amino acid position 4 (Table 5); nevertheless, the *cyc1-333* mutant lacks iso-1-cytochrome *c*. However, if the AUG codon at the normal position -1 in the *cyc1-333* mutant is inactivated by mutation, then translation of iso-1-

Table 5 Sequence with a deletion of nucleotide 8 or 9 and with various combinations of AUG codons at amino acid position −1 and at or near amino acid position 4

	−1　　　2　　　　　4	
CYC1	(Met)Thr-Glu-Phe-Lys-Ala- AUG ACU GAA UUC AAG GCC 　1　　　　　　　8	100%
CYC1-340a	(Met)Thr-Glu-Phe-*Met*-Ala- AUG ACU GAA UUC *AUG* GCC	100%
CYC1-13-S	(*Met*)Ala- *AUA* ACU GAA UUC *AUG* GCC	50%
cyc1-331	(Met)Thr-*Asp-Ser-Arg-Pro-* AUG ACU *GAU UCA AGG CCG*	0%
cyc1-333	(Met)Thr-*Asp-Ser-Trp-Pro-* AUG ACU GAU *UCA UGG CCG*	0%
CYC1-333-F	(*Met*)Ala- *AUA* ACU GA　UUC *AUG* GCC	50%

All amino acid sequences were directly determined experimentally except for those suggested to occur in the *cyc1-331* and *cyc1-333* mutants. The percentages of the normal amounts of iso-1-cytochrome *c* are indicated in the last column. The *cyc1-331* and *cyc1-333* mutants are listed as having 0% because no recognizable iso-1-cytochrome *c* is formed. The normal AAG or altered AUG codons corresponding to the normal amino acid position 4 are underlined. (See footnote to Table 3 for other designations.) (Adapted from Sherman and Stewart 1975; F. Sherman and J.W. Stewart, unpubl.)

cytochrome *c* initiates at the correctly phased AUG codon situated at the site corresponding to position 4. Furthermore, the amount of iso-1-cytochrome *c* in this mutant (*CYC1-333-F*) is equivalent to the amount in the *CYC1-13-S* mutant, which does not contain the nucleotide deletion. The results with the *cyc1-333* and *CYC1-333-F* mutants again reveal that the first AUG initiator codon sets the reading frame for translation.

MUTATIONS INFLUENCING EFFICIENCY OF TRANSLATION

The mutants initiating translation at abnormal sites have been used to evaluate the influence of adjacent sequences on the efficiency of translation. The nucleotides on both sides of the AUG codon and the distances between sites of normal and abnormal initiation particularly have been considered. The long forms of iso-1-cytochrome c (listed at the top of Table 6) all contain normal amounts of iso-1-cytochrome c. Thus, the adjacent nucleotides A, G, and C that are 3' to the AUG initiation codon appear equivalent, and there is no evidence that sequences following the initiation codon modify the efficiency of translation. In contrast, iso-1-cytochrome c initiating at amino

Table 6 Amount of iso-1-cytochrome c in mutants having alterations in the vicinity of the AUG initiation codon

	-2 -1 1 2 3 4 5 6 7 8 9 10 11	
CYC1	UUA AUA AUG ACU GAA UUC AAG GCC GGU UCU GCU AAG AAA GGU (Met)Thr-Glu-Phe-Lys-Ala-Gly-Ser-Ala-Lys-Lys-Gly-	100%
CYC1-13-A	UUA *AUG AUA* ACU GAA UUC AAG GCC GGU UCU GCU AAG AAA GGU *Met-Ile*-Thr-Glu-Phe-Lys-Ala-Gly-Ser-Ala-Lys-Lys-Gly-	100%
CYC1-133-A	UUA *AUG* AGG ACU GAA UUC AAG GCC GGU UCU GCU AAG AAA GGU *Met-Arg* Thr-Glu-Phe-Lys-Ala-Gly-Ser-Ala-Lys-Lys-Gly-	100%
CYC1-493-A	UUA *AUG* AAG ACU GAA UUC AAG GCC GGU UCU GCU AAG AAA GGU *Met-Lys*-Thr-Glu-Phe-Lys-Ala-Gly-Ser-Ala-Lys-Lys-Gly-	100%
CYC1-131-C	UUA *AUG* GUG ACU GAA UUC AAG GCC GGU UCU GCU AAG AAA GGU *(Met)Val* Thr-Glu-Phe-Lys-Ala-Gly-Ser-Ala-Lys-Lys-Gly-	100%
CYC1-51-C	UUA *AUG* CUG ACU GAA UUC AAG GCC GGU UCU GCU AAG AAA GGU *Met-Leu* Thr-Glu-Phe-Lys-Ala-Gly-Ser-Ala-Lys-Lys-Gly-	100%
CYC1-13-S	UUA AUA *AUA* ACU GAA UUC *AUG* GCC GGU UCU GCU AAG AAA GGU *(Met)*Ala-Gly-Ser-Ala-Lys-Lys-Gly-	50%
CYC1-341-D	UUA AUA *AUA* ACU *UAA* UUC *AUG* GCC GGU UCU GCU AAG AAA GGU *(Met)*Ala-Gly-Ser-Ala-Lys-Lys-Gly-	50%
CYC1-333-F	UUA AUA *AUA* ACU *GA* UUC *AUG* GCC GGU UCU GCU AAG AAA GGU *(Met)*Ala-Gly-Ser-Ala-Lys-Lys-Gly-	50%
CYC1-332-A	UUA AUA *AUA* ACU GAA AU^U_AC *AUG* GCC GGU UCU GCU AAG AAA GGU *(Met)*Ala-Gly-Ser-Ala-Lys-Lys-Gly-	100%
CYC1-240-B	UUA AUA *AUA* ACU GAA *AUG* GCC GGU UCU GCU AAG AAA GGU *(Met)*Ala-Gly-Ser-Ala-Lys-Lys-Gly-	100%
CYC1-31-A	Thr-Glu——— Lys-Ala-Gly-Ser-Ala-Lys-Lys-Gly-	100%
CYC1-31-B	Thr-Glu ——————— Ala-Gly-Ser-Ala-Lys-Lys-Gly-	100%
CYC1-31-I	Thr——————————— Gly-Ser-Ala-Lys-Lys-Gly-	100%
CYC1-31-K	Thr———————————————Ala-Lys-Lys-Gly-	75%

The percentages of the normal amounts of iso-1-cytochrome c are indicated in the last column. The AUG initiator codons are underlined. (See footnote to Table 3 for other designations.) (Adapted from Stewart et al. 1971; Stewart and Sherman 1973; Sherman and Stewart 1974; F. Sherman and J.W. Stewart, unpubl.)

acid position 4 is found at 50% the normal amount in the *CYC1-13-S* mutant, as well as in the two *CYC1-341-D* and *CYC1-333-F* mutants that differ at the site corresponding to amino acid position 2. However, normal amounts of the identical short form of iso-1-cytochrome *c* are found in the mutants *CYC1-332-A* and *CYC1-240-B*, which contain alterations to the left of the AUG initiation codon; the *CYC1-332-A* mutant contains one of the isoleucine codons, AUA, AUC, or AUU, at the site corresponding to amino acid position 3; the *CYC1-240-B* mutant lacks the normal codon at position 3 and has a GAA codon 5' to the initiator codon.

The lower amount of iso-1-cytochrome *c* in the *CYC1-13-S, CYC1-341-D, CYC1-333-F*, etc., mutants therefore is not due to instability or to post-translational deficiencies, nor is it simply due to the distance between the initiation site at position 4 and the normal site at position −1. The normal amount of iso-1-cytochrome *c* in the *CYC1-333-A* and *CYC1-240-B* mutants and the lower amount in the *CYC1-13-S, CYC1-341-D*, and *CYC1-333-F* mutants are most simply explained by the differences in the nucleotide and/or codon adjacent to the 5' side of the AUG initiator codon. The normal or near normal amounts of iso-1-cytochrome *c* in the *CYC1-332-A* (Table 6), *CYC1-345-H, CYC1-13-A, CYC1-345-C*, and *CYC1-345-J* mutants (Table 3) indicate that initiation of translation at numerous sites along the mRNA can occur with normal or near normal efficiency. Because the initiator codon can be relocated within an extended region, the secondary structure of mRNA does not appear to play a role in the selection of the site of initiation or in the efficiency of expression of an initiator codon. Consistent with this view is finding normal or near normal amounts of iso-1-cytochrome *c* in the series of *cyc1-31* revertants listed at the bottom of Table 6; the deletions in these mutants should be expected to disrupt any secondary structure in the vicinity of the normal initiator codon. However, these results do not exclude the possibility that certain secondary structures can inhibit or prevent the expression of initiator codons within other regions.

In summary, the normal amounts of iso-1-cytochrome *c* in the various mutants indicate that initiation along the mRNA can occur with approximately normal efficiency. The slightly lower amount of iso-1-cytochrome *c* in *CYC1-13-S* and certain other mutants with an initiation site at amino acid position 4 is most simply explained by the influence of the nucleotides adjacent to the 5' side of the AUG initiator codon. On the other hand, the substantial deficiency of iso-1-cytochrome *c* in the *CYC1-345-F* mutant (Table 3) may reflect a limitation of the protein that lacks the 8 amino-terminal residues. Although data on steady-state levels of mRNA are required before definitive statements can be made regarding translational efficiencies of various *CYC1* mRNAs, these results are quite suggestive. The analysis of additional mutants currently under way may define more completely the influence of adjacent sequences on the efficiency of translation.

cyc1-362 MUTATION IN THE LEADER REGION

Although the site of initiation of mRNA transcription has not been precisely determined, it is probably no more than 90 bp from the 5' side of the AUG initiator codon. Within the DNA sequence of this region, there are no ATG triplets. If translation can initiate at any site within the leader region of mRNA, then the generation of an out-of-phase AUG triplet should cause deficiency of iso-1-cytochrome *c*. As pointed out by Stiles et al. (1981), there are nine or less potential sites within the leader region in which AUG codons can be formed by single base-pair changes.

Although the vast majority of *cyc1* mutations have lesions within the translated region of the gene, detailed deletion mapping indicated that 2 out of 490 *cyc1* mutants contained alterations in adjacent regions. The exceptional mutants, *cyc1-512* (Zaret et al. 1980; Zaret and Sherman 1982) and *cyc1-362* (Stiles et al. 1981), contained, respectively, alterations in the nearby 3'- and 5'-untranslated regions. The rarity of *cyc1* mutations outside the translated region and the fact that almost all *cyc1* mutations consist of single base-pair changes suggest that translation as well as transcription is not appreciably affected when most single nucleotides in the untranslated regions are altered.

The nucleotide changes associated with the *cyc1-362* mutation, which causes deficiency of the iso-1-cytochrome *c* protein but not the iso-1-cytochrome *c* mRNA, have been determined by cloning and sequencing the appropriate DNA fragment. The *cyc1-362* mutation consists of two base-pair substitutions, resulting in an A → G change and a G → A change, respectively, 18 and 30 nucleotides in front of the AUG initiator codon, as shown in Table 7 (Stiles et al. 1981). The A → G substitution generates an AUG triplet that is 5' to the normal AUG initiation codon and that is out of phase with the normal reading frame. The deficiency of iso-1-cytochrome *c* in the *cyc1-362* mutant is most simply explained by assuming that translation initiates at the abnormal AUG codon and not at the normal site, similar to the mutants described in Tables 4 and 5.

Table 7 Nucleotide sequences of the region encompassing the *cyc1-362* mutation

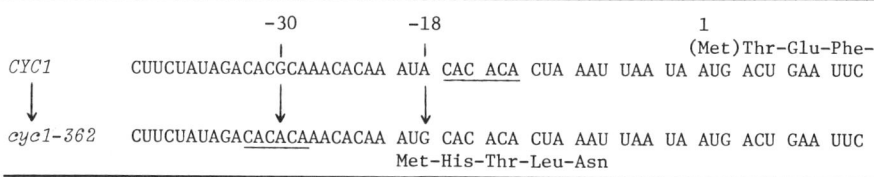

The nucleotide sequences from the normal strain (*CYC1*) and the *cyc1-362* mutant were determined experimentally. A hypothetical peptide translated from the *cyc1-362* mRNA is shown at the bottom. (Adapted from Stiles et al. 1981.)

The G → A substitution in the *cyc1-362* mutant generated a CACACA sequence that appears at approximately the same relative position in front of the normal initiation site in the normal *CYC1* mRNA (Table 7). In addition, this sequence appears at similar positions in the yeast mRNA of the glyceraldehyde-phosphate dehydrogenase gene (Holland and Holland 1979) and the *TRP1* gene (Tschumper and Carbon 1980) but not in mRNA of actin (Gallwitz and Sures 1980) or iso-2-cytochrome *c* (Montgomery et al. 1980). The analysis of *cyc1-362* revertants containing normal amounts of iso-1-cytochrome *c* should reveal whether this CACACA sequence plays a role in translation and in the *cyc1-362* deficiency.

CONTRASTS BETWEEN EUKARYOTES AND PROKARYOTES

The results with the iso-1-cytochrome *c* mutants revealed certain features of the translational process, which can be summarized as follows. (1) AUG is the only triplet capable of initiating translation in yeast and probably in all eukaryotes. As mentioned above, the use of solely AUG for initiation in eukaryotes and the use of both AUG and GUG in prokaryotes has been explained by the presence of the t^6A hypermodified nucleoside in eukaryotic initiator tRNA and its absence in prokaryotic initiator tRNA (Sherman et al. 1980a). (2) The site of initiation of translation can be relocated by mutations to sites preceding and following the normal site. (3) The most proximal AUG codon at the 5' end of the mRNA sets the reading frame and is the site of initiation of translation. (4) Reinitiation at internal AUG codons does not occur, even if they are separated from the normal initiator site by nonsense codons. (5) It appears as if the sequence 5' to the AUG initiator codon may slightly influence the efficiency of translation.

Certain conclusions reached in the iso-1-cytochrome *c* study and in other diverse studies indicate that prokaryotes and eukaryotes use fundamentally different initiation mechanisms by which the ribosomes recognize the initiator codon on the mRNA. Although the translational apparatus in both prokaryotes and eukaryotes consists of the same general array of ribosomes, mRNA, tRNA, and protein factors, some of these components exhibit differences that may reflect differences in the initiation process. Most striking, mRNA from eukaryotes but not prokaryotes is usually modified at the 5' end. Although formyl-methionyl tRNA ($tRNA^{fMet}$) is involved in the initiating process of prokaryotes, the special $tRNA^{Met}$ that is required for initiation in eukaryotes is apparently not formylated in vivo. In eukaryotes, translation cannot initiate after only a portion of the mRNA is transcribed, as it can in prokaryotes, because the mRNA first must be transported from its site of synthesis in the nucleus to the ribosomes in the cytoplasm.

Substantial evidence has been presented that base-pairing occurs between the 3' end of the 16S rRNA and a sequence, the so-called ribosomal binding site, located approximately 10 nucleotides to the 5' side of AUG or GUG initiator codons on prokaryotic mRNAs. This complementary pairing during initiation is believed to be critical for selection of the correct initiator codon (Shine and Dalgarno 1974; Steitz 1979). Direct evidence that basepairing of mRNA and rRNA is required for the ribosomal recognition of initiation sites has been provided by the examination of mutations within the ribosomal binding site of a gene in bacteriophage T7 (Dunn et al. 1978). In contrast to the situation with prokaryotic mRNA, ribosomal binding sites at fixed positions are not usually observed in eukaryotic mRNAs. In addition to the lack of fixed ribosomal binding sites, the flexibility with which AUG initiator codons can be relocated in *CYC1* mutants, the sequences of numerous eukaryotic mRNAs, and the apparent attachment of the ribosomal 40S subunit at or near the 5' end of mRNA all suggest that translation initiates at the AUG codon closest to the 5' terminus without regard to adjacent sequences (Sherman and Stewart 1975; Kozak 1978, 1981).

To explain these findings and the apparent migration of the 40S ribosomal subunit along the mRNA, a mechanism has been proposed in which the 40S ribosomal subunit, initiating tRNA^{Met} and associated factors, binds near or at the 5' end of the mRNA, and the resulting complex subsequently advances until the first AUG is encountered. At this AUG site the 60S subunit is joined to the complex and translation is initiated (Kozak 1978, 1981; Kozak and Shatkin 1978). A detailed and critical evaluation of this "scanning hypothesis" and of other alternative mechanisms has recently been presented by Kozak (1981). Although most findings are adequately explained by the scanning hypothesis, at least 9 out of over 70 eukaryotic mRNAs contain AUG triplets between the capping site and the AUG initiator site (Kozak 1981). These few exceptional sequences indicate that there are additional but unknown restrictions in the selection of the AUG codon for initiation of translation. Nevertheless, AUG initiation sites are always located in regions that are at or near the 5' terminus and are almost always the most proximal AUG codon. These hypothetical regions where translation can initiate have been referred to as "initiation regions" (Kozak 1981).

Another major distinction between prokaryotes and eukaryotes is the occurrence and absence, respectively, of polycistronic translation, i.e., the ability to produce more than a single polypeptide directly from a single mRNA molecule. It should be emphasized that the process for producing multiple peptide chains from polycistronic mRNAs is distinctly different from the process involving proteolytic cleavage of a single polypeptide chain into several distinct polypeptide chains. The proteolytic maturation of polypeptide chains after their release from ribosomes operates in diverse prokaryotic and eukaryotic organisms and in animal viruses and bacteriophages (Hershko and Fry 1975). Evidence for the exclusive or almost exclusive existence of monocistronic mRNA in eukaryotes comes from kinetic experiments where the rates of polypeptide chain synthesis on different size polysomes were determined in animal cells (Kuff and Roberts 1967) and in yeast (Petersen and McLaughlin 1973). In addition, polycistronic operons have never been revealed in any of the numerous genetic studies with eukaryotes. The genes for functionally related enzymes usually are not linked on the genetic map of yeasts or other fungi (King 1974). Even the gene clusters occasionally observed in eukaryotic organisms differ from those found in bacterial operons; most cases of enzymes specified by contiguous genes were shown to be due to distinct enzymic domains on single polypeptide chains (e.g., see Gaertner and Cole 1977; Paukert et al. 1977; Giles 1978; Keesey et al. 1979; Cooper et al. 1980). An organization, still different from a bacterial operon, is found for the *GAL7 GAL10 GAL1* gene cluster, which determines three distinct enzymes (Broach 1979) by way of three distinct transcripts (Hopper and Rowe 1978; St. John and Davis 1981; St. John et al. 1981). Even in vitro studies with polycistronic mRNA from bacteriophages indicate that eukaryotes are limited to translating a single polypeptide chain. Only the first 5' cistron in a polycistronic mRNA from bacteriophage λ was translated with a wheat germ system (Rosenberg and Paterson 1979). Although multiple peptides were observed when RNA bacteriophages and mRNA of bacteriophages T3 and T7 were translated with components of eukaryotic cells (Davies and Kaesberg 1973; Morrison and Lodish 1973; Anderson et al. 1976; Atkins et al. 1979), the cleavage of the RNA was recently shown to be required for initiation of internal cistrons when the bacteriophage R17 is translated with the wheat germ system (Kozak 1980).

The lack of reinitiation after termination in the mutants *cyc1-341*, *cyc1-599* (Table 4), and *cyc1-362* (Table 7) is consistent with the view that yeast is unable to translate polycistronic mRNA even when appropriate signals are available. In contrast to these results with yeast, reinitiation occurs in *E. coli* containing certain UAG mutations in the *lacI* gene (Platt et al. 1972; Ganem et al. 1973; Files et al. 1975; Steege 1977; Weber and Geisler 1978). Reinitiation at residue positions 23, 42, and 62 was activated by UAG mutations that preceded these sites. The shortest distance between the nonsense codon and the reinitiation codon was 3 bp, which is equivalent to the distance from the UAA codon at position 2 and the AUG codon at position 4 in the iso-1-cytochrome *c* mutants. Thus, the two systems are in some respects comparable. In addition, the studies with the *lacI* strains containing UAG mutations indicated that certain in-phase GUG codons do not serve as initiator codons and that the codons preferentially used for initiation were those appropriately positioned in front of sequences complementary to the 3' end of a 16S RNA (Steege 1977). The restriction of initiation sites in the *lacI* mutants and the apparent lack of restrictions in the *cyc1* mutants again point out the distinction between the initiation process in *E. coli* and yeast.

Distinctions between the initiation process in *E. coli* and yeast, which presumably reflect the differences between prokaryotes and eukaryotes, are schematically emphasized in Figure 2. Initiation of translation in prokary-

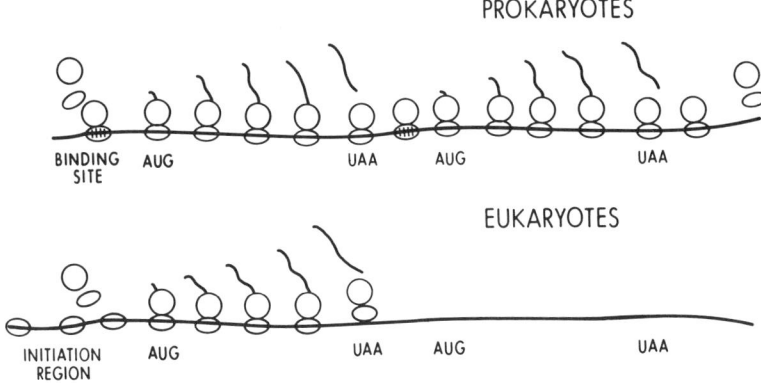

Figure 2 The proposed mechanisms for initiation of translation in prokaryotes and eukaryotes. Translation on prokaryotic mRNA initiates at AUG or GUG initiator codons 3' to the ribosomal binding sites. After translation and after peptide release at UAA, UAG, or UGA chain termination codons, the initiator complex is stabilized by a second ribosomal binding site, and translation of a second polypeptide chain is initiated at the second initiator codon. In contrast, small ribosomal subunits migrate from the 5' end of eukaryotic mRNA along the initiation region, until the first AUG codon is encountered. At this site, the large subunit and associated components complex and translation is initiated. At the site of UAA, UAG, or UGA chain termination codons, the polypeptide chain is terminated and released along with the ribosome. Reinitiation does not take place, and only a single polypeptide chain can be synthesized.

otes is confined to sites appropriately positioned with respect to the 3' side of ribosomal binding sites. After translation and termination with peptide release, the initiation of a second polypeptide chain takes place if a second initiator codon is again appropriately situated by a second ribosomal binding site. It can be suggested that the second binding sequence merely serves to stabilize the initiation complex and that the same ribosome or, at least, the same 30S subunit serves to translate an entire polycistronic mRNA. In addition, it could be suggested that there may be special instances where internal binding of entire ribosomes takes place.

On the other hand, translation of eukaryotic mRNA appears to be able to initiate at any site within an extended region, denoted in Figure 2 as the initiation region. As discussed above, initiation has been suggested to occur by migration of the 40S ribosomal subunit from the 5' end of the mRNA to the first AUG codon, as proposed by Kozak (1978, 1981). In contrast to the situation with prokaryotes, we suggest that the ribosome, along with the polypeptide, is released from the mRNA when chain termination signals are encountered. The lack of the 40S ribosomal subunit prevents initiation of a second polypeptide chain at a second AUG codon. Because UAA triplets have no effect if they are situated before initiator codons (see *CYC1-341-D* and *CYC1-599-D* sequences in Table 4), the chain termination codons do not cause release of 40S ribosomal subunits by themselves but rather cause release of the entire complex. The binding of release factors to both subunits is consistent with this requirement (Caskey 1980).

Although it remains to be seen whether the two proposed mechanisms are precisely as presented, it is clear that the overall properties of initiation of translation are markedly different in prokaryotes and eukaryotes. In conclusion, we wish to emphasize that the two distinct mechanisms for initiation of translation require that prokaryotes and eukaryotes also have two distinct methods for coordinately regulating sets of proteins.

ACKNOWLEDGMENTS

We thank Dr. M. Kozak (University of Pittsburgh) for providing her current papers prior to publication. The writing of this review and the cited unpublished experiments were supported in part by U.S. Public Health Service research grant GM-12702, from the National Institutes of Health, and in part by U.S. Department of Energy contract no. DE-AC02-76EV3490, at the University of Rochester, Department of Radiation Biology and Biophysics. This paper has been designated Report no. UR-3490-1959.

Note Added in Proof

Regarding the *cyc1-362* mutation in the leader region, nuclease S1 digestion of RNA-DNA duplexes indicates that there are multiple 5' ends of *CYC1* mRNAs and that abundant mRNAs initiate at approximately nucleotide positions −36, −44, and −60, and minor mRNAs initiate at approximately positions −29, −52, −69, and −92 (Faye et al. 1981). Thus, there is only one other site at position −31 in addition to the *cyc1-362* site at −18 that can form an out-of-phase AUG initiation codon (Table 7).

Recently, S. B. Baim and F. Sherman (unpubl.) constructed and analyzed a series of nested deletions 5' to the AUG initiation codon. The (−18)ACAC-ACA(−12) region, as well as other regions not encompassing major transcriptional start sites, can be deleted without any effect on iso-1-cytochrome *c* production. These, along with the other results presented in this paper, firmly establish that initiation of translation does not require a specific sequence in the 5'-untranslated region.

With regard to contrasts between eukaryotes and prokaryotes, from the results of experiments performed in vitro with ribosomal releasing factors and bacteriophage R17 as well as from other results, Ryoji et al. (1981) suggested that ribosomes are released from prokaryotic polycistronic mRNAs at each termination codon and that ribosomes independently bind to sites at each cistron. Whether prokaryotic ribosomes or at least the 30S subunit glide on intercistronic region, as shown in Figure 2 and discussed by others (see Ryoji et al. 1981), or whether they bind independently at each cistron as suggested by Ryoji et al. (1981) and others, it is still "clear that the overall properties of initiation of translation are markedly different in prokaryotes and eukaryotes."

To explain the approximately 10% exceptional cases where translation does not initiate at the AUG codons closest to the 5′ termini of eukaryotic mRNAs and to explain rare mRNAs, which seem to direct synthesis of two independently initiated proteins, Kozak (1981b) has modified the scanning hypothesis. AUG codons within the two favored sequences ANNAUGG and GNNAUGG are preferentially selected for initiation of translation even if they are preceded by AUG triplets and if the more 5′-AUG triplets are within other sequences such as UNNAUGA, CNNAUGA, and so on. However, so far the results with yeast indicate that translation initiates at the first AUG irrespective of the surrounding sequences. Apparently, initiation occurs at the first AUAAUGA sequence and not the second UUCAUGG sequence (*cyc1-341*, Table 4; *cyc1-133*, Table 5) or the second GCUAUGA sequence (*cyc1-599*, Table 4), and at the first CAAAUGC sequence and not the second AUAAUGA sequence (*cyc1-362*, Table 7). It remains to be seen whether translation in yeast will initiate at a preferred AUG site other than at the most 5′-proximal AUG codon. Also, the results in Table 6, showing that AUGA, AUGG, and AUGC sequences give rise to normal amounts of iso-1-cytochrome *c*, indicate that the nucleotide adjacent to the 3′ end of the AUG initiation codon does not influence efficiency of translation. However, as discussed above in Mutations Influencing Efficiency of Translation, sequences 5′ to the AUG initiation codon appear to modify the efficiency of translation.

In addition to the 16S mRNA of SV40 (Jay et al. 1981) and possibly other eukaryotic polycistronic mRNAs discussed by Kozak (1981a), the genomic segment A of the infectious pancreatic necrosis virus was suggested to give rise to a polycistronic mRNA capable of producing three proteins (Mertens and Dobos 1982). Possibly rare eukaryotic mRNAs may have more than one initiation region at distinct sites along a single molecule.

APPENDIX

The construction of mutant sequences by mutation and recombination was based upon the capacity of lactate medium to preferentially support the growth of strains containing functional iso-1-cytochrome *c* and of chlorolactate medium to preferentially support the growth of strains lacking either iso-1-cytochrome *c* or its activity (Sherman et al. 1974). Using similar techniques described for isolating *cyc1* mutants (Sherman et al. 1974, 1975), it was possible to obtain *cyc1* recombinants with desired sequences by crossing certain *CYC1* mutants that contained altered iso-1-cytochromes *c* and plating the sporulated cross on chlorolactate medium. The resistant colonies were analyzed genetically for *cyc1* defects, and the site of the lesions was determined by fine-structure mapping with defined *cyc1* tester strains. Likewise, *CYC1* recombinants could be obtained from two *cyc1* mutants by crossing, sporulating, and plating on lactate medium.

Most of the nucleotide sequences were ultimately deduced from the amino acid sequences of altered iso-1-cytochrome *c*. This could be directly determined with *CYC1* recombinants which, of course, contain iso-1-cytochrome *c*. On the other hand, the nucleotide sequences of the *cyc1* recombinants lacking iso-1-cytochrome *c* were deduced from the examination of intragenic revertants derived from the *cyc1* recombinants. In addition, some sequences were determined by sequencing DNA fragments that encompassed the alterations.

The mutant iso-1-cytochromes *c* were compared with normal iso-1-cytochrome *c* by peptide mapping of tryptic and chymotryptic digests. Iso-1-cytochrome *c* is sufficiently small and is of such a composition that the peptide maps reliably reveal alterations in the aminoterminal region and, in some instances, suggest the nature of changes. Finally, most alterations could be unambiguously determined by sequential Edman degradation of the protein, from the amino terminus through and beyond the altered regions.

Most of the mutant sequences were derived from the *cyc1-9*, *cyc1-13*, and *cyc1-183* mutants or their intragenic revertants shown in Table 8. The *cyc1-9* (Stewart et al. 1972), *cyc1-13* (Stewart et al. 1971), and *cyc1-183* (Sherman and Stewart 1973; Stewart and Sherman 1974) mutants and their revertants have been described previously. The construction of some of the *cyc1-239*, *cyc1-242*, *cyc1-331*, *cyc1-333*, *cyc1-341*, *cyc1-345*, *cyc1-346*, *cyc1-347*, *CYC1-340a*, and *CYC1-340d* recombinants (Table 9) and their intragenic revertants (Table 10) were briefly described by Sherman and Stewart (1975) and Sherman et al. (1980b) and are described in detail elsewhere (F. Sherman and J.W. Stewart, in prep.). The construction of still other *cyc1* and *CYC1* recombinants shown in Table 11 are described below.

In the selection scheme for the recombinants, assumptions first had to be made as to whether or not particular nucleotide sequences would give rise to functional iso-1-cytochrome *c*. The acquisition of strains either containing

Table 8 Mutational events leading from the normal gene *CYC1* to the deficient mutants *cyc1-9*, *cyc1-13*, and *cyc1-183* and the mutational events giving rise to intragenic revertants with altered iso-1-cytochromes *c*

CYC1	(Met)Thr-Glu-Phe-Lys-Ala-Gly-Ser-Ala-Lys-Lys-Gly- AUG ACU GAA UUC AAG GCC GGU UCU GCU AAG AAA GGU $\quad\quad\quad\;\;\Big\downarrow$ G → U	
cyc1-9	AUG ACU *UAA* UUC AAG GCC GGU UCU GCU AAG AAA GGU $\quad\quad\quad\quad\;\;\;\Big\downarrow$ A → U or C	
CYC1-9-S	AUG ACU *UA$_C^U$* UUC AAG GCC GGU UCU GCU AAG AAA GGU (Met)Thr-*Tyr*-Phe-Lys-Ala-Gly-Ser-Ala-Lys-Lys-Gly-	

CYC1	$\quad\quad\quad\quad$-1$\quad\quad\quad\quad\quad\;\;$4 (Met)Thr-Glu-Phe-Lys-Ala-Gly-Ser-Ala-Lys-Lys-Gly- AUG ACU GAA UUC AAG GCC GGU UCU GCU AAG AAA GGU $\;\;\Big\downarrow$ G → A	
cyc1-13	*AUA* ACU GAA UUC AAG GCC GGU UCU GCU AAG AAA GGU $\quad\quad\quad\quad\quad\quad\quad\;\;\Big\downarrow$ A → U	
CYC1-13-S	*AUA* ACU GAA UUC *AUG* GCC GGU UCU GCU AAG AAA GGU *(Met)*Ala-Gly-Ser-Ala-Lys-Lys-Gly-	

CYC1	$\quad\quad$-1$\;\;$1$\;\;\;$2$\;\;\;$3$\;\;\;$4$\;\;\;$5$\;\;\;$6$\;\;\;$7$\;\;\;$8$\;\;\;$9$\;$10$\;$11 (Met)Thr-Glu-Phe-Lys-Ala-Gly-Ser-Ala-Lys-Lys-Gly- AUG ACU GAA UUC AAG GCC GGU UCU GCU AAG AAA GGU $\quad\quad\quad\quad\quad\quad\quad\quad\quad\quad\quad\quad\quad\quad\quad\;\;$+A $\quad\quad\quad\quad\quad\quad\quad\quad\quad\quad\quad\quad\quad\;\;\downarrow\downarrow\downarrow\;\downarrow$	
cyc1-183	AUG ACU GAA UUC AAG GCC GGU UCU GCU AAG AAA *AGG U* $\quad\quad\quad\quad\quad\quad\quad\quad\quad\quad\quad\Big\lceil$ -G $\quad\quad\quad\quad\quad\quad\quad\quad\quad\quad\quad\downarrow\downarrow$	
CYC1-183-L	AUG ACU GAA UUC AAG *CCG GUU CUG CUA AGA* AAA GGU (Met)Thr-Glu-Phe-Lys-*Pro-Val-Leu-Leu-Arg*-Lys-Gly- $\quad\Big\lvert$ -A $\quad\;\;\downarrow\downarrow$	
CYC1-183-U	AUG ACU *GAU UCA AGG CCG GUU CUG CUA AGA* AAA GGU (Met)Thr-*Asp-Ser-Arg-Pro-Val-Leu-Leu-Arg*-Lys-Gly- $\Big\lvert$ -C $\downarrow$	
CYC1-183-T	AUG *AUG AAU UCA AGG CCG GUU CUG CUA AGA* AAA GGU X-*Met-Met-Asn-Ser-Arg-Pro-Val-Leu-Leu-Arg*-Lys-Gly-	

The amino acid residues and codons that differ from the normal are shown in italics. *X* in the *CYC1-183-T* sequence refers to an unknown blocking group. The methionine residues shown in parentheses are excised and not found in mature protein. (Adapted from Stewart et al. 1971, 1972; Stewart and Sherman 1973.)

Table 9 Formation of *cyc1* and *CYC1* mutants by recombination

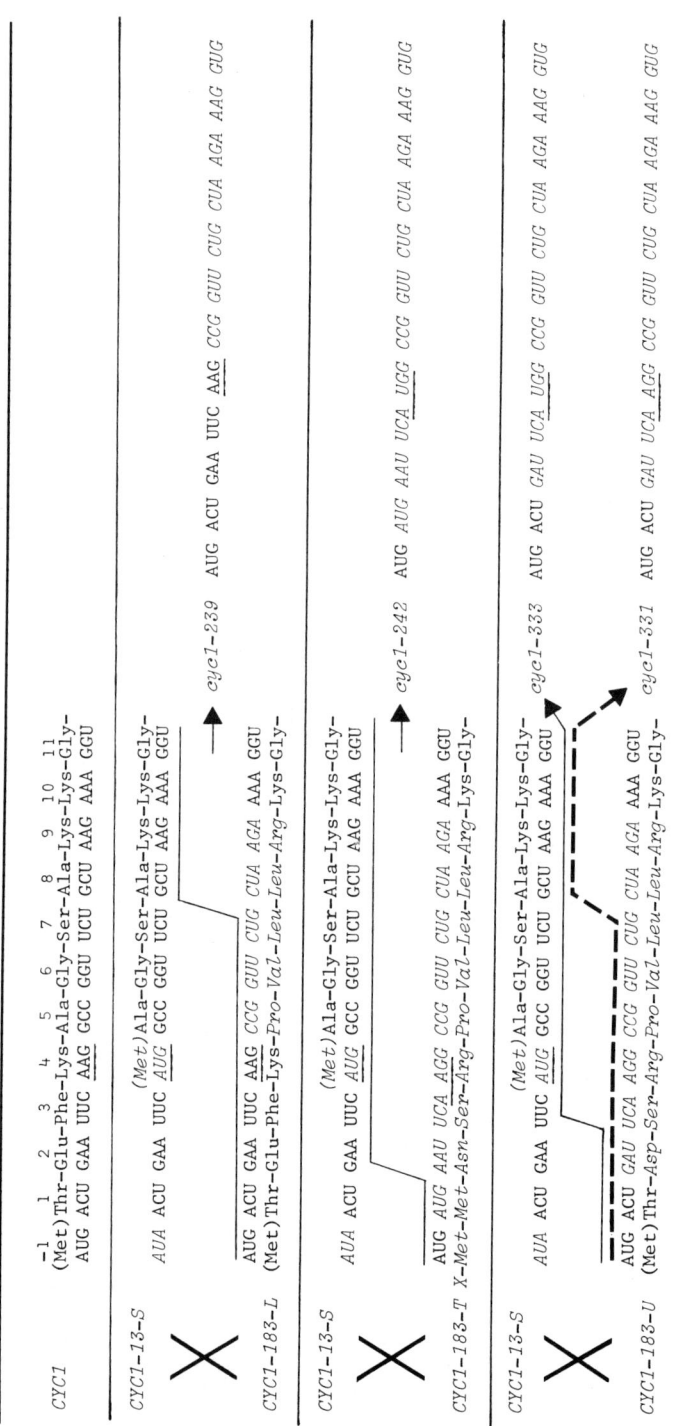

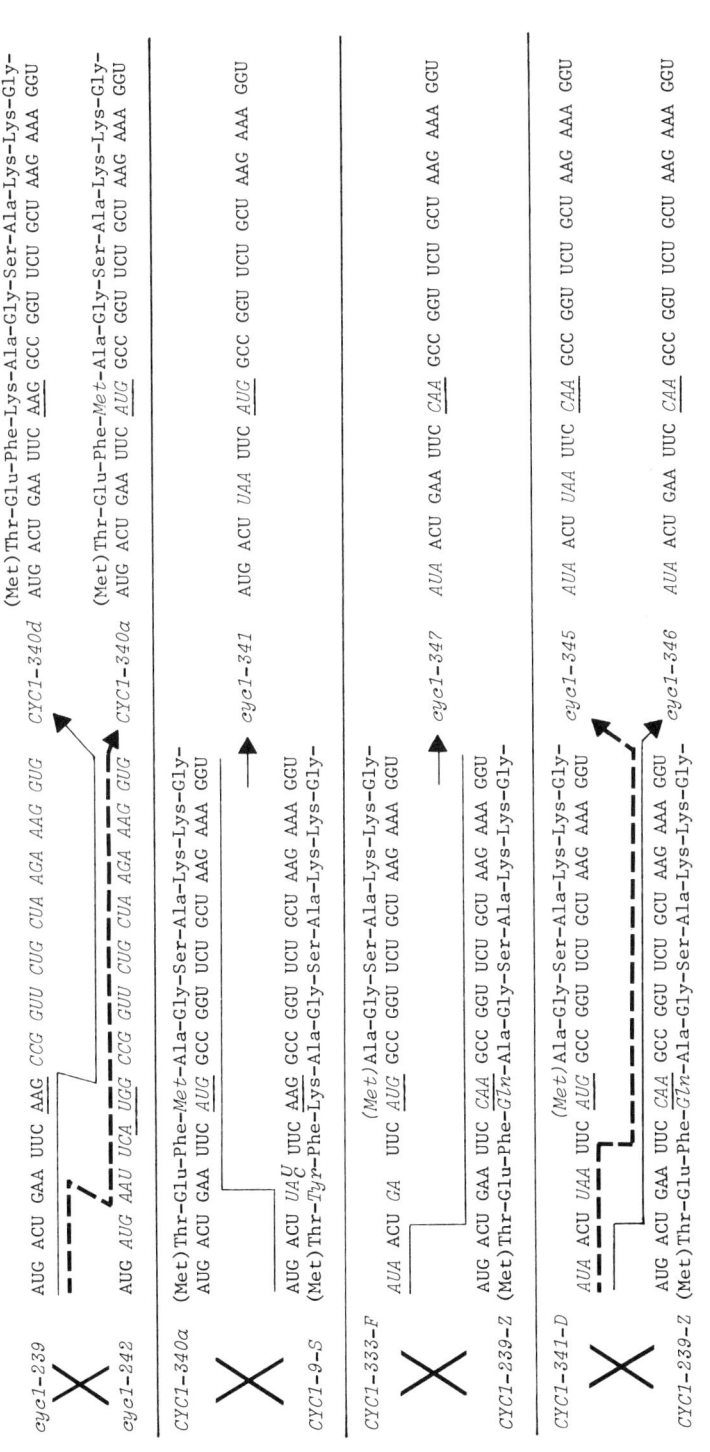

The normal AAG or altered AUG codons corresponding to position 4 are underlined. Other designations are denoted in the footnote to Table 8. (Data from Sherman and Stewart 1975, and in prep.)

Table 10 Intragenic revertants of *cyc1* recombinants

CYC1		1 1 2 3 4 5 6 7 8 9 10 11 (Met)Thr-Glu-Phe-Lys-Ala-Gly-Ser-Ala-Lys-Lys-Gly- AUG ACU GAA UUC AAG GCC GGU UCU GCU AAG AAA GGU

cyc1-239	AUG ACU GAA UUC AAG *CCG GUU CUG CUA AGA AAG GUG*
	                     +A or G
	                 ↓↓ ↓↓↓ ↓↓

CYC1-239-A	AUG ACU GAA UUC|AAG *CCG GUU CUG CUA|AG^A_G AA^A_G* GGU
	(Met)Thr-Glu-Phe-Lys-*Pro-Val-Leu-Leu-Arg*-Lys-Gly-
	             +N or A
	           ↓ ↓↓

CYC1-239-B	AUG ACU GAA UUC|AAG *CCG GUU CUG CUN AAG* AAA GGU
	(Met)Thr-Glu-Phe-Lys-*Pro-Val-Leu-Leu*-Lys-Lys-Gly-
	       +U or C
	     ↓

CYC1-239-Z	AUG ACU GAA *UU^U_C CAA* GCC GGU UCU GCU AAG AAA GGU (Met)Thr-Glu-Phe-*Gln*-Ala-Gly-Ser-Ala-Lys-Lys-Gly-

cyc1-242	AUG *AUG AAU UCA UGG CCG GUU CUG CUA AGA AAG GUG*
	     −UU
	    ↓
	AUG *AUG AAC AUG* GCC GGU UCU GCU AAG AAA GGU

CYC1-242-O	Met-Met-*Ile*-Met-Ala-Gly-Ser-Ala-Lys-Lys-Gly-

cyc1-331	AUG ACU *GAU UCA AGG CCG GUU CUG CUA AGA AAG GUG*
	              +N or A
	             ↓ ↓↓

CYC1-331-B	AUG ACU *GAU UCA AGG CCG|GUU CUG CUN* AAG AAA GGU
	(Met)Thr-*Asp-Ser-Arg-Pro-Val-Leu-Leu*-Lys-Lys-Gly-
	       +C, N or G
	     ↓↓↓ ↓↓

CYC1-331-K	AUG ACU *GAU UCA AGG CCN* GGU UCU GCU AAG AAA GGU (Met)Thr-*Asp-Ser-Arg-Pro*-Gly-Ser-Ala-Lys-Lys-Gly-

cyc1-333	AUG ACU *GAU UCA UGG CCG GUU CUG CUA AGA AAG GUG*
	                 +A or G
	               ↓↓ ↓↓↓ ↓↓

CYC1-333-A	AUG ACU *GAU UCA UGG CCG|GUU CUG CUA AG^A_G AA^A_G* GGU
	(Met)Thr-*Asp-Ser-Trp-Pro-Val-Leu-Leu-Arg*-Lys-Gly-
	         +N or G
	       ↓ ↓↓

CYC1-333-D	AUG ACU *GAU UCA UGG CCN* GGU UCU GCU AAG AAA GGU
	(Met)Thr-*Asp-Ser-Trp-Pro*-Gly-Ser-Ala-Lys-Lys-Gly-
	   G → A
	  ↓

CYC1-333-F	*AUA* ACU *GA* UUC *AUG* GCC GGU UCU GCU AAG AAA GGU *(Met)* Ala-Gly-Ser-Ala-Lys-Lys-Gly-

Table 10 (Continued)

	-2 -1 1 2 3 4 5 6 7 8 9 10 11
CYC1	(Met)Thr-Glu-Phe-Lys-Ala-Gly-Ser-Ala-Lys-Lys-Gly-
	AUA AUG ACU GAA UUC AAG GCC GGU UCU GCU AAG AAA GGU

cyc1-341	AUA AUG ACU *UAA* UUC *AUG* GCC GGU UCU GCU AAG AAA GGU
	⎨ ↓ U → C
CYC1-341-A	AUA AUG ⎮ACU *CAA* UUC *AUG* GCC GGU UCU GCU AAG AAA GGU
	(Met)Thr-*Gln*-Phe-*Met*-Ala-Gly-Ser-Ala-Lys-Lys-Gly-
	↙ G → A
CYC1-341-D	AUA *AUA* ACU *UAA* UUC *AUG* GCC GGU UCU GCU AAG AAA GGU
	(Met)Ala-Gly-Ser-Ala-Lys-Lys-Gly-

cyc1-346	AUA *AUA* ACU GAA UUC *CAA* GCC GGU UCU GCU AAG AAA GGU
	⎨ ↓ A → G
CYC1-346-A	AUA⎮AUG ACU GAA UUC *CAA* GCC GGU UCU GCU AAG AAA GGU
	(Met)Thr-Glu-Phe-*Gln*-Ala-Gly-Ser-Ala-Lys-Lys-Gly-
	↙ A → G
CYC1-346-B	*AUG AUA* ACU GAA UUC *CAA* GCC GGU UCU GCU AAG AAA GGU
	Met-*Ile*-Thr-Glu-Phe-*Gln*-Ala-Gly-Ser-Ala-Lys-Lys-Gly-

cyc1-347	AUA *AUA* ACU GAA UUC *CAA* GCC GGU UCU GCU AAG AAA GGU
	⎨ ↓ A → G
CYC1-347-B	AUA⎮AUG ACU GAA UUC *CAA* GCC GGU UCU GCU AAG AAA GGU
	(Met)Thr-Glu-Phe-*Gln*-Ala-Gly-Ser-Ala-Lys-Lys-Gly-
	↙ A → C
CYC1-347-A	*AUG AUA* ACU GAA UUC *CAA* GCC GGU UCU GCU AAG AAA GGU
	Met-*Ile*-Thr-Glu-Phe-*Gln*-Ala-Gly-Ser-Ala-Lys-Lys-Gly-

See footnote to Table 8 for designations. (Data from Sherman and Stewart 1975, and in prep.)

or not containing iso-1-cytochrome c was ultimately used to deduce whether translation either could or could not, respectively, effectively initiate in the particular sequences. The first series of desired sequences included those that contained various combinations of triplets at amino acid positions −1, 2, and 4. The desired codons at positions −1 and 4 were AUG; the desired alterations at position 2 were the nonsense codon UAA and a deletion of a single nucleotide. The construction of these sequences is briefly described

Table 11 Formation of the *cyc1-599* recombinant
and the *CYC1-599-A* and *CYC1-599-B* revertants

	-1 2 4 9	
CYC1	(Met)Thr-Glu-Phe-Lys-Ala-Gly-Ser-Ala-Lys-Lys-Gly- AUG ACU GAA UUC AAG GCC GGU UCU GCU AAG AAA GGU	100%
cyc1-345 ↓ UV *CYC1-345-F*	AUA ACU UAA UUC CAA GCC GGU UCU GCU AAG AAA GGU ↓ *Met*-Lys-Gly AUA ACU UAA UUC CAA GCC GGU UCU GCU *AUG* AAA GGU	0% 20%
✕ *CYC1-31-K* ↓ recomb.	AUG ACU ─────────────── GCU AAG AAA GGU (Met)Thr ─────────────── Ala-Lys-Lys-Gly-	75%

cyc1-599
↓ UV
CYC1-599-A
↓ EMS
CYC1-599-B
	AUG ACU UAA UUC CAA GCC GGU UCU GCU AUG AAA GGU
	     ↓
	AUG ACU CAA UUC CAA GCC GGU UCU GCU AUG AAA GGU
	(Met)Thr-*Gln*-Phe-*Gln*-Ala-Gly-Ser-Ala-*Met*-Lys-Gly-
	
	AUA ACU UAA UUC CAA GCC GGU UCU GCU AUG AAA GGU
	                                                      Met-Lys-Gly-
		0%
		
		100%
		
		20%

Amino acid residues 2 through 7 are deleted from the *CYC1-31-K* iso-1-cytochrome *c* (Sherman and Stewart 1973; Stewart and Sherman 1974). The percentages of the normal amounts of iso-1-cytochrome *c* are indicated in the last column. (See footnote to Table 8 for other designations.) (Data from F. Sherman and J.W. Stewart, unpubl.)

below. The role of these sequences in defining the initiation process is discussed in the previous sections.

The first *cyc1* recombinants were obtained from the cross *CYC1-13-S* × *CYC1-183-L* shown in Table 9. Theoretically, this cross could yield three types of *cyc1* recombinants; *cyc1-239* was shown to be the desired frameshift mutation (a deletion of the G from the AAG codon at position 4 or its equivalent, a deletion of the G from the GCC codon at position 5) by fine-structure mapping and by examination of intragenic revertants. The structures of iso-1-cytochromes *c* from 37 intragenic revertants, which have been reported previously (Stewart and Sherman 1974), unambiguously established the *cyc1-239* sequence that is presented in Tables 9 and 10. The mutational changes and the resulting sequences of three representative revertants are presented in Table 10, including the *CYC1-239-Z* revertant, which was used for generating other *cyc1* recombinants.

Similarly, three other recombinants, *cyc1-242, cyc1-331,* and *cyc1-333,* were obtained from the crosses shown in Table 9. The sites of the alterations in *cyc1* recombinants were first estimated by fine-structure mapping with *cyc1* tester strain. In addition, the *cyc1-331* and *cyc1-333* recombinants, which both have alterations at position 2, could be differentiated by the rates of ethylmethanesulfonate (EMS)-induced reversion, which will be discussed below. Finally, the expected nucleotide sequences were directly deduced from the amino acid sequences of revertant iso-1-cytochromes *c*, which are shown in Table 10 and are presented elsewhere (F. Sherman and J.W. Stewart, in prep.).

Two types of recombinants with functional iso-1-cytochromes *c* were expected from the cross *cyc1-239* × *cyc1-242*, as shown in Table 9. Only protein analysis could be used to differentiate between the one having normal iso-1-cytochrome *c* and the other having iso-1-cytochrome *c* with an abnormal residue of methionine at position 4. The first recombinant, *CYC1-340a*, proved to have the sequence with methionine 4; another recombinant from the cross *CYC1-340d* contained normal iso-1-cytochrome *c*.

One crucial recombinant, *cyc1-341*, was obtained from cross *CYC1-340a* × *CYC1-9-S*, as shown in Table 9. The *cyc1-341* sequence was verified from the structures of iso-1-cytochrome *c* in intragenic revertants that were induced with UV. The *cyc1-9* mutant, which contains a UAA codon at position 2 (Table 8), is highly revertible with UV, and most of the resulting revertants contain replacements of glutamine and occasionally leucine (Stewart et al. 1972; Lawrence et al. 1974; Sherman and Stewart 1974). Similarly, the *cyc1-341* mutant was demonstrated to be highly revertible with UV; a UV-induced revertant (*CYC1-341-A*) that was examined by sequencing contained glutamine 2 along with the expected methionine 4 (Table 5).

As discussed above, the *cyc1-341* and *cyc1-333* recombinants lack iso-1-cytochrome *c* because reinitiation apparently cannot occur at the AUG codon corresponding to position 4 when it is preceded by either a UAA or a frameshift mutation. However, if initiation can occur at this site when the AUG codon at position −1 is altered, then one would expect to uncover intragenic revertants lacking the four aminoterminal residues, similar to the iso-1-cytochrome *c* in the *CYC1-13-S* revertant (Table 8). This assumption was tested by reverting the *cyc1-333* and *cyc1-341* mutants with EMS, which shows a high degree of specificity for GC → AT transitions at least at some sites (Prakash and Sherman 1973). The two revertants induced with EMS, *CYC1-333-F* and *CYC1-341-D*, both contained the expected short form of iso-1-cytochrome *c* (Table 10), which suggests that the normal AUG codon was mutated to the isoleucine codon AUA. To verify the presence of the isoleucine codon, the *CYC1-333-F* and *CYC1-341-D* revertants were each crossed to the *CYC1-239-Z* mutant, and the *cyc1* recombinants were isolated and tested. The genetic analysis of the types of *cyc1* recombinants from each of the *CYC1-333-F* × *CYC1-239-Z* and *CYC1-341-D* × *CYC1-239-Z* crosses was aided by the presence of the CAA codon at position 4 instead of the normal AAG codon, which can mutate to AUG by a single base-pair change.

For example, the *CYC1-341-D* × *CYC1-239-Z* cross would be expected to yield three types of *cyc1* recombinants: one like the *cyc1-341* mutant, and the others like the *cyc1-345* and *cyc1-346* mutants shown in Table 9. The *cyc1-341* mutant is highly revertible by UV due to the UAA codon at position 2 and is revertible by EMS due to the AUG codon at position −1. The *cyc1-346* mutant would be expected to be UV-revertible but not EMS-revertible. In contrast, the *cyc1-345* mutant should not revert by a single base-pair change. Similarly three types of *cyc1* recombinants, distinguishable by UV and EMS revertibility, would also be expected from the *CYC1-333-F* × *CYC1-239-Z* cross; one of the types represented by *cyc1-347* should uniquely be UV-revertible but not EMS-revertible. The structures of iso-1-cytochrome *c* from intragenic revertants established the sequences of the *cyc1-346* and *cyc1-347* recombinants that were inferred from their rates of UV and EMS reversion. The expected presence of isoleucine codons at position −1 is shown by the revertants *CYC1-346-B* and *CYC1-347-A*, which both contain iso-1-cytochrome *c* with Met-Ile- appended at the amino terminus of the protein (Table 10). Because EMS preferentially produces GC → AT transitions and because the alterations occurred at position −1 in the *CYC1-333-F* and *CYC1-341-D* revertants, these isoleucine codons undoubtedly are AUA.

The additional sequences *cyc1-599*, *CYC1-599-A*, and *CYC1-599-B* were constructed by the procedures summarized in Table 11. The *cyc1-599* recombinant was derived from a cross of the *CYC1-345-F* (Sherman et al. 1980b) and *CYC-31-K* (Sherman and Stewart 1973; Stewart and Sherman 1974) mutants. As discussed above, for producing other *cyc1* revertants, the *CYC1-599-B* and *CYC1-599-A* revertants were produced by taking advantage of, respectively, the specific GC → AT action of EMS and the preferential action of UV on the UAA codon at position 2.

REFERENCES

Adams, J.M. 1968. On the release of the formyl group from nascent proteins. *J. Mol. Biol.* **33:** 571.

Anderson, C.W., J.F. Atkins, and J.J. Dunn. 1976. Bacteriophage T3 and T7 early RNAs are translated by eukaryotic 80S ribosomes: Active phage T3 coded *S*-adenosylmethionine cleaving enzyme is synthesized. *Proc. Natl. Acad. Sci.* **73:** 2752.

Atkins, J.F, J.A. Steitz, C.W. Anderson, and P. Model. 1979. Binding of mammalian ribosomes to MS2 phage RNA reveals an overlapping gene encoding a lysis function. *Cell* **18:** 247.

Broach, J.R. 1979. Galactose regulation in *Saccharomyces cerevisiae*: The enzymes encoded by the *GAL7,10,1* cluster co-ordinately controlled and separately translated. *J. Mol. Biol.* **131:** 41.

Caskey, C.T. 1980. Peptide chain termination. *Trends Biochem. Sci.* **5:** 234.

Cooper, T.G., C. Lam, and V. Turoscy. 1980. Structural analysis of the *dur* loci in *S. cerevisiae*: Two domains of a single multifunctional gene. *Genetics* **94:** 555.

Davis, J.W. and P. Kaesberg. 1973. Translation of virus mRNA: Synthesis of bacteriophage Qβ proteins in a cell-free extract from wheat embryo. *J. Virol.* **12:** 1434.

Dunn, J.J., E. Buzash-Pollert, and F.W. Studier. 1978. Mutations of bacteriophage T7 that affect initiation of synthesis of the 0.3 protein. *Proc. Natl. Acad. Sci.* **75:** 2741.

Faye, G., D.W. Leung, K. Tatchell, B.D. Hall, and M. Smith. 1981. Deletion mapping of sequences essential for in vivo transciption of the iso-1-cytochrome *c* gene. *Proc. Natl. Acad. Sci.* **78:** 2258.

Fiers, W., R. Contreras, F. Duerinck, G. Gaegeman, J. Merragaert, W. Min Jou, A. Raeymakers, G. Volckaert, M. Ysebaert, J. Van de Kerckhova, F. Nolf, and M. Van Montagu. 1975. A-protein gene of bacteriophage MS2. *Nature* **256:** 273.

Files, J.G., K. Weber, C. Coulondre, and J.H. Miller. 1975. Identification of the UUG codon as a translational initiation codon *in vivo*. *J. Mol. Biol.* **95:** 327.

Gaertner, F.H. and K.W. Cole. 1977. A cluster-gene: Evidence for one gene, one polypeptide, five enzymes. *Biochem. Biophys. Res. Comm.* **75:** 259.

Gallwitz, D. and I. Sures. 1980. Structure of a split yeast gene: Complete nucleotide sequence of the actin gene in *Saccharomyces cerevisiae*. *Proc. Natl. Acad. Sci.* **77:** 2546.

Ganem, D., J.H. Miller, J.G. Files, T. Platt, and K. Weber. 1973. Reinitiation of a *lac* repressor fragment at a codon other than AUG. *Proc. Natl. Acad. Sci.* **70:** 3165.

Giles, N.H. 1978. The organization, function and evolution of gene clusters in eucaryotes. *Amer. Nat.* **112:** 641.

Hershko, A. and M. Fry. 1975. Post-translational cleavage of polypeptide chains: Role in assembly. *Annu. Rev. Biochem.* **44:** 775.

Holland, J.P. and M.J. Holland. 1979. The primary structure of a glyceraldehyde-3-phosphate dehydrogenase gene from *Saccharomyces cerevisiae*. *J. Biol. Chem.* **254:** 9839.

Hopper, J.E. and L.B. Rowe. 1978. Molecular expression and regulation of the galactose pathway genes in *Saccharomyces cerevisiae*. *J. Biol. Chem.* **253:** 7566.

Jay, G., S. Normura, C.W. Anderson, and G. Khoury. 1981. Indentification of the SV40 agnogene product: A DNA binding protein. *Nature* **291:** 346.

Keesey, J.K., Jr., R. Bigelis, and G.R. Fink. 1979. The product of the *his4* gene cluster in *Saccharomyces cerevisiae*: A trifunctional polypeptide. *J. Biol. Chem.* **254:** 7427.

Kerwar, S.S., H. Weissbach, and G.G. Glenner. 1971. An aminopeptidase activity associated with brain ribosomes. *Arch. Biochem. Biophys.* **143:** 336.

King, R.C. 1974. *Handbook of genetics*, vol. 1. Plenum Press, New York.

Kolehmainen, L. and J. Mikola. 1971. Partial purification and enzymatic properties of an aminopeptidase from barley. *Arch. Biochem. Biophys.* **145:** 633.

Kozak, M. 1978. How do eucaryotic ribosomes select initiation regions in messenger RNA? *Cell* **15:** 1109.

―――. 1980. Binding of wheat germ ribosomes to fragmented viral mRNA. *J. Virol.* **35:** 748.

―――. 1981a. Mechanism of mRNA recognition by eukaryotic ribosomes during initiation of protein synthesis. *Curr. Top. Microbiol. Immunol.* **93:** 81.

―――. 1981b. Possible role of flanking nucleotides in recognition of the AUG initiator codon by eurkaryotic ribosomes. *Nucleic Acids Res.* **9:** 5233.

Kozak, M. and A.J. Shatkin. 1978. Migration of 40S ribosomal subunits on messenger RNA in the presence of edeine. *J. Biol. Chem.* **253:** 6568.

Kuff, E. and N.E. Roberts. 1967. *In vivo* labeling patterns of free polyribosomes: Relationship to tape theory of messenger ribonucleic acid function. *J. Mol. Biol.* **26:** 211.

Lawrence, C.W., F. Sherman, M. Jackson, and R.A. Gilmore. 1975. Mapping and gene conversion studies with the structural gene for iso-1-cytochrome *c* in yeast. *Genetics* **81:** 615.

Lawrence, C.W., J.W. Stewart, F. Sherman, and R. Christensen. 1974. The specificity and frequency of ultraviolet-induced reversion of an iso-1-cytochrome *c* ochre mutant in radiation-sensitive strains of yeast. *J. Mol. Biol.* **85:** 137.

Matheson, A.T., A.J. Dick, and F. Rollin. 1970. A ribosomal-bound aminopeptidase in *Escherichia coli* B: Substrate specificity. *Can. J. Biochem.* **48:** 1292.

Mertens, P.P.C. and P. Dobos. 1982. Messenger RNA of infectious pancreatic necrosis virus is polycistronic. *Nature* **297:** 243.

Montgomery, D.L., D.W. Leung, M. Smith, P. Shalit, G. Faye, and B.D. Hall. 1980. Isolation and sequence of the gene for iso-2-cytochrome *c* in *Saccharomyces cerevisiae*. *Proc. Natl. Acad. Sci.* **77**: 541.

Morrison, T.G. and H.F. Lodish. 1973. Translation of bacteriophage Qβ RNA by cytoplasmic extracts of mammalian cells. *Proc. Natl. Acad. Sci.* **70**: 315.

Paukert, J.L., G.R. Williams, and J.C. Rabinowitz. 1977. Formyl-methenyl-methylenetetrahydrofolate synthetase (combined): Correlation of enzymic activities with limited proteolytic degradation of the protein from yeast. *Biochem. Biophys. Res. Commun.* **77**: 147.

Petersen, N.S. and C.S. McLaughlin. 1973. Monocistronic messenger RNA in yeast. *J. Mol. Biol.* **81**: 33.

Platt, T., K. Weber, D. Ganem, and J.H. Miller. 1972. Translational restarts: AUG reinitiation of a *lac* repressor fragment. *Proc. Natl. Acad. Sci.* **69**: 897.

Prakash, L. and F. Sherman. 1973. Mutagen specificity: Reversion of iso-1-cytochrome *c* mutants of yeast. *J. Mol. Biol.* **79**: 65.

Prakash, L., J.W. Stewart, and F. Sherman. 1974. Specific induction of transitions and transversions of G · C base-pairs by 4-nitroquinoline-1-oxide in iso-1-cytochrome *c* mutants of yeast. *J. Mol. Biol.* **85**: 51.

Rosenberg, M. and B.M. Paterson. 1979. Efficient cap-dependent translation of polycistronic mRNAs is restricted to the first gene in the operon. *Nature* **279**: 696.

Ryoji, M., R. Berland, and A. Kaji. 1981. Reinitiation of translation from the triplet next to the amber termination codon in the absence of ribosome-releasing factor. *Proc. Natl. Acad Sci.* **78**: 5973.

St. John, T.P. and R.W. Davis. 1981. The organization and transcription of the galactose gene cluster of *Saccharomyces*. *J. Mol. Biol.* **152**: 285.

St. John, T.P., S. Scherer, M.W. McDonell, and R.E. Davis. 1981. Deletion analysis of the *GAL* gene cluster: Transcription from three promoters. *J. Mol. Biol.* **152**: 317.

Sherman, F. 1964. Mutants of yeast deficient in cytochrome *c*. *Genetics* **49**: 39.

Sherman, F. and J.W. Stewart. 1973. Mutations at the end of the iso-1-cytochrome *c* gene of yeast. In *The biochemistry of gene expression in higher organisms* (ed. J.K. Pollak and J.W. Lee), p. 56. Australian and New Zealand Books, Sydney.

———. 1975. The use of iso-1-cytochrome *c* mutants of yeast for elucidating the nucleotide sequences that govern initiation of translation. *FEBS Proc. Meet.* **38**: 175.

Sherman, F., G. McKnight, and J.W. Stewart. 1980a. AUG is the only initiation codon in eukaryotes. *Biochim. Biophys. Acta* **609**: 343.

Sherman, F., J.W. Stewart, and A.M. Schweingruber. 1980b. Mutants of yeast initiating translation of iso-1-cytochrome *c* within a region spanning 37 nucleotides. *Cell* **20**: 215.

Sherman, F., M. Jackson, S.W. Liebman, A.M. Schweingruber, and J.W. Stewart. 1975. A deletion map of *cyc1* mutants and its correspondence to mutationally altered iso-1-cytochromes *c* of yeast. *Genetics* **81**: 51.

Sherman, F., J.W. Stewart, M. Jackson, R.A. Gilmore, and J.H. Parker. 1974. Mutants of yeast defective in iso-1-cytochrome *c*. *Genetics* **77**: 255.

Sherman, F., J.W. Stewart, E. Margoliash, J. Parker, and W. Campbell. 1966. The structural gene for yeast cytochrome *c*. *Proc. Natl. Acad. Sci.* **55**: 1498.

Sherman, F., J.W. Stewart, J.H. Parker, E. Inhaber, N.A. Shipman, G.J. Putterman, R.L. Gardisky, and E. Margoliash. 1968. The mutational alteration of the primary structure of yeast iso-1-cytochrome *c*. *J. Biol. Chem.* **243**: 5446.

Shine, J. and L. Dalgarno. 1974. The 3'-terminal sequence of *Escherichia coli* 16S ribosomal RNA: Complementarity to nonsense triplets and ribosome binding sites. *Proc. Natl. Acad. Sci.* **71**: 1342.

Steege, D.A. 1977. 5'-terminal nucleotide sequences of *Escherichia coli* lactose repressor mRNA: Features of translational initiation and reinitiation sites. *Proc. Natl. Acad. Sci.* **74**: 4163.

Steitz, J.A. 1979. Genetic signals and nucleotide sequences in messenger RNA. In *Biological regulation and development* (ed. R. Goldberger), vol. I, p. 349. Plenum Press, New York.

Stewart, J.W. and F. Sherman. 1974. Yeast frameshift mutations identified by sequence changes in iso-1-cytochrome *c*. In *Molecular and environmental aspects of mutagenesis* (ed. L. Prakash et al.), p. 102. Thomas, Springfield.

Stewart, J.W., F. Sherman, N.A. Shipman, and M. Jackson. 1971. Identification and mutational relocation of the AUG codon initiating translation of iso-1-cytochrome *c* in yeast. *J. Biol. Chem.* **246:** 7429.

Stewart, J.W., F. Sherman, M. Jackson, F.L.X. Thomas, and N. Shipman. 1972. Demonstration of the UAA ochre codon in baker's yeast by amino acid replacements in iso-1-cytochrome *c*. *J. Mol. Biol.* **68:** 83.

Stiles, J.I., J.W. Szostak, A. Young, R. Wu, S. Consaul, and F. Sherman. 1981. DNA sequence of a mutation in the leader region of the yeast iso-1-cytochrome *c* mRNA. *Cell* **25:** 277.

Takeda, M. and R.E. Webster. 1968. Protein chain initiation and deformylase in *B. subtilis* homogenates. *Proc. Natl. Acad. Sci.* **60:** 1487.

Tschumper, G. and J. Carbon. 1980. Sequence of a yeast DNA fragment containing a chromosomal replicator and the *TRP1* gene. *Gene* **10:** 157.

Vogt, Y.M. 1970. Purification and properties of an aminopeptidase from *Escherichia coli*. *J. Biol. Chem.* **245:** 4760.

Weber, K. and N. Geisler. 1978. *lac* repressor fragments produced *in vivo* and *in vitro*: An approach to the understanding of the interaction of repressor and DNA. In *The operon* (ed. J.H. Miller and W.S. Reznikoff), p. 155. Cold Spring Harbor Laboratory, Cold Spring Harbor, New York.

Yoshida, A. and M. Lin. 1972. NH_2-terminal formylmethionine- and NH_2-terminal methionine-cleaving enzymes in rabbits. *J. Biol. Chem.* **247:** 952.

Zaret, K.S. and F. Sherman. 1982. DNA sequence required for efficient transcription termination in yeast. *Cell* **28:** 563.

Zaret, K.S., J. Kotval, and F. Sherman. 1980. A *cyc1* mutant of yeast that reverts by translocations and transpositions. *Genetics* **94:** s115.

Yeast Cell Wall and Cell Surface

Clinton E. Ballou
Department of Biochemistry, University of California,
Berkeley, California 94720

1. **Vegetative Cell Wall**
   A. Composition and Structure
   B. Immunochemistry
   C. Localization of Wall Components
2. **Changes in the Wall during Cell Growth and Division**
   A. Cell Growth
   B. Bud Formation
   C. Septation
   D. Protoplast Formation and Regeneration
3. **Changes in the Wall during Sporulation**
   A. Parent Cell Wall
   B. Spore Coat
   C. Spore Germination
4. **Role of the Cell Wall in Mating**
   A. Constitutive Agglutination Factors
   B. Pheromone-induced Changes in the Yeast Cell Wall
5. **Questions and Suggestions for New Directions in Research**

The structure and organization of the yeast cell wall and the nature of the cell surface have been investigated in terms of the composition of the wall; the structures of the wall components; the immunochemistry of the cell surface and binding of dyes, lectins, or specific antibodies; the degradative action of enzymes; and the visual evidence from scanning and thin-section electron microscopy. Most or all of these methods have been applied to the cells during vegetative growth, sporulation, and mating, and the analyses have been furthered by the availability in recent years of a variety of cell-wall mutants. A general conclusion from all of these studies is that yeast cell walls show less apparent organization than bacterial walls but that the yeast wall is as metabolically active and the cell surface as antigenically polymorphic as those of Gram-positive bacteria. To keep this review manageable, it will be limited in most instances to *Saccharomyces cerevisiae*, with only occasional reference to other yeasts. The aim is to treat the subject topics briefly and to direct the reader to several recent reviews in which leading references to the original literature can be found.

VEGETATIVE CELL WALL

Composition and Structure

Three components, glucan, mannoprotein (previously called mannan), and chitin, make up over 90% of the cell wall, whereas only small and variable amounts of lipid have been reported (Phaff 1971). When grown on hexadecane, however, the cell wall of *Candida tropicalis* increases in covalently linked fatty acid (Kappeli et al. 1978). Thus, cell-wall composition may be affected by the growth conditions, a conclusion supported by the observation that *S. cerevisiae* cells from a culture growing in logarithmic phase are more susceptible to the action of β-glucanase than those from a stationary-phase culture (Nečas 1971). Washed cell-wall preparations of some yeasts are reported to contain the enzymes invertase, acid phosphatase, endo-$\beta 1 \rightarrow 3$-glucanase, and proteases (Fleet and Phaff 1974).

Glucan

In *S. cerevisiae*, this polysaccharide is composed predominantly of $\beta 11 \rightarrow 3$-linked D-glucopyranose units (Manners et al. 1973a), with some branching via $\beta 1 \rightarrow 6$-linked D-glucopyranose units (Manners et al. 1973b). A part of the cell-wall glucan is soluble in water, but most of it is insoluble. The shape of a $\beta 1 \rightarrow 3$-glucan is such that the chain can form a long, twisted ribbon, one side of which is hydrophilic, owing to the concentration of hydroxyl groups, with the other side being hydrophobic, owing to the concentration of methine groups. It is possible, although not established, that two such chains could wind around each other with the hydrophobic surfaces in contact to form a double helix with exceptional rigidity. Such organized structures have been demonstrated in X-ray studies on $\beta 1 \rightarrow 3$-xylan, a closely related molecule (Atkins et al. 1969); and, if double helical chains of glucan are present in the yeast cell wall, the chains could slide along each other and allow the cell wall to expand. The alternative process for enlargement of the cell would involve enzymic cleavage of the preformed glucan chains, and the wall is known to contain glucanases capable of this action (see Changes in the Wall during Cell Growth and Division).

The arrangement of the glucan network within the matrix of the wall is unknown, although Bacon (1973) has presented a thoughtful discussion of the inferences one might draw about this organization from the results of enzymic, physical, and chemical treatments of the yeast cell. The material formed by regenerating protoplasts has a netlike appearance in electron micrographs (Nečas 1971) and seems to differ from the insoluble glucan that remains after extraction of the cell wall to remove the mannoprotein (Kreger and Kopecká 1973). The relevance of both observations to the organization of glucan in the wall has been considered in detail (Bartnicki-Garcia and McMurrough 1971).

No evidence suggests that the glucan is exposed on the cell surface. For some reason, it has not been possible to raise a $\beta 1 \rightarrow 3$-glucan antiserum, even with isolated insoluble glucan as the immunogen, so an immunochemical probe for glucan localization is not available.

Bulk Mannoprotein

Detailed structural analysis has been performed only on *S. cerevisiae* mannoproteins of the cell wall, periplasm, and cytoplasm; most of this section deals with a comparison of these substances (Ballou 1976). However, a brief discussion is also given of the immunochemistry of mannoproteins from several genera.

Figure 1 shows the structure of the carbohydrate part of *S. cerevisiae* X2180 bulk cell-wall mannoprotein. This is an average structure for the material obtained by extraction of the cells with hot citrate buffer followed by precipitation as the cetyl trimethylammonium-borate complex, and the purified external invertase from this same organism differs only in that it lacks oligosaccharides linked to serine or threonine.

Notable features of the structure in Figure 1 are the $\alpha 1 \rightarrow 6$-linked backbone to which are attached numerous sidechains in $\alpha 1 \rightarrow 2$ linkage and the differentiation of the polysaccharide chain into a core and outer chain. The core is distinguished by the fact that some sidechains are attached by $\alpha 1 \rightarrow 3$ linkage and that the first mannose unit is linked $\beta 1 \rightarrow 4$ to the di-*N*-acetylchitobiose unit. The core of yeast mannoproteins is identical in many of its structural features to similar oligosaccharides in mammalian glycoproteins (Kornfeld and Kornfeld 1976; Reading et al. 1978; Chapman and

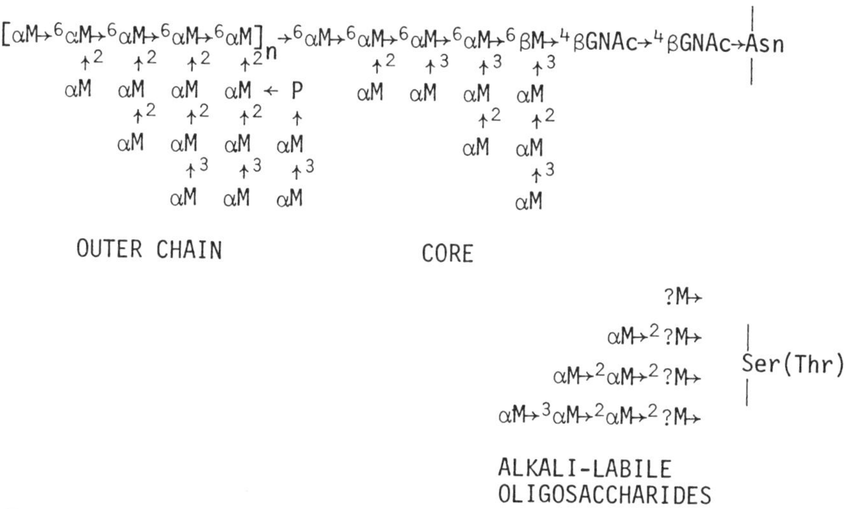

$$[\alpha M {\rightarrow} ^6 \alpha M {\rightarrow} ^6 \alpha M {\rightarrow} ^6 \alpha M {\rightarrow} ^6 \alpha M]_n {\rightarrow} ^6 \alpha M {\rightarrow} ^6 \alpha M {\rightarrow} ^6 \alpha M {\rightarrow} ^6 \alpha M {\rightarrow} ^6 \beta M {\rightarrow} ^4 \beta GNAc {\rightarrow} ^4 \beta GNAc {\rightarrow} Asn$$

(with branching sidechains of αM units at positions 2 and 3 as shown)

OUTER CHAIN   CORE

$$?M{\rightarrow}$$
$$\alpha M{\rightarrow}^2 ?M{\rightarrow}$$
$$\alpha M{\rightarrow}^2 \alpha M{\rightarrow}^2 ?M{\rightarrow} \quad Ser(Thr)$$
$$\alpha M{\rightarrow}^3 \alpha M{\rightarrow}^2 \alpha M{\rightarrow}^2 ?M{\rightarrow}$$

ALKALI-LABILE
OLIGOSACCHARIDES

Figure 1 Representative structures for *S. cerevisiae* mannoprotein carbohydrate components. (M) Mannose; (GNAc) *N*-acetylglucosamine; (Asn, Ser, Thr) amino acids.

Kornfeld 1979) (Fig. 2), and it is a reasonable hypothesis that this structure, which has so far been found only in eukaryotes, serves a function unique to such organisms.

The carbohydrate chains of the mannoproteins are the principal immunogens when whole yeast cells are injected into rabbits or goats, and with *S. cerevisiae* X2180, the terminal $\alpha 1 \rightarrow$ 3-linked mannose is immunodominant (Suzuki et al. 1968; Ballou 1970). Antiserum with this specificity can be used to enrich a mutagenized culture for cells with altered surface antigens by agglutinating the unchanged cells (Raschke et al. 1973). In this manner, a number of mannoprotein mutants (designated *mnn*) have been obtained (Fig. 3).

The *mnn1* mutant lacks terminal $\alpha 1 \rightarrow$ 3-linked mannose in all parts of the mannoprotein molecule (Fig. 3), and cell-free extracts are deficient in an activity that forms this structure, with GDP-mannose as the donor and $\alpha M \rightarrow {}^2\alpha M \rightarrow {}^2M$ as the acceptor (Nakajima and Ballou 1975). We have concluded that a single enzyme is involved that is able to utilize as mannosyl acceptors six structurally distinct, although similar, parts of the mannoprotein molecule. The broad pleiotrophy of this mutation is unusual, and our rationalization assumes an unusually broad specificity for the mannosyltransferase activity. A presumed wild-type strain of *S. cerevisiae* (A364A) has the same phenotype, and its properties appear to be controlled by the same locus that is affected in the *mnn1* mutant. This locus is of interest since it is the most tightly centromere-linked of all loci placed on the *S. cerevisiae* chromosome map thus far (Antalis et al. 1973).

A striking transformation of surface immunochemistry occurs in the *mnn1* mutant that is correlated with the change in the structure of the phosphodiester sidechain units (Rosenfeld and Ballou 1974). Whereas the

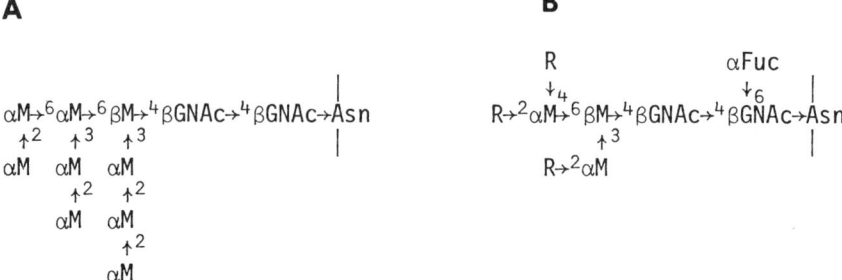

Figure 2 Mammalian asparagine-linked glycoprotein oligosaccharide units with structural analogies to yeast mannoprotein. (*A*) From a human immunoglobulin M myeloma protein (Chapman and Kornfeld 1979). (*B*) From vesicular stomatitis virus (VSV) glycoprotein in which R = αNeuNAc $\rightarrow {}^3$ βGal $\rightarrow {}^4$ βGNAc (Reading et al. 1978). (Fuc) fucose.

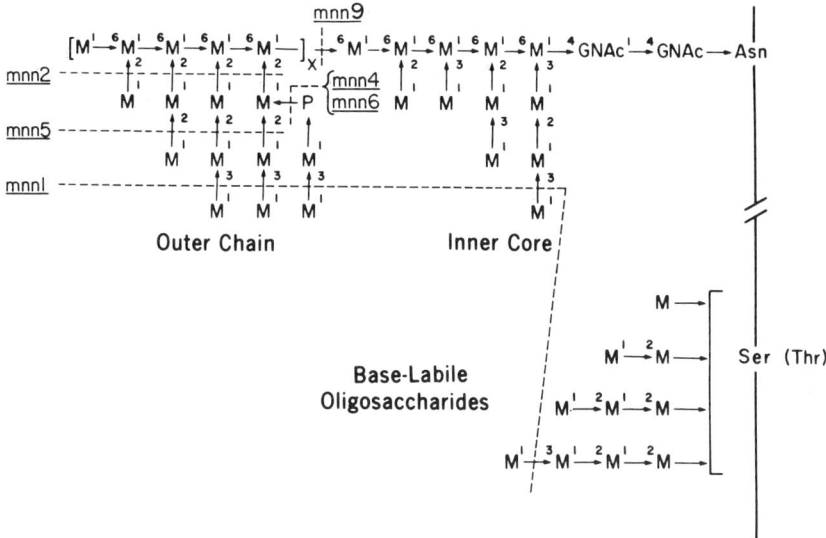

Figure 3 Mannoprotein alterations associated with *S. cerevisiae mnn* mutants. (----) Portion of the carbohydrate component that is lacking in each mutant.

$\alpha 1 \rightarrow$ 3-mannobiosylphosphate unit is not distinguished immunochemically from any other terminal $\alpha 1 \rightarrow$ 3-mannose, the α-mannosylphosphate unit that is exposed in the *mnn1* mutant becomes the immunodominant determinant of the cell surface. The phosphate in cell-wall mannoproteins is the major source of negative charge on the yeast cell surface, and its presence or absence is easily scored with the basic phthalocyanine dye, alcian blue (Friis and Ottolenghi 1970; Ballou 1975).

The *mnn2* mutant leads to formation of mannoprotein in which the outer chain is devoid of sidechains (Fig. 3), and this defect is associated with the absence of an enzyme activity that will transfer mannose from GDP-mannose to free mannose or $\alpha 1 \rightarrow$ 6-mannotriose to form an $\alpha 1 \rightarrow$ 2 linkage (Nakajima and Ballou 1975). This enzyme activity is called the $\alpha 1 \rightarrow$ 2-mannosyltransferase-I to distinguish it from a second transferase that is defective in the *mnn5* mutant (see below). One important feature of the *mnn2* mutant is that a portion of the polymannose chain linked to asparagine is unaffected, and this fact led to the discovery of the core oligosaccharide in yeast mannoproteins (Nakajima and Ballou 1974). Even though the core and the serine- and threonine-linked oligosaccharides possess terminal $\alpha 1 \rightarrow$ 3-linked mannose units in the *mnn2* mutant, the cells do not react with antiserum to such units. Rather, the important surface antigen of the *mnn2* mutant is the unbranched $\alpha 1 \rightarrow$ 6-linked outer chain. This suggests that the outer chain, which is about 100 sugar units long, effectively covers the other potential determinants so that they are inaccessible to antibody. This property has been exploited in the use of fluorescence-activated cell sorting for the isolation of a new mannoprotein mutant class (see below).

The *mnn3* locus originally was thought to affect specifically the $\alpha 1 \rightarrow 2$-mannosyltransferase-II that adds the second mannose in the outer-chain sidechains, but more recent studies have shown that this mutant is altered in both the outer chain and the oligosaccharides that are linked to serine and threonine (Cohen 1980). It now appears to be a regulatory mutation, and its phenotype could be rationalized as resulting from an altered availability of sugar nucleotide donor. To clarify this point, a new mutant selection was performed, and a strain designated *mnn5* was obtained that had the phenotype expected if the defect were limited to the $\alpha 1 \rightarrow 2$-mannosyltransferase-II. The mannoprotein from this mutant is altered only in the outer chain (Fig. 3), with the result that the sidechains consist of a single $\alpha 1 \rightarrow 2$-linked mannose unit (Cohen et al. 1980).

The *mnn4* mutant was first detected after mutagenesis of the *mnn1* strain and selection for a clone that no longer bound alcian blue (Ballou et al. 1973). The single mutant, recovered from this double mutant by genetic manipulation, also failed to bind the dye although the sidechains were otherwise wild type in nature. The *mnn4* locus has been placed on chromosome XI (Ballou 1975), and it is near another locus (*dbl1*) reported by Friis and Ottolenghi (1970) to control the dye-binding property of a wild-type *Saccharomyces* strain. The *mnn4* mutant lacks an enzyme activity found in the wild type that catalyzes transfer of mannosylphosphate units from GDP-mannose to $\alpha 1 \rightarrow 2$-mannotriose units to form the phosphodiester mannosylphosphorylmannotriose (Karson and Ballou 1978). Of the mannoprotein mutants so far obtained, *mnn4* is the only one to show dominance in the heterozygous diploid, and extensive studies suggest that it is probably a regulatory defect acting at the level of transcription. Although the property controlled by the *dbl1* locus was reported to be dominant (Friss and Ottolenghi 1970), subsequent analyses have given variable results (Ballou 1975). A recessive mutant, *mnn6*, with the same mannoprotein chemotype has been obtained, and this may represent the structural gene locus for the mannosylphosphate transferase in *S. cerevisiae*.

In the studies leading to the isolation of these first six mannoprotein mutants, it is notable that no mutant altered in synthesis of the $\alpha 1 \rightarrow 6$-linked backbone was obtained. It is a reasonable hypothesis that this was a consequence of the method of selection, which involved repetitive steps of growth and antibody agglutination, during which mutants that grew slowly or had poor viability would be lost. In a search for mutants altered in the mannoprotein backbone, a new procedure has been developed using a fluorescence-activated cell sorter. In this procedure, the *mnn2* strain served as the parent, and mutants were selected that failed to bind fluorescein-conjugated antibody Fab' fragments directed against the $\alpha 1 \rightarrow 6$-linked mannose units of the *mnn2* backbone (Ballou et al. 1980). This procedure was first applied for the isolation of yeast cell-wall mutants by Douglas and Ballou (1980), when they used fluorescein-conjugated wheat germ agglutinin for the selection of *Kluyveromyces lactis* mutants that lack terminal N-acetylglucosamine units in the mannoprotein outer chain (Smith et al. 1975; Douglas 1979). In the sorted cultures of the mutagenized *S. cerevisiae mnn2* strain, about 5% of the colonies that grew up had the desired phenotype in that they were not agglutinated by anti- $\alpha 1 \rightarrow$ 6-mannose serum but were agglutinated by anti- $\alpha 1 \rightarrow$ 3-mannose serum, as expected if the core of the mannoprotein were exposed by removal of the outer chain. Of 50 such isolates, 10 have been subjected to genetic analysis, yielding four complementation groups (designated *mnn7–mnn10*). Presumably, many more complementation groups with the same phenotype could be obtained from analysis of the other isolates.

Chemical studies suggest that the *mnn9* mutant makes mannoprotein with no outer chain (Fig. 3), and the other isolates are very similar in structure, perhaps differing slightly in the length of the backbone or in the structure of the core (Hashimoto et al. 1980). This mutant class is the first isolated in which important physiological defects are associated with an altered mannoprotein structure. The cells are clumpy, they tend to lyse even when grown with an osmotic stabilizer, and thin-section electron microscopy shows that the walls and septal regions are deformed (Ballou et al. 1980). The mutant cells mate reasonably well, but the homozygous diploids sporulate poorly, and heterozygous diploids that do sporulate usually have two deformed spores as though there were a postmeiotic defect in the formation of the spore wall.

The genetic lesions in the new mannoprotein mutants are not apparent from their properties. They probably do not concern the $\alpha 1 \rightarrow$ 6-mannosyltransferase that is involved in making the backbone, unless there are several such enzymes; but they could be defects that interfere with the action of this enzyme by altering the availability of donor substrate or the temporal or spatial relationships between enzyme and acceptor during the processing of the mannoprotein molecule. This latter possibility is enhanced by the fact that a form of invertase has been isolated from the *mnn2* mutant that has a carbohydrate component seemingly identical to that of the bulk mannoprotein of the *mnn9* mutant (Lehle et al. 1979). This invertase appears to be membrane-associated and is thought to be a precursor of the external form found in the periplasm (see below). This suggests, therefore, that the mannoproteins of the new mutant class could represent biosynthetic intermediates that are secreted from the cell in an incomplete form because of defects in the terminal stages of the pathway. Because the mechanism of mannoprotein synthesis and secretion in yeast is so poorly understood, it is not profitable to speculate further on the nature of the possible genetic defects. It should be emphasized, however, that although the lesions in the *mnn7–mnn10* class interfere with completion of the mannoprotein molecules, they do not prevent secretion of the products. All four mutants make and secrete wild-type levels of the external invertase, and the secreted invertase appears to be retained in the periplasm of the cell.

Structure of External Invertase

S. cerevisiae external invertase is a dimeric mannoprotein of 270,000 daltons, with 50% of the weight being mannose that is attached to the protein part as 18–20 asparagine-linked polysaccharide units, or 9–10 per protein subunit (Trimble and Maley 1977). The carbohydrate component can be released by endo-*N*-acetylglucosaminidase digestion to yield a protein subunit of 60,000 daltons and polysaccharide chains of two size classes (Tarentino et al. 1974). The structure of invertase made by the *S. cerevisiae mnn2* strain has been analyzed in further detail (Lehle et al. 1979) in order to define more precisely the nature of the carbohydrate component. Several important facts emerge. The invertase obtained from broken whole cells can be separated into a fraction that is soluble in 75% ammonium sulfate (S75) and one that is insoluble in the salt (P75). The latter is about 10% of the total and only a part of it is carbohydrate-free, with the remainder containing 36% mannose. This P75 invertase is not altered by digestion with endo-$\alpha 1 \rightarrow 6$-mannanase, which indicates that it lacks long outer chains, whereas digestion with endo-*N*-acetylglucosaminidase releases carbohydrate fragments containing 33 mannose units, corresponding to a core of 15 units with a short outer chain of 18 mannoses. The S75 invertase contains 53% mannose, and endomannanase digestion converts it to a form that is very similar to the P75 invertase; this suggests that it has exposed $\alpha 1 \rightarrow 6$-linked outer chain units that are readily removed by the endomannanase. Endo-*N*-acetylglucosaminidase digestion removes the asparagine-linked polysaccharide units of S75 invertase preferentially and produces a series of glycoprotein intermediates with variable numbers of small oligosaccharide core units of a size similar to those in the P75 invertase. The P75 invertase yields an identical pattern of glycoprotein intermediates by endo-*N*-acetylglucosaminidase digestion, which indicates that the carbohydrate chains of both forms of invertase differ in their accessibility to the endoglucosaminidase. The conclusion from this analysis is that about two thirds of the carbohydrate chains of the completed external invertase molecules consist of core oligosaccharides with very short outer chains and that the remaining one third of the chains have cores with long outer chains of about 100 mannose units. Moreover, the latter chains appear to be exposed on the surface of the protein because they are readily removed by endoglucosaminidase digestion, whereas the smaller units may be folded inside where they are protected from the action of the enzyme. If a similar organization applies to the mannoprotein molecules exposed on the cell surface, the cryptic nature of the core antigenic determinants in the *mnn2* mutant would be understandable.

Immunochemistry

The ability to discriminate between yeast strains on the basis of their immunochemical properties (Hasenclever and Mitchell 1964) was an early indicator of the now-established polymorphic nature of yeast mannoproteins (Ballou and Raschke 1974). Of some interest, however, are the facts that the carbohydrate portion of the mannoprotein in the intact wall is the principal immunogen and that some structures are much more immunogenic than others. In contrast, immunization with isolated soluble mannoprotein raises antibodies to both the protein and carbohydrate components. Schematic outer-chain structures for several yeasts are shown in Figure 4, and the immunodominant parts are indicated in Figure 5 by the curved lines encompassing portions of the molecules. Some of these determinants seem to be species-specific, with the $\alpha 1 \rightarrow 2$-N-acetylglucosamine being found only in the genus *Kluyveromyces*. On the other hand, terminal $\alpha 1 \rightarrow 3$-mannose occurs very widely as does the glycosylphoshate diester structure. The identification of yeast immunochemical determinants is still in its infancy and, as more mannoproteins are analyzed, one may predict the discovery of a structural complexity approaching that of the O antigens of enteric bacteria (Lüderitz et al. 1966).

Localization of Wall Components

This subject has been investigated by the use of specific antibodies, lectins, dyes, scanning and transmission electron microscopy, incorporation of

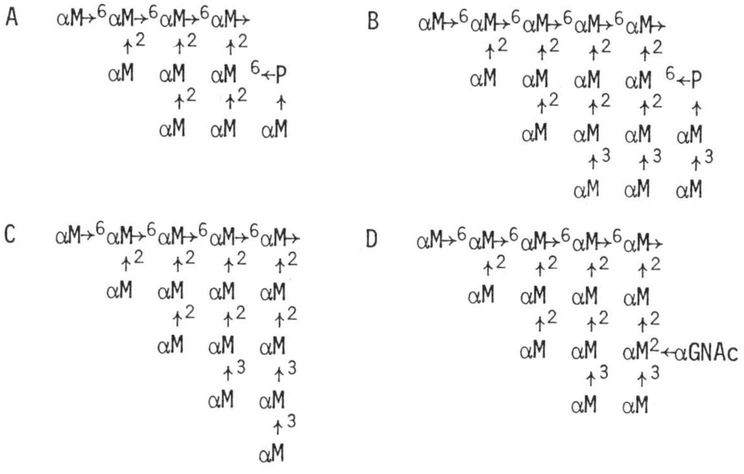

Figure 4 Representative structures for the polysaccharide component of yeast mannoproteins. (A) *Kloeckera brevis*, (B) *S. cerevisiae*, (C) *S. italicus*, (D) *K. lactis*.

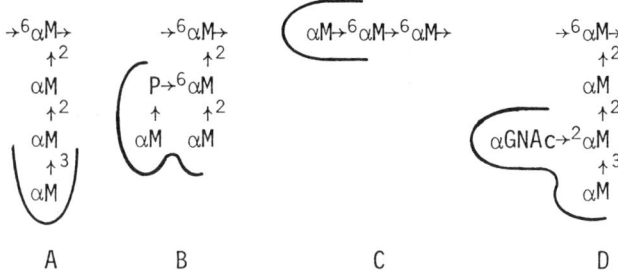

Figure 5 Partial structures of important yeast mannoprotein immunochemical determinants. (A) S. cerevisiae X2180 or the mnn4 mutant; (B) S. cerevisiae mnn1 mutant or S. cerevisiae A364A and K. brevis; (C) S. cerevisiae mnn2 mutant; (D) K. lactis. The region of antibody recognition is indicated by curved lines.

radioactivity, and enzymic digestion. Some of these studies are listed in Table 1. The results may be summarized by saying that, although there is some localization of wall components, there appears to be less organization in most yeast cell walls than in those of bacteria (Salton 1964).

CHANGES IN THE WALL DURING CELL GROWTH AND DIVISION

Owing to the limited knowledge concerning the organization of the yeast cell wall, there is relatively little that is well defined concerning the structural changes that occur during cell growth, bud formation, and septation (Cabib 1975). As a model, one would expect the need for a mechanism for wall expansion during growth, either through the weakening action of hydrolytic enzymes (β-glucanases) or through a physical stretching of associated glucan fibers. If the wall is cross-linked by protein disulfide bonds (Nickerson and Falcone 1956), then proteases or disulfide reductases may be important. On the initiation of bud formation, a localized outgrowth of the wall must occur (Farkas 1979), and at septation, the ordered deposition of chitin (Cabib et al. 1974) and mannoprotein is needed to close the space between the mother and daughter cells. As a final step, the two cells must separate. The enzymology and physiological control of some of these processes have been investigated, and these are dealt with in other sections of this paper. This part is concerned solely with the demonstrated changes in wall structure or organization during cell growth and division. An excellent general review of the literature prior to 1970 is provided by Bartnicki-Garcia and McMurrough (1971).

Cell Growth

S. cerevisiae cells are oval in shape, and several observations support the conclusion that new growth occurs at the ends of the cells. An inhibitor of glucan biosynthesis, 2-deoxyglucose, causes growing cells to lyse at the ends, presumably by interfering with the reconstruction of the autolyzed wall (Johnson 1968). Newly deposited cell-wall material can be detected by autoradiography after growth in the presence of tritiated wall precursors (Johnson and Gibson 1966), and the label shows an apical localization. In *Schizosaccharomyces pombe*, the cells of which have two clearly defined ends, localized deposition of new material may occur in a particular cell mainly at one end or at both ends (Johnson 1965).

Bud Formation

Bud formation can occur at any place on an *S. cerevisiae* cell as demonstrated by the observation that a specific cell can bud more than 20 times before it loses viability, after which most of the cell surface is covered with bud scars (Mortimer and Johnston 1959). Nevertheless, the initiation of any one bud does not occur randomly, and it involves the localized deposition of new wall material as demonstrated by autoradiography (Johnson 1965) and fluorescent probes (Tkacz and Lampen 1972). Presumably, the bud wall is also a site of enhanced enzyme activity, but this has been demonstrated only indirectly. Whereas new wall synthesis is continuous throughout the cell cycle, chitin synthesis (see below) occurs mainly during septation. Cells that are unable to form septa, owing to inability to synthesize chitin, continue to bud and undergo nuclear division with the result that long, beadlike strings of cells, each with a nucleus, are formed (Ballou et al. 1977). Electron microscopy of thin sections of such cells shows that the septal regions become progressively more deformed. Thus, neither bud initiation nor nuclear division is regulated by septation and cell separation.

Septation

The principal observable change in the wall during septation involves formation of the bud scar on the mother cell and the birth scar on the daughter cell (Cabib et al. 1974). The bud scar is composed of chitin and mannoprotein and, in *S. cerevisiae*, synthesis of chitin is regulated such that it is formed mainly at the time of septation (Cabib 1975). Various dyes and lectins bind selectively to the region of septation, which suggests that the chitin is exposed to such reagents. A similar regulation of glucosamine synthesis occurs during sporulation, and a glucosamine auxotroph that is unable to make glucosamine during sporulation is also defective in septation (Ballou et al. 1977). This suggests that at least some of the same enzymatic machinery is involved in both processes. The asymmetric nature of septation is a novel feature and, although the bud scar remains essentially unchanged during the life of the cell, the birth scar becomes less and less apparent.

Table 1 Localization of yeast cell-wall components

Conclusion of experiment	Probe	Detection method	Reference
Mannoprotein is located on the cell surface of S. cerevisiae	Ab Fl-Ab CG-Ab	AGG FM TEM	Hanseneclever and Mitchell (1964) Horisberger and Von-lanthem (1977)
Mannoprotein is located on the surface of some yeasts but not others	Fl-Con A Fl-WGA Fl-SBA Fl-PNA		Barkai-Golan and Sharon (1978)
Mannoprotein is located in the bud scar	CG-Ab	TEM	Horisberger and Von-lanthen (1977)
Mannoprotein is located on the protoplast surface	CG-Ab	SEM	Horisberger et al. (1976)
Chitin is located mainly in the bud scar	primulin	FM	Streiblova and Beran (1963)
	CG-WGA	TEM	Horisberger and Von-lanthen (1977)
Chitin is localized in a ring around the bud scar, sandwiched between two layers of glucan	glusulase β-glucanase and chitinase	LM LM	Cabib and Bowers (1971)
Chitinlike material is dispersed on the cell wall	CG-WGA	TEM	Horisberger and Von-lanthen (1977)
Chitin accumulates in the area of pheromone-stimulated growth of **a** cells	primulin	FM	Schekman and Brawey (1979)
Chitin accumulates over surface of cdc24 at nonpermissive temperature			Sloat and Pringle (1978)

Observation	Reagent	Technique	Reference
Fuzzy surface material accumulates in area of pheromone-stimulated growth on **a** cells	KMnO$_4$/OsO$_4$	TEM	Lipke et al. (1976)
Mannoprotein antigen shows enhanced reactivity in area of pheromone-stimulated growth of **a** cells	Fl-Ab	FM	Lipke (1976); Lipke and Ballou (1980)
	Fl-Con A	FM	Tkacz and MacKay (1979)
Cell wall is layered	TAPO/OsO$_4$	TEM	Djaczenko and Cassone (1971)
Free sulfhydryl groups accumulate in region of budding	[^{3}H]phenyl-HgCl	autoradiography	Robson and Stockley (1962)
Bud formation occurs by localized deposition of new wall material	[^{3}H]glucose	autoradiography	Johnson (1965)
	Fl-Ab	FM	May (1962)
	Fl-Con A	FM	Tkacz and Lampen (1972)
Receptor for killer toxin is located in the cell wall of *S. cerevisiae*	direct isolation	binding of ^{35}S-labeled toxin	Bussey et al. (1979)
Vegetative cell-wall glucan is accessible from the outside	β-glucanase	LM	Nečas (1971)
Reducing agents or proteases expose glucan	β-mercaptoethanol, proteases	LM	Scott (1980)
Spores are protected from β-glucanase attack by a surface layer	β-glucanase	TEM	Whelan and Ballou (1975); Kane and Roth (1974)

Abbreviations: Ab, specific antibody; AGG, agglutination; CG, colloidal-gold-conjugated; Con A, concanavalin A; Fl, fluorescein-conjugated; FM, fluorescence microscopy; LM, light microscopy; PNA, peanut agglutinin; SBA, soybean agglutinin; SEM, scanning electron microscopy; TAPO, Tris(l-aziridinyl)phosphine oxide; TEM, transmission electron microscopy; WGA, wheat germ agglutinin.

Scanning electron micrographs confirm the shape of the bud scar as inferred from light microscopy of cells and the chitinous bud-scar material isolated from cell walls. Whether the birth scar has a distinguishing chemical composition is unknown.

Protoplast Formation and Regeneration

Protoplast formation from *S. cerevisiae* generally has employed crude glucanase preparations such as strepzyme, glusulase (helicase), or zymolyase, in conjunction with a reducing agent such as mercaptoethanol (Nečas 1971). It has been demonstrated that a highly purified β-glucanase from *Oerskovia xanthineolytica* can form protoplasts from early log-phase cells of strain X2180 only in combination with a reducing agent or a protease (Scott 1980; Scott and Schekman 1980). From these minimum requirements with purified enzymes, it appears that the wall mannoprotein matrix must be loosened (by reduction of disulfide cross-links or cleavage of peptide bonds?) so that the glucanase can act effectively on the glucan network. Time-lapsed photomicrographs of the process reveal that, with appropriate osmolarity, the membrane-enclosed cytoplasm may move intact through the first small opening in the wall as though there were no adhesions to hold it in physical association with the remaining wall structure (shown by Dr. H. Weber at the 5th International Protoplast Symposium in Szeged, Hungary, 1979).

Surface antigens of yeast protoplasts cross-react with antisera to the vegetative cell surface (C. E. Ballou, unpubl.), which is attributable to the presence on protoplasts of glycoproteins with mannose-rich core oligosaccharide units. Cross-reacting oligosaccharides are also found on intracellular glycoproteins of *S. cerevisiae*, such as carboxypeptidase Y (R. E. Cohen and C. E. Ballou, unpubl.). It also has been reported, however, that antiserum to *Candida utilis* protoplast membranes will cross-react with *C. utilis* cell-wall fragments but not with the whole cell (Garcia Mendoza et al. 1968). In this latter instance, it is probable that the antibody was directed against a component of the membrane unrelated to mannoprotein and that the cross-reaction was due to membrane material adhering to the wall fragments.

According to Nečas (1979), protoplast regeneration occurs in three steps: (1) growth of protoplast, (2) regeneration of the cell wall, and (3) reversion, during which all cell functions are restored. Regeneration of the cell wall by *S. cerevisiae* protoplasts has two novel features. First, the wall macromolecules that are secreted must be physically held in the vicinity of the protoplast, otherwise they diffuse away (Nečas 1971). This suggests that the materials are released in a free form, not attached to the plasma membrane. Second, the first one or two generations of buds that are formed by the regenerated cell do not have the characteristic shape of the parent cell, and it takes several divisions before this morphology is recovered. The implication is that some information for determining shape is lost on protoplast formation, which is only slowly recovered by the regenerated cell. Whether this information resides in the wall, in the membrane, or in some interaction between the two is unknown (Nečas 1971; Farkaš 1979). Recent interest in fusion of yeast protoplasts with one another or with protoplasts from other organisms has led to standardized conditions for protoplast regeneration (Kuo and Yamamoto 1975). For a recent and more extensive review, see Peberdy (1979).

CHANGES IN THE WALL DURING SPORULATION

This section deals with changes that occur in the wall of the parent cell and of the developing spores during sporulation. More detailed discussions of meiosis and ascospore formation can be found in Fowell (1969) and Haber et al. (1975).

Parent Cell Wall

Because sporulation of *S. cerevisiae* can occur on potassium acetate alone, it is apparent that extensive recycling of the parent cell components is required to provide precursors for the synthesis of spore macromolecules. Thin-section electron microscopy of asci, however, does not suggest that the parent wall itself is extensively reutilized (Fowell 1969). Thus, the recycling is probably restricted mainly to the cell contents. The ascal wall, however, is disrupted when the spores begin to germinate and break out of the confines of the parent cell.

Spore Coat

Thin-section electron microscopy of *S. cerevisiae* ascospores shows that the spore wall is composed of two distinct layers: a thick inner wall that has the appearance of the parent vegetative wall and a thin outer layer (Fig. 6A). Disagreement exists concerning the composition of the outer wall (see Haber and Halvorson 1975), with some evidence suggesting it is lipid in nature and other evidence suggesting that it is proteinaceous. Interestingly, this layer is absent in spores produced by a glucosamine auxotroph (Whelan and Ballou 1975) (Fig. 6B), which suggests that this hexosamine may be an important component of the layer. The synthesis of this layer appears to be a late marker of sporulation, and glucosamine synthesis during sporulation occurs only after tetranucleate cells begin to appear in the culture. The spore wall is also reported to contain glucan and mannan (Kane and Roth 1974), although it is not established that these components are identical to those of the parent wall. Like the parent wall, however, the thick walls of the spores

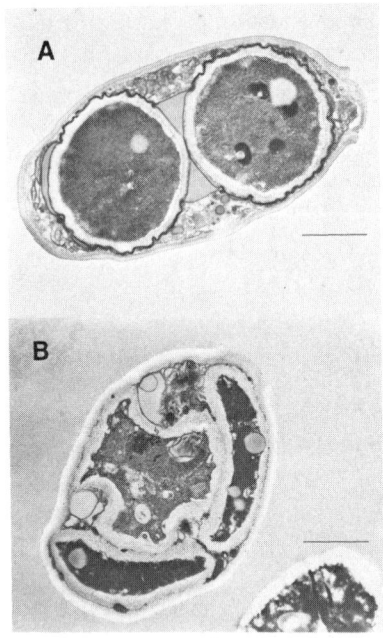

Figure 6 Thin-section electron micrographs of *S. cerevisiae* asci. (*A*) Strain X2180; (*B*) a glucosamine auxotroph. Note the absence of the dark-staining surface layer and the deformed walls in *B*. (Reprinted, with permission, from Whelan and Ballou 1975.) Bars represent 1 µM.

produced by the above glucosamine auxotroph are readily solubilized by the action of β-glucanases, which are unable to attack wild-type spores.

Some of the early developmental steps in the biogenesis of the *S. cerevisiae* ascospore wall have been elucidated through the analysis of mutants that are defective in the process (Esposito and Esposito 1975). These studies and other work (Lynn and Magee 1970) suggest that the inner thick wall of the spore is synthesized by the inner ascospore-delimiting membrane that becomes the plasmalemma, and that the outer thin coat is synthesized by the outer ascospore-delimiting membrane. It is postulated that the hydrophobic, glucanase-resistant coat of the spore is formed during spore maturation from the cytoplasmic material left over from the earlier meiotic processes (Fowell 1969).

Spore Germination

This process in *S. cerevisiae* presumably involves the disruption and loss of the glucanase-resistant surface layer (Fowell 1969) to yield almost immediately a haploid cell able to mate with a vegetative cell of opposite mating type. Wall changes associated with this outgrowth thus far have gone unstudied, except by electron microscopy (Hashimoto et al. 1959), which suggests that the spore wall becomes the cell wall of the germinated spore. This is consistent with the appearance (Whelan and Ballou 1975) and composition (Kane and Roth 1974) of the spore wall.

ROLE OF THE CELL WALL IN MATING

Sexual mating normally occurs between haploid cells of opposite mating type, and it may involve a specific agglutination of the cells and an exchange of pheromones that initiate changes in the wall, which lead to fusion. The agglutination may be strong and constitutive, as with *Hansenula wingei, Pichia amethionina,* and *S. kluyveri* (Burke et al. 1980); or it may be weak and become enhanced as a result of pheromone-induced changes, as with *S. cerevisiae* X2180 (Fehrenbacher et al. 1978). The constitutive agglutination factors can be isolated and some have been characterized, whereas the nature of the inducible factors is in dispute. The changes that the wall undergoes upon exposure to pheromone have been probed with antibodies, lectins, electron microscopy, and by direct analysis of the wall components (Brock 1965). Some of the results are discussed in this paper.

Constitutive Agglutination Factors

The two haploid mating types of the yeast *H. wingei*, 5 cells and 21 cells, agglutinate strongly when mixed (Crandall and Brock 1968). The agglutinative activity of the 5-cell mating type is stable to heat but is destroyed by mercaptoethanol, whereas the activity of the 21-cell mating type is destroyed by heat and is unaffected by reducing agents. Treatment of the 5-cell mating type with subtilisin releases a glycoprotein of 10^6 daltons that will agglutinate the 21-cell mating type and has been shown to have five to six binding sites (Taylor and Orton 1971; Yen and Ballou 1974). These binding sites are released by mercaptoethanol and are glycopeptides of 12,000 daltons that bind to the 21-cell mating type. Treatment of the 21-cell mating type with trypsin releases an acidic protein of 27,000 daltons that does not agglutinate the 5-cell mating type but does inhibit the agglutination of the 21-cell mating type by purified 5-agglutinin. This 21-cell factor is heat-labile, but its activity is not affected by mercaptoethanol. These results suggest the model in Figure 7 for the *H. wingei* sexual agglutination system.

Similar factors appear to be involved in the sexual agglutination of *P. amethionina* and *S. kluyveri* (Burke et al. 1980). The agglutinative activity of one haploid cell of each strain is stable to heat and is inactivated by mercaptoethanol, whereas the activity of the other haploid cell is destroyed by heat but is unaltered by the reducing agent. Although *S. kluyveri* does not mate well with *S. cerevisiae*, one of its haploid cells (16-cell mating type) responds to *S. cerevisiae* α-factor with a typical premating change in morphology (McCullough and Herskowitz 1979). By extrapolation, then, formation of the heat-stable agglutination factor appears to be under control of the **a**-mating-type locus, and the heat-labile recognition factor is under control of the α-mating-type locus. These distinctions may be significant in that the heat-stable agglutinin appears to be a passive site of recognition (a

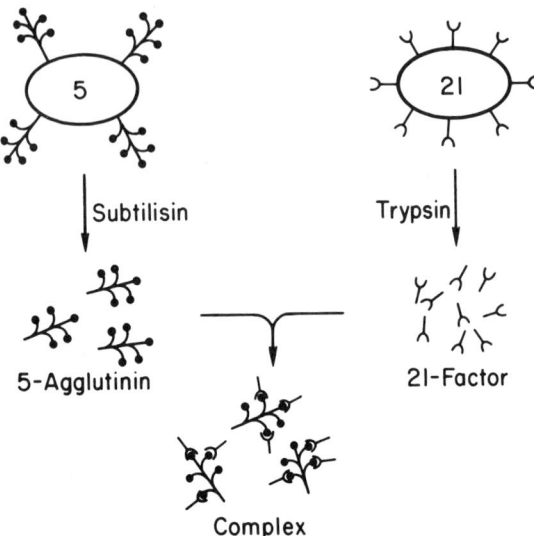

Figure 7 Model for sexual agglutination in *H. wingei*.

cognon), whereas the heat-labile factor is an active recognizer molecule (a cognor) (Burke et al. 1980). The interaction of these factors does not appear to be lectinlike because it is not inhibited by carbohydrate fragments of the 5-agglutinin.

The attempts to characterize agglutination factors from *S. cerevisiae* so far have been relatively unproductive, although material with specific activity has been obtained in soluble form (Shimoda et al. 1975).

Pheromone-induced Changes in the Yeast Cell Wall

The morphological change in *S. cerevisiae* **a** cells on exposure to α-cell supernatant (Levi 1956) and isolated α-factor (Duntze et al. 1970) had remained uninvestigated until Lipke et al. (1976) showed that there are significant changes in the cell surface and in the cell-wall polymers. Thin-section electron microscopy (Fig. 8) reveals a dramatic reorganization of the cell surface at the region of pheromone-induced cell enlargement, and this change is accompanied by an enhanced affinity for specific antibodies (Lipke 1976) and concanavalin A (Tkacz and MacKay 1979). Lipke and Ballou (1980) have obtained evidence that this surface change results in the enhanced expression of an antigen that is present at a low level on untreated **a** cells, but at a much higher level on α cells, and is not observed on diploid cells. Because antiserum to **a** cells that were not treated with pheromone binds preferentially to the disorganized cell surface of pheromone-treated **a** cells and the binding can be reversed by mannoprotein oligosaccharides, it is

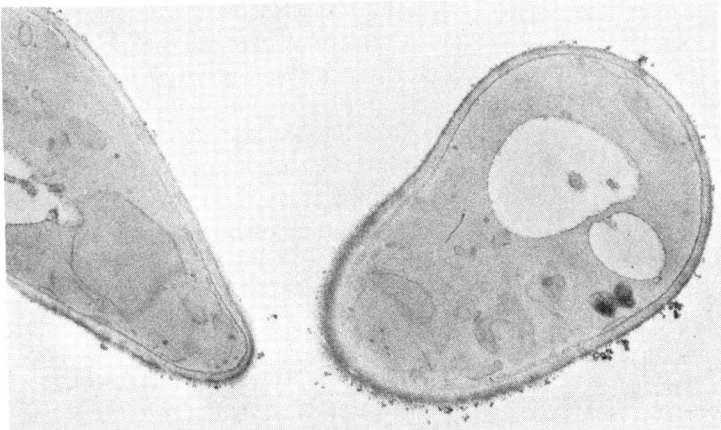

Figure 8 Thin-section electron micrograph of *S. cerevisiae* after exposure to α-factor. (Reprinted, with permission, from Lipke et al. 1976.) Magnification, 14,000×.

possible that this kind of polar binding of antibody is due as much to antigen accessibility as to the presence of any mating-specific antigen. If such an antigen is formed, it must be a part of the polysaccharide component of the mannoprotein, because none of the *mnn* mutants was able to express it (Lipke and Ballou 1980).

The kinds of biochemical changes one can analyze in yeast cell walls include the altered carbohydrate structure of mannoproteins glucan and chitin and changes in glucanase activity. Enzyme activities associated with mannoprotein or chitin synthesis have not been demonstrated in the *S. cerevisiae* cell wall, although a disulfide reductase may be involved in cross-linking mannoproteins (Nickerson and Falcone 1956). Lipke et al. (1976) found that *S. cerevisiae* **a** cells exposed to α-factor make more glucan and less mannoprotein than during vegetative growth, and the mannoprotein sidechains are shorter than normal as a consequence of a reduced expression of the α1 → 3-mannosyltransferase activity. If the morphogenesis is done in the presence of [^{3}H]mannose, there is an enhanced incorporation of the labeled mannose into an acid-labile structure, which suggests that the pheromone does affect some biosynthetic process. Such cells also become more susceptible to lysis by β1 → 3-glucanases, as though the glucan is more accessible to such hydrolytic enzymes. *S. cerevisiae* **a** cells, when exposed to α-factor also show an enhanced deposition of chitin in the region of pheromone-stimulated growth (Schekman and Brawley 1979). The physiological significance of this change is unknown, but it is associated with an increased activity and altered distribution of chitin synthetase within the cell.

Many studies have dealt with yeast glucanases in intracellular vesicles

(Matile et al. 1971), in the periplasm (Farkaš et al. 1973), in the cell wall (Fleet and Phaff 1974), and secreted in the medium (Farkaš et al. 1973). Two activities are most commonly observed: an exo-β-glucanase, assayable with the $\beta 1 \rightarrow$ 3-glucan laminarin and the $\beta 1 \rightarrow$ 6-glucan pustulan, and the endo-$\beta 1 \rightarrow$ 3-glucanase, assayable with periodate-oxidized laminarin that prevents action by the exoglucanase. No change in activity of any of these enzymes was detected in cell extracts during mating in *S. cerevisiae* (del Ray et al. 1979). A mutant of *S. cerevisiae* lacking the exo-β-glucanase activity is able to mate and grow normally (Santos et al. 1979), so this enzyme apparently is not involved in that process, although the endo-$\beta 1 \rightarrow$ 3-glucanase may be. Because the majority of the glucanase activity is not bound in the cell wall, it is not clear what role these enzymes have in metabolism of the wall glucan.

QUESTIONS AND SUGGESTIONS FOR NEW DIRECTIONS IN RESEARCH

The yeast cell surface would appear to be a poor analog for the cell surfaces of higher organisms, owing to its rigid cell-wall construction. In this sense, it parallels the Gram-positive bacteria more closely than lymphocytes. In other ways, however, the yeast cell wall may mimic more closely the intercellular matrix of plant and animal tissues, being composed of a structural glucan, which has similarities to cellulose, and a matrix of mannoproteins, which have their counterpart in the mucins, blood group substances, and immunoglobulins.

Clearly, there is much to be learned about the structure and organization of the yeast cell wall. The exact nature of the wall glucan is still uncertain and little is known of the way in which the molecules are assembled or associate to form the fibrous network. Although striking advances have been made in elucidating mannoprotein structures, we have only a hazy notion as to how the molecules are integrated into the wall. Are they cross-linked in any way? Does disulfide bonding play a role in stabilizing the wall? Do phosphodiester bonds have a function as cross-linkers?

What is the difference between the inner and outer surfaces of the wall? How does the inner surface of the wall relate to the outer surface of the plasmalemma? In this regard, Farkaš (1979) has postulated that cell growth is regulated by cognitive interaction between membrane and wall, possibly a parallel to the "contact inhibition of growth" observed in higher organisms.

Almost nothing is known of the way in which the wall is assembled. What determines the different shapes of the cells? What is the function of the polymorphism in the carbohydrate structure of the mannoproteins of different species? How are the changes in morphology regulated that occur during cell division and mating or during sporulation? What proteins are present in the wall and how do they change during these processes? What shape-determining information is lost during protoplast formation and how is it recovered during regeneration?

All proposed models for the yeast cell wall that attempt to picture cross-linking at the molecular level must be considered highly speculative. Phosphodiester cross-linking as an important structural element in *S. cerevisiae* (reviewed by Farkaš 1979) is untenable because all of the phosphate can be accounted for as mannosylphosphate units linked to position 6 of the mannoprotein sidechains, and mutants that lack phosphate in the wall are perfectly normal (reviewed by Ballou 1976). Disulfide cross-linking, on the other hand, seems well documented by the demonstrated effectiveness of mercaptoethanol in facilitating disruption of the wall fabric by a highly purified β-glucanase (Scott 1980; Scott and Schekman 1980). It is instructive that the role of the reducing agent can be replaced in this process by purified protease, which suggests that both agents act on wall proteins.

New directions for study of the yeast cell wall and cell surface should concentrate on an expanded use of mutants, and the use of the fluorescence-activated cell sorter for their selection will make this work much easier. More selective analysis of the individual mannoproteins of the wall, periplasm, and the plasmalemma should be fruitful, particularly if ways can be developed to remove the carbohydrate components completely without denaturation of the protein part. Dissection of the wall with enzymes may lead to a better understanding of its organization, but new techniques such as X-ray photoelectron spectroscopy (Millard et al. 1976) might be even more revealing. With the development of more selective staining procedures, electron microscopy should also help to resolve this structure.

Developmental changes associated with cell growth and division, with sporulation and spore germination, and during the mating reaction can all be investigated productively in yeasts. The unusual effect of glucosamine auxotrophy on spore-wall development provides a system for study of the synthesis of the glucan layer and surface coat as separate processes. In this regard, the turn-on of glucosamine synthesis at fixed times during cell division and sporulation are attractive processes for the study of gene regulation.

As a model for the study of cell-cell recognition, the yeast system is highly attractive. Several strains with parallel but non-cross-reacting factors are available, and reasonable amounts of pure material can be obtained without excessive effort. Moreover, the sexual agglutination factors from *H. wingei* have interesting properties that may find parallels in higher organisms, particularly in fertilization reactions that are unilateral interactions between unlike cells, although such unilateral models are less likely to be involved in adhesion between like cells.

Note Added in Proof

The core oligosaccharide structure of yeast mannoprotein (Fig. 1) has been refined (Zhang et al. 1982) and its identity with the oligosaccharides of higher eukaryotes, established. With regard to the defect in the *mnn2* mutant, it has been reported (Parodi 1979) that cell extracts of this strain make mannoprotein with the wild-type structure and that the extracts contain a normal $\alpha 1 \rightarrow$ 2-mannosyltransferase activity. This suggests that the *mnn2* mutation may be similar to that in *Kluyveromyces lactis*, in which an altered mannoprotein is secreted lacking terminal $\alpha 1 \rightarrow$ 2-linked *N*-acetyl-D-glucosamine units even though cell extracts contain the appropriate enzyme activity (Douglas and Ballou 1982).

REFERENCES

Antalis, C., S. Fogel, and C.E. Ballou. 1973. Genetic control of yeast mannan structure. Mapping the first gene concerned with mannan biosynthesis. *J. Biol. Chem.* **248:** 4655.

Atkins, E.D.T., K.D. Parker, and R.D. Preston. 1969. The helical structure of the β-1,3-linked xylan in some siphoneous green algae. *Proc. R. Soc. Lond. B* **173:** 209.

Bacon, J.S.D. 1973. The contribution of β-glucanases to the lysis of fungal cell walls. In *Yeast, mold and plant protoplasts* (ed. J.R. Villanueva et al.), p. 61. Academic Press, New York.

Ballou, C.E. 1970. A study of the immunochemistry of three yeast mannans. *J. Biol. Chem.* **245:** 1197.

———. 1976. Structure and biosynthesis of the mannan component of the yeast cell envelope. *Adv. Microb. Physiol.* **14:** 93.

Ballou, C.E. and W.C. Raschke. 1974. Polymorphism of the somatic antigen of yeast. *Science* **184:** 127.

Ballou, C.E., K.A. Kern, and W.C. Raschke. 1973. Genetic control of yeast mannan structure. Complementation studies and properties of mannan mutants. *J. Biol. Chem.* **248:** 4667.

Ballou, C.E., S.K. Maitra, J.W. Walker, and W.L. Whelan. 1977. Developmental defects associated with glucosamine auxotrophy in *Saccharomyces cerevisiae*. *Proc. Natl. Acad. Sci.* **74:** 4351.

Ballou, D.L. 1975. Genetic control of yeast mannan structure: Mapping genes *mnn2* and *mnn4* in *Saccharomyces cerevisiae*. *J. Bacteriol.* **123:** 616.

Ballou, D.L., R.E. Cohen, and C.E. Ballou. 1980. *Saccharomyces cerevisiae* mannoprotein mutants that make mannoproteins with a truncated carbohydrate outer chain. *J. Biol. Chem.* **255:** 5986.

Barkai-Golan, R. and N. Sharon. 1978. Lectins as a tool for the study of yeast cell walls. *Exp. Mycol.* **2:** 110.

Bartnicki-Garcia, S. and I. McMurrough. 1971. Biochemistry of morphogenesis in yeasts. In *The yeasts* (ed. A.H. Rose and J.S. Harrison), vol. 2, p. 441. Academic Press, New York.

Brock, T.D. 1965. Biochemical and cellular changes occurring during conjugation in *Hansenula wingei*. *J. Bacteriol.* **90:** 1019.

Burke, D., L. Mendonca-Previato, and C.E. Ballou. 1980. Cell-cell recognition in yeast: Purification of *Hansenula wingei* 21-cell sexual agglutination factor and comparison of the factors from three genera. *Proc. Natl. Acad. Sci.* **77:** 318.

Bussey, H., D. Saville, K. Hutchins, and R.G.E. Palfree. 1979. Binding of yeast killer toxin to cell wall receptor on sensitive *Saccharomyces cerevisiae*. *J. Bacteriol.* **140:** 888.

Cabib, E. 1975. Molecular aspects of yeast morphogenesis. *Annu. Rev. Microbiol.* **29:** 191.

Cabib, E. and B. Bowers. 1971. Chitin and yeast budding. Localization of chitin in yeast bud scars. *J. Biol. Chem.* **246:** 152.

Cabib, E., R. Ulane, and B. Bowers. 1974. A molecular model for morphogenesis: The primary septum of yeast. *Curr. Top. Cell. Regul.* **8:** 1.

Chapman, A. and R. Kornfeld. 1979. Structure of the high mannose oligosaccharides of a human IgM myeloma protein. I. The major oligosaccharides of the two high mannose glycopeptides. *J. Biol. Chem.* **254:** 816.

Cohen, R.E. 1980. "Structure of glycoprotein core oligosaccharides." Ph.D. thesis, University of California, Berkeley.
Cohen, R.E., L. Ballou, and C.E. Ballou. 1980. *Saccharomyces cerevisiae* mannoprotein mutants. Isolation of the *mnn5* mutant and comparison with the *mnn3* strain. *J. Biol. Chem.* **255**: 7700.
Crandall, M.A. and T.D. Brock. 1968. Molecular basis of mating in the yeast *Hansenula wingei*. *Bacteriol. Rev.* **32**: 139.
del Rey, F., T. Santos, I. Garcia-Acha, and C. Nombela. 1979. Synthesis of 1,3-β-glucanases in *Saccharomyces cerevisiae* during the mitotic cycle, mating, and sporulation. *J. Bacteriol.* **139**: 924.
Djaczenko, W. and A. Cassone. 1971. Visualization of new ultrastructural components of the cell wall of *Candida albicans* with fixatives containing TAPO. *J. Cell Biol.* **52**: 186.
Douglas, R.H. 1979. "Mannan biosynthesis in *Kluyveromyces lactis*." Ph.D. thesis, University of California, Berkeley.
Douglas, R.H. and C.E. Ballou. 1980. Isolation of *Kluyveromyces lactis* mannoprotein mutants by fluorescence-activated cell sorting. *J. Biol. Chem.* **255**: 5979.
Douglas, R. and C.E. Ballou. 1982. Purification of an α-N-acetylglucosaminyltransferase from the yeast *Kluyveromyces lactis* and a study of mutants defective in this enzyme activity. *Biochemistry* **21**: 1561.
Duntze, W., V. MacKay, and T.R. Manney. 1970. *Saccharomyces cerevisiae*: A diffusible sex factor. *Science* **168**: 1472.
Esposito, M.S. and R.E. Esposito. 1975. Mutants of meiosis and ascospore formation. *Methods Cell Biol.* **11**: 303.
Farkaš. V. 1979. Biosynthesis of cell walls of fungi. *Microbiol. Rev.* **42**: 117.
Farkaš, V., P. Biely, and S. Bauer. 1973. Extracellularβ-glucanases of the yeast *Saccharomyces cerevisiae*. *Biochem. Biophys. Acta* **321**: 246.
Fehrenbacher, G., K. Perry, and J. Thorner. 1978. Cell-cell recognition in *Saccharomyces cerevisiae*: Regulation of mating-specific adhesion. *J. Bacteriol.* **134**: 893.
Fleet, G.H. and J.H. Phaff. 1974. Glucanases in *Schizosaccharomyces*. Isolation and properties of the cell-wall associated β-(1 $\rightarrow$ 3)-glucanases. *J. Biol. Chem.* **249**: 1717.
Fowell, R.R. 1969. Sporulation and hybridization of yeasts. In *The yeasts* (ed. A.H. Rose and J.S. Harrison), vol. 1, p. 303. Academic Press, New York.
Friis, J. and P. Ottolenghi. 1970. The genetically determined binding of alcian blue by a minor fraction of yeast cell walls. *C.R. Trav. Lab. Carlsberg* **37**: 327.
Garcia Mendoza, C., M.D. Garcia Lopez, F. Uruburu, and J.R. Villanueva. 1968. Structural and immunological studies on the protoplast membrane of the yeast *Candida utilis*. *J. Bacteriol.* **95**: 2393.
Haber, J.E. and H.O. Halvoroson. 1975. Methods in sporulation and germination of yeasts. *Methods Cell Biol.* **11**: 45.
Haber, J.E., M.S. Esposito, P.T. Magee, and R.E. Esposito. 1975. Current trends in genetic and biochemical study of yeast sporulation. In *Spores VI* (ed. P. Gerhardt et al.), p. 132. American Society for Microbiology, Washington, D.C.
Hasenclever, H.F. and W.O. Mitchell. 1964. A study of yeast surface antigens by agglutination inhibition. *Sabouraudia* **3**: 288.
Hashimoto, C., R.E. Cohen, and C.E. Ballou. 1980. Characterization of phosphorylated oligomannosides from *Hansenula wingei* mannoprotein. *Biochemistry* **19**: 5932.
Hashimoto, J., S.F. Conti, and H.B. Naylor. 1958. Fine structure of microorganisms. III. Electron microscopy of resting and germinating ascospores of *Saccharomyces cerevisiae*. *J. Bacteriol.* **76**: 406.
Horisberger, M. and M. Vonlanthen. 1977. Location of mannan and chitin on thin sections of budding yeasts with gold markers. *Arch. Microbiol.* **115**: 1.
Horisberger, M., J. Rosset, and H. Bauer. 1976. Localization of mannan at the surface of yeast protoplasts by scanning electron microscopy. *Arch. Microbiol.* **109**: 9.

Johnson, B.F. 1965. Autoradiographic analysis of regional cell growth of yeasts. *Schizosaccharomyces pombe. Exp. Cell Res.* **39:** 613.

———. 1968. Lysis of yeast cell walls induced by 2-deoxyglucose at sites of glucan synthesis. *J. Bacteriol.* **95:** 1169.

Johnson, B.F. and E.J. Gibson. Autoradiographic analysis of regional cell growth of yeasts. III. *Saccharomyces cerevisiae. Exp. Cell Res.* **41:** 580.

Kane, S.M. and R. Roth. 1974. Carbohydrate metabolism during ascospore development in yeast. *J. Bacteriol.* **118:** 8.

Kappeli, O., M. Muller, and A. Fiechter. 1978. Chemical and structural alterations at the cell surface of *Candida tropicalis,* induced by hydrocarbon substrate. *J. Bacteriol.* **133:** 952.

Karson, E.M. and C.E. Ballou. 1978. Properties of a mannosylphosphate transferase in *Saccharomyces cerevisiae. J. Biol. Chem.* **253:** 6484.

Kornfeld, R. and S. Kornfeld. 1976. Comparative aspects of glycoprotein structure. *Annu. Rev. Biochem.* **45:** 217.

Kreger, D.R. and M. Kopecká. 1973. On the nature of the fibrillar nets formed by protoplasts of *Saccharomyces cerevisiae* in liquid media. In *Yeast, mold and plant protoplasts* (ed. J.R. Villanueva et al.), p. 117. Academic Press, New York.

Kuo, S.-C. and S. Yamamoto. 1975. Preparation and growth of yeast protoplasts. *Methods Cell Biol.* **11:** 169.

Lehle, L., R.E. Cohen, and C.E. Ballou. 1979. Carbohydrate structure of yeast invertase. *J. Biol. Chem.* **254:** 12209.

Levi, J.D. 1956. Mating reaction in yeast. *Nature* **177:** 753.

Lipke, P.N. 1976. "Biochemical studies on morphogenesis induced by a sex pheromone in *Saccharomyces cerevisiae.*" Ph.D. thesis, University of California, Berkeley.

Lipke, P.N. and C.E. Ballou. 1980. Altered immunochemical reactivity of *Saccharomyces cerevisiae* a-cells after α-factor-induced morphogenesis. *J. Bacteriol* **141:** 1170.

Lipke, P.N., A. Taylor, and C.E. Ballou. 1976. Morphogenic effects of α-factor on *Saccharomyces cerevisiae* a cells. *J. Bacteriol.* **127:** 610.

Lüderitz, O., A.M. Staub, and O. Westphal. 1966. Immunochemistry of O and R antigens of *Salmonella* and related Enterobacteriaceae. *Bacteriol. Rev.* **30:** 192.

Lynn, R.R. and P.T. Magee. 1970. Development of the spore wall during ascospore formation in *Saccharomyces cerevisiae. J. Cell Biol.* **44:** 688.

Manners, D.J., A.J. Masson, and J.C. Patterson. 1973a. The structure of a β-(1 → 3)-D-glucan from yeast cell walls. *Biochem. J.* **135:** 19.

Manners, D.J., A.J. Masson, J.C. Patterson, H. Bjorndal, and B. Lindberg. 1973b. The structure of a β-(1 → 6)-D-glucan from yeast cell walls. *Biochem. J.* **135:** 31.

Matile, P., M. Cortat, A. Wiemken, and A. Frey-Wyssling. 1971. Isolation of glucanase-containing particles from budding *Saccharomyces cerevisiae. Proc. Natl. Acad. Sci.* **68:** 636.

May, J.W. 1962. Sites of cell-wall extension demonstrated by the use of fluorescent antibody. *Exp. Cell Res.* **27:** 170.

McCullough, J. and I. Herskowitz. 1979. Mating pheromones of *Saccharomyces kluyveri*: Pheromone interactions between *Saccharomyces kluyveri* and *Saccharomyces cerevisiae. J. Bacteriol.* **138:** 146.

Millard, M.M., R. Scherrer, and R.S. Thomas. 1976. Surface analysis and depth profile composition of bacterial cells by X-ray photoelectron microscopy and oxygen plasma etching. *Biochem. Biophys. Res. Commun.* **72:** 1209.

Mortimer, R.K. and J.R. Johnston. 1959. Life span of individual yeast cells. *Nature* **183:** 1751.

Nakajima, T. and C.E. Ballou. 1974. Structure of the linkage region between the polysaccharide and protein parts of *Saccharomyces cerevisiae* mannan. *J. Biol. Chem.* **249:** 7685.

———. 1975. Yeast manno-protein biosynthesis: Solubilization and selective assay of four mannosyltransferases. *Proc. Natl. Acad. Sci.* **72:** 3912.

Nečas, O. 1971. Cell wall synthesis in yeast protoplasts. *Bacteriol. Rev.* **35:** 149.

———. 1979. Reversion of protoplasts. In *Abstracts of 5th International Protoplast Symposium*, Szeged, Hungary.
Nickerson, W.J. and G. Falcone. 1956. Enzymatic reduction of disulfide bonds in cell wall protein of baker's yeast. *Science* **124:** 318.
Parodi, A.J. 1979. Biosynthesis of yeast mannoproteins. Synthesis of mannan outer chain and of dolichol derivatives. *J. Biol. Chem.* **254:** 8343.
Peberdy, J.F. 1979. Fungal protoplasts: Isolation, reversion, and fusion. *Annu. Rev. Microbiol.* **33:** 21.
Phaff, H.J. 1971. Structure and biosynthesis of the yeast cell envelope. In *The yeasts* (ed. A.H. Rose and J.S. Harrison), vol. 2, p. 135. Academic Press, New York.
Raschke, W.C., K.A. Kern, C. Antalis, and C.E. Ballou. 1973. Genetic control of yeast mannan structure. Isolation and characterization of mannan mutants. *J. Biol. Chem.* **248:** 4660.
Reading, C.L., E.E. Penhoet, and C.E. Ballou. 1978. Carbohydrate structure of vesicular stomatitis virus glycoprotein. *J. Biol. Chem.* **253:** 5600.
Robson, J.E. and H.M. Stockley. 1962. Sulfhydryl metabolism of fungi grown in submerged culture. *J. Gen. Microbiol.* **28:** 57.
Rosenfeld, L. and C.E. Ballou. Genetic control of yeast mannan structure. Biochemical basis for the transformation of *Saccharomyces cerevisiae* somatic antigen. *J. Biol. Chem.* **249:** 2319.
Salton, M.R.J. 1964. *The bacterial cell wall*. Elsevier, New York.
Santos, T., F. del Rey, J. Conde, J.R. Villanueva, and C. Nombela. 1979. *Saccharomyces cerevisiae* mutant defective in exo-1,3-β-glucanase production. *J. Bacteriol.* **139:** 333.
Schekman, R. and V. Brawley. 1979. Localized deposition of chitin on the yeast cell surface in response to mating pheromone. *Proc. Natl. Acad. Sci.* **76:** 645.
Scott, J.H. 1980. "Enzymatic lysis of yeast cells." Ph.D. thesis, University of California, Berkeley.
Scott, J.H. and R. Schekman. 1980. Lyticase: Endoglucanase and protease activities that act together in yeast cell lysis. *J. Bacteriol.* **142:** 414.
Shimoda, C., S. Kitano, and N. Yanagishima. 1975. Mating reaction in *Saccharomyces cerevisiae*. VII. Effect of proteolytic enzymes on sexual agglutinability and isolation of crude sex-specific substances responsible for sexual agglutination. *Antonie van Leeuwenhoek J. Microbiol. Serol.* **41:** 513.
Sloat, B.F. and J.R. Pringle. 1978. A mutant of yeast defective in cellular morphogenesis. *Science* **200:** 1171.
Smith, W.L., T. Nakajima, and C.E. Ballou. 1975. Biosynthesis of yeast mannan. Isolation of *Kluyveromyces lactis* mannan mutants and a study of the incorporation of N-acetyl-D-glucosamine into the polysaccharide side chains. *J. Biol. Chem.* **250:** 3246.
Streiblova, E. and K. Beran. 1963. Demonstration of yeast scars by fluorescence microscopy. *Exp. Cell Res.* **30:** 603.
Suzuki, S., H. Sunayama, and T. Saito. 1968. Studies on the antigenic activity of yeasts. I. Analysis of the determinant groups of the mannan of *Saccharomyces cerevisiae*. *Jpn. J. Microbiol.* **12:** 19.
Tarentino, A.L., T.H. Plummer, Jr., and F. Maley. 1974. The release of intact oligosaccharides from specific glycoproteins by endo-β-N-acetylglucosaminidase H. *J. Biol. Chem.* **249:** 818.
Taylor, N.W. and W.L. Orton. 1971. Cooperation among the binding sites in the sex-specific agglutinin from the yeast *Hansenula wingei*. *Biochemistry* **10:** 2043.
Tkacz, J.S. and J.O. Lampen. 1972. Wall replication in *Saccharomyces* species: Use of fluorescein-conjugated concanavalin A to reveal the site of insertion. *J. Gen. Microbiol.* **72:** 243.
Tkacz, J.S. and V. MacKay. 1979. Sexual conjugation in yeast. Cell surface changes in response to the action of mating hormones. *J. Cell Biol.* **80:** 326.

Trimble, R.B. and F. Maley. 1977. Subunit structure of external invertase from *Saccharomyces cerevisiae*. *J. Biol. Chem.* **252**: 4409.

Whelan, W.L. and C.E. Ballou. 1975. Sporulation in D-glucosamine auxotrophs of *Saccharomyces cerevisiae*: Meiosis with defective ascospore wall formation. *J. Bacteriol.* **124**: 1545.

Yen, P.H. and C.E. Ballou. 1974. Partial characterization of the sexual agglutination factor from *Hansenula wingei* Y-2340 type 5 cells. *Biochemistry* **13**: 2428.

Zhang, W.-j., R.E. Cohen, and C.E. Ballou. 1982. *Saccharomyces cerevisiae* core oligosaccharide structure. *Fed. Proc.* **41**: 887.

The Secretory Process and Yeast Cell-surface Assembly

Randy Schekman and Peter Novick
Department of Biochemistry
University of California
Berkeley, California 94720

1. **Secretory Molecules and the Pattern of Cell-surface Growth**
   A. Secretory and Plasma-membrane Proteins
   B. Synthesis and Localization of Cell-wall Structural Polysaccharides
   C. Localization and the Pattern of Bud Growth
2. **Synthesis and Maturation of Secretory and Membrane Proteins**
   A. Synthesis of the Carbohydrate Portion
   B. Synthesis of the Protein Portion
   C. Organelles Implicated in the Secretory Process
3. **A Genetic Approach to the Secretory Process**
   A. Isolation and Characterization of *sec* Mutants
   B. Order of Events in the Secretory Pathway

INTRODUCTION

The goal of this paper is to review our knowledge of macromolecular secretion in *Saccharomyces cerevisiae* in terms of the synthesis and movement of exported proteins and carbohydrates, the membrane organelles that participate in this process, and the contribution that parts of this process make to the localization of organelle-associated proteins. Some of the material presented in this paper has been reviewed recently (Cabib 1975; Farkaš 1979; Byers 1981), and the chapters in this volume by Ballou, Henry, and Cooper are pertinent.

The mechanism of budding growth is relevant to this discussion. During most of the division cycle, cell-surface growth is restricted to the bud. Membrane enzymes, such as glucan synthetase, are responsible for much of this localized growth, and yet the plasma membrane is continuous and surrounds both the bud and mother portions of the cell. Thus, budding growth implies some mechanism for membrane differentiation and localization of surface components. Our attention has also been drawn by the connection between surface growth and secretion. These processes coincide temporally and spatially in a budding cell. We will present the evidence that supports a role for the secretory process in the formation of a bud.

It is tempting, though risky, to draw on principles derived from studies of other secretory processes. This paper includes some analogies and speculations that we hope will stimulate interest in yeast as a simple system for the study of eukaryotic secretion and cell-surface growth. Unfortunately, yeast has not been a popular source material for the cell biologists and biochemists who study secretion and, consequently, our knowledge may seem primitive in comparison to similar studies with bacteria or mammalian cells. Three factors may have contributed to this disapprobation. First, yeast cells are difficult to preserve for cytologic examination. The cell wall apparently excludes plastic embedding reagents. This problem can now be overcome by enzymatic removal of the cell wall after fixation (Byers and Goetsch 1975). Second, normal *S. cerevisiae* cells maintain a low concentration of intracellular precursors of secretion (Gascón and Lampen 1968; Novick and Schekman 1979) and, consequently, the secretory organelles are not as obvious in a cytologic examination as are those seen in higher exocrine cells or even other fungi (Grove et al. 1970). However, the pool of secretory organelles and intermediates can now be expanded with different temperature-sensitive yeast mutants (see below, A Genetic Approach to the Secretory Process). Third, rupture of the thick cell wall requires shear forces that also obliterate most of the internal organelles. However, relatively intact membranes can be isolated from spheroplasts prepared with cell-wall-degrading enzymes. Although crude zymolytic preparations such as Glusulase and Zymolyase (Cabib 1971; Kuo and Yamamoto 1975) are satisfactory for the extraction of internal structures, the isolation of native plasma membranes may require purified lytic enzymes, which are now available (Vrsanská et al. 1977; Scott and Schekman 1980).

Although these technical difficulties have caused problems in the past, we are convinced that many aspects of the secretory process may be pursued more easily in yeast than in other eukaryotic systems.

SECRETORY MOLECULES AND THE PATTERN OF CELL-SURFACE GROWTH
Secretory and Plasma-membrane Proteins

The yeast cell wall consists of a large number of secreted proteins, many of which are mannoproteins, and the structural polysaccharides glucan and chitin. A representation of the cell surface is shown in Figure 1. The particulars of this organization are discussed by Ballou (this volume). The secreted proteins are distributed among one or more of three extracytoplasmic spaces. Certain proteins, such as α-factor and killer toxin, are secreted through the cell wall into the growth medium (Woods and Bevan 1968; Duntze et al. 1970). Although these proteins are not glycosylated, culture supernatant samples contain a limited number of mannoproteins (Rogers et al. 1979). Endoglucanase and exoglucanase may be among the mannoproteins that are secreted into the medium (Farkaš et al. 1973). The outer

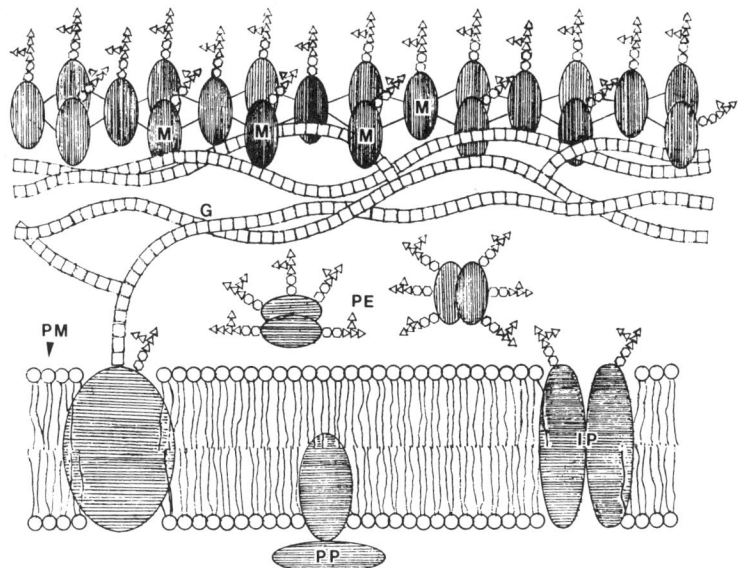

Figure 1 The yeast cell surface. (G) Glucan; (M) mannoprotein; (PE) periplasmic enzymes; (PM) plasma membrane; (IP) integral membrane protein; (PP) peripheral membrane protein.

surface of the wall contains a cross-linked network of mannoprotein, the average composition of which is about 10% protein and 90% carbohydrate (Ballou 1976). It is not known whether the protein in this material is a homogeneous species, nor is there evidence concerning its function or enzymatic composition. Mannoproteins are also found in a space (the periplasm) between the plasma membrane and the wall.

Most of the secreted enzymes are located somewhere between the plasma membrane and the surface of the wall. A list of the secreted enzymes that have been described is presented in Table 1. Invertase and the acid phosphatases have been studied most extensively, yet there is not agreement concerning the localization of these enzymes. On the basis of release by physical and autolytic disruption of yeast cells, Arnold has proposed that these enzymes are restricted to the periplasmic space (Arnold 1972a,b). However, a histochemical study has shown acid phosphatase activity in the periplasmic space and on the surface of the wall (Linnemans et al. 1977). Furthermore, some of the secreted invertase can be removed from intact cells by an alkaline protease activity or by mercaptoethanol (Scott and Schekman 1980). Perhaps the secreted enzymes are retained in two cell-wall locations by noncovalent trapping. The factors that determine cell-wall or extracellular location are unknown; the regulation of this distribution is likely to be of interest in the commercial use of yeast as a host in genetic engineering.

Table 1 Secreted Proteins

	Structural gene	Regulated by	Glycosylated	Extracellular location
Invertase[a]	SUC1–SUC6	hexose	+	cell wall
Acid phosphatase[b]	PHO5	Pi	+	cell wall
Acid phosphatase[b]	PHO3	constitutive	?	cell wall
α-Galactosidase[c]	MEL1	galactose	+	cell wall and medium
L-Asparaginase[d]	ASP3	nitrogen	+	cell wall
Exo-1,3-β-glucanase[e,f]	EXB1	constitutive?	–?	cell wall and medium
Endo-1,3-β-glucanase[f]	?	constitutive?	+	cell wall and medium
Pheromone peptidase[g]	?	?	?	medium
Pheromone (α, a factors)[g]	?	MAT locus	–	medium
Killer toxin[h]		KEX genes	–	medium

[a] References in this chapter.
[b] Toh-e et al. (1973).
[c] Kew and Douglas (1976).
[d] Dunlop et al. (1978); R. Roon (pers. comm.).
[e] Phaff (1979).
[f] Santos et al. (1979).
[g] Thorner (1981).
[h] Wickner (1981).

Invertase and acid phosphatase activities are also incorporated into the yeast vacuole (Meyer and Matile 1975; Wiemken 1975). Although it is not certain that these are the same gene products as the secreted enzymes, histochemical staining and thin-section electron microscopy have revealed that repressible acid phosphatase activity is present within the vacuole and at the cell surface (Linnemans et al. 1977; P. Novick and R. Schekman, unpubl.). Vacuoles also contain a large family of glycoproteins distinct from the secreted enzymes (Wiemken et al. 1979). A list of several vacuolar enzymes is in Appendix II of this volume. Vacuolar enzymes may originate in the same cellular organelle as secreted enzymes, and both groups may use the same mechanisms for their localization. This subject will be dealt with more extensively, below, in Synthesis and Maturation of Secretory and Membrane Proteins.

The best evidence suggests that yeast cells use a standard exocrine process for secretion of mannoproteins. Initial electron microscopy evidence supported an exocytic process in which secretory vesicles, derived from the endoplasmic reticulum, fused with the plasma membrane to deliver soluble mannoproteins to the cell surface (Plate 1[1]; Moor 1967). In subsequent studies, vesicles were partially purified (Matile et al. 1971; Cortat et al. 1972) and shown to contain endoglucanase activity, exoglucanase activity, and man-

[1] For Plates 1–5, see pp. 395–398.

nan. Pulse-label and autoradiographic studies have demonstrated that radioactive mannose incorporation into nascent mannoproteins occurs in the cell interior followed by transport of the completed molecules to the cell surface (Kosinova et al. 1974; Ruiz-Herrera and Sentandreu 1975). More recently, we have isolated a series of conditional secretory mutants that accumulate secretory enzymes enclosed within membrane-bounded organelles (Novick and Schekman 1979; Novick et al. 1980). The results with these mutants (see below, A Genetic Approach to the Secretory Process) are entirely consistent with an exocrine process in yeast.

An alternative secretory mechanism has been proposed that involves a direct transfer of invertase across the plasma membrane. This hypothesis attempts to account for the preponderance of nonglycosylated, cytoplasmic invertase found in spheroplasts. In one study, the physical rupture of spheroplasts released a form of glucanase that sediments (presumably vesicle-bound), whereas invertase was released in a soluble form (Meyer and Matile 1975). If the glucanase and invertase were both exported by an exocrine process, both should be in membrane-enclosed organelles within the cell. In these experiments, however, most of the invertase was a nonglycosylated species, which probably is not a precursor of the secreted enzyme (see below, Synthesis of the Protein Portion). In another investigation, trypsin treatment of spheroplasts blocked secretion and caused a transient accumulation of the nonglycosylated invertase (Holbein and Kirby 1977, 1979). Thus, if a protein that spanned the plasma membrane was involved in the export of secretory proteins, inactivation by trypsin might lead to the intracellular accumulation of secretory precursors. Although this result is consistent with a direct extrusion of invertase through the plasma membrane, it is not clear that the trypsin-induced, accumulated material can be secreted. In contrast, results with the thermosensitive secretory mutants demonstrate that invertase is glycosylated and enclosed within secretory organelles long before export through the plasma membrane (see below, A Genetic Approach to the Secretory Process).

Another alternative secretory mechanism, which is a variation on the direct extrusion scheme, suggests that infoldings of the plasma membrane are specialized for the extrusion of acid phosphatase during maturation of the mother portion of a cell wall (Linnemans et al. 1977). Infoldings of the plasma membrane that enclose acid phosphatase activity are seen under certain conditions of fixation. This hypothesis requires that glycosylation of secretory proteins can occur in the plasma membrane. Although partially purified plasma-membrane fractions contain some mannosyltransferase activity (Cortat et al. 1973), the most abundant and enriched source of activity is the endoplasmic reticulum (Marriot and Tanner 1979). In addition, it is possible that either the plasma-membrane infoldings are fixation artifacts or the putative infoldings are actually endoplasmic reticulum (ER), which in yeast is often seen in apposition to the plasma membrane.

Very little is known about transcriptional and translational regulation of secretory enzyme production, although the genetic loci for several secretory enzymes are known (Table 1). Conflicting claims have been made regarding the presence or absence of acid phosphatase mRNA in cells grown under repressing conditions. Although the data were not presented, Schweingruber and Schweingruber (1979) claimed that acid phosphatase mRNA was present in the same amount in repressed and derepressed cells where the enzyme specific activity differed by 20-fold. In the same study, acid phosphatase immunoreactive protein was repressed only twofold to threefold. Thus, under certain conditions, the major controls of acid phosphatase enzyme production are at translational and posttranslational stages. In contrast to these findings, large differences in the repressed/derepressed level of acid phosphatase mRNA have been found by Bostian et al. (1980). These findings were also made by translation in vitro in a wheat-germ extract; thus, there is no easy explanation for the discrepancy.

The same controversy exists for regulation of invertase expression. Chu and Maley (1980a,b) reported that mRNA extracted from both repressed and derepressed cells supported the production of invertase immunoreactive polypeptide in a wheat germ protein synthesis reaction. They did find, however, that invertase mRNA is absent in cells from a late logarithmic culture. On the basis of these two observations, Chu and Maley proposed a role for transcriptional and translational regulation of invertase production. However, by translation of electrophoretically resolved mRNA, Perlman has found a large increase in invertase mRNA synthesis during derepression (Perlman and Halvorson 1981). Recently, the derepressible acid phosphatase gene(s) and invertase structural gene (*SUC2*) have been cloned (Kramer and Anderson 1980; Carlson and Botstein 1982); and in both cases, a large increase in complementary mRNA is detected under derepressed conditions.

The formation of active invertase and acid phosphatase requires N-glycosidic bond formation concurrent with protein synthesis (Kuo and Lampen 1974). Mannoprotein synthesis is inhibited by tunicamycin, a drug that mimics N-acetylglucosamine (GlcNAc) and blocks the formation of the dolichyl-P-P-GlcNAc intermediate in core oligosaccharide assembly. Invertase and acid phosphatase enzyme synthesis are blocked by tunicamycin, although enzymatically inactive invertase polypeptide continues to be synthesized (J. Tkacz, pers. comm.). The requirement for glycosylation is not seen with the derepressible vacuolar alkaline phosphatase; this enzyme is made in active form in the presence of the drug (Bauer and Sigarlakie 1975; Onishi et al. 1979). Perhaps, only certain enzymes require a core oligosaccharide for correct folding during protein synthesis.

Secretion of invertase and acid phosphatase is localized to the bud portion of a dividing cell during most of the division cycle. Tkacz and Lampen used fluorescein isothiocyanate (FITC)-conjugated concanavalin A (Con A) to show that bulk mannan incorporation was restricted to the bud (Tkacz and Lampen 1972). They also used FITC antibody to show that secretion of invertase was similarly localized (Tkacz and Lampen 1973). These reagents recognize determinants that are located at the cell surface, and yet, as was discussed earlier, much of the secreted glycoprotein is in the periplasmic space and is probably inaccessible to large molecules. We have devised a histochemical stain for acid phosphatase that employs a small, wall-permeable dye and substrate that form a precipitate on the cell surface in the area of the secreted enzyme (Field and Schekman 1980). Acid phosphatase activity produced during derepressed synthesis on a phosphate-limited growth medium was localized initially to the bud (Plate 2). After two to three generations of phosphate-limited growth, most of the cells could be stained; when further phosphatase synthesis was repressed by growth in excess phosphate, dividing cells were produced in which the parent but not the bud could be stained. These experiments suggest that locally secreted enzyme is not free to diffuse about the entire cell surface.

Localized secretion is also observed during mating. Budding growth is interrupted in **a**-mating-type cells by a pheromone (α-factor) secreted by the opposite mating type; cell-surface growth continues in the presence of α-factor and produces a characteristic cell tip. Mannan and acid phosphatase secretion are restricted to this cell tip (Tkacz and Mackay 1979; Field and Schekman 1980).

The behavior of the division mutant *cdc24* (Hartwell et al. 1974) suggests that localized secretion is regulated by cell-cycle events. The mutant fails to bud at a nonpermissive temperature and instead expands symmetrically as a sphere. This altered pattern of surface growth is reflected by delocalized mannan incorporation (Sloat et al. 1981) and acid phosphatase secretion (Field and Schekman 1980).

Both bud-limited surface growth and localized secretion may be achieved by exocytosis of secretory vesicles (Plate 1). Several groups have suggested that fusion of vesicles with the inner surface of the bud may contribute to the assembly of the wall and the plasma membrane (Moor 1967; Cabib 1975; Farkaš 1979). This proposal requires that vesicle fusion sites that promote local growth will be available only on the bud portion of a dividing cell or the tip portion of a pheromone-treated cell and that these sites will be inactive during subsequent cell cycles. It is likely that the recognition sites, as components of a fluid plasma membrane, are confined to the growing portion of a cell by some mechanism that resists the tendency of lateral diffusion. The defect that results in delocalized growth of *cdc24* cells may affect the mechanism that organizes the vesicle recognition sites.

Another prediction of the secretory-vesicle–membrane-assembly hypothesis is that new plasma-membrane proteins and, perhaps, lipids will first appear on the bud. Localization techniques, such as those described for secretory enzymes, have not yet been applied to membrane proteins. This approach is limited by the small number of plasma-membrane enzymes that have been identified in yeast, which include chitin synthetase (Durán et al. 1975), vanadate-sensitive-Mg^{++} ATPase (Bowman et al. 1978; Willsky 1979), some unspecified mannosyltransferases (Cortat et al. 1973), adenylate cyclase (Liao and Thorner 1980), and glucan synthetase (Shematek and Cabib 1980; Shematek et al. 1980). None of these proteins has been purified to homogeneity from a *Saccharomyces* yeast strain, and no direct technique is available to localize a membrane enzyme in the plasma membrane of an intact yeast cell. An approach to this question might be to develop a visual label for one of the many inducible transport permeases in yeast (see Cooper [Transport in *Saccharomyces cerevisiae*], this volume).

Synthesis and Localization of Cell-wall Structural Polysaccharides

The structure and organization of the cell-wall polysaccharides are presented by Ballou (this volume), and the synthesis of these molecules has recently been reviewed by Farkaš (1979). In this paper we direct our attention to more recent contributions on the localization and regulation of chitin and glucan synthesis.

Unlike the glycosylation of secretory proteins, assembly of the structural polysaccharides probably occurs on the cell surface. No intracellular deposits of glucan and chitin have been found. Although the export of mannoprotein depends upon continuous protein synthesis, glucan and chitin synthesis proceed in the presence of cycloheximide (Elorza et al. 1976). This is not unexpected since neither glucan nor chitin contains protein, whereas mannan does. An analogous situation has been documented in the synthesis of cellulose by plants and by certain algae (Bowles and Northcote 1972; Montezinos and Brown 1976). Polymerization of monosaccharide units from a sugar-nucleotide substrate may occur on the inner surface of the plasma membrane and be coupled with extrusion of the polysaccharide through a transmembrane protein channel.

Cabib and his colleagues have proposed that the primary division septum in yeast is composed of chitin that is deposited into the cell wall by an integral plasma-membrane enzyme, chitin synthetase (Cabib and Farkaš 1971; Cabib et al. 1974; Durán et al. 1975). The enzyme is a zymogen, which can be activated by a variety of proteases, including trypsin and yeast protease B (Ulane and Cabib 1976). The activation site is probably on the inner surface of the plasma membrane, because trypsin activates the zymogen in a lysate but not in intact spheroplasts. Cabib has proposed that the zymogen is activated locally in the nascent division septum by a vesicle that somehow delivers a protease (or some other activating factor) to the inner surface of the plasma membrane (Cabib et al. 1974). Although it is reasonable to expect that active synthetase is restricted to the nascent division septum during budding growth, it is less obvious how the zymogen is distributed. Spheroplasts deposit chitin uniformly about the cell surface (Nečas 1971), and large plasma-membrane fragments prepared from Con-A-coated spheroplasts can be activated by trypsin to produce chitin all over the membrane surface (Durán et al. 1979). Although these observations suggest a uniform distribution of the zymogen, it is possible that the act of producing spheroplasts redistributes plasma-membrane components. Spheroplast formation with zymolytic enzymes may be analogous to trypsination of tissue culture cells, which is known to affect the distribution of surface markers (Burger 1969).

The timing and localization of chitin synthesis is regulated by cell-cycle events. During budding growth, chitin first appears as a ring surrounding an incipient bud (Hayashibe and Katohda 1973). The maximum rate of chitin synthesis occurs at some stage prior to septation. Although chitin is essential for normal septation (Bowers et al. 1974), division mutants blocked at various stages prior to septum formation nevertheless make chitin (Cabib and Bowers 1975; Sloat and Pringle 1978). No cell-cycle signal is known to precede the initiation of chitin synthesis. A very early G_1 block, α-factor-mediated arrest, does not block chitin synthesis. α-Factor treated cells have more chitin than budding cells; most of the chitin is found in the tip portion of an arrested cell (Schekman and Brawley 1979). The distribution of chitin is influenced in a different way by the *cdc24* mutant, which blocks division at an early stage and leads to nonpolar surface growth. In the absence of a division septum, *cdc24* cells deposit chitin uniformly over the growing surface (Sloat and Pringle 1978), much as is seen in regenerating spheroplasts. Thus, whereas chitin is normally restricted to the division septum, delocalization is possible during G_1-arrested growth; the *CDC24*-gene product and, perhaps, some zymolytic enzyme-sensitive structure are required for localization.

Chitin synthetase zymogen has been solubilized from yeast membranes by digitonin (DeRousset-Hall and Gooday 1975; Durán and Cabib 1978). A partially purified fraction has the same substrate and activation requirements as the membrane-bound enzyme and, in addition, is stimulated by phosphatidylserine (Durán and Cabib 1978). Although the solubilized and membrane-bound forms of the enzyme are activated by the yeast protease B, chain-termination mutants in the protease-B structural gene (*prb1*; Zubenko et al. 1979) have no effect on chitin synthesis in vivo. Some other activation mechanism, perhaps another protease, must operate in the cell. Complete purification of both the zymogen and an in-vivo-activated form of the synthetase might allow an indirect analysis of the activation mechanism. Another approach that has not yet been exploited successfully is the isolation of conditionally lethal chitin-synthesis mutants.

Glucan synthesis occurs along the bud portion of a dividing cell. Autoradiography of cells pulse-labeled with [^{3}H]glucose shows grain tracks restricted to the bud surface of dividing cells (Johnson and Gibson 1966). An inhibitor, 2-deoxyglucose, which interferes with hexose metabolism and blocks glucan synthesis (Kuo and Lampen 1972), causes cells to lyse at the bud surface (Johnson 1968). Under these conditions, glucanase-mediated hydrolysis may promote bud lysis. The localization of glucan synthesis has not been examined in cell-division mutants, although synthesis is stimulated somewhat during G_1-arrested growth in the presence of α-factor (Lipke et al. 1976).

In vitro synthesis of the β-1,3 glucan component has been reported (Shematek and Cabib 1980; Shematek et al. 1980). A substantial fraction of the in vivo rate of synthesis is seen in a reaction that requires either GTP or ATP for the polymerization of glucose from UDPG. The conditions for stimulation by the purine nucleotides differ: GTP stimulation is optimal at 10 μM, requires EDTA, and does not appear to involve cleavage of the β-γ-phosphate bond, whereas ATP stimulation is optimal at 1 mM, is inhibited by EDTA, and may involve transfer of the γ-phosphate to an alkaline-phosphatase-sensitive site on the synthetase. Shematek and Cabib (1980) have proposed a model in which glucan synthetase may be irreversibly activated by GTP or reversibly activated by ATP and some enzyme-bound small molecule that can be phosphorylated. This model provides an on-off regulatory feature that could explain activation of glucan synthetase in the bud and inactivation of the enzyme in the mother portion of a cell.

Mutational analysis has yet to be applied to studies of glucan synthesis and regulation. The fact that inhibition of glucan synthesis by 2-deoxyglucose causes cell lysis suggests that mutants that lyse in a nonpermissive growth condition might be candidates for those with a lesion in glucan synthesis. This phenotype is commonly found in collections of unselected temperature-sensitive mutants (Hartwell 1974).

Localization and the Pattern of Bud Growth

At this point, it is worth emphasizing similarities in the localization of structural polysaccharides and secretory mannoproteins. During the normal cell cycle, and probably also during G_1-arrested growth, glucan and mannan are incorporated into the same growth zone. Despite differences in the mechanism of assembly, these two processes may be connected by a common regulation. Farkaš has proposed that structural polysaccharide synthesis is regulated by a periplasmic inhibitor that allows transient synthesis in an area of newly assembled membrane (Farkaš 1979). Polysaccharide synthetase may be delivered to the plasma membrane in association with a secretory vesicle, become active when exposed to the appropriate environment, and then be inhibited by interaction with other molecules. Thus, an exocytic reaction may localize synthetases and their regulatory factors.

Even if these speculations are accepted, it is not clear how the secretory-membrane precursor vesicles are directed to an appropriate point for fusion with the plasma membrane. This problem may be divided into two parts:

transport of vesicles into the bud, and fusion of vesicles with the inner surface of the bud.

Vesicle transport may be achieved by targeted movement along the tracks of a cytoskeletal structure such as the microtubule. In *S. cerevisiae*, Byers has observed a group of microtubules that emanate from the spindle pole body on the surface of the nuclear membrane and project into the bud (Byers and Goetsch 1974). The microtubules are often seen in proximity to bud-localized vesicles. Microtubules have been implicated in secretory processes of many eukaryotic cells. Colchicine, a drug that causes microtubule depolymerization, affects some cells by inhibiting secretion of newly synthesized proteins (Chambaut-Guérin et al. 1978). In the fungus *Fusarium acuminatum*, the antitubule drug methyl benzimidazole-2-yl carbamate (MBC) causes branched and undirected hyphal growth and at the same time causes randomization, but not accumulation, of secretory vesicles that are normally restricted to the apex of a growing hypha (Grove et al. 1970; Howard and Aist 1980). Although the *cdc24* mutant of *S. cerevisiae* shows undirected growth, nuclear division, a process that requires microtubules, continues in this mutant. Thus, tubulin is not a likely target of the *cdc24* lesion. Actin, another cytoskeletal component, is present in yeast (Koteliansky et al. 1979; C. Greer and R. Schekman, in prep.). A function for yeast actin has not been defined because mutants are not yet available, and drugs that affect actin polymerization, such as cytochalasin and phalloidin, do not penetrate yeast cells (C. Greer and R. Schekman, unpubl.). In other cases, notably maize root tips (Mollenhauer and Mooré 1976) and *Fucus* embryos (Brawley and Quatrano 1979), cytochalasin causes the accumulation of secretory vesicles.

Alternatively, localization may be achieved by chemical or electrical currents that are directed with a unique polarity through a budding cell. Evidence for this phenomenon has been presented in studies of fertilized *Fucus* eggs by Jaffe and co-workers (1974). Jaffe has proposed that a negatively charged secretory vesicle might be propelled along a cation gradient going from the rhizoid (growing tip) to the thallus (opposite side) of a fertilized *Fucus* egg. A variety of data support this concept in diverse cells (Nuccitelli 1978; Woodruff and Telfer 1980); in *Neurospora*, a 60-mV electrical potential exists along the last 1 mm of a hypha (Slayman and Slayman 1962).

a-Mating-type *S. cerevisiae* cells form cell tips predominantly in the direction of a higher concentration of α-factor (J. Rine, pers. comm.). Thus, yeast cells may respond to or generate endogenous chemical gradients as a means of localized growth. According to this view, delocalized secretion observed with the *cdc24* mutant suggests that localization, but not movement, of secretory vesicles could be influenced by a transcellular current.

Transported secretory vesicles must fuse selectively with the bud plasma membrane. The vesicles do not appear to fuse with one another or with other organelles such as mitochondria. Vesicle recognition of the plasma membrane may be provided by a specific protein receptor, such as in the interaction of a mucocyst and a fusion rosette in the plasma membrane of *Tetrahymena* (Satir et al. 1973), or by interaction with a lipid bilayer of unique composition, such as in the fusion of Semliki Forest virus with phospholipid vesicles that contain cholesterol (White and Helenius 1980). In either case, the bud plasma membrane must be differentiated from the parent cell membrane with which it is connected. This situation is analogous to the pancreatic acinar cell, where exocytosis of zymogen granule contents is restricted to a small area of the cell surface, the lumenal membrane. The rest of the acinar cell surface, the basolateral membrane, is attached to but segregated from the lumenal membrane by a tight junction. This structure consists of a ring of adjacent intracellular particles that enclose the lumenal surface. The junction may serve to segregate the lumenal membrane, thus, allowing specific interaction and fusion with zymogen granules. Yeast cells possess a different structure that could serve the same function. Byers has described a filamentous ring that coils around the cytoplasmic surface of the plasma membrane in the region of the nascent division septum (Byers and Goetsch 1976a). This structure could restrict the diffusion of membrane proteins between the parent and bud surface membranes, while allowing passage of cytoplasmic components. Conditional mutants defective in this structure might have a phenotype similar to the *cdc24* mutant. The cytokinesis-defective *cdc* mutants (*cdc3, cdc10, cdc11,* and *cdc12;* Byers and Goetsch 1976b) fail to assemble this structure at the nonpermissive-growth temperature; these mutants may show delocalized surface growth. The potential for a genetic and biochemical approach should make yeast an ideal experimental system for studies of the molecular mechanism of localization and cell-surface differentiation.

SYNTHESIS AND MATURATION OF SECRETORY AND MEMBRANE PROTEINS
Synthesis of the Carbohydrate Portion

The mechanism of mannoprotein oligosaccharide synthesis has been examined extensively in yeast, and several recent reviews have dealt with this subject (Farkaš 1979; Parodi and Leloir 1979): Our treatment is brief and emphasizes the stages and localization of this process. The paper by Ballou (this volume) contains a description of the oligosaccharide structures.

The initial glycosylation of secretory proteins probably occurs during translocation of the polypeptide through the ER membrane. Two types of transfer occur at this stage. In one, a core oligosaccharide composed of $(GlcNAc)_2$-Man_9Glc_3 is transferred from a lipid carrier (dolichyl-P-P-oligosaccharide) to form an *N*-glycosidic linkage to asparaginyl residues on the nascent polypeptide (Parodi 1979b; Lehle 1980). A separate reaction transfers one mannose moiety from dolichyl-P-Man to form an *O*-glycosidic linkage to seryl or threonyl residues on the nascent protein (Sharma et al. 1974).

Dolichyl-oligosaccharide assembly is initiated by transfer of a GlcNAc-P from UDP-GlcNAc to dolichyl-P to form dolichyl-P-P-GlcNAc (Lehle and Tanner 1976). Another GlcNAc is added from a sugar-nucleotide donor, and further additions of mannose are provided by dolichyl-P-Man and possibly from GDP-Man (Lehle and Tanner 1978). Glucose is added from dolichyl-P-glucose, and then the entire unit is transferred to the nascent polypeptide. The presence of glucose on the dolichyl-P-P-oligosaccharide stimulates the rate of core oligosaccharide transfer by 20-fold. The glucose residues are removed shortly after transfer of the oligosaccharide to nascent secretory protein, and then the outer chain oligosaccharide is elaborated. This portion is built up from GDP-Man by a set of mannosyltransferases that are distinct from those responsible for inner-core oligosaccharide synthesis (Raschke et al. 1973; Parodi 1979a). O-glycosidically linked mannose is extended by transfer from GDP-Man to form mannobiose, mannotriose, and mannotetraose (Lehle et al. 1977). The final secreted form of a mannoprotein such as invertase is heterogeneous in carbohydrate content, which is due to a variable repetition of the outer chain (Lehle et al. 1979).

The core N- and O-glycosylation reactions occur in the ER. Partially purified ER is enriched in enzymes that transfer mannose and N-acetylglucosamine from sugar nucleotides to dolichylphosphates and in enzymes that transfer sugars from dolichylphosphate and dolichylpyrophosphate to endogenous protein acceptor (Marriot and Tanner 1979). The membrane fractions are probably rough ER, because the presence or absence of Mg^{++} changes the density of the material on sucrose gradients in a way characteristic of ribosome-bound or ribosome-free membranes. Recently, Welten-Verstegen et al. (1980) observed some core glycosylation in highly purified plasma-membrane fractions. Unfortunately, the specific activity of this material was not compared with purified ER fractions, and as the yield of plasma membrane was not quantified, it is not possible to exclude a contaminating fraction of ER as the source of glycosyl transferase.

Extension of the N-linked core oligosaccharide and O-linked mannose to produce the outer chains and the O-linked oligosaccharides, respectively, probably occurs in a Golgi-like organelle. Outer chain assembly has not yet been demonstrated in a membrane fraction distinct from the ER. Nevertheless, indirect evidence that supports this view has come from analysis of the pattern of glycosylation in secretory mutants that are blocked before and after the Golgi stage in the yeast secretory pathway (see below, A Genetic Approach to the Secretory Process).

The attachment of both core and outer chain sugars to protein is likely to occur on the cisternal surface of the ER and Golgi membranes, even though the enzymes that assemble the dolichyl-bound oligosaccharide are localized on and obtain their substrate on the cytoplasmic surface (Snider et al.

1980). Thus, organelles derived from the ER and Golgi are expected to have oligosaccharides on the membrane surface away from the cytoplasm. This is true of vacuoles and the plasma membrane.

Only the inner core appears to be necessary for secretion and cell viability. A new class of mannan structural mutants (*mnn7, mnn8, mnn9,* and *mnn10*) have been isolated that are missing most of the outer chain, yet these mutants grow, albeit slowly (Ballou et al. 1980). The *mnn7-mnn10* strains secrete a form of invertase that has a higher electrophoretic mobility than the enzyme produced by the parent strain (L. Ballou and C. E. Ballou, pers. comm.). These results suggest that a set of enzymes participate in the assembly of outer chain oligosaccharides on both bulk mannan and invertase and that outer chains are not required for export of mannoproteins. On the other hand, as discussed earlier, tunicamycin, a drug that prevents dolichyl-oligosaccharide synthesis, blocks invertase and acid phosphatase enzyme production (Kuo and Lampen 1974). The vacuolar enzymes carboxypeptidase Y and alkaline phosphatase do not depend on glycosylation for synthesis and proenzyme cleavage; both enzymes are formed in the presence of tunicamycin (Hasilik and Tanner 1978b; Onishi et al. 1979).

The oligosaccharide structure determined from analysis of bulk mannan seems to apply to specific secretory mannoproteins such as invertase (Lehle et al. 1979). However, the oligosaccharide associated with carboxypeptidase Y (Hasilik and Tanner 1978b), and probably also with alkaline phosphatase (Onishi et al. 1979), resembles the inner core. Although this difference may pertain to the localization of these enzymes, too few examples are available to draw a generalization. Some of the plasma-membrane outer surface components are glycosylated and probably contain mannose, because Con A binds to spheroplasts (Durán et al. 1975). The binding sites may be glycoprotein or glycolipid; however, in no case has an oligosaccharide structure been determined. Likewise, the inner surface of the vacuole contains Con-A binding material (Boller et al. 1976; Niedermeyer 1976). At least one vacuolar membrane protein has been identified; α-mannosidase is an integral membrane enzyme (Van der Wilden et al. 1973; D. Opheim, pers. comm.).

Synthesis of the Protein Portion

In contrast to the relatively advanced studies of glycoprotein oligosaccharide synthesis in yeast, synthesis of secretory and plasma-membrane polypeptides has been nearly ignored. Two problems have contributed to this sad state of affairs. First, only recently has it been possible to initiate and complete the synthesis of yeast proteins in vitro with mRNA, ribosomes, and soluble components all from a yeast extract (Gasior et al. 1979; Tuite et al. 1980). Second, ER fractions from yeast have not been used in vitro for the incorporation of membrane and secretory proteins such as have been reported for a variety of other systems. Yeast mitochondrial membranes, on the other hand, have been a popular source material for studies of assembly in vitro.

An alternative approach has been to characterize intracellular forms of cell-surface proteins in wild-type and mutant yeast strains. Although invertase export is completed within 5 minutes after synthesis (Novick et al. 1981), intracellular forms of the enzyme have been detected. We will consider two of these forms and their relationship to the secreted enzyme.

A cytoplasmic nonglycosylated form of invertase is found in constant amounts in cells that are repressed or actively producing the secreted enzyme (Gascón and Lampen 1968), although in fully derepressed cells the cytoplasmic form represents only 3% of the total invertase (Gascón and Lampen 1968). Both the secreted and nonglycosylated enzymes are dimers of a 60K polypeptide (Trimble and Maley 1977). The amino acid composition of the cytoplasmic and secreted forms were reported to differ (Gascón et al. 1968); however, more recent evidence suggests that the two proteins are produced by the same gene. In one study, a lesion resulting in a temperature-sensitive external enzyme also affected the stability of the cytoplasmic form (Mizunaga et al. 1981). The mutation mapped close to, or as part of, the *SUC1* locus, one of the six genes that allows sucrose fermentation in yeast. Chain-termination nonsense mutations, which abolish both the secreted and cytoplasmic invertases, have been isolated in the *SUC2* gene (Carlson et al. 1981). Most laboratory strains contain one of the *SUC* genes, which produce secreted forms of invertase that vary slightly in carbohydrate content (Abrams et al. 1978). *SUC1* strains produce a nonglycosylated form that migrates more slowly during native gel electrophoresis than the internal invertase from other *SUC* strains (Grossmann and Zimmerman 1979). Diploids containing *SUC1* and another *SUC* gene produce three electrophoretically distinct forms of cytoplasmic invertase. Subunit mixing between the two gene products would explain this observation. Thus, the cytoplasmic and secreted gene products appear to be allelic.

Cytoplasmic invertase is probably not an intermediate in the production of the secreted enzyme. The cytoplasmic form is not more abundant during derepression of external enzyme, nor does it accumulate when glycosylation is blocked by tunicamycin (Gallili and Lampen 1977). The constant level of cytoplasmic invertase is consistent with the presence of invertase mRNA in repressed cells. Distinct mRNA species could serve for production of the cytoplasmic and secreted forms, or one mRNA species could produce a low level of the cytoplasmic form and be converted to production of the external enzyme by some invertase-specific translational activation. The first of these two possibilities has recently been demonstrated (Perlman and Halvorson 1981; Carlson and Botstein 1982).

A variable and a low level of incompletely glycosylated forms of invertase have been observed in derepressed cells (Moreno et al. 1975; Holbein et al.

1976; Babczinski and Tanner 1978; Babczinski 1980). These forms are retained when cells are converted to spheroplasts, and their level declines when external invertase production ceases. Although these presumed intermediates have not been purified from wild-type cells, they appear to contain core oligosaccharide chains (Babczinski 1980). A discrete intermediate form is found in a fraction that sediments from cells lysed in the presence of Mg^{++} (Holbein et al. 1976; Babczinski and Tanner 1978). This form, which is detected by an enzyme activity stain, migrates in between the secreted and cytoplasmic forms on a polyacrylamide gel. The intermediate form appears prior to the formation of the mature external enzyme, and the production of both species is inhibited by tunicamycin (Babczinski 1980). The membrane-bound form of invertase appears to cosediment with ER. This association, which is also seen with intracellular transit forms of serum glycoproteins in rat liver (Redman and Cherian 1972), may play a role in the packaging of invertase for export. Intracellular forms of invertase that accumulate in secretory mutants are discussed below, in A Genetic Approach to the Secretory Process.

Intracellular forms of acid phosphatase have been localized by histochemical staining. The enzyme is present in ER, Golgi-like membranes, vacuoles, and bud-localized vesicles (Linnemans et al. 1977). The structures of intracellular transit forms of acid phosphatase have not been examined.

Among the few plasma-membrane enzymes that are known, only chitin synthetase has been examined for intracellular transit forms. Bartnicki-Garcia and his colleagues have proposed that chitin synthetase is delivered to the plasma membrane in vesicles (35–100 nm) called chitosomes (Ruiz-Herrera et al. 1975, 1977). These particles have been isolated from a variety of fungi, including yeast, although they may be derived from other membranes by breakage during lysis. If chitosomes are involved uniquely in the transport of chitin synthetase, purified chitosome membranes should contain a simple spectrum of proteins compared with plasma-membrane and other secretory organelle membranes. On the other hand, chitisomes may be exocytic vesicles that contain other proteins destined for the cell surface. A third possibility is that chitosomes are a mixed population comprising different specific exocytic vesicles.

We have investigated the distribution and activation of chitin synthetase during G_1-limited growth of **a**-mating-type cells treated with α-factor (Schekman and Brawley 1979). Under these conditions, cells make three times more chitin than during budding growth. Chitin synthetase activity is distributed equally between intracellular and plasma membranes; both membrane fractions have fivefold to tenfold more activity in pheromone-treated cells. Although chitin synthetase normally requires proteolytic activation, the plasma membrane of pheromone-treated cells shows a significant fraction of activated enzyme; intracellular membrane-bound synthetase is found exclusively in the zymogen form. Thus, the activation process is restricted to the plasma-membrane enzyme and fails to operate on intracellular, possibly transit, forms of synthetase.

Vacuole assembly may depend upon the synthesis and movement of its constituents through other secretory organelles. Thus, the structure of precursor forms of such enzymes as carboxypeptidase Y (61K) is relevant to a discussion of the secretory process. A 67K precursor of this enzyme, detected by immunoprecipitation from extracts of radioactively labeled cells, is seen in wild-type and protease-deficient *pep4-3* mutant strains (Hasilik and Tanner 1978a; Hemmings et al. 1981; Jones et al. 1981). The precursor is found initially in a fraction that sediments, perhaps membrane-bound, and in vivo conversion to the mature form occurs with a half-time of 6 minutes. Although normally glycosylated, the precursor is synthesized and processed in the absence of core oligosaccharide synthesis (Hasilik and Tanner 1978b). The precursor may be synthesized in association with the ER and transmitted to the vacuole for processing.

Either trypsin or the yeast protease B is capable of converting the carboxypeptidase Y precursor to a mature size in vitro. Protease-B mutants, however, make mature carboxypeptidase Y, so some other yeast protease must be capable of this processing (Zubenko et al. 1979). A large class of protease mutants, represented by at least 17 complementation groups, was isolated as having low levels of carboxypeptidase Y activity (Jones 1977; Jones et al. 1981). Two of the groups, *pep4* and *pep17*, are defective in the three major vacuolar proteases (protease A, B, and carboxypeptidase Y); *pep4* is also defective in the vacuolar alkaline phosphatase and RNase (Hemmings et al. 1981; Jones et al. 1981). The 67K precursor of carboxypeptidase Y accumulates in *pep4* strains, and trypsin treatment converts the accumulated proenzyme to a mature active form. The *PEP4*-gene product may be responsible for processing or transmission of carboxypeptidase Y and other enzymes to the vacuole. Localization of the accumulated precursor will reveal which of these possibilities is likely.

Organelles Implicated in the Secretory Process

Isolation and characterization of secretory organelles from yeast is another subject that has received only selective attention. In this section, we briefly review some of the isolation techniques and properties of the secretory organelles.

Endoplasmic Reticulum—Nuclear Membrane

The ER in yeast is not as extensive as is observed in exocrine cells of higher organisms. Stains used for electron microscopy of thin sections, such as lead citrate and uranyl acetate, which accentuate ribosomes, tend to obscure organelle details in wild-type yeast cells. Potassium-permanganate staining highlights membrane profiles at the expense of ribosome images. Thin sections of yeast cells show ER tubules closely associated with the inner surface of the plasma membrane. In favorable sections it appears that ribosomes are excluded from the cytoplasm between the ER and the plasma membrane. Occasionally, the ER appears connected to the nuclear membrane, with a lumen shared between the two. The high density of cytoplasmic ribosomes has made it difficult to visualize ER-bound ribosomes in thin sections of fixed cells. Such an association might be more apparent in thin sections of spheroplasts where the cytoplasm is diluted by swelling.

ER-enriched fractions have been obtained from lysates prepared by vigorous shearing of intact cells with glass beads (Matile et al. 1971) or blending of spheroplasts with glass beads in a Vortex mixer (Marriot and Tanner 1979). The ER is distinguished from other membranes by the presence of NADPH-cytochrome-c reductase, and by dolichyl-dependent glycosyltransferase activities (Marriot and Tanner 1979). When extracted in the absence of Mg^{++}, the reductase marker has a buoyant density of about 1.11 g/ml on Renografin density gradients (J. Tschopp and R. Schekman, unpubl.). On sucrose density gradients, the reductase marker equilibrates at 1.17 g/ml in the absence of Mg^{++}, and 1.2 g/ml with Mg^{++}; the difference is attributed to the absence or presence of bound ribosomes.

A connection between ER and the nuclear membrane is seen occasionally in electron microscopy of thin sections. Isolated mammalian nuclei retain shreds of ER, unless vigorous shear or mild detergents are included in the isolation protocol. Because of this continuity, certain enzyme markers may appear in both membranes. Thus, dolichyl-mediated glycosylation has been described in membranes isolated from yeast nuclei treated with Triton X-100 (Palamarczyk and Janczura 1977). Detergent treatment, however, may not be optimal for the evaluation of other enzyme markers, which could be solubilized by the isolation medium.

Golgi

Convincing profiles of Golgi membrane stacks have not appeared in thin sections of wild-type *S. cerevisiae*. Evidence for the existence of this organelle in yeast has come from our studies of secretory mutants (see below, A Genetic Approach to the Secretory Process). Golgi membranes have been documented in other fungi by thin-section and histochemical techniques (Grove et al. 1970; Howard 1981). Marker enzymes that have proved reliable for the detection of Golgi membranes in higher eukaryotes are thiamine pyrophosphatase and sugar-nucleotide-dependent glycosyltransferase.

Vacuole

A variety of yeast vacuole isolation procedures have been described, and this appears to be the one organelle, aside from the mitochondrion, that has been studied most intensively. Several reviews on the vacuole have appeared recently (Schwencke 1977; Matile 1978); our comments are restricted to considerations of how this organelle relates to the secretory process.

In stationary-phase cells and in certain yeast strains, the vacuole is the most prominent organelle. During exponential growth, the vacuole may be represented as smaller vesicles (300–1000 nm), which can be distinguished from other vesicles by the content of darkly staining or granular material (Figure 2).

Hypotonic lysis of spheroplasts from certain yeast strains allows a significant separation of low-density vacuoles from other secretory organelles. Major differences distinguish the composition of purified vacuolar and plasma-membrane fractions, suggesting no precursor-product relationship between the two membranes. Thus, membrane enzymes destined for the cell surface, such as the vanadate-sensitive ATPase (R. Davis, pers. comm.) and chitin synthetase (Cabib et al. 1973), are missing from the vacuole. Sterols such as ergosterol, which are enriched in the plasma membrane, are present in much lower amounts in the vacuolar membrane (Schwencke and de Robichon-Szulmajster 1976). Conversely, an α-mannosidase is present in the vacuolar membrane and absent in the plasma membrane (Van der Wilden et al. 1973).

The yeast vacuole is analogous to a mammalian lysosome in containing hydrolytic glycoproteins, many of which are derived from proenzyme forms (Jones et al. 1981). Unlike lysosomal enzymes, yeast vacuolar enzymes are not detected in the culture medium or in the cell wall during logarithmic growth; if they are secreted as inactive proenzymes, they may have escaped detection. A possible exception to this pattern is that during sporulation a vacuolarlike protease-A activity appears at the cell surface (Chen and Miller 1968). It is not clear, however, whether the protease activity is secreted or whether it is being released by lysed cells and binding adventitiously to intact cells.

Certain secretory enzymes are contained within the vacuole. Histochemical staining and electron microscopy of thin sections reveal repressible acid phosphatase in the vacuole (Linnemans et al. 1977; P. Novick and R. Schekman, unpubl.). Isolated vacuoles contain a form of invertase that appears to be as heavily glycosylated as the external enzyme; the two forms show a similar low electrophoretic mobility on polyacrylamide gels, in contrast to the cytoplasmic nonglycosylated enzyme (Meyer and Matile 1975). The presence of invertase in the vacuole has not been analyzed by histochemical techniques; thus, it is possible that the enzyme is a contaminant in purified vacuolar fractions.

The site of synthesis of soluble and membrane proteins of the vacuole is not known. For glycoproteins, the ER is the likely site of initial appearance. The proenzyme forms of these glycoproteins may be transferred directly to the vacuole or indirectly, via the Golgi. Separation of secretory proteins from those destined for the vacuole could occur either in the ER or in the Golgi body. Conversion of the proenzyme forms could occur at any of these three locations. Unlike the maturation of lysosomal enzymes in mammalian cells, the formation of active vacuolar alkaline phosphatase in yeast is not affected by blocking oligosaccharide synthesis with tunicamycin (Onishi et al. 1979). The cellular location of the active apoenzyme formed in the presence of tunicamycin is not known.

Secretory Vesicles

Secretory vesicles were discussed briefly, above, in Secretory and Plasma-membrane Proteins. The vesicles are about 60–80 nm in diameter and are localized to the bud during logarithmic growth but are not found in stationary-phase cells (Matile et al. 1971). The vesicles have been partially purified from extracts prepared by shaking log-phase cells with glass beads in an osmotically supporting medium (0.5 M mannitol). Two glucanase activities have been used as enzyme markers during fractionation, although other secretory enzymes such as invertase and acid phosphatase could also be used. The vesicles are reported to sediment after centrifugation at 55,000g for 20 min and to equilibrate at 1.14 g/ml in Urografin density gradients. As the vesicles are not abundant in wild-type cells, a homogeneous preparation has not been obtained. In this regard, the secretory mutants that accumulate vesicles at a nonpermissive temperature should be particularly valuable (see below, A Genetic Approach to the Secretory Process).

The secretory vesicle preparations that have been described could actually be vacuolar precursors. This possibility may be eliminated by demonstrating that the vesicles contain plasma-membrane marker enzymes and are missing vacuole-membrane enzymes. A vesicle fraction that contains acid phosphatase activity sediments together with chitin synthetase activity on a sucrose density gradient (Scott 1980).

Most other eukaryotic cells contain coated vesicles that are thought to mediate secretion and endocytosis (Goldstein et al. 1979). The protein coat of these vesicles consists of a cage-forming structural protein called clathrin (Pearse 1976). Bristle-coated vesicles have been seen in thin sections of several fungal species (Bracker et al. 1970), although neither coated vesicles nor clathrin have as yet been reported in yeast.

Plasma Membrane

The yeast plasma membrane has been the subject of numerous structural studies using electron microscopy. Some of the earlier work was discussed in a chapter on yeast cytology by Matile et al. (1969). More recently, improved freeze-fracturing techniques have revealed prominent intramembranous particles facing out on the cytoplasmic half of the bilayer (Sosinsky et al. 1980; Steere et al. 1980). These particles form semicrystalline arrays when cells are grown to stationary phase or starved for a carbon source. The function of these particles and aggregates is not known.

The literature abounds with yeast plasma-membrane purification procedures (Durán et al. 1975; Wehrli et al. 1975; Bussey et al. 1979; Welten-

Verstegen et al. 1980), none of which have been sufficiently documented to allow a degree of purification or homogeneity to be assessed. Satisfactory documentation requires a measure of the yield of a plasma-membrane marker and a demonstration that markers diagnostic of other membranes have been resolved and not merely inactivated. Part of the problem is that no enzyme marker has been shown to reside only in the plasma membrane. Thus, chitin synthetase, which has been used as a marker during plasma-membrane isolation, is also present on intracellular membranes. Another problem is that the ER in yeast is closely apposed to the inner surface of the plasma membrane, and the two may be connected at points. We have found that the ER is the most persistent contaminant in plasma-membrane preparations. Resolution of the two membranes requires sonication (J. Tschopp and R. Schekman, unpubl.).

Two useful purification steps have been developed. Coating of spheroplasts with Con A followed by hypotonic lysis generates large plasma-membrane sheets (Durán et al. 1975; Stroobant and Scarborough 1979). These sheets sediment at low speed and have a high buoyant density on Renografin gradients. Although these two steps generate partially purified plasma membranes, modifications to the basic procedure are required to remove other contaminating membranes.

Vanadate-sensitive ATPase activity is a convenient marker for the plasma membrane; the assay is easy and the enzyme is stable to several cycles of freeze and thaw. Mitochondrial contamination can be assayed by the oligomycin-sensitive ATPase; ER and nuclear membranes are assessed by the NADPH-cytochrome-c reductase; vacuolar membrane is marked by α-mannosidase.

No special components of the plasma membrane have been identified as essential for the secretory process. Presumably, some plasma-membrane molecules participate in exocytosis; their identification must await an in vitro membrane fusion reaction that approximates the cellular event.

A GENETIC APPROACH TO THE SECRETORY PROCESS
Isolation and Characterization of *sec* Mutants

Although much less is known about the secretory process in yeast than in mammalian cells—and in some ways there are greater technical problems in the use of yeast as an experimental system—the potential for a genetic approach may prove a crucial advantage. Mutants have already been useful in analyzing the structure and synthesis of secretory mannoproteins (see Ballou, this volume) and the synthesis and maturation of vacuolar glycoproteins (see above, Synthesis of the Protein Portion).

A large class of temperature-sensitive secretory mutants have been identified that are also blocked in bud growth, cell division, and the incorporation of a plasma-membrane permease activity (Novick and Schekman 1979; Novick et al. 1980). Secretory mutants are defined as those strains that fail to export active invertase and acid phosphatase but continue to synthesize protein at a nonpermissive temperature (permissive, 25°C; restrictive, 37°C). We have assumed that secretion of cell-wall components is necessary for cell viability. If secretion is coupled to cell-surface assembly, then a mutation blocking this process would be lethal. Thus, our analysis is restricted to conditional mutants, temperature-sensitive for growth.

The first secretory mutant was identified in an unselected collection of conditional mutants that were temperature-sensitive for growth. This mutant, called *sec1* (*sec*retory), accumulates eightfold more invertase at 37°C in an internal pool than does the wild-type strain. Acid phosphatase activity also accumulates at the expense of secretion. Cell growth and incorporation of a plasma-membrane sulfate permease activity stop abruptly at 37°C, whereas protein synthesis continues for several hours. Electron microscopy of mutant cells incubated at 37°C reveals a large increase in the number of membrane-bounded vesicles, which are distributed throughout the cytoplasm (Plate 3a). Histochemical staining and vesicle isolation (Plate 3b; W. Hansen and R. Schekman, unpubl.) reveal that the acid phosphatase and invertase activities are enclosed within the accumulated organelles. The number of vesicles decreases, and the accumulated enzymes are secreted when cells are returned to 25°C, whether or not protein synthesis is allowed to continue. The results with this mutant support a role for vesicles in secretion and cell-surface growth in yeast.

One other *sec* mutant was found in the collection of random temperature-sensitive mutants. This mutant, *sec2*, is nonallelic with *sec1*, yet it behaves identically at 37°C. It is possible that the accumulated vesicles are unable to fuse with the plasma membrane because of defective transport, membrane recognition, or exocytosis. The Sec phenotype could also appear in mutants blocked in the transfer of secretory proteins from the ER or the Golgi apparatus. Additional secretory mutants have been isolated in an attempt to evaluate the stages in the secretory pathway.

A selection for new secretory mutants relies on the fact that *sec1* cells become dense during incubation at 37°C. Henry et al. (1977) showed that during inositol starvation of an auxotrophic strain, net cell-surface growth stopped and cell mass increased. Starved cells could be resolved from normal cells on a Ludox density gradient. The same separation of normal and *sec* mutant cells is achieved on a Ludox gradient. The densest 1–2% of mutagenized cells from a culture incubated at 37°C are about 33-fold enriched in *sec* mutants. Thus far, in a screen of 18,500 colonies derived from dense cells, 220 *sec* mutants have been obtained.

The *sec* mutants are of two types. Class-A *sec* mutants (192 total) are like *sec1* and *sec2*, in that active secretory enzymes accumulate in an intracellular pool. Class-B *sec* mutants (27 total) do not secrete or accumulate active secretory enzymes, yet protein synthesis continues unabated at 37°C (S. Ferro and R. Schekman, unpubl.). Complementation group analysis has revealed 23 *sec* loci in the A group and 4 *sec* loci in the B group. The distribution of mutant alleles suggests that more of both classes could be found.

All *sec* mutants are also defective in bud growth and export of acid phosphatase and sulfate permease. Other export defects that have been identified are temperature-sensitive secretion of α-factor (*sec1* and *sec18;* D. Julius, unpubl.), blocked secretion of L-asparaginase (*sec1*; R. Roon, pers. comm.), blocked secretion of α-galactosidase (*sec1*; R. Buckholz, pers. comm.), and temperature-sensitive export of arginine permease (most of the mutants; P. Call and R. Schekman, unpubl.).

Many of the class-A *sec* mutants secrete a large fraction of the invertase that accumulates at 37°C when cells are returned to the permissive temperature. In most cases, the secretion of accumulated invertase is insensitive to cycloheximide. This implies that the affected gene product is reversibly inactivated by the temperature shift. The result demonstrates that ongoing protein synthesis is not essential for posttranslational transit of secretory enzymes and thus excludes the possibility that newly synthesized secretory protein forces the flow of the export process. None of the class-B *sec* mutants allows the production of normal amounts of active invertase from precursors that may be formed during derepression at 37°C when the mutants are returned to 25°C.

Perhaps the most dramatic feature of the class-A *sec* mutants is that they accumulate or exaggerate specific secretory organelles. Mutants in ten groups produce 80-100-nm vesicles at 37°C. As with *sec1*, these mutants accumulate vesicles throughout the cytoplasm. Mutants in another nine genes produce exaggerated ER (Plate 4). In these mutants, the ER lines the inner surface of the plasma membrane and winds through the cytoplasm where multiple connections with the nuclear membrane are seen. The lumen of both the ER and the nuclear membrane is wider than the corresponding wild-type structure. Due to the high density of ribosomes in the background, it has not been possible to determine whether the exaggerated ER is in the rough or smooth form. A third class of mutant, represented by two genes, produces a new organelle, which we have called the Berkeley body (Bb) (Plate 5a). The Bb, although varied in form, appears as a membrane-enclosed toroid or cup-shaped structure. These two forms may be related by perpendicular planes of sectioning. Evidence that the Bb is an altered Golgi body will be presented in the next section. Histochemical staining has shown that acid phosphatase accumulates in the lumen of the ER or Bb in representative mutants (Esmon et al. 1981).

The class-B *sec* mutants have not been studied in as much detail as those of class A. At least two of the class-B genes may operate in the earliest events in the secretory pathway. Mutants in these genes show temperature-sensitive incorporation of mannose into glycoprotein, yet indirect evidence suggests that these mutants are not defective in oligosaccharide synthesis. The affected gene products may be responsible for transmission of membrane and secretory proteins into and through the ER membrane. This possibility will be resolved when an in vitro protein synthesis reaction is performed with wild-type and mutant, soluble, ribosome, and ER-membrane fractions.

Order of Events in the Secretory Pathway

A simple technique exists for the ordering of events along a linear irreversible pathway in which distinct intermediates accumulate in different mutants. In this circumstance, a double mutant will accumulate the intermediate prior to the first block that is encountered. This analysis has been performed with a representative ER-accumulating strain (*sec18*), a Bb-accumulating strain (*sec7*), and all of the vesicle-accumulating strains (*sec1-sec6, sec8-sec10, sec15*). The ER phenotype is epistatic to the Bb and vesicle classes, and the Bb phenotype is epistatic to the vesicle class. Thus, by this analysis, the order of events is Er → Bb → vesicles → cell surface (Novick et al. 1981).

An independent line of evidence supports this order of events. Analysis of the extent of glycosylation of invertase accumulated at 37°C in the *sec* strains indicates at least two stages in oligosaccharide assembly (Esmon et al. 1981). The mutants that produce ER also accumulate a form of invertase that has only half as much carbohydrate as is found associated with the enzyme accumulated in the other *sec* mutants or on the secreted enzyme. The ER block also interrupts the synthesis of bulk mannan at 37°C. Under these conditions, about 50% of the total mannan synthesized is in the *O*-glycosidic linkage, and instead of the usual mannotriose and mannotetraose, the *O*-linked sugar is mannose and mannobiose. Other indirect analysis suggests that the ER-accumulated mannan is missing the outer chain carbohydrate. Thus, the *N*-glycosidically linked core oligosaccharide and the *O*-linked mannose are attached in the ER. Finally, upon return to 25°C, the mannan chains are extended, and nearly mature glycoproteins are secreted.

The epistatic influence of the ER phenotype is also seen in analysis of the form of invertase that accumulates in single and double *sec* mutants at 37°C. Invertase, immunoprecipitated from extracts of wild-type, vesicle-accumulating, and Bb-producing strains, behaves on SDS-polyacrylamide gels as a heterogenous, high-molecular-weight (100K-140K) glycoprotein. In contrast, the ER mutants reveal a more discrete 79K-83K species; this form also appears when an ER mutant is combined with a Bb or vesicle mutant.

The double mutant and glycosylation analyses indicate that the Bb occupies a position and role in the yeast secretory pathway equivalent to that of the Golgi body in mammalian exocrine cells. This view is strengthened by the appearance of typical stacks of Golgi lamellae in the *sec7* mutant when the cells are incubated in medium with low glucose (< 0.1%) at 37°C (Plate 5b). Secretion of accumulated invertase is inhibited by higher glucose concentrations in *sec7* but not in other *sec* mutants. Thus, Bbs produced by *sec7* at 37°C in 2% glucose are nonfunctional side products of the secretory pathway. The Golgi lamellae produced by this mutant at 37°C in low glucose are functional in that accumulated invertase is secreted efficiently when cells are returned to 25°C.

The results are consistent with the following model (Fig. 2). Secretory proteins enter the ER where the initial steps of glycosylation occur. Nine or more *SEC*-gene products are required to transport the secretory proteins to the Golgi where further glycosylation occurs. Two or more functions facilitate the packaging of nearly fully glycosylated proteins into vesicles, which are then transported into the bud where they fuse with the plasma membrane under the direction of ten additional gene products. The function of the *SEC*-gene products and the role of the secretory pathway in plasma-membrane and organelle assembly remain as challenging problems for future investigation.

Note Added in Proof

The role of the secretory process in plasma membrane and vacuole assembly has now been addressed directly by analysis of proteins exposed at the surface of the plasma membrane and by analysis of the maturation and transport of carboxypeptidase Y to the vacuole.

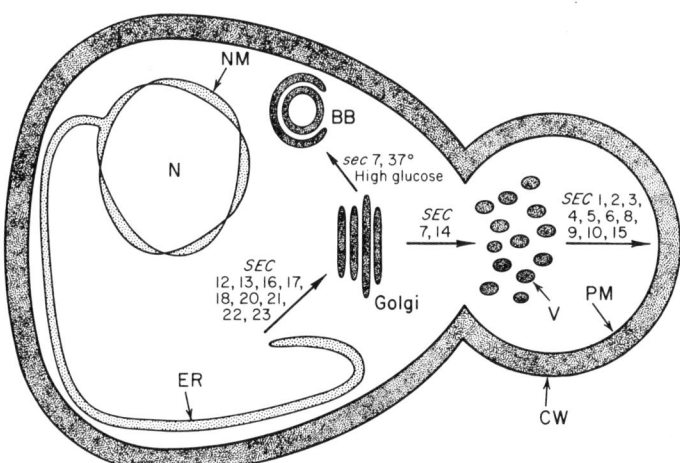

Figure 2 The yeast secretory pathway. (CW) Cell wall.

Plasma-membrane Assembly

The initial characterization of the *sec* mutants suggested a general block in secretion and cell-surface growth at the nonpermissive temperature. In addition to an immediate halt in bud growth, export of a number of secreted proteins (invertase, acid phosphatase, L-asparaginase, α-galactosidase, α-factor, killer toxin) and plasma-membrane permease activities ($SO_4^=$ permease, arginine permease, proline-specific permease) is blocked. A more general probe of surface assembly that was used to examine the export and turnover of macrophage plasma-membrane proteins (Kaplan et al. 1979) has now been adapted by Novick to examine surface assembly in yeast (Novick and Schekman, in prep.). Modification of cell-surface amino groups with trinitrobenzenesulfonate (TNBS) followed by precipitation with trinitrophenylated (TNP) antibody allows analysis of newly exported proteins. In this procedure, wild-type and mutant cells are labeled with protein synthesis precursors at 37°C and then tagged with TNBS at 0°C. Under these conditions, TNBS does not penetrate into the cell. Both secreted and plasma-membrane surface proteins are tagged in this procedure and can be examined separately: Secreted proteins are released when cells are converted to spheroplasts, and tagged membrane proteins are recovered in a sedimented fraction from lysed spheroplasts. Wild-type cells export distinct sets of membrane and secreted proteins, as revealed by SDS-gel electrophoresis of TNP-antibody precipitates. The major proteins in both fractions are not exported in *sec* mutant cells at 37°C, but they are at 24°C. These results suggest that the secretory process is responsible for the localization of most cell-surface proteins in yeast.

Vacuole Assembly

The yeast vacuole is a lysosomelike organelle that has a set of hydrolytic enzymes distinct from the secreted glycoproteins. Among the vacuolar enzymes, CPY has been studied in some detail. CPY is made as a 67-kD precursor containing four *N*-glycosidically linked oligosaccharides (Hasilik and Tanner 1978b). The aminoterminal 8-kD propeptide portion is cleaved under the direction of at least two gene products (*PEP4, PEP17*) that are required for the maturation of several other vacuolar enzymes (Hemmings et al. 1981). A *pep4* mutant accumulates enzymatically inactive pro-CPY in the vacuole, suggesting that maturation occurs after the precursor is localized (Stevens et al. 1982). If vacuolar proteins require the secretory pathway for transport, proenzyme forms will not be localized and processed at 37°C in the *sec* mutants. Stevens has shown that mutants blocked in movement from the ER or from the Golgi body accumulate pro-CPY in some place other than the vacuole, presumably in the accumulated organelle (Stevens et al. 1982). Upon return to the permissive temperature, the accumulated proenzyme forms become processed normally. Mutants that block after the

Golgi step have no effect on CPY localization. These results suggest that vacuolar and secretory proteins travel together from the ER to the Golgi body where sorting may occur. The results also rule out a secretion-recapture mechanism of localization such as has been suggested in mammalian lysosomal enzyme studies (Neufield et al. 1977).

REFERENCES

Abrams, B.B., R. Hackel, T. Mizunaga, and J.O. Lampen. 1978. Relationship of large and small invertases in *Saccharomyces*: Mutant selectively deficient in small invertase. *J. Bacteriol.* **135**: 809.

Arnold, W.N. 1972a. The structure of the yeast cell wall: Solubilization of a marker enzyme, β-fructofuranosidase, by the autolytic enzyme system. *J. Biol. Chem.* **247**: 1161.

———. 1972b. Location of acid phosphatase and β-fructofuranosidase within yeast cell envelopes. *J. Bacteriol.* **112**: 1346.

Babczinski, P. 1980. Partial purification, characterization and localization of the membrane-associated invertase of yeast. *Biochim. Biophys. Acta* **614**: 121.

Babczinski, P. and W. Tanner. 1978. A membrane-associated isozyme of invertase in yeast; precursor of the external glycoprotein. *Biochim. Biophys. Acta* **538**: 426.

Ballou, C.E. 1976. Structure and biosynthesis of the mannan component of the yeast cell envelope. *Adv. Microb. Physiol.* **14**: 93.

Ballou, L., R.E. Cohen, and C.E. Ballou. 1980. *Saccharomyces cerevisiae* mutants that make mannoproteins with a truncated carbohydrate outer chain. *J. Biol. Chem.* **255**: 5986.

Bauer, H. and E. Sigarlakie. 1975. Localization of alkaline phosphatase in *Saccharomyces cerevisiae* by means of ultrathin frozen sections. *Ultrastruc. Res.* **50**: 208.

Boller, T.P., M. Durr, and A. Wiemken. 1976. Asymmetric distribution of concanavalin A binding sites on yeast plasmalemma and vacuolar membrane. *Arch. Microbiol.* **109**: 115.

Bostian, K.A., J.M. Lemire, L.E. Cannon, and H.O. Halvorson. 1980. In vitro synthesis of repressible yeast acid phosphatase: Identification of multiple mRNAs and products. *Proc. Natl. Acad. Sci.* **77**: 4504.

Bowers, B., G. Levin, and E. Cabib. 1974. Effect of polyoxin D on chitin synthesis and septum formation in *Saccharomyces cerevisiae*. *J. Bacteriol.* **119**: 564.

Bowles, D.J. and D.H. Northcote. 1972. The sites of synthesis and transport of extracellular polysaccharides in the root tissues of maize. *Biochemistry* **130**: 1133.

Bowman, B.J., S.E. Mainzer, K.E. Allen, and C. Slayman. 1978. Effects of inhibitors on the plasma membrane and mitochondrial adenosine triphosphates of *Neurospora crassa*. *Biochim. Biophys. Acta* **512**: 13.

Bracker, C.E., C.E. Heintz, and S.N. Grove. 1970. Structural and functional continuity among endomembrane organelles in fungi. In *International Congress of Electron Microscopy* (ed. P. Favard), vol. III, p. 103. Société Francaise de Microscopie Electronique, Paris.

Brawley, S.H. and R.S. Quatrano. 1979. Sulfation of Fucoidin in *Fucus* embryos. IV. Autoradiographic investigations of Fucoidin sulfation and secretion during differentiation and the effect of cytochalasin treatment. *Dev. Biol.* **73**: 193.

Burger, M.M. 1969. A difference in the architecture of the surface membrane of normal and virally transformed cells. *Proc. Natl. Acad. Sci.* **62**: 994.

Bussey, H., D. Saville, M.R. Chevallier and G.H. Rank. 1979. Yeast plasma membrane ghosts. An analysis of proteins by two-dimensional gel electrophoresis. *Biochim. Biophys. Acta* **553**: 185.

Byers, B. 1981. Cytology of the yeast life cycle. In *The molecular biology of the yeast* Saccharomyces. I. *Life cycle and inheritance* (ed. J. Strathern et al.), p.59. Cold Spring Harbor Laboratory, Cold Spring Harbor, New York.

Byers, B. and L. Goetsch. 1974. Duplication of spindle plaques and integration of the yeast cell cycle. *Cold Spring Harbor Symp. Quant. Biol.* **38**: 123.

———. 1975. Behavior of spindles and spindle plaques in the cell cycle and conjugation of *Saccharomyces cerevisiae*. *J. Bacteriol.* **124**: 511.

———. 1976a. A highly ordered ring of membrane-associated filaments in budding yeast. *J. Cell Biol.* **69**: 717.

———. 1976b. Loss of the filamentous ring in cytokinesis-defective mutants of budding yeast. *J. Cell Biol.* **70**: 2a.

Cabib, E. 1971. Yeast spheroplasts. *Methods Enzymol.* **22**: 120.

———. 1975. Molecular aspects of yeast morphogenesis. *Annu. Rev. Microbiol.* **29**: 191.

Cabib, E. and B. Bowers. 1975. Timing and function of chitin synthesis in yeast. *J. Bacteriol.* **124**: 1586.

Cabib, E. and V. Farkaš. 1971. The control of morphogenesis: An enzymatic mechanism for the initiation of septum formation in yeast. *Proc. Natl. Acad. Sci.* **68**: 2052.

Cabib, E., R.E. Ulane, and B. Bowers. 1973. Yeast chitin synthetase. Separation of the zymogen from its activating factor and recovery of the latter in the vacuole fraction. *J. Biol. Chem.* **248**: 1451.

Cabib, E., V. Farkaš, R.E. Ulane, and B. Bowers. 1974. Yeast septum formation as a model system for morphogenesis. *Curr. Top. Cell. Regul.* **22**: 1.

Carlson, M. and D. Botstein. 1982. Two differentially regulated mRNAs with different 5' ends encode secreted and intracellular forms of yeast invertase. *Cell* **28**: 145.

Carlson, M., B.C. Osmond, and D. Botstein. 1981. Mutants of yeast defective in sucrose utilization. *Genetics* **98**: 25.

Chambaut-Guérin, A.-M., P. Muller, and B. Rossignol. 1978. Microtubules and protein secretion in rat lacrimal glands. Relationship between colchicine binding and its inhibitory effect on the intracellular transport of proteins. *J. Biol. Chem.* **253**: 3870.

Chen, A.W.-C. and J.J. Miller. 1968. Proteolytic activity of intact yeast cells during sporulation. *Can. J. Microbiol.* **14**: 957.

Chu, F.K. and F. Maley. 1980a. Growth phase dependence of invertase mRNA levels in yeast. *J. Biol. Chem.* **255**: 6387.

———. 1980b. The effect of glucose on the synthesis and glycosylation of the polypeptide moiety of yeast external invertase. *J. Biol. Chem.* **255**: 6392.

Cortat, M., P. Matile, and F. Kopp. 1973. Intracellular localization of mannan synthetase activity in budding baker's yeast. *Biochem. Biophys. Res. Commun.* **53**: 482.

Cortat, M., P. Matile, and A. Wiemken. 1972. Isolation of glucanase-containing vesicles from budding yeast. *Arch. Mikrobiol.* **82**: 189.

DeRousset-Hall, A. and G.W. Gooday. 1975. A kinetic study of a solubilized chitin synthetase preparation from *Coprinus cinereus*. *J. Gen. Microbiol.* **89**: 146.

Dunlop, P.C., G.M. Meyer, D. Ban, and R.J. Roon. 1978. Characterization of two forms of asparaginase in *Saccharomyces cerevisiae*. *J. Biol. Chem.* **253**: 1297.

Duntze, W., V. Mackay, and T.R. Manney. 1970. *Saccharomyces cerevisiae*: A diffusible sex factor. *Science* **168**: 1472.

Durán, A. and E. Cabib. 1978. Solubilization and partial purification of yeast chitin synthetase. Confirmation of the zymogenic nature of the enzyme. *J. Biol. Chem.* **253**: 4419.

Durán, A., B. Bowers, and E. Cabib. 1975. Chitin synthetase zymogen is attached to yeast plasma membrane. *Proc. Natl. Acad. Sci.* **72**: 3952.

Durán, A., E. Cabib, and B. Bowers. 1979. Chitin synthetase distribution on the yeast plasma membrane. *Science* **203**: 363.

Elorza, M.V., C.M. Lostau, J.R. Villaneuva, and R. Sentandreu. 1976. Cell wall synthesis regulation in *Saccharomyces cerevisiae*. Effect of RNA and protein inhibition. *Biochim. Biophys. Acta* **454**: 263.

Esmon, B., P. Novick, and R. Schekman. 1981. Compartmentalized assembly of oligosaccharides on exported glycoproteins in yeast. *Cell* **25**: 451.

Farkaš, V. 1979. Biosynthesis of cell walls of Fungi. *Microbiol. Rev.* **43:** 117.
Farkaš, V., P. Biely, and S. Bauer. 1973. Extracellular β-glucanases of the yeast, *Saccharomyces cerevisiae. Biochim. Biophys. Acta* **321:** 246.
Field, C. and R. Schekman. 1980. Localized secretion of acid phosphatase reflects the pattern of cell-surface growth in *Saccharomyces cerevisiae. J. Cell Biol.* **86:** 123.
Gallili, G. and J.O. Lampen. 1977. Large and small invertase and the yeast cell cycle pattern of synthesis and sensitivity to tunicamycin. *Biochim. Biophys. Acta* **475:** 113.
Gascón, S. and J.O. Lampen. 1968. Purification of the internal invertase of yeast. *J. Biol. Chem.* **243:** 1567.
Gascón, S., N.P. Neumann, and J.O. Lampen. 1968. Comparative study of the properties of the purified internal and external invertases of yeast. *J. Biol. Chem.* **243:** 1573.
Gasior, E., F. Herrera, I. Sadnik, C.S. McLaughlin, and K. Moldave. 1979. The preparation and characterization of a cell-free system from *Saccharomyces cerevisiae* that translates natural messenger ribonucleic acid. *J. Biol. Chem.* **254:** 3965.
Goldstein, J.L., R.G. Anderson, and M.S. Brown. 1979. Coated pits, coated vesicles, and receptor-mediated endocytosis. *Nature* **279:** 679.
Grossman, M.K. and F.K. Zimmerman. 1979. The structural genes of internal invertases in *Saccharomyces cerevisiae. Mol. Gen. Genet.* **175:** 223.
Grove, S.N., C.E. Bracker, and D.J. Morré. 1970. An ultrastructural basis for hyphal tip growth in *Pythium ultimum. Am. J. Bot.* **67:** 245.
Hartwell, L.H. 1974. *Saccharomyces cerevisiae* cell cycle. *Bacteriol. Rev.* **38:** 164.
Hartwell, L.H., J. Culotti, J.R. Pringle, and B.J. Reid. 1974. Genetic control of the cell division cycle in yeast. *Science* **183:** 46.
Hasilik, A. and W. Tanner. 1978a. Biosynthesis of the vacuolar yeast glycoprotein carboxypeptidase Y. Conversion of precursor into the enzyme. *Eur. J. Biochem.* **85:** 599.
Hasilik, A. and W. Tanner. 1978b. Carbohydrate moiety of carboxypeptidase Y and perturbation of its biosynthesis. *Eur. J. Biochem.* **91:** 567.
Hayashibe, M. and S. Katohda. 1973. Initiation of budding and chitin ring. *J. Gen. Appl. Microbiol.* **19:** 23.
Hemmings, B.A., G.S. Zubenko, A. Hasilik, and E.W. Jones. 1981. Mutant defective in processing of an enzyme located in the lysosome-like vacuole of *Saccharomyces cerevisiae. Proc. Natl. Acad. Sci.* **78:** 435.
Henry, S.A., K.D. Atkinson, A.I. Kolat, and M.R. Culbertson. 1977. Growth and metabolism of inositol-starved *Saccharomyces cerevisiae. J. Bacteriol.* **130:** 472.
Holbein, B.E. and D.K. Kidby. 1977. Effects of proteolytic enzymes on invertase secretion in sphaeroplasts of *Saccharomyces:* Inhibition by trypsin. *Can. J. Microbiol.* **23:** 202.
———. 1979. Trypsin-uncoupled synthesis and secretion of yeast invertase: Implications for the mechanism of secretion. *Can. J. Microbiol.* **25:** 528.
Holbein, B.E., C.N. Forsberg, and D.K. Kidby. 1976. A modified procedure for studying enzyme secretion in yeast spheroplasts: Sub-cellular distribution of invertase. *Can. J. Microbiol.* **22:** 989.
Howard, R.J. 1981. Ultrastructural analysis of hyphal tip cell growth in Fungi: Spitzenkorper, cytoskeleton, and endomembranes after freeze-substitution. *J. Cell Sci.* **48:** 89.
Howard, R.J. and J.R. Aist. 1980. Cytoplasmic microtubules and fungal morphogenesis: Ultrastructural effects of methyl benzimidazole-2-yl-carbamate determined by freeze-substitution of hyphal tip cells. *J. Cell Biol.* **87:** 55.
Jaffe, L.F., K.R. Robinson, and R. Nuccitelli. 1974. Local cation entry and self-electrophoresis as an intracellular localization mechanism. *Ann. N.Y. Acad. Sci.* **238:** 372.
Johnson, B.F. 1968. Lysis of yeast cell walls induced by 2-deoxyglucose at sites of glucan synthesis. *J. Bacteriol.* **95:** 1169.
Johnson, B.F. and E.J. Gibson. 1966. Autoradiographic analysis of regional cell wall growth of yeasts. *Exp. Cell Res.* **41:** 580.
Jones, E.W. 1977. Proteinase mutants of *Saccharomyces cerevisiae. Genetics* **85:** 23.

Jones, E.W., G.S. Zubenko, R.R. Parker, B.A. Hemmings, and A. Hasilik. 1981. Pleiotropic mutations of *S. cerevisiae* which cause deficiency for proteinases and other vacuole enzymes. In *Alfred Benzon Symp. 16. Molecular genetics in yeast* (ed. D. von Wettstein et al.), p. 182. Munksgaard, Copenhagen.

Kaplan, G., J.C. Unkeless, and Z.A. Cohn. 1979. Insertion and turnover of macrophage plasma membrane proteins. *Proc. Natl. Acad. Sci.* **76**: 3824.

Kew, O.M. and H.C. Douglas. 1976. Genetic co-regulation of galactose and melibiose utilization in *Saccharomyces*. *J. Bacteriol.* **125**: 33.

Kosinova, A., V. Farkaš, S. Machala, and S. Bauer. 1974. Site of mannan synthesis in yeast: An autoradiographic study. *Arch. Mikrobiol.* **99**: 255.

Koteliansky, V.E., M.A. Glukhova, W.V. Bejanian, A.P. Surguchov, and V.N. Smirnow. 1979. Isolation and characterization of actin-like protein from yeast *Saccharomyces cerevisiae*. *FEBS Lett.* **102**: 55.

Kramer, R.A. and N. Anderson. 1980. Isolation of yeast genes with mRNA levels controlled by phosphate concentration. *Proc. Natl. Acad. Sci.* **77**: 6541.

Kuo, S.-C. and J.O. Lampen. 1972. Inhibition by 2-deoxy-D-glucose of synthesis of glycoprotein enzymes by protoplasts of *Saccharomyces*: Relation to inhibition of sugar uptake and metabolism. *J. Bacteriol.* **111**: 419.

———. 1974. Tunicamycin—An inhibitor of yeast glycoprotein synthesis. *Biochem. Biophys. Res. Commun.* **58**: 287.

Kuo, S.-C. and S. Yamamoto. 1975. Preparation and growth of yeast protoplasts. *Methods Cell Biol.* **11**: 169.

Lehle, L. 1980. Biosynthesis of the core region of yeast mannoproteins. Formation of a glucosylated dolichol-bound oligosaccharide precursor, its transfer to protein and subsequent modification. *Eur. J. Biochem.* **109**: 589.

Lehle, L. and W. Tanner. 1976. The specific site of tunicamycin inhibition in the formation of dolichol-bound *N*-acetylglucosamine derivatives. *FEBS Lett.* **71**: 167.

———. 1978. Biosynthesis and characterization of large dolichol diphosphate-linked oligosaccharides in *Saccharomyces cerevisiae*. *Biochim. Biophys. Acta* **539**: 218.

Lehle, L., F. Bauer, and W. Tanner. 1977. The formation of glycosidic bonds in yeast glycoproteins. Intracellular localization of the reactions. *Arch Mikrobiol.* **114**: 77.

Lehle, L., R.E. Cohen, and C.E. Ballou. 1979. Carbohydrate structure of yeast invertase. Demonstration of a form with only core oligosaccharides and a form with completed chains. *J. Biol. Chem.* **254**: 12209.

Liao, H. and J. Thorner. 1980. Yeast mating pheromone factor inhibits adenylate cyclase. *Proc. Natl. Acad. Sci.* **77**: 1898.

Linnemans, W.A.M., P. Boer, and P.F. Elbers. 1977. Localization of acid phosphatase in *Saccharomyces cerevisiae*: A clue to cell wall formation. *J. Bacteriol.* **131**: 638.

Lipke, P.N., A. Taylor, and C.E. Ballou. 1976. Morphogenic effects of α-factor on *Saccharomyces cerevisiae* **a** cells. *J. Bacteriol.* **127**: 610.

Marriot, M. and W. Tanner. 1979. Localization of dolichyl phosphate- and pyrophosphate-dependent glycosyl transfer reactions in *Saccharomyces cerevisiae*. *J. Bacteriol.* **139**: 565.

Matile. P. 1978. Biochemistry and function of vacuoles. *Annu. Rev. Plant Physiol.* **29**: 193.

Matile, P., H. Moor, and C.F. Robinow. 1969. Yeast cytology. In *The yeasts* (ed. A.H. Rose and J.S. Harrison), vol. 1, p. 219. Academic Press, New York.

Matile, P., M. Cortat, A. Wiemken, and A. Frey-Wysling. 1971. Isolation of glucanase-containing particles from budding *Saccharomyces cerevisiae*. *Proc. Natl. Acad. Sci.* **68**: 636.

Meyer, J. and P. Matile. 1975. Subcellular distribution of yeast invertase isoenzymes. *Arch. Mikrobiol.* **103**: 51.

Mizunaga, T., J.S. Tkacz, L. Rodriguez, R.A. Hackel, and J.O. Lampen. 1981. Temperature-sensitive forms of large and small invertase in a mutant derived from a *SUC1* strain of *Saccharomyces cerevisiae*. *Mol. Cell. Biol.* **1**: 460.

Mollenhauer, H.H. and D.J. Morré. 1976. Cytochalasin B, but not colchicine, inhibits migration of secretory vesicles in root tips of maize. *Protoplasma* **87:** 39.

Montezinos, D. and R.M. Brown. 1976. Surface architecture of the plant cell: Biogenesis of the cell wall, with special emphasis on the role of the plasma membrane in cellulose biosynthesis. *J. Supramol. Struct.* **5:** 277.

Moor, H. 1967. Endoplasmic reticulum as the initiator of bud formation in yeast. *Arch. Mikrobiol.* **57:** 135.

Moreno, F., A.G. Ochoa, S. Gascón, and J.R. Villanueva. 1975. Molecular forms of yeast invertase. *Eur. J. Biochem.* **50:** 571.

Nečas, O. 1971. Cell wall synthesis in yeast protoplasts. *Bacteriol. Rev.* **35:** 149.

Neufeld, E.F., G.N. Sando, A.J. Garvin, and L.H. Rome. 1977. The transport of lysosomal enzymes. *J. Supramol. Struct.* **6:** 95.

Niedermeyer, W. 1976. The elasticity of yeast cell tonoplast related to its ultrastructure and chemical composition. II. Chemical and cytochemical investigations. *Cytobiologie* **13:** 380.

Novick, P. and R. Schekman. 1979. Secretion and cell-surface growth are blocked in a temperature-sensitive mutant of *Saccharomyces cerevisiae*. *Proc. Natl. Acad. Sci.* **76:** 1858.

Novick, P., S. Ferro, and R. Schekman. 1981. Order of events in the yeast secretory pathway. *Cell* **25:** 461.

Novick, P., C. Field, and R. Schekman. 1980. Identification of 23 complementation groups required for post-translational events in the yeast secretory pathway. *Cell* **21:** 205.

Nuccitelli, R. 1978. Ooplasmic segregation and secretion in the *Pelvetia* egg is accompanied by a membrane-generated electrical current. *Dev. Biol.* **62:** 13.

Onishi, H.R., J.S. Tkacz, and J.O. Lampen. 1979. Glycoprotein nature of yeast alkaline phosphatase: Formation of active enzyme in the presence of tunicamycin. *J. Biol. Chem.* **254:** 11943.

Palamarczyk, G. and E. Janczura. 1977. Lipid mediated glycosylation in yeast nuclear membranes. *FEBS Lett.* **77:** 169.

Parodi, A.J. 1979a. Biosynthesis of yeast mannoproteins: Synthesis of mannan outer chain and of dolichol derivatives. *J. Biol. Chem.* **254:** 8343.

———. 1979b. Biosynthesis of yeast glycoproteins. Processing of the oligosaccharides transferred from dolichol derivatives. *J. Biol. Chem.* **254:** 10051.

Parodi, A.J. and L.F. Leloir. 1979. The role of lipid intermediates in the glycosylation of proteins in the eukaryotic cell. *Biochim. Biophys. Acta* **559:** 1.

Pearse, B.M.F. 1976. Clathrin: A unique protein associated with intracellular transfer of membrane by coated vesicles. *Proc. Natl. Acad. Sci.* **73:** 1255.

Perlman, D. and H.O. Halvorson. 1981. Distinct repressible mRNAs for cytoplasmic and secreted yeast invertase are encoded by a single gene. *Cell* **25:** 525.

Phaff, H.J. 1979. A retrospective and current view on endogenous β-glucanases in yeast. In *Proceedings of the 5th International Protoplast Symposium,* July 9–14, Szeged, Hungary, p. 171. Akedemiai Kiado, Budapest.

Raschke, W.C., K.A. Kern, C. Antolis, and C.E. Ballou. 1973. Genetic control of yeast mannan structure. *J. Biol. Chem.* **248:** 4660.

Redman, C.M. and M.G. Cherian. 1972. The secretory pathways of rat serum glycoproteins and albumin. Localization of newly-formed proteins within the endoplasmic reticulum. *J. Cell Biol.* **52:** 231.

Rogers, D.T., D. Saville, and H. Bussey. 1979. *Saccharomyces cerevisiae* killer expression: Mutant *KEX*2 has altered secretory proteins and glycoproteins. *Biochem. Biophys. Res. Commun.* **90:** 187.

Ruiz-Herrera, J. and R. Sentandreu. 1975. Site of initial glycosylation of mannoproteins from *Saccharomyces cerevisiae*. *J. Bacteriol.* **124:** 127.

Ruiz-Herrera, J., E. López-Romera, and S. Bartnicki-Garcia. 1977. Properties of the chitin synthetase in isolated chitosomes from yeast cells of *Mucor rouxii*. *J. Biol. Chem.* **252:** 3338.

Ruiz-Herrera, J., V.O. Sing, V.J. van der Voude, and S. Bartnicki-Garcia. 1975. Microfibril assembly by granules of chitin synthetase. *Proc. Natl. Acad. Sci.* **72:** 2706.

Santos, T., F. del Rey, J. Conde, J.R. Villanueva, and C. Nombela. 1979. *Saccharomyces cerevisiae* mutant defective in exo-1,3-β-glucanase production. *J. Bacteriol.* **139:** 333.

Satir, B., C. Schooley, and P. Satir. 1973. Membrane fusion in a model system: Mucocyst secretion in *Tetrahymena*. *J. Cell Biol.* **56:** 153.

Schekman, R. and V. Brawley. 1979. Localized deposition of chitin on the yeast cell surface in response to mating pheromone. *Proc. Natl. Acad. Sci.* **76:** 645.

Schweingruber, M.E. and A.-M. Schweingruber. 1979. Posttranslational regulation of repressible acid phosphatase in yeast. *Mol. Gen. Genet.* **173:** 349.

Schwencke, J. 1977. Characteristics and integration of the yeast vacuole with cellular functions. *Physiol. Veg.* **15:** 491.

Schwencke, J. and H. de Robichon-Szulmajster. 1976. The transport of S-adenosyl-L-methionine in isolated yeast vacuoles and spheroplasts. *Eur. J. Biochem.* **65:** 49.

Scott, J.H. 1980. "Enzymatic lysis of yeast cells." Ph.d. thesis, University of California, Berkeley.

Scott, J.H. and R. Schekman. 1980. Lyticase: Endoglucanase and protease activities that act together in yeast cell lysis. *J. Bacteriol.* **142:** 414.

Sharma, C.B., P. Babczinski, L. Lehle, and W. Tanner. 1974. The role of dolichol monophosphate in glycoprotein biosynthesis in *Saccharomyces cerevisiae*. *Eur. J. Biochem.* **46:** 35.

Shematek, E.M. and E. Cabib. 1980. Biosynthesis of the yeast cell wall. II. Regulation of β-(1 → 3) glucan synthetase by ATP and GTP. *J. Biol. Chem.* **255:** 895.

Shematek, E.M., J.A. Bratz, and E. Cabib. 1980. Biosynthesis of the yeast cell wall. I. Preparation and properties of β-(1 → 3) glucan synthetase. *J. Biol. Chem.* **255:** 888.

Slayman, C.L. and C.W. Slayman. 1962. Measurement of membrane potentials in *Neurospora*. *Science* **136:** 876.

Sloat, B.F. and J.R. Pringle. 1978. A mutant of yeast defective in cellular morphogenesis. *Science* **200:** 1171.

Sloat, B.F., A. Adams, and J.R. Pringle. 1981. Roles of the *CDC24* gene product in cellular morphogenesis during the *Saccharomyces cerevisiae* cell cycle. *J. Cell. Biol.* **89:** 395.

Snider, M.D., L.A. Sultzman, and P.W. Robbins. 1980. Transmembrane location of oligosaccharide-lipid synthesis in microsomal vesicles. *Cell* **21:** 385.

Sosinsky, G., R. Schekman, and R. Glaeser. 1980. Enlarged crystalline patches on the plasma membrane of yeast protoplasts. In *Annual Proceedings of the Electron Microscopy Society of America* (ed. G.W. Bailey), p. 686. Claitor's, Baton Rouge.

Steere, R.L., E.F. Erbe, and J.M. Moseley. 1980. Prefracture and cold-fracture images of yeast plasma membranes. *J. Cell Biol.* **86:** 113.

Stevens, T., B. Esmon, and R. Schekman. 1982. Early stages in the yeast secretory pathway are required for transport of carboxypeptidase Y to the vacuole. *Cell* (in press).

Stroobant, P. and G.A. Scarborough. 1979. Large scale isolation and storage of *Neurospora* plasma membranes. *Anal. Biochem.* **95:** 554.

Thorner, J. 1981. Pheromonal regulation of development in *Saccharomyces cerevisiae*. In *The molecular biology of the yeast* Saccharomyces. I. *Life cycle and inheritance* (ed. J. Strathern et al.), p. 143. Cold Spring Harbor Laboratory, Cold Spring Harbor, New York.

Tkacz, J.S. and J.O. Lampen. 1972. Wall replication in *Saccharomyces* species: Use of fluorescein-conjugated concanavalin A to reveal the site of mannan insertion. *J. Gen. Microbiol.* **72:** 243.

———. 1973. Surface distribution of invertase on growing *Saccharomyces* cells. *J. Bacteriol.* **113:** 1073.

Tkacz, J.S. and V.L. Mackay. 1979. Sexual conjugation in yeast: Cell surface changes in response to the action of mating hormones. *J. Cell Biol.* **80:** 326.

Toh-e, A., Y. Ueda, S.-I. Kakimoto, and Y. Oshima. 1973. Isolation and characterization of acid phosphatase mutants in *Saccharomyces cerevisiae*. *J. Bacteriol.* **113:** 727.

Trimble, R.B. and F. Maley. 1977. Subunit structure of external invertase from *Saccharomyces cerevisiae. J. Biol. Chem.* **252**: 4409.
Tuite, M.F., J. Plesset, K. Moldave, and C.S. McLaughlin. 1980. Faithful and efficient translation of homologous and heterologous mRNAs in an mRNA-dependent cell-free system from *Saccharomyces cerevisiae. J. Biol. Chem.* **255**: 8761.
Ulane, R.E. and E. Cabib. 1976. The activating system of chitin synthetase from *Saccharomyces cerevisiae*. Purification and properties of the activating factor. *J. Biol. Chem.* **251**: 3367.
Van der Wilden, W., P. Matile, M. Schellenberg, J. Meyer, and A. Wiemken. 1973. Vacuolar membranes: Isolation from yeast cells. *Z. Naturforsch.* **28c**: 416.
Vrsanská, M., P. Biely, and Z. Krátký. 1977. Enzymes of the yeast lytic system produced by *Arthrobacter* GJM-1 bacterium and their role in the lysis of yeast cell walls. *Z. Allg. Mikrobiol.* **17**: 465.
Wehrli, E., C. Boehm, and G.F. Fuhrman. 1975. Yeast plasma membrane vesicles suitable for transport studies. *J. Bacteriol.* **124**: 1594.
Welten-Verstegen, G.W., P. Boer, and E. Steyn-Parvé. 1980. Lipid-mediated glycosylation of endogenous proteins in isolated plasma membrane of *Saccharomyces cerevisiae. J. Bacteriol.* **141**: 342.
White, J. and A. Helenius. 1980. pH-dependent fusion between the Semliki forest virus membrane and liposomes. *Proc. Natl. Acad. Sci.* **77**: 3273.
Wickner, R.B. 1981. Killer systems in *Saccharomyces cerevisiae*. In *The molecular biology of the yeast* Saccharomyces. I. *Life cycle and inheritance* (ed. J. Strathern et al.), p. 415. Cold Spring Harbor Laboratory, Cold Spring Harbor, New York.
Wiemken, A. 1975. Isolation of vacuoles from yeasts. *Methods Cell Biol.* **12**: 99.
Wiemken, A., M. Schellenberg, and K. Urech. 1979. Vacuoles: The sole compartments of digestive enzymes in yeast (*Saccharomyces cerevisiae*)? *Arch. Mikrobiol.* **123**: 23.
Willsky, G.R. 1979. Characterization of the plasma membrane Mg^{2+}-ATPase from the yeast, *Saccharomyces cerevisiae. J. Biol. Chem.* **254**: 3326.
Woodruff, R.I. and W.H. Telfer. 1980. Electrophoresis of proteins in intercellular bridges. *Nature* **286**: 84.
Woods, D.R. and E.A. Bevan. 1968. Studies on the nature of the killer factor produced by *Saccharomyces cerevisiae. J. Gen. Microbiol.* **51**: 115.
Zubenko, G.S., A.P. Mitchell, and E.W. Jones. 1979. Septum formation, cell division, and sporulation in mutants of yeast deficient in proteinase B. *Proc. Natl. Acad. Sci.* **76**: 2395.

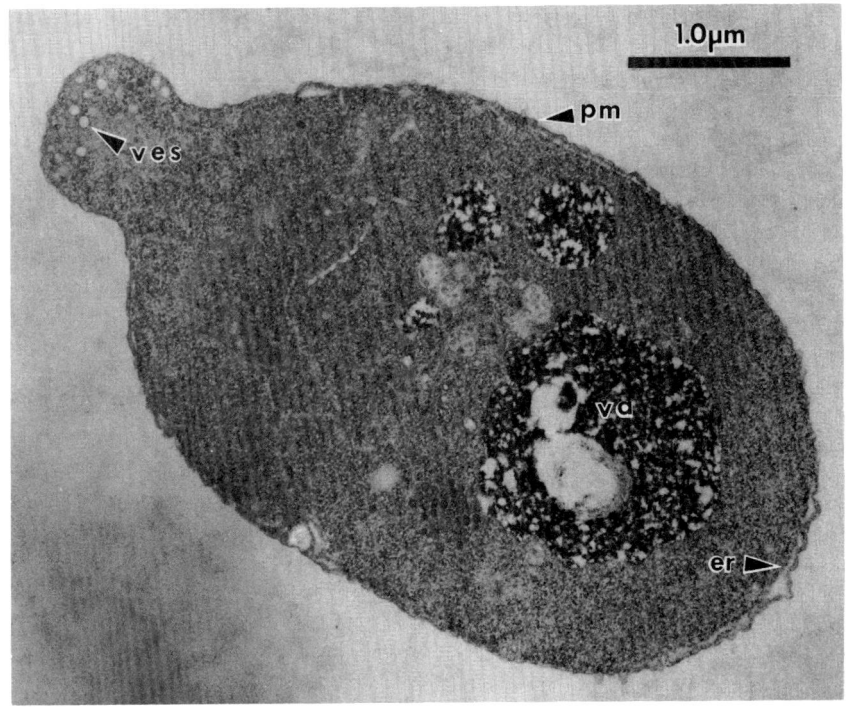

Plate 1 Budding cell. (ves) Vesicle; (va) vacuole; (er) endoplasmic reticulum; (pm) plasma membrane.

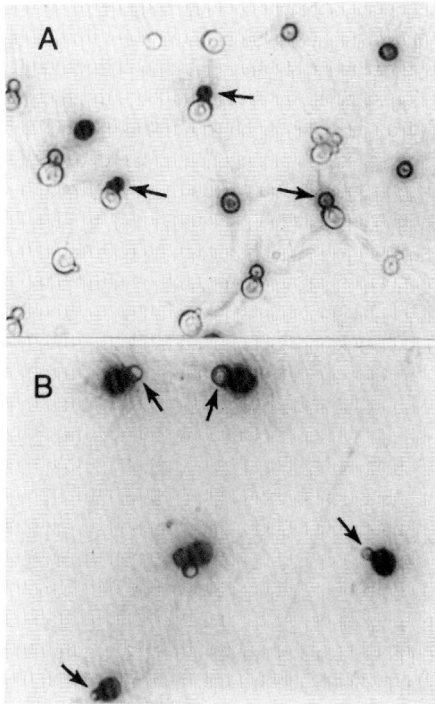

Plate 2 Acid phosphatase secreted into the bud. (*A*) Repressed cells were shifted to a phosphate-free medium for 5 hr, and a histochemical stain was then applied. (*B*) Derepressed cells were shifted to phosphate-supplemented medium for 2 hr and then stained (Field and Schekman 1980).

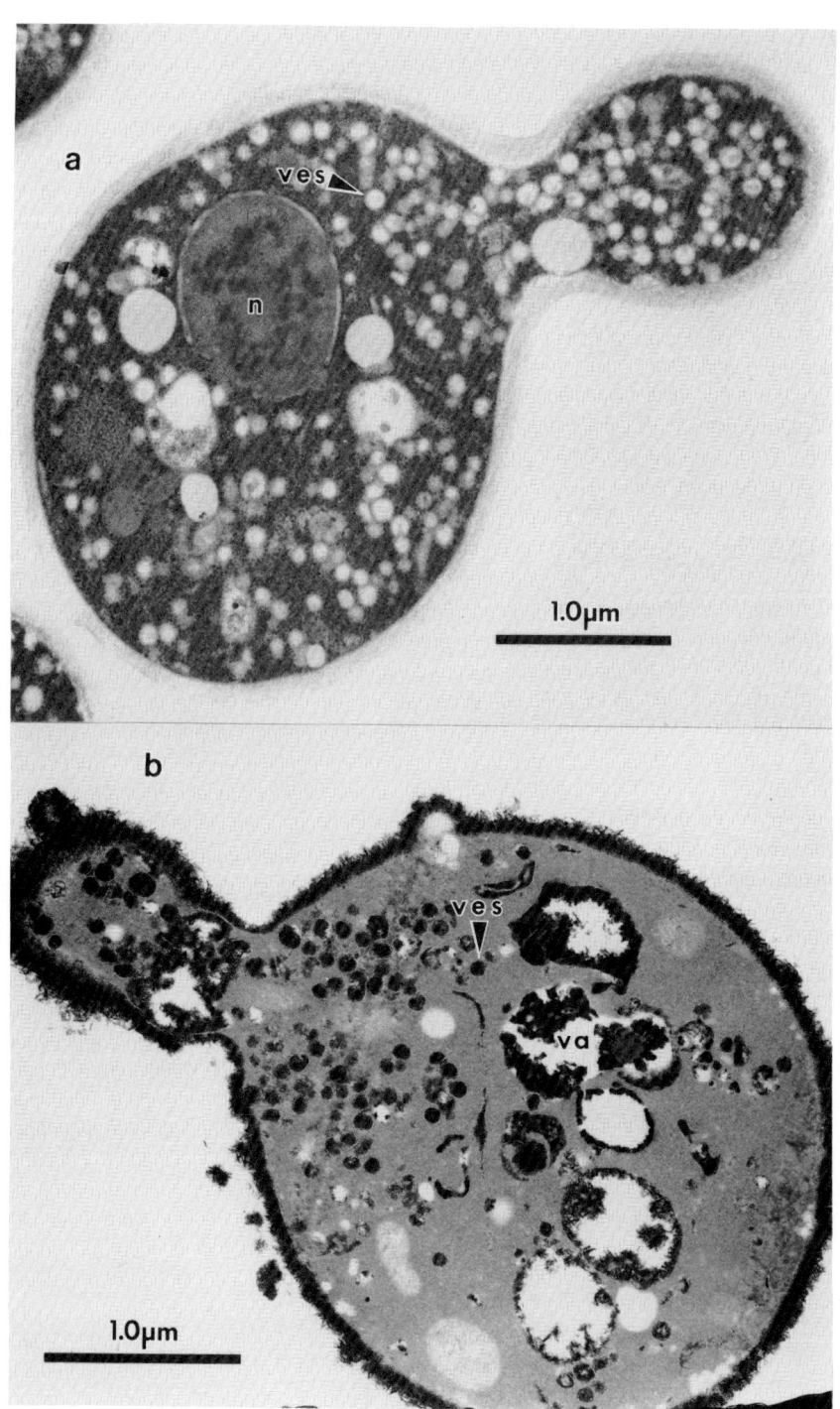

Plate 3 (See facing page for legend.)

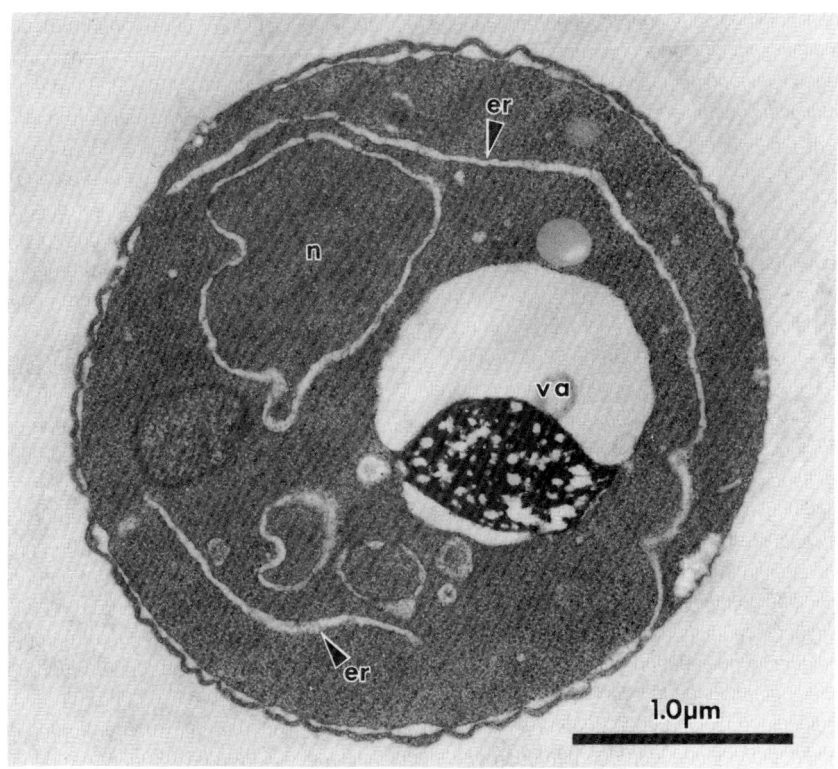

Plate 4 ER-accumulating *sec* mutant. A *sec18* strain was incubated for 2 hr at 37°C, and thin sections were examined by electron microscopy (Novick et al. 1980). (n) Nuclear membrane.

Plate 3 Vesicles accumulate in *sec1*. (*A*) The *sec1* strain was incubated for 3 hr at 37°C, and thin sections were examined by electron microscopy (Novick and Schekman 1979). (*B*) A *sec1 pho80* (acid phosphatase constitutive) strain was incubated for 1½ hr at 37°C. The cells were stained for phosphatase activity, and thin sections were examined by electron microscopy (Esmon et al. 1981).

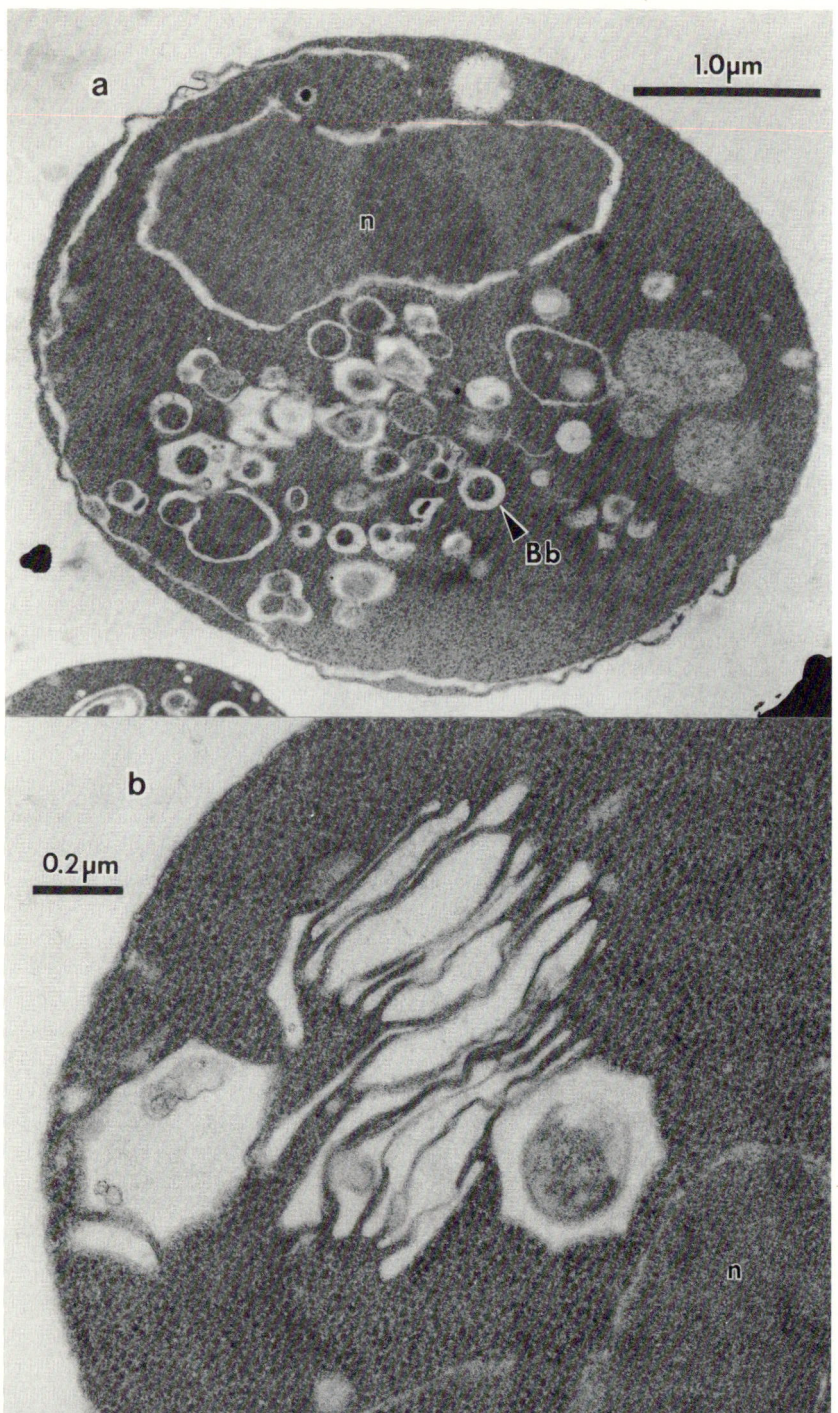

Plate 5 Golgi-like organelles accumulate in *sec7*. The mutant was incubated for 2 hr at 37°C in a medium containing 2% glucose (*a*) or 0.1% glucose (*b*), and thin sections were examined by electron microscopy (Novick et al. 1980; Novick et al. 1981).

Transport in *Saccharomyces cerevisiae*

Terrance G. Cooper
Department of Biological Sciences
University of Pittsburgh
Pittsburgh, Pennsylvania 15260

1. **Introduction**
   A. Types of Transport and Their Distinguishing Characteristics
   B. Studies of Transport in *S. cerevisiae*: A Perspective
2. **Transport of Carbohydrates in *S. cerevisiae***
   A. Constitutive Hexose Transport
   B. Inducible Galactose Transport
   C. Disaccharide Transport
   D. Inducible Maltose Transport
   E. α-MG Transport
   F. Uptake of Trehalose
   G. Uptake of Acyclic Polyols
   H. Acetate, Lactate, Glycerol, and Ethanol Uptake
3. **Transport of Nitrogenous Compounds**
   A. Catabolic Nitrogen Transport Systems
   B. Compounds of the Allantoin-degradative Pathway
   C. Other Catabolic Nitrogenous Compounds
   D. Anabolic Nitrogen Transport Systems
4. **Vitamin Transport**
   A. Thiamine Transport
   B. Biotin Transport
   C. Riboflavin Uptake
   D. Choline and Inositol Transport
5. **Ion Transport**
   A. Sulfate Transport
   B. Phosphate Transport
   C. Divalent Cation Transport
6. **Summary of Transport System Genetics**
7. **Metabolic Compartmentation and Vacuolar Transport Systems**
   A. Intracellular Compartmentation of Metabolites
   B. Vacuolar Transport Systems
   C. Labeling Difficulties Caused by Sequestered Metabolites
   D. Radioactive Labeling of Protein
   E. Radioactive Labeling of Nucleic Acids
8. **Regulation of Transport Activity**
   A. Control Exerted at Gene Expression
   B. Feedback Inhibition
   C. Transinhibition
   D. The Vacuole and Unidirectional Solute Transport
9. **Isolation and Characterization of Transport Components**
   A. Cell Envelope
   B. Plasma-membrane Vesicles and ATPase
   C. Metabolite-binding Proteins

INTRODUCTION

One of the most challenging problems for a cell is to regulate and integrate its metabolism. In the case of eukaryotic microorganisms, this must often be done in the face of changing internal needs and an uncertain external environment. Past studies addressing such control have focused on gene expression. Transport, on the other hand, has been cast as the means by which cells collect needed metabolites and precursors from their environment.[1] To be sure, metabolite accumulation is a central function of transport systems, but such a parochial view of transport in *Saccharomyces cerevisiae* is becoming increasingly unsatisfactory. It ignores the complex internal structure of a yeast cell and the complexity of the membranes forming these structures.

The following example illustrates the need for adopting a more dynamic view of transport processes and their role in metabolic control. For the past 20 years, induction and repression of enzyme synthesis have been thought to be the principal means of shifting from one metabolic mode to another; the molecular events associated with operation of the *lac* operon form the paradigm. The model involves switching on synthesis of an enzyme when it is needed and switching it off again when the need is past. This is exactly what is observed for the allantoin-degradative enzymes in *S. cerevisiae* (Cooper and Lawther 1973; Lawther and Cooper 1973; Whitney et al. 1973). If allantoin or one of its degradation products is added to a culture of yeast, synthesis of the allantoin-degradative enzymes is induced. If a better nitrogen source (e.g., asparagine or serine) is subsequently added to the medium, synthesis of these enzymes halts. At face value, a better example of controlled gene expression could not have been chosen. However, this mode of control works well only if the enzymes in question turn over rapidly. As shown in Figure 1, this is not the case. In terms of catalytic capacity, the cells remain competent to degrade allantoin for two or more generations after asparagine is added to the medium. However, a strikingly different result is observed if one monitors allantoin uptake following addition of asparagine. Uptake drops to undetectable levels in the first few minutes after asparagine addition, identifying this as the preeminent means of halting allantoin metabolism on addition of a better nitrogen source to the medium.

There is an increasing number of situations similar to the one just described, in which metabolism appears to be regulated by intercellular and intracellular metabolite compartmentation. Gene expression and substrate compartmentation appear to be coequal and interactive in their ability to

---

[1] To maintain this work at a reasonable size, the discussion has been confined, with few exceptions, to *Saccharomyces cerevisiae* and covers work largely appearing prior to October 1982. A broad literature concerning the mechanistic details of transport energetics has been dealt with only peripherally due to space limitations and the availability of excellent reviews elsewhere (Borst-Pauwels 1981; Eddy 1982).

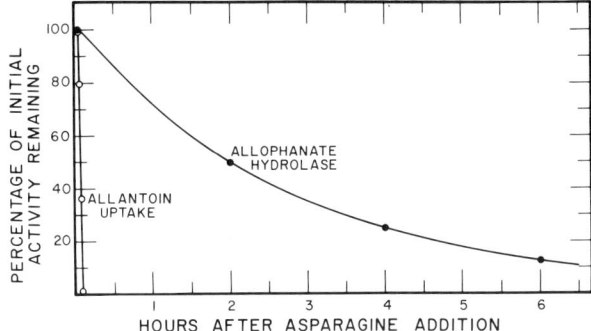

Figure 1 Allophanate hydrolase and allantoin transport activity following addition of asparagine to a derepressed culture of *S. cerevisiae*. Allophanate hydrolase activity is expressed on a per cell basis, since significant cell division occurred during the course of the experiment. (Data from T.G. Cooper and R.A. Sumrada, in prep.)

integrate acquisition, production, and utilization of metabolites. Hence, a rigorous understanding of metabolism and its control must be predicated on a thorough awareness of the pertinent transport systems as well.

Types of Transport and Their Distinguishing Characteristics

Three modes of transport are available for metabolite accumulation: simple diffusion, facilitated diffusion, and active transport. Simple chemical diffusion, which is rarely observed as a means of metabolite uptake in yeast, is governed by the law of mass action. As a result, intracellular concentrations of diffusible metabolites never exceed those in the surrounding medium, and the process is not saturable with respect to substrate concentration. There is also no effect of temperature or metabolic inhibitors.

Facilitated diffusion is a variation of simple diffusion in which a membrane carrier participates in the diffusion process. Here, as before, the metabolite is not concentrated intracellularly, nor is the process sensitive to metabolic inhibitors, i.e., the process is not dependent on an input of energy. Attainment of influx-efflux equilibrium is very rapid, often being attained in seconds. Participation of the membrane-carrier protein has several consequences. Since a membrane protein is involved, the uptake process is both temperature-dependent and saturable with respect to substrate concentration; i.e., substrate levels are reached beyond which the uptake rate is independent of substrate concentration. Binding of the transported compound is a prerequisite of facilitated diffusion and, hence, it may be competitively inhibited by structural analogs of the substrate. The carrier may also be inactivated by appropriate mutations. Facilitated diffusion systems are often, but not always, capable of high capacity uptake.

The last and most frequently observed form of uptake is active transport. Like facilitated diffusion, active transport is carrier-mediated. Therefore, the process is temperature-dependent, saturable with respect to substrate concentration, and sensitive to competitive inhibition by substrate analogs. Inactivation of uptake may also be brought about by appropriate mutations. The distinguishing characteristic of active transport is the requirement for energy input, which makes possible the accumulation of metabolites against steep concentration gradients. Metabolites may reach internal concentrations of 25,000–30,000 times greater than those in the surrounding medium. As a result of energy-coupling, active transport systems are sensitive to inhibitors of energy metabolism. In addition, influx-efflux equilibria are attained more slowly than for facilitated diffusion, often ranging from minutes to hours.

Studies of Transport in S. cerevisiae: A Perspective

Most past studies of transport in *S. cerevisiae* have been directed toward identifying the specificity of the carriers that mediate the uptake process or, put another way, the metabolites that share common carriers. Two basic approaches have been employed: one, biochemical, and the other, genetic. The kinetic parameters of many uptake systems have been thoroughly investigated. Here, specificity has been assessed by determining the effect of one compound on the uptake rate of another. Data are most commonly analyzed using some form of the Michaelis-Menton equation. The appearance of competitive inhibition has been taken as the diagnostic result of two compounds sharing a common carrier. Genetic analysis is a second means of defining carrier specificity. Mutants that are unable to accumulate one metabolite are isolated and tested for their ability to transport other compounds as well. In far too many instances, only one of the above approaches has been used to characterize a transport system, and ambiguous conclusions have been the result. Each method possesses limitations that can be overcome by using a combined biochemical and genetic attack.

The core of the biochemical approach is the appearance of competitive inhibition. However, the ability of a compound to prevent uptake does not in itself provide substantial evidence that the competitive ligand is transported into the cell. The method does not speak in a direct way to the mechanism behind the effect. A more serious limitation of this approach is observed when a metabolite can enter the cell using more than one carrier or transport process. If the apparent K_ms for the two processes or carriers differ by more than tenfold, the presence of two uptake systems will probably be detected kinetically, assuming that the uptake rates for the two systems are reasonably similar. If more than two systems mediate uptake, the situation is probably hopeless.

With mutants, one can detect and resolve the presence of multiple uptake systems with relative ease. One can also define, with reasonable certainty, the compounds transported by each system. However, major problems arise in the mechanistic interpretation of genetic results. Mutants are isolated and characterized on the basis of function. In many cases, mutations resulting in impaired transport are assumed or implied to give rise to defective forms of the carrier protein. However, in only a few cases is this conclusion rigorously supported experimentally. Transport is an enormously complex process that no doubt requires the participation of many gene products. Although mutant strains can certainly be used to delineate the minimum number of gene products required and which compounds require the defective gene products for uptake, precise explanation of the defect requires detailed biochemical analysis.

TRANSPORT OF CARBOHYDRATES IN *S. cerevisiae*

The uptake systems by which *S. cerevisiae* accumulates carbohydrates are summarized in Table 1. Each is briefly described in the subsequent sections.

Constitutive Hexose Transport

There is now broad consensus that glucose enters *S. cerevisiae* by way of a constitutive facilitated diffusion system. Much concern has been directed toward determining whether or not yeast possesses a phosphoenolpyruvate-phosphotransferase system similar to that in *Escherichia coli*; thus far, it apparently does not. Glucose uptake was studied using a series of analogs (Cirillo 1962). Table 2 provides a rough indication of the relative "affinities" of various carbohydrates for the hexose carrier (Kotyk and Janacek 1975), although the precise values and ordering of relative affinities differ somewhat from laboratory to laboratory (Cirillo 1968a; Heredia et al. 1968).

Since glucose and glucosamine were effective inhibitors of D-galactose and D-arabinose uptake, Burger et al. (1959) inferred that all four sugars used a common carrier. L-sorbose and D-xylose were also found to be nonmetabolizable glucose analogs that could inhibit glucose utilization by 80–85% (Kotyk 1967) and were competitive inhibitors of its uptake (Cirillo 1962, 1968a). Again, the existence of a common carrier was inferred.

Intracellular concentrations of both galactose and arabinose were always found to be below those in the medium, eliminating active transport of glucose from consideration. However, both influx and efflux of galactose

Table 1 Carbohydrate uptake systems in S. cerevisiae

Carbohydrate(s) transported	Mode of uptake		Control of uptake	Gene(s)	References[b]
	type	accumulation[a]			
Glucose, fructose mannose	facilitated	—	constitutive	—	1
Galactose	facilitated	0.85	inducible	GAL2	2
Sucrose	active		—	—	3
Maltose	active	15	inducible	—	4
α-methylglucoside	facilitated		constitutive	MGL2, MGL4	5
Trehalose	unclear		unknown		6
Acyclic polyols	unclear		unknown		7
Acetate, lactate glycerol, ethanol	unknown		unknown		

[a] Ratio of internal to external concentration of the metabolite at equilibrium.
[b] (1) Burger et al. (1959); Cirillo (1962; 1968a); Kotyk and Janacek (1975); Meredith and Romano (1977). (2) Cirillo (1968b); Kou and Cirillo (1970); Kou et al. (1970). (3) Santos et al. (1982); (4) Harris and Thompson (1961); Gorts (1969); Okada and Halvorson (1964); Serrano (1977). (5) Okada and Halvorson (1964a,b). (6) Okada and Halvorson (1964a); Kotyk and Michaljanicova (1979). (7) Maxwell and Spoerl (1971); Canh et al. (1975).

Table 2 Half-saturation constants of the hexose uptake system

Sugar	K_T (mM)
D-Glucose	5
D-Fructose	17
D-Mannose	27
2-Deoxy-D-glucose	4
L-Sorbose	21
D-Galactose	35
1,5-Anhydroglucitol	50
L-Glucose	100
D-Arabinose	115
D-Xylose	130
D-Allose	225
3-*O*-Methyl-D-glucose	250
L-Xylose	400
D-Ribose	600

Data from Kotyk and Janecek (1975).

possessed a Q_{10} greater than 2.4, which led to the conclusion that uptake was mediated by facilitated diffusion. Serrano and De la Fuente (1974) reported that xylose and arabinose uptake were linear for 30 seconds, whereas glucose, fructose, and mannose accumulated linearly for less than 5 seconds. They interpreted their kinetic data as suggesting the existence of two states of the carrier molecules, but alternative explanations are equally plausible.

Meredith and Romano (1977) followed uptake and phosphorylation of 2-deoxyglucose (K_m = 1 mM), yet another competitive inhibitor of glucose uptake (Cirillo 1968a). These investigators found that deoxyglucose could be concentrated as deoxyglucose-phosphate 30–42-fold. The phosphorylated sugar was always seen before the unphosphorylated form, but there was no concentration of free deoxyglucose. In addition, only unphosphorylated sugar was observed to exit from the cell. More recently, Franzusoft and Cirillo (1982) reported a set of pulse-labeling experiments carried out in mutant strains, each possessing only one of the enzymes presumed to phosphorylate hexoses (hexokinases PI and PII and a glucokinase). Again, 2-deoxyglucose first appeared in the sugar phosphate pool in a manner similar to that observed with wild-type cells. The investigators suggested that, "if transport associated phosphorylation does occur in baker's yeast," it was not a function specifically associated with any one of the kinases studied. In sum, the question remains open as to whether or not glucose is phosphorylated as it is transported.

From the data cited above, it cannot be claimed with absolute certainty that a single constitutive facilitated diffusion system mediates hexose uptake in *S. cerevisiae*. The question could be unambiguously settled, however, if addressed genetically.

Inducible Galactose Transport

Kotyk (1967) hypothesized the presence of two hexose carriers because he found that the carbohydrates whose uptake he assayed could be divided into two groups on the basis of their uptake rates and competition with glucose for uptake. For one group, initial uptake was rapid, but it declined with time. Members of the second group, including galactose, were taken up slowly from time zero. A lack of reciprocal competitive inhibition was also noted between glucose and the latter group of sugars, but the effect was not seen in all cases. Of greater significance, though not seemingly recognized at the time, was the observation that growth in the presence of galactose altered the uptake rate of the "slowly" transported (group-two) sugars.

Largely through the efforts of Cirillo and his colleagues (Cirillo 1968b; Kou and Cirillo 1970; Kou et al. 1970), galactose has been shown to enter the cell by way of an inducible facilitated diffusion system (ratio of intracellular to extracellular galactose, fucose, or arabinose < 0.85). The native inducer is apparently D-galactose (Cirillo 1968b; Azum et al. 1971). In uninduced cultures or induced *gal2* mutants (isolated by Douglas and Condie 1954), galactose enters the cell only as a poor analog of glucose. The sugar may be accumulated against a concentration gradient in galactose-1-phosphotransferase (*gal7⁻*) mutants. However, this result was shown to arise from entrapment of the phosphorylated sugar, since (1) all of the accumulated sugar was phosphorylated, and (2) free sugar (either galactose or its nonmetabolizable analogs) could not be accumulated in galactokinase (*gal1⁻*) strains. Uptake of L-arabinose and D-fucose is negligible in uninduced cells. The biochemical defect possessed by *gal2* transport mutants has not yet been identified, although these mutants have been hypothesized to possess a defective galactose carrier protein (Cirillo 1968b). Metabolism of galactose in *gal2* strains presumably occurs because galactose enters via the glucose-hexose system. Induction of the galactose uptake system has been shown to be affected by mutations in the *gal3*, *gal4*, and *gal80* loci in a manner similar to that of the galactose-degradative enzymes.

Matern and Holzer (1977) reported that on adding glucose to galactose-induced cultures, the K_m for galactose increased from 3.6 mM to 11 mM and that recovery of low K_m transport was inhibited by cycloheximide following removal of glucose. This was interpreted as indicating destruction or alteration of the carrier in response to glucose addition and the involvement of de novo synthesis in recovery.

Disaccharide Transport

It has been commonly believed that not all disaccharides fermented by *S. cerevisiae* are taken into the cell. Sucrose, e.g., was reported to be converted to glucose and fructose by invertase that is sequestered in the periplasmic space (Arnold 1972). The hexoses produced were then thought to be taken into the cell by the systems described above (De la Fuente and Sols 1962). This view has been recently challenged by Santos et al. (1982). Their multiphasic Eadie-Hoffstee plot suggests that multiple uptake systems are involved in sucrose uptake that is not inhibited by ATP metabolism inhibitors, such as iodoacetamide or cyanide. Uptake was found highly sensitive, however, to uncoupling agents, such as sodium azide and 2,4-dinitrophenol (DNP). Alkalinization of the medium during sucrose uptake raised the possibility that transport was mediated by a H^+/sucrose symport.

Inducible Maltose Transport

Maltose uptake appears to be mediated by an inducible active transport system (Harris and Thompson 1961; Okada and Halvorson 1964a; Serrano 1977). This conclusion is based on the observations that (1) maltose was concentrated up to 15-fold over extracellular levels, and (2) uptake was inhibited by the antimetabolites NaF, DNP, carbonylcyanide-m-chlorophenyl hydrazone (CCCP), and azide; somewhat surprisingly, iodoacetate did not appear to inhibit uptake. Increases in medium pH as a result of maltose uptake have been taken as an indication that protons (1 mole/mole of maltose) might be cotransported with maltose. Previously accumulated maltose has been shown to leave the cell only slowly with efflux and exchange exhibiting half-lives of greater than 100 minutes and 45 minutes, respectively. Although the maltose uptake system appears to be quite specific, α-methylglucoside (αMG) is a reasonably effective inhibitor at high concentrations. However, it is doubtful that this competitive ligand is transported into the cell by the maltose permease. Evidence for this conclusion will be given later.

Zimmermanna et al. (1973) isolated mutants unable to accumulate maltose (designated *dsf6, dsf7, dsf17,* and *dsf21* (*di*saccharide *f*ermentation). The mutants ferment sucrose, glucose, and galactose normally. None of the *DSF* loci were linked to known *MAL* genes. The isolation of four groups of mutants lacking uptake activity suggests that maltose uptake or its control may be quite complex.

The maltose uptake system has been shown to be induced up to 150-fold by pregrowth in the presence of maltose (Harris and Thompson 1961; Okada and Halvorson 1964a). The native inducer has not been rigorously identified nor have the events associated with induction. In contrast to the amino acid transport systems to be discussed later, maltose uptake is unaffected by the addition of cycloheximide to growing cultures even after 60 minutes of treatment. However, addition of 2% glucose was found to inhibit maltose uptake by 85-100% within 60 minutes. If cycloheximide was added along with glucose, inactivation was slowed by twofold (Gorts 1969). Upon removing glucose from the medium, maltose uptake is recovered. The prevention of recovery by cycloheximide raises the possibility that de novo synthesis may be involved in the recovery process.

αMG Transport

αMG transport has been characterized using the nonmetabolizable analog, α-thioethyl D-glucopyranoside (αTEG) as the major probe (Okada and Halvorson 1964a,b). In the absence of induction, uptake is mediated by a constitutively produced facilitated diffusion system with an apparent K_m of 50 mM. Uptake by this route is insensitive to inhibitors of energy metabolism, and no intracellular concentration of αTEG was observed. If, on the other hand, cells were pregrown in the presence of αMG or αTEG, a second active transport system appears. This system concentrates αMG and αTEG 10-150-fold over the external levels and is inhibited by azide, arsenite, and arsenate. The ability of a αTEG to serve as inducer is consistent with the suggestion that αMG is the native inducer. Induction can be largely prevented if 0.1 M acetate is also present. In addition, αTEG uptake capability is lost with a half-life of 60 minutes following suspension of previously induced cells in acetate medium. The molecular basis of both inhibitory effects remains obscure.

The *MGL2*-gene product is required for operation of both the facilitated diffusion and active transport systems. αMG is not transported by the inducible maltose uptake system, because *mgl2* mutants no longer accumulate αTEG or αMG and yet retain the ability to transport and ferment maltose. Since maltose is an effective inhibitor of αTEG and αMG uptake, it is reasonable to inquire whether maltose uptake can be mediated by the αMG transport systems as well as the maltose-specific system. The *dsf* mutants will be most useful in answering this question.

Uptake of Trehalose

Trehalose has been thought to be a storage sugar for yeast rather than a carbon source to be accumulated from the medium. However, Kotyk and Michaljanicova (1979) reported uptake of the disaccharide, and Okada and Halvorson (1964a) observed that trehalose inhibited uptake of αTEG.

Uptake of the sugar was found to be insensitive to addition of DNP, iodoacetamide, and cycloheximide, but was sensitive to uranyl ions. Perhaps the most interesting aspect of trehalose uptake is the fact that almost no efflux of preaccumulated disaccharide is observed on resuspension of preloaded cells in trehalose-free medium. This raises the possibility that accumulated sugar may be sequestered.

An inordinately large number of transport systems have been hypothesized to mediate trehalose uptake. The system mediating trehalose uptake has not been identified, for no genetic studies have been done.

Uptake of Acyclic Polyols

Acyclic polyols include such compounds as sorbitol, erythritol, mannitol, xylitol, ribitol, arabinitol, and galactinol. None of these carbohydrates is metabolized by *S. cerevisiae* (Canh et al. 1975). In a brief report (Maxwell and Spoerl 1971) mannitol was observed to accumulate threefold to fourfold. This is marginal at best, though the accumulation was shown to be sensitive to inhibition by DNP and iodoacetic acid. Of relevance is the observation that mannitol and sorbitol can be used as osmotic stabilizers for spheroplasts, which implies that cells do not accumulate these polyols *above* the extracellular concentrations.

Acetate, Lactate, Glycerol, and Ethanol Uptake

Though acetate, lactate, glycerol, and ethanol can serve as carbon sources for *S. cerevisiae,* their modes of entry have apparently been overlooked completely. This is indeed unfortunate, considering the central position they play in such processes as relief of glucose repression, sporulation, and inhibition of induction of the αMG transport system.

TRANSPORT OF NITROGENOUS COMPOUNDS

Two distinct classes of nitrogenous compounds are transported into a yeast cell. One group serves as nitrogen sources and its members are catabolized on entry into the cell. Transport associated with these metabolites possesses characteristics and patterns of control similar to those of the degradative enzyme systems they supply. The second group of compounds are biosynthetic precursors and intermediates that are incorporated more or less directly into macromolecules without first being catabolized. Here, the patterns of regulation are similar to those of the biosynthetic enzyme systems. In both cases, transport systems may be viewed as first steps in the catabolic or anabolic pathways they serve.

A thorough understanding of any enzyme system associated with nitrogen metabolism is grossly incomplete unless the routes by which pertinent metabolites enter the cell and become compartmentalized thereafter is known. In addition, pragmatically, mutants of transport functions at times exhibit properties characteristic of "control" mutants. Several examples of such mutants will be described in subsequent sections.

Catabolic Nitrogen Transport Systems

The catabolic uptake systems by which *S. cerevisiae* accumulates potential nitrogen sources are summarized in Table 3. Each one is also briefly described in the subsequent sections.

Compounds of the Allantoin-degradative Pathway

Allantoin Transport

Allantoin is accumulated against a concentration gradient by way of an active transport system exhibiting an apparent K_m of about 12 μM (Sumrada and Cooper 1977). This conclusion is based on the observation that (1) allantoin attained 7700-fold greater concentrations in cells than in the surrounding medium, (2) the rate of uptake plateaued with increasing time and allantoin concentrations if cell division was prevented, and (3) uptake did not occur in the absence of glucose or in the presence of arsenate, fluoride, DNP, or CCCP. The rate of accumulation was maximal at pH 5.2. Preloaded allantoin is not lost from cells that are resuspended in allantoin-free buffer and does not exchange with nonradioactive allantoin provided exogenously. However, treatment of the cells with nystatin released the preloaded allantoin, which could then be isolated and shown to be unaltered chemically. Subsequent studies demonstrated that this initial failure to observe efflux and exchange of preloaded allantoin derived from the extremely slow rates of these processes (Sumrada et al. 1978).

The allantoin permeation system appears to be induced by allantoin itself, unlike the enzymes of the allantoin-degradative pathway, which are induced by allophanate (Sumrada et al. 1978). Hydantoin and hydantoin acetate may be nonmetabolizable inducers, for they induce allantoin transport to 63% and 95% of the levels observed with allantoin, respectively (Sumrada et al. 1978). Some caution is necessary, however, because recent preliminary data indicate that the induction process may be more complex than originally conceived. *dal4* mutants are unable to accumulate allantoin (Cooper et al. 1979a).

Table 3 Catabolic uptake systems for nitrogenous compounds

Compound	Apparent K_m	Concentration (inside/ outside)	Mode of uptake	Control[a]	pH optimum	Efflux	Exchange	Inhibitors	Gene	Reference[b]
Allantoin	12 μM	7,700	active	ind, N-rep	5.2	very slow	no		DAL4	1
Allantoate	50 μM	15,000	active	const, N-rep	5.8	very slow	no		DAL5	2
Urea	14 μM	200	active	ind, N-rep	3.3	yes		formamide, acetamide	DUR3	3
Urea	2.5 mM		facilitated	const					DUR4	3
Oxalurate	1.2 mM	2–4	active	const	3.0	rapid	slow			4
Methylamine (ammonia)	0.25 mM	850	active	?	6.0–6.5				MEP2	5
Methylamine (ammonia)	2 mM		active	?	6.0–6.5				MEP1	5
Amino acids	variable		active	const, N-rep					GAP1	6
Proline	0.025 mM 13 mM			const, N-rep	4.0				PUT4	7
Asparagine									*[c]	8

[a] ind, inducible; N-rep, sensitive to nitrogen repression; const, constitutive.
[b] (1) Sumrada and Cooper (1977); Sumrada et al. (1978); Cooper et al. (1979a). (2) Morlion and Domnas (1962); Turoscy and Cooper (1979). (3) Cooper and Sumrada (1975); Sumrada et al. (1976). (4) Cooper et al. (1979b). (5) Roon et al. (1975 a,b; 1977a,b); Dubois and Grenson (1979). (6) Grenson et al. (1966, 1970); Darte and Grenson (1975); Rytka (1975); Roon et al. (1977b); Larrimore and Roon (1978); Larrimore et al. (1980). (7) Lasko and Brandriss (1981). (8) Dunlop and Roon (1975).
[c] *, mutant exists but locus is unnamed.

The allantoin transport system is subject to nitrogen repression and transinhibition when cells are provided with good nitrogen sources. Transinhibition of transport activity is also seen when cycloheximide is added to the culture. Amino acids, which accumulate on cessation of protein synthesis, are thought to bring about this inhibition because such inhibition is not observed in cultures that are starved for a nitrogen source prior to the addition of cycloheximide.

Allantoate Transport

Allantoate uptake is mediated by an energy-dependent active transport system with an apparent K_m value of 50 μM (Morlion and Domnas 1962; Turoscy and Cooper 1979). Cells were able to accumulate allantoate to greater than 15,000 times the extracellular concentration. The energy system driving uptake is probably quite different from that driving allantoin, urea, and oxalurate uptake. These latter systems were inhibited by arsenate, fluoride, DNP, and CCCP, whereas allantoate accumulation was sensitive to only DNP and CCCP. Efflux and exchange of preloaded allantoate did not occur at detectable levels and is similar in this way to the allantoin system (Turoscy and Cooper 1979).

It is probable that production of the allantoate-uptake system is constitutive, because no way has been found of increasing the activity, regardless of what compounds are added to the medium. More concretely, the system is also repressible and transinhibitable by up to 100-fold when cells are provided with readily used nitrogen sources such as glutamine, asparagine, or serine. One also observes inhibition of transport (transinhibition) on treating an unstarved culture with cycloheximide.

dal5 mutants are unable to accumulate allantoate. Uptake of allantoin was decreased by 60% in *dal5* strains. In a reciprocal experiment we found that allantoate uptake drops to 50% of the wild-type level in *dal4* mutants. Since the loss of either transport system by mutation results in total inability to accumulate one metabolite and a marked decrease in uptake of the other, perhaps the two transport systems share a common component or, otherwise, directly interact with one another (V. Turoscy and T.G. Cooper, in prep.).

Urea Transport

Urea transport occurs by two routes. The first mode is a low K_m (14 μM), energy dependent, active transport system (Cooper and Sumrada 1975). This conclusion was based on the observations that (1) urea attained 200-fold greater concentrations in cells than in the surrounding medium, (2) uptake did not occur in the absence of glucose or in the presence of arsenate, fluoride, DNP, or CCCP, and (3) uptake was absent in mutant strains. This transport system is induced by allophanate or oxalurate, native and gratuitous inducers, respectively, of the allantoin-degradative enzymes. Its production is also subject to nitrogen repression when cultures are grown with good nitrogen sources. Previously accumulated urea was lost from cells treated with metabolic inhibitors such as DNP (Cooper and Sumrada 1975; Sumrada et al. 1976).

dur3 mutants are defective in the active transport system for urea and are unable to accumulate urea at external concentrations of 0.25 mM or less (Sumrada et al. 1976). The locus containing these mutations appears to be linked to the centromere of chromosome VIII (V. Turoscy et al., in prep.).

At concentrations of urea in excess of 0.5 mM, urea is transported by a constitutively produced, facilitated diffusion system, which is insensitive to nitrogen repression (Cooper and Sumrada 1975). The apparent K_m of this system was approximated to be 2.5 mM. Mutant strains that were defective in the low K_m urea active transport system also exhibited a higher apparent K_m value for facilitated diffusion of urea and a decreased growth rate on 10 mM urea, indicating the possible existence of a functional relationship between the two modes of urea uptake (Sumrada et al. 1976). This behavior is similar in formal terms to the characteristics of *mgl2* mutants that have lost the capacity for both facilitated diffusion and active transport of αMG.

dur4 mutants are unable to carry out facilitated diffusion of urea. The *DUR4* locus appears to be tightly linked to the *DUR3* locus, which is responsible for some component of the active transport system (R. Sumrada and T.G. Cooper, in prep.).

Oxalurate Transport

Oxalurate, a gratuituous inducer of the allantoin-degradative enzymes, is concentrated twofold to fourfold in the cell by an energy-dependent, active transport system with an apparent K_m of 1.2 mM. Uptake is maximum at pH 3.0 and is inhibited by arsenate, fluoride, DNP, and CCCP. Efflux of preloaded oxalurate is rapid with a half-life of about 2 minutes. Exchange, on the other hand, is much slower with an apparent half-life of 25 minutes (Cooper et al. 1979b).

Since oxalurate is not a normal metabolite of *S.cerevisiae*, it is presumably transported by an uptake system associated with some other cellular constituent. Information presently available does not permit identification of this system or even certify that only one system performs this function. The system(s) associated with oxalurate transport appears to be constitutively produced and is insensitive to nitrogen repression. These characteristics raise the possibility that oxalurate uptake may be mediated by a biosynthetic transport system, because most, if not all, of the catabolic uptake systems are inducible and/or repressible.

Other Catabolic Nitrogenous Compounds

Ammonia Transport

Ammonia transport has been studied by Roon and his colleagues using the radioactive ammonia analog, methylamine (Roon et al. 1977b). Methylamine uptake is mediated by an active transport system with a pH optimum of 6.0–6.5, suggesting that it is the quaternary ammonium ion that is accumulated. This system, which exhibits an apparent K_m of 0.2 mM for methylamine, concentrates the ammonia analog 850–1000-fold over external levels. It requires glucose to be functional and is inhibited by CCCP, DNP, and Dio-9. Ammonium ion is a strong competitive inhibitor of uptake of methylamine, and metabolites such as arginine, urea, or asparagine, which are degraded to ammonia, are inhibitory. Uptake is also markedly decreased as a function of medium ionic strength. Methylamine accumulation was found to be decreased 80% by 0.2 M KCl compared to only 5–10% inhibition for alanine or aspartate uptake. In contrast to the allantoin and allantoate transport systems just described and the amino acid uptake systems to be described later, the rate of methylamine uptake was not observed to decrease following addition of cycloheximide to the medium. However, this observation must be viewed somewhat skeptically for the present, because cells used in the experiment were starved prior to addition of cycloheximide and, hence, were depleted of their internal amino acid and other nitrogen pools, the likely inhibitory agents for the former transport systems.

Roon observed that cell growth with glutamate as nitrogen source could be inhibited by high concentrations of methylamine and used this effect to isolate resistant mutants. These strains, initially designated *mea^r* and, later, *amt*, transported both methylamine and ammonia 20-fold more slowly than wild-type and were unable to grow with ammonia as sole nitrogen source. Revertants selected for ability to use ammonia as nitrogen source simultaneously recovered the ability to accumulate both ammonia and methylamine. The mutations in these strains were shown to be nonallelic with the *GAP*, *AAP*, or *APF* loci (see below).

More recently, it has been suggested that there are two, and possibly three, ammonia transport systems. This hypothesis was derived from the observation of a biphasic Lineweaver-Burke plot of methylamine uptake rates at increasing concentrations of methylamine and the isolation of additional methylamine-resistant mutants (Dubois and Grenson 1979). In this case, proline was used as the nitrogen source in place of glutamate. The methylamine-resistant strains isolated were found to contain two mutations, and the unlinked loci that contained them were designated *mep1* and *mep2*. Each mutation was found to eliminate one segment of the biphasic curve cited earlier. It was thus concluded that *MEP1* was needed for the high-capacity (V_{max} = 50 nmoles/min/mg of protein), high K_m (2 mM) system and that *mep2* was required for a lower-capacity (V_{max} = 20 nmoles/min/mg of protein), low K_m (0.25 mM) system. Methylamine resistance was reported to segregate with the *mep1* mutation. The investigators tentatively concluded that the system studied by Roon seemed to correspond to their *mep1* system, because it was linked with methylamine resistance. However, the K_m observed by Roon (0.25 mM) is much closer to the *MEP2* system than it is to

MEP1, which exhibits a K_m of 2 mM. A complementation test between the *amt* and *mep* mutations would clearly settle the question. If the kinetic data obtained by the two laboratories are carefully compared, another, perhaps provocative, interpretation can be suggested. The range of methylamine concentrations used by Roon was probably sufficiently broad to observe biphasic kinetic behavior if it was present (i.e., the break point in Dubois and Grenson's plot occurred at a concentration of ~0.1 mM, and the concentrations used by Roon ranged from 0.05 mM to >1 mM). It is possible that the strains Roon used did not possess the *MEP1* system. Since methylamine was inhibiting cell growth, probably through inhibition of glutamic dehydrogenase, any mutation that prevented its entry into the cell would confer resistance. Therefore, if only one system was present, mutation of the cognate genetic locus would yield methylamine resistance. Roon's mutants were derived from strain S288C, whereas those of Dubois and Grenson came from strain Σ1278b. Strain Σ1278b is far more sensitive to nitrogen repression when ammonia is used as nitrogen source than is strain S288C. It would be most interesting to test whether existence of a second high-capacity ammonia uptake system (*MEP1*) in Σ1278b compared with only one lower-capacity system in S288C was the reason for the increased sensitivity to ammonia repression observed for Σ1278b. It is important to note here that most, if not all, of the experiments studying nitrogen repression have been done with an ammonia concentration of 0.1% or 8 mM, which greatly exceeds the K_m value of even the high K_m system reported for strain Σ1278b.

Very little is known about the control of ammonia uptake. Roon et al. (1977a) reported that addition of 2 mM aspartate or alanine decreased methylamine uptake by 65% in a noncompetitive manner. Inhibition was not observed when the experiment was conducted in *gap* or *aap* mutants, which have lost the high-capacity, general amino acid uptake system (i.e., transport of alanine and aspartate was decreased ~50% in these strains). Roon noted a correlation between depletion of the alanine or aspartate from the medium and relief of their inhibitory effects on methylamine uptake. This led him to conclude that it was the extracellular concentration of amino acid that was most important. He suggested that the observed phenomena were consistent with those expected if the ammonia and amino acid uptake systems were competing for a limiting energy source or structural component needed for operation of both classes of uptake systems. However, it is equally reasonable to argue that the intracellular amino acid levels are the basis for inhibition. Although depletion of amino acids from the medium leads to relief of inhibition, it also leads to a decrease in the rapidly turning over intracellular amino acid pools. This is a particularly important consideration, because the cells used in Roon's experiments were severely nitrogen starved prior to beginning the measurements. Since cycloheximide was not present to prevent protein synthesis and both alanine and aspartate are reasonably good nitrogen sources, it is possible that internal metabolism and utilization of accumulated alanine and aspartate occurred at high rates. In this case, the internal pool would be a function of the external amino acid concentrations and their rates of entry.

Catabolic Amino Acid Transport

Amino acids serve as nitrogen sources or, alternatively, are amino-acylated and incorporated directly into protein. In view of this dual role, it is not surprising that *S. cerevisiae* possesses both catabolic and anabolic amino acid transport systems. The idea of a transport system that would accumulate all amino acids originated in the early work of Halvorson and Cohen (1958). Surdin et al. (1965) were first to isolate mutants pleiotropically defective in the uptake of many amino acids and to make a detailed study demonstrating the broad specificity of a general amino acid accumulation system. Later, another class of mutants was isolated by Grenson et al. (1970), and these strains are now generally accepted to be the ones defective in catabolic amino acid uptake. Although the functional significance of the mutants isolated by Surdin et al. was disputed and the strains largely ignored since then, their pleiotropic defectiveness in all forms of amino acid uptake necessitates their careful scrutiny and reevaluation. A meaningful understanding of catabolic and anabolic amino acid transport is probably not possible without this information.

General Amino Acid Transport

Visualization and characterization of the general amino acid transport (*GAP*) system has been largely based on studies of the *gap* mutants. The isolation of *gap* mutants was made possible by the availability of canavanine-resistant mutants (*can1* = *argp*), which lack the arginine-specific transport system (Grenson et al. 1966). When the *can1* mutants were grown in the absence of ammonia on glucose-proline medium, arginine uptake and, hence, canavanine sensitivity were reacquired (Grenson et al. 1970). The double mutant *can1 gap* was then isolated as resistant to canavanine on glucose-proline-canavanine medium. Mutations in the gap locus can be directly selected on glucose-proline medium as strains resistant to D-histidine using the procedures of Rytka (1975).

The rates at which a number of amino acids are transported into *gap* cells growing in glucose-proline medium are decreased severalfold to manyfold compared with wild type (Grenson et al. 1970; Roon et al. 1977b). A similar decrease in uptake rate in the wild type is noted when proline is replaced by ammonia as nitrogen source, leading to the conclusion that the general amino acid transport system is subject to nitrogen repression. The high capacity of the system and its sensitivity to nitrogen repression prompt categorizing it as a catabolic system. The control of this system is discussed in Nitrogen Metabolism in *Saccharomyces cerevisiae* (Cooper, this volume).

Various types of experiments have been used to delineate specifically the

GAP system. *GAP* apparently transports all amino acids, including D-amino acids and citrulline. Comparison of transport rates in wild type and *gap* cells indicate that glutamine, methionine, glycine (Grenson et al. 1970), tryptophan, alanine, aspartate (Randolph et al. 1973; Roon et al. 1977b), glutamate (Darte and Grenson 1975), and citrulline (Rytka 1975) and N-δ-chloroacetyl-L-ornithine (Larimore and Roon 1978; Larimore et al. 1980) are transported by the *GAP* system. (That citrulline is transported by *GAP* was inferred by Grenson et al. [1970], for *gap* mutants cannot use citrulline as a nitrogen source.) If repression of transport by growth on ammonia is taken as diagnostic of transport by *GAP*, then histidine, serine, and valine are transported by *GAP*. Sensitivity of *gap1 can1* and *gap1 lyp1* mutants to their respective analogs on glucose-proline medium implicates *GAP* in transport of arginine and lysine. Transport of D-amino acids seems to proceed through *GAP* (Rytka 1975). Resistance of *gap* mutants to D-stereoisomers of methionine, tryptophan, serine, arginine, and histidine reinforces previous inferences, and resistance to D-phenylalanine, D-leucine, and D-tyrosine adds the respective L-amino acids to the list of those transported by *GAP*. The fact that L-asparagine and L-threonine can relieve canavanine inhibition of a *can1* mutant growing on glucose-proline medium implies that these amino acids are also transported by *GAP*, although the relief might occur by transinhibition (see below). Transport of cysteine and isoleucine by this system have not been examined.

There has been general agreement in the past that proline is not transported by *GAP*. However, since all *gap* mutants that have been examined in any detail necessarily retained the ability to utilize proline as a nitrogen source, this conclusion must be viewed cautiously. The data of Lasko and Brandriss (1981) raise the clear possibility that proline is transported by the *GAP* system.

Control of the *GAP* system appears to be complex, with little general agreement with respect to the molecular events involved. Its production appears to be constitutive, a conclusion based exclusively on the absence of identifying an inducer. The predominant modulation of the transport system involves ammonia inhibition and catabolic nitrogen repression. In the initial report describing the *GAP* system, Grenson et al. (1970) concluded from kinetic data that ammonium ions inhibited the activity of the general amino acid permease but not its synthesis. Furthermore, it was found that ammonia inhibition of both arginine and tryptophan uptake was noncompetitive. Subsequently, Grenson and Hou (1972) found that ammonia inhibition of the uptake system was relieved in *gdha* mutants, which lack nicotinamide adenine dinucleotide phosphate (NADP)-dependent glutamate dehydrogenase (GDH). The possibility that relief of the ammonia effect of NADP-GDH minus cell resulted from their inability to produce glutamate was dismissed because addition of proline (a poor, nonrepressive nitrogen source) to the medium increased glutamate levels to those of wild type.

Roon et al. (1975a) reported that ammonia inhibition was restored if the experiment was repeated in a *gdha gdhCR* double mutant, which lacks anabolic NADP-dependent GDH activity due to mutation of the *gdha* locus but contains high levels of catabolic, NAD-dependent GDH as a result of the *gdhCR* mutation. The nearly normal growth rate of the double mutant in glucose-ammonia medium indicated that the increased levels of catabolic NAD-dependent GDH were able to replace the enzymatic function of the defective anabolic NADP-dependent enzyme. The simultaneous reacquisition of ability to use ammonia as a nitrogen source and for ammonia to inhibit gap-mediated amino acid transport in the double mutant led Roon to conclude that inhibition by ammonia resulted from its uptake into the cell and subsequent incorporation into the α-amino groups of glutamate and other amino acids. NADP-dependent GDH effected repression of *gap* by virtue of its catalytic capacity for incorporating ammonium ions into glutamate and not as a regulatory element.

Against the backdrop of uncertainty that surrounds our understanding of how the *GAP* system is regulated, it is useful to keep in mind that there do not yet exist clear data that would permit one to decide whether ammonia inhibition is exerted at the level of (1) transport activity, (2) synthesis of one or more components needed for operation of the transport system, or (3) both of the above. The isolation of recombinant plasmids containing the DNA sequences of the *gap* locus (T.G. Cooper, unpubl.) may provide one of the important tools in answering this question. This of course assumes that the *gap* locus codes for an element that functions only in the *GAP* system. No mutants have been reported in the *gap* locus that result in the alteration of the apparent K_m of the system for its various substrates.

Pleiotropic Mutations Altering Amino Acid and Ammonia Uptake

The isolation and characterization of two classes of mutant strains provide just a hint of the complexity that may well underlie operation of the ammonia and general amino acid uptake systems. Sorsoli et al. (1964) isolated a mutant (A3E90) whose growth was not inhibited by high levels of the methionine analog, ethionine, or the phenylalanine analog, *p*-fluorophenylalanine. Methionine uptake in this strain was decreased 20-fold as was that of leucine and serine. The pleiotropic nature of the recessive mutation was highlighted by the fact that uptake of a radioactive algal hydrolysate was depressed 40-fold. A similar mutation designated *aap* (*a*mino *a*cid *p*ermease) was described by Surdin et al. (1965). *aap* strains were shown to be resistant to the inhibitory effects of ethionine, canavanine, thienyl alanine, and *p*-fluorophenylalanine. Yet a third mutant with this phenotype was isolated by Grenson and Hennaut (1971). This mutant, designated *apf*, was resistant to growth inhibition by thiosine, norleucine, azetidine-2-carboxylate, cycloleucine, and cycloserine, in addition to the analogs mentioned above. This mutant, in contrast with the wild-type, had acquired the ability to use glycine as sole nitrogen source but had lost the ability to use proline. Growth was also slowed when glutamate, aspartate, valine, tryptophan, phenylalanine, or serine was provided as a nitrogen source. Growth of the *aap* or *apf* on glucose-citrulline medium distinguished it from strains carrying the *gap* mutation (Grenson et al. 1970).

Grenson and Hennaut (1971) reported that complementation was not observed when strains carrying mutations in the *aap* and *apf* loci were mated to one another. However, this result must be viewed with some caution, because Roon et al. (1977b) were unable to obtain mating between strain RA 68 (the *aap* strain of Grenson and Hennaut) with either **a** or α tester strains. Since the growth characteristics of strain RA 68 and the *aap/apf* diploid are identical, one cannot be sure, from the data provided, that mating occurred. The *aap* and *apf* mutations appear to result in loss of both general and specific amino acid transport systems. This suggests that the catabolic and anabolic transport systems share at least one common element, though they appear to be controlled quite differently. The possibility of a general membrane-associated lesion in these strains is unlikely, because active transport of uracil, uridine, cytosine, and adenine were found to be normal (Grenson and Hennaut 1971).

The second class of mutants was isolated by Roon et al. (1977b). During the isolation of *amt* mutants using resistance to methylamine as the selective agent, it was noticed that two different classes of strains appeared. In 20% of the mutants, ammonia uptake was lost, but amino acid uptake appeared normal. In the remaining 80%, designated *amt**, tryptophan, alanine, and proline transport systems were depressed by 70–97% in addition to loss of ammonia transport. The *amt** mutant grew poorly when glutamine was provided as sole nitrogen source, whereas *amt*⁻ strains grew normally. Surprisingly, the *amt* and *amt** mutations failed to complement each other. In heterozygous diploids the *amt* allele was shown to be dominant over *amt**. On reversion of *amt** to the plus phenotype for ammonia uptake and assimilation, the ability to transport proline and tryptophan was recovered. It is intriguing that the *amt** mutation inactivated the general amino acid permease and the unrelated proline system but apparently left the specific aspartate uptake system intact.

Proline Transport

Proline uptake is mediated by at least two transport systems in *S. cerevisiae*, as indicated by a biphasic Lineweaver-Burke plot (apparent K_ms of 0.025 mM and 13 mM) generated at varying external proline concentrations. In a genetic dissection of the problem, Lasko and Brandriss (1981) isolated two classes of mutants resistant to the proline analogs L-azetidine-2-carboxylate and 3,4-dehydro-DL-proline: One class possessed the characteristics of the *apf* or *aap* mutations described elsewhere, and the other class (*put4*) was unable to use proline as a nitrogen source when histidine was present in the medium. Proline uptake at an external concentration of 0.2 mM was reduced 15-fold in the second class of mutants. The low K_m component of the biphasic Lineweaver-Burke plot also disappeared in these mutants, leading to the conclusion that they had lost a low K_m proline transport system. *put4* mutants grew normally at high proline concentrations but not when the concentration was low again, pointing to the *PUT4*-gene product mediating low K_m proline active transport.

Inhibition of the high K_m system by histidine raised the possibility of a role for the general amino acid permease, which is known to transport a variety of amino acids, including histidine. The high K_m transport of proline "was not zero" in a *gap1* mutant. The investigators raised two possible explanations for this observation. It is possible that two high K_m transport systems participate in proline uptake: one is the *GAP1* product, and the other remains unidentified. Their second explanation seems more reasonable. The *gap1* mutant they used was the same one Grenson et al. (1970) selected using proline as sole nitrogen source. It is likely that such a selection scheme would identify only *gap1* mutants that retained proline transport activity. The question could be addressed unequivocally, as Lasko and Brandriss (1981) point out, by testing proline uptake in a *gap1* mutant isolated on a poor nitrogen source that is not an amino acid.

Asparagine Transport

Dunlop and Roon (1975) reported that asparagine uptake exhibited a pH optimum of 4.0. Mutant strains resistant to the growth inhibitory effects of D-asparagine were found to be deficient in asparagine uptake.

Anabolic Nitrogen Transport Systems

The anabolic uptake systems by which *S. cerevisiae* accumulates amino acids are summarized in Table 4.

Basic Amino Acid (Arginine) Permease

The basic amino acids, arginine (K_m = 10 μM), lysine, ornithine, canavanine, and probably histidine are accumulated by an active transport system, designated the basic amino acid permease. This conclusion is based on the demonstration that the above basic amino acids are competitive inhibitors of arginine uptake into cells growing in glucose-ammonia medium; the *GAP* system is inactive here due to ammonia inhibition (Grenson et al. 1966). Mutants of the basic amino acid transport system, designated *can1* (*argp1*), were isolated by selecting cells that were resisitant to canavanine (10 μg/ml) inhibition as described earlier. Arginine uptake was found to be decreased 26-fold in these strains. Lysine uptake was also decreased, but to a much

Table 4 Anabolic nitrogen transport systems: Amino acids

Compound	Apparent K_m	Concentration (inside/outside)	Mode of uptake	Control	pH optimum	Efflux	Exchange	Inhibitors	Gene	Reference[a]
Arginine	10 μM		active						CAN1	1, 8
Lysine	0.2 mM		active						CAN1	2, 8
Lysine	25 μM								LYP1	3
Histidine	20 μM					no	no		HIP1	4
Histidine	0.5 mM					no	no		CAN1?	4
Methionine	12 μM	56	active						METP	5
Methionine	0.8 mM		active							5
Cysteine	17		active							5
S-adenosyl-methionine	1.6–3.3 μM	active							SAP3	6
Glutamate	(see text)									
Leucine	0.05 mM, 1 mM	6	active					S-adenosyl-methionine		7

[a] (1) Grenson et al. (1966); Chan and Cossins (1976); Whelan et al. (1979). (2) Grenson (1966); Grenson et al. (1966). (3) Grenson (1966). (4) Crabeel and Grenson (1970). (5) Maw (1963); Gits and Grenson (1967). (6) Spence (1971); Murphy and Spence (1972); Petrotta-Simpson (1975). (7) Bussey and Umbarger (1970b); Ramos et al. (1975, 1980); Law and Ferro (1980). (8) Larimore and Roon (1978); Larimore et al. (1980).

smaller extent due to the operation of a second, lysine-specific transport system. Inhibition of transport by DNP, azide, and N-ethylmaleimide (NEM) supports the suggestion that uptake is energy-driven. However, no inhibition of arginine uptake was observed when cells were treated with cyanide or fluoride ions (Chan and Cossins 1976).

A detailed genetic map of the *can1* locus has been constructed (Whelan et al. 1979). Although a great deal of structural information is available for the locus, little is known about its control or that of its gene product. An increase of arginine uptake was reported following nitrogen starvation of wild-type cells (Chan and Cossins 1976). This effect could be eliminated by treating cells with cycloheximide just prior to starvation or preloading them with a variety of amino acids. Unfortunately, most, if not all, of these observations can be explained by derepression of the *GAP* system during nitrogen starvation and its feedback inhibition on preloading the cells with amino acids.

Lysine Transport

Kinetic data indicate that there are two separate modes of transport of lysine in *gap* mutants or in wild-type cells growing in glucose-ammonia medium. The basic amino acid permease (*CAN1*) mediates low-affinity (K_m = 0.2 mM) transport; the lysine-specific permease (*LYP1*) mediates high-affinity transport (K_m = 25 μM). In contrast to transport by the *CAN1* permease, lysine uptake by the *LYP1* permease is not inhibited by any of the other basic amino acids (Grenson 1966). Mutants lacking lysine uptake (*lyp1*, originally designated *lysp1*) were isolated as cells resistant to the growth-inhibitory action of the lysine analog, L-thiosine (10 μg/ml). Lysine uptake was decreased by 50-fold in these mutants, confirming that resistance resulted from the inability of thiosine to enter the cell (Grenson 1966).

Histidine Transport

Histidine appears to enter the cell by way of two biosynthetic transport systems, as evidenced by a biphasic Lineweaver-Burke plot of histidine uptake in cells growing in glucose-ammonia medium (Crabeel and Grenson 1970). Resolution of the plot yields apparent K_m values of 20 μM and 0.5 mM. Up to 2 hours elapse between addition of histidine to the medium and attainment of the influx-efflux equilibrium. No efflux or exchange of previously accumulated radioactive histidine was observed when cells were resuspended in fresh histidine-free medium or medium containing excess nonradioactive histidine. This led to the conclusion that histidine transport was irreversible. Alternatively, these observations might derive from a sequestration of histidine within a cellular organelle in a manner similar to that reported for allantoin and arginine (Sumrada and Cooper 1978a; Zacharski and Cooper 1978).

A mutant (*hisp1*, now *hip1*) lacking the low K_m (20 μM) histidine uptake system was isolated by identifying a histidine auxotroph that could not grow when supplemented with low concentrations of histidine. Histidine uptake was shown to be depressed tenfold in this mutant (Crabeel and Grenson 1970); the high K_m (0.5 mM) uptake system was not affected by the mutation. These investigators also observed that histidine uptake was decreased 15% when arginine or lysine was added to a wild-type strain. However, these amino acids had no effect when the experiment was repeated with a *can1* (*argp1*) *lyp1* double mutant. Although not interpreted in this manner by Crabeel and Grenson, this result raises the possibility that high K_m histidine uptake is mediated by the basic amino acid transport system (*CAN1*). The information that is necessary but missing is whether or not similar behavior is observed when the experiment is repeated using a strain that carries only the *can1* mutation. Histidine addition had no affect on lysine uptake in wild-type strains and hence was concluded not to be a substrate for the lysine permease.

Methionine Transport

Maw (1963) observed that inorganic sulfate and methionine were completely removed from the medium. Cystine, on the other hand, was not taken up at all, and cysteine was removed only slowly. All of the compounds Maw studied were accumulated by active transport against a concentration gradient: methionine, 56-fold; cysteine, 17-fold; ethionine, 12-fold; *S*-methyl cysteine, 31-fold; and *S*-ethyl cysteine, 21-fold. Uptake was inhibited by azide and DNP, indicating the action of an energy-coupling system associated with transport. Although these studies may be taken as a point of departure, they permit little rigorous characterization, because two and probably more systems were being assayed simultaneously.

In a systematic kinetic study, Gits and Grenson (1967) found evidence for two biosynthetic methionine transport systems in addition to the *GAP* catabolic system. The low K_m (12 μM) system was defective in mutants (*metp1*) that were resistant to very low concentrations of ethionine (0.28 mM) (Gits and Grenson 1967). These *metp1* mutants retained the high K_m (0.8 mM) system. The only compounds found to inhibit methionine uptake were ethionine and seleno-methionine, demonstrating the high specificity of these systems and reducing the possibility that the high K_m methionine uptake system is the result of a broadened specificity of some other system, as might be the case with histidine uptake.

S-*adenosylmethionine Transport*

The current understanding of *S*-adenosylmethionine transport is due largely to the efforts of Spence and his colleagues (Spence 1971; Murphy and Spence 1972; Petrotta-Simpson et al. 1975). *S*-adenosylmethionine accumulation is mediated by a low K_m (1.6–3.3 μM) active transport system with a pH optimum of 4.8. Uptake is inhibited by 20 mM azide and 10 mM cyanide, suggesting that accumulation is coupled to energy-generating reactions. S-adenosyl ethionine (K_i = 10 μM), S-adenosyl homocysteine (K_i = 0.2 mM), S-adenosyl methyl thiopropylamine (K_i = 1.4 μM), and S-inosyl methionine (K_i = 0.15 mM) were all found to be effective competitive inhibitors of S-adenosylmethionine uptake. Methionine, however, was without effect.

Mutant strains defective in S-adenosylmethionine transport have been isolated in two ways. In the first method, S-adenosylmethionine was used in place of methionine to permit growth of a methionine auxotroph. Mutants (*samp3*) in which S-adenosylmethionine was unable to perform this function were then isolated (Spence 1971). In the second method, an adenine auxotroph was plated in the presence of inhibitory levels of S-adenosyl ethionine. Those cells resistant to this condition were then scored for growth with S-adenosylmethionine serving as the source of adenine. Cells able to grow in the presence of adenine, but not in medium supplemented with S-adenosylmethionine, were those shown by subsequent biochemical studies to be defective in S-adenosylmethionine uptake. Mutations isolated in this manner failed to complement the *samp3* mutation isolated earlier. Two *samp* mutant alleles, *samp1* and *samp2*, were observed to complement each other, but no recombinants were observed among 100 randomly selected asci derived from a cross of strains harboring the two mutations. These observations led to the conclusion that the observed complementation was intragenic. Nakamura and Schlenk (1974) identified a strain of *S. cerevisiae* and growth conditions that permitted a fivefold increase in the amount of S-adenosylmethionine accumulated. These conditions have been useful for vacuolar accumulation studies.

Glutamate Transport

Glutamate uptake is suggested to occur by three different routes. Transport via the *GAP* system is evidenced by disappearance of the high K_m component of a biphasic Lineweaver-Burke plot when proline is replaced with ammonia as nitrogen source (Darte and Grenson 1975) and in *gap* mutants. The remaining two systems are ill defined largely because of difficulty encountered in resolving them either kinetically or genetically.

Glutamate uptake has been reported to involve the cotransport of three equivalents of H^+ and the loss of one to two equivalents of K^+ from the cell (Cockburn et al. 1975). However, in view of the existence of two, and possibly three, uptake systems operating simultaneously, the significance of this result is unclear at present.

Leucine Transport

Characterization of leucine uptake is limited. Transport appears to require glucose and was observed to be inhibited by DNP. However, only sixfold concentration of the amino acid was found. Valine and isoleucine were inhibitory, though their routes of uptake are totally obscure (Bussey and Umbarger 1970b; Ramos et al. 1975). Note, however, that leucine is probably transported by GAP (Rytka 1975). Law and Ferro (1980) also reported that S-adenosylmethionine inhibited leucine uptake by 70-75%. Inhibition was competitive and did not require the entry of S-adenosylmethionine into the cell. More recently, Ramos et al. (1980) have reported a kinetic analysis of leucine uptake in which they conclude the existence of two uptake systems possessing K_T (K_m) values of 0.05 mM and 1 mM, respectively. Dio-9, a known inhibitor of the plasma membrane Mg^{++}-ATPase, inhibited both uptake systems.

Peptide Transport

A limited number of dipeptides and tripeptides seem to be taken into the cell (Becker et al. 1973; Naider et al. 1974; Marder et al. 1977; Parker et al. 1980). A methionine auxotrophy could be fulfilled by supplementing the growth medium with any of the following peptides: methionine-methionine, methionine-methionine-methionine, or methionine-glycine-methionine-methionine. Transport of the homopolymeric methionine tripeptide was shown to be competitively inhibited by a polypeptide–polyethylene glycol conjugate (Naider et al. 1980). A similar set of leucine- and lysine-containing dipeptides and tripeptides could be used to supplement auxotrophies of these amino acids. Both dipeptides and tripeptides competed with one another for entry by the uptake system. A mutant isolated on the basis of its resistance to the dipeptide ethionine-alanine totally lacks the ability to accumulate dipeptides (Marder et al. 1978).

Cytosine and Purine Transport

The uptake systems for purine and cytosine uptake have been elucidated by the combined but independent efforts of R.A. Woods, F. Lacroute, M. Grenson, and their collaborators (Table 5). Two systems are potentially involved: one for uptake of adenine, hypoxanthine, and cytosine and a second for guanine and hypoxanthine. Pickering and Woods (1972) isolated mutants that were resistant to the inhibitory effects of 4-aminopyrazolo pyrimidine (APP), an analog of adenine. Assay of purine uptake in these strains revealed that adenine, hypoxanthine, and APP were accumulated at only 44%, 32%, and 0% of the wild-type rates, respectively. Guanine uptake, by comparison, was unaltered. The apparent K_m value for adenine was also increased from 0.12 µM to 2.6 µM as a result of mutating the $APP1$ locus. Hypoxanthine was an effective inhibitor of both adenine and guanine uptake in wild-type and APP-resistant mutants. Adenine, on the other hand, had no effect on guanine transport in either strain. These observations led Pickering and Woods (1972) to conclude that hypoxanthine could act as substrate for either the adenine or guanine uptake system.

Table 5 Anabolic nitrogen transport systems: Purines and pyrimidines

Compound	Apparent K_m	Concentration (inside/outside)	Mode of uptake	Control	pH optimum	Efflux	Exchange	Inhibitors	Gene	Reference[a]
Adenosine, guanine[b], hypoxanthine, cytosine	0.12 μM	1,900	active			rapid	rapid		APP1	1
Uracil	12 μM, 156 μM	40	active		4.3–4.5	rapid		dihydro-uracil	FUR4	2
Uridine	27 μM	10,000	active	const, N-rep	4.2–4.7	no	no		URD1	3
Ureido-succinate			active					*[c]	DAL5 UEP1	4

[a] (1) Lacroute (1966); Grenson (1969); Jund and Lacroute (1970); Pickering and Woods (1972); Polak and Grenson (1973). (2) Lacroute (1966); Jund and Grenson (1969); Lacroute (1970); Jund et al. (1977). (3) Grenson (1969). (4) Drillien and Lacroute (1972); Greth et al. (1977).
[b] Here, the data are controversial (see text).
[c] *, N-carbamyl-glutamate, -butyrate, -valine, -alanine, -glycine, and N-acetylglycine.

Lacroute (1966) isolated classes of mutants resistant to 5-fluorocytosine and designated one of the mutated loci *FCY2*. Strains carrying a mutation in this locus were impaired in cytosine uptake but not in its conversion to uracil (Jund and Lacroute 1970). Grenson (1969) reported isolation of a mutant that had lost the ability to use cytosine as sole nitrogen source. Cytosine deaminase, the enzyme that deaminates cytosine to yield ammonia plus uracil, was found at normal levels in the strain, leading her to conclude that defective uptake was responsible for the observed phenotype. The mutated locus was designated *cytp*. Cytosine and cytidine accumulation were depressed 47- and 14-fold, respectively, in the *cytp* mutants. Cytosine is the preferred substrate of the system, as evidenced by the fact that 0.1 mM cytidine was unable to inhibit cytosine uptake when the latter was present at a concentration of 0.02 mM. However, in the converse experiment, 0.02 mM cytosine was an effective inhibitor of cytidine (0.1 mM) uptake (Grenson 1966).

The fact that adenine and cytosine share a common transport system was found by Polak and Grenson (1973). They observed that exogenous adenine and hypoxanthine interfered with utilization of cytosine added as a supplement to the growth medium of a cytosine auxotroph. Conversely, growth of adenine auxotrophs was inhibited when cytosine was simultaneously added along with adenine as a medium supplement. Direct measurements of adenine and hypoxanthine uptake in $cytp^-$ strains demonstrated a reduction of 8.6- and 7.9-fold, respectively, in the accumulation rates of these purines compared with the wild-type. These investigators reported that Wood's wild type was able to use cytosine as sole nitrogen source, whereas the *app1* mutant had lost this ability. The *app1* mutant was also unable to complement mutations in the *cytp* locus.

More recently, Chevallier et al. (1975) have conducted a thorough biochemical characterization of cytosine transport in wild-type and mutant strains. Transport was shown to occur against a concentration gradient (1900-fold) and to require glucose. Addition of energy metabolism inhibitors such as DNP not only prevented cytosine accumulation but also resulted in loss of previously accumulated cytosine. Efflux and exchange of preloaded cytosine occurred at high rates. Also described were mutants exhibiting an altered K_m for their transport substrates—one of the few instances where this has been reported. However, the most provocative result of this work concerned the specificity of the uptake system. These investigators demonstrated that guanine competitively inhibited cytosine uptake and, conversely, that cytosine competitively inhibited guanine accumulation. Guanine uptake was reported to be "barely detectable" in the *fcy2* mutant. These observations are in marked contrast with those of Woods, who did not observe any interaction between the uptake systems for adenine and guanine. Chevallier et al. (1975) indicated that they did not investigate "the possible existence of other lower affinity systems." However, the substrate concentrations used by Woods were similar to those used by Chevallier et al. Unfortunately, the two observations cannot be rectified at the moment.

Uracil and Uridine Transport

Transport systems for the uptake of uracil and uridine are distinct from one another and from that of cytosine. Strains lacking uracil transport (*fur4*) were first isolated by Lacroute (1966; Jund and Lacroute 1970) as one of several classes of mutants that was resistant to the growth-inhibitory effects of the uracil analog 5-fluorouracil. Uracil accumulation is decreased by 11-fold in the mutant strains, whereas that of uridine is down only 1.5-fold. On the basis of these results, Lacroute designated the transport system as the pyrimidine permease with the implication that it mediated both uracil and uridine entry into the cell. Cytosine uptake was not affected in the *fur4* mutants, attesting to its entry via a separate system.

A similar class of mutants was subsequently isolated by Grenson (1969) and designated *urap*. She reported that a Lineweaver-Burke plot of uracil uptake was biphasic and that both components were lost in the *urap* mutants. Also isolated was a class of mutants that were resistant to 5-fluorouridine and tenfold depressed for uridine uptake. From these observations, Grenson concluded, in contrast to Lacroute (1966), that uridine and uracil uptake was mediated by two separate systems. The 50% decrease in uridine uptake observed in the *fur4* mutants, however, remains unexplained.

In a more recent report, Jund et al. (1977) have performed a biochemical analysis of uracil uptake in mutant strains that permit uptake measurements in the absence of metabolism. They found that uracil is concentrated 40-fold by two systems with a pH optimum of 4.3–4.5 and apparent K_m values of 12 μM and 156 μM. The degree of concentration, which is 48-fold less than that observed for cytosine possibly derives from a very high rate of uracil efflux. The half-life for loss of previously accumulated uracil is approximately 1 minute. Also reported was the isolation of two mutants that were able to transport sufficient uracil to fulfill an auxotrophy in the presence of inhibitory concentrations of dihydrouracil, a competitive inhibitor of uracil uptake. One of the mutations was dominant (*DHU2*), and the other was recessive (*dhu1*). However, both mutations resulted in a tenfold increase in uracil transport. Chevallier (1982) demonstrated a commensurate increase in *FUR4*-specific RNA in the latter mutant.

Ureidosuccinate Transport

Ureidosuccinate (USA), the first specific intermediate in pyrimidine biosynthesis, can support growth of *ura2* mutants, Uptake of USA is mediated by an active transport system operating with a pH optimum of 4.2–4.7 and a K_m value of 27 μM (Drillien and Lacroute 1972; Greth et al. 1977). This conclusion was based on an observed 10,000-fold concentration of USA over external levels and the sensitivity of this concentration to DNP, azide, and CCCP. Although uptake required energy input, maintenance of preloaded USA did not possess such a requirement. Neither efflux nor exchange of the metabolite was observed. A variety of competitive inhibitors were identified, including *N*-carbamyl-glutamate, -butyrate, -valine, -alanine, -glycine and *N*-acetylglycine. Allantoin is not a competitor of uptake.

Mutations lacking USA uptake (*urep1*, now *uep1* = *dal5*) were isolated by making use of the observation that high levels of USA are growth inhibitory. The majority of the resistant strains were found to be defective in USA accumulation; uptake was decreased approximately 15-fold. The *UEP1* locus was shown by a variety of tests to be different from *GAP1*. Other transport systems shown to be unrelated to USA uptake included proline, glutamate, and arginine.

The most striking characteristic of USA uptake is its susceptibility to nitrogen repression. USA was able to fulfill the *ura2* auxotrophy if the derepressing nitrogen source proline was used but could not perform this function when either glutamate or ammonia was provided instead. Lacroute also found that addition of cycloheximide to the medium resulted in complete inhibition of USA transport within 60 minutes, signaling the occurrence of transinhibition by amino acids that accumulate on cessation of protein synthesis (Greth et al. 1977). Both types of modulation cited above and the lack of efflux and exchange are identical to characteristics observed for the degradative nitrogen transport systems. Yet USA is a biosynthetic intermediate that is incapable of serving as a nitrogen source for *S. cerevisiae*, and the system mediating its transport seems to be quite specific.

The paradoxical regulatory characteristics of USA transport prompted us to search for degradative-transport systems whose substrates were structurally similar to USA. It quickly became apparent that USA was a nearly perfect analog of allantoate and suggested that USA was transported by way of the degradative allantoate transport system. The following observations confirm this hypothesis: (1) Growth of *dal5* mutants are not inhibited by USA, previously reported to be an exclusive property of *uep1* mutants; (2) *uep1* mutants were not able to use allantoate as sole nitrogen source or accumulate the compound within the cell; (3) mutations in the *UEP1* and *DAL5* loci were unable to complement one another, and no recombinant asci were observed on crossing these mutants. In sum, the data were interpreted as indicating that mutations in the *UEP1* and *DAL5* loci were allelic, thereby explaining the strange regulatory behavior of the USA transport system (V. Turoscy and T.G. Cooper, in prep.).

VITAMIN TRANSPORT

The uptake systems by which *S. cerevisiae* accumulates vitamins are summarized in Table 6.

Table 6 Vitamin transport systems

Compound	Apparent K_m	Concentration (inside/ outside)	Mode of uptake	Control	pH optimum	Efflux	Exchange	Inhibitors	Gene	Reference[a]
Thiamine	0.1 μM	10,000	active	repressible (by thiamine)	4.5			fatty acids, pyrathiamine	THP1	1
Biotin	0.3 μM	1,100	active	repressible (by biotin)	3.8		rapid	d-homobiotin, d-norbiotin		2
Riboflavin	15 μM	6			7.5					3
Choline	0.56 mM		active					β-methyl-choline	*[b]	4
Inositol	0.1 mM	35	active	repressible (by inositol)					*[b]	5

[a] (1) Iwashima et al. (1973, 1975, 1977); Iwashima and Nose (1976). (2) Rogers and Lichstein (1969a,b); Cicmanec and Lichstein (1974). (3) Perl et al. (1976). (4) Hosaka and Yamashita (1980). (5) Nikawa et al. (1982).
[b] *, Mutant exists, but locus is unnamed.

Thiamine Transport

Thiamine accumulation is mediated by a single active transport system that operates optimally at pH 4.5 and exhibits an apparent K_m value for thiamine of 0.1 μM (Iwashima et al. 1973). Thiamine is concentrated up to 10,000-fold over the external levels, reaching a plateau 30–40 minutes after its addition to the medium. Uptake is inhibited by arsenate, cyanide, DNP, and NEM. Fatty acids ranging in length from 2 carbon atoms to 6 carbon atoms are also inhibitory when added at high concentration (20 mM for caproate). Iwashima et al. (1973) reported that pyrithiamine ($K_i = 0.1$ μM), chloroethylthiamine, and dimethialium are far more potent inhibitors. More recently, two additional competitive inhibitors have been identified, which possess K_i values in the nanomolar range: thiamine disulfide ($K_i = 57$ nM) and O-benzoylthiamine ($K_i = 1.8$ nM) (Iwashima et al. 1977).

Resistance to the thiamine analog pyrithiamine was used to isolate strains that lacked thiamine uptake (Iwashima et al. 1975). The two mutants reported, PT-R1 and PT-R2, had lost 98.1% and 99% of their uptake activity, respectively. These strains were not characterized genetically and, hence, it is not known whether or not they are allelic.

Thiamine transport is reported to be repressed by pregrowth of cells in thiamine-containing medium (Iwashima and Nose 1976). However, feedback inhibition of thiamine uptake does not appear to function, since preloading the cells just before assay did not decrease uptake. Repression may be exerted by regulating a specific thiamine-binding protein (Iwashima and Nishimura 1979). This conclusion was derived from the observation that a specific thiamine-binding protein could be purified from the osmotic shock fluid of cells grown in thiamine-deficient medium but was not found when thiamine was included in the medium. However, more recent data (Iwashima et al. 1979; Nishimura et al. 1982) argue in a compelling way that this protein is not required for thiamine transport. These observations exemplify the difficulties associated with physically identifying those proteins that participate in transport processes.

Biotin Transport

Lichstein and his colleagues are primarily responsible for our understanding of biotin uptake in *S. cerevisiae*. The vitamin is transported by an active transport system with a pH optimum of 3.8 and a K_m value of 0.3 μM (Rogers and Lichstein 1969a). This conclusion is supported by the observations that (1) the vitamin is concentrated 1100-fold over extracellular levels, and (2) uptake is stimulated 13-fold by glucose and inhibited by azide or iodoacetate. Exchange between intracellular and extracellular biotin is very rapid with a half-life of 10–15 minutes. Several competitive inhibitors have been identified, including d-homobiotin, d,l-desthiobiotin, d-norbiotin, d-biotin sulfone, and biotinyl-p-nitrophenyl ester. The latter analog has been reported to progressively and irreversibly inactivate the transport system and hence has been claimed to be a good candidate as an affinity label for the biotin-binding protein (Becker et al. 1971). If binding of this compound turns out to be covalent, it may be useful as a radioactive probe for purification of the protein.

Control of the uptake system seems to be mediated by repression (Rogers and Lichstein 1969b; Cicmanec and Lichstein 1974). Transport activity is seen only when cells are grown in the presence of diminishingly small amounts of biotin (< 0.25 ng/ml). Also, if avidin is added to the culture medium, transport activity increases over an undetectable level. The increase can be prevented if cycloheximide is added simultaneously. Addition of cycloheximide, however, does not affect preexisting transport activity. Repression is distinguished from feedback inhibition by the observation that preloading derepressed cells with biotin does not prevent uptake of the vitamin. Therefore, by exclusion, inhibition of biotin transport following long-term growth of cells in the presence of the vitamin was concluded to result from repression.

Riboflavin Uptake

Perl et al. (1976) reported that uptake of riboflavin can only be observed in riboflavin-requiring mutants. The system exhibits a pH optimum of 7.5 and an apparent K_m of 15 μM. Although uptake appears to be carrier-mediated, it is difficult to distinguish whether the carrier is part of an active transport or facilitated diffusion system. Riboflavin seems to be concentrated sixfold, but accumulation was not inhibited by inhibitors of energy metabolism. Also glucose inhibited uptake rather than being stimulatory. It was hypothesized that the observed inhibition of total accumulation is derived from a glucose-stimulated efflux of the vitamin. It is argued that uptake is repressed, because it cannot be demonstrated in cells that can synthesize the vitamin. However, no experiments are reported that would distinguish the hypothesized repression from feedback inhibition.

Choline and Inositol Transport

Hosaka and Yamashita (1980) have characterized the choline transport system. Choline is accumulated linearly for only a short time (1–2 min at 30°C) and is largely phosphorylated on entry into the cell. The uptake system exhibits a K_m of 0.56 mM and depends on input of energy, as indicated by a requirement for glucose and its sensitivity to inhibitors such as DNP, arsenate, and cyanide. The system is highly specific with B-methyl choline as the only effective analog inhibitor. A mutant lacking choline transport was also reported.

Inositol transport was characterized in a similar manner (Nikawa et al. 1982). Accumulation (35-fold over medium levels) of this metabolite was mediated by an energy-dependent, active transport system possessing a K_m of 0.1 mM. Inositol uptake decreased over a 2-hour period following addition of high levels of inositol to the medium. Transport was reacquired following removal of inositol from the medium; the latter was completely inhibitable by treatment with cycloheximide, leading to the conclusion that production of the activity was repressible. Feedback inhibition of transport was not evaluated. A mutant exhibiting only 4% of the wild-type activity was also isolated.

ION TRANSPORT

The uptake systems by which *S. cerevisiae* accumulates inorganic ions are summarized in Table 7.

Sulfate Transport

Inorganic sulfate appears to enter the cell by way of two transport systems as evidenced by a biphasic Lineweaver-Burke plot of sulfate accumulation as a function of external concentration. The apparent K_m values derived from these data are 40 μM and 350 μM (Breton and Surdin-Kerjan 1977). Uptake is strongly dependent on temperature and decreases 33-fold in the absence of glucose (McCready and Din 1974). Recently, Roomans et al. (1979a) reported that sulfate uptake is accompanied by a net influx of three equivalents of H^+ and an efflux of one equivalent of K^+ for each sulfate ion taken up. Divalent cations were also found to be stimulatory, a characteristic shared with the phosphate uptake system. However, it is important to point out that these biochemical studies monitored sulfate uptake without distinguishing whether one or two systems were operating simultaneously.

McCready and Din (1974) reported that selenate and chromate were potent inhibitors of both sulfate accumulation and cell growth. Breton and Surdin-Kerjan (1977) isolated mutant strains that were resistant to the toxic effects of these compounds. When a single inhibitor was used, only strains with defects in the biosynthetic pathway were found. However, when both inhibitors were present simultaneously, a strain lacking sulfate transport was isolated. Genetic analysis revealed two mutations, a dominant one leading to chromate resistance and a recessive one leading to selenate resistance. The high K_m uptake system was totally absent in the selenate mutants, but both mutations were required for loss of the low K_m system.

Table 7 Ion transport systems

Compound	Apparent K_m	Concentration (inside/outside)	Mode of uptake	Control	pH optimum	Efflux	Exchange	Inhibitors	Gene	Reference[a]
Sulfate	40 μM 0.35 mM		active	repressible[b]					CHR1 SEL1	1
Phosphate	0.6 μM	100	active[c]		7.2				PHO84?	2
Phosphate	30 μM		active[d]		7.2				PHO84?	2
Divalent cations			active				no			3

[a](1) McCready and Din (1974); Breton and Surdin-Kerjan (1977). (2) Borst-Pauwels and Jager (1969); Huygen and Borst-Pauwels (1972); Ueda and Oshima (1975); Roomans and Borst-Pauwels (1977, 1979); Roomans et al. (1977). (3) Jennings et al. (1958); Rothstein et al. (1958); Fuhrmann and Rothstein (1968); Okorokov et al. (1977).
[b]Methionine or S-adenosylmethionine
[c]Na^+-dependent.
[d]H^+ cotransport.

Both mutations were shown to change the apparent K_m values for both systems, leading the investigators to hypothesize an interaction between the two systems. Their observations are similar in principle to those observed for the active transport and facilitated diffusion systems associated with urea and αMG transport.

Sulfate uptake is controlled in a coordinated manner with production of homocysteine synthase (Breton and Surdin-Kerjan 1977). The system appears to be repressed after growth of cells in medium containing either 1.5 mM methionine or 0.2 mM S-adenosylmethionine. Neither of these compounds, however, inhibits uptake when added simultaneously with sulfate. Feedback inhibition of both transport systems is seen when sulfate is added to the medium of preloaded cells. It has been hypothesized that both intracellular sulfate and adenosine 5'-phosphosulfate function as feedback inhibitors (Breton and Surdin-Kerjan 1977).

Phosphate Transport

Phosphate transport has been studied using both biochemical and genetic methods. Most of the biochemical studies, however, have employed commercial sources of yeast, making it difficult to interface the information gained from each approach. Phosphate uptake requires metabolism (Kamen and Spiegelman 1949) and is dependent on the presence of cations such as potassium and magnesium ions, which are cotransported (Schmidt et al. 1949). Although potassium ions were observed to be most effective for stimulating phosphate uptake, ammonium ions could adequately replace them. This raises the possibility that the third, rather ill-defined, high K_m, ammonia transport system mentioned by Dubois and Grenson (1979) might be associated with phosphate and/or sulfate uptake; the latter anion is also cotransported with a variety of cations.

Phosphate is concentrated over 100-fold within the cell by an active transport system exhibiting a K_m value of 0.4 mM (Goodman and Rothstein 1957). Concentration required a hexose that could be replaced by lactate or pyruvate only at the expense of a 16–25-fold loss in uptake rate. However, only a threefold to fourfold decrease in transport was observed when ethanol replaced the hexoses. The presence of potassium ions in the assay medium stimulated transport sevenfold and also shifted the pH optimum for transport from 6.5 to 4.8. "The effect of potassium [was] a specific one, for sodium was unable to stimulate the phosphate uptake" (Goodman and Rothstein 1957).

More recent studies have led to the conclusion that two mechanisms of phosphate transport exist at pH 7.2 (Borst-Pauwels and Jager 1969; Huygen and Borst-Pauwels 1972; Roomans and Borst-Pauwels 1977, 1979; Roomans et al. 1977). One route, which possesses a K_m of 30 μM, is independent of sodium ions, and the other, involving sodium-dependent cotransport, possesses a K_m of 0.6 μM. It was proposed that the sodium-phosphate symport system possessed one phosphate site and two sites with an affinity for Na^+ or Li^+ (Roomans and Borst-Pauwels 1977). The sodium-independent phosphate transport system, on the other hand, was reported to be accompanied by cotransport of three protons and the release of two potassium ions, the latter to maintain the charge balance within the cell (Cockburn et al. 1975). Borst-Pauwels and Peters (1977) reported that the rate of phosphate uptake by this system and its K_m value for phosphate were complex functions of medium pH. Proton–amino acid symports have similarly been proposed to be associated with uptake of neutral and basic amino acids (Eddy et al. 1970a,b,c; Eddy and Nowacki 1971; Seaston et al. 1973).

At the other end of the spectrum, Ueda and Oshima (1975) isolated a mutant strain producing repressible, extracellular acid phosphatase at constitutive levels even when phosphate was present in the growth medium. The mutations generating this phenotype were recessive, and the locus containing them was designated *PHOT* (now *PHO84*). Biochemical assay of *pho84⁻* strains revealed that phosphate accumulation was severely reduced, thus identifying transport deficiency as the basis for the observed phenotype; cells unable to transport phosphate derepressed acid phosphatase production in response to their starvation for phosphate. The latter situation is one of many in which mutations generating defects in transport or metabolism exhibit phenotypes similar to those expected of a mutated "regulatory" locus. In some of these cases, to seek the bizarre is to miss the obvious.

Three apparent inconsistencies surface upon correlating the diverse observations concerning phosphate transport. Goodman and Rothstein (1957) reported that sodium ions could not replace potassium ions as cosubstrates for phosphate uptake, whereas the experiments of Borst-Pauwels involved the exclusive use of sodium ions (Roomans and Borst-Pauwels 1979). The K_m values reported by the two groups differed by 13–670-fold, depending on whether the 30-μM or the 0.6-μM system is used for comparison. It is also difficult to see how a phosphate-transport-deficient mutant could have been isolated in the absence of selective pressure when phosphate could enter the cell by two separate routes. These inconsistencies are likely to be resolved only by detailed biochemical analysis of genetically defined strains.

Divalent Cation Transport

Manganese and magnesium ion uptake has been studied biochemically by Rothstein and his colleagues (Jennings et al. 1958; Rothstein et al. 1958); no genetic studies have been reported. Exogenous manganese was observed to rapidly bind to the cell surface. Bound manganese was easily exchanged, and binding occurred in the absence of glucose. The ion was absorbed more slowly into the cell. In this case, glucose was required, and absorbed manganese was nonexchangeable. The most striking characteristic of this transport system, however, was its absolute requirement for exogenous inorganic phosphate and its twofold stimulation by 3 mM K^+; 100 mM K^+ decreased manganese uptake by 90%. The capacity for manganese uptake was maintained for a short time after phosphate was removed from the medium. Sugar fermentation induced a very rapid loss of manganese uptake capacity following removal of phosphate, whereas respiration of ethanol or lactate had little effect. Inhibitor studies with arsenate revealed parallel effects on manganese and phosphate absorption, leading to the conclusion that uptake of manganese and magnesium involved a close association with phosphate uptake. Although a model was proposed (Jennings et al. 1958) involving a phosphorylated membrane carrier, no data have since appeared supporting its existence or describing its properties. More recently, Okorokov et al. (1977) reported that manganese uptake is inhibited by 25 μg/ml of oligomycin, 1 mM DNP, 1-50 mM deoxyglucose, and 1-10 mM azide.

Several additional divalent cations were reported to be accumulated via the divalent cation transport system (Fuhrmann and Rothstein 1968). Magnesium was most efficient, being capable of totally inhibiting manganese uptake when added at the same concentration. From crude competitive studies such as these, it was concluded that the approximate "relative affinities" of the various cations were Mg^{++}, Co^{++}, $Zn^{++} > Mn^{++} > Ni^{++} > Ca^{++} > Sr^{++}$. It was also found that two potassium ions were secreted for each divalent cation absorbed. Roomans et al. (1979b) have more recently studied the kinetics of calcium and strontium uptake. They observed a marked influence of cell-surface potential and phosphate on the transport of these cations.

Copper uptake has been studied only superficially. Uptake seems to be enhanced by glucose and reduced by low temperature. Unfortunately, the toxicity of the ion and lack of radioactive probes in past studies have made detailed characterization of the system almost impossible (Ross 1977).

SUMMARY OF TRANSPORT SYSTEM GENETICS

Previous sections summarize the characteristics of each transport system that has been described in the literature. However, it is also useful to organize some of the genetic information in a manner that permits it to be used in experiments not necessarily associated with transport per se. Therefore, mutations resulting in transport defects have been divided into two groups (Table 8). The first group of mutations give rise to recessive, resistant phenotypes for various growth inhibitors. The second group of

Table 8 Summary of S. cerevisiae transport system loci

Transport system	Locus	Resistance phenotype
Ammonia	amt	methylamine
	mep1	
	mep2	
General amino acid	gap	D-histidine
Basic amino acid	can1	L-canavanine
Lysine	lysp1	L-thiosine
Proline	put4	L-azetidine-2-carboxylate
		3,4-dehydro-DL-proline
Methionine	metp1	L-ethionine
S-adenosylmethionine	samp	S-adenosyl ethionine
Dipeptide	—	L-ethionyl-alanine
Cytosine	fcy2	5-fluorocytosine
Adenine	app	4-aminopyrazolo pyrimidine
Uracil	fur4	5-fluorouracil
Uridine	uridp	5-fluorouridine
Ureidosuccinate	dal5 (urep)	ureidosuccinate
Thiamine	pt-r162	pyrithiamine
Sulfate	—	selenate and chromate
Galactose	gal2	
Maltose	dsf6,	
	dsf7, dsf17,	
	dsf21	
α-Methyl glucoside	mg2	
Allantoin	dal4 (IX-R)	
Urea active transport	dur3	
Urea facilitated diffusion	dur4	
Histidine	hisp1	
Phosphate	pho84	

mutations are those that inactivate one or more transport systems but are not known to possess resistance to the inhibitory action of any compound. Transport-associated genes that have been isolated on plasmids include *CAN1* (Broach et al. 1979), *DAL4* (V. Turoscy and T.G. Cooper, in prep.), and *FUR4* (Chevallier 1982; Chevallier and Lacroute 1982).

METABOLIC COMPARTMENTATION AND VACUOLAR TRANSPORT SYSTEMS

Intracellular Compartmentation of Metabolites

If a growing culture of *S. cerevisiae* is harvested, washed, and resuspended in water, the cells are found to complete one and sometimes two cell-division cycles before growth fully stops. Although cell division is markedly slowed by such treatment, it occurs nonetheless. One recent explanation of this

Transport in S. cerevisiae 439

phenomenon hypothesized the existence of intracellularly sequestered metabolites that are drawn upon during times of stress (Sumrada and Cooper 1978a). Experimental support for the idea that metabolites could be sequestered, however, was first obtained by Svihla and Schlenk (1959, 1960), who demonstrated a striking production and intracellular accumulation of S-adenosylmethionine in cultures of *Candida utilis* provided with L-methionine or a combination of L-homocysteine and L-methylmethionine. The chromophoric nature of S-adenosylmethionine permitted its visualization in the cell and led these investigators to identify the vacuole as the site of S-adenosylmethionine sequestration. Similar studies were subsequently extended to *S. cerevisiae*, where an analogous situation was found (Svihla et al. 1963, 1964).

Making use of several early reports claiming that basic proteins made plant cell membranes permeable, Schlenk and co-workers (Schlenk and Dainko 1965; Schlenk and Zydek-Cwick 1970; Schlenk et al. 1970) disrupted the plasma membrane of yeast cells while leaving the tonoplast intact. This permitted the selective extraction of cytoplasmic and tonoplast metabolite pools and the presumptive intracellular localization of several sulfonium compounds, including S-adenosylmethionine. Wiemken and his collaborators applied similar techniques to studies of intracellular arginine and glutamate distribution in yeast (Wiemken and Nurse 1973). They selected cytochrome c as the basic protein for disrupting the cell membrane and observed that most of the soluble arginine was found in the vacuolar fraction (i.e., it was released after hypotonic treatment of the tonoplast). Glutamate, on the other hand, was largely cytoplasmic (i.e., it was released after treatment of cells with cytochrome c only). They concluded that as much as 80–90% of the soluble amino acids are located in the vacuole (Wiemken and Nurse 1972, 1973; Wiemken 1975). These conclusions were later supported by direct measurements of amino acids in purified vacuole preparations (Wiemken and Durr 1974). The latter preparations were made using gentle mechanical disintegration of spheroplasts, which avoided losses of low-molecular-weight substances resulting from stretching of the vacuolar membrane during osmotic lysis. These improved isolation procedures also permitted pulse-labeling experiments, demonstrating that cytoplasmic metabolites were more quickly labeled than their vacuolar counterparts. This was particularly true for glutamine and the basic amino acids. Using these and other methods, the intracellular distributions of several compounds have been determined. Zacharski and Cooper (1978) reported that greater than 95% of the intracellular allantoin and allantoate was localized in the tonoplast, whereas 70–80% of the intracellular urea was found in the cytoplasm. Okorokov and colleagues (Lichko and Okorokov 1976; Okorokov et al. 1977, 1978) concluded that up to 40% of the intracellular Mg^{++} and a major portion of Mn^{++} was within the vacuole. The monovalent cations, potassium, rubidium, and cesium, on the other hand, were found in the cytoplasm at concentrations about twofold higher than in the vacuole (Roomans and Seveus 1976). Finally, Indge (1968a,b) came to the tentative conclusion that about 40% of the acid-soluble phosphorus compounds (polyphosphates with up to 260 phosphate units per chain [Langen et al. 1962]) were associated with a vacuole-rich particulate fraction sedimenting at 2000g. More recently, Urech et al. (1978) and Durr et al. (1979) reported that over 98% of the cellular polyphosphates could be recovered from purified vacuoles.

Although a great deal of evidence points to the vacuole as the organelle containing many sequestered low-molecular-weight substances, some caution is appropriate. In several of the cases cited above, the procedures used essentially divide soluble metabolites into two fractions: those released by disrupting the cell membrane, and those that are lost by bursting all of the remaining osmotically sensitive organelles. Included here are nuclei, mitochondria, microbodies, and vacuoles, to name a few. Clearly, a role for the other cell compartments cannot be completely ruled out without quantitative partition data derived from highly purified and characterized preparations.

Vacuolar Transport Systems

Metabolite distribution experiments such as those cited above suggest that intracellular concentrations of arginine and allantoin reach 10–20 mM in logarithmically growing wild-type cultures. In spite of these high metabolite concentrations, arginase (whose inducer is arginine) and the allantoin-degradative enzymes (whose inducer is allophanate, a direct degradative product of allantoin) remain at low basal levels. The highly inducible behavior of these catabolic enzyme systems suggests that intracellular arginine and allantoin are very efficiently sequestered against enormous concentration gradients. Such efficient sequestration almost certainly requires the participation of highly specific transport systems. Studies of these transport systems have long been hampered by the lack of methods for isolating vacuoles. Pioneering experiments were first performed by Indge (1968a,b) and later by Durr et al. (1975). Polybasic substances, such as polylysine, disrupt the plasma membrane but not the vacuole. However, stretching of the vacuoles during this procedure prompted development of a gentle mechanical method of disrupting spheroplast preparations (Wiemken and Durr 1974). Vacuoles derived from cells lysed in this or alternative manners have then been purified by sedimentation velocity centrifugation. The vacuolar transport systems studied thus far are described below.

Boller et al. (1975) and Durr et al. (1976) reported that vacuoles accumulate arginine. However, there was no net increase in the intravacuolar concentration of arginine, leading to the conclusion that accumulation of radioactive arginine was the result of exchange across the vacuolar membrane rather than transport against a gradient. Consistent with this observation was the fact that no energy requirement could be demonstrated for the process. A K_m of 33 µM was observed for the vacuolar-exchange process compared to a K_m of 1.5 µM for arginine transport by analogous spheroplast preparations. Arginine transport in spheroplasts was inhibited by lysine, ornithine, histidine, and D-arginine, whereas vacuolar exchange was sensitive only to D-arginine.

S-adenosylmethionine was concentrated 170-fold by isolated vacuole preparations (Schwencke and de Robichon-Szulmajster 1976). Neither efflux nor exchange of intravacuolar S-adenosylmethionine could be demonstrated. An apparent K_m of 68 µM was observed for the vacuolar transport system compared to a K_m of 11 µM for the S-adenosylmethionine transport system of spheroplasts. Uptake was competitively inhibited by the S-adenosylmethionine analogs, S-adenosyl ethionine and S-adenosyl homocysteine, in both vacuole and spheroplast preparations. As was the case with the vacuolar arginine transport system, it was impossible to identify the energy source driving S-adenosylmethionine transport.

Nagy (1979) reported a vacuolar purine transport system. The isolated organelles used in his studies were able to accumulate guanosine twice as fast as inosine or hypoxanthine and about ten times as fast as adenosine. Adenine was not measurably accumulated. However, in no case was concentration of the substrate against a gradient observed. Preloaded purines were found to efflux from the vacuoles only slowly (4% after 7 min), but exchange of intravacuolar and extravacuolar purines was very rapid with a half-life of about 30 seconds.

In each of the above cases, one was unable to identify the energy source driving the transport processes. It is quite possible that the observed activities were uncoupled from their cognate energy-rich driving reactions. This view is supported by two reports describing an energy-driven basic amino acid transport system in purified vacuolar membrane vesicles (Kakinuma et al. 1981; Ohsumi and Anraku 1981). These investigators demonstrated that arginine was concentrated 40-fold in a Mg^{++}-ATP-requiring process that was sensitive to the ATPase inhibitor N,N'-dicyclohexylcarbodiimide (DCCD). Accumulated arginine was readily released on addition of the protonophore uncoupler SF6847. This and the demonstrated electrochemical potential difference of 170 mV across the vesicle membrane suggested that basic amino acid transport is due to a H^+/arginine antiport. Lysine, histidine, glutamine, tyrosine, isoleucine, and asparagine were also accumulated by the vesicle preparations to various degrees. Glutamate, aspartate, threonine, glycine, and proline, on the other hand, were not taken up by the vesicles. Arginine inhibition of lysine and histidine accumulation raised the possibility that a common carrier might mediate transport of all three basic amino acids.

Labeling Difficulties Caused by Sequestered Metabolites

Isotopic labeling of macromolecules in yeast is complicated considerably by the sequestered reserve metabolites discussed above. In prokaryotic organisms the internal and external pools of protein and nucleic acid precursors equilibrate rapidly. For example, uridine equilibrates within 12-15 seconds after its addition to a culture of *E. coli* growing at 37°C in minimal medium. For *S. cerevisiae*, 50-60 minutes were required for equilibrium to be established (Gross and Pogo 1974). Although no comprehensive studies of labeling procedures have been reported, a few widely scattered observations might prove useful.

Radioactive Labeling of Protein

The rapidity with which yeast proteins may be labeled is in part a function of the radioactive amino acid added to the culture and the amount of it contained in the cells being labeled. As shown in Table 9, the amounts of various amino acids differ markedly. In addition, intracellular amino acid levels vary widely as a function of medium composition (Chan and Cossins

Table 9 Intracellular pools of amino acids in cells grown on different nitrogen sources

Amino acid	Growth conditions[a]		
	rich	minimal ammonia	nitrogen free[b]
Lysine	2067	763	791
Histidine	171	71	47
Ammonia	1034	1136	1419
Arginine	259	390	47
Asparatate	172	82	70
Threonine	1069	—	64
Serine	4828	3516	195
Glutamate	1578	1451	335
Proline	129	53	trace
Glycine	690	624	126
Alanine	474	586	93
Cysteine	19	390	116
Valine	60	568	42
Methionine	121	71	42
Isoleucine	35	28	19
Leucine	35	32	42
Tyrosine	60	7	23
Phenylalanine	60	18	28

[a] Concentration, μmoles/10 g of cells.
[b] Cells were starved for 2 hr before measurements were made.
(Data from Chan and Cossins 1975.)

1976). If rapid labeling is desired, amino acids present in high concentrations should be avoided. Lysine, e.g., has been reported to require 70-80 minutes to reach isotopic equilibrium (Gross and Pogo 1974), whereas methionine requires less than 30 seconds. Methionine is probably the precursor of choice for labeling proteins in yeast.

Miller et al. (1979) made the interesting observation that methionine incorporation is enhanced 8-40-fold by addition of histidine or tyrosine, respectively, to the growth medium. This effect is dose-dependent up to a concentration of 20 µg/ml of tyrosine and appears to be at least somewhat specific since it was not observed when radioactive leucine was used in place of methionine (M.J. Miller, pers. comm.).

Radioactive Labeling of Nucleic Acids

Studies of DNA synthesis are often made difficult by the fact that specific labeling of yeast DNA is impossible. Thymine, the usual specific precursor of DNA synthesis, cannot be used for this purpose, because *S. cerevisiae* lacks thymidine kinase, the enzyme responsible for phosphorylating thymidine monophosphate (TMP) (Grivell and Jackson 1968), and fails to take up thymine or thymidine (Grenson 1969). However, various mutants with the ability to take up TMP have been isolated (for a discussion of these strains, see Brendel et al. [1975]; Schindler and Davies [1975]; Wickner [1975]) and may be useful for such studies. Labeling DNA with adenine, uracil, or uridine is possible, but isotopic equilibration is very slow. Hutchison and Hartwell (1967) observed a $2-2\frac{1}{2}$-hour lag between addition of [^{14}C]adenine to cells or spheroplasts and linear incorporation of radioactivity into RNA and DNA. Sumrada and Cooper (1978b) observed a similar lag of $3-3\frac{1}{2}$ hours before [^{14}C]uracil was linearly incorporated into DNA. Incorporation of radioactive uracil into RNA, on the other hand, was much more rapid, becoming linear in under 1 minute (Lawther et al. 1975).

An important set of labeling problems is highlighted by the experiments of Shulman et al. (1977). These investigators found a drastic drop in incorporation of radioactive precursors into RNA on starving cells for a required amino acid. Upon measuring the nucleotide pools, however, they observed an equally drastic decrease in specific activity of the RNA precursors. This prompted the conclusion that amino acid starvation markedly decreased uptake of the precursors which, in turn, accounted for much of the observed decrease in radioactive RNA synthesis. Related difficulties can also be encountered when inorganic phosphate, uracil, or uridine is used as the source of radioactive precursors. The precursor is not taken into the cell if a sufficient amount is already available either from (1) the medium, in the case of phosphate; or (2) in vivo synthesis, in the case of uracil and uridine. In view of the serious possibility for misinterpretation, it is useful to monitor the effects of experimental parameters on the uptake of macromolecular precursors as well as on the synthesis and turnover of the macromolecules themselves.

Aware of the problems inherent with attempts to rapidly label nucleic acids in general and mRNA in particular, Warner and his colleagues (1976) studied the kinetics of labeling of the S-adenosylmethionine pool. This was of interest, because Sripati et al. (1976) had shown that most, if not all, yeast mRNAs possess a 5′-terminal "cap" composed of either $m^7G(5')pppAp$ or $m^7G(5')pppGp$. Hence, mRNA could be labeled with [$methyl$-^{3}H]methionine. In addition, mRNA was shown to have only this one methyl group, which was not found to be exchangeable (Shulman et al. 1977). The S-adenosylmethionine pool was shown to reach half-maximal specific activity 30 seconds after radioactive methionine was added to the medium; maximum specific activity was attained between 5 minutes and 10 minutes. Upon chasing by the addition of excess nonradioactive methionine, the specific activity of S-adenosylmethionine dropped 50% in 20 seconds and 90% in 5 minutes. The precipitous drop in S-adenosylmethionine specific activity was found to result from exchange and a 3.5-fold expansion in the metabolite pool. Also, completely inhibiting protein synthesis either by addition of cycloheximide or starvation for an amino acid had only a slight effect on the rate at which the S-adenosylmethionine pool equilibrated. In view of these considerations, methionine represents a highly advantageous means of labeling mRNA for gross measurements. There is, however, one caveat. Only the 5′ terminus is labeled. The amount of label incorporated, therefore, represents the number of termini available for methylation (capping), not necessarily the number of full-length or functional mRNA molecules.

REGULATION OF TRANSPORT ACTIVITY

Control Exerted at Gene Expression

Transport systems may be regulated either at the level of their production or operation. For all cases reported thus far, controls on production appear to be exerted at the level of gene expression. They take the form of induction and/or repression. The phenomenology associated with nitrogen repression of transport systems has been discussed briefly already (see discussion of the *GAP* transport system) and is dealt with more broadly and in detail in Nitrogen Metabolism in *Saccharomyces cerevisiae* (Cooper, this volume). However, the molecular events involved in those processes are not well defined at present and hence will not be discussed further. Of greater concern here is the regulation of transport system activity. A good deal of inconsistency seems to surround the interpretation of past observations and the terms feedback and transinhibition. In view of this situation, it would be useful to clearly define these terms and the phenomena associated with them even though there presently exists too little rigorous data to permit construction of meaningful molecular models.

Feedback Inhibition

The salient characteristics of feedback inhibition are well illustrated by the following experimental results derived from studies of the major histidine biosynthetic transport system, *hip1* (Crabeel and Grenson 1970). If [^{14}C] histidine (20 μM) is provided to a wild-type culture, it accumulates within the cell at increasingly slower rates for 90–120 minutes. Thereafter, no further accumulation occurs, i.e., accumulation plateaus. The decreasing rate of histidine entry could result from (1) increased efflux of previously accumulated material or (2) inhibition of further uptake. The first explanation of inhibition (i.e., increased histidine efflux) is unlikely, because efflux of previously accumulated histidine cannot be demonstrated. This argues, by elimination, that further transport is in some way prevented. High intracellular levels of histidine could be visualized to prevent further uptake by occupying the histidine carrier, a molecule that might be envisioned to return to the external surface of the plasma membrane only when it is unoccupied. This is one simplified conceptualization of feedback inhibition. Consistent with this suggestion is the observation that preloading cells with nonradioactive histidine for increasing lengths of time prevented subsequent uptake of the radioactive amino acid to correspondingly greater degrees. Complete loss of uptake ability was observed in the same time as that required for radioactive histidine accumulation to plateau (90–120 min). Alternatively, when the cells were preloaded with nonradioactive methionine, arginine, serine, or lysine, little or no inhibition of histidine transport was observed, attesting to a high degree of specificity for the inhibitory effect. The essence of feedback inhibition is that the compound transported inhibits further transport activity from inside the cell and that inhibition exhibits the same specificity as the transport carrier. In other words, if a compound can be transported by a given system, it can also inhibit that system once its intracellular concentration is sufficiently high. Whether or not the same binding site is used to transport solute into the cell and to bring about feedback inhibition is unclear. However, a single site is rather strongly implied.

Feedback inhibition has been observed in every transport system tested so far, though it was not at first apparent in the allantoin (Sumrada and Cooper 1977) and α-amino isobutyric acid (Kotyk and Ríhová 1972) transport systems. In both instances, radioactive solute seemed to accumulate indefinitely. Resolution of this paradox depends on the fact that both allantoin and α-amino isobutyric acid are sequestered in the cell vacuole (Kotyk and Ríhová 1972; Zacharski and Cooper 1978). It is therefore reasonable to suggest that the vacuolar capacity of allantoin or α-amino isobutyric acid sequestration would have to be exceeded before feedback inhibition was observed. If the rate of solute transport into the cell was slow compared to the time needed to fill the vacuole, it would never fill up before the cells divided. Following this reasoning, Sumrada et al. (1978) predicted that if cell division could be prevented for sufficient time, the vacuole would have a chance to fill up, and allantoin accumulation would plateau as observed in other systems. This was found to be the case, although 20 hours were required to reach a steady state. It is probable that a similar result would be observed for α-amino isobutyric acid.

Transinhibition

Transinhibition is illustrated by the allantoin transport system. As shown in Figure 1, addition of asparagine to a culture accumulating allantoin halts further accumulation. In this case, however, the inhibiting species (asparagine) has no structural relation whatever to the transport system inhibited. It could be argued that asparagine repressed production of the transport system. Athough this model requires very rapid turnover of at least one allantoin transport system component, it is conceptually feasible. The model is discredited by the observation that preloading nitrogen-starved cells with proline (one of the least repressive nitrogen sources available) results in inhibition of allantoin transport (V. Turoscy and T.G. Cooper, in prep.). It might be alternatively argued that both proline and asparagine were metabolized to unspecified compounds structurally related to allantoin. Hence, they inhibited uptake by feedback inhibition. This explanation is also unlikely, because preloaded D-asparagine was observed to effectively inhibit allantoin transport even though it could not be metabolized (V. Turoscy and T.G. Cooper, in prep.). The essence of transinhibition is that the inhibitory compound shares little structural similarity with the substrate of the inhibited transport system, nor is the inhibitory compound transported by that system.

Roon et al. (1977a) suggested that transinhibition results from a reduction of the plasma membrane adenosine-triphosphate-generated proton-motive force when two solutes (methylamine and an amino acid, in their case) are being transported simultaneously, compared to the transport of either compound alone. This suggestion, in its present form, however, is somewhat inconsistent with data demonstrating that inhibition of protein synthesis resulted in severe inhibition of allantoin uptake in normal unstarved cultures but not in starved cultures. Here, materials already in the cell appear to have caused inhibition of transport. This effect has also been reported by others (Greth et al. 1977; Schulman et al. 1977).

In many past instances, the terms feedback inhibition and transinhibition have been used synonomously. Although the possibility exists that these phenomena are expressions of the same basic molecular events, the results discussed above point to important differences in specificity, which cannot easily be reconciled with the idea of a common mechanism. This is particularly true in the case of D-asparagine inhibiting allantoin transport. In view of these unresolved differences, it might be advantageous to use the terms feedback inhibition and transinhibition as they have just been described.

The Vacuole and Unidirectional Solute Transport

Several *S. cerevisiae* transport systems possess the striking feature of operating in an apparently unidirectional fashion (Crabeel and Grenson 1970; Kotyk and Ríhová 1972; Greth et al. 1977; Indge et al. 1977; Sumrada et al. 1978; Turoscy and Cooper 1979). Although there might be numerous advantages to such a model of operation, Cooper and his colleagues suggested that the observed irreversibility might be more apparent than real (Sumrada et al. 1978). This was confirmed by the observations that efflux and exchange of both allantoin and allantoate could be easily demonstrated when more sensitive procedures were used. These findings prompted a reevaluation of the question and led to the hypothesis that unidirectional transport might result from sequestration of the transported metabolites within the vacuole (Sumrada et al. 1978). This predicted that metabolites transported in a unidirectional manner should be largely sequestered, and those that crossed the plasma membrane reversibly should be localized in the cytoplasm. All of the metabolites thus far reported to be irreversibly transported have also been independently shown to be localized in the vacuole. Urea, on the other hand, is transported reversibly into the cell and is largely excluded from the vacuole (80% is found in the cytoplasmic fraction). Although these observations do not prove the hypothesis, they do materially support it.

To the extent that this hypothesis proves accurate, it will be necessary to reevaluate the kinetic parameters gathered about these transport systems. If, for the sake of argument, transport of these solutes into the cell is rapid and freely reversible, as has been found for many other metabolites, then many of the characteristics previously attributed to the irreversible transport systems would in reality be characteristics of the corresponding organelle transport apparatus. What these considerations point out is the fact that each of the sequestered metabolites participates in two sequential transport reactions

solute (external) ⟶ solute (cytoplasmic) ⟶ solute (sequestered)

and the need for future mutational separation and study of these systems.

ISOLATION AND CHARACTERIZATION OF TRANSPORT COMPONENTS

Cell Envelope

The yeast cell envelope occupies about 15% of the total cell volume and controls both the osmotic and permeability parameters of the cell (Arnold and Lacy 1977). The envelope appears to be composed of three layers in a manner somewhat analogous to that observed in prokaryotes. The heavy and rigid cell wall forms the outermost layer. Scherrer et al. (1974) measured the uptake exclusion threshold for *S. cerevisiae* cell walls and concluded that intact cells were permeated by polyethylene glycol molecules smaller than 700 m.w.; this corresponds to an Einstein-Stokes hydrodynamic radius (r_{ES}) of 0.42 nm. Interior to the cell wall is the periplasmic space, which has been approximated to be 35–45 Å thick (Arnold 1973). Here are thought to be located extracellular enzymes such as invertase and acid phosphatase and the metabolite-binding proteins associated with transport.

Innermost is the plasma membrane, the major permeability barrier of the cell. Hennaut et al. (1970) calculated the surface area and cell volumes for cells of increasing ploidy. The mean surface areas for haploid, diploid, triploid, and tetraploid cells were estimated to be 45 μm^2, 71 μm^2, 96 μm^2, and 114 μm^2, respectively. The volumes calculated for these strains were 29 μm^3, 55 μm^3, 83 μm^3, and 105 μm^3, respectively. From these measurements and the levels of transport activity observed, they concluded that the space for insertion of constitutively produced permeases into the cytoplasmic membrane was limited. However, the correlative nature of the evidence does not permit one to decide whether the relatively decreasing amounts of transport activity in cells of increasing ploidy were due to decreased space for these proteins in the membrane or relatively decreased rates of synthesis. The analysis is based on the assumption that all proteins exhibit a strict gene dosage controlled rate of expression; this may not be true in all cases.

There is very little information available concerning the physicochemical characteristics of yeast membrane proteins. Bussey et al. (1979) iodinated yeast cell ghosts and, on subjecting the subsequently solubilized proteins to two-dimensional gel electrophoresis, observed about 200 species. Four of the species, including a major one were periodic acid-Schiff (PAS) reagent positive, identifying them as glycoproteins. Santos et al. (1982) reported a similar study, suggesting that the main structural element was a glycoprotein with an apparent molecular weight of 27,500. The results of double label experiments led Loss et al. (1978) to conclude that membrane proteins do not turn over at measurable rates; the duration of their measurements was 3 hours.

Comparisons of plasma and vacuolar membranes reveal marked differences. Images of the plasmotic and exoplasmotic fracture faces derived by freeze-fracturing the two membranes are quite distinct (Kramer et al. 1978). The ratio of protein to lipid is significantly higher for the plasmalemma, and pronounced differences are observed in the composition of total lipids, phospholipids, and sterols derived from the two membranes. Boller et al. (1976) used fluorescently labeled concanavalin A to demonstrate that the exterior face of the plasmalemma and the interior face of the vacuole membrane both possessed large numbers of concanavalin-A-binding sites. This suggests that these surfaces may be protected by a high concentration of carbohydrate.

Plasma-membrane Vesicles and ATPase

Several attempts have been made to use yeast membrane vesicle preparations similar to those used by Kaback and his collaborators in prokaryotes (Christensen and Cirillo 1972; Fuhrmann et al. 1974, 1976; Wehrli et al. 1975). In one preparation, up to one third of the vesicles remained intact and exhibited osmotic behavior (Fuhrmann et al. 1976). Their surface charge density was found to be similar to that of intact cells with an isoelectric point of 3. Entry of glucose and galactose into these vesicles was inhibited by uranyl ions as expected from in vivo results. Vesicles derived from cells grown in galactose-containing medium accumulated the sugar to a significantly higher degree than did those derived from uninduced cells. However, the rate of entry was not increased in a commensurate manner. The yeast plasma-membrane Mg^{++} ATPase has been used as a marker enzyme ever since early attempts were made to purify this membrane (Matile et al. 1967; Nurminen et al. 1970), because it had been postulated to be involved in solute transport (Conway et al. 1950; Eddy and Indge 1962). Amino acid uptake by isolated vesicles has been recently reported by Merkel et al. (1980). Their preparations, purified by discontinuous sucrose gradient centrifugation, accumulated leucine and lysine. Addition of CCCP eliminated glycine uptake and reduced leucine accumulation 50%.

Christensen and Cirillo (1972) were among the first to report the presence of ATPase activity in a membrane vesicle preparation. However, a lack of homogeneity clouded the issue of whether or not the ATPase activity observed was derived from contaminating mitochondria. Subsequently, Fuhrmann et al. (1974) confirmed and extended these early observations by purifying their vesicles on sucrose density gradients. These vesicles possessed an oligomycin-insensitive ATPase with a pH optimum of about 7. They concluded that this ATPase was associated with divalent metal ion transport, because it was activated by these ions and exhibited the same inhibitor specificity (lanthanum- and carbodiimide-sensitive) as in vivo divalent metal ion transport. Similar studies have been performed by Ahlers et al. (1978) and Serrano (1978), who found the activity to be inhibited by p-chloromercuryphenyl sulfonate (a sulfhydryl reagent), fluoride, quercetin, and the antibiotic Dio-9. The enzyme is distinct from the Na, K ATPase since it is ouabain-resistant. Similarly, it is distinguished from the calcium ATPase by its lack of response to calcium stimulation. The major evidence that this ATPase acts in a manner similar to that of the mammalian plasma-membrane ion pumps is the isolation of a phosphoprotein intermediate after incubation with [γ-^{32}P]ATP (Willsky 1979; Foury et al. 1981; Malpartida and Serrano 1981a,b,c). Although two or more vanadate-sensitive species have been reported for membrane preparations, one protein with a molecular weight of 100,000 can be seen to be phosphorylated with purified enzyme preparations assayed at substrate levels approaching the K_m of the enzyme (millimolar levels). The phosphoprotein intermediates form at 0°C, are sensitive to hydroxylamine (indicative of an acyl phosphate bond), and can be detected within seconds of the addition of labeled substrate. Isolated enzyme has been reconstituted into liposomes and has been shown to catalyze a [^{32}P]ATP exchange reaction that was sensitive to proton ionophores and nigericin, a proton potassium exchanger (Malpartida and Serrano 1980, 1981a,b). The general consensus is that the yeast plasma-membrane ATPase is involved in the generation of a proton gradient that is needed for the transport of metabolites into the cell. Future studies will focus on the mechanism associated with formation of the proton gradient and control of the process (Sigler et al. 1981; De la Pena et al. 1982).

Metabolite-binding Proteins

The first attempts to isolate metabolite-binding proteins were reported by Haskovec and Kotyk (1969). These investigators obtained preliminary evidence for the binding of radioactive galactose to a crude membrane fraction. Bussey and Umbarger (1970a) made similar low-resolution observations of a leucine-binding component. A hexose-binding protein was extensively purified by Horak and Kotyk (1973). This preparation bound D-glucose, D-xylose, D-arabinose, and D-sorbose but was unable to bind L-arabinose or D-galactose. In other words, it possessed the same specificity as the constitutive hexose transport system. The lipoprotein behaved as a single species on native polyacrylamide gels and possessed an apparent molecular weight of 250,000 (determined by thin-layer chromatography using Sephadex G-150 as the matrix). The physicochemical parameters of this preparation, however, must be viewed skeptically, because it was not possible to separate the lipid from the protein.

All of the above studies were conducted with crude membrane preparations or solubilized total cell protein and, hence, it is appropriate to query whether or not the proteins were in fact associated with transport. An alternative approach to the problem was developed by Opekarova et al. (1975). They observed that suspending cells in water or dilute buffers resulted in loss of ability to transport arginine via the basic amino acid transport system. Transport ability was recovered after 2 hours, but recovery could be prevented by addition of cycloheximide. In the osmotic shock fluid, Opekarova and his collaborators found amounts of protein that were proportional to the transport activity lost by osmotic shock. Using the shock fluid as starting material, a 4500–4800-m.w. protein was purified. This protein possessed a dissociation constant for arginine of 0.38 mM and a K_i of 0.41 mM for lysine inhibition of this binding. However, the protein could also be purified from the osmotic shock fluid of *can1* mutants, which possess a defective form of the basic amino acid transport system. This observation led to the untested conclusion that Grenson's mutant must be defective in an integral protein needed for translocation.

Iwashima and Nishimura (1979) have reported the isolation of a thiamine-binding protein. The purified preparation is claimed to be electrophoretically pure and to possess an apparent molecular weight of 140,000. If shock fluid derived from cells grown in the presence and absence of thiamine is subjected to SDS-gel electrophoresis, the 140K species is found only when thiamine is omitted from the medium. This result is congruent with that predicted from in vivo studies. Unfortunately, more recent data have questioned this interpretation. Iwashima et al. (1979) found that a thiamine transport mutant (PT-R2) still contained the soluble thiamine-binding protein in amounts comparable to the wild-type strain. In addition, a second, membrane-bound thiamine-binding activity was discovered possessing an apparent K_d (0.17 μM) almost identical to the apparent K_m of the thiamine transport system in vivo. This activity was repressible by exogenous thiamine and was largely absent from the PT-R2 mutant strain. Finally, protoplasts that do not contain the soluble, thiamine-binding protein are able to transport thiamine normally (Nishimura et al. 1982). In sum, these data argue that the soluble 140-kD thiamine-binding protein is not an essential component of the transport system and point to the membrane-bound activity as an interesting candidate to mediate this function.

Woodward and Kornberg (1980) reported isolation of a periplasmic chloroacetyl-ornithine-binding protein with a molecular weight of 14,000. No chloroacetyl ornithine binding to cell exteriors was observed if (1) a *gap1* mutant was used in place of the wild type, (2) cells were grown under conditions leading to repression of *GAP1* transport system, or (3) substrates of the general amino acid permease were included along with the radioactive chloroacetyl ornithine. Unlike the protein described by Operkorova, this species was not released by osmotic shock but was solubilized on converting the cells to spheroplasts. In sum, these observations led Woodward and Kornberg to conclude that this species was the periplasmic-binding protein of the general amino acid permease. At least three other membrane-associated proteins (53K, 45K, 30K) were markedly decreased in the *gap1* mutant cells. The putative functions of these species and the basis for their absence in the mutant remains obscure.

Parlebas and Chevallier (1976) compared the pattern of membrane proteins separated from the wild-type by SDS-polyacrylamide gel electrophoresis and two mutants lacking cytosine permease activity. An 80-kD species was markedly decreased from both mutant preparations, leading the investigators to suggest that this protein might be responsible for cytosine transport.

SUMMARY

At the beginning of this paper, I described my view of transport in yeast as it exists today—a body of literature devoted largely to identifying transport systems that function in this organism. That task, though not finished, is reaching a point of diminishing return. The information already in hand points to new and more intriguing questions. A striking paradox is emphasized by the current descriptive literature. Many investigators have isolated many transport-defective mutants. Yet, in each case the mutations resulting in loss of any specific transport system all fall into a single complementation group. It is naive to argue that metabolite transport is mediated by single proteins. Therefore, it will be enormously interesting to sort out the network of elements that function in multiple transport systems. Some of these proteins may serve not only transport functions but structural roles as well. In the latter instance, mutant forms may generate lethality. Elements of the energy transduction system, already an area of high activity, will form a major subclass of these pleiotropically acting species. Synthesis and assembly of the membrane proteins is a second area of increasing interest. Here, a major beginning is well under way as described elsewhere in this volume (Schekman and Novick). Finally, there are the control mechanisms that integrate overall operation of the cellular transport and metabolite compartmentation systems. The magnitude and subtlety of control exerted on these systems is already clear from the superficial phenomenology. An understanding of the molecular mechanisms by which such regulation is achieved is a major objective for the immediate future.

ACKNOWLEDGMENTS

I express my gratitude to Professors Gail Willsky for her help with the ATPase section, Vince Cirillo for assistance with the carbohydrate uptake systems, and Dr. Roberta Sumrada for indexing both this chapter and the one on nitrogen repression. Thanks are also due to the editors and reviewers whose patience and probing comments materially improved the quality of this work. Finally, I thank my colleagues at the University of Pittsburgh for their understanding; their manuscripts were often delayed because I was occupied with this review.

REFERENCES

Ahlers, J., E. Ahr, and A. Seyfarth. 1978. Kinetic characterization of plasma membrane ATPase from *Saccharomyces cerevisiae*. *Mol. Cell. Biochem.* **22**: 39.
Arnold, W.N. 1972. Location of acid phosphate and β-fructofuranosidase within yeast cell envelopes. *J. Bacteriol.* **112**: 1346.
———. 1973. Volume and enzyme content of the periplasmic space in yeast. *Physiol. Chem. Physics* **5**: 117.
Arnold, W.N. and J.S. Lacy. 1977. Permeability of the cell envelope and osmotic behavior in *Saccharomyces cerevisiae*. *J. Bacteriol.* **131**: 564.
Azam, F., S.-C. Kuo, and V.P. Cirillo. 1971. Production of phenotypically epimeraseless yeast by L-arbinose. *J. Bacteriol.* **106**: 915.
Becker, J.M., F. Naider, and E. Katchalski. 1973. Peptide utilization in yeast. Studies on methionine and lysine auxotrophs of *Saccharomyces cerevisiae*. *Biochim. Biophys. Acta* **291**: 388.
Becker, J.M., M. Wilchek, and E. Katchalski. 1971. Irreversible inhibition of biotin transport in yeast by biotinyl-p-nitrophenyl ester. *Proc. Natl. Acad. Sci.* **68**: 2604.
Boller, T., M. Durr, and A. Wiemken. 1975. Characterization of a specific transport system for arginine in isolated yeast vacuoles. *Eur. J. Biochem.* **54**: 81.
———. 1976. Asymmetric distribution of concanavalin A sites on yeast plasmalemma and vacuolar membrane. *Arch. Microbiol.* **109**: 115.
Borst-Pauwels, G.W.F.H. 1981. Ion transport in yeast. *Biochim. Biophys. Acta* **650**: 88.
Borst-Pauwels, G.W.F.H. and S. Jager. 1969. Inhibition of phosphate and arsenate uptake in yeast monoiodoacetate, fluoride, 2,4-dinitrophenol and acetate. *Biochim. Biophys. Acta* **172**: 399.
Borst-Pauwels, G.W.F.H. and P.J.H. Peters. 1977. Effect of the medium pH and the cell pH upon the kinetical parameters of phosphate uptake by yeast. *Biochim. Biophys. Acta* **466**: 488.
Brendel, M., W.W. Fath, and W. Laskowski. 1975. Isolation and characterization of mutants of *Saccharomyces cerevisiae* able to grow after inhibition of dTMP synthesis. *Methods Cell Biol.* **11**: 287.
Breton, A. and Y. Surdin-Kerjan. 1977. Sulfate uptake in *Saccharomyces cerevisiae*: Biochemical and genetic study. *J. Bacteriol.* **132**: 224.
Broach, J.R., J.N. Strathern, and J.B. Hicks. 1979. Transformation in yeast: Development of a hybrid cloning vector and isolation of the *CAN1* gene. *Gene* **8**: 121.
Burger, M., L. Hejmova, and A. Kleinzeller. 1959. Transport of some mono- and disaccharides into yeast cells. *Biochem. J.* **71**: 233.
Bussey, H. and H.E. Umbarger. 1970a. Biosynthesis of the branched-chain amino acids in yeast: A leucine-binding component and regulation of leucine uptake. *J. Bacteriol.* **103**: 277.
———. 1970b. Biosynthesis of branched-chain amino acids in yeast—A trifluoroleucine resistant mutant with altered regulation of leucine uptake. *J. Bacteriol.* **103**: 286.
Bussey, H., D. Saville, M.R. Chevallier, and G.H. Rank. 1979. Yeast plasma membrane ghosts: An analysis of proteins by two-dimensional gel electrophoresis. *Biochim. Biophys. Acta* **553**: 185.
Canh, D.S., J. Horak, A. Kotyk, and L. Ríhová. 1975. Transport of acyclic polyols in *Saccharomyces cerevisiae*. *Folia Microbiol.* **20**: 320.
Chan, P.Y. and F.A. Cossins. 1976. General properties and regulation of arginine transporting systems in *Saccharomyces cerevisiae*. *Plant and Cell Physiol.* **17**: 341.
Chevallier, M.-R. 1982. Cloning and transcriptional control of a eucaryotic permease gene. *Mol. Cell. Biol.* **2**: 977.
Chevallier, M.-R. and F. Lacroute. 1982. Expression of the cloned uracil permease gene of *Saccharomyces cerevisiae* in a heterologous membrane. *EMBO J.* **1**: 375.

Chevallier, M.-R., R. Jund, and F. Lacroute. 1975. Characterization of cytosine permeation in *Saccharomyces cerevisiae. J. Bacteriol.* **122:** 629.
Christensen, M.S. and V.P. Cirillo. 1972. Yeast membrane vesicles: Isolation and general characteristics. *J. Bacteriol.* **110:** 1190.
Cicmanec, J.F. and H.C. Lichstein. 1974. Biotin uptake by cold-shocked cells, spheroplasts, and repressed cells of *Saccharomyces cerevisiae*: Lack of feedback control. *J. Bacteriol.* **119:** 718.
Cirillo, V.P. 1962. Mechanism of glucose transport across the yeast cell membrane. *J. Bacteriol.* **84:** 485.
―――. 1968a. Relationship between sugar structure and competition for the sugar transport system in baker's yeast. *J. Bacteriol.* **95:** 603.
―――. 1968b. Galactose transport in *Saccharomyces cerevisiae*. I. Nonmetabolized sugars as substrate and inducers of the galactose transport system. *J. Bacteriol.* **95:** 1727.
Cockburn, M., P. Earnshaw, and A.A. Eddy. 1975. The stoichiometry of the absorption of protons with phosphate and L-glutamate by yeasts of the genus *Saccharomyces. Biochem. J.* **146:** 705.
Conway, E.J., T.G. Brady, and E. Carton. 1950. Biological production of acid and alkali 2. A redox theory for the process in yeast with application to the production of gastric acidity. *Biochem. J.* **47:** 369.
Cooper, T.G. and R.P. Lawther. 1973. Induction of the allantoin degradative enzymes in *Saccharomyces cerevisiae* by the last intermediate of the pathway. *Proc. Natl. Acad. Sci.* **70:** 2340.
Cooper, T.G. and R. Sumrada. 1975. Urea transport in *Saccharomyces cerevisiae. J. Bacteriol.* **121:** 571.
Cooper, T.G., M. Gorski, and V. Turoscy. 1979a. A cluster of three genes responsible for allantoin degradation in *Saccharomyces cerevisiae. Genetics* **92:** 383.
Cooper, T.G., J. McKelvey, and R. Sumrada. 1979b. Oxalurate transport in *Saccharomyces cerevisiae. J. Bacteriol.* **139:** 917.
Crabeel, M. and M. Grenson. 1970. Regulation of histidine uptake by specific feedback inhibition of two histidine permeases in *Saccharomyces cerevisiae. Eur. J. Biochem.* **14:** 197.
Darte, C. and M. Grenson. 1975. Evidence for three glutamic acid transporting systems with specialized physiological functions in *Saccharomyces cerevisiae. Biochem. Biophys. Res. Commun.* **67:** 1028.
De la Fuente, G. and A. Sols. 1962. Transport of sugars in yeast. II. Mechanisms of utilization of disaccharides and related glycosides. *Biochim. Biophys. Acta* **56:** 49.
De la Pena, P., F. Barros, S. Gascon, S. Ramos, and P.S. Lazo. 1982. The electrochemical proton gradient of *Saccharomyces*. The role of potassium. *Eur. J. Biochem.* **123:** 447.
Douglas, H.C. and F. Condie. 1954. The genetic control of galactose utilization in *Saccharomyces. J. Bacteriol.* **68:** 662.
Drillien, R. and F. Lacroute. 1972. Ureidosuccinic acid uptake in yeast and some aspects of its regulation. *J. Bacteriol.* **109:** 203.
Dubois, E. and M. Grenson. 1979. Methylamine/ammonia uptake systems in *Saccharomyces cerevisiae*: Multiplicity and regulation. *Mol. Gen. Genet.* **175:** 67.
Dunlop, P.C. and R.J. Roon. 1975. L-asparagine of *Saccharomyces cerevisiae*: An extracellular enzyme. *J. Bacteriol.* **122:** 1017.
Durr, M., T. Boller, and A. Wiemken. 1975. Polybase induced lysis of yeast spheroplasts. *Arch. Microbiol.* **105:** 319.
―――. 1976. Action of proteinases on the arginine transport system of purified vacuoles from *Saccharomyces cerevisiae. Biochem. Biophys. Res. Commun.* **73:** 193.
Durr, K. Urech, T. Boller, A. Wiemken, J. Schwencke, and M. Nagy. 1979. Sequestration of arginine by polyphosphate in vacuoles of yeast (*Saccharomyces cerevisiae*). *Arch. Microbiol.* **121:** 169.

Eddy, A.A. 1982. Mechanism of solute transport in selected eukaryotic micro-organisms. *Adv. Microbiol. Physiol.* **23**:2.

Eddy, A.A. and K.J. Indge. 1962. The dependence of amino acid transport in yeast on both energy supply and intracellular potassium ions. *Biochem. J.* **82**:15p.

Eddy, A.A. and J.A. Nowacki. 1971. Stoichiometrical proton and potassium ion movements accompanying absorption of amino acids by yeast *Saccharomyces carlsbergensis*. *Biochem. J.* **122**:701.

Eddy, A.A., K. Backen, and J. Nowacki. 1970a. Translocation of protons and alkalimetal cations accompanying uptake of neutral amino acids by yeast. *Biochem. J.* **116**:34.

Eddy, A.A., K. Backen, and G. Watson. 1970b. Concentration of amino acids by yeast cells depleted of adenosine-triphosphate. *Biochem. J.* **120**:853.

Eddy, A.A., K.J. Indge, K. Backen, and J.A. Nowacki. 1970c. Interactions between potassium ions and glycine transport in yeast *Saccharomyces carlsbergensis*. *Biochem. J.* **120**:845.

Foury, F., A. Amory, and A. Goffeau. 1981. Large-scale purification and phosphorylation of a detergent-treated adenosine triphosphatase complex from plasma membrane of *Saccharomyces cerevisiae*. *Eur. J. Biochem.* **119**:395.

Franzusoff, A. and V.P. Cirillo. 1982. Uptake and phosphorylation of 2-deoxy-D-glucose by wild-type and single-kinase strains of *Saccharomyces cerevisiae*. *Biochim. Biophys. Acta* **688**:295.

Fuhrmann, G.F. and A. Rothstein. 1968. The transport of Zn^{2+}, CO^{2+} and Ni^{2+} into yeast cells. *Biochim. Biophys. Acta* **163**:325.

Fuhrmann, G.F., C. Boehm, and A.P.R. Theuvenet. 1976. Sugar transport and potassium permeability in yeast plasma membrane vesicles. *Biochim. Biophys. Acta* **433**:583.

Fuhrmann, G.F., E. Wehrli, and C. Boehm. 1974. Preparation and identification of yeast plasma membrane vesicles. *Biochim. Biophys. Acta* **363**:295.

Gits, J.J. and M. Grenson. 1967. Multiplicity of the amino acid permeases in *Saccharomyces cerevisiae*. III. Evidence for a specific methionine-transporting system. *Biochim. Biophys. Acta* **135**:507.

Goodman, J. and A. Rothstein. 1957. The active transport of phosphate into the yeast cell. *J. Gen. Physiol.* **40**:915.

Gorts, C.P.M. 1969. Effect of glucose on the activity and the kinetics of the maltose-uptake system and of α-glucosidase in *Saccharomyces cerevisiae*. *Antonie Leeuwenhoek. J. Microbiol.* **35**:233.

Greasham, R.L. and A.G. Moat. 1973. Amino acid transport in a polyaromatic amino acid auxotroph of *Saccharomyces cerevisiae*. *J. Bacteriol.* **115**:975.

Grenson, M. 1966. Multiplicity of the amino acid permeases in *Saccharomyces cerevisiae*. II. Evidence for a specific lysine-transporting system. *Biochim. Biophys. Acta* **127**:339.

———. 1969. The utilization of exogenous pyrimidines and the recycling of uridine-5'-phosphate derivatives in *Saccharomyces cerevisiae*, as studied by means of mutants affected in pyrimidine uptake and metabolism. *Eur. J. Biochem.* **11**:249.

Grenson, M. and C. Hennaut. 1971. Mutation affecting activity of several distinct amino transport systems in *Saccharomyces cerevisiae*. *J. Bacteriol.* **105**:477.

Grenson, M. and C. Hou. 1972. Ammonia inhibition of the general amino acid permease and its suppression in NADPH-specific glutamate dehydrogenaseless mutants of *Saccharomyces cerevisiae*. *Biochem. Biophys. Res. Commun.* **48**:749.

Grenson, M., C. Hou, and M. Crabeel. 1970. Multiplicity of the amino acid permeases in *Saccharomyces cerevisiae*. IV. Evidence for a general amino acid permease. *J. Bacteriol.* **103**:770.

Grenson, M., M. Mousset, J.M. Wiame, and J. Bechet. 1966. Multiplicity of the amino acid permeases in *Saccharomyces cerevisiae*. I. Evidence for a specific arginine-transporting system. *Biochim. Biophys. Acta* **127**:325.

Greth, M.L. M.R. Chevallier, and F. Lacroute. 1977. Ureidosuccinic acid permeation in *Saccharomyces cerevisiae*. *Biochim. Biophys. Acta* **465**:138.

Grivell, A.R. and J.F. Jackson. 1968. Thymidine kinase: Evidence for its absence from *Neurospora crassa* and some other micro-organisms and the relevance of this to the specific labeling of deoxyribonucleic acid. *J. Gen. Microbiol.* **54**: 307.

Gross, K.J. and O. Pogo. 1974. Control mechanism of ribonucleic acid synthesis in eukaryotes. *J. Biol. Chem.* **249**: 568.

Halvorson, H.O. and G.N. Cohen. 1958. Incorporation des amino-acides endogenes dans les proteines de la levure. *Ann. Inst. Pasteur* **95**: 73.

Harris, G. and C.C. Thompson. 1961. The uptake of nutrients by yeasts. III. The maltose permease of a brewing yeast. *Biochim. Biophys. Acta* **52**: 176.

Haskovec, C. and A. Kotyk. 1969. Attempts at purifying the galactose carrier from galactose-induced baker's yeast. *Eur. J. Biochem.* **9**: 343.

Hennaut, C., F. Hilger, and M. Grenson. 1970. Space limitation for permease insertion in the cytoplasmic membrane of *Saccharomyces cerevisiae*. *Biochem. Biophys. Res. Commun.* **39**: 666.

Heredia, C.F., A. Sols, and G. De la Fuente. 1968. Specificity of the constitutive hexose transport in yeast. *Eur. J. Biochem.* **5**: 321.

Horak, J. and A. Kotyk. 1973. Isolation of a glucose-binding lipoprotein from yeast plasma membrane. *Eur. J. Biochem.* **32**: 36.

Hosaka, K. and S. Yamashita. 1980. Choline transport in *Saccharomyces cerevisiae*. *J. Bacteriol.* **143**: 176.

Hutchinson, H.T. and L.H. Hartwell. 1967. Macromolecule synthesis in yeast spheroplasts. *J. Bacteriol.* **94**: 1697.

Huygen, P.L.M. and G.W.F.H. Borst-Pauwels. 1972. The effect of N,N'-dicyclohexylcarbodiimide on anaerobic and aerobic phosphate uptake by baker's yeast. *Biochim. Biophys. Acta* **283**: 234.

Indge, K.J. 1968a. The isolation and properties of the yeast cell vacuole. *J. Gen. Microbiol.* **51**: 441.

———. 1968b. Polyphosphates of the yeast cell vacuole. *J. Gen. Microbiol.* **51**: 447.

Indge, K., A. Seaston, and A.A. Eddy. 1977. The concentration of glycine by *Saccharomyces uvarum*: Role of the main vacuole and conditions leading to the explosive absorption of the amino acid. *J. Gen. Microbiol.* **99**: 243.

Iwashima, A. and H. Nishimura. 1979. Isolation of a thiamine-binding protein from *Saccharomyces cerevisiae*. *Biochim. Biophys. Acta* **577**: 217.

Iwashima, A. and Y. Nose. 1976. Regulation of thiamine transport in *Saccharomyces cerevisiae*. *J. Bacteriol.* **128**: 855.

Iwashima, A., H. Nichino, and Y. Nose. 1973. Carrier-mediated transport of thiamine in baker's yeast. *Biochim. Biophys. Acta* **330**: 222.

Iwashima, A., H. Nishimura, and Y. Nose. 1979. Soluble and membrane-bound thiamine-binding proteins from *Saccharomyces cerevisiae*. *Biochim. Biophys. Acta* **557**: 460.

Iwashima, A., Y. Wakabayashi, and Y. Nose. 1975. Thiamine transport mutants of *Saccharomyces cerevisiae*. *Biochim. Biophys. Acta* **413**: 243.

———. 1977. Inhibition of thiamine transport in *Saccharomyces cerevisiae* by thiamine disulfides. *J. Bacteriol.* **131**: 1013.

Jennings, D.H., D.C. Hooper, and A. Rothstein. 1958. The participation of phosphate in the formation of a "carrier" for the transport of Mg^{++} and Mn^{++} ions into yeast cells. *J. Gen. Physiol.* **41**: 1019.

Jund, R. and F. Lacroute. 1970. Genetic and physiological aspects of resistance to 5-fluoropyrimidines in *Saccharomyces cerevisiae*. *J. Bacteriol.* **102**: 607.

Jund, R., M.R. Chevallier, and F. Lacroute. 1977. Uracil transport in *Saccharomyces cerevisiae*. *J. Membr. Biol.* **36**: 233.

Kakinuma, Y., Y. Ohsumi, and Y. Anraku. 1981. Properties of H^+-translocating adenosine triphosphatase in vacuolar membranes of *Saccharomyces cerevisiae*. *J. Biol. Chem.* **256**: 10859.

Kamen, M.D. and S. Spiegelman. 1949. Studies on the phosphate metabolism of some unicellular organisms. *Cold Spring Harbor Symp. Quant. Biol.* **13:** 151.
Kotyk, A. 1967. Properties of the sugar carrier in baker's yeast. II. Specificity of transport. *Folia Microbiol.* **12:** 121.
Kotyk, A. and K. Janacek. 1975. *Cell membrane transport*, p. 341. Plenum Press, New York.
Kotyk, A. and D. Michaljanicova. 1979. Uptake of trehalose by *Saccharomyces cerevisiae*. *J. Gen. Microbiol.* **110:** 323.
Kotyk, A. and L. Říhová. 1972. Transport of α-aminoisobutyric acid in *Saccharomyces cerevisiae*. Feedback control. *Biochim. Biophys. Acta* **288:** 380.
Kou, S.-C. and V.P. Cirillo. 1970. Galactose transport in *Saccharomyces cerevisiae*. III. Characteristics of galactose uptake in transferaseless cells: Evidence against transport-associated phosphorylation. *J. Bacteriol.* **103:** 679.
Kou, S.-C., M.S. Christensen, and V.P. Cirillo. 1970. Galactose transport in *Saccharomyces cerevisiae*. II. Characteristics of galactose uptake and exchange in galactokinaseless cells. *J. Bacteriol.* **103:** 671.
Kramer, R., F. Kopp, W. Niedermeyer, and G.F. Fuhrmann. 1978. Comparative studies of the structure and composition of the plasmalemma and the tonoplast in *Saccharomyces cerevisiae*. *Biochim. Biophys. Acta* **507:** 369.
Lacroute, E.F. 1966. "Regulation de la chaine de biosynthese de l'uracile chez *Saccharomyces cerevisiae*." Ph.D. thesis, University of Paris.
Langen, P., E. Liss, and K. Lohmann. 1962. Art, bildung und umsatz der polyphosphate der hefe. In *Acides Ribonucleiques et Polyphosphates, Colloques Internationaux du CNRS*, Strasbourg. p. 603.
Larimore, F.S. and R.J. Roon. 1978. Possible site-specific reagent for the general amino acid transport system of *Saccharomyces cerevisiae*. *Biochemistry* **17:** 431.
Larimore, F.S., I. Kuist, P.M. Korkowski, and R.J. Roon. 1980. Transport and metabolism of N-δ-chloroacetyl-L-ornithine by *Saccharomyces cerevisiae*. *Arch. Biochem. Biophys.* **204:** 34.
Lasko, P.F. and M.C. Brandriss. 1981. Proline transport in *Saccharomyces cerevisiae*. *J. Bacteriol.* **148:** 241.
Law, R.E. and A.J. Ferro. 1980. Inhibition of leucine transport in *Saccharomyces* by S-adenosylmethione. *J. Bacteriol.* **143:** 427.
Lawther, R. and T.G Cooper. 1973. Induction of the allantoin degradative enzymes by allophanic acid, the last intermediate of the pathway. *Biochem. Biophys. Res. Commun.* **52:** 137.
Lawther, R.P., S.L. Phillips, and T.G. Cooper. 1975. Lomofungin inhibition of allophanate hydrolase synthesis in *Saccharomyces cerevisiae*. *Mol. Gen. Genet.* **137:** 89.
Lichko, L.P. and L.A. Okorokov. 1976. The compartmentalization of magnesium and phosphate ions in *Saccharomyces carlsbergenesis* cells. *Dokl. Akad. Nauk. U.S.S.R.* **227:** 756.
Loss, R., R. Jund, and F. Lacroute. 1978. Studies on membrane protein turnover in growing yeast cells. *FEBS Lett.* **86:** 285.
Malpartida, F. and R. Serrano. 1980. Purification of the yeast plasma membrane ATPase solubilized with a novel Zwitterionic detergent. *FEBS Lett.* **111:** 69.
———. 1981a. Proton translocation catalysed by the purified yeast plasma membrane ATPase reconstituted in liposomes. *FEBS Lett.* **131:** 351.
———. 1981b. Reconsitution of the proton-translocating adenosine triphosphatase of yeast plasma membrane. *J. Biol. Chem.* **256:** 4175.
———. 1981c. Phosphorylated intermediate of the ATPase from the plasma membrane of yeast. *Eur. J. Biochem.* **116:** 413.
Marder, R., J.M. Becker, and F. Naider. 1977. Peptide transport in yeast: Utilization of leucine- and lysine-containing peptides in *Saccharomyces cerevisiae*. *J. Bacteriol.* **131:** 906.
Marder, R., B. Rose, J.M. Becker, and F. Naider. 1978. Isolation of a peptide transport-deficient mutant of yeast. *J. Bacteriol.* **36:** 1174.

Matern, H. and H. Holzer. 1977. Catabolite inactivation of the galactose uptake system in yeast. *J. Biol. Chem.* 252:6399.

Matile, P., H. Moor, and K. Muhlethaler. 1967. Isolation and properties of plasmalemma in yeast. *Arch. Mikrobiol.* 58:201.

Maw, G.A. 1963. The uptake of some sulphur-containing amino acids by a brewer's yeast. *J. Gen. Microbiol.* 31:247.

Maxwell, W.A. and E. Spoerl. 1971. Mannitol uptake by *Saccharomyces cerevisiae*. *J. Bacteriol.* 105:753.

McCready, R.G.L. and G.A. Din. 1974. Active sulfate transport in *Saccharomyces cerevisiae*. *FEBS Lett.* 38:361.

Meredith, S.A. and A.H. Romano. 1977. Uptake and phosphorylation of 2-deoxy-D-glucose by wild type and respiration-deficient baker's yeast. *Biochim. Biophys. Acta* 497:745.

Merkel, G.J., F. Naider, and J.M. Becker. 1980. Amino acid uptake by *Saccharomyces cerevisiae* plasma membrane vesicles. *Biochim. Biophys. Acta* 595:109.

Miller, M.J., N. Xuong, and E.P. Geiduschek. 1979. A response of protein synthesis to temperature shift in the yeast *Saccharomyces cerevisiae*. *Proc. Natl. Acad. Sci.* 76:5222.

Morlion, M. and A. Domnas. 1962. Uptake and use of allantoin and allantoic acid by yeasts. *Natuurwet. Tijdschr.* 44:100.

Murphy, J.T. and K.D. Spence. 1972. Transport of S-adenosylmethionine in *Saccharomyces cerevisiae*. *J. Bacteriol.* 109:499.

Nagy, M. 1979. Studies on purine transport and on purine content in vacuoles isolated from *Saccharomyces cerevisiae*. *Biochim. Biophys. Acta* 558:221.

Naider, F., J.M. Becker, and E. Katzir-Katchalski. 1974. Utilization of methione-containing peptides and their derivatives by a methionine-requiring auxotroph of *Saccharomyces cerevisiae*. *J. Biol. Chem.* 249:9.

Naider, F., S.A. Khan, D.D. Parker, and J.M. Becker. 1980. Inhibition of oligopeptide transport in *S. cerevisiae* by a peptide-poly (ethylene glycol) conjugate. *Biochem. Biophys. Res. Commun.* 95:1187.

Nakamura, K.D. and F. Schlenk. 1974. Active transport of exogenous S-adenosylmethionine and related compounds into cells and vacuoles of *Saccharomyces cerevisiae*. *J. Bacteriol.* 120:482.

Nikawa, J.-I., T. Nagumo, and S. Yamashita. 1982. Myo-inositol transport in *Saccharomyces cerevisiae*. *J. Bacteriol.* 150:441.

Nishimura, H., K. Sempuku, and A. Iwashima. 1982. Thiamine transport in *Saccharomyces cerevisiae* protoplasts. *J. Bacteriol.* 150:960.

Nurminen T., E. Oura, and H. Soumalainen. 1970. The enzymic composition of the isolated cell wall and plasma membrane of baker's yeast. *Biochem. J.* 116:61.

Ohsumi, Y. and Y. Anraku. 1981. Active transport of basic amino acids driven by a proton motive force in vacuolar membrane vesicles of *Saccharomyces cerevisiae*. *J. Biol. Chem.* 256:2079.

Okada, H. and H.O. Halvorson. 1964a. Uptake of α-thioethyl D-glucopyranoside by *Saccharomyces cerevisiae*. I. The genetic control of facilitated diffusion and active transport. *Biochim. Biophys. Acta* 82:538.

———. 1964b. Uptake of α-thioethyl D-glucopyranoside by *Saccharomyces cerevisiae*. II. General characteristics of an active transport system. *Biochim. Biohys. Acta* 82:547.

Okorokov, L.A., S.B. Petrikevich, L.P. Lichko, and E.V. Melnikova. 1978. Magnesium pool in cell vacuoles of *Saccharomyces cerevisiae*. *Izv. Akad. Nauk. U.S.S.R.* 5:791.

Okorokov, L., L. Lichko, V. Kadomtseva, V. Kholodenko, V. Titovsky, and I. Kulaev. 1977. I. Energy-dependent transport of manganese into yeast cells and distribution of accumulated ions. *Eur. J. Biochem.* 75:373.

Opekarova, M., A. Kotyk, J. Horak, and V.P. Kholodenko. 1975. Isolation and properties of an arginine-binding protein from *Saccharomyces cerevisiae*. *Eur. J. Biochem.* 59:73.

Parker, D.D., F. Naider, and J.M. Becker. 1980. Separation of peptide transport and hydrolysis in trimethionine uptake by *Saccharomyces cerevisiae*. *J. Bacteriol.* **141**: 1066.

Parlebas, N. and M.R. Chevallier. 1976. Biochemical studies on the cytosine permease of *Saccharomyces cerevisiae*. *FEBS Lett.* **65**: 327.

Perl, M., E.B. Kearney, and T.P. Singer. 1976. Transport of riboflavin into yeast cells. *J. Biol. Chem.* **251**: 3221.

Petrotta-Simpson, T.F., J.E. Talmadege, and K.D. Spencer. 1975. Specificity and genetics of *S*-adenosylmethionine transport in *Saccharomyces cerevisiae*. *J. Bacteriol.* **123**: 516.

Pickering, W.R. and R.A. Woods. 1972. The uptake and incorporation of purines by wild-type *Saccharomyces cerevisiae* and a mutant resistant to 4-aminopyrazolo-(3,4-d)pyrimidine. *Biochim. Biophys. Acta* **264**: 45.

Polak, A. and M. Grenson. 1973. Evidence for a common transport system for cytosine, adenine and hypoxanthine in *Saccharomyces cerevisiae* and *Candida albicans*. *Eur. J. Biochem.* **32**: 276.

Ramos, E.H., L.C. De Bongioanni, and A.O.M. Stoppani. 1980. Kinetics of L-(^{14}C) leucine transport in *Saccharomyces cerevisiae*. Effect of energy coupling inhibitors. *Biochim. Biophys. Acta* **599**: 214.

Ramos, E.H., L.C. De Bongioanni, M.L. Claise, and A.O.M. Stoppani. 1975. Energy requirements for the uptake of L-leucine by *Saccharomyces cerevisiae*. *Biochim. Biophys. Acta* **394**: 470.

Rogers, T.O. and H.C. Lichstein. 1969a. Characterization of the biotin transport system in *Saccharomyces cerevisiae*. *J. Bacteriol.* **100**: 557.

———. 1969b. Regulation of biotin transport in *Saccharomyces cerevisiae*. *J. Bacteriol.* **100**: 565.

Roomans, G.M. and G.W.F.H. Borst-Pauwels. 1977. Interaction of phosphate with monovalent uptake in yeast. *Biochim. Biophys. Acta* **470**: 84.

———. 1979. Interactions of cations with phosphate uptake in *Saccharomyces cerevisiae*. Effects of surface potential. *Biochem. J.* **178**: 521.

Roomans, G.M. and L.A. Seveus. 1976. Subcellular localization of diffusible ions in the yeast *Saccharomyces cerevisiae*: Quantitative microprobe analysis of thin freeze-dried sections. *J. Cell Sci.* **21**: 119.

Roomans, G.M., F. Blasco, and G.W.F.H. Borst-Pauwels. 1977. Cotransport of phosphate and sodium by yeast. *Biochim. Biophys. Acta* **467**: 65.

Roomans, G.M., G.A.J. Kuypers, A.P.R. Theuvenet, and G.W.F.H. Borst-Pauwels. 1979a. Kinetics of sulfate uptake by yeast. *Biochim. Biophys. Acta* **551**: 197.

Roomans, G.M., A.P.R. Theuvenet, T.P.R. Van Den Berg, and G.W.F.H. Borst-Pauwels. 1979b. Kinetics of Ca^{+2} and Sr^{+2} uptake by yeast-effects of pH, cations and phosphate. *Biochim. Biophys. Acta* **551**: 187.

Roon, R.J., F. Larimore, and J.S. Levy. 1975a. Inhibition of amino acid transport by ammonium in *Saccharomyces cerevisiae*. *J. Bacteriol.* **124**: 325.

Roon, R.J., J.S. Levy, and F. Larimore. 1977a. Negative interactions between amino acid and methylamine/ammonia transport systems of *Saccharomyces cerevisiae*. *J. Biol. Chem.* **252**: 3599.

Roon, R.J., G.M. Meyer, and F.S. Larimore. 1977b. Evidence for a common component in kinetically distinct systems of *Saccharomyces cerevisiae*. *Mol. Gen. Genet.* **158**: 185.

Roon, R.J., H.L. Even, P. Dunlop, and F.L. Larimore. 1975b. Methylamine and ammonia transport in *Saccharomyces cerevisiae*. *J. Bacteriol.* **122**: 502.

Ross, I.S. 1977. Effect of glucose on copper uptake and toxicity in *Saccharomyces cerevisiae*. *Trans. Br. Mycol. Soc.* **69**: 77.

Rothstein, A., A. Hayes, D. Jennings, and D. Hooper. 1958. The active transport of Mg^{++} and Mn^{++} into the yeast cell. *J. Gen. Physiol.* **41**: 585.

Rytka, J. 1975. Positive selection of general amino acid permease mutants in *Saccharomyces cerevisiae*. *J. Bacteriol.* **121**: 562.

Santos, E., F. Leal, and R. Sentandreu. 1982. The plasma membrane of *Saccharomyces cerevisiae* molecular structure and asymmetry. *Biochim. Biophys. Acta* **685:** 329.

Santos, E., L. Rodriguez, M.V. Elorza, and R. Sentandreu. 1982. Uptake of sucrose transport by *Saccharomyces cerevisiae*. *Arch. Biochem. Biophys.* **216:** 652.

Scherrer, R., L. Louden, and P. Gerhardt. 1974. Porosity of the yeast cell wall and membrane. *J. Bacteriol.* **118:** 534.

Schindler, D. and J. Davies. 1975. Inhibitors of macromolecular synthesis in yeast. *Methods Cell Biol.* **12:** 17.

Schlenk, F. and J.L. Dainko. 1965. Action of ribonuclease preparations on viable yeast cells and spheroplasts. *J. Bacteriol.* **89:** 428.

Schlenk, F. and C.R. Zydek-Cwick. 1970. Enzymatic activity of yeast cell ghosts produced by protein action on the membranes. *Arch. Biochem.* **138:** 220.

Schlenk, F., J.L. Dainko, and G. Svihla. 1970. The accumulation and intracellular distribution of biological sulfonium compounds in yeast. *Arch Biochem. Biophys.* **140:** 228.

Schmidt, G., L. Hecht, and S.J. Thannhauser. 1949. The effect of potassium ions on the absorption of orthophosphate and the formation of metaphosphate by baker's yeast. *J. Biol. Chem.* **178:** 733.

Schwencke, J. and H. de Robichon-Szulmajster. 1976. The transport of S-adenosyl-l-methionine in isolated yeast vacuoles and spheroplasts. *Eur. J. Biochem.* **65:** 49.

Seaston, A., C. Inkson, and A.A. Eddy. 1973. The absorption of protons with specific amino acids and carbohydrates by yeast. *Biochem. J.* **134:** 1031.

Serrano, R. 1977. Energy requirements for maltose transport in yeast. *Eur. J. Biochem.* **80:** 97.

———. 1978. Characterization of the plasma membrane ATPase of *Saccharomyces cerevisiae*. *Mol. Cell Biochem.* **22:** 51.

Serrano, R. and G. De la Fuente. 1974. Regulatory properties of the constitutive hexose transport in *Saccharomyces cerevisiae*. *Mol. Cell. Biochem.* **5:** 161.

Shulman, R.W., C.E. Sripati, and J.R. Warner. 1977. Noncoordinated transcription in the absence of protein synthesis in yeast. *J. Biol. Chem.* **252:** 1344.

Sigler, K., A. Knotkova, and A. Kotyk. 1981. Factors governing substrate-induced generation and extrusion of protons in the yeast *Saccharomyces cerevisiae*. *Biochem Biophys. Acta* **643:** 572.

Sorsoli, W.A., K.D. Spencer, and L.W. Parks. 1964. Amino acid accumulation in ethionine-resistant *Saccharomyces cerevisiae*. *J. Bacteriol.* **88:** 20.

Spence, K.D. 1971. Mutation of *Saccharomyces cerevisiae* preventing uptake of S-adenosylmethionine. *J. Bacteriol.* **106:** 325.

Sripati, C.E., Y. Groner, and J.R. Warner. 1976. Methylated, blocked 5′ termini of yeast mRNA. *J. Biol. Chem.* **251:** 2898.

Sumrada, R. and T.G. Cooper. 1977. Allantoin transport in *Saccharomyces cerevisiae*. *J. Bacteriol.* **131:** 839.

———. 1978a. Control of vacuole permeability and protein degradation by the cell cycle arrest signal in *Saccharomyces cerevisiae*. *J. Bacteriol.* **136:** 234.

———. 1978b. Basic amino acid inhibition of cell division and macromolecular synthesis in *Saccharomyces cerevisiae*. *J. Gen. Microbiol.* **108:** 45.

Sumrada, R., M. Gorski, and T.G. Cooper. 1976. Urea-transport-defective strains of *Saccharomyces cerevisiae*. *J. Bacteriol.* **125:** 1048.

Sumrada, R., C.A. Zacharski, V. Turoscy, and T.G. Cooper. 1978. Induction and inhibition of the allantoin permease in *Saccharomyces cerevisiae*. *J. Bacteriol.* **135:** 498.

Surdin, Y., W. Sly, J. Sire, A.M. Bordes, and H. De Robichon-Szulmajster. 1965. Proprietes et controle genetique du systeme d'accumulation des acides amines chez *Saccharomyces cerevisiae*. *Biochim. Biophys. Acta* **107:** 546.

Svihla, G. and F. Schlenk. 1959. Localization of S-adenosylmethionine in *Candida utilis* by ultraviolet microscopy. *J. Bacteriol.* **78:** 500.

———. 1960. S-adenosylmethionine in the vacuole of *Candida utilis*. *J. Bacteriol.* **79:** 841.

Svihla, G., J.L. Dainko, and F. Schlenk. 1963. Ultraviolet microscopy of purine compounds in the yeast vacuole. *J. Bacteriol.* **85**: 399.

———. 1964. Ultraviolet microscopy of the vacuole of *Saccharomyces cerevisiae* during sporulation. *J. Bacteriol.* **88**: 449.

Turoscy, V. and T.G. Cooper. 1979. Allantoate transport in *Saccharomyces cerevisiae*. *J. Bacteriol.* **140**: 971.

Ueda, Y. and Y. Oshima. 1975. A constitutive mutation, *phoT*, of the repressible acid phosphatase synthesis with inability to transport inorganic phosphate in *Saccharomyces cerevisiae*. *Mol. Gen. Genet.* **136**: 255.

Urech, K., M. Durr, T. Boller, and A. Wiemken. 1978. Localization of polyphosphate in vacuoles of *Saccharomyces cerevisiae*. *Arch. Microbiol.* **116**: 275.

Warner, J.R., S.A. Morgan, and R.W. Shulman. 1976. Kinetics of labeling of the *S*-adenosylmethionine pool of *Saccharomyces cerevisiae*. *J. Bacteriol.* **125**: 887.

Wehrli, E., C. Boehm, and G.F. Fuhrmann. 1975. Yeast plasma membrane vesicles suitable for transport studies. *J. Bacteriol.* **124**: 1594.

Whelan, W.L., E. Gocke, and T.R. Manney. 1979. The Can1 locus of *Saccharomyces cerevisiae*: Fine structure analysis and forward mutation rates. *Genetics* **91**: 35.

Whitney, P.A., T.G. Cooper, and B. Magasanik. 1973. The induction of urea carboxylase and allophanate hydrolase in *Saccharomyces cerevisiae*. *J. Biol. Chem.* **248**: 6203.

Wickner, R.B. 1975. Mutants of *Saccharomyces cerevisiae* that incorporate deoxythymidine-5′-monophosphate into deoxyribonucleic acid *in vivo*. *Methods Cell Biol.* **11**: 295

Wiemken, A. 1975. Isolation of vacuoles from yeasts. *Methods Cell Biol.* **12**: 99.

Wiemken, A. and M. Durr. 1974. Characterization of amino acid pools in the vacuolar compartment of *Saccharomyces cerevisiae*. *Arch. Microbiol.* **101**: 45.

Wiemken, A. and P. Nurse. 1972. Isolation and characterization of the amino-acid pools located within the cytoplasm and vacuoles of *Candida utilis*. *Planta* **109**: 293.

———. 1973. The vacuole as a compartment of amino acid pools in yeast. In *Proceedings of the Third International Spec. Symposium Yeasts Otaniemi/Helsinki*, Part II, p. 331.

Willsky, G.R. 1979. Characterization of the plasma membrane Mg^{2+}-ATPase from the yeast, *Saccharomyces cerevisiae*. *J. Biol. Chem.* **254**: 3326.

Woodward, J.R. and H.L. Kornberg. 1980. Membrane proteins associated with amino acid transport by yeast (*Saccharomyces cerevisiae*). *Biochem. J.* **192**: 659.

Zacharski, C.A. and T.G. Cooper. 1978. Metabolite compartmentation in *Saccharomyces cerevisiae*. *J. Bacteriol.* **135**: 490.

Zimmermann, F.K., N.A. Khan, and N.R. Eaton. 1973. Identification of new genes involved in disaccharide fermentation in yeast. *Mol. Gen. Genet.* **132**: 29.

Suppression in the Yeast
Saccharomyces cerevisiae

Fred Sherman
Department of Radiation Biology and Biophysics
University of Rochester School of Medicine and Dentistry
Rochester, New York 14642

1. UAA and UAG Suppressors
2. UGA Suppressors
3. Frameshift Suppressors
4. Missense Suppressors
5. Ribosomal and Presumptive Ribosomal Suppressors
6. Extrachromosomal ψ^+ Determinant
7. Phenotypic Suppression
8. Modifiers of Suppressors
9. Discriminating Suppressible Alleles
10. Problems Working with Suppressors
11. Concluding Remarks

INTRODUCTION

It can be said that significant advancements in determining the molecular basis of suppression have been made so far only with the prokaryotes *Escherichia coli* and *Salmonella typhimurium* and with the eukaryotic yeasts *Saccharomyces cerevisiae* and *Schizosaccharomyces pombe*. With notable and important exceptions, the properties of suppressors from yeast are identical or similar to the properties of suppressors from bacteria. In fact, the conceptual framework for the work on yeast suppressors has been based on the earlier and more extensive work on bacterial suppressors. In this brief review, the suppressors of the yeast *S. cerevisiae* are tabulated, and their properties are summarized. The chromosomal map positions of many of the suppressors can be found on the genetic map that is presented in Appendix I of this volume.

Suppressors in yeast, as well as in prokaryotes, have proved to be important for probing the translational process, for investigating tRNA biosynthesis, and for providing a convenient means of deducing the nucleotide sequences of suppressible mutations. Investigations with bacteria have shown that extragenic information suppression is primarily mediated by altered tRNAs or altered ribosomes. Another class of suppressors, the polarity suppressors, has altered components required for proper transcriptional termination, and these types of suppressors are not to be expected to occur in yeast or other eukaryotes. Altered tRNAs with suppressor activity can have base changes within or outside anticodons, or they can have bases lacking proper modifications. Prokaryotic tRNA suppressors have been reviewed recently by Körner et al. (1978), Smith (1979), and Ozeki (1980). General reviews on suppression with some mention of yeast have appeared (Hartman and Roth 1973; Steege and Söll 1979). Hawthorne and Leupold (1974) have presented a useful review on yeast suppressors that unfortunately contains some interpretations that proved to be incorrect. Sherman et al. (1979) have recently summarized the results that were used to establish the nonsense codons in *cyc1* mutants and the use of these *cyc1* mutants for determining the amino acid residues inserted by UAA and UAG suppressors. The genetic studies of the suppressors from the yeast *S. pombe* (Hawthorne and Leupold 1974) will not be covered in this paper; although it should be mentioned that certain *S. pombe* suppressors were recently demonstrated to act on UGA mutations, and their activity was shown to be mediated by serine and leucine tRNA, which was mutated at their anticodons (Kohli et al. 1979; Rafalski et al. 1979; Wetzel et al. 1979).

Over the years, yeast suppressors have been characterized, and the molecular bases of their action have been investigated in the following ways: by genetic analysis of suppressors; by examining the pattern of suppression of groups of undefined markers, as well as defined *cyc1* mutations and presumably defined nutritional markers; by determining the amino acids inserted into iso-1-cytochrome *c* at the site corresponding to defined *cyc1* mutations; by the in vitro translation of Qβ RNA and globin mRNA with purified tRNA components; by examining the altered chromatographic position of tRNAs; by determining the DNA sequence of cloned tRNA genes; by sequencing tRNA molecules; by examining the translational properties of ribosomes and ribosomal components; and by examining the altered electrophoretic positions of ribosomal proteins. Presented in Table 1 is a somewhat idealized summary of the types of yeast suppressors. The summary was prepared not only by considering the results of the various investigations of

Table 1 Idealized summary of informational suppression in yeast

I. Genotypic suppression
   A. Altered tRNA or presumably altered tRNA
      1. Nonsense suppressors
         a. UAA
         b. UAG
         c. UGA
      2. Frameshift suppressors
      3. Missense suppressors
   B. Altered ribosomes or presumably altered ribosomes
   C. ψ^+ extrachromosomal determinant
II. Phenotypic suppression

yeast suppressors but also by considering analogies with prokaryotic suppressors.

It is now accepted that dominant and recessive suppressors should be denoted, respectively, by three uppercase or three lowercase letters, followed by a locus designation, e.g., *SUP4, SUF1, sup35, suf11*, etc. In some instances UAA suppressors and UAG suppressors are further designated, respectively, by o and a following the locus. For example, *SUP4*-o refers to suppressors of the *SUP4* locus that insert tyrosine residues at UAA sites; *SUP4*-a refers to suppressors of the same *SUP4* locus that insert tyrosine residues at UAG sites. The corresponding wild-type locus coding for the normal tyrosine tRNA and lacking suppressor activity can be referred to as sup4$^+$. Thus, the nomenclature describing suppressor and wild-type alleles in yeast is unrelated to the bacterial nomenclature. For example, an ochre *E. coli* suppressor that inserts tyrosine residues at both UAA and UAG sites is denoted as su_4^+, and the wild-type locus coding for the normal tyrosine tRNA and lacking suppressor activity can be referred to as Su_4, su_4^-, or *supC*.

UAA AND UAG SUPPRESSORS

The most extensively studied yeast suppressors are those that act on either UAA or UAG mutations. The UAA and UAG suppressors examined with the iso-1-cytochrome *c* system are listed in Table 2. Evidence that suppression is mediated by mutationally altered tRNA molecules comes from different lines of investigation of representatives of these UAA and UAG suppressors. Strains containing the UAG suppressors *SUP6*-a, *SUP-RL1*-a (Gesteland et al. 1976), *SUP7*-a, and *SUP16*-a (Capecchi et al. 1975), or the UAA suppressors *SUP4*-o (Gesteland et al. 1976) and *SUP16*-o, *SUP17*-o, and *SUP19*-o (B. Ono et al.; C. Waldron et al.; both unpubl.), were shown to contain tRNA molecules capable of translating either UAA or UAG codons in Qβ RNA and globin mRNA with in vitro protein-synthesizing systems. Nucleotide sequencing of suppressor tRNAs has revealed base changes at the anticodon for the UAG suppressors *SUP5*-a (Piper et al. 1976) and *SUP-RL1*-a (Piper 1978; also see Etcheverry et al. 1979). Similarly, DNA sequencing of the *SUP4*-o gene encoding the suppressor tRNA has revealed a base-pair change at the anticodon (Goodman et al. 1977). By assuming that all eight of the tyrosine-inserting suppressors are determined by tyrosine tRNA genes, the eight genetic loci could be related to one of the eight *Eco*RI restriction fragments that hybridize to tyrosine tRNA (Olson et al. 1979a,b; P. Philippsen et al. [pers. comm.], cited in Olson et al. 1979b). Investigations in progress in several laboratories also indicate that other UAA and UAG suppressors listed in Table 2 have mutationally altered tRNA molecules with suppressor activity.

In Table 3, the UAA and UAG suppressors have been placed into groups

Table 2 UAA and UAG suppressors examined with the iso-1-cytochrome c system

Locus	Amino acid replacement UAA	Amino acid replacement UAG	Efficiency	Chromosome arm	Possible equivalents	References[a] insertion	References[a] map position
SUP2	tyrosine	tyrosine	high	IV right		1,2	3,4
SUP3	tyrosine	n.t.	high	XV left		1	3,4
SUP4	tyrosine	tyrosine	high	X right		1,2	3,4
SUP5	tyrosine	tyrosine	high	(frag. 8)		1,2	3,4
SUP6	tyrosine	tyrosine	high	VI right		1,5	3,4
SUP7	tyrosine	tyrosine	high	X left		1,5	3,4
SUP8	tyrosine	n.t.	high	XIII right		1	3,4
SUP11	tyrosine	tyrosine	high	VI right		1,2	3,4
SUP16	serine		moderate	XVI right	SUQ5, SUP15	6,7	7
SUP17	serine		moderate	IX left		7	7
SUP19	serine		low	V right	SUP20	8	8
SUP22	serine		low	IX left		8	8
SUP26	leucine		very low	XII right		9	7
SUP27	leucine		very low	IV right		9	9
SUP28	leucine		very low	unknown		9	9
SUP29	leucine		very low	X left	SUP30	9	9
SUP32	leucine		very low	unknown		9	9
SUP33	leucine		very low	unknown		9	9
SUP-RL1		serine	high	III right	SUP61	10	10
SUP52		leucine	low	X left	SUP51	11	11
SUP53		leucine	low	III left		12	13
SUP54		leucine	low	VII left		12	13
SUP55		leucine	low	unknown		12	13
SUP56		leucine	low	unknown		12	13

The efficiencies of suppression were estimated from the levels of iso-1-cytochrome c in suppressed cyc1 strains and from the pattern of suppression of nutritional markers. n.t. indicates not tested.

[a] (1) Gilmour et al. (1971); (2) Liebman et al. (1976); (3) Hawthorne and Mortimer (1968); (4) Mortimer and Hawthorne (1973); (5) Sherman et al. (1973); (6) Liebman et al. (1975); (7) Ono et al. (1979a); (8) B. Ono et al. (in prep.); (9) Ono et al. (1979b); (10) Brandriss et al. (1976); (11) Liebman et al. (1976); (12) S.W. Liebman et al. (in prep.); (13) C.R. Reed and S.W. Liebman (1979, and in prep.).

Table 3 Major groups of UAA and UAG suppressors examined with the iso-1-cytochrome c system

Types	Loci	No. of loci	Insertion	References
(a) *UAA suppressors*				
Recessive lethals	none	—	—	
High efficiency	SUP2, etc.	8	tyrosine	Gilmour et al. (1971)
Moderate efficiency	SUP16, SUP17	2	serine	Ono et al. (1979a)
Low efficiency	SUP19, etc.	2 or more	serine	B. Ono et al. (in prep.)
Very low efficiency	SUP26, etc.	6	leucine	Ono et al. (1979b)
	others	unknown	unknown	
(b) *UAG suppressors*				
Recessive lethal	SUP-RL1	1	serine	Brandriss et al. (1976)
	others	unknown	unknown	
High efficiency	SUP2, etc.	8	tyrosine	Liebman et al. (1976)
Low efficiency	SUP52, etc.	5 or more	leucine	S.W. Liebman et al. (1977, and in prep.)
	others	unknown	unknown	

Summary of the data presented in Table 2.

on the basis of similar properties. One conclusion is that suppressors having more or less the same pattern of suppression of nutritional markers and similar efficiencies of action all cause the insertion of the same amino acid. Although the *SUP52* suppressor, a leucine-inserting UAG suppressor, could not be distinguished from the tyrosine-inserting UAG suppressors on the basis of suppression of nutritional markers, the *SUP52* suppressor had a distinctly lower efficiency of action on the *cyc1-179* allele (Liebman et al. 1977). However, there are slight variations in the efficiencies of suppression of certain suppressors within a single group. Among the tyrosine-inserting UAA suppressors, *SUP11* appears to act with a slightly lower efficiency. The serine-inserting UAA suppressors appear to have various efficiencies of suppression (B. Ono et al. 1979a, and in prep.).

The analysis of extensive numbers of revertants should reveal the maximum number of loci that can yield suppressors with particular characteristic patterns of suppression. Thus, the genetic analyses indicate that there are only eight loci yielding tyrosine-inserting suppressors, and each of these loci can yield both UAA suppressors and UAG suppressors (Table 3). Similarly, there appear to be only six leucine-inserting UAA suppressors. These suppressors undoubtedly arise by single base-pair changes, and the suppressors within a group appear to arise from redundant genes that determine the same tRNA; each of the genes within a group appears to be expressed at approximately the same level. The incomplete number of analyzed revertants and the limited number of mutations of certain loci suggest that there may be more low-efficiency UAA suppressors that insert serine and more low-efficiency UAG suppressors that insert leucine.

Up to now, only UAA and UAG suppressors causing replacement of tyrosine, serine, and leucine have been uncovered. It is not known why suppressors inserting glutamine, lysine, glutamic acid, and tryptophan have not been observed, since tRNA molecules altered at the anticodon of these tRNAs should be able to suppress UAA and UAG codons. Suppressors acting solely on UAG codons, found in both yeast and bacteria, arise by single base changes of the anticodon of tRNAs, which result in properties consistent with the "wobble" hypothesis. The yeast suppressors *SUP5*-a and *SUP-RL1*-a were shown to contain CUA anticodons similar to the anticodons of amber suppressors from *E. coli* (Piper et al. 1976; Piper 1978). However, in contrast to UAG suppressors, the ochre suppressors acting on both UAG and UAA codons are observed only in bacteria, and the suppressors acting solely on UAA codons are observed only in yeast. Although the exact nature of the changes in tRNAs of yeast UAA suppressors is unknown, the alterations are believed to be a uridine derivative at the wobble position of the anticodon of at least some of the tyrosine tRNAs; the DNA sequence of the cloned *SUP4* gene established that the unmodified anticodon in the suppressor tRNA is UUA (Goodman et al. 1977). The UAA-specific suppressors could be attributed to anticodons having the sequence SUA, as first suggested by Gilmore et al. (1971), where S could be a 2-thio-5-carboxymethyluridine derivative (Yoshida et al. 1971) or possibly a 5-carboxymethyluridine derivative (Kuntzel et al. 1975). Assuming that the properties of known tRNAs apply to suppressors, then suppressors with SUA anticodons would decode only the UAA nonsense codon. The tRNAs of the leucine-inserting and at least some of the serine-inserting UAA suppressors also could have SUA anticodons that were formed by single base-pair changes. However, it is not yet clear whether some of the serine-inserting UAA suppressors were generated by anticodon mutation of a serine tRNA, in which the normal IGA anticodon was changed to IUA; the IUA anticodon would be expected to decode the UAA nonsense codon and the two UAU and UAC tyrosine codons, but not the UAG nonsense codon. Also, it should not be excluded that certain UAA suppressors or other yeast suppressors arose by mutations external to the anticodon.

In addition to the UAA and UAG suppressors listed in Table 2, there is the UAA suppressor *SUP25*-o on the right arm of chromosome XI and the UAG suppressor *SUP50*-a on fragment 6 (Hawthorne and Mortimer 1968; Mortimer and Hawthorne 1973); neither of these suppressors was uncovered in investigations with the iso-1-cytochrome *c* and neither was examined for the amino acid it inserts.

UGA SUPPRESSORS

Presumptive UGA suppressors were first identified on the basis of suppression of the *ade5,7-143* allele, which was not suppressed by UAA or UAG suppressors but could be converted by mutation to a UAA-suppressible allele (Hawthorne 1976). Subsequent characterizations of the suppressors included their suppression of presumptive UGA alleles that were derived by mutation from UAA alleles. In addition, UAA suppressors could be converted by mutation to suppressors that acted on these UGA alleles (Hawthorne 1976). The 15 UGA suppressors isolated by suppression of UGA alleles are listed in Table 4 (Hawthorne 1981).

The identity and mode of action of the UGA suppressors were inferred indirectly by assuming that the formation and interconversion of the suppressors occurred by single base-pair substitutions at the sites corresponding to anticodons of tRNA (Hawthorne 1976). Recently a UGA-suppressing tRNA was identified on the basis of readthrough of the normal UGA terminator of rabbit globin mRNA in an in vitro protein-synthesizing system (Gesteland and Wills 1979); the UGA suppressor used in the in vitro experiment suppressed at least some markers that are suppressed by the UGA suppressors listed in Table 4.

Although the tRNA corresponding to the UGA suppressors has not been determined, there appear to be close linkages or equivalences in the loci for the serine-inserting UAA suppressor *SUP19* with the UGA suppressor

Table 4 Presumptive UGA suppressors

Class I		Class II			
locus	chromosome arm	locus	chromosome arm	locus	chromosome arm
SUP85	V right	SUP71	V right	SUP76	VII right
SUP86	XII right	SUP72	II right	SUP77	VII right
SUP87	II right	SUP73	X left	SUP78	XIII right
SUP88	IV right	SUP74	X left	SUP79	XIII left
SUP90	IX left	SUP75	II left	SUP80	IV right

Class-II suppressors cause less deleterious effects and do not suppress some of the markers that are suppressed by class-I suppressors, probably because their efficiency of action is less. (Adapted from Hawthorne 1976, 1981.)

SUP85, for the leucine-inserting UAA suppressor *SUP26* with the UGA suppressor *SUP86*, and for the leucine-inserting UAA suppressor *SUP27* with the UGA suppressor *SUP88*.

FRAMESHIFT SUPPRESSORS

Certain mutations in yeast were inferred to be frameshift because their properties were similar to the properties of frameshift mutations in *S. typhimurium* and *E. coli*. The mutations produced polar effects in the *HIS4* locus; they reverted by internal suppressors; they were not suppressed by UAA, UAG, or UGA suppressors; and they were induced by ICR-170, an acridine half-mustard compound similar to the mutagen that produces frameshift mutations in *S. typhimurium* (Culbertson et al. 1977, 1980; Cummins et al. 1980). The suppressors that were obtained as revertants of the frameshift mutations could be assigned to the 11 distinct loci listed in Table 5. These frameshift suppressors could be divided on the basis of suppression of either the so-called group-II mutations (e.g., *his4-519*) or the so-called group-III mutations (e.g., *his4-713*). The *SUF5* mutant contains a glycyl-1-tRNA with an altered chromatographic position; the *SUF1*, *SUF4*, and *SUF6* mutants have glycyl-3-tRNA with reduced glycine iso-accepting activities. These results suggest that the gene product of the *SUF5* locus is a mutationally altered glycyl-1-tRNA that translates frameshift mutations; similarly, the *SUF1, SUF4*, and *SUF6* suppressors are believed to have mutations in three redundant loci that determine glycyl-3-tRNA (Culbertson et al. 1977).

MISSENSE SUPPRESSORS

Missense suppressors in yeast have been identified on the basis of their suppression of certain mutations that appear to be missense and by the lack

Table 5 Frameshift suppressors

Group II		Group III	
locus	chromosome arm	locus	chromosome arm
SUF1	unknown	*SUF2*	III right
SUF3	unknown	*SUF7*	XIII left
SUF4	unknown	*SUF8*	VIII right
SUF5	XV right	*SUF9*	VI right
SUF6	unknown	*SUF10*	XIV left
		suf11	unknown

Group-II suppressors act on the *his4-519*, etc., alleles, and the group-III suppressors act on the *his4-713* allele. (Adapted from Culbertson et al. 1977, 1981; Cummins et al. 1980.)

of their suppression of nonsense mutations. The suppressors *SUH1* and *SUH2*, previously denoted as *SUP-H1* and *SUP-H2*, respectively, suppress the osmotic remedial allele *his2-1* but do not suppress other *his2* alleles or nutritional markers containing UAA or UAG mutations (Gorman and Gorman 1971). The *SUH1* locus is on the right arm of chromosome XII (see the chromosome map in Appendix I). The *ilv1-83* and *ilv1-51* alleles are each suppressed by suppressors that do not suppress the UAA mutation *ilv1-1* nor the markers presumably containing UAG or UGA mutations (Gundelach 1973). *SUP101* suppresses the missense alleles *trp5-18* and *trp5-67* but not other *trp5* alleles nor other UAA or UAG alleles (Singh and Manney 1974). The molecular basis for the action of these missense suppressors is not known, although it has been suggested that the suppression is mediated by mutationally altered tRNA analogous to *E. coli* missense suppressors.

RIBOSOMAL AND PRESUMPTIVE RIBOSOMAL SUPPRESSORS

A class of suppressors, often referred to as omnipotent suppressors (Hawthorne and Leupold 1974), differ in the following ways from the suppressors that are mediated by mutationally altered tRNAs: They have a broad pattern of suppression, apparently acting on certain UAA, UAG, and UGA alleles, as well as alleles not containing nonsense mutations; independent suppressors arising from the same locus can have different suppressing patterns; high proportions of the suppressors are recessive; suppressor activities are often temperature sensitive; and the growth of the suppressor strains can be temperature sensitive. Suppressors that are usually recessive have been reported to arise from two loci designated *sup1* and *sup2* (or s_1 and s_2) by Inge-Vechtomov and Andrianova (1970), *sup45* and *sup35*, by Hawthorne and Leupold (1974), and *supQ* and *supP,* by Gerlach (1975). The map position of the suppressors near *lys2* on chromosome II suggests that *sup1* (Ter-Avanesian and Inge-Vechtomov 1974), *sup45* (Hawthorne and Mortimer 1968), and *supQ* (Gerlach 1975) are equivalent; similarly, *sup35* and *supP*, which are located on chromosome IV, and, possibly, *sup2* may be equivalent. In addition, there are other loci, including *SUP46*, that can give rise to the so-called omnipotent suppressors (B. Ono et al., pers. comm.).

The kinetics of incorporation of radioactive amino acids in a strain containing the recessive suppressor *sup2*, a suppressor that also prevents growth at higher incubation temperatures, demonstrated that protein synthesis ceased when the strain was shifted to the restrictive temperature of 36°C (Smirnov et al. 1974). Furthermore, studies of Mg^{++} concentration optima of in vitro amino acid incorporation using heterologous ribosomes indicated that the 60S ribosomal subunit was the defective component (Smirnov et al. 1976). The accumulation of radioactive polypeptide bound in vivo to the ribosome and the shift of the in vitro Mg^{++} concentration with puromycin suggested that the defect in protein synthesis is the lack of release of peptidyl tRNA (Smirnov et al. 1974, 1976; Surguchov et al. 1980a). The 60S ribosomal subunit from the *sup2* suppressor is associated with the presence of the protein L30, which is completely or almost completely absent in the ribosomes from the normal parental strain (Smirnov et al. 1978). Although these results do not clearly define the product of the *sup2* gene, they suggest that peptide chain release is impaired and that suppression at the permissive temperature may be due to partial readthrough of nonsense codons in unreleased peptides. Ribosomes from a *sup1* suppressor exhibited an abnormally higher rate of translation errors in a cell-free system programmed with poly(U) (Surguchov et al. 1980b). In addition, the *sup1* ribosomes had an increased tendency to dissociate into subunits. It was suggested that the gene product of *sup1* was a component of the ribosome.

Similarly, extensive investigations of the dominant suppressor *SUP46* indicated that the suppression is caused by translation errors mediated by mutationally altered ribosomes. The *SUP46* mutation maps on chromosome II, but at a site distinct from the *sup45* locus (Ono et al. 1981). The *SUP46* suppressor acts on certain but not all UAA, UAG, and UGA alleles (Chattoo et al. 1979; B. Ono et al., in prep.) and very efficiently on two *lys2* alleles that apparently do not contain nonsense mutations (Chattoo et al. 1979). The *SUP46* suppressor causes the insertion of serine residues in iso-1-cytochrome *c* at a site corresponding to a UAA mutation (Ono et al. 1981). The growth of *SUP46* strains is unusually sensitive to paromomycin (Masurekar et al. 1979, 1981), an antibiotic that causes efficient phenotypic suppression of a wide range of mutations and that stimulates mistranslation (see below, Phenotypic Suppression). In vitro misreading with components from the *SUP46* strain occurred at higher basal level and was stimulated with paromomycin to a higher degree than misreading with components from a normal related strain. The levels of misreading with mixed components and with reconstituted heterologous ribosomes established that the alteration was in the 40S ribosomal subunit of the *SUP46* strain (Masurekar et al. 1979, 1981). An altered electrophoretic position of the protein S11 of the 40S ribosomal subunit was revealed by two-dimensional gel electrophoresis (J. Ishiguro et al. 1981). Thus, it appears as if the mutationally altered ribosomal protein S11 causes the ribosome to misread at an abnormally high level and to suppress certain nonsense and other mutations. Consistent with this interpretation was the finding that all *lys2* alleles suppressible by *SUP46* were also suppressible by paromomycin (Chattoo et al. 1979), an agent that is believed to stimulate misreading (see below).

Further systematic investigations should reveal whether loci other than *sup35*, *sup45*, and *SUP46* can yield these types of "ribosomal" suppressors and which suppressors are associated with mutationally altered ribosomes causing abnormalities in the translational process.

EXTRACHROMOSOMAL ψ^+ DETERMINANT

A non-Mendelian genetic determinant, ψ^+, was first identified by its ability to modify the expression of a nonsense suppressor (Cox 1965, 1971). Tetrad analysis (Cox 1965) and heterokaryon analysis (Fink and Conde 1977) demonstrated that the ψ^+ determinant does not reside on a chromosome or in the nucleus. Genetic analyses have shown that the ψ^+ factor is not associated with mitochondrial DNA (Young and Cox 1972) and probably not with double-stranded RNA that controls the killer phenotype (McCready et al. 1977). High frequencies of $\psi^+ \rightarrow \psi^-$ mutations can be induced by growing cells in media made hypertonic with KCl or ethylene glycol, and such growth conditions are not believed to induce chromosomal mutations (Singh et al. 1979a). Although the molecular identity of ψ^+ and its site of action is not known, it appears to be a self-replicating extranuclear determinant that affects the translational process.

Comparisons of ψ^+ and ψ^- strains indicate that the ψ^+ determinant weakly suppresses certain UAA mutations; ψ^+ causes the production of approximately 1% of the normal level of iso-1-cytochrome c in a strain containing the UAA mutation *cyc1-72*; it clearly suppresses the UAA mutation *trp5-48* allele in at least some genetic backgrounds (Liebman and Sherman 1979). Formally speaking, the ψ^+ factor can be considered a weak UAA suppressor whose efficiency is modified by chromosomal genes. It is possible that the suppressor activity of ψ^+ is due to the enhanced expression of certain normal tRNAs that, by themselves, have no detectable suppressor activity.

A more pronounced effect is the ability of the ψ^+ determinant to enhance the expression of UAA suppressors and certain frameshift suppressors. Approximately tenfold higher efficiencies of action of UAA suppressors are observed in ψ^+ strains in comparison with ψ^- strains (Liebman and Sherman 1979). Similarly, ψ^+ produces efficiencies of action of certain frameshift suppressors (Culbertson et al. 1977) threefold to tenfold higher. ψ^+ enhances the activity of all of the UAA suppressors that have been examined. The activity of the tyrosine-inserting UAA suppressors is increased to such a level in ψ^+ strains that inviability or retardation of growth is observed. By using *cyc1* mutants, ψ^+ was directly demonstrated to increase the level of suppression of all four serine-inserting UAA suppressors (Liebman and Sherman 1979; Ono et al. 1979a, and in prep.). The recovery of leucine-inserting UAA suppressors in ψ^+ strains but not in ψ^- strains suggests that ψ^+ increases the activity of all six of these UAA suppressors (Ono et al. 1979b). In contrast, there was no detectable difference of suppression between ψ^+ and ψ^- strains containing the UAG suppressor *SUP7*-a (Liebman and Sherman 1979); also, the pattern of suppression of UAG markers by UAG suppressors appears the same in ψ^+ and ψ^- strains.

The ψ^+ enhancement of suppression of frameshift suppressors appears to be restricted to the *SUF1*, *SUF4*, and *SUF6* suppressors, which are associated with glycyl-3-tRNA; ψ^+ does not affect the remaining eight frameshift suppressors listed in Table 5 (Cummins et al. 1981).

The levels of suppression in ψ^+ *SUP*-o strains are considerably greater than the sum of the levels in the individual ψ^- *SUP*-o strains and ψ^+ strains, thus indicating that the combination produces synergistic activities far greater than the individual activities. Because *SUP16*-o inserts serine in both ψ^+ and ψ^- strains (Liebman et al. 1975), it appears that the ψ^+ determinant enhances the activity of the *SUP16*-o suppressor rather than vice versa. Because ψ^+ itself does not suppress frameshift mutation, its action on the specific class of frameshift suppressor is clearly the enhancement of *SUF* activities. The results with the nonsense and frameshift suppressors indicated that certain mutationally altered tRNAs are expressed at a higher level in ψ^+ strains than in ψ^- strains.

In addition to acting as a weak UAA suppressor and enhancing the activity of UAA suppressors and certain frameshift suppressors, the ψ^+ determinant enhances the phenotypic suppression of UAA, UAG, UGA, and possibly other mutations (Palmer et al. 1979a). The aminoglycoside paromomycin phenotypically suppresses a variety of different types of mutation, probably by stimulating misreading (see below). Finding threefold to sevenfold increases in paromomycin-induced phenotypic suppression in ψ^+ strains in comparison with ψ^- strains indicates that the ψ^+ determinant effects are not limited to mutationally altered forms of tRNA having suppressor capacity.

PHENOTYPIC SUPPRESSION

The reversal of mutant phenotypes by environmental conditions, referred to as phenotypic suppression, can occur by distinctly different mechanisms. In yeast, the phenotypic reversal of mutant characters by hypertonic media is believed to occur by restoring the activity of mutationally altered proteins that have elevations of their osmotic pressure optima (Mortimer and Hawthorne 1969). Recently, diverse types of mutations were shown to be phenotypically suppressed by the aminoglycoside antibiotic paromomycin and, to a lesser degree, by neomycin, etc. (Chattoo et al. 1979; Palmer et al. 1979b; Singh et al. 1979b). Because these antibiotics stimulate mistranslation in cell-free systems containing ribosomes from yeast (Palmer et al. 1979b; Singh et al. 1979b) as well as other eukaryotes, it is believed that the phenotypic suppression involves the misreading of the mutant codons. In one study it was demonstrated that paromomycin could phenotypically suppress 14 of the 17 alleles that contained UAA, UAG, or UGA mutations in different genes (Palmer et al. 1979a). A comprehensive examination involving 358 *lys2* mutants indicated that paromomycin suppressed 13% of the mutants; 74% of the mutants suppressible by paromomycin were also suppressible by UAA, UAG, or UGA suppressors. Thus, not all but most alleles suppressible by paromomycin contain nonsense mutations, and not all nonsense mutations are suppressible by paromomycin. As noted above, all alleles suppressible by the ribosomal suppressor *SUP46* were suppressible by paromomycin. Apparently, efficient suppression by misreading involves not only the mutant codon but also the adjacent regions.

MODIFIERS OF SUPPRESSORS

In addition to the non-Mendelian determinant ψ^+, which affects suppressors as described above, there are chromosomal mutations extrinsic to suppressor loci that can reduce or enhance the expression of suppressors. The mutations that reduce or eliminate the expressions of suppressors are referred to as antisuppressors, and the corresponding mutant genes have been denoted as either *asu, mod*, or *sin*. Mutations enhancing the expression of suppressors have been referred to as allosuppressors, with the corresponding gene symbol denoted as *sal*; in the case of frameshift suppressors, allosuppressors have been referred to as up-frameshift-suppressor mutations, with the corresponding gene symbol denoted as *upf*. In addition to defined chromosomal mutations affecting suppressors, differences in the expression of suppressors are often observed in the meiotic progeny from different crosses, which indicates that laboratory strains commonly contain different, undefined genetic backgrounds capable of influencing suppressors. In addition to the ψ^+ determinant and chromosomal mutations, the ρ^-

mitochondrial mutation has been suggested to modify the expression of certain UAA suppressors (Ono et al. 1979a); also, the ρ^- mutation appears to be associated with the recovery of the antisuppressor *sin1* (Gorman and Gorman 1971).

A systematic investigation of antisuppressors acting on the UAA suppressor *SUP11* resulted in the identification of eight distinct antisuppressor loci, *ASU1-ASU8* (McCready and Cox 1973). The *asu1-asu8* antisuppressors reduced the activities of the UAA suppressors *SUP11, SUP2,* and *SUP16* but had no observable effect on the UAG suppressor *SUP11* (McCready and Cox 1976). Similarly, mutations in seven distinct antisuppressor loci, *mod1-mod6* and *moda*, were uncovered on the basis of reversion of the UAA suppressor *SUP7* (J. Gorman et al. [unpubl.], cited in Laten et al. 1978). So far the primary defect of only the *mod5* antisuppressor has been identified; tRNAs from the *mod5* antisuppressor contain only 1.5% of the normal level of isopentenyladenosine (Laten et al. 1978).

Although no systematic study has been made of antisuppressors that act on UAG suppressors, a recessive antisuppressor of the UAG suppressor *SUP8* has been observed (Liebman and Sherman 1976). Also, it has been possible to mutate UAG suppressors so that they become sensitive to the dominant antisuppressor *SIN2*, which preexists in many laboratory stocks (Hawthorne and Leupold 1974); *sin1*, an antisuppressor of the missense suppressors *SUH1* and *SUH2* (Gorman and Gorman 1971), and *asu9*, an antisuppressor of omnipotent suppressors (S. W. Liebman and M. Cavenagh, pers. comm.), have been isolated.

Recessive mutations at five distinct loci have been shown to give rise to the allosuppressors *sal1-sal5*, which increase expression of the UAA suppressor *SUP16* (Cox 1977). Similar to the action of the ψ^+ determinant, these allosuppressors increase the expression of the tyrosine-inserting UAA suppressors to a level that causes lethality or reduced viability.

In addition, the allosuppressors have other interesting but unexplained properties. ψ^+ *sal3* and ψ^+ *sal4* strains, as well as haploid strains carrying both *sal3* and *sal4* allosuppressors, are inviable. All *sal3* mutants are cold-sensitive, i.e., they do not grow or grow poorly at 12°C. The temperature-sensitive allosuppressor *sal4-2*, when in combination with *sal3* mutations, acts as a weak UAA suppressor at permissive temperature (Cox 1977).

Another group of *sal* allosuppressors, which may be different than *sal-1-sal5*, increases the expression of the weak UAA suppressor *SUP20* (or *SUP19*) and the mutated UAG suppressor *SUP5* when present with the antisuppressor *SIN2* (Hawthorne and Leupold 1974). These allosuppressors by themselves act as weak suppressors and cause tyrosine-inserting UAA suppressors to suppress UAG mutations.

An allosuppressor that increases the expression of serine- and leucine-inserting UAA suppressors did not influence the type of insertion; the

SUP17 suppressor caused the insertion of serine both in the presence and absence of the allosuppressor (Ono et al. 1979a,b).

Linkage results suggested allosuppressors could be alleles of the omnipotent suppressors *sup35* and *sup45* (Hawthorne and Leupold 1974). However, two-dimensional polyacrylamide gel electrophoresis of ribosomal proteins did not reveal any differences between allosuppressors or normal strains (Waldron and Cox 1978).

Eighteen allosuppressors acting on frameshift suppressors were obtained by increasing the action of *SUF1* on the weakly suppressible frameshift mutation *his4-38* (Culbertson et al. 1981). Genetic analysis indicated that the allosuppressors could be assigned to two loci, *UPF1* and *UPF2*; *upf1* increased the expression of *SUF1, SUF4, SUF5,* and *SUF6*; whereas *upf2* increased the expression of *SUF1, SUF3, SUF4,* and *SUF6*.

DISCRIMINATING SUPPRESSIBLE ALLELES

The effective suppression of markers is determined by several factors, including the nature of the suppressible codon, the acceptibility of the amino acid inserted at the mutant site, the position of the mutant codon within the locus, and the level of expression of the suppressor. There are a number of useful nutritional markers that are highly suppressible by a wide range of suppressors, and there are other markers that exhibit particularly restricted patterns of suppression. Restrictions in effective suppression can be due to either insufficiencies in the levels of gene products or the insertion of amino acids incompatible with the function of gene products. Representatives of markers highly suppressible by UAA, UAG, or UGA suppressors are listed in Table 6, along with markers that are suppressible by ribosomal suppressors or paromomycin, but not by nonsense suppressors. Certain markers that exhibit restricted patterns of suppression are listed in Table 7. Thus, the

Table 6 Examples of highly suppressible markers

	Specific suppressors			Ribosomal suppressor (*SUP46*)	Phenotypic suppressor (paromomycin)
	UAA	UAG	UGA		
leu2-1, trp5-48,[a] *ilv1-2*	+	0	0	+	+
met8-1, aro7-1, trp1-1	0	+	0	+	+
leu2-2, lys2-101, his4-166	0	0	+	+	+
lys2-187	0	0	0	++	++
lys2-12	0	0	0	0	+

[a]In some genetic backgrounds, the *trp5-48* marker is suppressed by ψ^+. Further information can be found in Gilmour (1967), Hawthorne and Leupold (1974), Liebman et al. (1976), B. Ono et al. (1979a,b, and unpubl.), Palmer et al. (1979a,b), Liebman and Sherman (1979), and Chattoo et al. (1979).

Table 7 Discriminating suppressible markers

	UAA suppressors[a]		
	tyrosine inserters	serine inserters	leucine inserters
leu2-1, etc.	++	++	++
ura4-1	+	0	+
his4-1176	0	+	0
lys2-52[b]	++	0	0
lys2-255[a]	0	+	++

	UGA suppressors[c]	
	class I	class II
leu2-2	+	+
ade5,7-143	+	0

	Frameshift suppressors[d]	
	group II	group III
his4-519	+	0
his4-713	0	+

[a] See Table 2.
[b] The properties of *lys2* markers were determined in diploid strains and were not verified by examining haploid segregants.
[c] See Table 4.
[d] See Table 5.

alleles listed in Tables 6 and 7 are of use in the selection and initial characterization of classes of suppressors. In addition, a number of discriminating *lys2* alleles were uncovered by testing heterozygous diploid strains, including markers that appear to respond preferentially to different UAG suppressors (Chattoo et al. 1979); however, the suppressibility of most of these *lys2* markers has not been verified by examining meiotic segregants containing the appropriate suppressors.

PROBLEMS WORKING WITH SUPPRESSORS

A few precautions should be taken when working with suppressors or in the interpretation of suppressibilities. In particular, one major difficulty is their instabilities. Strains containing highly efficient UAA and UAG suppressors, class-I UGA suppressors, or certain ribosomal suppressors grow poorly on nutrient medium, whereas normal or nearly normal growth rates are observed when these strains lose the suppressors or when the suppressors are mutated to forms having lower levels of expression. In addition, the frameshift suppressors *SUF1*, *SUF4*, and *SUF6* are highly unstable, whereas

SUF3 and *SUF5* are relatively stable. Although the loss of suppressors can be detected conveniently if suppressible markers are present in the strain, it is sometimes difficult to detect when a suppressor has mutated to a state having a lower efficiency of expression. For example, high-efficiency and low-efficiency UAG suppressors could not be distinguished by the suppression of nutritional markers, and the determination of the degree of suppression relied on the level of iso-1-cytochrome *c* in strains containing suppressible *cyc1* markers (Liebman and Sherman 1976).

Care should be taken in purifying suppressor strains by subcloning on selective media. In attempting to retrieve rare cells still containing a particular suppressor, one can inadvertently isolate new and possibly different suppressors.

In addition to inhibiting vegetative growth, suppressors often completely or partially inhibit sporulation (Liebman and Sherman 1976; Rothstein et al. 1977). Confusion can arise in a genetic analysis if there was preferential sporulation of cells that lacked the suppressor or contained a suppressor with a lower level of expression.

Although tRNA suppressors act with high specificity, the lack of suppression does not establish that the mutation is not nonsense. Some UAA and UAG mutations at certain sites are barely suppressible even when the appropriate amino acid replacements at these sites are compatible with functional gene products. For example, the UAA mutation *cyc1-9* is barely suppressible by tyrosine-inserting and serine-inserting UAA suppressors, although tyrosine and serine replacements at the *cyc1-9* site lead to fully functional iso-1-cytochromes *c* (Liebman et al. 1975). Also, it should be noted that there are rare instances of alleles being suppressed by both UAA and UAG suppressors; the *leu1-101* (Hawthorne and Leupold 1974) and the *lys2-59* (Chattoo et al. 1979) markers appear to be suppressed both by certain UAA and certain UAG suppressors.

CONCLUDING REMARKS

In the past and probably in the future, the most common use of yeast tRNA suppressors is to provide a convenient genetic means for deducing the nature of suppressible mutations. In addition, there are a number of other important uses of yeast suppressors that have not been stressed in this paper. Temperature-sensitive suppressors provide a means of recovering conditional mutants (Rasse-Messenguy and Fink 1973). Yeast has been and will continue to be a source for suppressor tRNAs that can be used to identify and investigate mutations in higher eukaryotes by in vitro suppression (Capecchi et al. 1975, 1977; Gesteland et al. 1976, 1977; Pelham 1978; Philipson et al. 1978; Chang et al. 1979; Cremer et al. 1979; Secher et al. 1979). Also, by employing recombinant DNA procedures, suppressor tRNA genes from yeast may provide a means of examining suppression in intact cells of other eukaryotic cells not amenable to conventional genetic manipulations. It has already been demonstrated that a tyrosine tRNA gene from yeast probably can be expressed in monkey cells (Goff and Berg 1979). Discussed elsewhere in this volume is the role of yeast suppressors for detecting tRNA functions and for determining the sites and regions required for the function, the maturation, and the processing of tRNAs as well as for investigating the function and requirement of intervening sequences. One aspect of yeast suppressors that only recently has been appreciated is the potential use of ribosomal suppressors, allosuppressors, and antisuppressors for perturbating the ribosome in order to investigate the relationship of defined mutational lesions to the fidelity of the translational process.

ACKNOWLEDGMENTS

The writing of this paper and some of the cited unpublished experiments was supported in part by U.S. Public Health Service research grant GM-12702 from the National Institutes of Health, and in part by U.S. Department of Energy contract no. DE-ACO2-76EVO3490 at the University of Rochester, Department of Radiation Biology and Biophysics. This paper has been designated Report no. UR-3490-1828.

Note Added in Proof
UAA and UAG Suppressors

Recently, Ono et al. (1981) and Broach et al. (1981) demonstrated that the serine-inserting UAA suppressors *SUP16, SUP17,* and *SUP19* (Table 2) encode the iso-accepting species of $tRNA^{Ser}_{UCA}$, and these three suppressors represent the only three genes for this tRNA species (Broach et al. 1981; Ono et al. 1981). DNA sequencing established that *SUP16* has the expected nucleotide change at the anticodon (Broach et al. 1981) and that RNA sequencing of the tRNA established the corresponding $G \rightarrow U$ substitution, but the mutant anticodon lacks a modified uridine found in the wobble position of the normal tRNA and contains instead another modification in or near the anticodon (Waldron et al. 1981). The mechanism of suppression of the serine-inserting suppressor *SUP22* is unknown (Ono et al. 1981). The recessive lethality of the serine-inserting suppressor *SUP-RL1* (or *SUP61*) (Tables 1 and 2) was shown to be due to the loss of $tRNA^{Ser}_{UCG}$, which has a unique decoding function (Etcheverry et al. 1982).

Frameshift Suppressors

The numbers and specificities of the frameshift suppressors (Table 5) have been extended, and members of the following groups have been identified (Culbertson et al. 1982; Gaber and Culbertson 1982). Group II, class 1: *SUF1, SUF3, SUF4, SUF5, SUF6,* and *SUF17*; group II, class 2: *SUF15,*

SUF16, and *SUF18-SUF25*; group III, class 1: *SUF2* and *SUF10*; group III, class 2: *SUF7, SUF8, SUF9*, and *SUF11*; and a group with a wide range of specificity: *suf12, suf13*, and *suf14*. The chromosomal positions of most of these suppressor loci have been determined (see Appendix I). Donahue et al. (1981) demonstrated with DNA sequencing that the group-II markers *his4-38* and *his4-519* contained, respectively, mutations of the glycine codons GGU → GGGU and GGG → GGGG, whereas the group-III markers contained mutations of the proline codons CCU → CCCU. A group-II suppressor, *SUF16*, was shown to encode an altered glycine tRNA with a GCCC anticodon instead of the normal GCC anticodon (R. F. Gaber and M. R. Culbertson, unpubl.), whereas a group-III suppressor, *SUF2*, was shown to encode an altered proline tRNA with a GGGA anticodon instead of the normal GGA anticodon (C. M. Cummins et al., unpubl.). Thus, group-II and group-III suppression of 4-base codons is mediated, respectively, by altered glycine tRNA and altered proline tRNA through mRNA-tRNA interactions. The frameshift suppressors having a wide range of specificity are believed to misread a range of codons and may be equivalent or similar to the ribosomal suppressors (Culbertson et al. 1982).

Ribosomal and Presumptive Ribosomal Suppressors

Certain alleles of the *sup1* and *sup2* suppressors, which appear equivalent to the *sup45* and *sup35* suppressors, respectively, were found to have properties similar to some of the properties of the *SUP46* ribosomal suppressor described by Masurekar et al. (1981). Certain *sup1* and *sup2* suppressors are associated with ribosomal misreading in vitro and ribosomal abnormalities (Surguchov et al. 1980a,b, 1981b,c); the growth inhibition and phenotypic suppression by paromomycin is enhanced in some *sup1* strains (Surguchov et al. 1981a); certain *sup1* and *sup2* strains are more sensitive to osmotic pressure, to temperatures above and below the growth optimum, and are unable to utilize nonfermentable carbon sources (Ter-Avanesyan et al. 1982); in addition, the growth of some *sup45* suppressors is more sensitive to paromomycin and to osmotic media (Liebman and Cavenagh 1980). Eleven antisuppressors of the two *sup35* and *sup45* suppressors could be assigned to the two loci *ASU9* and *ASU10*. The recessive *asu9* antisuppressors reduced the action of *sup35* and *sup45*, whereas the dominant antisuppressor *ASU10* reduced the action of only *sup35* (Liebman and Cavenagh 1980; Liebman et al. 1980). The antisuppressor *asu9* caused increased sensitivity to paromomycin and an association of an extra protein at the 40S ribosomal subunit (Liebman and Cavenagh 1980, 1981). An antisuppressor *asu11*, which reduces the action of *SUP46*, was found to have an altered S27 ribosomal protein, indicating that *asu11* encodes the S27 protein whose altered form corrects the abnormal S11 protein in *SUP46* (Ishiguro 1981). Thus, the *SUP46* and its antisuppressor *asu11* encodes ribosomal proteins. Although the primary lesions in the *sup35* and *sup45* suppressors remain to be determined, these suppressors and at least some of their antisuppressors directly or indirectly alter the ribosome.

REFERENCES

Brandriss, M.C., J.W. Stewart, F. Sherman, and D. Botstein. 1976. Substitution of serine caused by a recessive lethal suppressor in yeast. *J. Mol. Biol.* **102**:467.

Broach, J.R., L.R. Friedman, and F. Sherman. 1981. Correspondence of yeast UAA suppressors to cloned $tRNA_{UCA}^{Ser}$ genes. *J. Mol. Biol.* **150**:375.

Capecchi, M.R., S.H. Hughes, and G.M. Wahl. 1975. Yeast super-suppressors are altered tRNAs capable of translating a nonsense codon *in vitro*. *Cell* **6**:269.

Capecchi, M.R., R.A. Van der Haar, N.E. Capecchi, and M.M. Sveda. 1977. The isolation of a suppressible nonsense mutant in mammalian cells. *Cell* **12**:371.

Chang, J.C., G.F. Temple, R.F. Trecartin, and Y.W. Kan. 1979. Suppression of the nonsense mutation homozygous β^0 thalassaemia. *Nature* **281**:602.

Chattoo, B.B., E. Palmer, B. Ono, and F. Sherman. 1979. Patterns of genetic and phenotypic suppression of *lys2* mutations in the yeast *Saccharomyces cerevisiae*. *Genetics* **93**:67.

Cox, B.S. 1965. A cytoplasmic suppressor of super-suppressor in yeast. *Heredity* **20**:505.

———. 1971. A recessive lethal super-suppressor mutation in yeast and other ψ phenomena. *Heredity* **26**:211.

———. 1977. Allosuppressors in yeast. *Genet. Res.* **30**:187.

Cremer, K.J., M. Bodemer, W.P. Summers, W.C. Summers, and R.F. Gesteland. 1979. *In vitro* suppression of UAG and UGA mutants in the thymidine kinase gene of herpes simplex virus. *Proc. Natl. Acad. Sci.* **76**:430.

Culbertson, M.R., R.F. Gaber, and C.M. Cummins. 1982. Frameshift suppression in *Saccharomyces cerevisiae*. V. Isolation and genetic properties of non-group-specific suppressors. *Genetics* (in press).

Culbertson, M.R., K.M. Underbrink, and G.R. Fink. 1980. Frameshift suppression in *Saccharomyces cerevisiae*. II. Genetic properties of group II suppressors. *Genetics* **95**:833.

Culbertson, M.R., L. Charmas, M.T. Johnson, and G.R. Fink. 1977. Frameshifts and frameshift suppressor in *Saccharomyces cerevisiae*. *Genetics* **86**:745.

Cummins, C.M., R.F. Gaber, M.R. Culbertson, R. Mann, and G.R. Fink. 1980. Frameshift suppression in *Saccharomyces cerevisiae*. III. Isolation and genetic properties of group III suppressors. *Genetics* **95**:855.

Donahue, T.F., P. Farabaugh, and G.R. Fink. 1981. Suppressible glycine and proline four base codons. *Science* **212**:455

Etcheverry, T., D. Colby, and C. Guthrie. 1979. A precursor to a minor species of yeast $tRNA^{Ser}$ contains an intervening sequence. *Cell* **18**:11.

Etcheverry, T., M. Salvato, and C. Guthrie. 1982. Recessive lethality of yeast strains carrying the *SUP61* suppressor results from the loss of a tRNA with a unique decoding function. *J. Mol. Biol.* (in press).

Fink, G.R. and J. Conde. 1977. Studies on *KAR1*, a gene required for nuclear fusion in yeast. In *International cell biology, 1976-1977* (ed. B.R. Brinkley and K.R. Porter), p. 414. Rockefeller University Press, New York.

Gaber, R.F. and M.R. Culbertson. 1982. Frameshift suppressors in *Saccharomyces cerevisiae*. IV. New suppressors among spontaneous co-revertants of the group II *his4-206* and *leu2-3* frameshift mutations. *Genetics* (in press).

Gerlach, W.L. 1975. Genetic properties of some amber-ochre suppressors in *Saccharomyces cerevisiae*. *Mol. Gen. Genet.* **138**:53.

Gesteland, R.F. and N. Wills. 1979. Use of yeast suppressors for identification of adenovirus nonsense mutants. In *Nonsense mutations and tRNA suppressors* (ed. J.E. Celis and J.D. Smith), p. 277. Academic Press, New York.

Gesteland, R.F., N. Wills, J.B. Lewis, and T. Grodzicker. 1977. Identification of amber and ochre mutants of the human virus Ad⁺ND1. *Proc. Natl. Acad. Sci.* **74:** 4567.

Gesteland, R.F., M. Wolfner, P. Grisafi, G. Fink, D. Botstein, and J.R. Roth. 1976. Yeast suppressors of UAA and UAG nonsense codons work efficiently *in vitro* via tRNA. *Cell* **7:** 381.

Gilmour, R.A. 1967. Super-suppressors in *Saccharomyces cerevisiae*. *Genetics* **56:** 641.

Gilmour, R.A., J.W. Stewart, and F. Sherman. 1971. Amino acid replacements resulting from super-suppression of nonsense mutants of iso-1-cytochrome *c* from yeast. *J. Mol. Biol.* **61:** 157.

Goff, S.P. and P. Berg. 1979. Construction, propagation and expression of simian virus 40 recombinant genomes containing the *Escherichia coli* gene for thymidine kinase and a *Saccharomyces cerevisiae* gene for tyrosine transfer RNA. *J. Mol. Biol.* **133:** 359.

Goodman, H.M., M.V. Olson, and B.D. Hall. 1977. Nucleotide sequence of a mutant eukaryotic gene: The yeast tyrosine-inserting ochre suppressor *SUP4-o*. *Proc. Natl. Acad. Sci.* **74:** 5423.

Gorman, J.A. and J. Gorman. 1971. Genetic analysis of a gene required for the expression of allele-specific missense suppression in *Saccharomyces cerevisiae*. *Genetics* **67:** 337.

Gundelach, E. 1973. Suppressor studies on *ilv1* mutants of *Saccharomyces cerevisiae*. *Mutat. Res.* **29:** 25.

Hartman, P.E. and J.R. Roth. 1973. Mechanism of suppression. *Adv. Genet.* **17:** 1.

Hawthorne, D.C. 1976. UGA mutations and UGA suppressors in yeast. *Biochimie* **58:** 179.

———. 1981. UGA suppressors in yeast. In *Alfred Benzon Symp. 16, Molecular genetics in yeast* (ed. D. von Wettstein et al.), p. 291. Munksgaard, Copenhagen.

Hawthorne, D.C. and U. Leupold. 1974. Suppressor mutations in yeast. *Curr. Top. Microbiol. Immunol.* **64:** 1.

Hawthorne, D.C. and R.K. Mortimer. 1968. Genetic mapping of nonsense suppressors in yeast. *Genetics* **60:** 735.

Inge-Vechtomov, S.G. and V.M. Andrianova. 1970. Recessive supersuppressors in yeast. *Genetika* **6:** 103.

Ishiguro, J. 1981. Genetic and biochemical characterization of antisuppressor mutants in the yeast *Saccharomyces cerevisiae*. *Curr. Genet.* **4:** 197

Ishiguro, J., B. Ono, M. Masurekar, C.S. McLaughlin, and F. Sherman. 1981. Altered ribosomal protein S11 from the *SUP46* suppressor of yeast. *J. Mol. Biol.* **147:** 391.

Kohli, J., T. Kwong, F. Altruda, D. Söll, and G. Wahl. 1979. Characterization of a UGA suppressing serine tRNA from *Schizosaccharomyces pombe* with the help of a new *in vitro* assay system for eukaryotic suppressor tRNAs. *J. Biol. Chem.* **254:** 1546.

Körner, A., S.I. Feinstein, and S. Altman. 1978. tRNA-mediated suppression. In *Transfer RNA* (ed. S. Altman), p. 105. MIT Press, Cambridge, Massachusetts.

Kuntzel, B., J. Weissenbach, R.E. Wolff, T.D. Tumaitis-Kennedy, B.G. Lane, and G. Dirheimer. 1975. The presence of the methylester of 5-carboxymethyl uridine in the wobble position of the anticodon of tRNA$_{III}^{Arg}$ from brewer's yeast. *Biochimie* **57:** 61.

Laten, H., J. Gorman, and M. Bock. 1978. Isopentenyladenosine deficient tRNA from an antisuppressor mutant of *Saccharomyces cerevisiae*. *Nucleic Acids Res.* **5:** 4329.

Liebman, S.W. and M. Cavenagh. 1980. An antisuppressor that acts on omnipotent suppressors in yeast. *Genetics* **95:** 49.

Liebman, S.W. and M.M. Cavenagh. 1981. 40S ribosomal protein from a *Saccharomyces cerevisiae* antisuppressor mutant exhibiting a unique 2D gel pattern. *Curr. Genet.* **3:** 27.

Liebman, S.W. and F. Sherman. 1976. Inhibition of growth by amber suppressors in yeast. *Genetics* **81:** 233.

———. 1979. Extrachromosomal ψ^+ determinant suppresses nonsense mutations in yeast. *J. Bacteriol.* **139:** 1068.

Liebman, S.W., M.M. Cavenagh, and L.N. Bennett. 1980. Isolation and properties of an antisuppressor in *Saccharomyces cerevisiae* specific for an omnipotent suppressor. *J. Bacteriol.* **143:** 1527.

Liebman, S.W., F. Sherman, and J.W. Stewart. 1976. Isolation and characterization of amber suppressor in yeast. *Genetics* **82**: 251.

Liebman, S.W., J.W. Stewart, and F. Sherman. 1975. Serine substitutions caused by an ochre suppressor in yeast. *J. Mol. Biol.* **94**: 595.

Liebman, S.W., J.W. Stewart, J.H. Parker, and F. Sherman. 1977. Leucine insertion caused by a yeast amber suppressor. *J. Mol. Biol.* **109**: 13.

Masurekar, M., E. Palmer, B. Ono, J.M. Wilhelm, and F. Sherman. 1981. Misreading of the ribosomal suppressor *SUP46* due to an altered 40S subunit in yeast. *J. Mol. Biol.* **147**: 381.

McCready, S.J. and B. Cox. 1973. Antisuppressors in yeast. *Mol. Gen. Genet.* **124**: 305.

———. 1976. Suppressor-specificity of antisuppressors in yeast. *Genet. Res.* **28**: 129.

McCready, S.J., B.S. Cox, and C.S. McLaughlin. 1977. The extrachromosomal control of nonsense suppression in yeast: An analysis of the elimination of [psi^+] in the presence of a nuclear gene PNM^-. *Mol. Gen. Genet.* **150**: 265.

Mortimer, R.K. and D.C. Hawthorne. 1969. Yeast genetics. In *The yeasts* (ed. A.H. Rose and J.S. Harrison), vol. 1, p. 385. Academic Press, New York.

———. 1973. Genetic mapping in *Saccharomyces*. IV. Mapping of temperature-sensitive genes and use of disomic strains in localizing genes. *Genetics* **74**: 33.

Olson, M.V., K. Loughney, and B.D. Hall. 1979a. Identification of the yeast DNA sequences that correspond to specific tyrosine-inserting nonsense suppressor loci. *J. Mol. Biol.* **132**: 387.

Olson, M.V., B.D. Hall, J.R. Cameron, and R.W. Davis. 1979b. Cloning of the yeast tyrosine transfer RNA genes in bacteriophage lambda. *J. Mol. Biol.* **127**: 285.

Ono, B., J.W. Stewart, and F. Sherman. 1979a. Yeast UAA suppressors effective in ψ^+ strains: Serine-inserting suppressors. *J. Mol. Biol.* **128**: 81.

———. 1979b. Yeast UAA suppressors effective in ψ^+ strains: Leucine-inserting suppressors. *J. Mol. Biol.* **132**: 507.

———. 1981. Serine insertion caused by the ribosomal suppressor *SUP46* in yeast. *J. Mol. Biol.* **147**: 373.

Ono, B.-I., N. Wills, J.W. Stewart, R.F. Gesteland, and F. Sherman. 1981. Serine inserting UAA suppression mediated by yeast $tRNA^{Ser}$. *J. Mol. Biol.* **150**: 361.

Ozeki, H. 1980. Genetics of nonsense suppressor tRNAs in *Escherichia coli*. In *Transfer RNA: Biological aspects* (ed. D. Söll et al.), p. 341. Cold Spring Harbor Laboratory, Cold Spring Harbor, New York.

Palmer, E., J. Wilhelm, and F. Sherman. 1979a. Variation of phenotypic suppression due to the ψ^+ and ψ^- extrachromosomal determinants in yeast. *J. Mol. Biol.* **128**: 107.

———. 1979b. Phenotypic suppression of nonsense mutants in yeast by aminoglycoside antibiotics. *Nature* **277**: 148.

Pelham, R.B. 1978. Leaky UAG termination codon in tobacco mosaic virus RNA. *Nature* **272**: 469.

Piper, P.W. 1978. A correlation between a recessive lethal amber suppressor mutation *S. cerevisiae* and an anticodon change in minor serine tRNA. *J. Mol. Biol.* **122**: 217.

Piper, P.W., M. Wasserstein, F. Engbaek, K. Kaltoft, J.E. Celis, J. Zeuthen, S. Leibman, and F. Sherman. 1976. Nonsense suppressors of *Saccharomyces cerevisiae* can be generated by mutation of the tyrosine tRNA anticodon. *Nature* **262**: 757.

Philipson, L., P. Anderson, V. Olshevsky, R. Weinberg, D. Baltimore, and R. Gesteland. 1978. Translation of MULV and MSV RNAs in nuclease-treated reticulocyte extract: Enhancement of the *gag-pol* polypeptide with yeast suppressor tRNA. *Cell* **13**: 189.

Rafalski, A., J. Kohli, P. Agris, and D. Söll. 1979. The nucleotide sequence of a UGA suppressor serine tRNA from *Schizosaccharomyces pombe*. *Nucleic Acids Res.* **6**: 2683.

Rasse-Messenguy, F. and G.R. Fink. 1973. Temperature-sensitive nonsense suppressors in yeast. *Genetics* **75**: 459.

Reed, C.R. and S.W. Liebman. 1979. New amber suppressors in *Saccharomyces cerevisiae*. *Genetics* **91**: s102.

Rothstein, R.J., R.E. Esposito, and M.E. Esposito. 1977. The effect of ochre suppression on meiosis and ascospore formation in *Saccharomyces*. *Genetics* **85**: 35.

Secher, D.S., R.G.C. Cotton, N.J. Cowan, R.F. Gesteland, and C. Milstein. 1979. Structural gene mutants of immunoglobulin heavy chains. In *Nonsense mutations and tRNA suppressors* (ed. J.E. Celis and J.D. Smith), p. 285. Academic Press, New York.

Sherman, F., B. Ono, and J.W. Stewart. 1979. The use of the iso-1-cytochrome c system for investigating nonsense mutants and suppressor in yeast. In *Nonsense mutations and tRNA suppressors* (ed. J.E. Celis and J.D. Smith), p. 133. Academic Press, New York.

Sherman, F., S.W. Liebman, J.W. Stewart, and M. Jackson. 1973. Tyrosine substitution resulting from suppression of amber mutants of iso-1-cytochrome c in yeast. *J. Mol. Biol.* **78**: 157.

Singh, A. and T.R. Manney. 1974. Suppression of two missense alleles of the *TRP5* locus of *Saccharomcyes cerevisiae*. *Genetics* **77**: 661.

Singh, A., C. Helms, and F. Sherman. 1979a. Mutation of the non-Mendelian suppressor, ψ^+, in yeast by hypertonic media. *Proc. Natl. Acad. Sci.* **76**: 1952.

Singh, A., D. Ursic, and J. Davies. 1979b. Phenotypic suppression and misreading in *Saccharomyces cerevisiae*. *Nature* **277**: 146.

Smirnov, V.N., V.G. Kreier, L.V. Lizlova, A.M. Andrianova, and S.G. Inge-Vechtomov. 1974. Recessive super-suppressors in yeast. *Mol. Gen. Genet.* **129**: 105.

Smirnov, V.N., A.P. Surguchov, V.V. Smirnov, Y.V. Berestetskaya, and S.G. Inge-Vechtomov. 1978. Recessive nonsense-suppression in yeast: Involvement of 60S ribosomal subunit. *Mol. Gen. Genet.* **163**: 87.

Smirnov, V.N., A.P. Surguchov, E.S. Fominykh, L.V. Lizlova, T.V. Saprygina, and S.G. Inge-Vechtomov. 1976. Recessive nonsense-suppression in yeast: Further characterization of a defect in translation. *FEBS Lett.* **66**: 12.

Smith, J.D. 1979. Suppressor RNAs in prokaryotes. In *Nonsense mutations and tRNA suppressors* (ed. J.E. Celis and J.D. Smith), p. 109. Academic Press, New York.

Steege, D.A. and D.G. Söll. 1979. Suppression. In *Biological regulation and development* (ed. R.F. Goldberger), vol. 1, p. 433. Plenum Press, New York.

Surguchov, A.P., E.M. Pospelova, and V.N. Smirnov. 1981a. Synergistic action of genetic and phenotypic suppression of nonsense mutations in yeast *Saccharomyces cerevisiae*. *Mol. Gen. Genet.* **183**: 197

Surguchov, A.P., Y.V. Berestetskaya, V.N. Smirnov, M.D. Ter-Avanesyan, and S.G. Inge-Vechtomov. 1981b. Ribosomal mutants in eukaryotes. The use of antibiotics for the study of recessive suppression in yeast. *FEMS Lett.* **12**: 381.

Surguchov, A.P., E.S. Fominykch, Y.V. Berestetskaya, V.N. Smirnov, and S.G. Inge-Vechtomov. 1980a. Recessive nonsense-suppression in yeast *Saccharomyces cerevisiae*. The study of 80S ribosomes accumulated in suppressor strain under non-permissive condition. *Mol. Gen. Genet.* **177**: 675.

Surguchov, A.P., E.S. Fominykch, V.N. Smirnov, M.D. Ter-Avanesyan, L.N. Mironova, and S.G. Inge-Vechtomov. 1981c. Further characterization of recessive suppression in yeast: Isolation of the cold-sensitive mutant of *Saccharomyces cerevisiae* defective in the assembly of 60S ribosomal subunit. *Biochim. Biophys. Acta.* **654**: 149.

Surguchov, A.P., Y.V. Berestetskaya, E.S. Fominykch, E.M. Pospelova, V.N. Smirnov, M.D. Ter-Avanesyan, and S.G. Inge-Vechtomov. 1980b. Recessive suppression in yeast *Saccharomyces cerevisiae* is mediated by a ribosomal mutation. *FEBS Lett.* **111**: 175.

Ter-Avanesyan, M.D. and S.G. Inge-Vechtomov. 1974. Location of mutations on the chromosome II of yeast *Saccharomyces cerevisiae*. *Genetika* **10**: 117.

Ter-Avanesyan, M.D., J. Zimmermann, S.G. Inge-Vechtomov, A.B. Sudarikov, V.N. Smirnov, and A.P. Surguchov. 1982. Ribosomal recessive suppressors cause a respiratory deficiency in yeast *Saccharomyces cerevisiae*. *Mol. Gen. Genet.* **185**: 319.

Waldron, C. and B.S. Cox. 1978. Ribosomal proteins of yeast strains carrying mutations which affect the efficiency of nonsense suppression. *Mol. Gen. Genet.* **159**: 223.

Waldron, C., B.S. Cox, N. Wills, R.F. Gesteland, P.W. Piper, D. Colby, and C. Guthrie. 1981. Yeast ochre suppressor *SUQ5-ol* is an altered tRNA$_{UCA}^{Ser}$. *Nucleic Acids Res.* **9**: 3077.

Wetzel, R., J. Kohli, F. Altruda, and D. Söll. 1979. Identification and nucleotide sequence of the *sup8-e* UGA-suppressor leucine tRNA from *Schizosaccharomyces pombe*. *Mol. Gen. Genet.* **172**: 221.

Yoshida, M., K. Takeishi, and T. Ukita. 1971. Structural studies on a yeast glutamic acid tRNA specific to GAA codon. *Biochim. Biophys. Acta* **228**: 153.

Young, C.S.H. and B.S. Cox. 1972. Extrachromosomal elements in a supersuppression system of yeast. II. Relations with other extrachromosomal elements. *Heredity* **28**: 189.

Organization and Expression of tRNA Genes in *Saccharomyces cerevisiae*

Christine Guthrie
Department of Biochemistry and Biophysics
University of California, San Francisco
San Francisco, California 94143

John Abelson
Department of Chemistry
University of California, San Diego
La Jolla, California 92093

1. **Gene Number and Organization**
   A. Number of tRNA Species
   B. Number and Distribution of tRNA Genes
2. **Intervening Sequences in tRNA Genes**
3. **Genetic Analyses**
   A. Rationale and Strategy
   B. Fine-structure Analysis of the *SUP4* Locus
   C. Effects of Intragenic Mutations on RNA Processing
   D. Unlinked Antisuppressor Mutations
4. **Enzymology of Splicing**
5. **Other Steps in tRNA Gene Expression**
6. **Speculations**

INTRODUCTION

Transfer RNAs have traditionally been among the favorite tools of molecular biologists. By virtue of their small size and relative abundance, tRNAs have long been the RNA molecules of choice for determination of primary, secondary, and tertiary structure. Their pivotal and multifaceted roles in cellular metabolism have made them the focus of biochemists' attention for more than three decades; as a result, probably as much is known about their functional interactions as about any other class of macromolecules. The pièce-de-résistance of their appeal stems from their amenability to genetic analysis. Once converted to a suppressor, a tRNA gene is no longer anonymous; its distinctive phenotype is readily identified and easily tracked. Given the appropriate genetic background, phenotypic loss of suppression can be used to identify each of the elements essential for transcription, maturation, and activity of the tRNA gene product.

The power of these complementary approaches has already been amply demonstrated in both phage and bacterial systems. The ultimate goal of knowing not only the physical relationships among each of the nucleotides in the three-dimensional structure, but also of understanding how each of these structural features relates to the synthesis and function of the mature tRNA, is now conceivably within reach. As the details of this picture have become more refined, the need for comparable information in eukaryotes has become more apparent. Although contemporary methodology has now made it possible to rapidly accumulate primary sequence data from a variety of higher organisms, results of this kind are of only limited value. Similarly, although the techniques of in vitro "genetics" enable one to mutate cloned genes more or less at will, the power of such tour-de-force approaches is severely diminished if the biological consequences can be ascertained only in heterologous or in vitro systems.

Even a casual consideration of these constraints makes the unique appeal of yeast immediately obvious. *Saccharomyces cerevisiae* and *Schizosaccharomyces pombe* are the only eukaryotic organisms in which the existence of nonsense suppressors has been both genetically and biochemically verified (see Sherman, this volume). In addition to the ease and sophistication of genetic manipulations in yeast, the low complexity of the genome makes it feasible to identify a particular gene of interest among as few as several thousand clones. These attractions are highlighted by the opportunity to assay the biological activity of the cloned gene, with or without in vitro alteration, following its reintroduction into the genome by transformation (for further information, see Botstein and Davis [this volume]). Surely no other eukaryote is accessible to such a complete repertoire of experimental strategies.

In this paper we have focused on the application of these approaches to questions of particular current interest. These include (1) the chromosomal distribution of tRNA genes with respect to one another and to other functionally related loci, (2) the structure of the primary transcriptional unit and, specifically, the relationship of the promoter to the structural gene, (3) the biological significance of intervening sequences in a small subset of tRNA genes, as well as (4) elucidation of the enzymatic mechanism(s) by which they are removed, and (5) the maturation pathways of spliced versus colinear tRNA genes. Whenever appropriate, we have tried to assess key similarities and distinctions with respect to the structure, organization, and expression of yeast tRNA genes as compared with their prokaryotic or higher eukaryotic counterparts. Finally, we have correlated available data on codon usage with gene copy number and tRNA abundance in an attempt to discern plausible relationships between the utilization of a given tRNA gene product and its specific mode of biosynthesis.

We have confined our attention to nuclear tRNA genes in *S. cerevisiae* in this paper; mitochondrial tRNAs are discussed elsewhere (see Dujon 1981).

GENE NUMBER AND ORGANIZATION

Number of tRNA Species

In 1966, Crick proposed a set of rules that predicted the likely pairings allowed between codon and anticodon in the third or "wobble" position of the codon. By this hypothesis, which considered evidence from in vitro codon-binding experiments and the three to four tRNA sequences known at the time, fewer than 61 tRNAs would be required to translate the 61 "sense" codons since, in many cases, a single tRNA species could recognize more than one codon. From the more than 177 tRNA sequences now reported (Gauss and Sprinzl 1981), it is clear that codon-anticodon wobble does occur but is somewhat more restricted in yeast, and probably in all eukaryotes, than had been thought previously.

As summarized in Table 1, at present there is sequence information for 28 distinct *S. cerevisiae* tRNA species. In two cases, $tRNA^{Ile}_{AUA}$ and $tRNA^{Gln}$, only the sequence of the gene is known. The only amino acid acceptor for which sequence data are lacking is asparagine. This information allows us to propose a revised set of wobble rules, which are given in Table 2. By extension of these empirically derived modifications of the wobble rules, we can predict the approximate number and nature of tRNA species likely to be present in yeast.

Codons ending in U or C are recognized by G in all cases where there are only two codons in the set. Thus, in the case of $tRNA^{Cys}$, $tRNA^{His}$, $tRNA^{Phe}$, and $tRNA^{Tyr}$, a single tRNA species is used to read both codons. When there are three or four codons in the set, U and C are usually recognized by I. (An exception to this rule is seen in $tRNA^{Gly}$, in which case the wobble base is G.) In the original wobble hypothesis, I was postulated to recognize A, as well as U and C. Recent genetic evidence, however, argues against this expanded recognition. In *S. pombe* there are two genes for a $tRNA^{Ser}$ that recognize UCA; haploid spores in which both of these genes have been converted to nonsense suppressors are inviable (Munz et al. 1981). Assuming that *S. pombe*, like *S. cerevisiae* and higher eukaryotes, contains a major $tRNA^{Ser}$ species with the anticodon IGA, these data suggest that I cannot decode A in vivo.

Another departure from the original wobble hypothesis is the finding that codons ending in A are decoded exclusively by tRNA species with a modified U in the wobble position of the anticodon and that these tRNAs do not appear to read the synonymous codon ending in G. Six tRNAs containing a modified U have been sequenced in yeast: $tRNA^{Arg}_3$ (5-methoxycarbonylmethyluridine [mcm^5U]), $tRNA^{Glu}_3$ and $tRNA^{Lys}_2$ (5-methoxycarbonylmethyl-2-thiouridine [mcm^5s^2U]), $tRNA^{Ser}_{UCA}$ (5-carbamoylmethyluridine [ncm^5U]), $tRNA^{Val}$, and $tRNA^{Leu}$. Each of these cases involves codon sets in which A and G specify the same amino acid. In four instances ($tRNA^{Leu}_2$, $tRNA^{Ser}_{UCA}$, $tRNA^{Val}$, and $tRNA^{Leu}$), a second isoacceptor species has been identified

Table 1 Yeast tRNAs: Codon usage, tRNA sequences, gene sequences, and copy number

Amino acid	Codon	Codon usage	Anticodons predicted	found tRNA	found DNA	tRNA Genes intervening sequences (length, bp)	number of genes
Ala	GCU	208	IGC	IGC			$\geq 6^{14}$
	GCC	76					
	GCA	3	U*GC				
	GCG	2	CGC				
Arg	CGU	7	ICG	ICG			
	CGC	0					
	CGA	0	U*CG				
	CGG	0	CCG				
	AGA	85	U*CU	U*CU	TCT[1,2]	none	$\geq 8^{14}$
	AGG	2	CCU				
Asn	AAU	5	GUU				
	AAC	101					
Asp	GAU	51	GUC	GUC	GTC[1]	none	$\geq 12^{14}$
	GAC	106					
Cys	UGU	22	GCA	GCA			
	UGC	1					
Gln	CAA	61	U*UG		TTG[3]	none	
	CAG	4	CUG				
Glu	GAA	150	U*UC	U*UC	TTC[23]	none	
	GAG	7	CUC				
Gly	GGU	214	ICC	GCC			$\geq 6^{16}$
	GGC	6					
	GGA	3	U*CC				
	GGG	2	CCC				
His	CAU	11	GUG	GUG			6^{17}
	CAC	57					
Ile	AUU	75	IAU				
	AUC	71					
	AUA	1	U*AU		TAT[4]	(60)	
Leu	UUA	14	U*AA	ZAA[20]			$\geq 5^{14}$
	UUG	170	CAA	CAA	CAA[5,6]	(32)	$\geq 10^{14}$
	CUU	3	IAG				
	CUC	1					
	CUA	7	U*AG	UAG			
	CUG	0	CAG				
Lys	AAA	15	U*UU	U*UU			
	AAG	184	CUU	CUU			
Met$_m$	AUG	56	CAU	CAU	CAT[24]	none	
Met$_i$	AUG		CAU	CAU	CAT[18]	none	$\geq 8^{18}$

Table 1 (Continued)

Amino acid	Codon	Codon usage	Anticodons predicted	Anticodons found tRNA	Anticodons found DNA	tRNA Genes intervening sequences (length, bp)	tRNA Genes number of genes
Phe	UUU	12	GAA	G_mAA	GAA^7	(18,19)	10^{19}
	UUC	74					
Pro	CCU	16	IGG		AGG^{16}	none	
	CCC	1					
	CCA	85	U*GG				
	CCG	0	CGG				
Ser	UCU	103	IGA	IGA	AGA^8	none	11^8
	UCC	94					
	UCA	6	U*GA	$U*GA^{21}$	$TGA^{9,10}$	none	$3^{9,10}$
	UCG	1	CGA	CGA	CGA^{10}	(19)	$1^{10,11}$
	AGU	5	GCU				
	AGC	1					
Thr	ACU	75	IGU	IGU			$\geq 4^{14}$
	ACC	75					
	ACA	8	U*GU				
	ACG	4	CGU				
Trp	UGG	27	CCA	C_mCA	$CCA^{5,12}$	(34)	$\geq 4^{12}$
Tyr	UAU	10	GUG	$G\Psi A$	GTA^{13}	(14)	8^{22}
	UAC	76					
Val	GUU	121	IAC	IAC			
	GUC	97					
	GUA	1	U*AC	N*AC			
	GUG	2	CAC	CAC			$\geq 7^{14}$

The codon usage is the total number of times each codon appears in ten different yeast genes: Glyceraldehyde-3-phosphate dehydrogenase (GPDH) genes pGap491 and pGap63 (Holland and Holland 1979, 1980); alcohol dehydrogenase (ADH) I (Bennetzen and Hall 1982a); histone genes H2B1 and H2B2 (Wallis et al. 1980); iso-1-cytochrome c (Smith et al. 1979); iso-2-cytochrome c (Montgomery et al. 1980); actin (Ng and Abelson 1980); enolase genes peno 8 and peno 46 (Holland et al. 1981). The table is derived from, and is an extension of, the compilation made by Bennetzen and Hall (1982b). The asymmetry of codon usage is most striking in the case of genes that are strongly expressed, i.e., GPDH, ADH, and enolase. Table 2 was used as the basis for predicting the anticodons. The anticodon sequences found are from the tabulation of tRNA sequences by Gauss and Sprinzl (1981), except as indicated in the footnotes. The underlined anticodon sequences are those from the major isoacceptors (Bennetzen and Hall 1982b). In some cases, anticodon sequences are also given from the result of DNA sequencing. The gene number for any particular species has been determined by hybridization using the techniques of Southern (1975); or in those cases where the number is approximate ($\geq$), it is the number of independent clones that have been found. References (indicated by superscript numbers): (1) Schmidt et al. (1980); (2) J. Villanueva (pers. comm.); (3) G. Tschumper and J. Carbon (pers. comm.); (4) R.C. Ogden et al. (unpubl.) (see Fig. 2); (5) Kang et al. (1980); (6) Venegas et al. (1979); (7) Valenzuela et al. (1978); (8) Page and Hall (1981); (9) Broach et al. (1981); (10) Olson et al. (1981); (11) Etcheverry et al. (1982); (12) J. Zalidvar et al. (pers. comm.); (13) Goodman et al. (1977); (14) Beckmann et al. (1977); (15) B. Hall (pers. comm.); (16) M. Culbertson (pers. comm.); (17) G. Fink (pers. comm.); (18) E. Gonzalez et al. (pers. comm.); (19) R. Bishop (pers. comm.); (20) The anticodon sequence of this tRNA was determined by Piper and Wasserstein (1977). Z is an unknown modified pyrimidine. (21) Etcheverry et al. (1979), Waldron et al. (1981); (22) Olson et al. (1977), (1979); (23) A. Eigel et al. (pers. comm.); (24) J. Olah and H. Feldmann (pers. comm.).

Table 2 Wobble rules for third position codon-anticodon pairing in yeast tRNAs

	Anticodon	
Codon	Crick (1966)	proposed
U	A, G, or I	G or I
C	G or I	G or I
A	U or I	$\overset{*}{U}{}^{a}$
G	C or U	C

$^a\overset{*}{U}$ indicates a modified uridine residue (see text).

that contains a C in the wobble position of the anticodon; in two of these cases, namely UCN (serine) and GUN (valine), yeast contains three tRNA species to decode a block of four codons. In other words, although the original rules predict that it should take only two tRNAs to translate all four UCN serine codons, yeast contains a tRNA species with a CGA anticodon in addition to those with IGA and UGA anticodons. The implication that wobble decoding is restricted is consistent with the long-standing observation that yeast ochre suppressors, unlike their prokaryotic counterparts, recognize UAA but not UAG nonsense codons (see Sherman, this volume).

The rules proposed in Table 2 can thus be summarized as follows. A single tRNA is required to recognize sets of two codons ending in U or C (as in the original wobble hypothesis). Two different tRNAs are required to recognize sets of two codons ending in A and G, and three different tRNAs are required to recognize a block of four codons. Since wobble pairings may be unpredictably affected by other specific anticodon modifications (or lack thereof), these rules may not hold in all cases. Indeed, a major isoacceptor of tRNA^{Leu} isolated from *S. cerevisiae* (Randerath et al. 1979) contains an unmodified uridine residue in the wobble position and apparently translates all six leucine codons in vitro. We can thus expect occasional idiosyncrasies in the pattern of codon recognition, but overall the rules are consistent. It is therefore likely that yeast will prove to have 46 different tRNAs, give or take a few.

Number and Distribution of tRNA Genes

Hybridization studies have shown that there are approximately 360 tRNA genes in yeast (Schweizer et al. 1969; Feldman 1976). Thus, the average reiteration frequency is on the order of eight genes per tRNA species. A number of interesting questions can be posed regarding the organization and properties of these gene families. Are the members of each family clustered, or are they dispersed in the genome? Is the repeated unit the tRNA structural gene per se, or is there flanking sequence information that is common to all members of a set? Are all genes within an isoaccepting family expressed at the same level, or is expression modulated depending on chromosome position or time in the cell cycle? Is the number of genes in a particular set directly related to the requirements for that tRNA, i.e., to codon usage? The answers to many of these questions are already available.

Certain general features of gene organization have emerged from cloning experiments. In one study (Beckmann et al. 1977), 175 of 4000 clones containing random *Hin*dIII fragments of yeast DNA were shown to contain tRNA genes. The 4000 independent clones, each containing an average size fragment of 2.5×10^6 daltons, comprise approximately one haploid equivalent of the yeast genome ($\sim 10^{10}$ daltons). The results suggest that there is, in general, relatively little clustering; on the average, the 175 clones should each contain two tRNA genes. Although extremely powerful for fine-structure mapping, the resolution afforded by studies of this type is of limited value for determining higher-order clustering. Thanks to the extensive genetic analysis of nonsense suppressors, however, we have definitive information about the organization of several tRNA gene families. Table 3 summarizes available data for those nonsense suppressor loci shown to encode tRNA products by one or more criteria, including specific amino acid insertion in vivo, suppression of chain termination in vitro, or (in most cases) direct sequence analysis.

The best-studied isoaccepting family is the set of genes coding for $tRNA^{Tyr}$. Both tyrosine codons (UAU and UAC) are read by a single tRNA species. This tRNA is transcribed from eight genes, each of which can be mutated to an ochre or amber suppressor. The results of genetic mapping (Hawthorne and Mortimer 1968; Mortimer and Hawthorne 1973) show that the eight $tRNA^{Tyr}$ loci are distributed among six linkage groups; in those instances where two tRNA genes are found on a common chromosome, they occupy different arms. Each of these genes has also been localized to a unique restriction fragment and cloned (see Tables 1 and 3). In no case is there evidence for the presence of any other tRNA gene within several kilobases of DNA. As described below, the $tRNA^{Tyr}$ genes have been and will continue to be the subject of intense genetic study.

The serine-specific tRNAs are the products of a more diverse set of genes. There are six codons for serine, UCN, and AG_C^U. The major isoacceptor $tRNA_2^{Ser}$ has the anticodon IGA and presumably recognizes the codons UCU and UCC (see above). Hybridization studies suggest that there are at least 11 $tRNA_2^{Ser}$ genes; 3 of these genes have been sequenced (Page and Hall 1981). No two $tRNA_2^{Ser}$ genes are found on a common restriction fragment, although in one case it is possible that there is a second tRNA gene within 0.6 kb. The only direct linkage information for any of these tRNAs comes from the fortuitous identification of the *CYC1* gene on one of

Table 3 Chromosomal distribution of suppressor tRNA genes

Amino acid	Suppressor (UAG)	Suppressor (UAA)	Chromosome arm	Reported linkage[a]	Sequence data RNA	Sequence data DNA	References[b] map	References[b] RNA	References[b] DNA
Serine	SUP-RLI[c]		IIIR	rad18	+	+	1	2	3
		SUP61[c]	IIIR	rad18	+	(+)[c]	4	5	
		SUP16	XVIR	aro7	+	+	6	7	3
		SUP17	IXL	lys11	(+)[d]	(+)[e]	6		
		SUP19	VR	his1	(+)[d]	(+)[e]	8		
Leucine		SUP26	XIIIR	ura4	–	–	6		
		SUP27	IVR	rad57	–	–	9		
		SUP29[f]	XC	met3	–	–	9		
	SUP52[f]		XC	met3	–	–	9		
	SUP53		IIIL	leu2	–	–	11		
	SUP54		VIIL	trp5	–	–	11		
Tyrosine	SUP2	SUP2	IVR	lys4	(+)[g]	(+)[g]	12,13		
	SUP3	SUP3	XVL	spd1	+	+	12,13		
	SUP4	SUP4	XR	rad7, cdc8	+	+	12,13	14	15
	SUP5	SUP5	F8	gal80	(+)[g]	(+)[g]	12,13	16	17
	SUP6	SUP6	VIR	met10	(+)[g]	+	12,13		
	SUP7	SUP7	XL	let5	(+)[g]	(+)[g]	12		18
	SUP8	SUP8	XIIIR	mak27, rna1	(+)[g]	(+)[g]	12,13		
	SUP11	SUP11	VIC	cly3	(+)[g]	(+)[g]	12,13		19

[a] Data from Mortimer and Schild (1980).
[b] References: (1) Brandriss et al. (1975, 1976); (2) Piper (1978); (3) Olson et al. (1981); (4) Hawthorne and Leupold (1974); (5) Etcheverry et al. (1982); (6) Ono et al. (1979a); (7) Waldron et al. (1981); (8) Ono et al. (1981); (9) Ono et al. (1979b); (10) Liebman et al. (1976); (11) Piper et al. (1976); (12) Hawthorne and Mortimer (1968); (13) Mortimer and Hawthorne (1973); (14) Piper and R. Davis (pers. comm.); (15) R. Rothstein (pers. comm.); (16) Knapp et al. (1978); (17) Goodman et al. (1977); (18) P. Phillipsen and R. Davis (pers. comm.); (19) S. Clark and P. Berg (pers. comm.).
[c] The allelism of SUP61 and SUP-RLI and their common derivation from the sup^+ tRNA$^{Ser}_{UCG}$ gene can be inferred from Etcheverry et al. (1982).
[d] The tRNAs have been shown to be active in in vitro suppression (Ono et al. 1981).
[e] Correspondence of suppressor DNA and wild-type tRNA$^{Ser}_{UCA}$ genes from which they derive (Broach et al. 1981).
[f] SUP29 and SUP52 may be closely linked but distinct tRNA^{Leu} genes; see Ono et al. (1979b).
[g] Cloned fragments that hybridize to tRNA^{Tyr} have been correlated with the respective suppressor loci by restriction mapping (Olson et al. 1977, 1979).

the cloned fragments; the two loci are separated by about 2 kb. A second tRNA^{Ser} species contains a modified U in the wobble position of the anticodon and is presumed to decode UCA (Etcheverry et al. 1979, and in prep.). This isoacceptor is encoded by three genes, each of which has been uniquely correlated with a nonsense suppressor locus (Broach et al. 1981); these loci are distributed among three chromosomes. A third tRNA^{Ser}, which differs in only three nucleotides from tRNA$^{\text{Ser}}_{\text{UCA}}$, recognizes the codon UCG. This tRNA is coded for by a single gene (Etcheverry et al. 1979, 1982). The allelic recessive lethal suppressors *SUP-RL1* (Brandriss et al. 1975, 1976) and *SUP61* (Hawthorne and Leupold 1974) derived from this gene map to chromosome III. Little is known about the tRNA that recognizes AGU and AGC.

Although there is less information about the leucine-inserting suppressors, results of genetic mapping again show that the six identified loci are distributed among five chromosomes. (There is, however, reason to suspect that two tRNA^{Leu} genes might be closely linked to one another; see the footnote to Table 3.)

To date, the only clear-cut evidence for tRNA juxtaposition comes from the cloning experiments described above (Beckmann et al. 1977). Of 175 clones, 4 contain genes for tRNA$^{\text{Arg}}_{3}$ and tRNA^{Asp}. DNA sequencing of the tRNA$^{\text{Arg}}_{3}$ and tRNA^{Asp} genes in two of these clones (Schmidt et al. 1980) showed that they are indeed tightly linked to one another. In both clones, tRNA$^{\text{Arg}}_{3}$ is located 5' proximal to the tRNA^{Asp} gene, and they are separated by an identical spacer of ten nucleotides. The 5'-flanking sequences are distinct in the two clones. As described below (see Other Steps in tRNA Gene Expression), there is reason to believe that the two tRNA genes comprise a single transcriptional unit. (A similar example of linkage has been described in *S. pombe* [Mao et al. 1980]. Here, a tRNA$^{\text{Ser}}_{\text{UCG}}$ gene is separated by 7 bp from a tRNA^{Met} gene; it is also probable that these two tRNAs arise via a single dimeric precursor.) In an independent analysis of tRNA gene organization in *S. cerevisiae*, a clone containing only a tRNA$^{\text{Arg}}_{3}$ gene has been identified (J. Villanueva et al., pers. comm.). The absence of a linked tRNA^{Asp} gene was confirmed by restriction mapping and DNA sequence analysis. Remarkably, however, the ten nucleotides immediately adjacent to the 3' end of the tRNA$^{\text{Arg}}_{3}$ coding sequence are *identical* to the 10-bp spacer separating tRNA$^{\text{Arg}}_{3}$ from tRNA^{Asp} in the two clones described above. The 5'- and 3'-flanking sequences are otherwise divergent. Certainly, this finding raises intriguing questions about the evolution of this multigene family.

A general pattern of tRNA gene organization thus emerges. A particular tRNA species is encoded by a repeated set of identical or very similar genes. These genes are not linked but rather are found on different chromosomes. The 17 nonsense suppressors shown in Table 3 are distributed over 11 chromosomes. Thus, there is little evidence for the clustering either of isoaccepting species or of tRNA genes per se. With possibly few exceptions, then, the overall distribution of tRNA genes may in fact be close to random. This organization is distinctly unlike that seen in Drosophila, where tRNA genes are typically found in clusters containing multiple copies of several different tRNA genes; here, cloning experiments suggest an average spacing of two to four tRNAs per kilobase within a cluster (Yen and Davidson 1980; Robinson and Davidson 1981).

The significance of the dispersed arrangement of tRNA genes in yeast is not clear. It is in striking contrast to the arrangement of 5S rRNA genes which, like the tRNA genes, are transcribed by RNA polymerase III (see Warner, this volume). One obvious distinction that derives from this type of organization is that the number of genes in a tandemly repeated set can expand or contract by unequal crossing-over. When a set of identical genes is dispersed among different chromosomes, the number of genes in that set would tend to remain fixed.

It is clear from the data in Table 1 that the number of genes varies substantially for different sets. There also appears to be a correlation between gene number and codon usage. This seems especially true for protein-coding genes expressed at high levels (see Bennetzen and Hall 1982b). This relationship is particularly striking in the case of serine. As described above, there are 11 genes for the major isoacceptor $tRNA_2^{Ser}$ (UC_C^U), 3 for $tRNA_{UCA}^{Ser}$, and a single gene for $tRNA_{UCG}^{Ser}$. The demand for these tRNAs is proportionately asymmetrical; based on sequence data for ten yeast proteins, 94% of all serine codons are read by $tRNA_2^{Ser}$ and 3%, by $tRNA_{UCA}^{Ser}$. In 210 opportunities, the UCG codon appears only once.

In general, this correlation probably extends as well to the relative abundance in the cell of different isoacceptors. For example, quantitations of the amount of $tRNA^{Ser}$ species agree well with the 3-to-1 ratio of $tRNA_{UCA}^{Ser}$ to $tRNA_{UCG}^{Ser}$ genes (Etcheverry et al. 1979, 1982). It is more difficult to determine whether all genes within a given set are expressed at the same level. In the case of $tRNA^{Tyr}$, estimates based on the amount of suppressor anticodon relative to the wild-type sequence are consistent with the roughly equivalent expression of eight genes (see, e.g., Colby et al. 1981). Yet the level of suppression in SUP11 strains is markedly lower than that of the other seven suppressors, which raises the possibility that this tRNA is transcribed (and/or processed) at a lower efficiency. Similarly, the three $tRNA_{UCA}^{Ser}$ genes are expressed at different levels, as indicated by the efficiency of suppression of the corresponding SUP alleles, even though the genes encode identical tRNAs (Broach et al. 1981; Ono et al. 1981). The biochemical basis for these differences may soon be revealed by RNA and DNA sequence analysis.

INTERVENING SEQUENCES IN tRNA GENES

In 1977, Goodman et al. found that the $tRNA^{Tyr}$ gene is not colinear with the mature product but instead contains an intervening sequence of 14 bp located immediately adjacent and to the 3' side of the anticodon. Three different tRNA^{Tyr} genes sequenced in that study each had a 14-bp interruption; the three intervening sequences differed at only a single position. In addition, the sequence of an ochre suppressor allele (*SUP4*) of one of the three genes was determined, and the expected anticodon change from GTA to TTA was observed. The presence of the identical intron in this cloned fragment provided good proof that a functioning gene had been sequenced. Subsequently, intervening sequences of 18 bp or 19 bp were also found in three tRNA^{Phe} genes (Valenzuela et al. 1978). In both sets of genes, the insertion occurs at an identical position relative to the coding sequences: one base removed from the 3' side of the anticodon. The intervening sequences within each set of tRNA genes are closely related, but the tyrosine- and phenylalanine-specific sequences bear no obvious relationship to each other.

From these findings, the existence of precursors to tRNA^{Tyr} and tRNA^{Phe} containing the intervening sequence was predicted. Presumably, the mature tRNA would be derived from the precursor by the process of splicing, in which the intervening sequence is clipped out and the resulting ends religated to give the mature product. To look for these intermediates, advantage was taken of the discovery by Hopper et al. (1978) that tRNA precursors accumulate in a mutant strain of yeast, ts136. Isolated by Hutchison et al. (1969), this mutant carries a temperature-sensitive lesion in the *RNA1* gene. The mutant is presumed to be defective in the transport of RNA from nucleus to cytoplasm. At the nonpermissive temperature, the 35S rRNA precursor accumulates (Hopper et al. 1978), the appearance of mRNA in the cytoplasm is halted and poly(A)-containing RNA accumulates in the nucleus (Shiokawa and Pogo 1974), and a particular subset of tRNA precursors accumulates (Knapp et al. 1978).

As shown in Figure 1, these tRNA precursors can be fractionated by two-dimensional polyacrylamide gel electrophoresis. The precursor-specific spots were initially identified by hybridization of the eluted RNAs to the set of *E. coli* recombinant plasmid clones described above (Beckmann et al. 1977). Six of the RNAs (spots indicated in Fig. 1) hybridized to clones containing genes for tRNA^{Tyr}, tRNA^{Phe}, tRNA$_3^{Leu}$, tRNA$_{UCG}^{Ser}$, tRNA^{Trp}, and tRNA$_{AUA}^{Ile}$. These identifications have been subsequently confirmed by RNA sequence analysis of the precursors (Knapp et al. 1978; O'Farrell et al. 1978; Etcheverry et al. 1979; Kang et al. 1980) and DNA sequence analysis of clones containing genes for tRNA^{Trp} (Kang et al. 1980), tRNA$_3^{Leu}$ (Venegas et al. 1979; Kang et al. 1980), tRNA$_{UCG}^{Ser}$ (Olson et al. 1981), and tRNA$_{AUA}^{Ile}$ (R.C. Ogden et al., unpubl.). The nucleotide sequences of the six precursor RNAs are presented in Figure 2.

Four other RNAs (designated 9, 11, 12, and 13 in Fig. 1) hybridize to clones containing tRNA genes of unknown specificity. The identity of these precursors is currently under investigation, but it is already clear that the intermediates that accumulate at the nonpermissive temperature in ts136 comprise a distinct subset of the possible tRNA precursors. There is, e.g., no

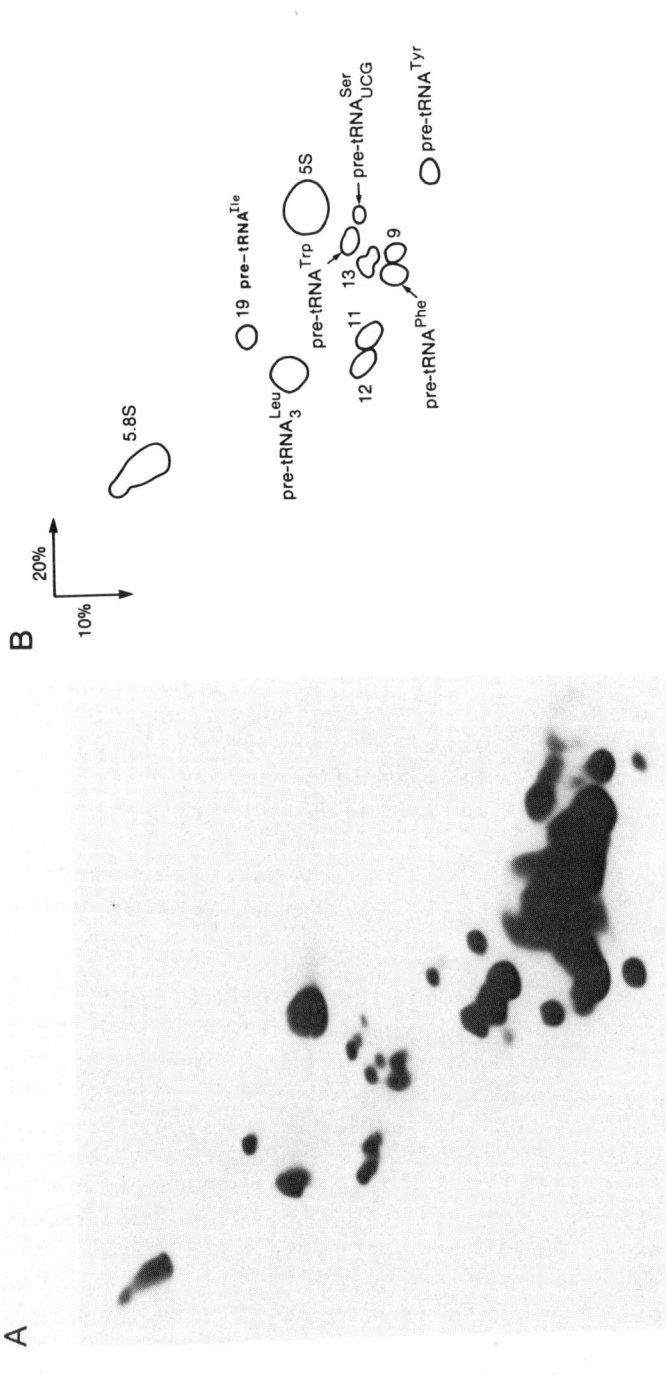

Figure 1 Fractionation of tRNA precursors accumulated in strain ts136. ^{32}P-labeled RNA was prepared using a temperature-sensitive diploid, M304, that is homozygous for the *rna1* locus. A typical preparation (cf. Knapp et al. 1978; Etcheverry et al. 1979) is purified by electrophoresis in a 10% polyacrylamide–4 M urea gel, followed by electrophoresis in a second dimension using a 20% polyacrylamide–4 M urea gel. (*A*) Autoradiography revealed the positions of the different precursor tRNAs. (*B*) Diagrammatic representation of the ^{32}P-labeled precursor tRNAs; identities were determined as described in the text. The sequences of some of these precursors are shown in Fig. 2.

accumulation of the expected dimeric precursor to tRNA$_3^{Arg}$-tRNA^{Asp}. Indeed, earlier investigations by Fraden et al. (1975), using in vivo pulse-labeling of yeast cell cultures, showed an accumulation of approximately 27 putative tRNA precursors that could be processed to 4S-sized material after incubation with a crude yeast extract (Blatt and Feldmann 1973).

Three lines of evidence suggest that the precursors found in *rna1* strains at the nonpermissive temperature are those containing intervening sequences. First, as can be seen in Figure 2, the six molecular species for which sequence information is available all contain intervening sequences. These sequences range in size from 14 nucleotides (tRNA^{Tyr}) to 60 nucleotides (tRNA^{Ile}) and, where evidence has been obtained, they have been shown to be faithful transcripts of the intervening sequence in the gene. Second, since all ten precursors can be matured in vitro by the purified splicing enzyme (see below, Other Steps in tRNA Gene Expression), it appears that the four precursors whose sequences are still unknown also contain intervening sequences. Third, at least nine tRNA genes that do not contain introns have been sequenced (see Table 1). Precursors for these tRNAs have not been detected in ts136. It is thus our working hypothesis that precursors to all of the tRNAs whose genes contain intervening sequences accumulate as a consequence of the *rna1* mutation.

Those precursors that accumulate at the nonpermissive temperature have already been partially matured, since they have the same 5' and 3' termini as do their mature tRNA counterparts. In particular, the presence of the CCA$_{OH}$ sequence at the 3' terminus indicates that these precursors are intermediates between the primary transcription product and mature molecules, because in no case has a eukaryotic tRNA gene been shown to encode these nucleotides.

Whatever the *rna1* defect may be, characterization of the accumulated precursors has obviously facilitated the identification and sequencing of those tRNA genes that are not colinear with their products. But why some tRNA genes contain intervening sequences and others do not remains a complete mystery. Attempts to "explain" the presence of introns in these particular tRNA genes have been numerous but uniformly unsatisfactory. The paradox is most clearly seen in the case of the tRNA^{Ser} genes. As we have described above, there are at least 15 genes that code for tRNAs recognizing the UCN codons. Of these, only one, the tRNA$^{Ser}_{UCG}$ gene, contains an intervening sequence (Etcheverry et al. 1979; Olson et al. 1981; Page and Hall 1981). Moreover, this is the *only* gene in yeast capable of decoding this rarely used triplet (Etcheverry et al. 1982). The tRNA$^{Ser}_{UCG}$ sequence otherwise differs from that of tRNA$^{Ser}_{UCA}$ in only three positions. This fact places a significant constraint on theories for the role of the intervening sequence in evolution and in the expression of tRNA genes.

It is thus of interest to consider the occurrence of intervening sequences in the tRNA genes of other eukaryotes. Introns have been found in a tRNA$^{Ser}_{UCG}$

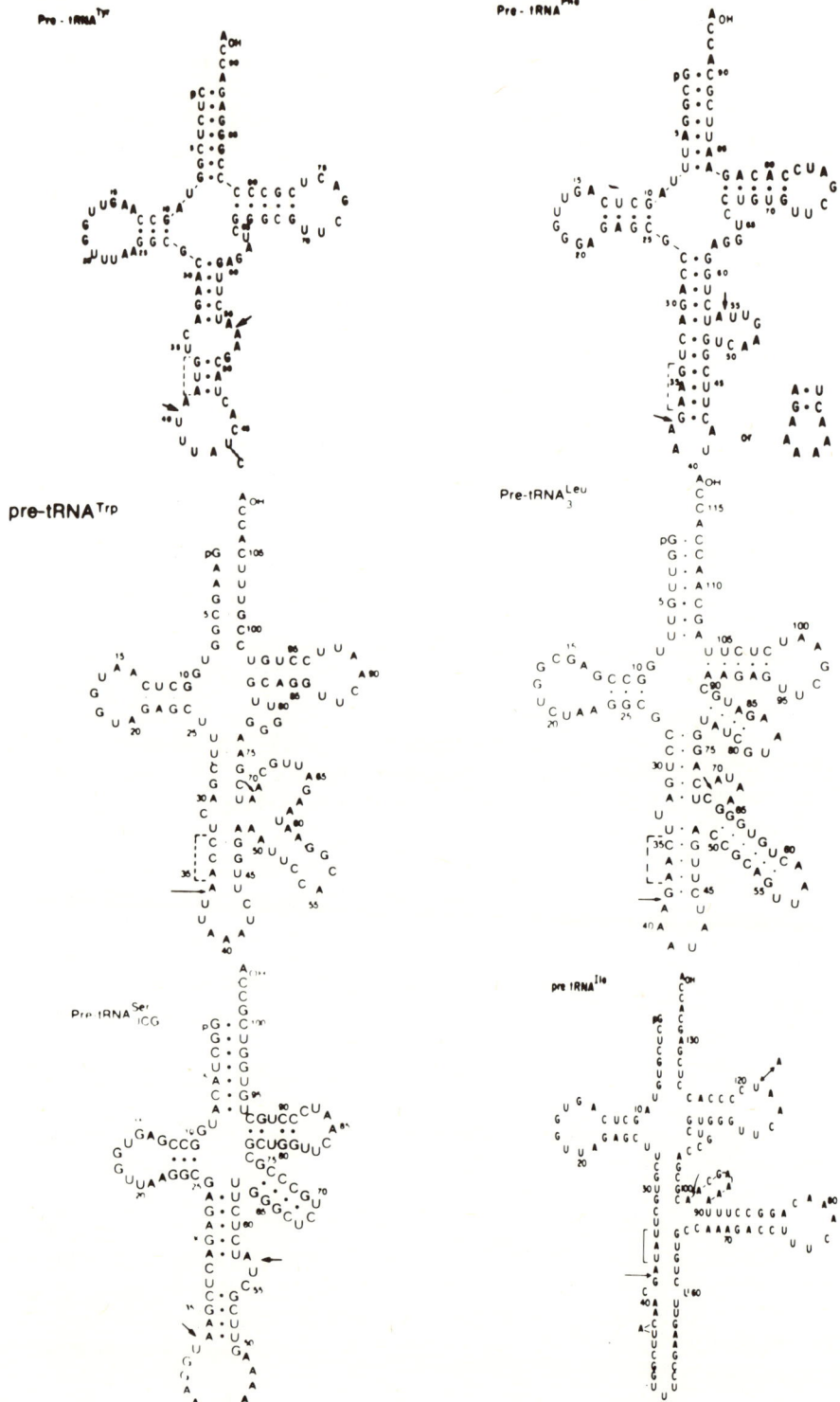

gene from *S. pombe* (Mao et al. 1980), a tRNA^{Phe} gene from *Neurospora* (Selker and Yanofsky 1980), two tRNA$^{Leu}_{UUG}$ genes from *Drosophila* (Robinson and Davidson 1981), and a tRNA^{Tyr} gene from *Xenopus* (Muller and Clarkson 1980). From these observations several generalizations can be made about spliced tRNAs from other eukaryotes.

1. So far, all of the tRNA genes found to contain intervening sequences have spliced counterparts in yeast.
2. Not all tRNA genes that are spliced in yeast are spliced in other eukaryotes. For example, colinear tRNA^{Phe} genes have been sequenced from *S. pombe* (Kohli et al. 1980) and from *Xenopus* (Muller and Clarkson 1980). Thus, tRNA^{Phe} genes in some organisms contain intervening sequences, and in other organisms they do not.
3. In most cases, the nucleotide sequences of the introns in other eukaryotic tRNAs bear little resemblance to those found in the corresponding yeast tRNA genes. Although they are generally similar in length, the most impressive correlation is in the location of the intervening sequence relative to the anticodon. The placement of the sequence is immediately adjacent to the anticodon in the *Xenopus* tRNA^{Tyr} gene; however, in all other examples it is one nucleotide removed.

From the apparent conservation of intron location within tRNA genes, it might be predicted that there is a common splicing mechanism that takes advantage of this positional information. In Figure 2, the sequences of precursors to tRNA^{Tyr} and tRNA$^{Leu}_3$ are drawn in secondary structures that have maximized base-pairing, as determined by calculating the most favorable free energy, using the rules derived by Tinoco et al. (1973) and Borer et al. (1974). The acceptor stem and the dihydro-U and TΨC stems and loops appear as they are found in the tRNA cloverleaf structure. The anticodon stem is intact, but it is augmented by a second helical region. The structures of the precursors to tRNA$^{Ser}_{UCG}$, tRNA^{Trp}, and tRNA$^{Ile}_{AUA}$ were derived simply by inspection and were chosen to maximize similarities with the first set; free energy minima for alternative structures have not been determined.

As an initial test of these structural models, nuclease sensitivity has been used to probe the conformation of two of the precursors accumulated in ts136. In its native conformation, only the anticodon loop of a tRNA mole-

Figure 2 (See facing page) Nucleotide sequences of six spliced tRNA precursors. The nucleotide sequences of six precursors that accumulate at the nonpermissive temperature in ts136 are drawn in secondary structures which, in the case of tRNA^{Tyr} and tRNA$^{Leu}_3$, have maximized base-pairing according to the rules of Tinoco et al. (1973) and Borer et al. (1974). Free energy calculations for the other structures have not been determined (see text). The anticodons appear in brackets. Sites of excision and ligation are indicated by arrows. The data are from Knapp et al. (1979); Etcheverry et al. (1979); and Kang et al. (1980).

cule is a target for the single-stranded specific nuclease S1 (Harada and Dahlberg 1975). Under comparable conditions, the anticodon region in the precursor to tRNA^{Tyr} is completely protected from digestion; the observed cleavage sites lie instead within the two predicted looped regions in the extended anticodon stem (O'Farrell et al. 1978). Analogous results were obtained with the precursor to tRNA$^{Ser}_{UCG}$ (Etcheverry et al. 1979). Although these data are consistent with the proposed structures, they do not exclude alternative models.

There is, in fact, at least one reason to challenge the current hypothesis. The predicted requirements for complementarity between anticodon and intervening sequence should not be satisfied in the case of most nonsense suppressors. As a consequence of the G → U mutation in the anticodon of *SUP-RL1* (Piper 1978; Olson et al. 1981), e.g., the proposed helical region encompassing the anticodon in the precursor to tRNA$^{Ser}_{UCG}$ would almost certainly exist as a large, non-base-paired loop; the same is true for the 2-nucleotide substitution in the *SUP61* ochre derivative (Etcheverry et al. 1982). Yet these precursors appear to be processed with efficiency and fidelity comparable to the wild-type molecule (T. Etcheverry and C. Guthrie, unpubl.). Thus, even if the proposed structures do exist in the cell, they cannot be an absolute prerequisite for splicing.

Clearly, much more data are required to identify the salient structural features of intervening sequences. Since precursor conformation is likely to play a major role in determining the kinetics and specificity of tRNA splicing (see below), this information is crucial if we hope to understand the biological significance of this otherwise paradoxical gene structure. The application of most conventional techniques for structural analysis is currently precluded by the limited yields of these precursors. Fortunately, there are alternative resources already at hand. Genetics has traditionally been a powerful ally of physical biochemistry; among its many virtues is the unique capacity to specifically distinguish those structural alterations that are of functional consequence to the living organism. As discussed below, this strategy can be exploited with particular effectiveness in yeast.

GENETIC ANALYSES
Rationale and Strategy

Genetics has been used extensively to identify the requirements for the production of a functional tRNA species. In most of these studies, the parental strain contains a tRNA gene that has been converted to a nonsense suppressor, so that loss of function of a specific tRNA can be monitored phenotypically by following suppression of nonsense mutations in several different genes. Two classes of mutations can be isolated: (1) lesions in unlinked genes coding for enzymes involved in the transcription and maturation of tRNA and (2) mutations within the structural gene for the suppressor tRNA itself. At least in principle, mutations in the latter class could be of two types: those that specifically affect the function of a normally synthesized tRNA and those that interfere with one or more steps in the synthesis of the mutant tRNA gene product. Distinguishing these alternatives requires direct biochemical analysis.

The most commonly used strategy for mutant isolation is that first employed by McCready and Cox (1973), in which mutations that reduce or abolish the activity of a suppressor tRNA can be selected directly by including in the parental strain the *can1-100* allele, an ochre mutation in the gene for arginine permease. Since the parental strain makes a functional arginine permease, it is sensitive to the toxic arginine analog, canavanine. Mutations that abolish the ability of the suppressor tRNA to suppress the *can1-100* allele render the cell resistant to canavanine. These can be readily distinguished from a second, nonsuppressible mutation in the *can1* gene by screening for the simultaneous loss of suppression of nonsense mutations in other genes carried in the parental strain.

Many studies to date have focused on mutational analysis of the *SUP4* locus, which encodes a nonsense suppressor derived from one of the eight tRNA^{Tyr} genes. This locus was initially chosen because of high suppression efficiency; its appeal more recently is based on the availability of DNA sequence information and on the opportunity to probe the structure and possible function of an intervening sequence.

Fine-structure Analysis of the *SUP4* Locus

As described above, the identification of presumptive antisuppressor mutations is based on the simultaneous loss of suppression of one or more additional nonsense alleles. When a large number of suppressible markers are included in the parental strain, several phenotypic classes can be identified (McCready and Cox 1973; Laten et al. 1978; Kurjan et al. 1980). It is presumed that the apparent gradient of suppressor inactivation (see Table 4) reflects the relative amount of suppressor activity required at each of the ochre sites. As shown in Table 4, the majority of mutants retain no detectable suppressor function. Virtually all mutations giving rise to this "tight" phenotype can be shown by complementation and tetrad analysis to lie within, or closely linked to, the *SUP4* locus (see footnote to Table 4).

It is of interest, for several reasons, to obtain more precise mapping data than are afforded by these techniques. Fine-structure analysis of a biochemically accessible locus allows the opportunity for a precise correlation between physical and genetic maps. In addition to providing information about genetic recombination per se, the availability of a reliable genetic map would make it possible to readily identify mutations in specific regions of the molecule of particular interest for biochemical characterization.

Fine-structure maps of the *SUP4* locus have been obtained by Rothstein (1979) and by Kurjan et al. (1980) by taking advantage of the ability to easily

Table 4 Phenotypic and genotypic distribution of antisuppressor mutations

Parental genotype	SUP4	can1-100	ade2-1	leu5-2	lys1-1	trp5-48	No./class	Percent of total	Percent of linked
Parental phenotype	Sup	s	+	+	+	+			
Mutant phenotype									
I		r	±	+	+	+	10	(12)	0
II		r	−	±	+	+	9	(11)	40
III		r	−	−	±	+	11	(14)	
IV		r	−	−	−	+	7	(9)	100
V		r	−	−	−	−	44	(54)	100
						Total	81	(100)	

Mutants were selected from haploid strains by spreading ~10^7 cells/plate on minimal medium supplemented with adenine, histidine, lysine, and tryptophan and containing 80 mg/liter of canavanine. Plates were incubated 3–5 days at 30°C. Independently arising resistant colonies were repurified and scored for suppression of auxotrophic suppressible alleles by replica-plating to selective media; ± indicates partial growth. To identify mutations within the SUP4 locus, mutants were crossed to SUP4 and sup4⁻ strains homozygous for all suppressible alleles. Mutations within the suppressor gene are semidominant; mutant diploids are either pink (with SUP4) or red (with sup4⁻) due to incomplete suppression of the ade2-1 allele. In many cases, linkage was confirmed by tetrad analysis. (Data are from D. Colby and C. Guthrie [in prep.] and Colby et al. [1981].)

detect rare meiotic recombination events between two antisuppressor mutations (or between an antisuppressor and the sup^+ anticodon) that will regenerate a functional suppressor gene product. Both investigators have shown that approximately 25% of all mutations at the *SUP4* locus are deletions, which are presumably promoted by the presence of repeated "δ" sequences flanking the $tRNA^{Tyr}$ gene. Kurjan et al. (1980) were able to assign 69 of 100 independent intragenic antisuppressor mutations to ten clusters; using data obtained both from pairwise crosses of mutants and from crosses of mutants and wild type (sup^+), an internally consistent map was generated. To correlate the genetic map with a physical map, Kurjan and co-workers (1980) determined the DNA sequence of *SUP4* genes cloned from 32 of these mutant strains. Twenty-six independent nucleotide alterations were identified, and the positions were all in excellent agreement with predictions based on meiotic fine-structure mapping.

An examination of the data of Kurjan et al. (1980) suggests two important conclusions. Assuming that the 98 mutations studied are a statistically random sample of genetic target sites, they can be taken to define the *SUP4* functional unit. All of the mutations sequenced, which included the extreme 5' and 3' alleles, fall within the tRNA structural gene. In other words, it appears as if no simple mutation outside of the gene alters a function essential for transcription. This finding is at least compatible with the provocative notion that RNA polymerase-III "promoters" occur within the structural gene (see below, Other Steps in tRNA Gene Expression). The second conclusion derives from an analysis of the relative distribution of mutations. Although 66 mutations are more or less randomly distributed throughout the 75 nucleotides comprising the structural or "coding" sequences, only three mutants were found to have lesions within the 14-nucleotide intervening sequence. Moreover, all three contained the identical mutation: an A → U transversion 4 nucleotides from the 5' splice junction. Since a disproportionately large number of mutations presumed to lie within or near the intron were chosen for sequence analysis, it is clear that this portion of the molecule does not constitute a functional target comparable with its physical size.

Since removal of the intervening sequence is required for generation of a functional product, these results suggest that single base substitutions within the intron rarely, if ever, prevent splicing of the precursor tRNA. To determine whether any of the sequence alterations identified affect the expression of the mutant gene product, as opposed to its function per se, requires biochemical analysis of RNA synthesis, as described below.

Effects of Intragenic Mutations on RNA Processing

Two genetic approaches have been taken to identify the specific elements of precursor sequence and/or structure that determine accurate and efficient recognition by the RNA processing machinery. In the first case, mutants isolated by the above strategy were characterized biochemically in order to identify those in which unspliced precursor accumulates at the expense of the mature suppressor tRNA product. Quantitative comparisons of the amount of in vivo ^{32}P-labeled suppressor-specific species can be made by exploiting the G → U mutation in the *SUP4* anticodon (Goodman et al. 1977); with the elimination of this T1-sensitive RNase site, the anticodon of the mature and precursor forms of the suppressor tRNA are each contained within a unique T1 oligonucleotide, which is easily distinguished from the seven wild-type copies by homochromatography (cf. Colby et al. 1981).

Using this approach, four mutants that exhibit elevated (threefold to tenfold) accumulation of precursor relative to the *SUP4* parent have been identified (Colby et al. 1981; D. Colby and C. Guthrie, in prep.). In most cases, there is a simultaneous reduction in the amount of suppressor-specific product detected in the mature tRNA^{Tyr} population. Each of the mutations map at distinct sites within the *SUP4* locus and, consistent with the presence of residual suppressor tRNA, they are all phenotypically "leaky" (classes II-IV, Table 4). In one such mutant (strain a122), the level of mutant precursor is increased at least sevenfold, although there is only a 30-50% reduction in the steady-state level of suppressor tRNA. Interestingly, the mutation in strain a122 has been shown to be an A → G transition at the 5'-splice junction (Colby et al. 1981). Thus, despite the fact that this nucleotide change involves the phosphodiester bond that is cleaved in the excision of the intervening sequence, the mutant precursor is nonetheless recognized with the same fidelity as is the wild-type substrate. It is not yet understood why this precursor is an inefficient substrate in the splicing reaction. A comparison of probable secondary structures (Fig. 3b) with those of other spliced precursors (Fig. 2) suggests that the presence of a base pair at the 3' side of this cleavage site (in 30-50% of the molecules at equilibrium) may correlate with the reduced efficiency of splicing.

The nucleotide alteration in strain a122 agrees with its location (immediately adjacent to the anticodon) on the fine-structure map (see Colby et al. 1981). From the relative map positions of the other mutations associated with elevated amounts of the unspliced precursor, it thus appears that at least the majority of such lesions lie outside (both 5' and 3' to) the intervening sequence. DNA sequence analysis of the cloned mutant genes is currently in progress. It is likely that many of these alterations will correspond to those identified by Kurjan et al. (1980) as mutations that confer leaky phenotypes. (Such is the case with the a122 A → G transition, which appeared twice in their analysis of 32 independently derived mutants.) These substitutions that confer leaky phenotypes are indicated in boldface, underlined type in Figure 3a. As has been observed previously in comparable analyses of prokaryotic suppressor tRNAs (see McClain 1977; Smith 1979; Guthrie 1980), such alterations occur in loops as well as stems in the clover-

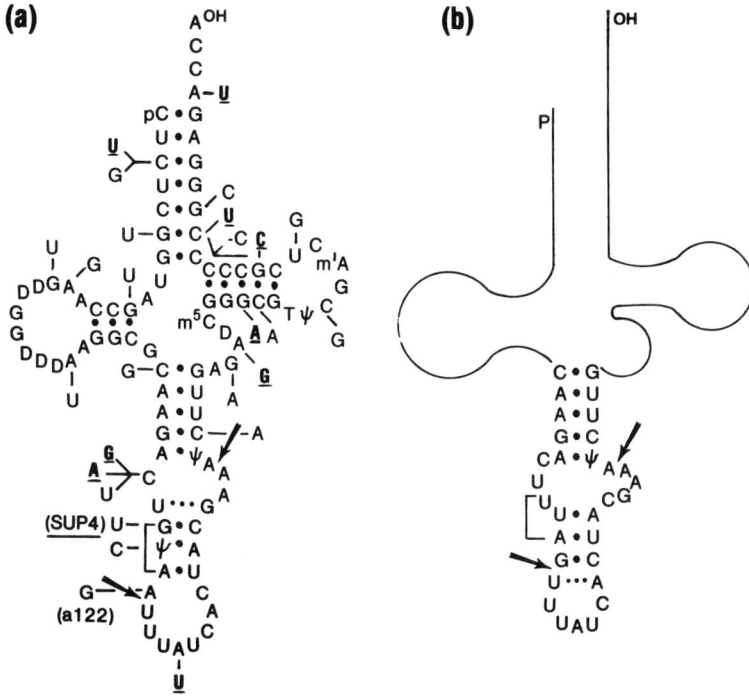

Figure 3 Sequenced alterations in suppressor mutants. The alterations in 32 of the mutants mapped in Fig. 4 were determined by DNA sequence analysis of the cloned genes (Kurjan et al. 1980). The nucleotide changes are indicated in *a*. Those lesions that inactivate the suppressor phenotype incompletely are shown in boldface, underlined type. The secondary structure shown is that predicted for the wild-type precursor to tRNA^{Tyr} (see text and Fig. 2). The anticodon is indicated by brackets. Arrows identify the sites of excision and ligation. The A → G transition mutation 3' to the anticodon has also been identified by RNA sequence analysis in an independent study (Colby et al. 1981); mutant strain a122 appears to be defective in the excision of the intervening sequence in vivo. A possible explanation for the decreased efficiency of this processing step is the alteration in secondary structure *(b)*.

leaf structure; in the former case, the mutations correspond almost exclusively to positions occupied by so-called "invariant" residues, consistent with predictions based on the three-dimensional structure of yeast tRNA^{Phe} (see, e.g., Rich and RajBhandary 1976). Thus, these mutations appear to exert their effect by destabilizing secondary and tertiary interactions involved in specifying the overall conformation of those portions of the precursor destined to become the product of the processing reaction (see Abelson 1979).

A complementary approach to classical genetic analyses of the role of the intervening sequence in tRNA splicing is afforded by recombinant DNA technology. On the assumption that gross alterations within the intron would reveal conformation criteria not identified by single nucleotide mutations, Johnson et al. (1980) inserted a synthetic 21-nucleotide fragment corresponding to *Escherichia coli* lac operator DNA near the middle of the 32-bp intervening sequence of tRNA$_3^{Leu}$. The effects of this alteration were assessed relative to the unmodified gene by comparing the transcription and processing of the cloned DNAs in *Xenopus* germinal vesicle extracts (see Other Steps in tRNA Gene Expression). From RNA sequence analysis of the products it was concluded that both genes are efficiently transcribed and processed in this system. Moreover, the precursors containing normal or mutant intervening sequences were also shown to be accurately and efficiently processed in yeast extracts. Although these results must still be reproduced in vivo, it is nonetheless striking that a perturbation of this extent has no appreciable effect, in either homologous or heterologous systems, on the specificity of splicing in vitro. Presumably not all insertions (or deletions) would have comparably benign consequences. It is tempting to speculate that this particular alteration is well tolerated by virtue of its inherent conformational symmetry. As can be seen in Figure 4, it is conceivable that the altered precursor differs from the wild-type structure primarily in the relative length of a helical arm that is normally present within the intervening sequence of the precursor to tRNA$_3^{Leu}$.

All of the available data are thus fundamentally consistent and point to the conclusion that the intervening sequence per se plays only a minor and relatively "passive" role in its own removal. The specificity of splicing is largely independent of the absolute number of nucleotides within, or the precise sequence of nucleotides at the boundaries of, a given intron. Rather, the fidelity of the reaction appears to be determined by recognition of conformational elements comprised of tRNA-specific sequences within the precursor. The efficiency of splicing may also depend on features of secondary and tertiary structure common to those in mature tRNA.

It is also interesting to consider these results in the light of experiments that emphasize the conservation of sequences across splice junctions in mRNA precursors and the importance of maintaining the integrity of these boundaries. That is, although the relationship of intervening sequences in tRNA and mRNA precursors is of particular interest from the standpoint of evolution, the mechanisms by which these precursor-specific sequences are removed may be as distinctive from one another as are the structures and functions of the spliced products. The resolution of this important issue will ultimately require an enzymological analysis. On the basis of comparable experience in prokaryotes, however, it is likely that genetics will also be an extremely valuable tool in estimating the total enzymological complexity of eukaryotic tRNA processing, as well as in identifying possible overlapping functions in tRNA or rRNA biosynthetic pathways (see, e.g., Gegenheimer and Apirion 1978; McClain 1979; Lund et al. 1980).

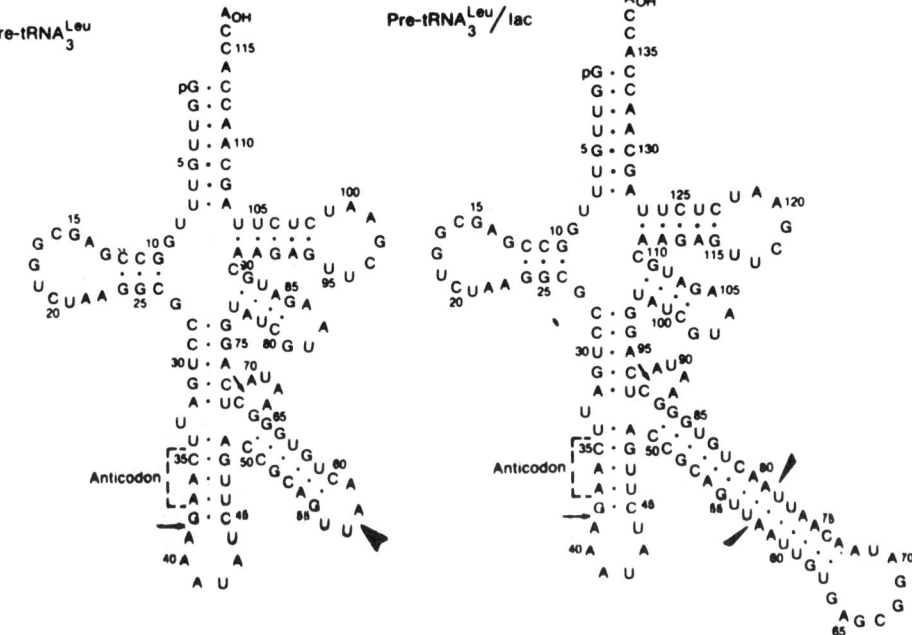

Figure 4 Secondary structure models of the normal and modified precursor to tRNA$_3^{Leu}$. The intervening sequence, normally excised at the sites indicated by small arrows, was modified by the insertion of a 21-bp fragment corresponding to *E. coli* lac operator DNA into the *Hpa*I site (large arrowhead) of the cloned tRNA gene (Johnson et al. 1980). Both genes are transcribed and accurately spliced in vitro (see text).

Unlinked Antisuppressor Mutations

A general discussion of extragenic mutations that affect the expression of suppressor tRNAs can be found elsewhere in this volume (see Sherman). Here, we summarize the current status of antisuppressor mutants that are associated with identifiable biochemical phenotypes.

Using a selection procedure similar to that described above (see footnote to Table 4), R. M. Bock and his colleagues (Laten et al. 1978) recently identified six loci that give rise to antisuppressors for the tyrosine-inserting ochre suppressor *SUP7*. To date, mutations in only a single complementation group, *mod5*, have been found to have a detectable biochemical phenotype; tRNAs synthesized in this antisuppressor strain contain only 1.5% of the normal level of isopentenyladenosine (i^{6}A). Apart from a moderate reduction in ochre-suppressor efficiency, however, there is no appreciable effect of this mutation on cell growth rate or on tRNA gene expression. That is, the absence of this hypermodified nucleotide 3' to the anticodon apparently lowers the efficiency of the tRNA^{Tyr} suppressor without detectably altering the activity of the wild-type tRNA molecules that normally contain this modification.

Attempts to isolate mutations in enzymes essential for tRNA synthesis or maturation have taken into account the likelihood that such lesions might only be recovered as conditional lethals. To date, however, the only tRNA-defective mutants identified in such screens appear to be temperature-sensitive only for suppression (Hopper et al. 1980; A. Hopper, pers. comm.). The best characterized of these mutations is *los1-1* (Hopper et al. 1980). When analyzed by pulse-labeling at 35°C, *los1* mutants show a pattern of tRNA precursor accumulation very similar to that seen in the *rna1* mutant ts136. In contrast to *rna1*, however, the *los1-1* mutation does not appear to affect synthesis of rRNA or of mRNA (as inferred from protein synthetic capacity). Mutations at the *rna1* and *los1* loci complement one another and segregate independently in meiosis.

Although the specific identities and structures of the precursors accumulated as a result of the *los1-1* mutation have not been determined, when incubated with wild-type extracts, these molecules are converted to mature-size products via half-molecule intermediates that comigrate with comparable species from ts136. Despite these indications that *LOS1* encodes a component essential for tRNA splicing, attempts to demonstrate a temperature-dependent defect in mutant extracts have so far been unsuccessful (A. Hopper, pers. comm.).

Twenty other independently isolated *los* mutants show no noticeable defects in tRNA biosynthesis at the nonpermissive temperature; nor, as in the case of mutations at five out of six *mod* loci (Laten et al. 1978), do they show a detectable decrease in the levels of 13 different modified nucleosides (A. Hopper, pers. comm.). More sensitive biochemical assays will be necessary to reveal whether these strains are defective in the expression or in the function of the suppressor tRNA gene, and at what step(s). Though some of these experiments may lie far in the future, with the current rate of progress on purification of the splicing enzymes, it should soon be possible to characterize mutants such as *los1* by direct biochemical complementation.

ENZYMOLOGY OF SPLICING

The precursors accumulated at the nonpermissive temperature in *rna1* strains have provided substrates for the assay of splicing in vitro. Peebles et al. (1979) have isolated an activity from a wash of yeast ribosomes that is capable of removing the introns from all ten of the precursors accumulated in ts136 strains. As illustrated in Figure 5 for tRNA^{Phe}, the extent of precursor utilization, with the concurrent appearance in excellent yield of mature

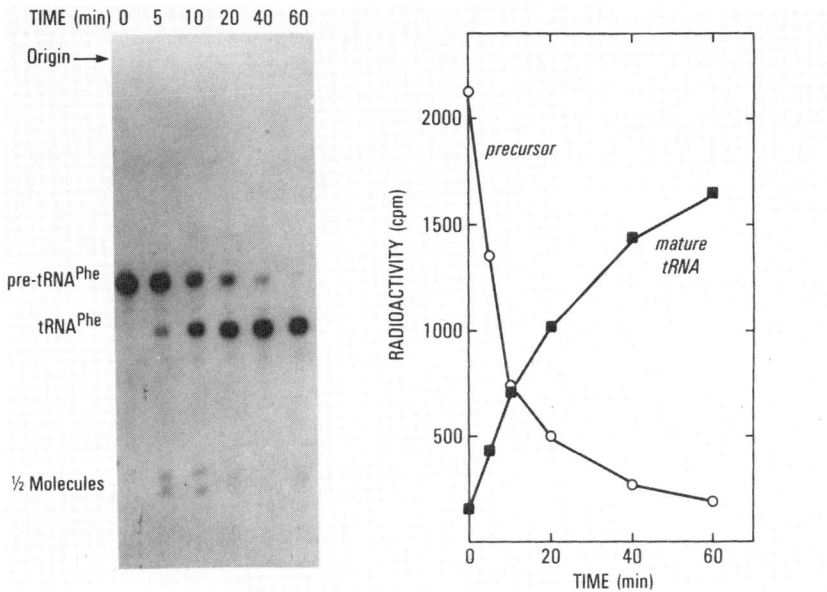

Figure 5 Time course of the in vitro splicing reaction. The precursor to tRNA^{Phe}, synthesized in vivo in the presence of ^{32}P at the nonpermissive temperature in strain ts136, was incubated with a ribosomal wash fraction from wild-type yeast (Peebles et al. 1979) for various times. The products were separated by polyacrylamide gel electrophoresis and identified by fingerprint analyses.

tRNA (96% of the theoretical maximum in the case presented here), suggests that nearly all of the substrate was competent for processing and that, even in the crude extract, neither random degradation nor abortive splicing pathways consume a significant fraction of the RNA precursor.

At early times one observes the transient appearance of smaller RNAs with the mobility of half-tRNA-sized molecules. From kinetic analyses it is probable that these halves, generated by excision of the intervening sequence (see below), are true intermediates in the splicing reaction. Further dissection of this process was made possible by the discovery that only half of the reaction occurs in the absence of ATP. The precursor is cut correctly, excising the intervening sequence and leaving a gapped tRNA molecule. All three of the polynucleotide products of the endonuclease reaction can be separated by acrylamide gel electrophoresis. The half molecules have been shown to be substrates for the second step in the splicing process, formation of a phosphodiester bond between the two half tRNAs. The ligation requires ATP at an optimal concentration of approximately 1 mM.

Subsequent characterization of the ligase reaction has revealed several interesting features. The first was determined by characterization of the structure of the products of endonucleolytic cleavage. The intervening sequence is excised as a discrete, linear polynucleotide with 5'-hydroxyl and 3'-phosphate termini (Knapp et al. 1979). Similar analyses of the gapped tRNA products reveals that in each case the endonuclease reaction produces 3'-phosphate and 5'-hydroxyl termini, which are therefore the substrates for the ensuing ligase reaction. Apparently, the 3'-phosphate group is absolutely required for the ligase reaction since 5'-half molecules, from which the 3'-phosphate has been selectively removed, are no longer substrates in the ATP-dependent joining reaction (Knapp et al. 1979). This finding was totally unanticipated, as all known DNA ligases as well as the T4 RNA ligase work in the opposite manner, joining a 5'-phosphate to a 3'-hydroxyl group. The specific mechanism used to activate the 3'-phosphate group remains to be elucidated. This information is now of utmost importance, since the unusual specificity of the tRNA ligase may well prove to be a common feature central to all RNA splicing mechanisms.

The results of the above analyses also provided valuable insights into the question of recognition specificity. By sequencing the half-molecule intermediates, the precise sites of excision could be unambiguously determined. As we have already pointed out, in the six precursors for which sequence data are available (Fig. 2), excision occurs at the identical location relative to the anticodon. This positional information could presumably determine the specificity of endonucleolytic cleavage and would thus account for the apparent insensitivity of the reaction to either the size or nucleotide composition of the intron, or to the specific sequences at the splice junctions. The prediction that recognition is based, rather, on features of secondary and tertiary structure common to all tRNA precursors is supported by several in vitro observations. The first is the finding that only complementary half molecules are substrates in the joining reaction. All attempts to ligate heterologous halves have so far been unsuccessful (C. Peebles and J. Abelson, unpubl.). Secondly, mature tRNA selectively inhibits the ligation reaction (Peebles et al. 1979). The inhibition is not tRNA species specific, however; e.g., pure yeast $tRNA^{Phe}$ inhibits equally the processing of precursors to $tRNA^{Tyr}$ and $tRNA^{Phe}$.

The latter result could be taken to suggest that there is only a single enzyme required for the splicing of all ten tRNA precursors. Although there is no preliminary biochemical evidence for activities that distinguish among the six precursors shown in Figure 2, extensive purification will be required before this possibility can be critically evaluated. The ability to separate the splicing reaction into two stages has substantially facilitated progress in this direction. The endonuclease is assayed by separating the cleavage products from the precursor by gel electrophoresis; similarly, the ligase is assayed by electrophoretic separation of the half-molecule substrates from the intact and slower-moving tRNA product. This type of analysis has yielded good evidence that the two stages of the reaction are carried out by functionally distinct and physically separable enzyme activities. Differential fractionation occurs, e.g., when crude extracts are chromatographed on heparin agarose; the endonuclease passes through the column, and the ligase is retained. Reconstitution of the entire splicing reaction requires the presence of both fractions (C. Peebles and J. Abelson, unpubl.). Both activities have been purified further, but not yet to homogeneity.

It will be of particular interest to determine whether either of these enzymatic functions is associated with the nuclear membrane. This possibility was initially prompted by the description of ts136 as defective in the transport of RNA from nucleus to cytoplasm. Although the nature of the *rna1* defect still remains obscure, there is now widespread belief that nuclear association is a general property of RNA splicing.

Direct evidence for a functional or physical relationship between transport and splicing, however, has been conspicuously lacking. It is perhaps this aspect of the splicing problem that provides the greatest challenge for the future. To understand this process it will be necessary to localize the various RNA components in the nucleus and to study the requirements for RNA transport per se. The advantage of the yeast system in studying these questions will inevitably be the application of genetics to the problem. In the interim, preliminary indications for the intracellular location and compartmentalization of tRNA processing activities have been provided by exploiting several heterologous systems. As discussed below, these results suggest that the enzymology of tRNA splicing has been fundamentally conserved throughout most of eukaryotic evolution.

OTHER STEPS IN tRNA GENE EXPRESSION

To date, the intervening-sequence-containing precursors accumulated in the *rna1* mutant ts136 are the only authentic in vivo intermediates in eukaryotic tRNA synthesis to be characterized. Since they are already mature at both 5' and 3' termini, it is also clear that there must be other processing steps yet to be detected. For example, the absence of the CCA sequence in eukaryotic tRNA genes argues that this ubiquitous 3' terminus must be added posttranscriptionally by tRNA nucleotidyltransferase. Furthermore, the ts136 intermediates contain some but not all of the modified bases observed in the mature tRNA. Modifications in the dihydro-U and TΨC loops are usually present in high molar yields (Kang et al. 1979), whereas modified bases in the anticodon loop are generally not observed. For example, i^6A in $tRNA^{Tyr}$, the Y base in $tRNA^{Phe}$, and $2'$-O-methyl modifications are not found. One interesting exception to this rule is that low yields of Ψ have been detected in the anticodon of the $tRNA^{Tyr}$ precursors.

It is probable that many of these maturation steps will be analogous to those seen in bacteria. Much of our knowledge about tRNA biosynthesis in

E. coli has come from the use of mutants defective in various processing steps. As a result, the expression of prokaryotic tRNA genes is now understood in considerable detail (see Abelson 1979; Guthrie 1980; Altman 1981). It is useful to review briefly the most salient of these features insofar as they might apply to the eukaryotic situation.

The tRNA processing reactions are carried out by a small number of enzymes that recognize common structural features in a large number of substrates. Transcription is directed by a promoter sequence located 5' to the tRNA coding sequence; the tRNA promoter sequences bear a strong resemblance to other prokaryotic promoters. The de novo transcript is matured by the action of several nucleases. The best characterized of these is RNase P, which is responsible for the 5'-terminal maturation of all tRNAs in *E. coli* and bacteriophage T4. The mode of action of this endonuclease has been of particular interest since the intriguing discovery that highly purified enzyme preparations contain one or more RNA molecules that appear to be essential for in vitro cleavage activity (Kole and Altman 1979; Guthrie and Atchison 1980). The enzymes that mature the 3' ends of tRNAs are generally less well characterized; they include at least one endonuclease (Fukada and Abelson 1980) as well as RNase D (Cudney and Deutscher 1980). The latter enzyme removes nucleotides from the 3' terminus of precursors that contain extra residues distal to the CCA sequence, generally encoded in bacterial tRNA transcripts (in distinct contrast to the situation in eukaryotic tRNA genes). As with RNase P, RNase D exhibits a high degree of conformational specificity and degrades denatured substrates. Several T4 precursors contain other residues in place of the CCA sequence; these transcripts, which are thus likely to be more similar to eukaryotic precursors, require the action of one or more distinct exonucleases as well as of tRNA nucleotidyltransferase (see McClain 1977). Finally, although nucleotide modifications are frequently made on the precursors, the 2'-*O*-methyl guanosine enzyme requires a 5'- and 3'-matured substrate.

As we have discussed above in Genetic Analyses, the tools for a comparable genetic dissection of yeast tRNA biosynthesis are at hand. For the time being, however, information about tRNA biosynthesis in eukaryotes is available only from heterologous systems. Of these, the one that has been exploited to the best advantage is the *Xenopus* oocyte. Either the cloned tRNA genes are injected directly into the giant germinal vesicle, or the germinal vesicles are manually dissected from the oocyte and disrupted to give a nuclear extract that contains both transcription and processing activities. In both of these systems, it has been possible to demonstrate specific transcription of yeast tRNA genes and maturation of the RNA transcripts. RNA polymerase III is presumably responsible for this transcription since synthesis is resistant to low levels of α-amanitin. The expression of a number of eukaryotic tRNA genes has now been studied in this way; in the case of yeast, these include the genes for $tRNA^{Trp}$ (Ogden et al. 1979), $tRNA_3^{Leu}$

(Johnson et al. 1980), tRNA$_{UCA}^{Ser}$ (J. Broach and R. Cortese, pers. comm.), tRNA$_{UCG}^{Ser}$ (M. Salvato and C. Guthrie, in prep.), tRNA^{Asp}-tRNA$_3^{Arg}$ (Schmidt et al. 1980), and tRNA^{Tyr} (DeRobertis and Olson 1979; Melton et al. 1980). Among these examples, tRNA^{Tyr} has been studied in greatest detail.

The plasmid pYT-C contains the tRNA^{Tyr} gene, which corresponds to the *SUP8* locus (Goodman et al. 1977). When this plasmid is injected into the germinal vesicle of the *Xenopus* oocyte, there is preferential transcription of the tRNA gene; the four major products have nucleotide lengths of 108, 104, 92, and 78 (DeRobertis and Olson 1979). The latter product corresponds to mature tRNA^{Tyr}, and the 92-nucleotide precursor is virtually indistinguishable from the intron-containing precursor observed in vivo in strain ts136. The 104-nucleotide intermediate carries precursor-specific residues at both 5' and 3' termini and still contains the 14-bp intervening sequence. The largest product (108 nucleotides) is presumed to be the primary transcript, since it is labeled by [γ-^{32}P]ATP. The site of termination has not been precisely identified but probably occurs within an oligo-T sequence immediately adjacent to the coding sequence. If these assumptions are correct, transcription is initiated at a position 19 nucleotides upstream of the mature tRNA terminus. The CCA sequence is not encoded and is lacking in both the 108- and 104-nucleotide RNAs.

These experiments were extended by Melton et al. (1980) to ascertain the order and intracellular location of the various maturation events. The 108- and 104-nucleotide RNAs do not leave the *Xenopus* oocyte nucleus, and mature tRNA is found only in the cytoplasm. However, the 92-nucleotide precursor is found in both cytoplasm and nucleus. It is currently thought that this distribution may be an artifact of the high level of expression from the vast excess of injected genes, since this precursor can be spliced when it is reinjected into the nucleus but not when it is introduced directly into the cytoplasm. The removal of the intervening sequence may thus be the rate-limiting step, at least in this heterologous situation, and the appearance of unspliced precursor in the cytoplasm could conceivably represent nonspecific "leakage" that results from the anomalously high concentration of these molecules in the nucleus. In any case, the splicing enzymes seem to be exclusively located in the nuclear compartment.

More recently, DeRobertis et al. (1981) have attempted to localize these activities more specifically by manually isolating the nuclear envelope away from the gellike structure formed by the nuclear contents in the presence of high Mg^{++}. Neither excision nor ligation activity could be demonstrated with the isolated nuclear envelope, although the nuclear contents appear to retain both of these functions. Taken at face value, these results argue against the type of membrane-associated splicing models considered above; on the other hand, alternative explanations of these experiments cannot yet be ruled out.

Further evidence that tRNA maturation entails a highly ordered series of specific molecular transformations is seen in the progression of nucleoside modifications. The majority of these take place in the nucleus, although some reactions are presumed to be cytoplasmic since modifications such as i^{6}A and 2'-O-methyl guanosine are present in the mature tRNA but absent in the unspliced 92-nucleotide precursor. The nuclear enzymes also appear to exhibit a high degree of substrate specificity, since certain (m^{1}A, m^{5}C, Ψ) modifications are present on the 104-nucleotide precursor, and others (e.g., D) are first detected in the 92-nucleotide product.

The expression of other yeast tRNA genes is likely to occur by a similar pathway, which is summarized in Figure 6. Yet it is also clear that certain characteristics, such as the number and structure of the specific intermediates, will be unique to a given tRNA gene or gene family. Of particular concern in this regard is the question of which specific features of sequence and/or structure are used to define the transcriptional unit. The thinking in this area is dominated by the surprising results of D. Brown and his colleagues (Bogenhagen et al. 1980; Sakonju et al. 1980), which demonstrated that transcriptional specificity in the case of the *X. borealis* 5S gene is determined by coding sequences (~ +50 to +80) in the interior of the gene. A specific transcriptional factor recognizes this same region and presumably guides RNA polymerase III to a start site at the beginning of the gene (Engelke et al. 1980).

Comparably convincing data for the transcription of tRNA genes are not yet available, but several conclusions can be stated with reasonable certainty. Genetic surgery has been performed on a number of cloned eukaryotic tRNA genes to remove or pare down substantially the 5'-flanking region. Since these genes can still be transcribed in vitro, albeit with variable efficiencies, it is argued that the coding portion of tRNA genes must also play a role in determining the specificity of initiation by RNA polymerase III (Garber and Gage 1979; Kressmann et al. 1979; Telford et al. 1979; DeFranco et al. 1980). At the same time, there is good evidence that 5'-flanking sequences are also required for selective transcription. Sprague and her colleagues (1980) have shown that a truncated gene for tRNA^{Ala} from *Bombyx mori* is completely unable to support specific transcription in silk-

Figure 6 Steps in the expression of a tRNA^{Tyr} gene. The pathway for the expression of a cloned tRNA gene—one of the eight tRNA^{Tyr} genes—is shown schematically. The results were obtained by injecting DNA into germinal vesicles of *Xenopus* oocytes; the isolated RNA products were fractionated by polyacrylamide gel electrophoresis and identified by fingerprinting using nearest-neighbor analysis (deRobertis and Olson 1979; Melton et al. 1980). Specific transcription occurs upstream of the mature tRNA terminus; the transcript is matured via a series of discrete intermediates. Removal of the intervening sequence (IVS) takes place only in the nucleus, whereas the mature tRNA is found exclusively in the cytoplasm. For other details, see text.

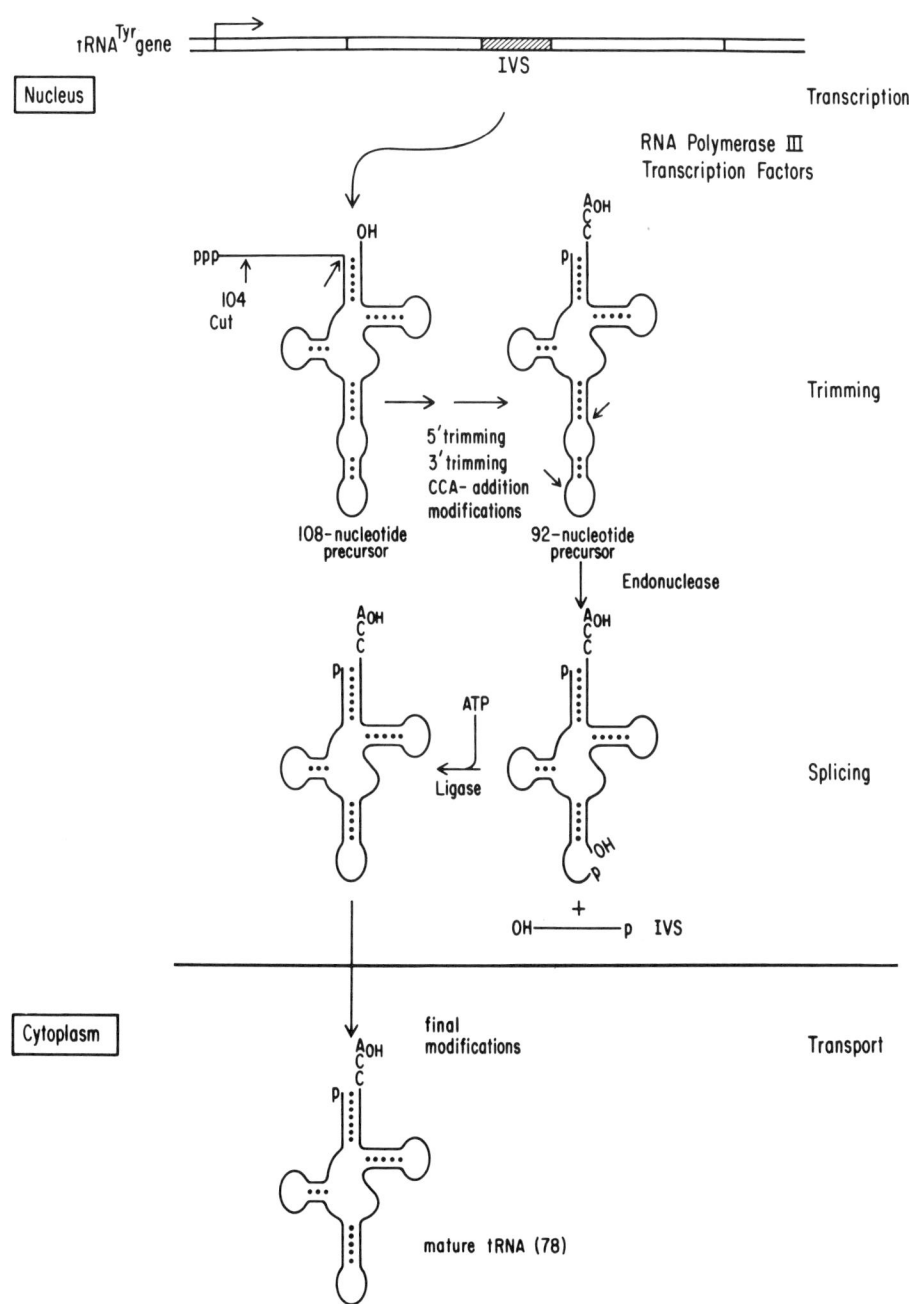

Figure 6 (See facing page for legend.)

worm extracts, although it is still able to bind a presumptive "specificity factor." These experiments thus argue that both coding and flanking (>11 nucleotides 5' to the start site) sequences are recognized by RNA polymerase III.

In the case of yeast tRNA genes, the bulk of available evidence also points to the involvement of 5' sequences in signaling the site of initiation. It is clear that flanking sequences differ considerably among otherwise identical members of several tRNA gene families. If the specific start site of transcription were dictated wholly by elements within the structural gene, one would then expect to find primary transcripts of identical length within a gene family. Yet, comparison of the transcription products from several distinct tRNA^{Tyr} genes injected into *Xenopus* oocytes suggests that the initial transcripts differ both in length and sequence (DeRobertis and Olson 1979). Similar results have been obtained for the three tRNA$^{Ser}_{UCA}$ genes (J.R. Broach and R. Cortese, pers. comm.), which, unlike the tRNA^{Tyr} gene family, lack intervening sequences. The primary transcript from the tRNA$^{Ser}_{UCG}$ gene, which does contain an intron but otherwise differs from the tRNA$^{Ser}_{UCA}$ sequence at only three positions, is also initiated at a different site with respect to the mature tRNA terminus (M. Salvato and C. Guthrie, in prep.). Furthermore, it would appear from the results of Johnson et al. (1980) that the center of the tRNA$^{Leu}_3$ gene cannot be of crucial importance in determining transcriptional specificity in the *Xenopus* system, since insertion of a 21-bp sequence into that region did not affect the location of the start site or the extent of transcription. Perhaps of greater importance, these experiments indicate that the absolute distance from any conserved sequence in the 3' half of the molecule to sequences in the 5' half or in the leader is not an important feature in determining transcriptional specificity.

On the other hand, as is reviewed in greater detail elsewhere (see Sentenac and Hall, this volume), there is at least suggestive evidence that structural gene sequences can influence the efficiency and/or specificity of initiation in the transcription of yeast tRNAs. Koski et al. (1980) have analyzed the transcriptional properties of 29 *SUP4*-defective mutants (Kurjan et al. 1980) in extracts of *Xenopus* tissue-culture cells. These extracts were shown to carry out transcription (but not processing) of the tRNA^{Tyr} gene. In the majority of cases, the length and amount of primary transcripts were comparable to those for the control *SUP4* gene. However, each of two mutations within the dihydro-U loop (see Fig. 3) resulted in the enhanced production of shortened transcripts, and mutation of the invariant C residue in the TΨC loop appeared to virtually abolish the transcriptional competence of the gene. Since these experiments fail to distinguish between alterations that affect the efficiency of initiation or termination per se rather than the stabilities of the altered products in vitro, these observations must be interpreted somewhat cautiously.

Taking all of these results into account, one is led to the general conclusion that both the selectivity and efficiency of transcription are influenced by features of sequence and/or structure that reside adjacent to, as well as within, the tRNA structural gene. The specific definition of a "consensus" RNA polymerase-III promoter, however, at least in the case of yeast tRNA genes, must await further genetic and biochemical analysis. Certainly, it will be necessary to assess the physiological relevance of conclusions that are based solely on results from heterologous in vitro systems. The need for this type of confirmation cannot be overemphasized. A particular case in point is the demonstration by Sprague et al. (1980) that 5' truncation of the *Bombyx* tRNA$_2^{Ala}$ gene, an alteration that eliminates all detectable transcriptional activity in *homologous* extracts, permits transcription with efficiency and fidelity comparable to the intact gene when the cell-free system is derived from *Xenopus* germinal vesicles. A reasonable interpretation of this discrepancy is that the specific molecular elements required for selective transcription by RNA polymerase III have been fundamentally conserved throughout eukaryotic evolution, whereas the ability to modulate these signals in response to particular developmental or physiological demands has been variously exploited. (These possibly idiosyncratic differences might be nonetheless profitably exploited for the identification and purification of specific regulatory factors.)

The analysis of posttranscriptional steps in gene expression can be seen in a similar light. Using a cell-free extract derived from HeLa cells, Standring et al. (1981) have recently demonstrated specific transcription and processing using a yeast tRNA$_3^{Leu}$ gene. Excision of the intervening sequence is accurate, as is the ligation of the resultant half molecules. However, in marked contrast to the tRNA$_3^{Leu}$ precursors observed in ts136 (Kang et al. 1979) or in *Xenopus* oocytes (Johnson et al. 1980), the splicing reaction in HeLa extracts occurs without prior 5' or 3' maturation. The molecular basis for this phenomenon remains to be determined. One plausible explanation is that these extracts are simply deficient in trimming activity but enriched for excision-ligase function; presumably the 5',3'-immature precursor is normally an inefficient substrate for splicing but can be recognized accurately in the absence of competing reactions. The observations are, in any case, intriguing and of definite interest at least from a purely enzymological standpoint. At the same time, they underscore the potential pitfalls inherent in in vitro approaches.

In summary, it is as yet premature to predict the extent to which the events observed in heterologous systems will turn out to accurately reflect the in vivo situation. As a first step in this direction, Hopper and Kurjan (1981) have recently presented presumptive evidence for the primary transcripts from several tRNA gene families in vivo. Using Northern hybridization procedures, three tRNA^{Tyr}-specific bands were identified in addition to components with mobilities appropriate for the mature tRNA and the intervening-sequence-containing precursor. The largest species has an estimated length (108 nucleotides) comparable to that observed for the initial transcript in *Xenopus*. Only a single precursor (102 nucleotides) to tRNA$_2^{Ser}$ was observed, whereas probes for tRNA$_{UCA}^{Ser}$ and tRNA$_{UCG}^{Ser\ 2}$ detected a species the size of the unspliced tRNA$_{UCG}^{Ser}$ precursor as well as a larger product (120 nucleotides). Also analyzed were 22 of the *SUP4*-defective mutants described earlier (Kurjan et al. 1980), including many of the mutations whose transcriptional properties had been characterized using *Xenopus* extracts as well (Koski et al. 1980). In all cases, the precursor-specific hybridization patterns were qualitatively and quantitatively indistinguishable from those of wild-type, despite the fact that this survey included five mutations that result in the production of markedly shorter transcripts in the heterologous in vitro system. Conceivably, these truncated products are also produced in vivo but are more rapidly degraded. Alternatively, these products may be the result of aberrant reactions unique to the in vitro system. It is encouraging nonetheless that the general features of the pathway delineated in Figure 6 are likely to hold for the case of wild-type gene expression.

Hopefully, the kinds of information needed to confirm and extend these preliminary conclusions will soon be forthcoming. Of particular value will be the biochemical exploitation of mutations like *rna1* and *los1*.

SPECULATIONS

The data described in the preceding sections have provided at least partial answers to many of the questions posed in the Introduction. In most cases, however, these answers have been primarily descriptive; as such, they fail to satisfy the deeper craving of biologists to interpret their knowledge within some broader, teleologically plausible context. In anticipation of reaching that destination, we will close by taking this opportunity to speculate from our current vantage point, somewhere in midstream.

We have seen that the physical organization of tRNA genes in yeast is characterized by the distinct absence of clustering. The interchromosomal scattering of identical genetic elements presumably serves to fix the size of a multigene family. In keeping with this interpretation, the number of genes in a particular tRNA gene family generally seems proportionate with the requirement for the cognate decoding capacity. Given the fact that certain codons are employed in marked preference to their degenerate synonyms, it is not so surprising that the copy number of the appropriate tRNA gene is correspondingly higher than that of less frequently required isoacceptors. Rather, the question is why either of these asymmetries exists.

As recently summarized by Hall (Bennetzen and Hall 1982b), the complex pattern of codon preference can be described by a few general rules. (1) In cases of threefold to fourfold degeneracy, C and U appear in the wobble position with comparable frequency, A and G, almost never; (2) in twofold pyrimidine degeneracy, only C is used; and (3) in twofold purine degeneracy, either A or G is used exclusively for any given pairwise option. Furthermore, the preferred codon-anticodon pairings are invariably those that allow perfect Watson-Crick complementarity. In accounting for the exceptional cases, a more comprehensive rule emerges: Choices are made to maximize similarities in codon-anticodon binding strengths for different amino acids. For example, the use of UAC rather than UAU, despite the fact that the single tRNA^{Tyr} recognizes both, avoids 100% A + U base-pairing and thus promotes "binding energy homeostasis" (Grosjean et al. 1978).

Over all, this hypothesis seems to account rather nicely for the "coevolution" of the 20 most abundant tRNAs and the 25 "preferred triplets," since, as Hall points out (Bennetzen and Hall 1982b), the degree of codon bias exhibited by a particular mRNA is in turn related to the abundance of that protein product. For example, under conditions of maximal translation, degenerate mutations would result in premature termination at the site read by a minor tRNA isoacceptor and thus be strongly selected against. Yet, is it clear that the complementary contention—that codon usage in mRNAs expressed at low level could safely approach random—is sufficient to explain the maintenance of these 20 minor tRNA genes throughout evolution?

An mRNA carries more information than the order and identity of amino acids dictated by the nucleotide sequence. For example, the signals for processing, transport, and stability of a given mRNA are likely to be specified at least in part by conformational features encoded within the same primary structure. In all likelihood, then, codon degeneracy is exploited for these dual roles. Conceivably, mRNA characteristics such as intrinsic stability will be shown to constitute primary targets for the regulation of gene expression; this mechanism may account for the temporally modulated expression of certain adenovirus gene products, for example (Wilson and Darnell 1981).

Until recently, attempts to account for the differential expression of eukaryotic genes have focused almost exclusively on transcriptional models. When models for translational control have been considered, they have been directed exclusively at those factors that would influence the rate of initation; it has traditionally been assumed that under a given set of normal growth conditions, the rate of elongation ("step time") is constant and invariant. Recently, Bossi and Roth (1980) have provided elegant and persuasive evidence that the efficiency of translation is in fact strongly influenced by the codon composition within an mRNA. These results, enforced by a large body of earlier data, argue that the speed of translation can vary by an order of magnitude, depending on the local context in which a particular codon is read.

This expanded concept of translational modulation has profound implications and certainly prompts us to reconsider the role of minor isoaccepting tRNAs in light of the opportunity for this additional regulatory strategy.

Proper placement of a "nonpreferred triplet" would specifically set the rate of translation in one environment, whereas in another the identical codon might be translated readily. Minor isoaccepting tRNA species could thus be essential factors in determining the particular pattern of gene expression in temporally or developmentally regulated programs. This hypothesis predicts specific variation in minor codon usage and, perhaps, in isoacceptor tRNA abundance as well; changes of both these types have in fact been observed. A particularly well-studied case is the appearance of a novel tRNA^{Ala} species in the posterior silk gland at the time that enormous amounts of fibroin are synthesized (Sprague et al. 1977). Other examples are numerous and include the recent observation that arginine residues are preferentially encoded by AG$_G^A$ in β-globin and immunoglobulins but by the normally nonpreferred CGN triplets in both histones and a number of polypeptide hormones (Grantham et al. 1980).

Context may affect more than the rate of elongation per se. Fox and colleagues (Fox and Weiss-Brummer 1980) have recently shown that the interplay of tRNA structure and local mRNA sequence can also have profound consequences on the fidelity of translation. A similar phenomenon of ribosomal frameshifting was induced simply by altering the ratio of several normal tRNA species in vitro (Atkins et al. 1979). These results suggest that it may be important for the cell to maintain the proper balance between two structurally related but functionally nonidentical tRNA species.

The bulk of the evidence overwhelmingly supports the interpretation that evolution has exploited the tremendous regulatory potential provided by close to the full spectrum of tRNAs. The high correlation between gene copy number and codon usage presumably provides a primary mechanism for ensuring the translatability of the genome. It also seems likely that it is important to maximize the flexibility of this coordination, so that the tRNA population is ideally adapted to the specific transcriptional needs of the cell in a given milieu. Control at the transcriptional level is an obvious device but difficult to evaluate in light of our current ignorance about the molecular signals required for promotion of yeast tRNA genes. We have previously noted the suggestion that all eight loci for tRNA^{Tyr} are not necessarily transcribed with equal efficiency.

A more intriguing possibility is the potential opportunity to exploit posttranscriptional processing as a site of differential regulation. Possible support for models of this type is provided by the curious observation that an essential tRNA gene encoded by phage T4 can be processed by two alternative pathways utilizing a common dimeric precursor. One of these routes results in the preferential underproduction of this minor tRNA species (Fukada and Abelson 1980; Guthrie and Scholla 1980). Interestingly, the case in point is the isoleucine isoacceptor that specifically reads the rare AUA codon. You will recall that in yeast the gene for tRNA$^{Ile}_{AUA}$ contains an intervening sequence and, moreover, there is reason to think that this particular precursor may normally be processed at low efficiency relative to other spliced precursors.

This speculation has certain obvious appeal in that it provides a plausible role for the presence of introns in some tRNA genes. By this interpretation, the requirement for the splicing of precursors to other minor isoacceptors such as $tRNA^{Ser}_{UCG}$ might provide one method for modulating the production of these tRNAs in accord with their translational demand. Until we understand much more about the enzymology of splicing, it is difficult to predict the detailed mechanisms by which this novel biosynthetic mode might be specifically coordinated with specialized functional roles of these rare isoacceptors.

For the present, we can only say that the presence of intervening sequences uniquely distinguishes the expression of these genes from their noninterrupted counterparts; whether the requirement for splicing has biologically useful consequences for transcription, processing, or transport, or whether any of these are in turn directly related to the function of these particular tRNAs, are among today's most challenging questions in molecular biology.

From our current vantage point, then, these and other questions raised in the preceding discussion seem likely to determine the next step, if not the final destination. Considering the wealth of experimental resources that can be marshaled toward these ends, many of these answers are probably close at hand.

ACKNOWLEDGMENTS

We thank our colleagues for providing data in advance of publication. We express our special gratitude to D. Colby, E. Lundburg, D. Maloney, R. Parker, and K. Wilson for their generous help in the preparation of this manuscript.

Cited work by the authors was supported by grants from the National Institutes of Health (GM-21119 to C. G. and CA-10984 to J.A.), the National Science Foundation (PCM-7912081 to C. G.), and the American Cancer Society (NP-302 to J. A.).

Note Added in Proof

With regard to the discussion on the enzymology of splicing, further experiments by C. Peebles and J. Abelson (unpubl.) have revealed that the purified (but not yet homogeneous) splicing endonuclease from yeast initially produces 5'-hydroxyl and 2'-3'-cyclic phosphate termini. It has been shown that half-tRNA-sized molecules with 2'-3'-cyclic phosphate termini at the splice junction are substrates for subsequent joining by the ligase. In this reaction the 2'-3'-cyclic phosphate is shifted to the 2' position; the phosphate in the 3'-5'-phosphodiester bond is derived from the γ phosphate of ATP (C. Greer et al., unpubl.). This mechanism is identical to that of an RNA ligase activity in wheat germ extracts (M. Konarska et al. 1981).

Preliminary experiments have provided evidence for a phosphodiesterase associated with the yeast splicing ligase, which converts the 2′-3′-cyclic phosphate to a 2′-phosphomonoester. The 2′-phosphomonoesters could be substrates, albeit poor ones for joining. This may explain the phosphatase sensitivity of tRNA substrates observed earlier (Knapp et al. 1979).

REFERENCES

Abelson, J. 1979. RNA processing and the intervening sequence problem. *Annu. Rev. Biochem.* **48**: 1035.

Altman, S. 1981. Transfer RNA processing enzymes. *Cell* **23**: 3.

Atkins, J.F., R.F. Gesteland, B. Reid, and C. Anderson. 1979. Normal tRNAs promote ribosomal frameshifting. *Cell* **18**: 1119.

Beckmann, J., P. Johnson, and J. Abelson. 1977. Cloning of yeast transfer RNA genes in *Escherichia coli*. *Science* **196**: 205.

Bennetzen, J.L. and B.D. Hall. 1982a. The primary structure of the *Saccharomyces cerevisiae* gene for alcohol dehydrogenase I. *J. Biol. Chem.* **257**: 3018.

———. 1982b. Codon selection in yeast. *J. Biol. Chem.* **257**: 3026.

Blatt, B. and H. Feldmann. 1973. Characterization of precursors to tRNA in yeast. *FEBS Lett.* **37**: 129.

Bogenhagen, D.F., S. Sakonju, and D.D. Brown. 1980. A control region in the center of the 5S RNA gene directs specific initiation of transcription. II. The 3′ border of the region. *Cell* **19**: 27.

Borer, P., B. Dengler, I. Tinoco, and O. Uhlenbeck. 1974. Stability of ribonucleic acid double-stranded helices. *J. Mol. Biol.* **86**: 843.

Bossi, L. and J. Roth. 1980. The influence of codon context on genetic code translation. *Nature* **286**: 123.

Brandriss, M., L. Soll, and D. Botstein. 1975. Recessive lethal amber suppressors in yeast. *Genetics* **79**: 551.

Brandriss, M., J. Stewart, F. Sherman, and D. Botstein. 1976. Substitution of serine caused by a recessive lethal suppressor in yeast. *J. Mol. Biol.* **102**: 467.

Broach, J.R., L. Friedman, and F. Sherman. 1981. Correspondence of yeast UAA suppressors to cloned tRNA$^{Ser}_{UCA}$ genes. *J. Mol. Biol.* **150**: 375.

Colby, D., P. Leboy, and C. Guthrie. 1981. Yeast tRNA precursor mutated at a splice junction is correctly processed *in vivo*. *Proc. Natl. Acad. Sci.* **78**: 415.

Crick, F. 1966. Codon-anticodon pairing: The wobble hypothesis. *J. Mol. Biol.* **19**: 548.

Cudney, H. and M.P. Deutscher. 1980. Apparent involvement of ribonuclease D in the 3′ processing of tRNA precursors. *Proc. Natl. Acad. Sci.* **77**: 837.

DeFranco, D., O. Schmidt, and D. Söll. 1980. Two control regions for eukaryotic tRNA gene transcription. *Proc. Natl. Acad. Sci.* **77**: 3365.

DeRobertis, E.M. and M.V. Olson. 1979. Transcription and processing of cloned yeast tRNA genes microinjected into frog oocytes. *Nature* **278**: 137.

DeRobertis, E.M., P. Black, and K. Nishikura. 1981. Intranuclear location of the tRNA splicing enzyme. *Cell* **23**: 89.

Dujon, B. 1981. Mitochondrial genetics and functions. In *The molecular biology of the yeast Saccharomyces. I. Life cycle and inheritance* (ed. J. Strathern et al.), p.505. Cold Spring Harbor Laboratory, Cold Spring Harbor, New York.

Engelke, D.R., S.Y. Ng, B.S. Shastry, and R.G. Roeder. 1980. Specific interaction of a purified transcription factor with an internal control region of 5S RNA gene. *Cell* **19**: 717.

Etcheverry, T., D. Colby, and C. Guthrie. 1979. A precursor to a minor species of yeast tRNA^{Ser} contains an intervening sequence. *Cell* **18:** 11.

Etcheverry, T., M. Salvato, and C. Guthrie. 1982. Recessive lethality of yeast strains carrying the *SUP61* suppressor results from loss of a tRNA with a unique decoding function. *J. Mol. Biol.* **168:** 599.

Feldmann, H. 1976. Arrangement of transfer-RNA genes in yeast. *Nucleic Acids Res.* **3:** 2379.

Fox, T. and B. Weiss-Brummer. 1980. Leaky +1 and −1 frameshift mutations at the same site in a yeast mitochondrial gene. *Nature* **288:** 60.

Fraden, A., H. Gruhl, and H. Feldmann. 1975. Mapping of yeast tRNAs by two-dimensional electrophoresis on polyacrylamide gels. *FEBS Lett.* **50:** 185.

Fukada, K. and J. Abelson. 1980. The DNA sequence of a T4 tRNA gene cluster. *J. Mol. Biol.* **139:** 377.

Garber, R.L. and L.P. Gage. 1979. Transcription of a cloned *Bombyx mori* tRNA$_1^{Ala}$ gene: Nucleotide sequence of the tRNA precursor and its processing *in vitro*. *Cell* **18:** 817.

Gauss, D. and M. Sprinzl. 1981. Compilation of tRNA sequences. *Nucleic Acids Res.* **9:** r1.

Gegenheimer, P. and D. Apirion. 1978. Processing of rRNA by RNase P: Spacer tRNAs are linked to 16S rRNA in an RNase P RNase III mutant of *E. coli*. *Cell* **15:** 527.

Goodman, H.M., M.V. Olson, and B.D. Hall. 1977. Nucleotide sequence of a mutant eukaryotic gene: The yeast tyrosine-inserting suppressor *SUP4-o*. *Proc. Natl. Acad. Sci.* **74:** 5423.

Grantham, R., C. Gautier, M. Gouy, R. Mercier, and A. Pave. 1980. Codon catalog usage and the genome hypothesis. *Nucleic Acids Res.* **8:** r49.

Grosjean, H., D. Sankoff, W. MinJou, W. Fiers, and R. Cedergren. 1978. Bacteriophage MS2 RNA: A correlation between the stability of the codon:anticodon interaction and the choice of code words. *J. Mol. Evol.* **12:** 113.

Guthrie, C. 1980. Folding up a transfer RNA molecule is not simple. *Q. Rev. Biol.* **55:** 335.

Guthrie, C. and R. Atchison. 1980. Biochemical characterization of RNase P: A tRNA processing activity with protein and RNA components. In *Transfer RNA: Biological aspects* (ed. D. Söll et al.), p. 83. Cold Spring Harbor Laboratory, Cold Spring Harbor, New York.

Guthrie, C. and C. Scholla. 1980. The asymmetric maturation of a dimeric tRNA precursor. *J. Mol. Biol.* **139:** 349.

Harada, F. and J.E. Dahlberg. 1975. Specific cleavage of tRNA by nuclease S$_1$. *Nucleic Acids Res.* **2:** 865.

Hawthorne, D.C. and U. Leupold. 1974. Suppressors in yeast. *Curr. Top. Microbiol. Immunol.* **64:** 1.

Hawthorne, D.C. and R.K. Mortimer. 1968. Genetic mapping of nonsense suppressors in yeast. *Genetics* **60:** 735.

Holland, J.P. and M.J. Holland. 1979. The primary structure of a glyceraldehyde-3-phosphate dehydrogenase gene from *Saccharomyces cerevisiae*. *J. Biol. Chem.* **254:** 9839.

―――. 1980. Structural comparison of two nontandemly repeated yeast glyceraldehyde-3-phosphate dehydrogenase genes. *J. Biol. Chem.* **255:** 2596.

Holland, M.J., J.P. Holland, G.P. Thill, and K.A. Jackson. 1981. The primary structure of two yeast enolase genes. *J. Biol. Chem.* **256:** 1385.

Hopper, A. and J. Kurjan. 1981. tRNA synthesis: Identification of *in vivo* precursor tRNAs from parental and mutant yeast strains. *Nucleic Acids Res.* **9:** 1019.

Hopper, A.K., F. Banks, and V. Evangelidis. 1978. A yeast mutant which accumulates precursor tRNAs. *Cell* **14:** 211.

Hopper, A.K., L.D. Schultz, and R.A. Shapiro. 1980. Processing of intervening sequences: A new yeast mutant which fails to excise intervening sequences from precursor tRNAs. *Cell* **19:** 741.

Hutchison, H.T., L.H. Hartwell, and C.S. McLaughlin. 1969. Temperature-sensitive yeast mutant defective in ribonucleic acid production. *J. Bacteriol.* **99:** 807.

Johnson, J.D., R. Ogden, P. Johnson, J. Abelson, P. Dembeck, and K. Itakura. 1980. Transcription and processing of a yeast tRNA gene containing a modified intervening sequence. *Proc. Natl. Acad. Sci.* **77:** 2564.

Kang, H.S., R.C. Ogden, and J. Abelson. 1980. Two yeast tRNA genes containing intervening sequences. *Miami Winter Symp.* **17**: 317.

Kang, H., R. Ogden, G. Knapp, C. Peebles, and J. Abelson. 1979. Structure of yeast tRNA precursors containing intervening sequences. In *Eukaryotic gene regulation* (ed. R. Axel et al.), p. 69. Academic Press, New York.

Knapp, G., R.C. Ogden, C.L. Peebles, and J. Abelson. 1979. Splicing of yeast tRNA precursors: Structure of the reaction intermediates. *Cell* **18**: 37.

Knapp, G., J.S. Beckmann, P.F. Johnson, S.A. Fuhrman, and J. Abelson. 1978. Transcription and processing of intervening sequences in yeast tRNA genes. *Cell* **14**: 221.

Kohli, J., J. Mao, and D. Soll. 1980. Transfer RNA and nonsense suppression in *Schizosaccharomyces pombe*. In *RNA polymerase, tRNA and ribosomes* (ed. S. Osawa et al.), p. 233. University of Tokyo Press, Tokyo.

Kole, R. and S. Altman. 1979. Reconstitution of RNase P activity from inactive RNA and protein. *Proc. Natl. Acad. Sci.* **76**: 3795.

Konarska, M., W. Filipowicz, H. Domdey, and H.J. Gross. 1981. Formation of a 2'-phosphomonoester, 3'-5'-phosphodiester linkage by a novel RNA ligase in wheat germ. *Nature* **293**: 112.

Koski, R.A., S.G. Clarkson, J. Kurjan, B.D. Hall, and M. Smith. 1980. Mutations of the yeast $SUP4$ tRNA^{Tyr} locus: Transcription of the mutant gene *in vitro*. *Cell* **22**: 415.

Kressman, A., H. Hofstetter, E. DiCapua, R. Grosschedl, and M.L. Birnstiel. 1979. A tRNA gene of *Xenopus laevis* contains at least two sites promoting transcription. *Nucleic Acids Res.* **7**: 1749.

Kurjan, J., B.D. Hall, S. Gillam, and M. Smith. 1980. Mutations at the yeast $SUP4$ tRNA^{Tyr} locus: DNA sequence changes in mutants lacking suppressor activity. *Cell* **20**: 701.

Laten, H., J. Gorman, and R.M. Bock. 1978. Isopentenyladenosine deficient tRNA from an antisuppressor mutant of *Saccharomyces cervisiae*. *Nucleic Acids Res.* **5**: 4329.

Liebman, S.W., F. Sherman, and J.W. Stewart. 1976. Isolation and characterization of amber suppressors in yeast. *Genetics* **82**: 251.

Lund, E., J.E. Dahlberg, and C. Guthrie. 1980. Processing of spacer tRNAs from ribosomal RNA transcripts of *E. coli*. In *Transfer RNA: Biological aspects* (ed. D. Söll et al.), p. 123. Cold Spring Harbor Laboratory, Cold Spring Harbor, New York.

Mao, J., O. Schmidt, and D. Söll. 1980. Dimeric transfer RNA precursors in *S. pombe*. *Cell* **21**: 509.

McClain, W.H. 1977. Seven terminal steps in a biosynthetic pathway leading from DNA to transfer RNA. *Acct. Chem. Res.* **10**: 418.

———. 1979. A role for ribonuclease III in synthesis of bacteriophage T4 transfer RNAs. *Biochem. Biophys. Res. Commun.* **86**: 718.

McCready, S. and B. Cox. 1973. Antisuppressors in yeast. *Mol. Gen. Genet.* **124**: 305.

Melton, D.A., E.M. DeRobertis, and R. Cortese. 1980. Order and intracellular location of the events involved in the maturation of a spliced tRNA. *Nature* **284**: 143.

Montgomery, D.L., D.W. Leung, M. Smith, P. Shalit, G. Faye, and B.D. Hall. 1980. Isolation and sequence of the gene for iso-2-cytochrome *c* in *Saccharomyces cerevisiae*. *Proc. Natl. Acad. Sci.* **77**: 541.

Mortimer, R.K. and D. Hawthorne. 1973. Genetic mapping in *Saccharomyces*. IV. Mapping of temperature-sensitive genes and use of disomic strains in localizing genes. *Genetics* **74**: 33.

Mortimer, R.K. and D. Schild. 1980. The genetic map of *Saccharomyces cerevisiae*. *Microbiol. Rev.* **44**: 519.

Muller, F. and S.G. Clarkson. 1980. Nucleotide sequence of genes coding for tRNA^{Phe} and tRNA^{Tyr} from a repeating unit of *X. laevis* DNA. *Cell* **19**: 345.

Munz, P., U. Leupold, P. Agris, and J. Kohli. 1981. *In vivo* decoding rules studied in *Schizosaccharomyces pombe* are at variance with *in vitro* data. *Nature* **294**: 187.

Ng, R. and J. Abelson. 1980. Isolation and sequence of the gene for actin in *Saccharomyces cerevisiae*. *Proc. Natl. Acad. Sci.* **77**: 3912.

O'Farrell, P.Z., B. Cordell, P. Valenzuela, W.J. Rutter. and H.M. Goodman. 1978. Structure and processing of yeast precursor tRNAs containing invervening sequences. *Nature* **274**: 438.
Ogden, R.C., J.S. Beckmann, J. Abelson, H.S. Kang, D. Söll, and O. Schmidt. 1979. In vitro transcription and processing of yeast precursor tRNAs containing intervening sequences. *Cell* **17**: 399.
Olson, M.V., B. Hall, J. Cameron, and R. Davis. 1979. Cloning of the yeast tyrosine tRNA genes in bacteriophage lambda. *J. Mol. Biol.* **127**: 285.
Olson, M.V., D.L. Montgomery, A.K. Hopper, G.S. Page, F. Horodyski, and B.D. Hall. 1977. Molecular characterization of the tyrosine tRNA genes in yeast. *Nature* **267**: 639.
Olson, M.V., G. Page, A. Sentenac, P. Piper, M. Worthington, R. Weiss, and B. Hall. 1981. Only one of two closely related yeast suppressor tRNA genes contains an intervening sequence. *Nature* **291**: 464.
Ono, B.J., W. Stewart, and F. Sherman. 1979a. Yeast UAA suppressors effective in ψ^+ strains: Serine-inserting suppressors. *J. Mol. Biol.* **128**: 81.
——. 1979b. Yeast UAA suppressors effective in ψ^+ strains: Leucine-inserting suppressors. *J. Mol. Biol.* **132**: 507.
Ono, B., N. Wills, J. Stewart, R. Gesteland, and F. Sherman. 1981. Serine inserting UAA suppression mediated by yeast tRNA^{Ser}. *J. Mol. Biol.* **150**: 361.
Page, G. and B. Hall. 1981. Characterization of the yeast tRNA^{Ser} gene family: Genomic organization and DNA sequence. *Nucleic Acids Res.* **9**: 921.
Peebles, C.L., R.C. Odgen, G. Knapp, and J. Abelson. 1979. Splicing of yeast tRNA precursors: A two stage reaction. *Cell* **18**: 27.
Piper, P. 1978. A correlation between a recessive lethal amber suppressor mutation in *S. cerevisiae* and an anticodon change in a minor serine tRNA. *J. Mol. Biol.* **122**: 217.
Piper, P.W. and M. Wasserstein. 1977. Separation of *Saccharomyces cereivisae* tRNAs on two-dimensional polyacrylamide gels as applied to investigations on the mutational alterations of tRNA that produce nonsense suppressors. *Eur. J. Biochem.* **80**: 103.
Piper, P.W., M. Wasserstein, F. Engbaek, K. Kaltoft, J.E. Celis, J. Zeuthen, S. Liebman, and F. Sherman. 1976. Nonsense suppressors of *Saccharomyces cerevisiae* can be generated by mutation of the tyrosine tRNA anticodon. *Nature* **262**: 757.
Randerath, E., R. Gupta, L. Chia, S. Chang, and K. Randerath. 1979. Yeast tRNA$^{Leu}_{UAG}$. Purification, properties and determination of the nucleotide sequence by radioactive derivative methods. *Eur. J. Biochem.* **93**: 79.
Reed, C.R. and S.W. Liebman. 1979. New amber suppressors in *Saccharomyces cerevisiae*. *Genetics* **91**: s102.
Rich, A. and U.L. RajBhandary. 1976. Transfer RNA: Molecular structure, sequence and properties. *Annu. Rev. Biochem.* **45**: 805.
Robinson, R.R. and N. Davidson. 1981. Analysis of a *Drosophila* tRNA gene cluster: Two tRNA^{Leu} genes contain intervening sequences. *Cell* **23**: 251.
Rothstein, R. 1979. Deletions of a tyrosine tRNA gene in *S. cerevisiae*. *Cell* **17**: 185.
Sakonju, S., D.F. Bogenhagen, and D.D. Brown. 1980. A control region in the center of the 5S RNA gene directs specific initiation of transcription. I. The 5' border of the region. *Cell* **19**: 13.
Schmidt, O., J. Mao, R. Ogden, J. Beckmann, H. Sakano, J. Abelson, and D. Söll. 1980. Dimeric tRNA precursors in yeast. *Nature* **287**: 750.
Schweizer, E., C. MacKechnie, and H.O. Halvorson. 1969. The redundancy of ribosomal and transfer RNA genes in *Saccharomyces cerevisiae*. *J. Mol. Biol.* **40**: 261.
Selker, E. and C. Yanofsky. 1980. A phenylalanine tRNA gene from *Neurospora crassa*. Conservation of secondary structure involving an intervening sequence. *Nucleic Acids Res.* **8**: 1033.
Shiokawa, K. and A.O. Pogo. 1974. The role of cytoplasmic membranes in controlling the transport of nuclear messenger RNA and initiation of protein synthesis. *Proc. Natl. Acad. Sci.* **71**: 2658.

Smith, J.D. 1979. Suppressor tRNAs in prokaryotes. In *Nonsense mutations and tRNA suppressors* (ed. J.E. Celis and J.D. Smith), p. 109. Academic Press, New York.

Smith, M., D.W. Leung, S. Gillam, and C.R. Astell. 1979. Sequence of the gene for iso-1-cytochrome *c* in *Saccharomyces cerevisiae*. *Cell* **16**: 753.

Southern, E.M. 1975. Detection of specific sequences among DNA fragments separated by gel electrophoresis. *J. Mol. Biol.* **98**: 503.

Sprague, K.U., O. Hagenbüchle, and M.C. Zuniga. 1977. The nucleotide sequence of two silk gland alanine tRNAs: Implications for fibroin synthesis and for initiator tRNA structure. *Cell* **11**: 561.

Sprague, K.U., D. Larson, and D. Morton. 1980. 5' flanking sequence signals are required for activity of silkworm alanine tRNA genes in homologous *in vitro* transcription systems. *Cell* **22**: 171.

Standring, D., A. Venegas, and W. Rutter. 1981. A yeast tRNA^{Leu} gene is transcribed and spliced in a HeLa cell extract. *Proc. Natl. Acad. Sci.* **78**: 5963.

Telford, J.L., A. Kressman, R.A. Koski, R. Grosschedl, F. Muller, S.G. Clarkson, and M.L. Birnsteil. 1979. Delimitation of a promoter for RNA polymerase III by means of a functional test. *Proc. Natl. Acad. Sci.* **76**: 2590.

Tinoco, I., Jr., P.N. Borer, B. Dengler, M.D. Levine, O.C. Uhlenbeck, D.M. Crothers, and J. Gralla. 1973. Improved estimation of secondary structure in ribonucleic acids. *Nat. New Biol.* **246**: 40.

Valenzuela, P., A. Venegas, F. Weinberg, R. Bishop, and W.J. Rutter. 1978. Structure of yeast phenylalanine tRNA gene: An intervening DNA segment within the region coding for the tRNA. *Proc. Natl. Acad. Sci.* **75**: 190.

Venegas, A., M. Quiroga, J. Zaldivar, W.J. Rutter, and P. Valenzuela. 1979. Isolation of yeast tRNA^{Leu} genes: DNA sequence of a cloned tRNA$_3^{Leu}$ gene. *J. Biol. Chem.* **254**: 12306.

Waldron, C., B. Cox, N. Willis, R. Gesteland, P. Piper, D. Colby, and C. Guthrie. 1981. Yeast ochre suppressor *SUQ5-ol* is an altered tRNA$_{UCA}^{Ser}$ gene. *Nucleic Acids Res.* **9**: 3077.

Wallis, J.W., L. Hereford, and M. Grunstein. 1980. Histone H2B genes of yeast encode two different proteins. *Cell* **22**: 799.

Wilson, M.C. and J.E. Darnell. 1981. Control of messenger RNA concentration by differential cytoplasmic half-life. *J. Mol. Biol.* **148**: 231.

Yen, P. and N. Davidson. 1980. The gross anatomy of a tRNA gene cluster at region 42A of the *Drosophila melanogaster* chromosome. *Cell* **22**: 137.

The Yeast Ribosome: Structure, Function, and Synthesis

Jonathan R. Warner
Departments of Biochemistry and Cell Biology
Albert Einstein College of Medicine
Bronx, New York 10461

1. **The Yeast Ribosome**
   A. Ribosomal RNA
   B. Ribosomal Proteins
2. **Activity of the Yeast Ribosome**
   A. Ribosomes and Polyribosomes
   B. Cell-free Translation in Yeast
   C. Effects of Antibiotics on Protein Synthesis
   D. Mutants involving the Protein Synthetic Apparatus
3. **Synthesis of the Yeast Ribosome**
   A. Transcription of rRNA Genes
   B. Processing of the rRNA Transcript
   C. Synthesis of Ribosomal Proteins and Their Assembly into Ribosomes
   D. Genes for Ribosomal Proteins
   E. Mutants Affected in Ribosome Synthesis
4. **Regulation of the Synthesis of the Yeast Ribosome**
   A. Dependence on Growth Rate
   B. Ribosome Synthesis and the Cell Cycle
   C. Dependence on Protein Synthesis
   D. Effects of Temperature Shift
   E. Mutants Defective in the Regulation of Ribosome Synthesis
   F. Conclusions
5. **Summary and Prospects**
   A. Protein Synthesis
   B. Regulation

INTRODUCTION

The ribosome holds our fascination for a number of reasons. It is a complex organelle, yet one that is comprehensible in terms of the number and stoichiometry of its constituent molecules. Its synthesis involves major interactions of nucleus and cytoplasm. It is ubiquitous in living cells, and as the site of protein synthesis, it plays a central role in cell growth and maintenance. The regulation of its synthesis is complex, not only because of the number of components involved, but also because it is closely tied to cell growth. Perhaps more important, the ribosome has been highly conserved during evolution and is the prime example of a "housekeeping enzyme." A comparison of the regulation of ribosome synthesis in prokaryotes, in simple eukaryotes, and in complex eukaryotes may provide insight into the way regulation itself has evolved.

This paper consists of three parts: (1) the mature ribosome; (2) what is known of the way it works in protein synthesis, which for yeast is precious little; and (3) ribosome synthesis, including a considerable body of data concerning its regulation in yeast and a hypothesis to account for some of the data. Mitochondrial ribosomes are discussed by Dujon (1981).

THE YEAST RIBOSOME

Ribosomal RNA

Saccharomyces is typical of eukaryotes in that it has four species of RNA in its ribosome: 25S, 5.8S, and 5S in the 60S subunit, and 18S in the 40S subunit. The properties of these molecules are summarized in Table 1. The sequence of 5S RNA from *S. carlsbergensis* (Hindley and Page 1972) and that of 5S (Miyazaki 1974) and 5.8S RNAs (Rubin 1973) from *S. cerevisiae* have been determined. The complete sequence of 18S RNA was recently published (Rubtsov et al. 1980).

Ribosomal Proteins

A number of groups (Kruiswijk and Planta 1974; Gorenstein and Warner 1976; Grankowski et al. 1976; Ishiguro 1976; Zinker and Warner 1976; Otaka and Kobata 1978) have analyzed yeast ribosomal proteins by two-dimensional electrophoresis using two gel systems: basic urea–acidic urea (Kaltschmidt and Wittman 1970) and acidic urea–SDS (Mets and Bogorad 1974). Most groups agree that the 40S subunit contains 30 ± 5 proteins, and the 60S subunit contains 40 ± 5 proteins. Unfortunately, the agreement ends there. Due presumably to different methods of preparation and to

Table 1 Characteristics of rRNA of *Saccharomyces*

RNA	Nucleotides	Methyl groups	Pseudouridines	Comments
25S	3360[a]	43[b]	32[c]	
18S	1650[a]	24[b]	14[c,d]	
5.8S	158[e]	0[f]	1[e]	hydrogen-bonded to 25S
5S	121[g]	0[f]	1[h]	a primary transcript

[a] Philippsen et al. (1978).
[b] Klootwijk and Planta (1974).
[c] Klootwijk and Planta (1973).
[d] Also contains one residue of 3-[3-amino-3-carboxypropyl]-1-methyl pseudouridine (Maden et al. 1975).
[e] Rubin (1973); minor species of 164 and 165 nucleotides are also found (Rubin 1974).
[f] Udem and Warner (1972).
[g] Hindley and Page (1972), *S. carlsbergensis*; Miyazaki (1974), *S. cerevisiae*.
[h] *S. carlsbergensis* 5S RNA has no ψ.

different modifications of standard gel systems, it is at present impossible to provide an unambiguous numbering system for the proteins. Nevertheless, I have attempted to correlate the various numbering systems. Because it is based on Kaltschmidt and Wittman analysis, in analogy to the system adopted for *Escherichia coli* (Wittman et al. 1971) and for mammalian cells (McConkey et al. 1979), I now propose that all workers provisionally adopt the numbering system first suggested by Kruiswijk et al. (1974, 1978c). Figures 1, 2, and 3 show pH 8.6 × pH 4.5 gel patterns from our laboratory (Zinker and Warner 1976), in which I have attempted to label the proteins with the Kruiswijk-Planta numbers. A more complete comparison is presented in Tables 2 and 3. A major difficulty is that many workers (e.g., Gorenstein and Warner 1976), including the one laboratory that has purified a number of the proteins (Higo and Otaka 1979; Itoh et al. 1979), use modifications of the SDS gel system (Fig. 4). Recently, however, Otaka and Osawa (1980) have established a substantial correlation between the systems (Tables 2 and 3). Furthermore, Wejksnora (in Warner and Gorenstein 1978a), using as a third dimension the analysis of proteolytic digests of proteins separated on two-dimensional gels, has cross-identified a number of proteins as indicated in Tables 2 and 3. As more proteins are purified and as cloned genes for ribosomal proteins become available (see below), it will be possible to give definitive names to more of the proteins separated in the SDS system. For the remainder of this paper I have used the Kruiswijk-Planta numbers wherever possible. For proteins analyzed on SDS gels where the equivalent is not known, I have used brackets, e.g., protein [39].

I have made one change. The protein originally named L1 (Kruiswijk and Planta 1974) now seems not to be a true ribosomal protein. However, the

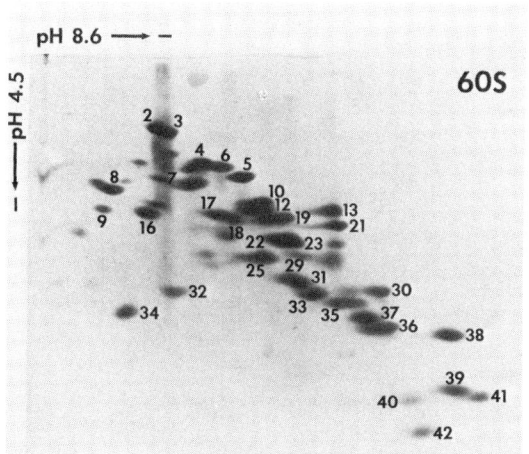

Figure 1 Basic proteins of the 60S subunit were displayed on a Kaltschmidt-Wittman (1970) gel according to Zinker and Warner (1976). The numbers identifying each protein are those of Kruiswijk and Planta (1974) (see Table 2).

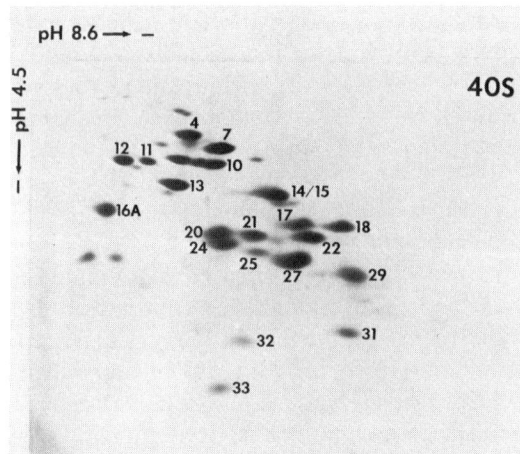

Figure 2 Basic proteins of the 40S subunit (see legend to Fig. 1 and Table 3).

protein that binds 5S RNA (Nazar et al. 1979) was not seen on the original gels because it is insoluble in 8 M urea between pH 9 and pH 5. I propose that this protein be designated L1 to avoid complicating the numbering system. By size, charge, and characteristic insolubility, L1 corresponds to proteins [3] and [4], which have the same peptide map and are presumably identical (Warner and Gorenstein 1978a).

Like ribosomal proteins from other organisms, most of the yeast proteins are small and basic, i.e., pI > 8.6. A few are more acidic (Tables 2 and 3); L44 and L45 are among the most acidic proteins in the cell. Removal of L44

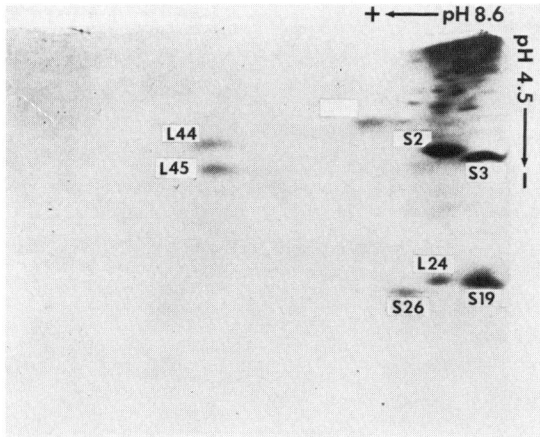

Figure 3 Acidic proteins of the 60S and 40S subunits (see legend to Fig. 1 and Tables 2 and 3).

and L45 causes the ribosome to lose most of its EF2-dependent GTPase activity, providing evidence that those proteins play a role analogous to the acidic L7/L12 of *Escherichia coli* (Sanchez-Madrid 1979b). The proteins have molecular weights of 8,000–45,000, all but a few being less than 30,000 (Kruiswijk et al. 1978c; Otaka and Kobata 1978; Higo and Otaka 1979; Itoh et al. 1979). Most of the structural data available have been obtained by Otaka and co-workers, who have purified a number of proteins from both the large (Itoh et al. 1979) and small (Higo and Otaka 1979) subunits and have commenced sequence determination (Itoh and Wittman-Liebold 1978; Itoh et al. 1980). Much of the sequence of L1 has also been determined (Nazar et al. 1979).

Several yeast ribosomal proteins are phosphorylated. In particular, S10, because of its rapid and reversible phosphorylation and its location on two-dimensional gels, is almost certainly the yeast analog of protein S6 of mammalian cells (Zinker and Warner 1976; Otaka and Kobata 1978). As in mammalian cells (Thomas et al. 1979), its degree of phosphorylation depends on the growth conditions. In rapidly growing cells it is sufficiently phosphorylated to cause satellite spots, as seen in Figure 2. The phosphorylated acidic proteins L44 and L45 (Zinker and Warner 1976; Grankowski et al. 1977; Kruiswijk et al. 1978a) are clearly analogous to the phosphorylated proteins EL7 and EL12 of *Artemia* (Amons et al. 1977). Phosphorylated S2 runs in the same place as a phosphorylated hamster protein in pH 5 × SDS gels (Schubart et al. 1977). These observations suggest that the phosphorylation of eukaryotic ribosomal proteins is highly conserved during evolution. Other phosphorylated proteins have been seen by some authors and not others (cf. Becker-Ursic and Davies 1976; Zinker and Warner 1976; Kruiswijk et al. 1978a; Otaka and Kobata 1978). Kruiswijk et al. find L9 and L30 phosphorylated, but only at one to two orders of magnitude less than S10. Several proteins of the 60S subunit, in particular, L15, are methylated (Cannon et al. 1977; Kruiswijk et al. 1978c). S10 and S15 (Hernandez et al. 1978) or S31 and S32 (Kruiswijk et al. 1978a) are methylated to a somewhat lesser extent. Whether this represents a strain difference or a methodological difference is not clear.

The stoichiometry of ribosomal proteins is difficult to determine experimentally, as a rather tedious series of papers on *E. coli* ribosomes demonstrates. If one washes with too little salt, nonribosomal proteins remain; with too much, ribosomal proteins are lost. Only by careful study of the recovery from *E. coli* of purified labeled ribosomal proteins was Hardy (1975) able to conclude that there was one copy of each except for L7/L12, of which there are four. Kruiswijk et al. (1978b) have some preliminary measurements suggesting that the majority of the proteins on yeast ribosomes are present in equimolar amounts. Nevertheless, the obvious differences in staining intensity of the spots in Figures 1–4 are cause for some concern.

Table 2 Compilation of the names of the 60S proteins

Kruiswijk and Planta (1974)	Zinker and Warner (1976)[a]	Grankowski et al. (1976)	Ishiguro (1976)	Gorenstein and Warner (1976)[b]	Otaka and Osawa (1980)	Comments[c]
(L1)[d]				[3], [4]	YL3; YL4	acidic[e], insoluble in 8 M urea[e,f], binds 5S RNA[e]
L2	1	1	2	[2]	YL2	
L3	2	1	1	[1][g]	YL1	trichodermin resistant[h]
L4	3	4	5	[6][g]	YL5	
L5	6	6	7	[8]	YL6	
L6	4	5	8	[11]	YL8	
L7	5	8	8		YL9	
L8	8	7	10	[25]	YL11	
L9	7	12	8		YL7	
L10	11	13	11	[16]	YL13	
L11	?	?	12		not listed	
L12	12	13	11		YL12	
L13	16	14	13		YL10	
L15	9	10	19		YL23	methylated
L16	10	11	18	[39][g]	YL22	
L17	13	15	17		YL16	
L18	14	16	16		YL18	
L19	15	18	15		YL19	
L20	15	18	14		YL17	
L21	17	19	21	[22]	YL15	
L22	18	22	22	[33]	YL20	
L23	18	23	22		YL14	
L24	37	21	26		YL31	acidic

L25	19	24	25		YL25
L26	?	?	23		YL28
L29	20	28	27		YL24
L30	24	29	28		YL21
L31	21	30	29		YL29
L32	22	26	30	[44][f]	YL38
L33	23	31	31		YL33
L34	25	27	32		YL36
L35	26	32	33	[73][f]	YL36
L36	28	33	34		YL34
L37	27	34	35	[47][f]	YL36
L38	29	35	36		YL26
L39	30	36	38		YL39
L40	32	?	?	may not be real	not listed
L41	31	37	37		YL27
L42	33	38	39		YL42
L43	runoff	39	40		YL40
L44	35	2	runoff		YL44 } acidic; phosphorylated;
L45	36	3	runoff		probably forms of the same protein

[a] There remains some ambiguity in the assignment of L31, L33, L35, and L37.
[b] Proteins numbered according to the SDS gel system (see Fig. 4).
[c] Only attributions that I consider unambiguous have been included (see text).
[d] (L1) is not the protein so designated by Kruiswijk and Planta (1974) but the 5S binding protein described by Nazar et al. (1979).
[e] Nazar et al. (1979).
[f] Zinker and Warner (1976).
[g] Confirmed by analysis of products of cloned genes.
[h] Fried and Warner (1981).
[i] Stocklein and Piepersberg (1980).

cycloheximide resistant[i]

535

Table 3 Compilation of the names of the 40S proteins

Kruiswijk and Planta (1974)	Zinker and Warner (1976)	Grankowski et al. (1976)	Ishiguro (1976)	Gorenstein and Warner (1976)[a]	Otaka and Ogata (1980)	Comments[b]
S2	27	1	4	[14]	YS8	acidic, phosphorylated
S3	28	2	3	[13]	YS3	acidic
S4	1	4	2	[12]	YS5	acidic
S5	?	3	5		not listed	
S6	?	5	?		YS7	
S7	2	6	6	[5][c]	YS6	
S10	5,6,7	8	7	[9][c]	YS4	phosphorylated; = S6 of mammals
S11	4	7	8	[40][c]	YS10	
S12	3	7	9	[30]	YS10	
S13	8	9	11	[21]	YS11	SUP46 protein[d]
S14	9	13	10	[19]	YS9	probably the same protein
S15	9	14	10			
S16	10	10	14	[55][d]	YS13	probably the same protein
S16A	10	21	14			
S17	11	16	13		YS15	

	runoff				
S18	12	17	12	YS12	
S19	29	11	18	YS13	acidic
S20	13	15	17	YS17	
S21	14	19	16	YS19	
S22	15	22	15	YS20	
S24	16	23	19	YS22	
S25	19	24	20	YS21	
S26	30	20	24	YS25	acidic
S27	20	26	21	YS18	
S28	21	27	25	YS14 ⎫	probably the same protein
S29	21	29	25		
S30	?	31	?		not listed
S31	22	33	26	YS23	
S32	23	32	?	YS28	
S33	24	34	27	YS27	
S34		35	28	YS24	

[41]
[42]
[52]
[50]
[61]
[37]
[45]

[a] Proteins numbered according to the SDS gel system (see Fig. 4).
[b] Only attributions that I consider unambiguous are included (see text).
[c] Confirmed by analysis of products of cloned genes.
[d] Ishiguro et al. (1981).

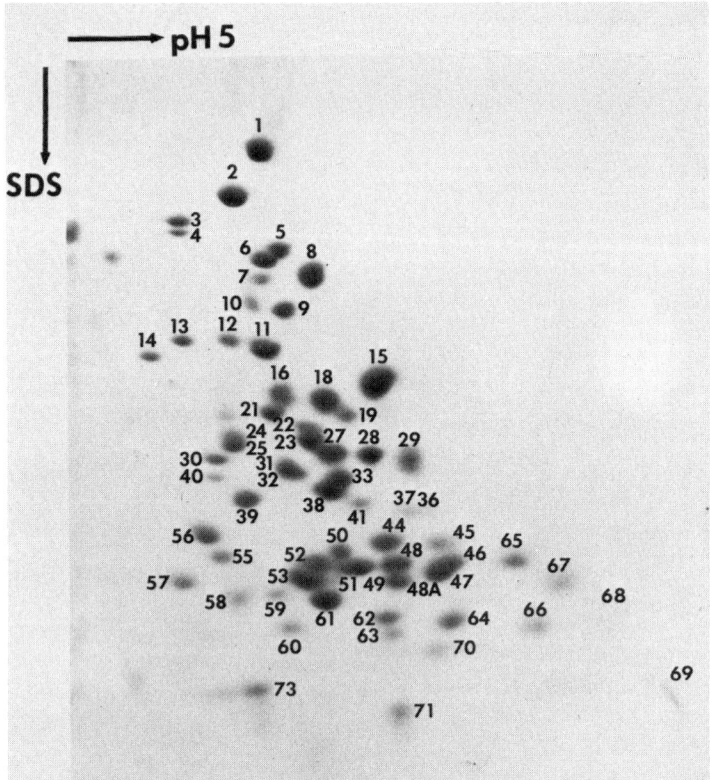

Figure 4 Total ribosomal proteins. Proteins of the 60S and 40S subunits were displayed on a two-dimensional gel run at pH 5 in the first dimension and in the presence of SDS in the second dimension (Gorenstein and Warner 1976). (Figure courtesy of C. Gorenstein.)

A few other facts should be recorded. Protein [61] is in fact two proteins, one from each subunit (P. Wejksnora, in Warner and Gorenstein 1978a). Sanchez-Madrid and Ballestra (1979) and Sanchez-Madrid et al. (1979a) have identified Ax, a phosphorylated, acidic protein, immunologically distinct from L44 and L45, which in some conditions can be bound firmly to the 60S subunit. Whether it should be classed as a ribosomal protein or a "factor" is unclear.

ACTIVITY OF THE YEAST RIBOSOME

The process of protein synthesis in yeast and the ribosome itself is essentially identical to those of higher eukaryotes by all criteria examined. Where yeast is perhaps unique is in the potential use of classical and modern genetic techniques to probe the mechanisms of eukaryotic protein synthesis. This section deals with our current state of understanding of protein synthesis in yeast, with an emphasis on those areas where mutants are available or genetic analysis appears promising.

Ribosomes and Polyribosomes

Mature ribosomes are distributed between polyribosomes and free ribosomes, as a function of the culture conditions. Although stationary cells have only a fraction of their ribosomes in polyribosomes, rapidly growing cells have more than 90% as polysomes, 5-8% as monosomes, and the remainder as subunits (Hartwell and McLaughlin 1969). A portion of the latter represents immature subunits (Udem and Warner 1973).

The high percentage of polysomes, as well as the fact that there is a lag in the decay of polysomes after mRNA synthesis has been inhibited, suggests that ribosomes, rather than mRNA, are limiting for protein synthesis. In sporulating cells, few of the ribosomes are in polysomes (Kraig and Haber 1980). However, if one adds cycloheximide, an inhibitor of elongation, the bulk of the ribosomes are recruited into the polysomes (Frank and Mills 1978); this suggests that during sporulation, it is the initiation of protein synthesis that is limiting, rather than the supply of mRNA.

Careful analysis of the radioactivity associated with nascent chains on polyribosomes led Peterson and McLaughlin (1973) to conclude that there are few, if any, polycistronic mRNAs in yeast. This conclusion is consistent with the elegant experiments of Sherman and Stewart (1975), who demonstrated genetically that a ribosome cannot reinitiate polypeptide synthesis after passing a termination codon.

As with other eukaryotes, it is difficult to dissociate from polyribosomes those subunits carrying a nascent polypeptide chain. Magnesium concentrations of 10^{-5} M or less or KCl concentrations of 0.5 M are required (Zinker and Warner 1976). The method of dissociation has some effect on the protein composition of the subunit, e.g., the phosphorylated protein S10 is found only on 40S subunits prepared in 0.5 M KCl but is distributed between 60S and 40S subunits prepared in 10^{-5} M $MgCl_2$ (Zinker and Warner 1976). Barbacid and Vazquez (1975), by studying the binding affinity of anisomycin for the ribosome, have suggested that there are substantial changes in conformation between polyribosomes, runoff ribosomes, and salt-washed ribosomes.

Cell-free Translation in Yeast

For a number of years, vigorous but unsuccessful (and mostly unpublished) attempts were made to develop authentic cell-free translation in extracts of yeast. Although extracts that could carry out chain elongation or translation of poly(U) were studied by a number of groups (Richter and Klink 1971;

Gallis and Young 1975; Skogerson and Wakatama 1976), no initiation of polypeptides using natural mRNA could be accomplished. Recently, however, Gasior et al. (1979a,b) have used the rather bizarre procedure of sedimenting most of the polyribosomes to prepare a 100,000g supernatant that contains ribosomes and all other factors necessary to translate exogenous mRNA. The extract is dependent on added mRNA and is sensitive to 7-methyl-GMP. Analysis of the products on two-dimensional O'Farrell gels demonstrates that the extract synthesizes authentic polypeptides (Tuite et al. 1980).

Using partial reactions on washed ribosomes, Skogerson and Wakatama (1976) have identified three factors required for poly(U)-directed polyphenylalanine synthesis. Two of them correspond to elongation factors 1 and 2 described for other eukaryotic systems. The third, a more active ribosome-dependent GTPase than EF2, is absolutely required for polyphenylalanine synthesis. It will be interesting to see whether yeast in fact has a distinctive elongation factor or whether an analogous factor will be found in other eukaryotic systems.

Effects of Antibiotics on Protein Synthesis

A wide range of antibiotics affect protein synthesis in cell-free extracts of yeast (reviewed by Vazquez 1979). However, the intact yeast cell is refractory to most. This discussion is restricted to four antibiotics, which can inhibit the growth of cells and which may prove useful in the genetic analysis of yeast ribosomes. Cycloheximide is an inhibitor of elongation, although with some effects on initiation and termination, all perhaps the result of an inhibition of translocation (Vazquez 1979). Analysis of mutants resistant to the antibiotic suggests that it acts on the 60S subunit (Rao and Grollman 1967; Jimenez et al. 1972; Stocklein and Piepersberg 1980). Cryptopleurine may also affect translocation (Vazquez 1979), although mutants resistant to this antibiotic appear to have an altered 40S subunit (Skogerson et al. 1973; Grant et al. 1974). All such mutations are allelic, but they fall into two classes, one of which is cross-resistant to emetine and tubulosine (Sanchez et al. 1977). Trichodermin and the related trichothecins are an interesting family of antibiotics. Although all are implicated in the inhibition of peptide bond formation (Vazquez 1979) some, such as verrucarin A, are effectively inhibitors of initiation since they block the formation of only the first few peptide bonds. Others, such as trichodermin, may be more effective on the termination step than on the elongation steps (Stafford and McLaughlin 1973). A single molecule of radioactive trichodermin binds to the 60S subunit of wild-type cells but not to that of trichodermin-resistant cells (Barbacid and Vazquez 1974; Wei et al. 1974). Paromomycin causes misreading in *Saccharomyces* (Sherman, this volume), much as streptomycin does in *E. coli*, and is able to suppress several types of nonsense mutations (Palmer et al. 1978).

Mutants Involving the Protein Synthetic Apparatus

A major impetus to studies on the molecular biology of yeast has been the promise of genetic analysis of the processes of transcription and translation in a eukaryote. The original search for temperature-sensitive mutants by Hartwell and McLaughlin (reviewed in Hartwell 1970) was designed with this in mind. A number of mutants were isolated that had temperature-sensitive aminoacyl-tRNA synthetases, e.g., isoleucyl (Hartwell and McLaughlin 1968a) and methionyl (McLaughlin and Hartwell 1969). Other mutants were deficient in either elongation (Hartwell and McLaughlin 1968b) or initiation (Hartwell and McLaughlin 1969) of protein synthesis. These mutants are now amenable to study using the in vitro system described above. However, no temperature-sensitive mutations affecting ribosome function have been isolated. This is not surprising, for none have been found in *E. coli*, even using sophisticated selection procedures. In *E. coli* mutants with altered ribosomal proteins, the assembly of the ribosome is far more temperature sensitive than its function (see below).

The genes causing resistance at the ribosome level to cycloheximide (*cyh2*), to trichodermin (*tcm1*), and to cryptopleurine (*cry1*) have been mapped, respectively, to chromosome VII (Mortimer and Hawthorne 1966), chromosome XV (Grant et al. 1976), and chromosome III (Skogerson et al. 1973; reviewed by McLaughlin 1974). The latter has attracted attention because of its tight linkage with the mating-type locus.

Although it has been difficult to demonstrate electrophoretically altered ribosomal proteins in such mutants, Stocklein and Piepersberg (1980) have found an allele for *cyh2* that yields an altered form of protein L29. Fried and Warner (1981) have isolated a recombinant DNA clone containing *tcm1* and find it to carry the gene for protein L3. The protein responsible for cryptopleurine resistance has yet to be identified. All three resistance mutations are qualitatively recessive, presumably because the sensitive ribosomes clog the polyribosomes in the presence of the drug. Nevertheless, heterozygous diploids carrying either *cyh2* or *tcm1* can grow in the presence of low concentrations of the appropriate drug.

Certain omnipotent suppressor genes have been isolated, which cause increased misreading (Sherman, this volume). These resemble the *ram* mutants of *E. coli*, in which an altered ribosomal protein, S4, also causes miscoding (Gorini 1974). Ishiguro et al. (1981) have found that one yeast ribosomal protein, probably S13, is altered in the omnipotent suppressor *SUP46*.

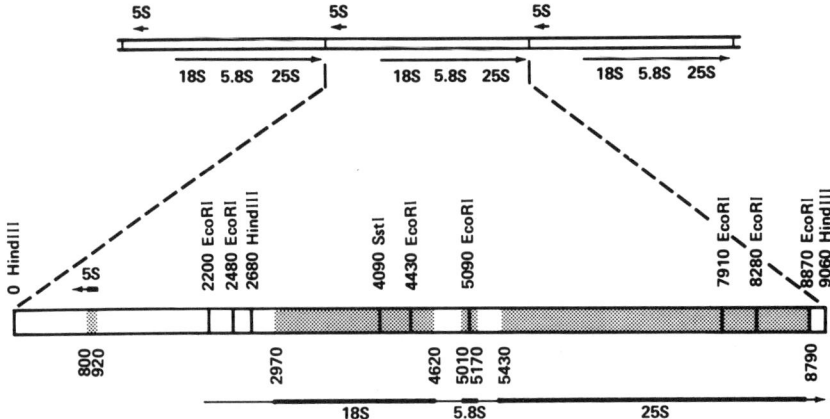

Figure 5 rDNA repeat unit. (Reprinted, with permission, from Philippsen et al. 1978.)

SYNTHESIS OF THE YEAST RIBOSOME

Transcription of rRNA Genes

It was first observed by Cramer et al. (1976) that the genes for rRNA are highly repetitive and tightly linked. It now appears that all 100-120 rRNA genes form a single tandem array on chromosome XII (Petes 1979) (see Fig. 5). These genes are discussed in detail by Fangman and Zakian (1981).

By restriction map (Bell et al. 1977; Kramer et al. 1978) and R-loop (Philippsen et al. 1978) analysis of the cloned rDNA repeat unit, the arrangement and orientation of the mature rRNA sequences have been established, as shown in Figure 5. This has been substantiated by extensive sequence analysis of the repeat unit, now approaching 25% of the total (Bell et al. 1977; Maxam et al. 1977; Valenzuela et al. 1977; Skryabin et al. 1978; Bayev et al. 1979; Rubtsov et al. 1980).

Two transcripts are synthesized from the rRNA genes. (Transcription is discussed by Sentenac and Hall [this volume]). 5S RNA is transcribed in the leftward direction, as shown in Figure 5 (Maxam et al. 1977). Since it has a 5′ triphosphate (Hindley and Page 1972), is colinear with its gene, and has no detectable methyl groups (Udem and Warner 1972), 5S RNA appears to undergo no posttranscriptional modifications. In contrast, 18S, 5.8S, and 25S RNAs are all derived from a single transcript synthesized in the rightward direction. Originally described as 3.1×10^6 daltons (42S) (Retel and Planta 1970), it now seems that the transcript is less than 7000 nucleotides in length (35S)[1] (Udem and Warner 1972; Trapman and Planta 1975b). The

---

[1]Unfortunately for the reader, workers in this area have not agreed on the names of the RNAs. By author's prerogative I use the nomenclature developed in our laboratory. It is related to that of some other authors by the following glossary: 35S = 37S; 27S = 29S; 25S = 26S; 20S = 18S; 18S = 17S.

recent finding of pppApUp... at the 5′ end of this molecule confirms that it is the initial transcript (Hadjiolov et al. 1978; Klootwijk et al. 1979). Klemenz and Geiduschek (1980) have shown that the 5′ end of the 35S precursor RNA is 48 nucleotides upstream from the left-most EcoRI site in Figure 5, i.e., at about 2150 on that map. The 3′ end of 35S is just 7 nucleotides downstream from the 3′ end of 25S RNA (G.M. Veldmann, pers. comm.).

Processing of the rRNA Transcript

After transcription is complete, approximately 60 methyl groups are added to the 35S RNA (Udem and Warner 1972; Klootwijk and Planta 1974), mostly at the 2′ OH of ribose, but some on the bases as well. Furthermore, most of the pseudouridines are formed (Brand et al. 1979). Subsequently, a number of processing steps occur which, by kinetics analysis, seem to be as follows (Sillevis Smitt et al. 1972; Udem and Warner 1972; Trapman and Planta 1975b):

$$35S \longrightarrow 32S \begin{array}{l} \longrightarrow 20S \longrightarrow 18S \\ \\ \longrightarrow 27S \longrightarrow 25S \\ \searrow 7S \longrightarrow 5.8S \end{array}$$

Little 32S is observed in normal cells, suggesting that it is rapidly cleaved to form the relatively stable 27S and 20S intermediates. 27S RNA is trimmed at the 3′ end and at the 5′ end to produce the 25S species and the 7S precursor to 5.8S RNA (Helser and McLaughlin 1975; Trapman and Planta 1975a). The final 5.8S molecule is heterogeneous at the 5′ end (Rubin 1974) due to nonuniform processing (Bayev et al. 1979). The 5′ end of the 5.8S RNA is involved (Nazar and Stitz 1980) in its rather firm hydrogen bonding to 25S RNA (Udem et al. 1971). All of these events occur in the nucleus. The 20S precursor, however, appears as a 40S ribonucleoprotein in the cytoplasm (Udem and Warner 1973; Trapman et al. 1975), where the 3′ end is trimmed (deJonge et al. 1977a) and two dimethyl adenines are formed, 19 and 20 nucleotides from the 3′ end (deJonge et al. 1977b). Then, and only then, can the particle participate in protein synthesis (Udem and Warner 1973).

Altogether, approximately 2000 nucleotides are disposed of during the processing of 35S RNA to the mature species. Few if any of those nucleotides are modified by methylation or pseudouridylation. R-loop analysis using both electron microscopy and nuclease S1 demonstrate that no splicing events occur during the processing (Phillipsen et al. 1978). In contrast, splicing is required for the maturation of the mitochondrial rRNA of some strains of yeast (see Dujon 1981). In examining the sequences on either side of the mature rRNAs, Veldman et al. (1980) find no regions of complementarity, such as those that exist in *E. coli*: they conclude that an RNase III type of enzyme is not likely to be involved in the maturation of rRNA in yeast.

Synthesis of Ribosomal Proteins and Their Assembly into Ribosomes

Our conception of the synthesis of ribosomal proteins and their assembly into ribosomes is derived largely from work on mammalian cells. Craig and Perry (1971) showed that ribosomal proteins of mouse L cells were synthesized on soluble cytoplasmic polyribosomes, from which they migrate rapidly to the nucleolus where they assemble with ribosomal precursor RNA to form ribonucleoprotein particles; these are the precursors of ribosomes (for review, see Warner 1974).

Such experiments are far more difficult in yeast. The nuclei are small, fragile, and difficult to purify. Nevertheless, some progress has been made. Nearly half the nucleus is composed of a "dark crescent," which has fibrillar and granular components resembling a mammalian nucleolus (Matile et al. 1969; Sillevis Smitt et al. 1973). After a brief pulse of [^{3}H]uracil, labeled 35S and 27S ribosomal precursor RNAs appear to be concentrated in the dark crescent, as demonstrated by autoradiography (Sillevis Smitt et al. 1972) and by partial fractionation of the nuclei (Sillevis Smitt et al. 1973). Thus, *S. cerevisiae* has an authentic nucleolus which, because of the cell's small content of DNA and its masive synthesis of ribosomes, dominates the nucleus.

From the nucleus, Trapman et al. (1975) have isolated precursor particles of 90S, 66S, and 43S, which contain ribosomal precursor RNA and are enriched for certain ribosomal proteins (Kruiswijk et al. 1978b), suggesting a possible order of assembly. Other ribosomal proteins, such as, L9, L44, and L45, are exchangeable, i.e., they appear on ribosomes in the absence of new ribosome synthesis (Zinker and Warner 1976; Kruiswijk et al. 1978b). Whereas the 90S and 66S particles are largely in the nucleus (Trapman et al. 1976), the final steps of maturation of the 40S particle occur in the cytoplasm (Udem and Warner 1973; Trapman et al. 1975). The presence of newly formed ribosomal proteins is essential for the maturation of yeast ribosomal precursor RNA. In the presence of cycloheximide (Mayo et al. 1968; Taber and Vincent 1969; Udem and Warner 1972) or in certain mutants in which the synthesis of ribosomal proteins is suppressed (Warner and Udem 1972; Shulman and Warner 1978), no intermediate or mature species of rRNA is produced in spite of substantial transcription of 35S precursor RNA.

Genes for Ribosomal Proteins

As mentioned above, recessive genes conferring resistance at the ribosome level to cycloheximide, to trichodermin, and to cryptopleurine have been mapped to three separate chromosomes. This suggests that the genes for yeast ribosomal proteins are dispersed throughout the genome. Gorenstein and Warner (1979) have shown that no ribosomal proteins were made on polycistronic mRNAs nor were they derived from polyprotein precursors. On the other hand, from the kinetics of UV inactivation, Mager et al. (1977) suggested substantial clustering of the genes for ribosomal proteins. Data from this laboratory, however (C. G. Gorenstein, unpubl.), suggest that UV inactivation of ribosomal protein synthesis may reflect a control mechanism responsive to inactivation of rRNA genes.

The advent of gene cloning techniques has clarified the situation. Woolford et al. (1979) have isolated clones containing genes for four ribosomal proteins: [39], [51], [52], and [63], none of which are linked. Fried and Warner (1981) have cloned the gene for trichodermin resistance, *tcm1*, and found that it codes for protein L3. It is on a 13.5-kb fragment of DNA that has no other ribosomal proteins. Fried et al. (1981) have cloned the genes for 14 other ribosomal proteins: S7, S10, S11, S16A, L4, L16, L32, L37, [10], [24], [28], [29], [58], and [63]. Although the average size of these clones is about 8.5 kb, in only one case are two ribosomal protein genes in the same clone. Interestingly, those two appear to be adjacent. With the exception of L3, none of the cloned ribosomal protein genes have been placed on the yeast genetic map; nevertheless, it is clear that they are essentially unlinked. The genes for all ribosomal proteins examined appear to be unique, except protein [39], for which Woolford et al. (1979) have identified two unlinked genes for protein [39].

Mutants Affected in Ribosome Synthesis

Although yeast has a number of disadvantages for the study of ribosome biosynthesis, e.g., the difficulty in preparing clean nuclei in high yield and the rapid rate of processing of ribosomal precursor RNA, it does hold the promise of a genetic analysis of the process. A number of mutants now available may lead to some insight.

Ursic and Davies (1979) have isolated a cold-sensitive mutant, *dip1*, in which RNA polymerase transcribing the rRNA genes often appears to terminate prematurely, leading to 14S and smaller RNA molecules instead of the usual 35S precursor. A substantial amount of the rRNA species that do accumulate are abnormal. Further analysis of this mutant is awaited with interest.

At least three groups have reported mutants defective in the processing of precursor RNA. Andrew et al. (1976) identified a temperature-sensitive mutant, ts351, which is almost totally defective in the reaction 27S → 25S at the nonpermissive temperature. Both 27S RNA and newly formed 60S ribosomal proteins are rapidly degraded (Gorenstein and Warner 1977). Venkov and Vasileva (1979) have isolated mutants in which the processing of 35S precursor RNA is substantially slower than normal. Carter et al. (1980) have reported an unusual mutant, in which the 20S → 18S step occurs very slowly, apparently because the 40S ribonucleoprotein containing 20S RNA is retained within the nucleus (Carter and Cannon 1980). Surprisingly, this phenotype is only observed in cells carrying the trichodermin-resistant allele of *TCM1*, the gene for a 60S protein.

Some years ago, Bayliss and Ingraham (1974) demonstrated that a mutant causing sensitivity to streptomycin at the ribosome level also rendered the cells cold sensitive because of improper assembly of ribosomal proteins on the 18S rRNA. Finally, there are a number of mutants defective in the regulation of ribosomal protein synthesis (Hartwell et al. 1970), which will be discussed below.

REGULATION OF THE SYNTHESIS OF THE YEAST RIBOSOME

Ribosome synthesis is comprised of two parallel but interrelated sets of reactions:

$$XTP \longrightarrow 35S \dashrightarrow 25S + 5.8S + 18S$$

$$XTP \longrightarrow \sum_{1}^{70} mRNA_{ribosomal\ proteins} \dashrightarrow \sum_{1}^{70} ribosomal\ proteins$$

As a rule, the yeast cell synthesizes the various components of the ribosome in approximately correct stoichiometry under a variety of environmental conditions, leading to different rates of ribosome synthesis. Thus, these two sets of reactions are coordinately controlled, making it a delightfully complex system in which to study regulation by perturbing ribosome synthesis. Discussed below are changes in growth rate and temperature shift, amino acid starvation, protein synthesis inhibition, mutants defective in protein synthesis or rRNA synthesis and maturation, and their contribution to understanding the coordinate regulation of the synthesis of the ribosomal components.

Dependence on Growth Rate

The concentration of ribosomes within the yeast cell seems to be correlated with the growth rate (Waldron and Lacroute 1975), although less dramatically than is the case in *E. coli*. Immediately following a shift-up, there is a substantial burst of total RNA synthesis (Wehr and Parks 1969), perhaps after a short lag (Waldron 1977), in spite of a temporary inhibition of protein synthesis. In cells growing on ethanol as a carbon source, the concentration of ribosomes (per unit of protein) is half that of a culture growing on glucose. When glucose is added to such a culture, there is an explosive increase in the synthesis of ribosomal components (Kief and Warner 1981). The rate of rRNA synthesis increases by 40% within 5 minutes, doubles within 15 minutes, and reaches a steady state characteristic of the new growth medium within 30 minutes. The rates of synthesis of more than 30 individual ribosomal proteins increase 2.5-fold within 15 minutes and reach 3.5-fold in 60 minutes.

On the other hand, when cells are induced to undergo sporulation by depriving them of a nitrogen source, the relative synthesis of individual ribosomal proteins declines to about 20% of the vegetative level within 60 minutes (Pearson and Haber 1980). Since total protein synthesis also declines, the absolute rate of synthesis of ribosomal proteins is reduced by more than 90%. Total RNA synthesis undergoes a similar decline (Hopper et al. 1974). Thus, yeast can coordinate the synthesis of ribosomal components over a 20- to 50-fold range.

Ribosome Synthesis and the Cell Cycle

Although there have been reports that many enzymes are synthesized in discontinuous fashion through the cell cycle (for review, see Mitchison 1971), this has not been confirmed by measurement of the synthesis of a large number of individual proteins (Elliott and McLaughlin 1978). In any case, it is now clear that both rRNA (Sogin et al. 1974; Elliott and McLaughlin 1979) and each of the ribosomal proteins (Elliott et al. 1979) are synthesized continuously throughout the cell cycle.

On the other hand, Johnston and Singer (1978) find that cells are blocked at the G_1 stage of the cell cycle by certain chelating agents or methionine analogs (Singer et al. 1978), which seem to inhibit rRNA synthesis but not protein synthesis. They suggest that the "production of ribosomal precursor RNA may...be an important regulatory event for progression through G_1." This provocative idea deserves further experimental analysis.

Dependence on Protein Synthesis

A number of groups have studied the effects of amino acid starvation and inhibition of protein synthesis on the synthesis of rRNA to determine whether yeast has a "stringent response" comparable to that of *E. coli*, i.e., in response to the deprivation of an amino acid, the cell rapidly shuts off the synthesis of both rRNA and ribosomal protein. deKloet (1966) was the first to show that rRNA synthesis was inhibited by cycloheximide. Wehr and Parks (1969) showed that RNA accumulation stopped some time after cells were deprived of an amino acid. Interestingly, methionine starvation had a much more rapid effect on RNA accumulation than did tryptophan starvation. An analogous observation in HeLa cells was shown to be due to lack of methylation of ribosomal precursor RNA, which was then inactive as a substrate for processing enzymes (Vaughan et al. 1967).

However, studies that attempted to distinguish between the effects of amino acid starvation and inhibitors of protein synthesis reached contradictory conclusions. Roth and Dampier (1972) and Oliver and McLaughlin (1977) found little difference in transcription whether the cells were deprived of an amino acid or treated with cycloheximide. Gross and Pogo (1974, 1976), on the other hand, suggested that cycloheximide caused phenotypic reversion of amino acid starvation, restoring some measure of rRNA synthesis, much as chloramphenicol relaxes the stringent response in *E. coli*. Shulman et al. (1977) found that, although cycloheximide appears to permit increased incorporation of labeled precursors into RNA during amino acid starvation, the effect is almost entirely at the level of labeling the pools of nucleotide precursors. When RNA is labeled with ^{3}H-labeled [CH$_3$]methionine, whose incorporation into the *S*-adenosylmethionine pool is relatively unaffected by amino acid starvation, it is found that transcription of rRNA is inhibited more than 80% after cells are deprived of tyrosine, whether or not cycloheximide is present. mRNA and tRNA continue to be synthesized at almost normal levels. Any ribosomal precursor RNA that is synthesized in the absence of protein synthesis is fully methylated. Similar inhibition of rRNA synthesis is observed at the restrictive temperature in temperature-sensitive mutants with lesions in various parts of the protein synthetic pathway (McLaughlin et al. 1969; Oliver and McLaughlin 1977). On the other hand, in certain temperature-sensitive mutants unable to synthesize ribosomal proteins (Hartwell et al. 1970; Gorenstein and Warner 1976), the transcription of 35S ribosomal precursor RNA continues almost normally (Shulman and Warner 1978), although the molecules are not processed to mature species.

The uniform response of yeast to various modes of protein synthesis inhibition suggests that the inhibition of rRNA transcription characteristic of the stringent response is a direct consequence of insufficient protein synthesis, not of the lack of ribosomal proteins specifically. This stringent response extends to ribosomal proteins whose mRNAs are specifically depleted in cells deprived of tyrosine (Warner and Gorenstein 1978b). The occurrence of a stringent response in yeast suggests the possibility of a molecular signal, which has yet to be found, and leads to the hope for a relaxed mutant, which may at long last have been isolated (J. Stevens and C. McLaughlin, pers. comm.).

Effects of Temperature Shift

It has recently become clear that the phenomenon of heat shock, first described in *Drosophila* (Ritossa 1962), has a counterpart in both yeast (Miller et al. 1979) and bacteria (Lemaux et al. 1978). When yeast cells are subjected to a temperature upshift, even if the higher temperature will support growth, there is a substantial change in the distribution of proteins synthesized. In particular, the synthesis of proteins involved in the protein synthetic apparatus is substantially inhibited (Gorenstein and Warner 1976; Warner and Gorenstein 1977). If exponentially growing cells of *Saccharomyces* are shifted from 23°C to 36°C, the synthesis of each of the ribosomal proteins declines rapidly to about 30% of normal after 15 minutes. The synthesis then begins to recover, reaching nearly normal levels in 60–90 minutes (Gorenstein and Warner 1976; Warner and Gorenstein 1977). During this time, the synthesis of rRNA continues at nearly normal levels, although its processing to mature species is slowed (Udem and Warner 1972; Warner and Udem 1972). Cell-free translation of RNA isolated from such cells suggests that the effect is largely due to limitations in the availability of mRNA for ribosomal proteins (Warner and Gorenstein 1977). The magnitude of the effect is dependent on the magnitude of the temperature shift. No effect is seen when the temperature is shifted down.

To determine whether protein synthesis is involved in bringing about or reversing the temperature shift effect, the experiment described in Table 4 was carried out. In the presence of cycloheximide, the level of mRNA for ribosomal proteins declines less than in the control. Although this result may be due to the mRNA decaying more slowly in the presence of cyclohex-

Table 4 Effects of cycloheximide on temperature shock

Ribosomal protein number	Template activity, 20 min / template activity, 0 min		Template activity, 60 min / template activity, 20 min	
	control	+ cycloheximide (0–20 min)	control	+ cycloheximide (20–60 min)
[1]	0.39	0.71	3.7	2.2
[2]	0.27	0.72	3.0	2.2
[5]	0.20	0.70	4.8	3.7
[6]	0.22	0.74	5.1	4.4
[8]	0.32	0.76	4.2	2.6
[28]	0.20	0.68	5.7	3.4
[39]	0.31	0.87	4.2	3.4

Total RNA was prepared from cells growing exponentially at 23°C and subjected to the following treatment: none; 36°C for 20 min; 36°C and cycloheximide (100 μg/ml) for 20 min; 36°C for 60 min; 36°C for 60 min and cycloheximide (100 μg/ml) from 20 to 60 min. The RNA was translated in a wheat germ extract, and the synthesis of individual ribosomal proteins, compared to the synthesis of total protein, was determined (Warner and Gorenstein 1977).

imide, it could imply that protein synthesis plays some role in the temperature shock effect. The right-hand portion of Table 4 shows that the recovery of the ability to make ribosomal proteins is largely independent of protein synthesis.

Two possible explanations for the effect of temperature were explored. Since a shift from 23°C to 36°C causes an instantaneous doubling of protein synthesis, the resulting drain on the aminoacyl-tRNA pool might lead to an increase in nonacylated tRNA, which could signal a shutdown of the synthesis of ribosomal proteins. To test this hypothesis, we carried out a shift-up experiment on cells exposed to 0.1 µg/ml of cycloheximide for 10 minutes. Under such conditions, there can be no drain on aminoacyl tRNA, since protein synthesis at 23°C is roughly half the control rate and increases very little on shifting to 36°C. Nevertheless, 15 minutes after the shift, the synthesis of ribosomal proteins was inhibited by 60-70% (J. Warner, unpubl.). Clearly, the sudden increase in the rate of protein synthesis does not bring about the inhibition of ribosomal protein synthesis. An alternative locus of regulation was suggested by the observation that a temporary accumulation of Magic Spot, ppGpp, occurs when yeast is shifted from 23°C to 36°C (Pao et al. 1977). However, in ρ^0 cells or in cells treated with tetracycline, there is no accumulation of ppGpp after a temperature shift (Pao et al. 1977). Nevertheless, under those conditions, the normal inhibition of ribosomal protein synthesis is observed. In short, there is no current molecular or biological explanation for the effect of temperature shift.

Mutants Defective in the Regulation of Ribosome Synthesis

Among the 400 temperature-sensitive mutants isolated by Hartwell (1967) were 23 whose phenotypes involved the inability to accumulate RNA at the restrictive temperature. These mutants, which fall into ten complementation groups, designated *rna2-rna11*, were shown to be deficient in the synthesis of rRNA and its processing into ribosomes (Hartwell et al. 1970). It has now become clear that the primary effect of these mutations is to block the synthesis of nearly all of the ribosomal proteins at the restrictive temperature (Gorenstein and Warner 1976), primarily by blocking the synthesis of their mRNAs (Hereford and Rosbash 1977; Warner and Gorenstein 1977). Although the synthesis of almost all ribosomal proteins declines to 5-10% of normal, the synthesis of 35S ribosomal precursor RNA continues normally for some time, but it is not processed and is rapidly degraded (Shulman and Warner 1978). The kinetics of decline of ribosomal protein synthesis on temperature shift, either in the wild-type or in the *rna2-rna11* temperature-sensitive mutants, suggest that the synthesis of mRNA for ribosomal proteins ceases almost immediately after the shift, and the available mRNA decays with its characteristic half-life (Gorenstein and Warner 1976). Thus, it seems likely that temperature-sensitive mutants defective in genes *rna2–rna11* are simply unable to reverse the usually temporary effect of heat shock. It should be noted, however, that the synthesis of a few ribosomal proteins, e.g., [16], [25], and [39], is refractory to the effects of such temperature-sensitive mutations. Surprisingly, in strains carrying a temperature-sensitive mutant in *rna2*, the synthesis of these proteins is protected from the usual inhibition accompanying temperature shift (Gorenstein 1977; Warner and Gorenstein 1977).

A temperature-sensitive mutation in another gene, *rna1*, is thought to block transport of mRNA to the cytoplasm (Hutchinson et al. 1969), as well as the processing of tRNA precursors (Hopper et al. 1978). This mutation also has a strange sparing effect on ribosomal protein synthesis after a temperature shift: No decline is seen during the first 10 or 15 minutes (Gorenstein and Warner 1976: cf. Figs. 4 and 5). A similar sparing effect has been noticed for other temperature shock proteins (Miller et al. 1979). No explanation for the effect has been proposed.

On the other hand, when cells are undergoing sporulation, a temperature upshift has no effect on the rate of ribosomal protein synthesis, even in strains homozygous for *rna2*, which suggests that a different regulatory regime is in effect during sporulation (Pearson and Haber 1980). Nevertheless, stains bearing the *rna2–rna11* mutations are temperature sensitive in some step of sporulation. Pearson (1979) has recently isolated two mutations that permit sporulation in such stains. One of these, *srp1*, a recessive suppressor of *rna2*, prevents the rapid shutoff of ribosomal protein synthesis after a temperature shift, but permits a slow one, with a half-time of decay of about 4 hours rather than 10–15 minutes. A strain of genotype *rna2 srp1* is temperature sensitive for growth. Another mutant, *SRN1*, a dominant suppressor of *rna2*, *rna6*, and *rna8*, permits growth at 36°C by preventing any shutoff of ribosomal protein synthesis after a temperature shift (N. J. Pearson and J. Haber, in prep.).

The effect of temperature shock is so broadly distributed throughout living things that it must reflect an important biological principle. The availability of the temperature-sensitive mutants and their second-site suppressors may make possible its molecular and genetic analysis.

Conclusions

The mass of data just presented can best be digested by considering the three levels at which a cell must regulate the synthesis of the components of the ribosome: (1) regulation that leads to the synthesis of equal numbers of all of the ribosomal proteins, (2) regulation that leads to balanced synthesis of the RNA and protein portions of the ribosome, and (3) regulation that leads to increased or decreased synthesis of ribosomes in response to changes in environmental conditions.

It is absolutely clear from the data presented above that yeast has developed a way to coordinate the synthesis of its ribosomal proteins, in spite of its genes being scattered over the genome. Except in strains carrying temperature-sensitive mutants in *rna2–rna11*, no ribosomal protein is synthesized in amounts substantially greater or smaller than the other ribosomal proteins, even under a variety of experimental conditions. This suggests that some common factor regulates the synthesis of mRNAs for all the ribosomal proteins. Furthermore, in view of potential variations in the lifetime of mRNA and the efficiency of its translation, it seems likely that excess ribosomal proteins exert a feedback repression on their own synthesis. Such an effect seems to occur in *E. coli* (Fallon et al. 1979; Lindahl and Zengel 1979) and should be amenable to testing in yeast using cloned genes for ribosomal proteins.

Among the experiments described above, in no instance are ribosomal proteins synthesized in substantial excess over rRNA. These observations suggest that the synthesis of rRNA is a necessary, though not sufficient, requirement for the synthesis of ribosomal proteins or their mRNAs. Two additional findings support this suggestion. In cells deprived of an essential amino acid, the decline of ribosomal protein synthesis lags behind the decline of rRNA synthesis by just about the lifetime of mRNA (Warner and Gorenstein 1978b). Secondly, in repair-deficient cells subjected to UV radiation, the dose required for half-maximal inhibition of synthesis of each of the ribosomal proteins is the same as it is for the rRNA and is substantially less than that for other cell proteins (C. Gorenstein and J. R. Warner, unpubl.). I suggest that the inhibition of rRNA synthesis by UV has the secondary effect of inhibiting synthesis of mRNA for ribosomal proteins.

After a temperature shock, as well as in the temperature-sensitive mutants of genes *rna2–rna11*, the situation is somewhat different. The transcription of rRNA continues uncoupled from the synthesis of ribosomal proteins or their mRNAs. Thus, neither the transcription nor the processing of ribosomal precursor RNA is sufficient to cause the synthesis of mRNA for ribosomal proteins. These data, together with the mutants isolated by Pearson (1979), which suppress the effects of both the temperature shock and the temperature-sensitive mutants, suggest that some positive effector is essential for the synthesis of mRNA for ribosomal proteins.

I propose that the synthesis of this effector, called Q, is dependent on the transcription of ribosomal precursor RNA. Since Q is unstable, its level in the cell depends on a balance between its synthesis and its degradation (or excretion). This proposal predicts that changes in environmental conditions, such as amino acid deprivation or nutritional shift-up, act primarily at the level of transcription of rRNA. This affects the synthesis of Q and, subsequently, its concentration and leads to changes in the synthesis of mRNA for ribosomal proteins. I suggest further that a temperature shock causes a rapid loss of Q. Only when it accumulates anew can mRNA synthesis resume. The recessive mutant *srp1* might act by preventing the loss of Q. Furthermore, perhaps the temperature-sensitive lesions in genes *rna2-rna11* prevent the synthesis of Q in spite of continued transcription of rRNA.

Note that Q, as a positive effector, plays a role opposite that of Magic Spot. Q is probably not related directly to protein synthesis, since protein synthesis continues normally under conditions in which Q should change markedly. Q is probably not related to the processing of ribosomal precursor RNA, e.g., a fragment of the transcribed spacer, since processing continues after a temperature shift, albeit at a declining rate. Nevertheless, this hypothesis, although by no means unique, makes a number of predictions about the concentrations of such a hypothetical molecule under different physiological conditions and will, I hope, provoke some experiments.

SUMMARY AND PROPSECTS

Protein Synthesis

Our level of understanding of eukaryotic protein synthesis is largely due to work on mammalian and plant systems. The details of yeast protein synthesis are only now, with the development of a complete in vitro system, coming to light. Nevertheless, it may well be yeast that will lead to our next level of understanding of protein synthesis, beyond the descriptive to the analytical, where one can ask the role of each of the polypeptides of the complex initiation factors, or determine the parts of the ribosome that contribute to the accuracy of the decoding process. Only yeast has the rich resources of classical genetics, coupled with the modern techniques of gene cloning and transformation, which can facilitate such analysis.

Regulation

It must be emphasized that we have not firmly established how yeast cells regulate any aspect of ribosome synthesis. No relaxed mutants have been reported; no regulatory mutations exist in individual rRNA or ribosomal protein genes. Nevertheless, the future looks promising. The number of experimental approaches that alter the synthesis of ribosomal components is increasing, and the methods for analyzing them are improving. The cloned ribosomal protein genes can be studied in detail and their transcription products, measured accurately. Finally, the ability to transform yeast with a cloned gene for a ribosomal protein should make possible a more penetrating analysis of these essential cell proteins.

ACKNOWLEDGMENTS

This research has been supported by National Institutes of Health grants RO1-GM-25532 and P30-CA-13330, American Cancer Society grant NP-72-

J, and National Institutes of Health fellowships to H. Fried (F32-GM-07165) and N. Pearson (F32-GM-07486). I am grateful to H. Fried, N. Pearson, and D.R. Kief for numerous discussions, to C. Gorenstein for permission to use Figure 4, to my many colleagues who provided preprints and reprints, and to Grace Sullivan for her tireless efforts. This review was completed in March, 1980.

REFERENCES

Amons, R., A.J. van Agthoven, W. Pluijms, W. Moller, K. Higo, T. Itoh, and S. Osawa. 1977. A comparison of the amino-terminal sequence of the L7/L12-type proteins of *Artemia salina* and *Saccharomyces cerevisiae*. *FEBS Lett.* **81**: 308.

Andrew, C., A.K. Hopper, and B.D. Hall. 1976. A yeast mutant defective in the processing of 27 rRNA precursor. *Mol. Gen. Genet.* **144**: 29.

Barbacid, M. and D. Vazquez. 1974. Binding of (acetyl-^{14}C)-trichodermin to the peptidyl transferase centre of eukaryotic ribosomes. *Eur. J. Biochem.* **44**: 437.

———. 1975. Ribosome changes during translation. *J. Mol. Biol.* **93**: 449.

Bayev, A.A., K.G. Skryabin, V.M. Zaharyev, A.S. Krayev, and P.M. Rubtsov. 1979. Yeast ribosomal genes. In *Macromolecules in the functioning cell* (ed. F. Salvatore et al.), p. 3. Plenum Press, New York.

Bayliss, F.T. and J.L. Ingraham. 1974. Mutation in *Saccharomyces cerevisiae* conferring streptomycin and cold sensitivity by affecting ribosome formation and function. *J. Bacteriol.* **118**: 319.

Becker-Ursic, D. and J. Davies. 1976. *In vivo* and *in vitro* phosphorylation of ribosomal proteins by protein kinases from *Saccharomyces cerevisiae*. *Biochemistry* **15**: 2289.

Bell, G.I., L.J. DeGennaro, D.H. Gelfand, R.J. Bishop, P. Valenzuela, and W.J. Rutter. 1977. Ribosomal RNA genes of *Saccharomyces cerevisiae*. *J. Biol. Chem.* **252**: 8118.

Brand, R.C., J. Klootwijk, C.P. Sibom, and R.J. Planta. 1979. Pseudouridylation of yeast ribosomal precursor RNA. *Nucleic Acids Res.* **7**: 121.

Cannon, M., D. Schindler, and J. Davies. 1977. Methylation of proteins in 60S ribosomal subunits from *Saccharomyces cerevisiae*. *FEBS Lett.* **75**: 187.

Carter, C.J. and M. Cannon. 1980. Maturation of ribosomal precursor RNA in *Saccharomyces cerevisiae*. A mutant with a defect in both the transport and terminal processing of the 20S species. *J. Mol. Biol.* **143**: 179.

Carter, C.J., M. Cannon, and A. Jimenez. 1980. A trichodermin-resistant mutant of *Saccharomyces cerevisiae* with an abnormal distribution of native ribosomal subunits. *Eur. J. Biochem.* **107**: 173.

Craig, N. and R. Perry. 1971. Persistent cytoplasmic synthesis of ribosomal proteins during the selective inhibition of ribosomal RNA synthesis. *Nat. New Biol.* **229**: 75.

Cramer, J.H., F.W. Farrally, and R.H. Rownd. 1976. Restriction endonuclease analysis of ribosomal DNA from *Saccharomyces cerevisiae*. *Mol. Gen. Genet.* **148**: 233.

deJonge, P., J. Klootwijk, and R.J. Planta. 1977a. Terminal nucleotide sequences of 17S ribosomal RNA and its immediate precursor 18S RNA in yeast. *Eur. J. Biochem.* **72**: 361.

———. 1977b. Sequence of the 3′ terminal 21 nucleotides of yeast 17S ribosomal RNA. *Nucleic Acids Res.* **4**: 3655.

deKloet, S.R. 1966. Ribonucleic acid synthesis in yeast. The effect of cycloheximide on the synthesis of ribonucleic acid in *Saccharomyces cerevisiae*. *Biochem. J.* **99**: 566.

Dujon, B. 1981. Mitochondrial genetics and functions. In *The molecular biology of the yeast Saccharomyces. I. Life cycle and inheritance* (ed. J. Strathern et al.), p. 505. Cold Spring Harbor Laboratory, Cold Spring Harbor, New York.

Elliott, S.G. and C.S. McLaughlin. 1978. The rate of macromolecular synthesis through the cell cycle of the yeast *Saccharomyces cerevisiae*. *Proc. Natl. Acad. Sci.* **75**: 4384.
———. 1979. Regulation of RNA synthesis in yeast. III. Synthesis during the cell cycle. *Mol. Gen. Genet.* **169**: 237.
Elliott, S.G., J.R. Warner, and C.S. McLaughlin. 1979. Synthesis of ribosomal proteins during the cell cycle of the yeast *Saccharomyces cerevisiae*. *J. Bacteriol.* **137**: 1048.
Fallon, A.M., C.S. Jinks, M. Yamamoto, and M. Nomura. 1979. Regulation of ribosomal protein synthesis in *Escherichia coli* by selective mRNA inactivation. *Proc. Natl. Acad. Sci.* **76**: 3411.
Fangman, W.L. and V.A. Zakian. 1981. Genome structure and replication. In *The molecular biology of the yeast* Saccharomyces. I. *Life cycle and inheritance* (ed. J. Strathern et al.), p. 27. Cold Spring Harbor Laboratory, Cold Spring Harbor, New York.
Frank, K.R. and D. Mills. 1978. Ribosome activity and degradation in meiotic cells of *Saccharomyces cerevisiae*. *Mol. Gen. Genet.* **160**: 59.
Fried, H. and J.R. Warner. 1981. Cloning of the yeast gene for trichodermin resistance and ribosomal protein L3. *Proc. Natl. Acad. Sci.* **78**: 238.
Fried, H.M., N.J. Pearson, C.H. Kim, and J.R. Warner. 1981. The genes for fifteen ribosomal proteins of *Saccharomyces cerevisiae*. *J. Biol. Chem.* **256**: 10176.
Gallis, B.M. and E.T. Young. 1975. Endogenous messenger RNA directed polypeptide chain elongation in a cell-free system from yeast: *Saccharomyces cerevisiae*. *J. Bacteriol.* **122**: 719.
Gasior, E., F. Herrera, C.S. McLaughlin, and K. Moldave. 1979a. The analysis of intermediary reactions involved in protein synthesis in a cell-free extract of *Saccharomyces cerevisiae* that translates natural messenger ribonucleic acid. *J. Biol. Chem.* **254**: 3970.
Gasior, E., F. Herrera, I. Sadnik, C.S. McLaughlin, and K. Moldave. 1979b. The preparation and characterization of a cell-free system from *Saccharomyces cerevisiae* that translates natural messenger ribonucleic acid. *J. Biol. Chem.* **254**: 3965.
Gorenstein, C. 1977. "Regulation of ribosomal protein synthesis in yeast." Ph.D. thesis, Albert Einstein College of Medicine, New York, New York.
Gorenstein, C.G. and J.R. Warner. 1976. Coordinate regulation of the synthesis of eukaryotic ribosomal proteins. *Proc. Natl. Acad. Sci.* **73**: 1547.
———. 1977. Synthesis and turnover of ribosomal proteins in the absence of 60S subunit assembly in *Saccharomyces cerevisiae*. *Mol. Gen. Genet.* **157**: 327.
———. 1979. The monocistronic nature of ribosomal protein genes in yeast. *Curr. Genet.* **1**: 9.
Gorini, L. 1974. Streptomycin and misreading of the genetic code. In *Ribosomes* (ed. M. Nomura et al.), p. 791. Cold Spring Harbor Laboratory, Cold Spring Harbor, New York.
Grankowski, N., W. Kudlicki, and E. Gasior. 1977. Phosphorylation of some acidic ribosomal proteins by cyclic AMP independent kinases from yeast. In *Symposium on translation of natural and synthetic polynucleotides* (ed. A. Legocki), p. 317. Poznan, Poland.
Grankowski, N., W. Kudlicki, E. Palen, and E. Gasior. 1976. Proteins of yeast ribosomal subunits: Number and general properties. *Acta Biochim. Pol.* **23**: 341.
Grant, P.G., L. Sanchez, and A. Jimenez. 1974. Cryptopleurine resistance: Genetic locus for a 40S ribosomal component in *Saccharomyces cerevisiae*. *J. Bacteriol.* **120**: 1308.
Grant, P.G., D. Schindler, and J.E. Davies. 1976. Mapping of trichodermin resistance in *Saccharomyces cerevisiae*: A genetic locus for a component of the 60S ribosomal subunit. *Genetics* **83**: 667.
Gross, K.J. and A.O. Pogo. 1974. Control mechanism of ribonucleic acid synthesis in eukaryotes. *J. Biol. Chem.* **249**: 568.
———. 1976. Control of ribonucleic acid synthesis in eukaryotes. 2. The effect of protein synthesis on the activities of nuclear and total DNA-dependent RNA polymerase in yeast. *Biochemistry* **15**: 2070.
Hadjiolov, A.A., M.D. Dabeva, K.P. Dudov, K.C. Gajdardjieva, O.I. Georgiev, N. Nikolaev, and B.B. Stoyanova. 1978. Control of ribosome RNA processing in eukaryotes. *FEBS Proc. Meet.* **51**: 319.

Hardy, S.J.S. 1975. The stoichiometry of the ribosomal proteins of *Escherichia coli. Mol. Gen. Genet.* **140**: 253.

Hartwell, L.H. 1967. Macromolecular synthesis in temperature sensitive mutants of yeast. *J. Bacteriol.* **93**: 1662.

———. 1970. Biochemical genetics of yeast. *Annu. Rev. Genet.* **4**: 373.

Hartwell, L.H. and C.S. McLaughlin. 1968a. Mutants of yeast with temperature-sensitive isoleucyl-tRNA synthetases. *Proc. Natl. Acad. Sci.* **59**: 422.

———. 1968b. Temperature-sensitive mutants of yeast exhibiting a rapid inhibition of protein synthesis. *J. Bacteriol.* **96**: 1664.

———. 1969. Mutants of yeast apparently defective in the initiation of protein synthesis. *Proc. Natl. Acad. Sci.* **62**: 468.

Hartwell, L.H., C.S. McLaughlin, and J.R. Warner. 1970. Identification of ten genes that control ribosome formation in yeast. *Mol. Gen. Genet.* **109**: 42.

Helser, T.L. and C.S. McLaughlin. 1975. Small ribonucleic acid molecules produced during ribosome biosynthesis in *Saccharomyces cerevisiae. J. Biol. Chem.* **250**: 2003.

Hereford, L. and M. Rosbash. 1977. Regulation of a set of abundant messenger RNA sequences. *Cell* **10**: 463.

Hernandez, F., M. Cannon, and J. Davies. 1978. Methylation of proteins in 40S ribosomal subunits from *Saccharomyces cerevisiae. FEBS Lett.* **89**: 271.

Higo, K. and E. Otaka. 1979. Isolation and characterization of fourteen ribosomal proteins from small subunits of yeast. *Biochemistry* **18**: 4191.

Hindley, J. and S.M. Page. 1972. Nucleotide sequence of yeast 5S ribosomal RNA. *FEBS Lett.* **26**: 157.

Hopper, A.K., F. Banks, and V. Evangelidis. 1978. A yeast mutant which accumulates precursor tRNAs. *Cell* **14**: 211.

Hopper, A.K., P.T. Magee, S.K. Welch, M. Friedman, and B.D. Hall. 1974. Macromolecular synthesis and breakdown in relation to sporulation and meiosis in yeast. *J. Bacteriol.* **119**: 619.

Hutchison, H.T., L.H. Hartwell, and C.S. McLaughlin. 1969. A temperature sensitive yeast mutant defective in RNA production. *J. Bacteriol.* **99**: 807.

Ishiguro, J. 1976. Proteins of yeast cytoplasmic ribosomes. *Mol. Gen. Genet.* **145**: 73.

Ishiguro, J., B.I. Ono, M. Masurekai, C.S. McLaughlin, and F. Sherman. 1981. Altered ribosomal protein S11 from the *SUP46* suppressor of yeast. *J. Mol. Biol.* **147**: 391.

Itoh, T. and B. Wittman-Liebold. 1978. The primary structure of protein 44 from the large subunit of yeast ribosomes. *FEBS Lett.* **96**: 399.

Itoh, T., K. Higo, and E. Otaka. 1979. Isolation and characterization of twenty-three ribosomal proteins from large subunits of yeast. *Biochemistry* **18**: 5787.

Itoh, T., K. Higo, E. Otaka, and S. Osawa. 1980. Studies on the primary structures of yeast ribosomal proteins. In *Genetics and evolution of RNA polymerase, tRNA and ribosomes* (ed. S. Osawa et al.), p. 609. University of Tokyo Press, Tokyo.

Jimenez, A., B. Littlewood, and J. Davies. 1972. In *Molecular mechanisms of antibiotic action on protein synthesis and membranes* (ed. E. Munoz et al.), p. 292. Elsevier, Amsterdam.

Johnston, G.C. and R.A. Singer. 1978. RNA synthesis and control of cell division in the yeast *S. cerevisiae. Cell* **14**: 951.

Kaltschmidt, E. and H.G. Wittman. 1970. Ribosomal proteins VII: Two dimensional polyacrylamide gel electrophoresis for fingerprinting of ribosomal proteins. *Anal. Biochem.* **36**: 401.

Kief, D.R. and J.R. Warner. 1981. Coordinate control of synthesis of ribosomal ribonucleic acid and ribosomal proteins during nutritional shift-up in *Saccharomyces cerevisiae. Mol. Cell. Biol.* **1**: 1007.

Klemenz, R. and E.P. Geiduschek. 1980. The 5' terminus of the precursor ribosomal RNA of *Saccharomyces cerevisiae. Nucleic Acids Res.* **8**: 2679.

Klootwijk, J. and R.J. Planta. 1973. Modified sequences in yeast ribosomal RNA. *Mol. Biol. Rep.* **1**: 187.

———. 1974. Analysis of the methylation sites in yeast ribosomal RNA. *Eur. J. Biochem.* **39:** 325.
Klootwijk, J., P. deJonge, and R.J. Planta. 1979. The primary transcript of the ribosomal repeating unit in yeast. *Nucleic Acids Res.* **6:** 27.
Kraig, E. and J.E. Haber. 1980. Messenger ribonucleic acid and protein metabolism during sporulation of *Saccharomyces cerevisiae. J. Bacteriol.* **144:** 1098.
Kramer, R.A., P. Philippsen, and R.W. Davis. 1978. Divergent transcription in the yeast ribosomal RNA coding region as shown by hybridization to separated strands and sequence analysis of cloned DNA. *J. Mol. Biol.* **123:** 405.
Kruiswijk, T. and R.J. Planta. 1974. Analysis of the protein composition of yeast ribosomal subunits by two dimensional acrylamide gel electrophoresis *Mol. Biol. Rep.* **1:** 409.
Kruiswijk, T., J.T. DeHay, and R.J. Planta. 1978a. Modification of yeast ribosomal proteins: Phosphorylation. *Biochem J.* **175:** 213.
Kruiswijk, T., R.J. Planta, and W.H. Mager. 1978b. Quantitative analysis of the protein composition of yeast ribosomes. *Eur. J. Biochem.* **83:** 245.
Kruiswijk, T., A. Kunst, R.J. Planta, and W.H. Mager. 1978c. Modification of yeast ribosomal proteins: Methylation. *Biochem. J.* **175:** 221.
Lemaux, P.G., S.L. Herendeen, P.L. Block, and F.C. Neidhardt. 1978. Transient rates of synthesis of individual polypeptides in *E. coli* following temperature shifts. *Cell* **13:** 427.
Lindahl, L. and J.M. Zengel. 1979. Operon-specific regulation of ribosomal protein synthesis in *Escherichia coli. Proc. Natl. Acad. Sci.* **76:** 6542.
Maden, B.E.H., J. Forbes, P. deJonge, and J. Klootwijk. 1975. Presence of a hypermodified nucleotide in HeLa cells 18S and *Saccharomyces carlsbergensis* 17S ribosomal RNAs. *FEBS Lett.* **59:** 60.
Mager, W.H., J. Retel, R.J. Planta, H.P.M.G. Bollen, C.H.F.V. deRegt, and H. Hoving. 1977. Transcriptional units for ribosomal proteins in yeast. *Eur. J. Biochem.* **78:** 575.
Matile, P., H. Moor, and C.F. Robinow. 1969. Yeast cytology. In *The yeasts* (ed. A.H. Rose and J.S. Harrison), p. 219. Academic Press, New York.
Maxam, A., R. Tizard, K.G. Skryabin, and W. Gilbert. 1977. Promoter region for yeast 5S ribosomal RNA. *Nature* **267:** 643.
Mayo, V.S., B.A.G. Andrean, and S.R. deKloet. 1968. Effects of cycloheximide and 5-fluorouracil on the synthesis of ribonucleic acid in yeast. *Biochim. Biophys. Acta* **169:** 297.
McConkey, E.H., H. Bielka, J. Gordon, S.M. Lastick, A. Lin, K. Ogata, J.-P. Reboud, J.A. Traugh, R.R. Traut, J.R. Warner, H. Welfle and I.G. Wool. 1979. Proposed uniform nomenclature for mammalian ribosomal proteins. *Mol. Gen. Genet.* **169:** 1.
McLaughlin, C.S. 1974. Yeast ribosomes: Genetics. In *Ribosomes* (ed. M. Nomura et al.), p. 815. Cold Spring Harbor Laboratory, Cold Spring Harbor, New York.
McLaughlin, C.S. and L.H. Hartwell. 1969. A mutant of yeast with a defective methionyl tRNA synthetase. *Genetics* **61:** 557.
McLaughlin, C.S., P.T. Magee, and L.H. Hartwell. 1969. Role of isoleucyl tRNA in RNA synthesis and enzyme repression in yeast. *J. Bacteriol.* **100:** 579.
Mets, L. and L. Bogorad. 1974. Two-dimensional polyacrylamide gel electrophoresis: An improved method for ribosomal proteins. *Anal. Biochem.* **57:** 200.
Miller, M.J., N.-H. Xuong, and E.P. Geiduschek. 1979. A response of protein synthesis to temperature shift in the yeast *Saccharomyces cerevisiae. Proc. Natl. Acad. Sci.* **76:** 5222.
Mitchison, J.M. 1971. *The biology of the cell cycle.* Cambridge University Press, Cambridge, England.
Miyazaki, M. 1974. Studies on the nucleotide sequences of pseudouridine containing 5S RNA from *Saccharomyces cerevisiae. J. Biochem.* **75:** 1407.
Mortimer, R. and D. Hawthorne. 1966. Genetic mapping in *Saccharomyces. Genetics* **53:** 165.
Nazar, R.N. and T.O. Sitz. 1980. Role of the 5' terminal sequence of the RNA binding site of yeast 5.8S rRNA. *FEBS Lett.* **115:** 71.

Nazar, R.N., M. Yaguchi, G.E. Willick, C.F. Rollin, and C. Roy. 1979. The 5S binding protein from yeast (*Saccharomyces cerevisiae*). *Eur. J. Biochem.* **102**: 573.
Oliver, S.G. and C.S. McLaughlin. 1977. The regulation of RNA synthesis in yeast. I. Starvation experiments. *Mol. Gen. Genet.* **154**: 145.
Otaka, E. and K. Kobata. 1978. Yeast ribosomal proteins: I. Characterization of cytoplasmic ribosomal proteins by two-dimensional gel electrophoresis. *Mol. Gen. Genet.* **162**: 259.
Otaka, E. and S. Osawa. 1980. Correlation of several nomenclatures for yeast ribosomal proteins: A proposal of standard nomenclature. *Mol. Gen. Genet.* **181**: 176.
Palmer, E., J.M. Wilhelm, and F. Sherman. 1978. Phenotypic suppression of nonsense mutants in yeast by aminoglycoside antibiotics. *Nature* **277**: 148.
Pao, C.C., J. Paietta, and J.A. Gallant. 1977. Synthesis of guanosine tetraphosphate (Magic Spot I) in *Saccharomyces cerevisiae*. *Biochem. Biophys. Res. Commun.* **74**: 314.
Pearson, N. 1979. "Regulation of ribosomal protein synthesis during sporulation in *Saccharomyces cerevisiae*." Ph.D. thesis, Brandeis University, Waltham, Massachusetts.
Pearson, N.J. and J.E. Haber. 1977. Changes in regulation of ribosome synthesis during different stages of the life cycle of *Saccharomyces cerevisiae*. *Mol. Gen. Genet.* **158**: 81.
―――. 1980. Changes in regulation of ribosomal protein synthesis during vegetative growth and sporulation of *Saccharomyces cerevisiae*. *J. Bacteriol.* **143**: 1411.
Peterson, N. and C.S. McLaughlin. 1973. Monocistronic messenger RNA in yeast. *J. Mol. Biol.* **81**: 33.
Petes, T.D. 1979. Yeast ribosomal DNA genes are located on chromosome XII. *Proc. Natl. Acad. Sci.* **76**: 410.
Philippsen, P., M. Thomas, R.A. Kramer, and R.W. Davis. 1978. Unique arrangement of coding sequences for 5S, 5.8S, 18S and 25S ribosomal RNA in *Saccharomyces cerevisiae* as determined by R-loop and hybridization analysis. *J. Mol. Biol.* **123**: 387.
Rao, S.S. and A.P. Grollman. 1967. Cycloheximide resistance in yeast: A property of the 60S ribosomal subunit. *Biochem. Biophys. Res. Commun.* **29**: 696.
Retel, J. and R.J. Planta. 1970. On the mechanism of the biosynthesis of ribosomal RNA in yeast. *Biochim. Biophys. Acta* **224**: 458.
Richter, D. and F. Klink. 1971. Isolation of peptide chain elongation factors from the yeast *Saccharomyces cerevisiae*. *Methods Enzymol.* **206**: 349.
Ritossa, F. 1962. A new puffing pattern induced by temperature shock and DNA in *Drosophila*. *Experientia* **18**: 571.
Roth, R.M. and C. Dampier. 1972. Dependence of ribonucleic acid synthesis on continuous protein synthesis in yeast. *J. Bacteriol.* **109**: 773.
Rubin, G.M. 1973. The nucleotide sequence of *Saccharomyces cerevisiae* 5.8S ribosomal ribonucleic acid. *J. Biol. Chem.* **248**: 3860.
―――. 1974. Three forms of the 5.8S ribosomal RNA species in *Saccharomyces cerevisiae*. *Eur. J. Biochem.* **41**: 197.
Rubtsov, P.M., M.M. Musakhanov, V.M. Zakharyev, A.S. Krayev, K.G. Skryabin, and A.A. Bayev. 1980. The structure of the yeast ribosomal RNA genes. I. The complete nucleotide sequence of the 18S ribosomal RNA gene from *Saccharomyces cerevisiae*. *Nucleic Acids Res.* **8**: 5779.
Sanchez, L., D. Vazquez, and A. Jimenez. 1977. Genetics and biochemistry of cryptopleurine resistance in the yeast *Saccharomyces cerevisiae*. *Mol. Gen. Genet.* **156**: 319.
Sanchez-Madrid, F. and J.P.G. Ballesta. 1979. An acidic protein associated to ribosomes of *Saccharomyces cerevisiae*. Changes during cell cycle. *Biochem. Biophys. Res. Commun.* **91**: 643.
Sanchez-Madrid, F., P. Conde, D. Vazquez, and J.P.G. Ballesta. 1979a. Acidic proteins from *Saccharomyces cerevisiae* ribosomes. *Biochem. Biophys. Res. Commun.* **87**: 381.
Sanchez-Madrid, F., R. Reyes, P. Conde, and J.P.G. Ballesta. 1979b. Acidic ribosomal proteins from eukaryotic cells: Effect on ribosomal functions. *Eur. J. Biochem.* **98**: 409.

Schubart, U.K., S. Shapiro, N. Fleischer, and O.M. Rosen. 1977. Cyclic adenosine 3′:5′-monophosphate-mediated insulin secretion and ribosomal protein phosphorylation in a hamster islet cell tumor. *J. Biol. Chem.* **252:** 92.
Sherman, F. and J.W. Stewart. 1975. The use of iso-1-cytochrome *c* mutants of yeast for elucidating the nucleotide sequences that govern initiation of translation. *FEBS Proc. Meet.* **38:** 175.
Shulman, R.W. and J.R. Warner. 1978. Ribosomal RNA transcription in a mutant of *S. cerevisiae* defective in ribosomal protein synthesis. *Mol. Gen. Genet.* **161:** 221.
Shulman, R.W., C.E. Sripati, and J.R. Warner. 1977. Non-coordinated transcription in the absence of protein synthesis in yeast. *J. Biol. Chem.* **252:** 1344.
Sillevis Smitt, W.W., J.M. Vlak, I. Molenaar, and T.H. Rozijn. 1973. Nucleolar function of the dense crescent in the yeast nucleus. A biochemical and ultrastructural study. *Exp. Cell Res.* **80:** 313.
Sillevis Smitt, W.W., J.M. Vlak, R. Schiphof, and T.H. Rozijn. 1972. Precursors of ribosomal RNA in yeast. *Exp. Cell Res.* **71:** 33.
Singer, R.A., G.C. Johnston, and D. Bedard. 1978. Methionine analogues and cell division regulation in the yeast *Saccharomyces cerevisiae*. *Proc. Natl. Acad. Sci.* **75:** 6083.
Skogerson, L. and E. Wakatama. 1976. A ribosome-dependent GTPase from yeast distinct from elongation factor 2. *Proc. Natl. Acad. Sci.* **73:** 73.
Skogerson, L., C. McLaughlin, and E. Wakatama. 1973. Modification of ribosomes in cryptopleurine-resistant mutants of yeast. *J. Bacteriol.* **116:** 818.
Skryabin, K.S., A.M. Maxam, and T.D. Petes. 1978. Location of the 5.8S rRNA gene of *Saccharomyces cerevisiae*. *J. Bacteriol.* **134:** 306.
Sogin, S.J., B.L.A. Carter, and H.O. Halvorson. 1974. Changes in the rate of ribosomal RNA synthesis during the cell cycle of *Saccharomyces cerevisiae*. *Exp. Cell Res.* **89:** 127.
Stafford, M.E. and C.S. McLaughlin. 1973. Trichodermin, a possible inhibitor of the termination process of protein synthesis in yeast. *J. Cell. Physiol.* **82:** 121.
Stocklein, W. and W. Piepersberg. 1980. Altered ribosomal protein L29 in cycloheximide-resistant strain of *Saccharomyces cerevisiae*. *Curr. Genet.* **1:** 177.
Taber, R.L. and W.S. Vincent. 1969. The synthesis and processing of ribosomal RNA precursor molecules in yeast. *Biochim. Biophys. Acta* **186:** 317.
Thomas, G., M. Siegmann, and J. Gordon. 1979. Multiple phosphorylation of ribosomal protein S6 during transition of quiescent 3T3 cells into early G_1, and cellular compartmentalization of the phosphate donor. *Proc. Natl. Acad. Sci.* **76:** 3952.
Trapman, J. and R.J. Planta. 1975a. On the biosynthesis of 5.8S ribosomal RNA in yeast. *FEBS Lett.* **57:** 26.
———. 1975b. Detailed analysis of the ribosomal RNA synthesis in yeast. *Biochim. Biophys. Acta* **414:** 115.
Trapman, J., J. Retel, and R.J. Planta. 1975. Ribosomal precursor particles from yeast. *Exp. Cell Res.* **90:** 95.
Tuite, M.F., J. Plesset, K. Moldave, and C.S. McLaughlin. 1980. Faithful and efficient translation of homologous and heterologous mRNAs in an mRNA dependent cell free system from *S. cerevisiae*. *J. Biol. Chem.* **255:** 8761.
Udem, S.A. and J.R. Warner. 1972. Synthesis and processing of ribosomal RNA in *Saccharomyces cerevisiae*. *J. Mol. Biol.* **65:** 227.
———. 1973. The cytoplasmic maturation of a ribosomal precursor RNA in yeast. *J. Biol. Chem.* **248:** 1412.
Udem, S.A., K. Kaufman, and J.R. Warner. 1971. Small ribosomal ribonucleic acid species of *Saccharomyces cerevisiae*. *J. Bacteriol.* **105:** 101.
Ursic, D. and J. Davies. 1979. A cold-sensitive mutant of *Saccharomyces cerevisiae* defective in ribosome processing. *Mol. Gen. Genet.* **175:** 313.
Valenzuela, P., G.I. Bell, A. Venegas, E.T. Sewell, F.R. Masiarz, L.J. DeGennaro, F. Wein-

berg, and W.J. Rutter. 1977. Ribosomal RNA genes of *Saccharomyces cerevisiae*. II. Physical map and nucleotide sequence of the 5S ribosomal RNA gene and adjacent intergenic regions. *J. Biol. Chem.* **252:** 8126.

Vaughan, M.H., R. Soeiro, J.R. Warner, and J.E. Darnell. 1967. Effects of methionine deprivation on ribosome synthesis in HeLa cells. *Proc. Natl. Acad. Sci.* **53:** 1527.

Vazquez, D. 1979. *Inhibitors of protein synthesis.* Springer-Verlag, New York.

Veldman, G.M., R.C. Brand, J. Klootwijk, and R.J. Planta. 1980. Some characteristics of processing sites in ribosomal precursor RNA of yeast. *Nucleic Acids Res.* **8:** 2907.

Venkov, P.V. and A.P. Vasileva. 1979. *Saccharomyces cerevisiae* mutants defective in the maturation of ribosomal RNA. *Mol. Gen. Genet.* **173:** 203.

Waldron. C. 1977. Synthesis of ribosomal and transfer ribonucleic acid in yeast during a nutritional shift up. *J. Gen. Microbiol.* **98:** 215.

Waldron, C. and F. Lacroute. 1975. Effect of growth rate on the amounts of ribosomal and transfer ribonucleic acids in yeast. *J. Bacteriol.* **122:** 855.

Warner, J.R. 1974. The assembly of ribosomes in eukaryotes. In *Ribosomes* (ed. M. Nomura et al.), p. 461. Cold Spring Harbor Laboratory, Cold Spring Harbor, New York.

Warner, J.R. and C.G. Gorenstein. 1977. Synthesis of eukaryotic ribosomal proteins *in vitro*. *Cell* **11:** 201.

―――. 1978a. The ribosomal proteins of *Saccharomyces cerevisiae*. *Methods Cell Biol.* **20:** 45.

―――. 1978b. Yeast has a true stringent response. *Nature* **275:** 338.

Warner, J.R. and S.A. Udem. 1972. The effects of *ts* mutations in ribosome synthesis in yeast. *J. Mol. Biol.* **65:** 243.

Wehr, C.T. and L.W. Parks. 1969. Macromolecular synthesis in *Saccharomyces cerevisiae* in different growth media. *J. Bacteriol.* **98:** 458.

Wei, C., I.M. Campbell, C.S. McLaughlin, and M.H. Vaughan. 1974. Binding of trichodermin to mammalian ribosomes and its inhibition by other 12, 13-epoxytrichothecenes. *Mol. Cell. Biochem.* **3:** 215.

Wittmann, H.G., G. Stoffler, I. Hindennach, C.G. Kurland, L. Randall-Hazelbauer, E.A. Birge, M. Nomura, E. Kaltschmidt, S. Mizushima, R.R. Traut, and T.A. Bickle. 1971. Correlation of 30S ribosomal proteins of *Escherichia coli* isolated in different laboratories. *Mol. Gen. Genet.* **111:** 327.

Woolford, J.L., L.M. Hereford, and M. Rosbash. 1979. Isolation of cloned DNA sequences containing ribosomal protein genes from *Saccharomyces cerevisiae*. *Cell* **18:** 1247.

Zinker, S. and J.R. Warner. 1976. The ribosomal proteins in *Saccharomyces cerevisiae*: Phosphorylated and exchangeable proteins. *J. Biol. Chem.* **251:** 1799.

Yeast Nuclear RNA Polymerases and Their Role in Transcription

André Sentenac
Service de Biochimie
C.E.N. Saclay
91191 Gif-sur-Yvette Cedex, France

Benjamin Hall
Department of Genetics
University of Washington
Seattle, Washington 98195

1. **Introduction and Terminology**
   A. Three Forms of Yeast Nuclear RNA Polymerases
   B. Nomenclature
2. **Purification of Enzymes A, B, and C**
   A. Yeast Strain
   B. Growth Conditions
   C. Cell Disruption
   D. Enzyme Alterations during Purification
   E. Large-scale Separation of the Three RNA Polymerases
   F. Small-scale Purification
3. **Molecular Properties and Polypeptide Composition**
   A. Polypeptide Composition
   B. Are the Polypeptide Components of a Given Enzyme Distinct Gene Products?
4. **Structural and Phylogenetic Relationships of the Three Forms of RNA Polymerase**
   A. Common Subunits
   B. Evolution of RNA Polymerase Genes
5. **Functional Role of Individual Polypeptide Chains**
   A. Partial Enzyme Dissociation
   B. Probes for Subunit Function
   C. Mutants
   D. Enzyme-A-associated RNase-H Activity
   E. Satellite Proteins
6. **General Catalytic Properties**
   A. Substrates and Reactions
   B. Inhibitors
7. **Regulation of RNA Polymerases**
8. **Function of RNA Polymerases**
9. **In Vitro Transcription Specificity**
   A. RNA Polymerase I
   B. RNA Polymerase II
   C. RNA Polymerase III
10. **DNA Elements Involved in Transcription Initiation and Termination**
    A. Transcription Initiation by RNA Polymerase II
    B. Transcription Termination by RNA Polymerase II and Poly(A) Addition

C. Specification of Transcription Initiation by RNA Polymerase III
D. Antisuppressor Mutants at the *SUP4* Locus
E. Transcription Termination by RNA Polymerase III
11. **Conclusions**
A. Yeast Transcription Compared to That in Other Eukaryotes
B. Future Prospects

INTRODUCTION AND TERMINOLOGY

The first eukaryotic RNA polymerase to be extensively purified was a yeast RNA polymerase by Frederick et al. (1969).

At the same time, the more definitive general description of eukaryotic RNA polymerase was initiated by the solubilization and resolution of the mammalian enzymes (Roeder and Rutter 1969). Afterward, there were many investigations to demonstrate the multiplicity of DNA-dependent RNA polymerases in different eukaryotic cells, including yeast.

Now, after more than a decade of research, the existence in eukaryotes of three forms of nuclear RNA polymerases is well established, on the basis of structural and, more recently, functional criteria. Studies on yeast RNA polymerases have considerably contributed to our knowledge of the eukaryotic transcription machinery. For practical reasons, among eukaryotic RNA polymerases, the yeast enzymes have been the most extensively studied at the structural level. On the other hand, specificity studies with yeast RNA polymerases have lagged behind their structural analysis, due primarily to the lack of simple defined templates, such as the viral DNAs that function in animal cells and the absence of information on in vivo transcripts. The advent of gene cloning in yeast has now largely compensated for this difficulty. The purpose of this paper is, first, to present the available information on the structure of the three forms of nuclear RNA polymerases from *Saccharomyces cerevisiae* and then to discuss various aspects of their function, including features of the DNA structure that govern transcription specificity.

Three Forms of Yeast Nuclear RNA Polymerases

The strategy that originally succeeded in the resolution of three distinct enzyme activities from animal cell nuclei involved high salt extraction of proteins and sonication (Roeder and Rutter 1969). When applied to yeast whole cells disrupted by glass beads, the same method allowed the separation of three main chromatographic forms of RNA polymerase on DEAE-Sephadex (Ponta et al. 1971; Adman et al. 1972). Occasionally, additional minor peaks of activity were resolved (Adman et al. 1972), although not consistently (Tipper 1973; Schultz and Hall 1976).

The nuclear origin of these activities is not as well documented as that for the corresponding animal enzymes because yeast nuclei are not easily prepared. To the extent that enzymes have been extracted from purified yeast nuclear preparations, the best indications are that all three of the activities observed in whole-cell extracts are present in yeast nuclei (Ponta et al. 1971; Sebastian et al. 1973a; Schultz 1978; Wandzilak and Benson 1978).

Nomenclature

The literature on mammalian RNA polymerases is frequently confusing to scientists working outside this field because there are two different systems of enzyme nomenclature. The designations I, II, and III were originally proposed by Roeder and Rutter (1969), primarily on the basis of enzyme chromatographic properties, whereas Chambon (1974) and the majority of European laboratories have used the system A, B, and C, on the basis of differential α-amanitin sensitivity of the three RNA polymerases. Fortunately, for yeast RNA polymerases, the two systems are congruent, and one can deduce from comparisons of protein subunit patterns of enzymes I, II, and III (Hager et al. 1976) with A, B, and C (Sentenac et al. 1976) that yeast RNA polymerase A is I, B is II, and C is III. In this paper we adhere to the A, B, C system for describing subunit structure and other protein properties. Each polypeptide component is identified by a letter corresponding to the enzyme from which it derives and by a subscript corresponding to its molecular weight $\times 10^{-3}$. We use the I, II, III system in describing the activities associated with the three proteins.

PURIFICATION OF ENZYMES A, B, AND C

It is not the purpose of this paper to describe in detail purification methods, but it may be of interest to comment on some specific problems concerning the selection of the yeast strain and methods for cell disruption, as well as possible alterations of enzymes during purification. A procedure is cited for purification of each of the three RNA polymerases from *S. cerevisiae*.

Yeast Strain

The selection of the most appropriate yeast strain has not been considered as an important parameter in most of the early work and even in more recent experiments. So, a variety of ill-defined strains obtained from commercial sources have been used (Hager et al. 1977; Valenzuela et al. 1978a; Wandzilak and Benson 1978). It would be advisable to agree on a standard strain and growth conditions, so that in interpreting varying results, one would at least exclude the possibility of strain variations or growth conditions. Fortunately, there were no apparent differences in the polypeptide pattern of homologous enzymes obtained from different *Saccharomyces* strains (see Fig. 1). On the other hand, enzymes from different yeast species can differ considerably in terms of number and molecular weight of their polypeptide components (Fig. 1). In view of the fact that the largest subunit of enzyme B is readily proteolyzed during purification (see below), a protease-deficient strain would be most helpful. Using a strain with the *pep4-3* mutation (pleitropic A⁻, B⁻, C⁻ proteinases and RNase deficient [Zubenko and Jones 1981]) has improved the recovery of unproteolyzed enzyme B. Other related strains could even be more favorable. Furthermore, for specificity studies, RNase-deficient strains such as *pep4-3* would be preferred.

Growth Conditions

There are reasons for growing the cells in a rich medium and collecting them in the exponential phase of growth to optimize the yield of enzymes and reproducibility of the purification. In cases where the cells are of commercial origin, they may be energy-source-limited to less than 10% of their maximal growth rate (Hager et al. 1977). As discussed later in the text, the number of RNA polymerase-A molecules per cell varies with the growth rate and is much higher when cells are grown in a rich medium. On the other hand, the cell content in enzyme B appears independent of the growth conditions. Furthermore, growth into stationary phase may induce the production of undesirable proteases, nucleases, or other inhibitors. By using synthetic

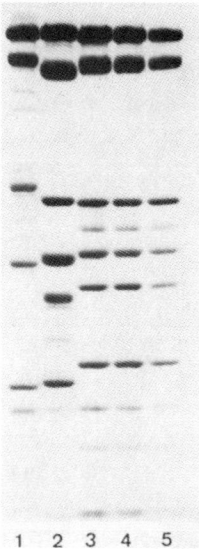

Figure 1 RNA polymerase A was purified from different yeast species, and the polypeptide composition of these enzymes was compared by electrophoresis on a polyacrylamide slab gel with SDS: *(1)Schizosaccharomyces pombe*; *(2) Candida tropicalis;* *(3) S. carlsbergensis;* *(4) S. cerevisiae* 4094B; *(5) S. cerevisiae* FL100. The molecular weights of the polypeptides from enzyme A of *S. cerevisiae* 4094B are given in Table 2.

templates to minimize the interference with nucleases, it was found that the activity of RNA polymerase B was identical in extracts from exponential or stationary-phase cells, whereas the activity of A + C enzymes was a little higher in exponentially growing cells (Ruet et al. 1978).

Cell Disruption

Solubilization of animal RNA polymerases requires sonication of the cell extract in high salt. Since yeast cells are relatively difficult to break, harsh mechanical treatments have to be used. The two methods most commonly used are (1) high-pressure disruption of the cell suspension in a Manton-Gaulin homogenizer (Dezélée et al. 1972) and (2) high-pressure disruption of the frozen cells in an Eaton press (Valenzuela et al. 1976d). The first method can handle almost unlimited amounts of yeast cells in a continuous way. The disruption of the frozen material is more gentle. In fact, intact yeast nuclei can be obtained with a modified French pressure cell (Sebastian et al. 1973a). Other methods like spheroplasting, grinding with alumina, and shaking with glass beads can be used for small-scale purifications.

Enzyme Alterations during Purification

Two kinds of enzyme structural alterations were encountered during purification: (1) proteolysis of subunits and (2) dissociation of some polypeptides. The first instance reported was the dissociation of two polypeptides from enzyme A during phosphocellulose chromatography or electrophoresis (Huet et al. 1975; Valenzuela et al. 1976d). The simplified form of the enzyme, A*, exhibits a number of altered properties (Huet et al. 1975). The same phenomenon occurs with enzyme B, which gives rise, upon DEAE-Sephadex chromatography or polyacrylamide gel electrophoresis, to an enzyme variant called B*, devoid of two polypeptide components (Dezélée et al. 1976; Ruet et al. 1980). Similarly, enzyme C loses three components upon electrophoresis on polyacrylamide gels (Valenzuela et al. 1976c). In all of these cases, the complete enzyme and its simplified form can be separated by chromatography or gel electrophoresis. The second type of alteration is due to in vitro proteolysis, especially in the case of enzyme B. The enzyme initially obtained by large-scale purification was the proteolyzed form (Dezélée et al. 1972; Valenzuela et al. 1976c; Hager et al. 1977). This in vitro proteolysis can be partly prevented by controlling the pH of the extract, by adding inhibitors of serine proteases (Dezélée et al. 1976), and by using a yeast strain with the *pep4-3* mutation.

Large-scale Separation of the Three RNA Polymerases

Purification methods are now available for obtaining milligram amounts of pure enzyme A in a few steps (Buhler et al. 1974; Huet et al. 1975; Valenzuela et al. 1976d), enzyme B (Dezélée and Sentenac 1973; Dezélée et al. 1976), as well as enzyme C (Valenzuela et al. 1976b; Wandzilak and Benson 1978) in good yield. Some methods allow the simultaneous purification of the three forms of enzymes (Hager et al. 1977; Valenzuela et al. 1978a) but include more steps per enzyme, which can be detrimental to enzyme integrity (Hager et al. 1977). Separation of the three enzymes can be achieved by chromatography on DEAE-Sephadex but not on DEAE-cellulose, which does not resolve enzymes A and C. These two enzymes are well separated on denatured DNA-cellulose or phosphocellulose (Table 1).

Table 1 Properties that distinguish between RNA polymerases A, B, and C

	RNA polymerase form		
	A	B	C
General catalytic properties			
specific activity	750	3000	1500
best homopolymer template	$d(I-C)_n$	$d(I-C)_n$, $(rC)_n$	$d(A-T)_n$
divalent cation requirement	Mg^{++} or Mn^{++}	Mn^{++}	Mg^{++} or Mn^{++}
ammonium sulfate optima (M)	0.025 (low)	0.05 (intermediate)	0.05–0.25 (biphasic)
Sensitivity to inhibitors			
α-amanitin (μg/ml for 50% inhibition)	300 (form A*) 600 (form A)	1(form B_I, B_{II})	insensitive
3′ dATP (K_i)	> 3 μM	0.3 μM	3 μM
Chromatographic properties (elution by ammonium sulfate) (M)			
DEAE-Sephadex	0.15 (A) 0.13 (A*)	0.22 ($B*_I$) 0.26 (B_I,B_{II})	0.3
DEAE-cellulose	0.12 (A,A*)	0.28	0.12
denatured DNA-cellulose (M KCl)	0.25	0.18	0.45
phosphocellulose	0.25 (A) 0.19 (A*)	0.12	0.17

Specific activity is given as nmoles of UMP incorporated per hour per milligram of protein, with a mixture of native and denatured calf thymus DNA as template, Mn^{++} ions, and 0.05 M ammonium sulfate (data from Hager et al. [1977]). The elution positions indicated are approximate, as they varied in the different reports. For references, see text. RNA polymerase forms A*, B_I, B_{II}, and B_I* are defined in Table 2.

Small-scale Purification

Methods for purification of yeast RNA polymerases from small amounts of cells are also available to investigate regulation of enzyme levels as a function of growth rate along the cell cycle or during physiological transitions. First, the three enzymes can be separated from a crude extract on DEAE-Sephadex, and their activity, quantitated (Adam et al. 1972; Gross and Pogo 1974). This approach is complicated by the fact that the activity measurements may be influenced by the presence of varying interfering components (proteases, nucleases, phosphatases, stimulatory factors), which themselves can be affected by the physiological transition under study. More recently, small-scale purification procedures have been developed for all three forms of RNA polymerase, starting from 30 g of spheroplasted cells (Bell et al. 1977). In addition, ^{32}P- or ^{35}S-labeled RNA polymerases A or B can be obtained in a homogeneous form starting from 2 g of whole cells (Huet et al. 1975; Buhler et al. 1976). Immunoprecipitation has also been used to obtain enzyme A (Buhler et al. 1976), as well as enzyme B$_I$ (Sentenac et al. 1978). The enzymes are pure enough for their polypeptide content to be visualized by SDS-gel electrophoresis. Immunodetection could be an excellent method to investigate variations in enzyme levels. However, because the three RNA polymerases are immunologically related (see below), it is necessary to use antibodies to unrelated subunits or to separate the enzymes by a preliminary purification step.

MOLECULAR PROPERTIES AND POLYPEPTIDE COMPOSITION

Yeast RNA polymerases are large multimeric proteins of high molecular weight, in the range of 400,000–600,000. Enzyme A sediments faster than the B enzyme (Dezélée et al. 1972; Ponta et al. 1972). The enzyme has a $S_{20,w}$ value of 16.2, consistent with a molecular weight of 650,000 (Valenzuela et al. 1976d). Enzyme B and *Escherichia coli* RNA polymerase can be co-sedimented in a glycerol gradient and assayed independently by virtue of their differential sensitivities to α-amanitin and rifampicin. The yeast enzyme sediments faster than the bacterial holoenzyme (M_r 480,000), at about 13S–17S at high ionic strength. The sedimentation coefficient increases significantly in low salt at 21S–24S, suggesting some protein aggregation (Dezélée et al. 1972). By electrophoresis in polyacrylamide gels of increasing porosity, the molecular weight of RNA polymerase B$_I$ (native, nonproteolyzed) was estimated as 465,000 ± 10% (Dezélée et al. 1976), which is smaller than that of the bacterial holoenzyme. This apparent discrepancy is explained by the dissociation of two polypeptides from the enzyme during electrophoresis (M_r 32,000 and M_r 16,500) (Dezélée et al. 1976). There is little data on enzyme C. It is reported to sediment, in glycerol gradient at high ionic strength, with an apparent molecular weight of 380,000 (Wandzilak and Benson 1978).

All three forms of RNA polymerase contain tightly bound zinc atoms, as shown by microwave excitation spectrometry and by inhibition by derivatives of 1,10-phenanthroline, chelating agents specific for zinc. Enzyme A contains 2 atoms of zinc (Auld et al. 1976), enzyme B contains 1 atom of zinc (Lattke and Weser 1976), and enzyme C contains 2 tightly bound zinc atoms (Wandzilak and Benson 1978).

Polypeptide Composition

The polypeptide content of the purified enzymes has been analyzed by polyacrylamide gel electrophoresis in the presence of SDS. All enzyme preparations have a complex polypeptide content, always with two high-molecular-weight components in the range of 130,000 to 200,000 and a collection of smaller proteins (Dezélée et al. 1972; Ponta et al. 1972; Buhler et al. 1974; Valenzuela et al. 1976c). Since the complexity of each of the three yeast enzymes is higher than that of the unique bacterial RNA polymerase, in terms of number and size of the putative subunits, a great deal of effort has been directed toward demonstrating that all of these polypeptides are part of the enzyme structure. Indeed, most of the polypeptides cosediment with the enzyme activity and migrate with the enzyme by electrophoresis in nondenaturing polyacrylamide gel. The polypeptide composition of the enzymes is essentially based on these two operational criteria (Table 2).

Enzyme A

Enzyme A comprises 13 polypeptides: A_{190}, A_{135}, A_{49}, A_{43}, A_{40}, $A_{34.5}$, A_{27}, A_{23}, A_{19}, $A_{14.5}$, A_{14}, $A_{12.2}$, and A_{10} (Huet et al. 1975). The fact that an identical polypeptide content was independently reported by Valenzuela et al. (1976b) supports the view that all of these components are bona fide subunit candidates. However, this argument is weak inasmuch as the two purification procedures differ only in the final steps. On the other hand, using a different purification strategy, Hager et al. (1977) found a simpler molecular structure, missing polypeptides A_{49} and $A_{44.5}$. Additionally, immunological techniques have been used to identify the polypeptides that are an integral part of the enzyme structure. When enzyme A is immunoprecipitated with antibodies to its largest subunit, A_{190}, all of the polypeptides are coprecipitated with A_{190}, with the exception of $A_{34.5}$ (Huet et al. 1975). This component, as well as A_{49}, is not as closely associated with the enzyme as the other polypeptides. Hence, a simplified form of enzyme A can be resolved by electrophoresis (Huet et al. 1975; Valenzuela et al. 1976d) or by chromatography on phosphocellulose (Huet et al. 1975). This enzyme variant, depleted of A_{49} and $A_{34.5}$, is called A*. Chelators with a high affinity for zinclike lomofungin and dithizone (Ruet et al. 1975), as well as mild urea treatment (Huet et al. 1977) or acid pH (Bull et al. 1981b), favor the dissociation of enzyme A into A*.

Table 2 Polypeptides of RNA polymerases A, B, and C from *Saccharomyces cerevisiae*

RNA polymerase form					
A	A*	B_I	B_{II}	B_I*	C
		220	220		
190	190				
			185		
					160
		150	150	150	
135	135				
					128
					82
					53
49					
		44.5	44.5	44.5	
43	43				
40	40				40
					37
34.5					
					34
		32	32		
27	27	27	27	27	27
23	23	23	23	23	23
19	19				19
		16	16		
14.5	14.5	14.5	14.5	14.5	14.5
14	14				
		12.6	12.6	12.6	
12.2	12.2				
					11
10	10	10	10	10	

Numbers indicate molecular weights $\times 10^{-3}$ as determined by electrophoresis on polyacrylamide gel in the presence of SDS. The data for enzyme A and A* are from Huet et al. (1975), data for enzyme B are from Dezélée et al. (1976), and data for enzyme C are from Valenzuela et al. (1978a). In some cases, the values differ slightly from previously reported values. For further details and references, see text.

Enzyme B

The following polypeptides cosediment with the nonproteolyzed form B_I: B_{220}, B_{150}, $B_{44.5}$, B_{32}, B_{27}, B_{23}, $B_{16.5}$, $B_{14.5}$, $B_{12.6}$, and B_{10} (Dezélée et al. 1976; Bell et al. 1977). The same components migrate with the enzyme by electrophoresis, except for B_{32} and $B_{16.5}$, which dissociate during electrophoresis (Dezélée et al. 1976). By analogy with the A → A* transition, the simplified form of enzyme B is called B*. The dissociation of these two polypeptides is accentuated by mild urea treatment, following which enzymes B and B* can be resolved by chromatography on DEAE-Sephadex (Ruet et al. 1980). The proteolyzed form B_{II} is altered at the level of two polypeptides: B_{220} gives rise by proteolysis to B_{185}, whereas $B_{44.5}$ is often replaced by B_{40} (Dezélée et al. 1976; Sentenac et al. 1976). As mentioned previously, in the early studies, only the B_{II} form was found. The $B_I \rightarrow B_{II}$ conversion can be reproduced by a protease extract from *S. cerevisiae* (Dezélée et al. 1976). Immunoprecipitation of enzyme B from a yeast crude extract yields exclusively form B_I in spite of the fact that the antibodies were directed against form B_{II} (proteolyzed) (Sentenac et al. 1978). Therefore, it is likely that form B_{II} has no physiological significance.

Enzyme C

Variable polypeptide compositions have been reported for enzyme C. The invariant polypeptide content of enzyme C purified on sucrose gradient is C_{160}, C_{128}, C_{82}, C_{40}, C_{34}, C_{27}, C_{23}, C_{19}, $C_{14.5}$, and C_{11} (Valenzuela et al. 1976b,c; Hager et al. 1977). Other polypeptides (C_{53}, $C_{40.5}$, and C_{37}) are not found consistently (Valenzuela et al. 1976c; Hager et al. 1976; 1977) and have sometimes been considered contaminants (Valenzuela et al. 1976b). Like enzymes A and B, enzyme C is resolved into two components by electrophoresis under nondenaturing conditions (Valenzuela et al. 1976b). One component retains the full complement of polypeptides, and the other lacks C_{82}, C_{37}, and C_{34}. It is not known whether this enzyme heterogeneity preexists or is induced by electrophoresis.

Molar ratios of the different polypeptides within each enzyme have been derived solely from Coomassie-blue staining of the protein bands in the polyacrylamide gels. Although these data should be considered as preliminary, it is nonetheless reassuring that the ratio of a given polypeptide chain relative to either of the largest two polypeptides is always close to one. The only exceptions are ABC_{27}, which is possibly represented twice (Hager et al. 1976), and B_{220} and $B_{44.5}$, which can be proteolyzed to varying degrees (Dezélée et al. 1976).

Additional data on the biochemical properties of different polypeptides of yeast RNA polymerases have been accumulated. The majority of the small polypeptides are acidic, with isolectric points in the range of 5. AB_{27} and A_{49} are exceptions, with isoelectric points of 9 (Buhler et al. 1976). The large subunits are slightly acidic with isoelectric points in the range of 6. The proteolyzed subunit B_{185} is less acidic than B_{220} (Sentenac et al. 1976).

Five polypeptide chains are phosphorylated in vivo in enzyme A to varying degrees: A_{190}, A_{43}, $A_{34.5}$, A_{23}, and A_{19} (Bell et al. 1976, 1977; Buhler et al. 1976). The ^{32}P label is recovered as phosphoserine and phosphothreonine after partial acid hydrolysis. Enzyme B is phosphorylated in vivo on subunits B_{220}, B_{23} and, possibly, $B_{44.5}$ (Buhler et al. 1976). The phosphorylation of subunit B_{220} was evidenced by two-dimensional gel electrophoresis and by immunoprecipitation of B_I enzyme from a ^{32}P-labeled cell extract (Sentenac et al. 1978). Enzyme C is phosphorylated in vivo on polypeptides C_{23} and C_{19} and also possibly on C_{53} (Bell et al. 1977).

This simple compilation of polypeptide molecular weight is very instructive, and several conclusions can be drawn at this point. (1) It is obvious at the molecular level that the transcription system in the yeast nucleus like that in animal cells, contains three distinct RNA polymerases (Roeder 1976). (2) Calculated from their polypeptide composition, assuming the presence of one polypeptide of each kind, the molecular weights of enzymes A, B_I, and C are 600,000, 540,000, and 390,000, respectively, with an error margin in each case of 10%. (3) This molecular complexity suggests that each form of RNA polymerase may be extensively regulated.

Are the Polypeptide Components of a Given Enzyme Distinct Gene Products?

This question arises in view of the enormous complexity of all three RNA polymerases, which comprise altogether 30 components (see Table 2). There is a possibility that some polypeptides derive from larger components by partial proteolysis. It has also been suggested that the multiplicity of eukaryotic RNA polymerases could result from the maturation of one unique enzyme. Except for the existence of common subunits (discussed in the following section), the available biochemical and immunological evidence does not support this view. Antibodies to the different polypeptide components of yeast RNA polymerase A react specifically, with minor exceptions, with the corresponding subunit (Buhler et al. 1980). The fingerprint patterns of several polypeptides of enzyme A or B are different (Buhler et al. 1976). Furthermore, the large subunits of enzyme B, B_{220} and B_{150}, yield a different pattern of peptides after partial proteolysis, whereas the same technique demonstrates that the B_{185} subunit derives from B_{220} (Ruet et al. 1980). There remains the possibility that some of the small components stem from the rapid, total and specific cleavage of a larger polypeptide. Obviously, such cleavage products would carry different immunological determinants and yield different fingerprints. This possibility appears unlikely for enzymes A and B, since they retain the same polypeptide composition when rapidly isolated from a crude extract by immunoprecipitation (Buhler et al. 1976; Sentenac et al. 1978). Therefore, it seems reasonable to conclude that the different polypeptide components of a given enzyme are distinct proteins and the products of different genes. Taking into account the common subunits, this gives the impressively large total of 22–25 distinct polypeptides in the three isoforms of RNA polymerase.

How the genes for each form of RNA polymerase are organized within the nuclear genome is a matter of speculation. To gain some insight into this problem, one can take advantage of the natural variation in the molecular structure of RNA polymerase A as seen in Figure 1. Some subunits of enzyme A from *S. cerevisiae* and *S. douglasii* differ slightly in molecular weight (A_{49}, A_{43}, A_{40}, $A_{34.5}$, A_{19}, and $A_{14.5}$), and therefore can be identified by electrophoresis on polyacrylamide gel (Fig. 2A). RNA polymerase A isolated from the diploid hybrid, derived by crossing the two species, contains all of the subunits characteristic of the two parents. One meiotic segregant (4203-sp52) contained a hybrid RNA polymerase A with five of the polymorphic polypeptides coming from *S. douglasii* and two from *S. cerevisiae* (the other cognate polypeptides could not be distinguished on the basis of molecular weight). In three successive backcrosses with *S. cerevisiae*, the genes for the *S. douglasii* subunits were found to recombine. One tetrad is shown in Figure 2B. RNA polymerase A from a strain disomic for chromosome XV showed the two polypeptides A_{43} characteristic of the two parents. This approach, although limited to those subunits exhibiting interspecies differences, might allow the identification of the chromosomes that harbor the genes for several of the RNA polymerase subunits. The results already suggest that the genes for several of the RNA polymerase-A subunits are located on different chromosomes (Riva et al. 1982).

STRUCTURAL AND PHYLOGENETIC RELATIONSHIPS OF THE THREE FORMS OF RNA POLYMERASE

Common Subunits

The presence of common subunits in yeast enzymes A, B, and C is well established. It was originally suggested by the finding of polypeptide chains having identical molecular weight in enzymes A and B (Buhler et al. 1974). Hildebrandt et al. (1973) were the first to show the antigenic homology of A and B RNA polymerases. This observation was confirmed using purified and well-characterized enzymes A and B (Buhler et al. 1976). Other authors did not detect a cross-reactivity between enzymes A and B but only between enzymes A and C (Valenzuela et al. 1976b). These different results probably reflect variations in the antibody preparations.

The existence of common subunits is now based on a variety of criteria. On the basis of identical molecular weight, the putative common subunits are AC_{40}, ABC_{27}, ABC_{23}, AC_{19}, and $ABC_{14.5}$ (see Table 2) (Buhler et al.

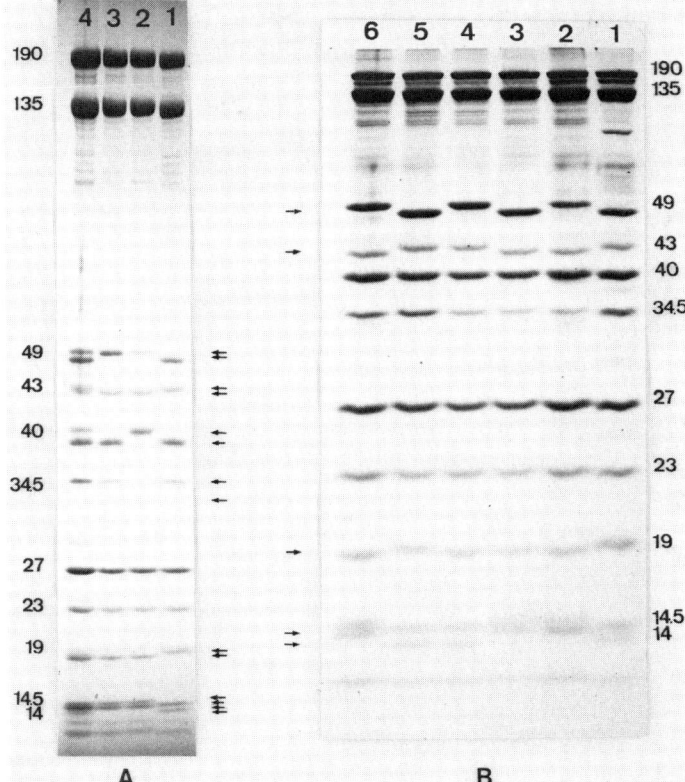

Figure 2 Polypeptide pattern of RNA polymerase A from *S. cerevisiae, S. douglasii,* interspecific hybrid, a segregant *(A),* and tetrad analysis *(B).* *(A)* Subunits of RNA polymerase A were resolved by high-resolution electrophoresis on a slab gel with a gradient of polyacrylamide (8-15%) in the presence of SDS. (→ǀ) Polypeptide variants. The polypeptides of *S. cerevisiae* enzyme A are identified by their molecular weight × 10^{-3}. (*1*) *S. cerevisiae;* (*2*) *S. douglasii;* (*3*) segregant 4203-sp52; (*4*) hybrid 4203. (*B*) RNA polymerase A was isolated from cells of a viable tetrad and analyzed by electrophoresis in a 15% polyacrylamide gel in the presence of SDS. The tetrad was isolated from a cross of *S. cerevisiae* × 4292-2D. (*1*) *S. cerevisiae;* (*2*) spore 4369-1A; (*3*) spore 4369-1B; (*4*) spore 4369-1C; (*5*) spore 4369-1D; (*6*) spore 4292-2D. In the tetrad, one can follow the segregation of A_{49}, A_{43}, A_{19}, $A_{14.5}$, and A_{14} (→). Note a case of gene conversion for the A_{19} genes.

1976; Valenzuela et al. 1976b). Among them, the following polypeptides also have an identical isoelectric point, as determined by two-dimensional mapping of the dissociated enzymes: AB_{27} (9.15), AB_{23} (4.5), and $AB_{14.5}$ (4.6) (Buhler et al. 1976). These results were confirmed and extended to enzyme C by comparing the electrophoretic mobility of the polypeptides in urea. The subunits that comigrated were AC_{40}, AB_{27}, AB_{23}, AC_{19}, and $ABC_{14.5}$, whereas C_{27} could not be detected and C_{23} migrated differently than AB_{23} (Valenzuela et al. 1976b). The fingerprint patterns of the ^{35}S-labeled polypeptides are indistinguishable in the case of AB_{27}, AB_{23}, and $AB_{14.5}$ (Buhler et al. 1976). Furthermore, ABC_{23}, as well as AC_{19}, is phosphorylated in vivo (Bell et al. 1977). Inhibition studies with antibodies to individual subunits from RNA polymerase A confirmed the structural relationship of A_{40} and C_{40}; A_{23}, B_{23}, and C_{23}; and A_{19} and C_{19} (Buhler et al. 1980). The common polypeptides from RNA polymerases A, B, and C can be identified, after transfer to a membrane, with antibodies to native RNA polymerase A or B (Fig. 3) or with antibodies to the individual components of enzyme A or B, using a spot-immunodetection method (Fig. 4). On the basis of these antibody-binding data, the immunological relationship of the following subunits is firmly established: AC_{40}, ABC_{27}, ABC_{23}, AC_{19}, and $ABC_{14.5}$. The charge difference between AB_{23} and C_{23}, observed by electrophoresis in urea, probably arises from a different degree of phosphorylation (Valenzuela et al. 1976b) or some other covalent modification; this might also be the case for C_{27}, which does not comigrate with A_{27}.

Evolution of RNA Polymerase Genes

In addition to the common subunits that are strongly recognized by antibodies, there is a small but discrete reaction at the level of the large subunits of enzymes A and B and, to a lesser extent, of enzyme C (Buhler et

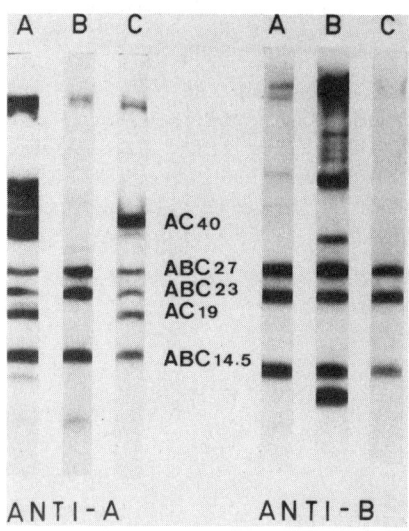

Figure 3 Immunological relationship of RNA polymerases A, B, and C from S. cerevisiae. Subunits of RNA polymerases A, B, and C were separated by electrophoresis on a polyacrylamide slab gel with SDS, transferred on a nitrocellulose membrane, and incubated with antibodies to native RNA polymerase A (anti-A) or to RNA polymerase B (anti-B). The polypeptide-immunoglobulin complexes were revealed by ^{125}I-labeled Protein A, and the membrane was subjected to autoradiography. A strong antibody-binding signal is found at the level of common subunits AC_{40}, ABC_{27}, ABC_{23}, AC_{19}, and $ABC_{14.5}$. Weaker reactions are visible on the largest subunits. For further comments, see text.

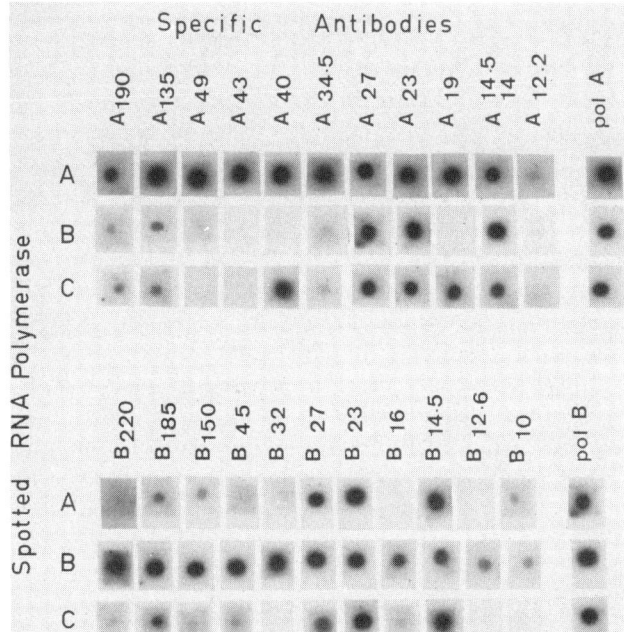

Figure 4 Immunological relatedness of the three forms of yeast RNA polymerase was investigated with antibodies to the isolated subunits of enzyme A or B. Purified RNA polymerase A, B, or C was spotted on nitrocellulose filters and challenged with the different antibodies. Bound antibodies were revealed by ^{125}I-labeled Protein A. Antibodies are identified by the name of the subunit used as antigen.

al. 1980). These data, obtained with antibodies to the native enzymes, were confirmed by using antibodies to each of the large subunits of RNA polymerase A or B (Huet et al. 1982) (see Fig. 4). These observations suggest that the large polypeptides in each form of enzyme evolved from a common ancestral pair of proteins to meet different functional and regulatory requirements. The enzymes are located in different parts of the nucleus and probably interact with different factors or chromatin components. The large subunits are likely to be the targets of these interactions. This may explain the very small residual immunological relationship of the homologous subunits in the three enzymes.

The immunological relationship of RNA polymerase A from yeast species was explored by using antibodies to *S. cerevisiae* enzyme A (Buhler et al. 1980). There was a marked conservation of immunologically related sites in the largest subunits of enzyme A from *Candida tropicalis, Endomycopsis fibuligera, Schizosaccharomyces pombe,* and *S. cerevisiae*. On the other hand, only weak cross-reactions were detected with the small subunits. These preliminary studies indicate that all of the subunits did not evolve at the same pace. A stronger degree of structural and functional constraint must be exerted on the largest subunits (Riva et al. 1982).

The immunological relationship of form B RNA polymerase from yeast, plant, insect, crustacea, and mammal was explored using the collection of antibodies to the components of the yeast enzymes (Huet et al. 1982). There was a general cross-reaction of all B enzymes with antibodies to the large subunits of yeast enzyme B (B_{220} and B_{150}). A small cross-reaction was observed with antibodies to the large components of yeast enzyme A (A_{190} and A_{135}). All of the B enzymes also shared a few immunological determinants with one of the common polypeptides (ABC_{23}). In addition to these general cross-reactions, other specific antibodies react with wheat germ enzyme B or the *Drosophila* enzyme. None of the antibodies reacts with *E. coli* RNA polymerase. However, interestingly, a strong cross-reaction is obtained with an archaebacterium, *Sulfolobus acidocaldaria*. The unique RNA polymerase of that microorganism is recognized by antibodies to subunits A_{190}, A_{135}, A_{40}, and B_{220}. A weak cross-reaction was also noted with the common subunit ABC_{23}. The archaebacterial enzyme was also recognized by one monoclonal antibody specific for A_{190} (J. Huet and W. Zillig, unpubl.). These findings illustrate the structural evolution of the multiple forms of eukaryotic RNA polymerases from an ancestral multimeric protein.

FUNCTIONAL ROLE OF INDIVIDUAL POLYPEPTIDE CHAINS

At this point a comment should be made concerning the term subunit, which is widely but improperly used when referring to the polypeptide chains associated with eukaryotic RNA polymerases. This term is taken in its very broadest sense. It refers to the various polypeptides that are found associated with the enzyme molecule. This definition differs markedly from the concept of subunit, which has been well defined biochemically, functionally, and genetically for the *E. coli* RNA polymerases. To be qualified as a true subunit, a given polypeptide chain should be shown to be required for transcription. It will be seen that this has not been established for most polypeptides.

Partial Enzyme Dissociation

Solely on the basis of structural grounds, certain polypeptide chains emerge as good candidates for subunit status. First, the two large polypeptide chains found in all eukaryotic RNA polymerases are reminiscent of the β' and β pair in the bacterial enzyme. The common polypeptides that copurify with the three RNA polymerases ABC_{27}, ABC_{23}, and $ABC_{14.5}$, or with only A and C RNA polymerases AC_{40} and AC_{19}, are also probably part of the fundamental enzyme that catalyzes the basic steps of RNA synthesis. The other polypeptides fall into two groups. One small group is always found closely associated with the RNA polymerases: $A_{12.2}$, $B_{12.6}$, and $B_{44.5}$. The second group is made up of polypeptides that can be dissociated under certain conditions: $A_{49} + A_{34.5}$ (Huet et al. 1975), $B_{32} + B_{16.5}$ (Dezélée et al. 1976), and $C_{82} + C_{37} + C_{34}$ (Valenzuela et al. 1976c), or those that were not consistently obtained when using different purification procedures: $A_{49} + A_{43}$, (Hager et al. 1977), C_{53} (Bell et al. 1977), $C_{53} + C_{37}$ (Valenzuela et al. 1976b), and A_{23} (Valenzuela et al. 1976a). In some specific cases, it has been possible to compare the catalytic properties of the complete enzyme to that of enzyme deficient for individual polypeptides.

The loss of polypeptides A_{49} and $A_{34.5}$ severely impairs the activity of RNA polymerase A. Enzyme A is practically inactive on native DNA at ionic strength above 0.1 (Huet et al. 1976b), whereas at low salt concentration it transcribes poly[d(A-T)] as efficiently as the complete enzyme A (Huet et al. 1975, 1976b). The two polypeptides possibly participate in DNA binding (Sentenac et al. 1976). Kinetic studies suggest that they play a role in DNA binding and chain elongation (Cooper and Quincey 1979a). Interestingly, enzyme A* is twice as sensitive to α-amanitin as enzyme A (Huet et al. 1975). An enzyme A deprived of both A_{49} and A_{43}, but retaining $A_{34.5}$, is also half as active as the complete enzyme (Hager et al. 1977). Therefore, it could be that the critical polypeptide is A_{49}. It is clear that although the loss of the polypeptides A_{49}, A_{43}, and $A_{34.5}$ markedly alter the properties of enzyme A, they are not essential to RNA synthesis in vitro.

A good correlation has been found between the relative abundance in the enzyme of polypeptides A_{43} and A_{23} and the activity of RNA polymerase A (Valenzuela et al. 1976a). Also, the presence of A_{23} in all three enzymes argues for its essentiality to enzyme activity. However, to be fully convincing, a reconstruction experiment should be performed as a control for the above experiment to substantiate that loss of A_{23} and A_{43} is a cause rather than a result of enzyme inactivation.

No drastic impairment in the activity of enzyme B is brought about by the removal of B_{32} and $B_{16.5}$ (Dezélée et al. 1976). These two proteins are probably bound together, since they comigrate by electrophoresis in spite of their different isoelectric points (Dezélée et al. 1976). The deficient enzyme, B*, is half as active as the complete enzyme, with poly(rC) as template, and has a lower affinity for denatured DNA (Ruet et al. 1980). These observations are not sufficient to ascribe or deny a function to these polypeptides. However, clearly they are not mandatory for RNA synthesis in vitro.

As far as enzyme C is concerned, it would be most interesting to investigate the catalytic properties of the enzyme lacking C_{82}, C_{37}, and C_{34}. This has not been done because this enzyme variant has only been observed after polyacrylamide gel electrophoresis (Valenzuela et al. 1976c).

The results described above were obtained by taking advantage of the spontaneous dissociation of some components. A more systematic approach in identifying the essential subunits and the accessory proteins requires reconstitution of active enzyme from isolated polypeptides. Unfortunately, each RNA polymerase, up to now, has resisted all attempts to carry out the elegant dissociation-reconstitution experiments similar to those done with bacterial RNA polymerase (Zillig et al. 1976).

Probes for Subunit Function

RNA polymerase A is inhibited by pyridoxal 5'-phosphate through formation of a Schiff base at a few lysyl residues (Martial et al. 1975). The subunits that react with the aldehyde are predominantly A_{190}, A_{49}, and $A_{34.5}$ and, to a lesser extent, A_{135} (Valenzuela et al. 1978b). DNA prevents binding of the reagent to the four polypeptides, whereas nucleoside triphosphates drastically decrease its binding to A_{190}. These results suggest that A_{190} contains the nucleotide binding site involved in enzyme activity and also participates in DNA binding, perhaps in conjunction with A_{135}, A_{49}, and $A_{34.5}$ (Valenzuela et al. 1978b).

Antibodies to isolated subunits can also be used as a probe of polypeptide function. DNA protects enzyme A from inhibition by antibodies to subunits A_{190}, A_{23}, and A_{19}. The protection by DNA against anti-A_{23} is effective for all three enzymes. A monoclonal antibody to subunit A_{135} is also much less inhibitory when enzyme A is preincubated with DNA (J. Huet, unpubl.). Taken together, all of these results suggest that the template interacts with many polypeptides in enzyme A. However, it is likely that the catalytic site is located on the large subunits.

Mutants

The isolation of mutants with altered RNA polymerases would be of great help to correlate loss of a function in vivo and in vitro with a structural defect. Unfortunately, the search for mutants has not been very successful. Yeast cells are not permeable to α-amanitin, and there is no straightforward means of selecting for RNA polymerase mutants (see below, Inhibitors). One mutant (*rpoB1*), deficient in RNA polymerase-B activity in vitro, was isolated though a selection procedure involving thiolutin treatment (Winsor et al. 1979). However, it has not been established that the drug contributed to the isolation of the mutant strain. The enzyme from the *rpoB1* mutant is defective in RNA chain initiation and elongation reactions (Ruet et al. 1980). Enzyme-DNA binding is comparatively much less affected. The B_{220} subunit in the *rpoB1* mutant has an altered peptide map, indicating that the polymerase lesion in this mutant resides in the largest subunit. The structural alteration of B_{220} resulting from this mutation also brings about the spontaneous dissociation of B_{32} and $B_{16.5}$. These observations confirm an essential structural role of the largest polypeptide (Ruet et al. 1980). Unfortunately, the *rpoB1* mutant has no detectable in vivo phenotype and it remains as yet the only yeast strain affected in one specific RNA polymerase. Thermosensitive mutants have been described that may be impaired in components essential for RNA synthesis. A dominant thermosensitive mutant (RNA-12), impaired in RNA biosynthesis, was selected in *S. cerevisiae*. However, measurements of RNA synthesis in isolated nuclei, or by purified RNA polymerases A and B, showed no significant difference between the mutant and the wild type (Lacroute 1975). The thermosensitivity of a common cofactor for all three RNA polymerases, dispensable in vitro, would be a possible explanation of the results. Thonart et al. (1976) isolated three interesting thermosensitive mutants substantially defective in the synthesis of all RNA species at the nonpermissive temperature (37°C). Since synthesis of all RNA species is affected, it is possible that the mutations are in genes encoding the common subunits. In support of this hypothesis is the observation that after growth at 37°C for a few hours, the mutants retain only a very low residual activity of RNA polymerases A, B, and C, as detected following chromatography on DEAE-Sephadex. In addition, a significant increase in the thermosensitivity of RNA polymerase activities was found in partially purified extracts from the three mutants (Thonart et al. 1976). However, this observation could not be confirmed with purified RNA polymerases A and B (P. Thonart and A. Ruet, unpubl.).

Enzyme-A-associated RNase-H Activity

The persistence of an RNase-H activity in homogeneous preparations of yeast RNA polymerase A has been observed. This ribonuclease, which degrades specifically RNA-DNA hybrids, copurifies with enzyme A through various fractionation procedures involving ion-exchange chromatography, sedimentation in glycerol gradients at high ionic strength, and polyacrylamide gel electrophoresis (Huet et al. 1976a, 1977). In fact, several observations suggest the presence of two distinct RNase-H activities. The nuclease produces a mixture of mononucleotides and dinucleotides with 5'-phosphate end. Two distinct RNase-H species can be renatured after electrophoresis of RNA polymerase A in an SDS-polyacrylamide gel (Iborra et al. 1979). They comigrate with A_{49} and A_{40}.

The role of a ribonuclease-H activity associated with RNA polymerase A remains obscure. RNA polymerase-B preparations have no RNase-H activity, so it is not a general feature of all yeast RNA polymerases. One could speculate that this nuclease activity is involved in termination of transcription to release the RNA product, or that it removes transient RNA-DNA hybrids required for initiation of replication or functions as part of a proofreading system. At any rate, one cannot disregard the possibility that the associated RNases H are unrelated proteins that happen to have a strong binding affinity for the RNA polymerase A.

Satellite Proteins

A 37,000-dalton protein (P37), has been purified from yeast cells on the basis of its ability to stimulate RNA polymerase B (Sawadogo et al. 1980a). A complex of P37 and RNA polymerase B was isolated by sedimentation through a glycerol gradient and by polyacrylamide gel electrophoresis under nondenaturing conditions. Under both conditions, enzyme B containing bound P37 has a drastically increased specific activity. The P37 protein binds preferentially to RNA polymerase B_1 (the nonproteolyzed enzyme form). The complex, which can be dissociated at high ionic strength, has a dissociation constant of 5×10^{-8} M. P37 stimulates specifically both yeast RNA polymerases A and B and protects both enzymes from inhibition by antibodies to A_{23} (which is common to A, B, and C RNA polymerases [Sawadogo et al. 1980b]). This suggests that P37 could be involved in the regulation of all three RNA polymerase activities. Since P37 stimulates RNA synthesis from preinitiated ternary transcription complexes but does not stimulate trinucleotide synthesis, it is likely that it acts at the level of RNA chain elongation (Sawadogo et al. 1981).

GENERAL CATALYTIC PROPERTIES

Substrates and Reactions

Enzyme B resembles the RNA polymerase from *E. coli* and other sources in its absolute requirement for a DNA template, a divalent cation, and the complementary nucleoside triphosphates. The two fundamental differences are a marked preference for a denatured template and for Mn^{++} ions (Frederick et al. 1969). On linear double-stranded DNA, enzyme B initiates preferentially at nicks (Lescure et al. 1981a). With denatured DNA, the specific activity of yeast enzyme B is similar to that of *E. coli* RNA polymerase on the same native template. The opposite behavior of the two enzymes is very striking (Dezélée and Sentenac 1973). The enzyme carries the exchange of pyrophosphate with a relatively low efficiency with Mg^{++} ions, executes the synthesis of poly(rA) on denatured DNA by a slippage mechanism, like the bacterial enzyme, and transcribes an alternated template with high fidelity (Frederick et al. 1969; Dezélée and Sentenac 1973). Among synthetic templates, poly[d(I-C)] is the most efficiently transcribed (Dezélée et al. 1974b). Homopolymer pairs are transcribed asymmetrically, the pyrimidine strand acting as template. Interestingly, poly(rC) is very actively transcribed (Dezélée et al. 1974b), and this property was used to detect specifically enzyme B in a yeast crude extract (Ruet et al. 1978). Enzyme B carries out a very efficient synthesis of trinucleotides (abortive reaction) on supercoiled DNA and even on linear DNA in a dinucleotide-primed reaction (Lescure et al. 1981b). Yeast RNA polymerase B binds to supercoiled SV40 DNA, but the complex is extremely unstable and decays in a few seconds in the presence of heparin or denatured DNA as competitor (Huet et al. 1976b). However, stable heparin-resistant complexes were found after a short pulse of RNA synthesis primed with a dinucleotide (Lescure et al. 1981b). It could be that priming with a dinucleotide favored chain initiation. The particularly active synthesis of trinucleotides on a supercoiled template can also be explained by the localized melting of DNA induced by supercoiling.

The mean rate of chain elongation at 30°C on denatured DNA is 6-7 nucleotides per second (Dezélée and Sentenac 1973). On supercoiled DNA, the initial rate of RNA elongation is closer to the in vivo estimates (Lacroute 1973) of 30 nucleotides per second at 30°C (B. Lescure, unpubl.). On denatured DNA, long RNA chains are made (Frederick et al. 1969), whereas elongation is very limited on a supercoiled template (Lescure et al. 1981a).

Catalytic properties of enzymes A and C have not been investigated as extensively. RNA polymerase A transcribes denatured and native DNA almost equally well. However, initiation occurs preferentially on nicks or gaps present in the native template (Dezélée et al. 1974a). The specific activity of the purified enzyme on native yeast DNA depends on the method of purification. The structural basis for this variation is not understood, as there was no apparent difference in the polypeptide content of the purified enzymes (Holland et al. 1977). With native DNA, the product is made of two populations of RNA that differ in size (Buhler 1974). Enzyme A catalyzes an active pyrophosphate exchange reaction with poly[d(A-T)] at very low concentrations of ATP, the initiator nucleotide (Sentenac et al. 1976). On the other hand, trinucleotide synthesis is very limited (P. Cottrelle, unpubl.).

Very little is known about the catalytic properties of enzyme C. Enzyme C shows a typical biphasic salt stimulation profile and is most active with poly[d(A-T)] and poly[d(I-C)] (Adman et al. 1972; Schultz and Hall 1976). These properties allowed the specific assay of enzyme C in yeast crude extracts using poly[d(I-C)] as template at high ionic strength (Ruet et al. 1978). The properties that distinguish RNA polymerases A, B, and C are summarized in Table 1.

Inhibitors

Inhibitors of yeast RNA polymerases act either at the level of the template or of the enzymes. There is little information on the effect of DNA binding inhibitors on the activity of the different forms of yeast RNA polymerases.

Such a comparative study might help disclose differences in their chain elongation properties. The binding of histones to denatured DNA restricted its template activity for RNA polymerase B by decreasing RNA chain initiation, whereas chain elongation by preinitiated complexes remained possible. In contrast, poly-L-lysine blocked elongation (Karagyozov et al. 1978).

The second class of compounds, which interact with RNA polymerases, permits a more ready discrimination between the three enzymes. Of these inhibitors, α-amanitin has been the most extensively studied. Yeast enzyme B, which is the most sensitive (Dezélée 1970; Adman et al. 1972; Schultz and Hall 1976; Valenzuela et al. 1976c), is 1000-fold less sensitive than the mammalian enzyme (see Table 1) (Roeder 1976).

Also, in contrast with the corresponding mammalian enzyme, yeast enzyme A can be inhibited by high concentrations of the mycotoxin (Huet et al. 1975; Schultz and Hall 1976; Valenzuela et al. 1976c). The removal of two polypeptides, A_{49} and $A_{34.5}$, enhances its sensitivity to the inhibitor (Huet et al. 1975). On the other hand, enzyme C is probably totally resistant. In different studies its sensitivity to α-amanitin concentrations of 2 mg/ml varied from 5% to 25% (Schultz and Hall 1976; Valenzuela et al. 1976c). Kinetic studies indicate that α-amanitin binds reversibly in a 1:1 stoichiometry to enzyme A or B (Sawadogo 1981) and blocks chain elongation (Dezélée et al. 1970) as well as trinucleotide synthesis. Therefore, it is likely that the drug interferes with phosphodiester bond formation rather than with the translocation process. The mechanism by which α-amanitin blocks chain elongation is not known. The toxin-binding subunit has not yet been identified. The proteolyzed enzyme B_{II} has the same sensitivity to the toxin as the native form B_I (Dezélée et al. 1976). In the mammalian enzyme B, the amatoxin receptor is the second largest subunit (Brodner and Wieland 1976).

A variety of drugs active on bacterial RNA polymerase are ineffective on the yeast enzymes. Yeast RNA polymerases are not inhibited by rifampicin (Ponta et al. 1971; Adman et al. 1972). Enzyme B is resistant to streptolydigin and streptovaricin (Dezélée and Sentenac 1973). Several derivatives of rifampicin inhibit the three enzymes at rather high concentrations (Adman et al. 1972; Di Mauro et al. 1974), but these compounds have a large spectrum of nonspecific binding to a variety of proteins (Chambon 1974).

Among the classical inhibitors of bacterial RNA polymerase, polyanions like heparin (Lescure et al. 1981b), polyphosphate (Ponta et al. 1974), poly(rI) (Dezélée et al. 1974b), Cibacron blue (Bull et al. 1981a), and tRNA (Sawadogo 1981) bind to and inactivate free RNA polymerase. RNA polymerases A (Cooper and Quincey 1979b) and B (Lescure et al. 1981b) in preinitiated complexes are resistant to heparin. tRNA probably competes for the template binding site on RNA polymerase B and, like α-amanitin, inhibits enzyme B more strongly than enzyme A (Sawadogo 1981). However, the binding sites of the two inhibitors are not related. Pyridoxal phosphate similarly inhibits mammalian and yeast RNA polymerases, which suggests a similarity in their active sites (Martial et al. 1975).

Substrate analogs could be potentially interesting tools to inhibit selectively one form of RNA polymerase and select for mutants. Cordycepin triphosphate (3'-dATP) inhibits the three yeast enzymes to different extents (Horowitz et al. 1976). It would be interesting to confirm these results with purified enzymes and compare their responses to other nucleotide analogs. Thiolutin inhibits RNA synthesis in vivo and the activity of yeast RNA polymerases A, B, and C in vitro (Jiminez et al. 1973; Tipper et al. 1973). Thiolutin has been used in attempts to select RNA polymerase mutants (Lacroute 1975; Winsor et al. 1979).

Another group of inhibitors of RNA synthesis in yeast are chelators. Lomofungin and 8-hydroxyquinoline function analogously by inhibiting in vivo synthesis of high-molecular-weight rRNA and poly(A)-containing RNA more than that of 5S and tRNA (Fraser and Creanor 1974; Johnston and Singer 1978). Lomofungin and 8-hydroxyquinoline are able to chelate Mn^{++} and Zn^{++} ions more efficiently than Mg^{++} at pH 5.8 (Ruet et al. 1975). These results suggest a difference in the nature or concentration of the divalent cation required by enzyme C. It is also possible that the chelators act by chelating the zinc associated with the polymerases. Another chelator, with high affinity for zinc, 1,10-phenanthroline, is a potent inhibitor of all three forms of yeast RNA polymerases (Auld et al. 1976; Lattke and Weser 1977; Wandzilak and Benson 1978). This result has been taken as strong evidence that the intrinsic zinc atoms of the enzyme are essential to the catalytic process.

REGULATION OF RNA POLYMERASES

Sebastian et al. (1973b) investigated the effect of growth rate on RNA polymerase content of yeast cells grown under controlled steady-state conditions in a chemostat. RNA polymerases were solubilized and separated by DEAE-Sephadex chromatography. On the basis of the enzyme activity recovered under these conditions, it appeared that fast-growing cells contained proportionately more RNA polymerase A than slow-growing cells, whereas the activity of RNA polymerase B was almost independent of growth rate. Other authors confirmed this observation and also found that shifting cells from minimal to rich medium led to a selective increase in the RNA polymerase-A activity, assayed on DEAE-Sephadex (Carter and Dawes 1975). It is unfortunate that enzyme C was not observed in these experiments. As discussed previously, activity measurements in crude fractions may not be representative of RNA polymerase content. For a direct estimation of the amount of RNA polymerases, enzymes A and B were selectively immunoprecipitated from crude extracts and quantitated (J.M. Buhler, unpubl.). In a rich medium (generation time, 1½ hr) the number of RNA polymerase molecules per diploid cell was 27,000 and 14,000 for enzymes A and B, respectively. In a poor medium (generation time, 2½ hr), the number of RNA polymerases A and B per cell was 14,000 and 11,000, respectively. These figures correspond to the enzymes solubilized by 0.3 M ammonium sulfate. Therefore, it is likely that there is indeed a variation in the number of RNA polymerase-A molecules with the growth rate. As the rate of rRNA and tRNA accumulation increases with the growth rate (Waldron and Lacroute 1975; Waldron 1977), synthesis of new RNA polymerase molecules may be required for the production of stable RNA species. It would be most interesting to know whether the synthesis of RNA polymerases A and C is coordinated. There is no information yet on the regulation of enzyme C level. On the other hand, it appears that the synthesis of enzymes A and B is regulated independently, and this has important implications, especially in view of the fact that the three forms of enzyme share common subunits.

Independent regulation of RNA polymerases A and B was also observed during the cell cycle (Sebastian et al. 1974; Carter and Dawes 1975). On the basis of the recovery of RNA polymerase activity on DEAE-Sephadex, the results are compatible with a continuous synthesis of enzyme A and a stepwise synthesis of enzyme B occurring after DNA synthesis. There is no information on enzyme C. Again, the results should be confirmed by measuring directly the number of enzyme molecules per cell rather than their activity.

In *S. cerevisiae* deprived of an amino acid, rRNA synthesis is drastically reduced (Gross and Pogo 1974; Shulman et al. 1977). Cycloheximide treatment induces the same effect. Glucose starvation also produces an abrupt interruption of RNA synthesis in spheroplasts. Gross and Pogo (1974) investigated the effect of starvation and cycloheximide treatment on the amount and distribution of RNA polymerase activities isolated by DEAE-Sephadex chromatography or present in nuclear extracts. Neither amino acid or glucose starvation nor cycloheximide treatment affected the peak pattern of the solubilized enzymes.

There is the possibility that transcriptive function of yeast RNA polymerases can be regulated by phosphorylation. The same polypeptides that are phosphorylated in vivo were found to be phosphorylated in vitro by a yeast protein kinase, independent of cyclic AMP (cAMP) (Bell et al. 1976, 1977). In vivo phosphorylation of two polypeptides, A_{23} and A_{19} (out of five phosphorylated in enzyme A), has been found to be unaffected by the presence of cycloheximide (Buhler et al. 1976). Attempts to alter RNA polymerase-A activity by phosphatase treatment have failed (Bell et al. 1977). However, this does not rule out a physiological role of this enzyme modification. The important phosphate groups may not have been removed by the phosphatase or, alternatively, the enzyme assay may not have been selective enough to detect a change in transcription function.

FUNCTION OF RNA POLYMERASES

The existence of three forms of yeast nuclear RNA polymerases with distinct molecular structure and different properties suggested that each enzyme has a specialized function in transcription of the genetic information. This was convincingly shown, in the case of the animal enzymes, by taking advantage of their different α-amanitin sensitivity (see Roeder 1976). By analogy with the animal enzymes, it should be possible to ascribe a priori the same function to the corresponding yeast enzymes. However, this has been precluded by the problem of establishing a good correspondence between the animal and yeast enzymes. The α-amanitin sensitivity of yeast RNA polymerases is very much different from that of animal enzymes. The most sensitive of the yeast RNA polymerases, enzyme B, is about 1000-fold less sensitive than animal enzyme B. In addition, the relationship between the relative elution position and the relative α-amanitin sensitivity of yeast polymerases is different than that of the vertebrate enzymes. Thus, more direct means have been required to determine the transcriptional function of each of the yeast RNA polymerases. This goal has been approached through the use of specific inhibitors or of mutants affected in one form of enzyme and through demonstration of the transcriptional specificity of the enzymes in vitro.

The only characterized mutant available is *rpoB1*. Unfortunately, despite its marked deficiency in RNA polymerase B in vitro, the mutant showed no significant impairment in growth and RNA biosynthesis (Winsor et al. 1979). It is possible that the enzyme alteration is compensated for in vivo through interactions with chromatin components.

Schultz (1978) investigated the effect of α-amanitin on RNA synthesis by isolated yeast nuclei. The synthesis of low-molecular-weight RNA species was resistant to very high concentrations of the toxin, which is a characteristic property of RNA polymerase C. Yeast nuclei also synthesized some heterodisperse RNA that was resistant to α-amanitin at very high concentrations. Assuming that the nuclear and the purified enzyme C responded similarly to the mycotoxin, the results strongly suggested that RNA polymerase C is responsible for the synthesis of 5S RNA, pre-4S transcripts, as well as some higher-molecular-weight species. The main transcription product made by isolated nuclei of *S. carlsbergensis*, in the presence of enough α-amanitin (20 μg/ml) to block enzyme B, is ribosomal precursor RNA as judged by its size and its characteristic methylation in the presence of *S*-adenosylmethionine (Klootwijk et al. 1976). Low-molecular-weight RNA species, resistant to low concentrations of α-amanitin, were also described in this study. Both enzymes A and C are active under these conditions.

There is a relationship between the growth rate, the rRNA content of the cells, and the level of RNA polymerase-A activity recovered by DEAE-cellulose chromatography (Sebastian et al. 1973b; Carter and Dawes 1975). This correlation suggests that enzyme A is responsible for the synthesis of rRNA.

Among the three enzymes, RNA polymerase A is the most efficient in making RNA primers required for initiation of in vitro DNA synthesis by yeast DNA polymerase I (Plevani and Chang 1977, 1978). Short RNA initiators are made by purified RNA polymerase A on single-stranded DNA, and uridine 5'-monophosphate (UMP) appears at the highest frequency at the RNA-DNA junction (Plevani and Chang 1978). These in vitro studies suggest a role for RNA polymerase A in DNA replication. Alternatively, these findings may simply reflect the tendency of RNA polymerase A to make short RNA chains on denatured DNA (whereas RNA polymerase B synthesizes long products) and to terminate transcription on dA-rich sequences (Martin and Tinoco 1980). However, the presence of associated RNase-H activity within RNA polymerase A lends some support to the first hypothesis. With the availability of in vitro replication systems (Jazwinski and Edelman 1979; Kojo et al. 1981), the involvement of one of the known RNA polymerases in initiation of DNA synthesis may be investigated using specific antibodies.

In conclusion, with the available circumstantial evidence on the function of RNA polymerases, it seems reasonable to infer, by analogy with what is known of the animal enzymes, that RNA polymerase B is responsible for the synthesis of mRNA: It is the most sensitive of the three forms of enzyme to α-amanitin and is immunologically related to the animal RNA polymerase B (Huet et al. 1982). RNA polymerase C is structurally related to the animal enzyme C and makes 5S RNA and tRNA in nuclei and in reconstituted systems (see below). Enzyme-C function, however, may well not be limited to the synthesis of small stable RNA. Although there is no evidence on the function of RNA polymerase A, one is left with the likely hypothesis that it makes rRNA.

IN VITRO TRANSCRIPTION SPECIFICITY

Since this discussion henceforth will be concerned with the activity of yeast RNA polymerases rather than their structure, we will use the designations RNA polymerases I, II, and III rather than A, B, and C.

RNA Polymerase I

The experiments that have been reported on in vitro transcription by yeast polymerase were all done prior to the availability of cloned rDNA templates. Total yeast chromosomal DNA was used as template for these early experiments (Hollenberg 1973; Cramer et al. 1974). By using competitive hybridization against 18S + 25S RNA to determine the amount of rRNA transcribed in vitro, Hollenberg (1973) concluded that transcription of the genome occurred in an apparently random fashion. Cramer et al. (1974) found that a fraction of yeast DNA enriched in ribosomal genes (γ DNA) was preferentially transcribed from the in vivo sense strand. In subsequent experiments, Van Keulen et al. (1975) and Holland et al. (1977) reported strongly selective transcription (10–25% of total RNA) of the ribosomal sequences in total yeast nuclear DNA.

In all of these studies, however, the sites of initiation of transcription have not been mapped. This could be done more easily with recombinant plasmids containing the ribosomal transcription unit as template. However, using such a simplified template, the specific transcription of ribosomal genes could not be reproduced consistently (Tekamp et al. 1979). At this time, the question of RNA polymerase-I specificity remains unanswered until the in vitro transcripts are shown to correspond to the initiation sites of the in vivo rRNA precursor (Bayev et al. 1980; Klemenz and Geiduschek 1980).

RNA Polymerase II

Study of the transcription specificity of purified yeast RNA polymerase II has been hampered by its low activity on intact DNA duplexes and its requirement for Mn^{++} ions (known to catalyze anomalous reactions by the bacterial RNA polymerase and nonspecific initiation events). RNA polymerase B transcribes single-stranded templates most efficiently, yielding a nonrandom pattern of transcription. Defined RNA products were visualized as discrete RNA-DNA hybrid bands following nuclease S1 treatment and agarose gel electrophoresis (Nagamine et al. 1981). On supercoiled DNA, RNA polymerase B catalyzes an active trinucleotide synthesis with a dinucleotide as primer and one nucleoside triphosphate. Using the best combination of dinucleotide and substrate, the specificity of initiation was investigated with a recombinant plasmid containing *ADC1*, the structural gene for yeast alcohol dehydrogenase I (ADHI). The enzyme bound to and initiated mostly within the yeast DNA insert (Lescure et al. 1981b). A very small number of chains was initiated as evidenced by electrophoresis of the oligonucleotides synthesized under substrate limitation. One preponderant transcript was initiated far upstream from the origin of the *ADC1* gene. One minor start was located at position –35 from the AUG initiation codon, which corresponds to an in vivo transcription start identified by nuclease S1 mapping of the RNA (Lescure et al. 1981a). On the other hand, when the template was truncated DNA, initiation occurred predominantly at nicks introduced by the restriction enzyme, as first observed in the case of wheat germ RNA polymerase II (Lewis and Burgess 1980).

These observations underscore the importance of the structure of the template during transcription. Clearly, further work is needed to establish whether purified enzyme B retains a basal level of specificity. Recently, a crude reconstituted transcription system has been developed from animal cells, which allows the accurate transcription of several viral and cloned animal genes (Weil et al. 1979; Manley et al. 1980; Wasylyk et al. 1980). However, the mechanisms that make the promoter available to recognition by RNA polymerase II are unknown. Probably because of the presence of nucleases, which degrade the transcripts and also the property of yeast RNA polymerase B to end-label RNA preexisting in yeast cell extracts, it has been difficult to develop a yeast in vitro transcription system similar to the animal one. Recent progress along these lines has been made by C. Parker (pers. comm.), using a system extracted from yeast nuclei.

RNA Polymerase III

Of the three yeast RNA polymerases, only enzyme III has been clearly shown to initiate and terminate in vitro transcription correctly. In early experiments, the yeast nuclear system, which transcribes 5S and pre-4S RNA, was depleted of RNA polymerase III by acid treatment. The addition of purified RNA polymerase III to these enzyme-depleted nuclei restored transcription of 5S and pre-4S molecules (Tekamp et al. 1979).

In recent experiments that employed as template purified plasmid DNA containing cloned yeast tRNA genes, accurate transcription initiation and termination were obtained using several different extraction procedures (Klekamp and Weil 1982; R.A. Koski et al. 1982). These extracts have high activity for pre-tRNA and 5S RNA synthesis; this activity exhibits a high degree of insensitivity to α-amanitin, as expected for RNA polymerase III. In the presence of 150 mM KCl, posttranscriptional processing of pre-tRNA molecules occurs, with the generation of apparently mature-sized 4S RNA molecules.

The protein components of the crude extract have been resolved by DEAE-Sephadex chromatography into two essential fractions, both of which are required in addition to purified RNA polymerase III, in order to achieve transcription of either tRNA genes or 5S rRNA genes (Klekamp and Weil 1982). On the basis of differential inhibition of 5S and pre-tRNA transcription by added 5S RNA, the investigators conclude that one of the active fractions contains both a tRNA-specific and a 5S-RNA-specific transcription factor.

DNA ELEMENTS INVOLVED IN TRANSCRIPTION INITIATION AND TERMINATION

Transcription Initiation by RNA Polymerase II

In higher eukaryotes, as in yeast, studies of the sequence requirements for transcription initiation by RNA polymerase II have not succeeded in localizing essential promoter sequences to any one or two small regions comparable to the Pribnow box and −35 sequences of *E. coli* genes (Siebenlist et al. 1980). Sequence comparisons between vertebrate RNA polymerase-II transcription units disclose that they possess, with few exceptions, a TATAAA-like sequence about 30–35 bp before the cap site. For in vitro transcription initiation, this sequence was shown to be essential (Mathis and Chambon 1981), whereas in vivo studies of its role suggest that the TATAAA element may act primarily to determine which one of several alternative initiation sites is actually used (Benoist and Chambon 1981). Farther upstream, a consensus sequence occurs at −73 to −92; alteration of sequences in this region and beyond frequently produce major effects upon the level of gene expression (Grosschedl and Birnstiel 1982).

There are indications of several types that the structural determinants for RNA polymerase-II transcription initiation may be different in yeast than in cells of higher animals. An early indication of this difference was the inability of purified yeast RNA polymerase II to complement the transcription activity of the mammalian in vitro system from which RNA polymerase II had been removed (Wasylk et al. 1980). Secondly, there is a difference between yeast and animal genes with regard to the position of TATAAA relative to the transcription start. Whereas the majority of sequenced yeast polymerase-II transcription units have TATAAA elements somewhere within the 200 bp of sequence preceding the ATG initiation codon, distances typically as large as 100 bp intervene between yeast TATAAAs and the corresponding mRNA 5′ end (Table 3), in contrast to the distance of 30 bp that generally separates these elements for mammalian genes (Breathnach and Chambon 1981).

In yeast, as in *E. coli,* the mutant screening and selection procedures for cell clones lacking a particular gene function should permit the identification of genetic lesions affecting all levels of gene function, including transcription initiation. Curiously, however, there are few if any reported cases from conventional mutant screening studies of simple up or down promoter mutants resulting from point genetic alterations. Rather, the major observed genetic events that alter the level of gene activity are either insertions of the transposable element Ty1 or some other type of gross DNA alteration (Errede et al. 1980; Williamson et al. 1981). The reasons for the relative dearth of simple promoter mutations are not clear; perhaps this indicates that yeast RNA polymerase-II promoter sequences consist of some simple DNA element, e.g., an AT-rich sequence of a certain approximate length,

Table 3 TATAAA sequence upstream from yeast mRNA start points

Yeast gene	TATA-like sequence	Distance from T to the major mRNA 5' end	Reference
PYK1	AATTATAAATAC	~180	Russell (1982); T. Alber (pers. comm.)
TPI	GAATATAAAGGG	150	Russell (1982); Alber and Kawasaki (1982)
ADR2	ACATATAAATAG	107	Young et al. (1981)
G3PDH (pGAP491)	GTATATAAAGAA	101–103	Holland and Holland (1980); A. Musti and R. Kramer (pers. comm.)
G3PDH (pGAP63)	ATATATAAAGGT	80	A. Musti and R. Kramer (pers. comm.); Holland and Holland (1980)
PHO5	GTATATAAGCGC	60, 65	G. Thill and R. Kramer (pers. comm.)
PGK	ATATATAAACTT	100–120	Dobson et al. (1982)
ADC1	AAGTATAAATAG	90	Bennetzen and Hall (1982)
CYC1	GTATATAAAACT	64–87	Faye et al. (1981)
HIS3	TTATATAAAGTA	42	Struhl (1982)
MATα1	TCATATGAAACA	57, 71, 79	Nasmyth et al. (1981)
MATα2	ATATATAAAGGA	37–53	Nasmyth et al. (1981)
MATa1	CTGTATAAAACT	60–70	Nasmyth et al. (1981)
S. pombe ADH	TGGTATAAATAG	47 in S. pombe 95 in S. cerevisiae	Russell (1982)

All yeast genes are included for which RNA 5'-end data are available.

which is not profoundly disrupted by single base-pair substitutions or deletions.

Most of the existing information about structural alterations that affects the ability of yeast genes to be transcribed by RNA polymerase II comes from experiments in which mutant genes were constructed in vitro. Through the use of the yeast transformation system (Beggs 1978; Hinnen et al. 1978), the altered genes can be reintroduced into yeast cells either by integration at a chromosomal site or within an autonomously replicating plasmid. By correlating the phenotype of the yeast transformant with the DNA sequence alteration in the mutant gene, functions can be ascribed to various regions of the DNA.

For the yeast *CYC1* gene, such experiments have identified as functional the regions lying 247-670 bp, 139-242 bp, and 75-100 bp upstream from the translation start. Deletions spanning each of these regions produce decreased function of the gene, measured either by lowered in vivo formation of a fusion protein (Guarente and Ptashne 1981) or lowered levels of *CYC1* mRNA (Faye et al. 1981). In addition to the sequence elements mentioned, a promoter function can also be ascribed to sequences lying between −99 and −139. Deletion of this region, which includes TATAAAA, produces a shift in the position of the major 5' end of *CYC1* mRNA. Another in vitro deletion experiment establishes that sequences within the *CYC1*-coding region are not required for transcription initiation. After removal of the region between +8 and +251 and religation of the remainder of *CYC1*, unimpaired starting specificity and total template activity are retained, indicating that the promoter lies elsewhere than within the gene (Faye et al. 1981).

For the *ADC1* gene, coding for the yeast alcohol dehydrogenase found in glucose-grown cells (ADHI), in vitro deletions have been constructed that define both the 5' and 3' extremities of the promoter region. These establish that the region between −33 and −170 is adequate to promote transcript formation in vivo (Hitzeman et al. 1981; D. Beier and E.T. Young, unpubl.). Attachment of non-ADH-coding genetic material by fusions in the range of −5 to −33 results in transcription of the adjoining segment with 5'-transcript ends beginning at positions −28 and −38.

A detailed analysis of sequences required for transcription initiation of the yeast *HIS3* gene has been carried out by generating deletion mutations in the 5'-flanking sequence and then testing the mutant *his3* genes for their ability to function in vivo (Struhl 1981, 1982). Two DNA elements essential for *HIS3* transcription were made evident by this analysis: the sequences from −112 to −155 and from −32 to −52. The latter region contains only two GC base pairs and includes the sequence TATAAA, which has been implicated as a phasing signal for polymerase-II transcription starts in higher eukaryotes (Breathnach and Chambon 1982).

For the three yeast genes considered, there appear to be no simple generalizations that explain why particular sequences are required for transcription initiation. The complexities seen, particularly for *CYC1*, may be expained in part by the nature of the lesions produced in each deletion mutant. Each deletion not only eliminated certain sequences, it also created novel junction sequences and established new distance relationships between existing blocks of DNA sequence. Future experiments to define within more narrow limits the locations of RNA polymerase-II promoter regions may need to employ mutagenesis procedures that generate smaller-scale lesions in DNA. Along these lines, one apparent promoter mutation in *MATα* (Tatchell et al. 1981) was produced by the linker insertion technique developed by Heffron et al. (1979). An intensive mutagenic analysis of the 5′-flanking region of yeast genes using this procedure should permit definitive identification of the regions essential for transcription initiation.

Transcription Termination by RNA Polymerase II and Poly(A) Addition

All yeast cytoplasmic mRNAs studied so far, including histone mRNAs, are polyadenylated at their 3′ ends. Comparisons of DNA sequences near the poly(A) addition site may give some insight into the termination mechanism, although it is not certain that transcription termination and poly(A) addition take place at the same point within the sequence; for some mammalian viruses, transcription termination occurs farther downstream, with subsequent 3′-terminal processing prior to poly(A) addition (Fitzgerald and Shenk 1981). Analogous to the conserved AATAAA sequence that occurs in mammalian genes 18–32 nucleotides before the poly(A) addition site, many yeast genes have a similarly placed AT-rich sequence (Bennetzen and Hall 1982). A deletion that removes sequences encoding the 3′ end of the yeast *CYC1* transcript prevents normal termination and polyadenylation of the *CYC1* mRNA and permits readthrough "wrong-strand" transcription of the *CYC1* gene sequence (Zaret and Sherman 1982).

Specification of Transcription Initiation by RNA Polymerase III

The most rapid progress in locating sequences that act in eukaryotic gene transcription has been made for genes transcribed by RNA polymerase III. This has occurred because the small size of these genes facilitates their manipulation and structural study, because in vitro RNA polymerase-III systems transcribe with higher specificity and activity than those for RNA polymerases I and II, and because of the great degree of evolutionary conservation of RNA polymerase-III behavior for enzymes obtained from different phyla. Consequently, the principles of RNA polymerase-III specificity deduced in *Xenopus* systems appear to apply to yeast cells and vice-versa.

Results of the deletion analyses that Brown and co-workers carried out on the *Xenopus laevis* and *X. borealis* 5S genes provide a useful paradigm for approaching other RNA polymerase-III transcription units (Bogenhagen et al. 1980; Sakonju et al. 1980). These investigators found that the retention of sequences 50–80 bp downstream from the 5S start point was absolutely required for transcription initiation. The element or elements within this region were named the "control region" to distinguish them from more conventionally located 5'-distal prokaryotic promoter elements. To a first approximation, any sequence may be substituted before position 50 or after position 83 without destroying the ability of the 5S control region to mediate transcription initiation 50 bp upstream of the 5' end of the control region. As the first major result in RNA polymerase-III promoter analysis, this surprising finding raised a number of questions about the detailed way in which the control region functions in specific cases. Many of these have already been answered.

1. What DNA-protein interaction does the +50 and +83 sequence of a *Xenopus* 5S gene mediate? A sequence-specific DNA binding protein, called transcription-factor A, found in stage-III *Xenopus* oocytes, binds to this region very tightly, protecting much of it from DNase-I digestion (Engelke et al. 1980). Transcription-factor A is specific for 5S genes; however, it also binds to 5S RNA molecules both in vivo and in vitro.
2. Do tRNA and other RNA polymerase-III genes have internal control regions that recognize protein factors? For the *X. laevis* tRNA^{MetA} (Hofstetter et al. 1981) and adenovirus 2 VA-RNA$_\text{I}$ genes (Fowlkes and Shenk 1980; Guilfoyle and Weinmann 1981), deletion experiments have identified sequences between +8 and +58 as absolutely essential for transcription initiation. The sequence at the boundaries of this region (D arm and TΨ arm) appear to be essential for transcription activity of the gene. These are the most highly conserved sequences among different eukaryotic tRNA molecules. The tRNA–VA-RNA$_\text{I}$ consensus sequences of these two regions have only slight homology to the 5S gene control region (Koski et al. 1980). Furthermore, the protein factors that activate transcription of the tRNA and the 5S genes can be separated by fractionation (Segall et al. 1980), implying that the central regions of the two types of genes bind different protein molecules.
3. Are the internal sequences that comprise the "control region" the only ones that normally affect promoter strength for genes transcribed by RNA polymerase III? For 5S RNA genes, this question cannot easily be answered, because these genes typically occur in tandemly repeated arrays with identical spacer sequences separating them. For tRNA genes,

however, it is quite often the case that duplicate genes for the same tRNA isoacceptor occur at different chromosomal loci with flanking sequences that are divergent for the different genes. There is evidence of two types, which suggests that these flanking sequences can affect the level of gene activity. In their in vivo strength of suppression, the eight tRNA^{Tyr} suppressor genes of yeast differ markedly from one another (Hawthorne and Leupold 1974; Rothstein 1977), even though all eight genes have identical sequences in the putative control region within the gene. Therefore, the structural determinant that controls the actual level of tRNA^{Tyr} produced by each locus (either by affecting transcription or RNA processing) must lie outside this control region. Similary, there are large differences in in vitro transcription template activity between different iso-coding copies of the *Xenopus* tRNA^{Met} and *Drosophila* tRNA^{Lys} genes. It has been shown in each of these cases that certain naturally occuring 5'-flanking sequences are associated with a high rate of in vitro transcription, and others, with a low rate. When hybrid combinations of these genes are constructed as, e.g., 5'-flanking sequence of A plus coding sequence of B, the rate of transcription is determined by the 5'-flanking region present (Clarkson et al. 1981; DeFranco et al. 1981). On the basis of these two types of data it appears that, for tRNA genes, the internal control region sequences provide only a coarse level of control of promoter activity, with the fine adjustment for a particular locus being supplied by the particular 5'-flanking sequence that is present. In this respect it is interesting to note (Olson et al. 1979, 1981) that the sequence regions flanking yeast tRNA genes exhibit a great degree of interstrain DNA sequence polymorphism between related yeasts. Much of this variation results from the insertion of a repeated DNA element, σ (del Rey et al. 1982; Sandmeyer and Olson 1982). It appears likely that the presence of an inserted σ sequence may specifically affect the rate of transcription of the tRNA genes near it.

Antisuppressor Mutants at the *SUP4* Locus

A collection of mutants at the yeast *SUP4* locus have been used to probe the relative contribution of different regions of the tRNA gene structure to the transcription activity of the gene (Koski et al. 1980, 1982). Each *SUP4*-gene mutation was picked up originally as a red canavanine-resistant adenine-requiring cell clone originating from a *SUP4 can1-100 ade2-1* parent strain that was phenotypically Ade$^+$ canavanine-sensitive and white in colony color. After intragenic mapping was performed to identify separate sites of mutation, representative mutant alleles of the *SUP4* gene were cloned in plasmid pBR322 and sequenced to determine the base-pair alteration responsible for loss of suppression (Kurjan et al. 1980). The collection of such mutant changes so far observed is shown in Figure 5.

The availability of this set of mutant *SUP4* genes in combination with an accurate RNA polymerase-III in vitro transcription system (Koski et al. 1980, and in prep.) has allowed specific tests to be made for the involvement of particular base pairs within the gene in transcription initiation. Recombinant plasmids containing *SUP4*-o genes bearing each one of the second-site mutations shown in Figure 5 were tested for their ability to act as template for pre-tRNA^{Tyr} transcript formation in either an *X. laevis* kidney cell, S100 (Koski et al. 1980), or a cell-free transcription system that uses concentrated yeast whole cell extract (Koski et al. 1982).

The size and amount of pre-RNA^{Tyr} transcription product made on each mutant DNA template was assayed by gel electrophoresis. A strong down-promoter effect was noted for two mutations, G56 and U56, affecting the C residue of the DNA sequence encoding GTΨCGA.

The possibility that other point mutations may have weaker down-promoter effects cannot be ruled out on the basis of these data, since the experiments measure RNA accumulation over a long period, rather than rate of transcription initiation. Recently, a more sensitive in vitro test of promoter strength has been devised by D. Allison (pers. comm.). By measuring the ability of various mutant *SUP4* genes to compete with a tRNA^{Arg} gene for limiting transcription factors, he has shown that mutations U10, U14, A54, and C57 also decrease the efficiency of the *SUP4* promoter.

Together with the deletion experiments mentioned earlier, these studies of mutant tRNA^{Tyr} gene transcription seem capable of precisely defining the regions of a tRNA gene that are essential for transcription initiation. Most of this required DNA sequence lies within two noncontiguous blocks encoding the beginning of the D arm and middle of the TΨ arm of a tRNA molecule (for review, see Hall et al. 1982). Paradoxically, these highly conserved sequences are separated from one another by a distance that varies greatly from one tRNA gene to another.

The essential promoter sequences within tRNA genes are presumed to function analogously to those of 5S RNA genes by binding a transcription factor that subsequently may act to position RNA polymerase III at the starting point for transcription. The isolation and characterization of these transcription factors and the definition of their relationship to the RNA polymerase-III intragenic promoter remain as one of the major unsolved but approachable problems in this area of research.

Transcription Termination by RNA Polymerase III

Both 5S rRNA genes and tRNA genes in yeast uniformly possess two structural features potentially related to transcription termination: extensive two-

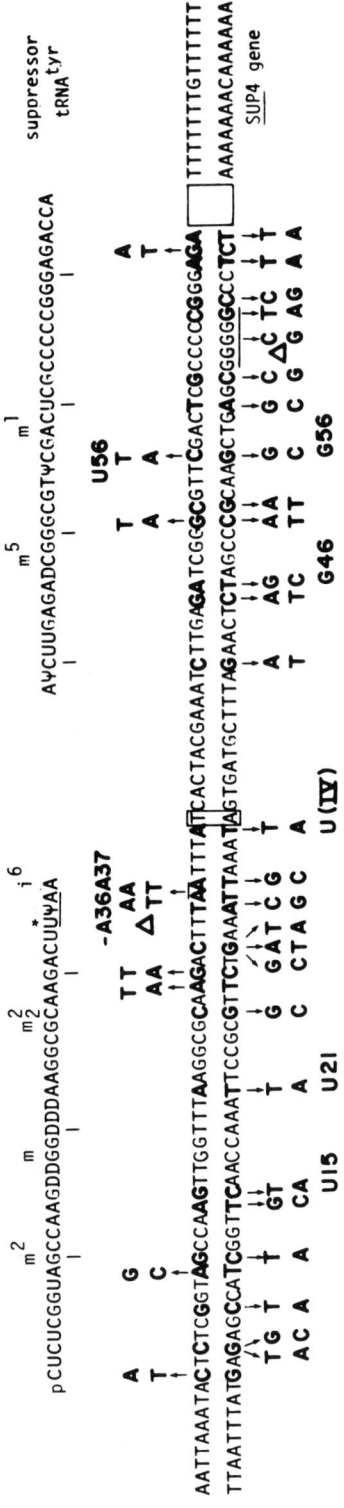

Figure 5 Sequence alterations in SUP4-o genes with second-site-mutations. Positions where mutants have been isolated are shown in boldface type, with previously reported mutational alterations (Kurjan et al. 1980). Each mutation is named according to the position in the tRNA and the nucleotide alteration, as described by Celis (1980).

fold symmetries within the gene and a cluster of Ts on the + strand opposite the RNA 3' terminus. These two features would permit a termination mechanism analogous to *E. coli* ρ-independent termination (Farnham and Platt 1981). Evidence for the involvement of T clusters in transcription termination by yeast and *X. laevis* RNA polymerase III was obtained by the in vitro transcription of mutant *SUP4* genes, -A36A37 and U(IV). These two mutations create T tracts of lengths 6 and 5 nucleotides, respectively, within the gene. In each case, transcription terminates at the site of the mutation (Koski et al. 1980, 1982).

CONCLUSIONS

Yeast Transcription Compared to That in Other Eukaryotes

In recent years, yeast has increasingly been used as a model system for the molecular genetic analysis of various functional macromolecules and cellular processes that are common to all eukaryotes. The suitability of yeast as a model system for eukaryotic transcription appears to be much greater for RNA polymerase-III- than for RNA polymerase-II-mediated transcription.

With regard to RNA polymerase-III enzymes, factors, and recognition sequences, there seems to be a large degree of similarity between all eukaryotic systems. Some effects of 5'-flanking sequences of tRNA genes upon RNA polymerase-III transcription have exhibited species-specificity (Sprague et al 1980) but, by and large, all eukaryotic tRNA genes (as well as many prokaryotic tRNA genes) are transcribed by all eukaryotic RNA polymerase-III systems. A particularly striking example of this is the range of effects on transcription exhibited by a set of mutations within the yeast $SUP4$ tRNA^{Tyr} gene. Each of them affects transcription initiation and/or termination in the same way in a yeast as in an *X. laevis* RNA polymerase-III transcription system. The conclusion, therefore, is that for RNA polymerase III and its associated components, yeast is a model eukaryote that is quite representative. The considerable advantages of suppressor, antisuppressor, and allosuppressor mutations can, in principle, be used in yeast to dissect apart the components of the eukaryotic RNA polymerase-III system.

General principles governing the behavior of RNA polymerase II in recognizing yeast genes are hard to come by; each gene seems to follow its own special logic. For example, the relationship of yeast TATAAA sequences to transcription start points is variable from gene to gene and is certainly very different than the relationship (30-35 bp from TATAAA to start) that prevails in vertebrates. A corollary of this would be that, generally speaking, RNA polymerase-II promoter sequences for animal genes will not function properly in yeast. This is indicated by the failure of the herpesvirus thymidine kinase promoter to function in yeast (Kiss et al. 1982) and by the failure in general to obtain complementation of yeast mutations by *Drosophila* genes in shotgun pools (Henikoff et al. 1981). The presence of a serendipitous yeast RNA polymerase-II promoter within an open reading frame of *Drosophila* DNA explains the special case of *ADE8*, for which *Drosophila* complementation was successful (C. Furlong and S. Henikoff, unpubl.).

At one level—the complementation of structural gene mutations with heterologous DNA in a shotgun pool—it appears that *S. pombe* promoters do function in *S. cerevisiae*. Many *S. pombe* genes have been cloned in this way (Russell 1982; F. Lacroute, pers. comm.). On closer examination, however, it is evident that the RNA polymerase-II recognition systems are markedly different in these two yeasts. A single alcohol-dehydrogenase gene from *S. pombe* has been cloned by complementation of an *S. cerevisiae* alcohol-dehydrogenase mutant strain. Subsequently, the *S. pombe* alcohol-dehydrogenase gene has been sequenced and its transcripts have been mapped, both in *S. pombe* and in *S. cerevisiae*.

From the sequence GTATAAATAG, which is common to the *S. pombe* alcohol dehydrogenase gene and the *S. cerevisiae ADC1* gene, transcripts from the *S. pombe* gene start 47 bp downstream when the gene is in *S. pombe* but 95 bp downstream when the gene is in *S. cerevisiae*. The latter figure agrees well with the distance between GTATAAATAG and the *ADC1* and *ADR1* starts in *S. cerevisiae* (Russell 1982). Thus, it appears that the spacing element or "arm," which phases starts relative to TATAAA, has a different length in *S. pombe* than in *S. cerevisiae*.

For RNA polymerase-II starting specificity, then, the provisional result seems to be that each organism does it in a different way. The general rules, if they exist, will only become apparent later.

There is a small amount of evidence that RNA polymerase-II termination (plus polyadenylation) recognizes different signals in yeast than in mammalian cells. Both for the rabbit β-globin (Beggs et al. 1980) and for the rat growth hormone (A. Barta and G. Ammerer, unpubl.) genes expressed in yeast, transcripts were terminated and polyadenylated at sites within introns. Thus, it appears that yeast and mammals use different signals to specify 3' ends of transcripts.

Future Prospects

Elucidation of the real subunit structure of yeast RNA polymerase requires the demonstration of a functional role in transcription for each polypeptide. This may be approached by dissociation-reconstitution experiments in vitro and by isolation of mutants. Yeast appears the best organism to carry these studies compared to other eukaryotic organisms, which are either not amenable to genetic studies or do not lend themselves to large-scale purification of RNA polymerases.

A number of questions arise concerning the genes coding for RNA polymerase subunits: how they are organized within the genome, and how their expression is regulated in order to obtain the polymerase subunits at the appropriate molar ratio. A second set of questions derive from the observation that the levels of RNA polymerases I and II appear to be regulated independently. Although there is no information on the variation of enzyme-III level compared to enzyme I, it is possible that the synthesis of these two enzymes is coordinated, since they are responsible for the synthesis of rRNA. Should the expression of the structural genes for the three enzymes be under distinct controls, there arises the problem of the genes for the common subunits. Is there one single gene for each common polypeptide or duplicated genes under distinct regulatory controls? One approach for answering some of these questions would be to clone the structural genes for RNA polymerase subunits, which is certainly not a simple task.

Reconstituted in vitro transcription will hopefully soon be available for RNA polymerases II and III as for the animal RNA polymerases. A number of yeast genes have been cloned and characterized, and regulation of transcription will then be accessible to in vitro experiments. Among the yeast mutants deficient in RNA synthesis, some may be affected at the level of essential factors. In that case, their isolation will be facilitated by complementation of extracts from mutant cells in vitro.

The enormous molecular complexity of yeast RNA polymerases probably reflects the variety of regulatory controls that are exerted on these enzymes through interaction with various chromatin components and specificity factors. It is likely that the purified enzymes are only the visible part of the iceberg and that much remains to be done at the genetic and biochemical levels to unravel the essential parts of the transcription system together with its regulatory controls.

ACKNOWLEDGMENTS

We thank our co-workers, past and present, and colleagues in other laboratories who have made available the results of their research for presentation in this paper.

REFERENCES

Adman, R., L.D. Schultz, and B.D. Hall. 1972. Transcription in yeast: Separation and properties of multiple RNA polymerases. *Proc. Natl. Acad. Sci.* **69**: 1702.

Alber, T. and G. Kawasaki. 1982. Nucleotide sequence of the triose phosphate isomerase gene of *Saccharomyces cerevisiae. J. Mol. Appl. Genet.* (in press).

Auld, D.S., I. Atsuya, C. Campino, and P. Valenzuela. 1976. Yeast RNA polymerase I: A eucaryotic zinc metalloenzyme. *Biochem. Biophys. Res. Commun.* **69**: 548.

Bayev, A.A., O.J. Georgiev, A.A. Hadjiolov, M.B. Kermekchiev, N. Nikolaev, K.G. Skryabin, and V.M. Zakharyev. 1980. The structure of the yeast ribosomal RNA genes. 2. The nucleotide sequence of the initiation site for ribosomal RNA transcription. *Nucleic Acids Res.* **8:** 4919.

Beggs, J.D. 1978. Transformation of yeast by a replicating hybrid plasmid. *Nature* **275:** 104.

Beggs, J.D., J. van den Berg, A. van Ooyen, and C. Weissman. 1980. Abnormal expression of chromosomal rabbit β-globin gene in *Saccharomyces cerevisiae*. *Nature* **283:** 835.

Bell, G.I., P. Valenzuela, and W.J. Rutter. 1976. Phosphorylation of yeast RNA polymerases. *Nature* **261:** 429.

———. 1977. Phosphorylation of yeast DNA-dependent RNA polymerases *in vivo* and *in vitro*. *J. Biol. Chem.* **252:** 3082.

Bennetzen, J.L. and B.D. Hall. 1982. The primary structure of the *Saccharomyces cerevisiae* gene for alcohol dehydrogenase I. *J. Biol. Chem.* **257:** 3018.

Benoist, C. and P. Chambon. 1981. *In vivo* sequence requirements of the SV40 early promoter region. *Nature* **290:** 304.

Bogenhagen, D.F., S. Sakonju, and D.D. Brown. 1980. A control region in the center of the 5S RNA gene directs specific initiation of transcription. II. The 3' border of the region. *Cell* **19:** 27.

Brodner, O.G. and T. Wieland. 1976. Identification of the amatoxin-binding subunit of RNA polymerase B by affinity labeling experiments. Subunit B3—The true amatoxin receptor protein of multiple RNA polymerase B. *Biochemistry* **15:** 3484.

Breathnach, R. and P. Chambon. 1981. Organization and expression of eucaryotic split genes coding for proteins. *Annu. Rev. Biochem.* **50:** 349.

Buhler, J.M., A. Sentenac, and P. Fromageot. 1974. Isolation, structure, and general properties of yeast ribonucleic acid polymerase A (or I). *J. Biol. Chem.* **249:** 5963.

———. 1976. The presence of phosphorylated subunits in yeast RNA polymerases A and B. *FEBS Lett.* **71:** 37.

Buhler, J.M., J. Huet, K.E. Davies, A. Sentenac, and P. Fromageot. 1980. Immunological studies of yeast nuclear RNA polymerases at the subunit level. *J. Biol. Chem.* **255:** 9949.

Bull, P., H. MacDonald, and P. Valenzuela. 1981a. The interaction of yeast RNA polymerase I and Cibacron blue F3GA. *Biochim. Biophys. Acta* **653:** 368.

Bull, P., C. Campino, G.I. Bell, A. Venegas, and P. Valenzuela. 1981b. The effect of pH on the structure and activity of yeast RNA polymerase I. *Arch. Biochem. Biophys.* **209:** 637.

Carter, B.L.A. and I.W. Dawes. 1975. Synthesis of two DNA-dependent RNA polymerases in yeast. *Exp. Cell. Res.* **92:** 253.

Celis, J.E. 1980. Collection of mutant tRNA sequences. *Nucleic Acids Res.* **8:** r23.

Chambon, P. 1974. Eukaryotic RNA polymerases. In *The enzymes* (ed. P.D. Boyer), vol 10, p. 261. Academic Press, New York.

Clarkson, S.G., R.A. Koski, J. Corlet, and R.A. Hipskind. 1981. Influence of 5' flanking sequences on tRNA transcription *in vitro*. *ICN-UCLA Symp. Mol. Cell. Biol.* **23:** 463.

Cooper, C.S. and R.V. Quincey. 1979a. The role of subunits in yeast DNA-dependent ribonucleic acid polymerase A. *Biochem. J.* **181:** 301.

———. 1979b. Convolution analysis of transcription by yeast DNA-dependent ribonucleic acid polymerase A. *Biochem. J.* **177:** 825.

Cramer, J.H., J. Sebastian, R.H. Rownd, and H.O. Halvorson. 1974. Transcription of *Saccharomyces cerevisiae* ribosomal DNA *in vivo* and *in vitro*. *Proc. Natl. Acad. Sci.* **71:** 2188.

DeFranco, D., S. Sharp, and D. Söll. 1981. Identification of regulatory sequences contained in the 5'-flanking region of *Drosophila* lysine tRNA genes. *J. Biol. Chem.* **256:** 12424.

del Rey, F., T.F. Donahue, and G.R. Fink. 1982. *sigma*, a repetitive element found adjacent to tRNA genes of yeast. *Proc. Natl. Acad. Sci.* **79:** 4138.

Dezélée, S. and A. Sentenac. 1973. Role of DNA · RNA hybrids in eukaryotes: Purification and properties of yeast RNA polymerase B. *Eur. J. Biochem.* **34:** 41.

Dezélée, S., A. Sentenac, and P. Fromageot. 1970. Study on yeast RNA polymerase. Effect of α-amanitin and rifampicin. *FEBS Lett.* **7:** 220.
———. 1972. Role of DNA · RNA hybrids in eucaryotes. I. Purification of yeast RNA polymerase B. *FEBS. Lett.* **21:** 1.
———. 1974a. Role of deoxyribonucleic acid-ribonucleic acid hybrids in eucaryotes. Study of the template requirements of yeast RNA polymerases and nature of the RNA product. *J. Biol. Chem.* **249:** 5971.
———. 1974b. Role of deoxyribonucleic acid-ribonucleic acid hybrids in eucaryotes. Synthetic ribo- and deoxy-ribopolynucleotides as templates for yeast RNA polymerase B (or II). *J. Biol. Chem.* **249:** 5978.
Dezélée, S., F. Wyers, A. Sentenac, and P. Fromageot. 1976. Two forms of RNA polymerase B in yeast: Proteolytic conversion *in vitro* of enzyme B_I into B_{II}. *Eur. J. Biochem.* **65:** 543.
Di Mauro, E., M. Mezzina, and M. Arca. 1974. *In vitro* inhibition of *Saccharomyces cerevisiae* RNA polymerase by rifamycin derivatives. *Arch. Biochem. Biophys.* **164:** 765.
Dobson, M.J., M.F. Tuite, N.A. Roberts, A.J. Kingsman, S.M. Kingsman, R.E. Perkins, S.C. Conroy, B. Dunbar, and L.A. Fothergill. 1982. Conservation of high efficiency promoter sequences in *Saccharomyces cerevisiae*. *Nucleic Acids Res.* **10:** 2625.
Engelke, D.R., S.Y. Ng, B.S. Shastry, and R.G. Roeder. 1980. Specific interaction of a purified transcription factor with an internal control region of 5S RNA genes. *Cell* **19:** 717.
Errede, B., T.S. Cardillo, F. Sherman, E. Dubois, J. Deschamps, and J.-M. Wiame. 1980. Mating signals control expression of mutations resulting from insertion of a transposable repetitive element adjacent to diverse yeast genes. *Cell* **22:** 427.
Farnham, P.J. and T. Platt. 1981. Rho-independent termination: Dyad symmetry in DNA causes polymerase to pause during transcription *in vitro*. *Nucleic Acids Res.* **9:** 563.
Faye, G., D.W. Leung, K. Tatchell, B.D. Hall, and M. Smith. 1981. Deletion mapping of sequences essential for *in vivo* transcription of the iso-1-cytochrome *c* gene. *Proc. Natl. Acad. Sci.* **78:** 2258.
Fitzgerald, M. and T. Shenk. 1981. The sequence 5'-AAUAAA-3' forms part of the recognition site for polyadenylation of late SV40 mRNAs. *Cell* **24:** 251.
Fowlkes, D.M. and T. Shenk. 1980. Transcriptional control regions of the adenovirus VA_I RNA gene. *Cell* **22:** 405.
Fraser, R.S.S. and J. Creanor. 1974. Rapid and selective inhibition of RNA synthesis in yeast by 8-hydroxyquinoline. *Eur. J. Biochem.* **46:** 67.
Frederick, E.W., U. Maitra, and J. Hurwitz. 1969. The role of deoxyribonucleic acid in ribonucleic acid synthesis. The purification and properties of RNA polymerase from yeast: Preferential utilization of denatured DNA as template. *J. Biol. Chem.* **244:** 413.
Gross, K.J. and A.O. Pogo. 1974. Control mechanism of ribonucleic acid synthesis in eucaryotes. The effect of amino acid and glucose starvation and cycloheximide on yeast DNA-dependent RNA polymerase. *J. Biol. Chem.* **249:** 568.
Grosschedl, R. and M.L. Birnstiel. 1982. Delimitation of far upstream sequences required for maximal *in vitro* transcription of an H2A histone gene. *Proc. Natl. Acad. Sci.* **79:** 297.
Guarente, L. and M. Ptashne. 1981. Fusion of *Escherchia coli* lacZ to the cytochrome *c* gene of *Saccharomyces cerevisiae*. *Proc. Natl. Acad. Sci.* **78:** 2199.
Guilfoyle, R. and R. Weinmann. 1981. Control region for adenovirus VA RNA transcription. *Proc. Natl. Acad. Sci.* **78:** 3378.
Hager, G.L., M.J. Holland, and W.J. Rutter. 1977. Isolation of ribonucleic acid polymerases I, II, and III from *Saccharomyces cerevisiae*. *Biochemistry* **16:** 1.
Hager, G.L., M. Holland, P. Valenzuela, F. Weinberg, and W.J. Rutter. 1976. RNA polymerases and transcriptive specificity in *Saccharomyces cerevisiae*. In *RNA polymerase* (ed. R. Losick and M. Chamberlin), p. 745. Cold Spring Harbor Laboratory, Cold Spring Harbor, New York.
Hall, B.D., S.G. Clarkson, and G.P. Tocchini-Valentini. 1982. Transcription initiation of eukaryotic tRNA genes. *Cell* **29:** 3.

Hawthorne, D.C. and U. Leupold. 1974. Suppressor mutations in yeast. *Curr. Top. Microbiol. Immunol.* **64**: 1.

Heffron, F., M. So, and B.J. McCarthy. 1978. In vitro mutagenesis of a circular DNA molecule by using synthetic restriction sites. *Proc. Natl. Acad. Sci.* **75**: 6012.

Henikoff, S., K. Tatchell, B.D. Hall, and K.A. Nasmyth. 1981. Isolation of gene from *Drosophila* by complementation in yeast. *Nature* **289**: 33.

Hildebrandt, A., J. Sebastian, and H.O. Halvorson. 1973. Yeast nuclear RNA polymerase I and II are immunologically related. *Nat. New Biol.* **246**: 73.

Hinnen, A., J.B. Hicks, and G.R. Fink. 1978. Transformation of yeast. *Proc. Natl. Acad. Sci.* **75**: 1929.

Hitzeman, R.A., F.E. Hagie, H.L. Levine, D.V. Goeddel, G. Ammerer, and B.D. Hall. 1981. Expression of a human gene for interferon in yeast. *Nature* **293**: 717.

Hofstetter, H., A. Kressman, and M.L. Birnstiel. 1981. A split promoter for a eukaryotic tRNA gene. *Cell* **24**: 573.

Holland, J.P. and M.J. Holland. 1980. Structural comparison of two nontandemly repeated yeast glyceraldehyde-3-phosphate dehydrogenase genes. *J. Biol. Chem.* **255**: 2596.

Holland, M.J., G.L. Hager, and W.J. Rutter. 1977. Transcription of yeast DNA by homologous RNA polymerase I and II: Selective transcription of ribosomal genes by RNA polymerase I. *Biochemistry* **16**: 16.

Hollenberg, C.P. 1973. Ribosomal ribonucleic acid synthesis by isolated yeast ribonucleic acid polymerases. *Biochemistry* **12**: 5320.

Horowitz, B., B.A. Goldfinger, and J. Marmur. 1976. Effect of cordycepin triphosphate on the nuclear DNA-dependent RNA polymerases and poly(A) polymerase from the yeast *Saccharomyces cerevisiae*. *Arch. Biochem. Biophys.* **172**: 143.

Huet, J., A. Sentenac, and P. Fromageot. 1982. Spot-immunodetection of conserved determinants in eukaryotic RNA polymerases. *J. Biol. Chem.* **257**: 2613.

Huet, J., J.M. Buhler, A. Sentenac, and P. Fromageot. 1975. Dissociation of two polypeptide chains from yeast RNA polymerase A. *Proc. Natl. Acad. Sci.* **72**: 3034.

———. 1977. Characterization of ribonuclease H activity associated with yeast RNA polymerase A. *J. Biol. Chem.* **252**: 8848.

Huet, J., F. Wyers, J.M. Buhler, A. Sentenac, and P. Fromageot. 1976a. Association of RNase H activity with yeast RNA polymerase A. *Nature* **261**: 431.

Huet, J., S. Dezélée, F. Iborra, J.M. Buhler, A. Sentenac, and P. Fromageot. 1976b. Further characterization of yeast RNA polymerases. Effect of subunit removal. *Biochimie* **58**: 71.

Iborra, F., J. Huet, B. Breant, A. Sentenac, and P. Fromageot. 1979. Identification of two different RNase H activities associated with yeast RNA polymerase A. *J. Biol. Chem.* **254**: 10920.

Jazwinski, S.M. and G.M. Edelman. 1979. Replication in vitro of the 2 micron DNA plasmid of yeast. *Proc. Natl. Acad. Sci.* **76**: 1223.

Jimenez, A., D.J. Tipper, and J. Davies. 1973. Mode of action of thiolutin, an inhibitor of macromolecular synthesis in *Saccharomyces cerevisiae*. *Antimicrob. Agents Chemother.* **3**: 729.

Johnston, G.C. and R.A. Singer. 1978. RNA synthesis and control of cell division in the yeast *S. cerevisiae*. *Cell* **14**: 951.

Karagyozov, L.K., M.A. Valkanov, and A.A. Hadjiolov. 1978. Transcription of DNA-histone complexes by yeast RNA polymerase B. *Nucleic Acids Res.* **5**: 1907.

Kiss, G.B., R.E. Pearlman, K.V. Cornish, J.D. Friesen, and V.L. Chan. 1982. The herpes simplex virus thymidine kinase gene is not transcribed in *Saccharomyces cerevisiae*. *J. Bacteriol.* **149**: 542.

Klekamp, M.S. and P.A. Weil. 1982. Specific transcription of homologous class III genes in yeast soluble cell-free extracts. *J. Biol. Chem.* **257**: 8432.

Klemenz, R. and P.E. Geiduschek. 1980. The 5' terminus of the precursor ribosomal RNA of *Saccharomyces cerevisiae*. *Nucleic Acids Res.* **8**: 2679.

Klootwijk, J., R.J. Planta, and J.P. Bakker. 1976. Synthesis and methylation of ribosomal RNA in isolated yeast nuclei. *J. Microsc. Biol. Cell.* **26:** 91.

Kojo, H., B.D. Greenberg, and A. Sugino. 1981. Yeast 2 micron plasmid DNA replication in vitro: Origin and direction. *Proc. Natl. Acad. Sci.* **78:** 7261.

Koski, R.A., D.S. Allison, M. Worthington, and B.D. Hall. 1982. An *in vitro* RNA polymerase III system from *S. cerevisiae:* Effects of deletions and point mutations upon *SUP4* gene transcription. *Nucleic Acids Res.* (in press).

Koski, R.A., S.G. Clarkson, J. Kurjan, B.D. Hall, and M. Smith. 1980. Mutations of the yeast *SUP4* tRNA^{Tyr} locus: Transcription of the mutant genes *in vitro*. *Cell* **22:** 415.

Kurjan, J., B. D. Hall, S. Gillam, and M. Smith. 1980. Mutations at the yeast *SUP4* tRNA^{Tyr} locus: DNA sequence changes in mutants lacking suppressor activity. *Cell* **20:** 701.

Lacroute, F. 1973. RNA and protein elongation rates in *Saccharomyces cerevisiae*. *Mol. Gen. Genet.* **25:** 319.

Lacroute, F., J. Huet, and F. Exinger. 1975. Dominant and semi-dominant mutations leading to thermosensitivity of ribonucleic acid biosynthesis in *Saccharomyces cerevisiae*. *J. Bacteriol.* **122:** 847.

Lattke, H. and U. Weser. 1976. Yeast RNA-polymerase B: A zinc protein. *FEBS Lett.* **65:** 288.

———. 1977. Functional aspects of zinc in yeast RNA-polymerase B. *FEBS Lett.* **83:** 297.

Lescure, B., J. Bennetzen, and A. Sentenac. 1981a. *In vitro* transcription of yeast alcohol dehydrogenase I gene by homologous RNA polymerase B. Selective initiation and discontinuous elongation on a supercoiled template. *J. Biol. Chem.* **265:** 11018.

Lescure, B., V. Williamson, and A. Sentenac. 1981b. Efficient and selective initiation by yeast RNA polymerase B in a dinucleotide-primed reaction. *Nucleic Acids Res.* **9:** 31.

Lewis, M.K. and R.R. Burgess. 1980. Transcription of simian virus 40 DNA by wheat germ RNA polymerase II. *J. Biol. Chem.* **255:** 4928.

Manley, J.L., A. Fire, A. Cano, P.A. Sharp, and M.L. Gefter. 1980. DNA-dependent transcription of adenovirus genes in a soluble whole-cell extract. *Proc. Natl. Acad. Sci.* **77:** 3855.

Martial, J., J. Zaldivar, P. Bull, A. Venegas, and P. Valenzuela. 1975. Inactivation of rat liver RNA polymerases I and II and yeast RNA polymerase I by pyridoxal 5'-phosphate. Evidence for the participation of lysyl residues at the active site. *Biochemistry* **14:** 4907.

Martin, F.H. and I. Tinoco. 1980. DNA · RNA hybrid duplex containing oligo (dA:rU) sequences are exceptionally unstable and may facilitate termination of transcription. *Nucleic Acids Res.* **8:** 2295.

Mathis, D.J. and P. Chambon. 1981. The SV40 early region TATA box is required for accurate *in vitro* initiation of transcription. *Nature* **290:** 310.

Nagamine, Y., J. Bennetzen, A. Sentenac, and P. Fromageot. 1981. Single-stranded DNA transcription by yeast RNA polymerase B. *Biochim. Biophys. Acta* **656:** 220.

Nasmyth, K.A., K. Tatchell, B.D. Hall, C. Astell, and M. Smith. 1981. A position effect in the control of transcription at yeast mating-type loci. *Nature* **289:** 244.

Olson, M.V., K. Loughney, and B.D. Hall. 1979. Identification of the yeast DNA sequences that correspond to specific tyrosine-inserting nonsense suppressor loci. *J. Mol. Biol.* **132:** 387.

Olson, M.V., G.S. Page, A. Sentenac, P.W. Piper, M. Worthington, R.B. Weiss, and B.D. Hall. 1981. Only one of two closely related yeast suppressor tRNA genes contains an intervening sequence. *Nature* **291:** 464.

Plevani, P. and L.M.S. Chang. 1977. Enzymatic initiation of DNA synthesis by yeast DNA polymerases. *Proc. Natl. Acad. Sci.* **74:** 1937.

———. 1978. Initiation of enzymatic DNA synthesis by yeast RNA polymerase I. *Biochemistry* **17:** 2530.

Ponta, H., U. Ponta, and E. Wintersberger. 1971. DNA-dependent RNA polymerases from yeast, partial characterization of three nuclear enzyme activities. *FEBS Lett.* **18:** 204.

———. 1972. Purification and properties of DNA-dependent RNA polymerases from yeast. *Eur. J. Biochem.* **29:** 110.

Ponta, H., U. Ponta, V. Kraft, and E. Wintersberger. 1974. Transcription of yeast DNA *in vitro*: Preparation of yeast DNA which is used as template by the purified DNA-dependent RNA polymerases A and B from yeast. *Eur. J. Biochem.* **46:** 473.

Riva, M., J.-M. Buhler, A. Sentenac, P. Fromageot, and D.C. Hawthorne. 1982. Natural variation in RNA polymerase A. Formation of a mosaic RNA polymerase A in a meiotic segregant from an interspecific hybrid. *J. Biol. Chem.* **257:** 4570.

Roeder, R.G. 1976. Eucaryotic nuclear RNA polymerases. In *RNA polymerase* (ed. R. Losick and M. Chamberlin), p. 285. Cold Spring Harbor Laboratory, Cold Spring Harbor, New York.

Roeder, R.G. and W.J. Rutter. 1969. Multiple forms of DNA-dependent RNA polymerase in eucaryotic organisms. *Nature* **224:** 234.

Rothstein, R. 1977. A genetic fine structure analysis of the suppressor 3 locus in *Saccharomyces*. *Genetics* **85:** 55.

Ruet, A., A. Sentenac, and P. Fromageot. 1978. A specific assay for yeast RNA polymerases in crude cell extracts. *Eur. J. Biochem.* **90:** 325.

Ruet, A., J.C. Bouhet, J.M. Buhler, and A. Sentenac. 1975. On the mode of action of Lomofungin, an inhibitor of RNA synthesis in yeast. *Biochemistry* **14:** 4651.

Ruet, A., A. Sentenac, P. Fromageot, B. Winsor, and F. Lacroute. 1980. A mutation of the B_{220} subunit gene affects the structural and functional properties of yeast RNA polymerase B *in vitro*. *J. Biol. Chem.* **255:** 6450.

Russell, P. 1982. "Gene structure and expression in the fission yeast *Schizosaccharomyces pombe*." Ph.D. thesis, University of Washington, Seattle.

Sakonju, S., D.F. Bogenhagen, and D.D. Brown. 1980. A control region in the center of the 5S RNA gene directs specific initiation of transcription. I. The 5' border of the region. *Cell* **19:** 13.

Sandmeyer, S.B. and M.V. Olson. 1982. Insertion of a novel repetitive element at the same position in the 5'-flanking regions of two dissimilar yeast tRNA genes. *Proc. Natl. Acad. Sci.* (in press).

Sawadogo, M. 1981. On the inhibition of yeast RNA polymerases A and B by tRNA and α-amanitin. *Biochem. Biophys. Res. Commun.* **98:** 261.

Sawadogo, M., A. Sentenac, and P. Fromageot. 1980a. Interaction of a new polypeptide with yeast RNA polymerase B. *J. Biol. Chem.* **255:** 12.

———. 1980b. Similar binding site for P_{37} factor on yeast RNA polymerases A and B. *Biochem. Biophys. Res. Commun.* **96:** 258.

Sawadogo, M., B. Lescure, A. Sentenac, and P. Fromageot. 1981. Native DNA transcription by yeast RNA polymerase—P_{37} complex. *Biochemistry* **20:** 3542.

Schultz, L.D. 1978. Transcriptional role of yeast deoxyribonucleic acid-dependent ribonucleic acid polymerase III. *Biochemistry* **17:** 750.

Schultz, L.D. and B.D. Hall. 1976. Transcription in yeast: α-amanitin sensitivity and other properties which distinguish between RNA polymerases I and III. *Proc. Natl. Acad. Sci.* **73:** 1029.

Sebastian, J., M.M. Bhargava, and H.O. Halvorson. 1973a. Nuclear deoxyribonucleic acid-dependent ribonucleic acid polymerases from *Saccharomyces cerevisiae*. *J. Bacteriol.* **114:** 1.

Sebastian, J., F. Mian, and H.O. Halvorson. 1973b. Effect of the growth rate on the level of the DNA-dependent RNA polymerase in *Saccharomyces cerevisiae*. *FEBS Lett.* **34:** 159.

Sebastian, J., I. Takano, and H.O. Halvorson. 1974. Independent regulation of the nuclear RNA polymerases I and II during the yeast cell cycle. *Proc. Natl. Acad. Sci.* **71:** 769.

Segall, J., T. Matsui, and R.G. Roeder. 1980. Multiple factors are required for the accurate transcription of purified genes by RNA polymerase III. *J. Biol. Chem.* **255:** 11986.

Sentenac, A., J.M. Buhler, A. Ruet, J. Huet, P. Iborra, and P. Fromageot. 1978. Eukaryotic RNA polymerases. In *Gene expression* (ed. B.F.C. Clark et al.), p. 187. Pergamon Press, Oxford, England.

Sentenac, A., S. Dezélée, F. Iborra, J.M. Buhler, J. Huet, F. Wyers, A. Ruet, and P.

Fromageot. 1976. Yeast RNA polymerases. In *RNA polymerase* (ed. R. Losick and M. Chamberlin), p. 763. Cold Spring Harbor Laboratory, Cold Spring Harbor, New York.

Shulman, R.W., C.E. Sripati, and J.R. Warner. 1977. Non-coordinated transcription in the absence of protein synthesis in yeast. *J. Biol. Chem.* **252**: 1344.

Siebenlist, U., R.B. Simpson, and W. Gilbert. 1980. *E. coli* RNA polymerase interacts homologously with two different promoters. *Cell* **20**: 269.

Sprague, K.V., D. Larson, and D. Morton. 1980. 5′ flanking sequence signals are required for activitity of silkworm alanine tRNA genes in homologous *in vitro* transcription systems. *Cell* **22**: 171.

Struhl, K. 1981. Deletion mapping a eukaryotic promoter. *Proc. Natl. Acad. Sci.* **78**: 4461.

———. 1982. A eukaryotic promoter contains at least two distinct elements. *Proc. Natl. Acad. Sci.* (in press).

Tatchell, K., K.A. Nasmyth, B.D. Hall, C. Astell, and M. Smith. 1981. *In vitro* analysis of the mating-type locus in yeast. *Cell* **27**: 25.

Tekamp, P.A., P. Valenzuela, T. Maynard, G.I. Bell, and W.J. Rutter. 1979. Specific gene transcription in yeast nuclei and chromatin by added homologous RNA polymerases I and III. *J. Biol. Chem.* **254**.: 955.

Thonart, P., J. Bechet, F. Hilger, and A. Burny. 1976. Thermosensitive mutations affecting ribonucleic acid polymerases in *Saccharomyces cerevisiae*. *J. Bacteriol.* **125**: 25.

Tipper, D.J. 1973. Inhibition of yeast ribonucleic acid polymerases by Thiolutin. *J. Bacteriol.* **116**: 245.

Valenzuela, P., G.I. Bell, and W.J. Rutter. 1976a. The 24,000 dalton subunits and the activity of yeast RNA polymerases. *Biochem. Biophys. Res. Commun.* **71**: 26.

Valenzuela, P., G.I. Bell, F. Weinberg, and W.J. Rutter. 1976b. Yeast DNA-dependent RNA polymerases I, II and III. The existence of subunits common to the three enzymes. *Biochem. Biophys. Res. Commun.* **71**: 1319.

———. 1978a. Isolation and assay of eucaryotic DNA-dependent RNA polymerases. *Methods Cell Biol.* **19**: 1.

Valenzuela, P., G.L. Hager, F. Weinberg, and W.J. Rutter. 1976c. Molecular structure of yeast RNA polymerase III: Demonstration of the tripartite transcriptive system in lower eucaryotes. *Proc. Natl. Acad. Sci.* **73**: 1024.

Valenzuela, P., F. Weinberg, G.I. Bell, and W.J. Rutter. 1976d. Yeast DNA-dependent RNA polymerase I: A rapid procedure for the large scale purification of homogeneous enzyme. *J. Biol. Chem.* **251**: 1464.

Valenzuela, P., P. Bull, J. Zaldivar, A. Venegas, and J. Martial. 1978b. Subunits of yeast RNA polymerase I involved in interactions with DNA and nucleotides. *Biochem. Biophys. Res. Commun.* **81**: 662.

Van Keulen, H., R.J. Planta, and J. Retel. 1975. Structure and transcription specificity of yeast RNA polymerase A. *Biochim. Biophys. Acta* **395**: 179.

Waldron, C. 1977. Synthesis of ribosomal and transfer ribonucleic acids in yeast during a nutritional shift-up. *J. Gen. Microbiol.* **98**: 215.

Waldron, C. and F. Lacroute. 1975. Effect of growth rate on the amounts of ribosomal and transfer ribonucleic acids in yeast. *J. Bacteriol.* **122**: 855.

Wandzilak, T.M. and R.W. Benson. 1978. *Saccharomyces cerevisiae* DNA-dependent RNA polymerase III: A zinc metalloenzyme. *Biochemistry* **17**: 426.

Wasylyk, B., C. Kedinger, J. Corden, O. Brison, and P. Chambon. 1980. Specific *in vitro* initiation of transcription on conalbumin and ovalbumin genes and comparison with adenovirus-2 early and late genes. *Nature* **285**: 367.

Weil, P.A., D.S. Luse, J. Segall, and R.G. Roeder. 1979. Selective and accurate initiation of transcription at the AD_2 major late promoter in a soluble system dependent on purified RNA polymerase II and DNA. *Cell* **18**: 469.

Williamson, V.M., E.T. Young, and M. Ciriacy. 1981. Transposable elements associated with constitutive expression of yeast alcohol dehydrogenase II. *Cell* **23**: 605.

Winsor, B., F. Lacroute, A. Ruet, and A. Sentenac. 1979. Isolation and characterisation of a strain of *Saccharomyces cerevisiae* deficient in *in vitro* RNA polymerase B (II) activity. *Mol. Gen. Genet.* **173**: 145.

Young, T., V. Williamson, A. Taguchi, M. Smith, A. Sledziewski, D. Russell, J. Osterman, C. Denis, D. Cox, and D. Beier. 1982. The alcohol dehydrogenase genes of the yeast, *Saccharomyces cerevisiae*: Isolation, structure and regulation. In *Genetic engineering of microorganisms for chemicals* (ed. A. Hollaender et al.), p. 335. Plenum Press, New York.

Zaret, K.S. and F. Sherman. 1982. DNA sequence required for efficient transcription termination in yeast. *Cell* **28**: 563.

Zillig, W., P. Palm, and A. Heil. 1976. Function and reassembly of subunits of DNA-dependent RNA polymerase. In *RNA polymerase* (ed. R. Losick and M. Chamberlin), p. 101. Cold Spring Harbor Laboratory, Cold Spring Harbor, New York.

Zubenko, G.S. and E.W. Jones. 1981. Protein degradation, meiosis and sporulation in proteinase deficient mutants of *Saccharomyces cerevisiae*. *Genetics* **97**: 45.

Principles and Practice of Recombinant DNA Research with Yeast

David Botstein
Department of Biology
Massachusetts Institute of Technology
Cambridge, Massachusetts 02139

Ronald W. Davis
Department of Biochemistry
Stanford University School of Medicine
Stanford, California 94305

1. **Technology**
   A. Acquisition and Amplification of Yeast Genes in *E. coli*
   B. Isolation of Particular Genes
   C. Return of Cloned Genes to Yeast: DNA Transformation
2. **Uses of Recombinant DNA Technology in Yeast Molecular Biology Research**
   A. Characterization of Cloned Genes
   B. Uses of Cloned Genes in Genetic Analysis
   C. Alteration of Cloned Genes
   D. Identifying Cloned DNA Segments with Yeast Genes
3. **Conclusion**

INTRODUCTION

Recombinant DNA research shows great promise in furthering understanding of yeast biology by making possible the analysis and manipulation of yeast genes not only in the test tube but also in yeast cells. The intent of this paper is to summarize the major features of current technology and to outline some implications for the immediate future of yeast molecular biology. There now exist simple and general methods for isolating and amplifying virtually any yeast gene, although these methods generally require an intermediate step in *Escherichia coli*. Powerful and exquisitely sensitive hybridization methods have been developed that allow direct physical analysis of any chromosomal region containing a gene that has been molecularly cloned. Most importantly, it is now possible to return to yeast, by transformation with DNA, cloned genes using a variety of selectable marker systems developed for this purpose. These technological advances have combined to make feasible truly molecular as well as classical genetic manipulation and analysis in yeast.

The many uses to which these new tools have already been put have recently been reviewed by Olson (1981), and the results that have been obtained are to be found scattered in many of the papers in this volume. The biological problems that have been most effectively addressed by recombinant DNA technology are ones that have the structure and organization of individual genes as their central issue. Thus, the reader will find elsewhere in this volume that many signal advances in the understanding of such diverse subjects as the switching of mating types, the maturation of tRNAs, and the maintenance of the reiterated rDNA have resulted from the use of recombinant DNA methods. However, it seems quite clear that gene structure issues are only the beginning, and it is hoped that this paper will help to identify the immediate prospects for use of this technology for studies of gene function, regulation, and evolution as well.

TECHNOLOGY

Acquisition and Amplification of Yeast Genes in *E. coli*

At the present time, virtually all systems for isolating and amplifying yeast genes require the initial isolation of in vitro recombinants in *E. coli* host-vector systems. The primary reasons for this are practical: *E. coli* systems offer high efficiency of transformation, good amplification with simple methods, and powerful hybridization screening systems. In addition to the practical necessity, however, the cloning of yeast genes in *E. coli* has some advantages in principle. These include the fact that the DNAs of bacterial hosts and vectors in general are not homologous to yeast probes, which facilitates hybridization screening. For example, a yeast sequence extracted from *E. coli* requires minimal purification to be maximally useful as a hybridization probe; the same sequence isolated from yeast would be useful in direct proportion to its purity. The possibility of maintaining cloned genes in two alternative host environments allows the experimenter options in carrying out manipulations and analysis of the cloned DNA, taking advantage of the unique properties of each host and its mutants. For instance, the availability of recombination-deficient *E. coli* strains makes easier the preservation of repeated DNA sequences.

Most yeast gene isolations thus begin with the insertion of yeast DNA into an *E. coli* vector system. Since the total size of the yeast genome is small (~ 15,000 kbp), it is usually not necessary to purify the DNA of interest prior to the cloning. Total yeast DNA is usually cloned into the *E. coli* host vector, producing a collection of recombinant organisms referred to as a "shotgun," "library," "pool," or "bank." Two main systems of cloning yeast DNA in *E. coli* have been used. One involves a self-replicating circular plasmid DNA encoding one or more markers (often antibiotic-resistance determinants) that permit selection for the plasmid to be applied. The other involves the insertion of yeast DNA into bacteriophage λ. Each of these systems has advantages and drawbacks; many laboratories have found it useful to employ both systems, exploiting each in turn as circumstances dictate.

Cloning with Bacteriophage λ

The genome of λ is a duplex linear DNA molecule with homologous single-stranded "cohesive" ends. The central third of the λ chromosome is dispensable; λ derivatives in which this region is replaced with other DNAs are viable phages fully capable of lytic growth. However, simple deletion of the central region results in inviability when the deletion removes more than about 26% of the DNA: There is a minimum DNA size required for successful packaging into the phage head (Parkinson and Huskey 1971; Thomas et al. 1974).

Most λ vector systems consist of phages in which most of the central region is flanked by a pair of restriction endonuclease targets that do not occur elsewhere in the vector genome (Murray and Murray 1974; Thomas et al. 1974; deWet et al. 1980; Karn et al. 1980). Digestion with the appropriate restriction endonuclease results in the production of three DNA fragments: a left end, a right end (both of which contain λ genes essential for viability), and a central fragment (containing no essential genes). The restriction endonuclease used must be one that produces cohesive single-stranded ends 4 bases or more in length. When a molar excess of foreign DNA digested with the same restriction endonuclease is added to this mixture of fragments and the mixture is subjected to the action of DNA ligase, complete phage genomes are reassembled, as shown in Figure 1. Other arrangements of DNA (e.g., two left ends joined, left ends attached to foreign DNA without right ends, and so on) are also produced, but these are not capable of giving rise to viable phages. Most importantly, simple joining of the left and right ends of λ (a relatively frequent event) does not result in a viable phage when (as is the case with most vectors) the arms together amount to less than 74% of wild-type λ DNA in length.

One might anticipate from Figure 1 that two bimolecular reactions are required to produce viable λ DNA molecules from the mixture of cleaved restriction fragments. This is not the case in reality, since the λ cohesive ends (12 nucleotides in length; Wu and Taylor 1971) are generally stably annealed before joining of the restriction endonuclease ends occurs. The λ cohesive ends are often deliberately annealed before the addition of foreign DNA and DNA ligase. The actual reaction that produces viable λ DNA molecules is essentially bimolecular. The accompanying nonproductive reactions are the unimolecular joining of the ends of a single foreign DNA fragment forming a circle, the essentially unimolecular circularization of the already annealed left and right arms of λ, and the bimolecular joining of two restriction fragments of the foreign DNA, among others. The ratios among the above products are affected by the absolute and relative concentrations of DNA ends. For example, the ratio of bimolecular to unimolecular reaction prod-

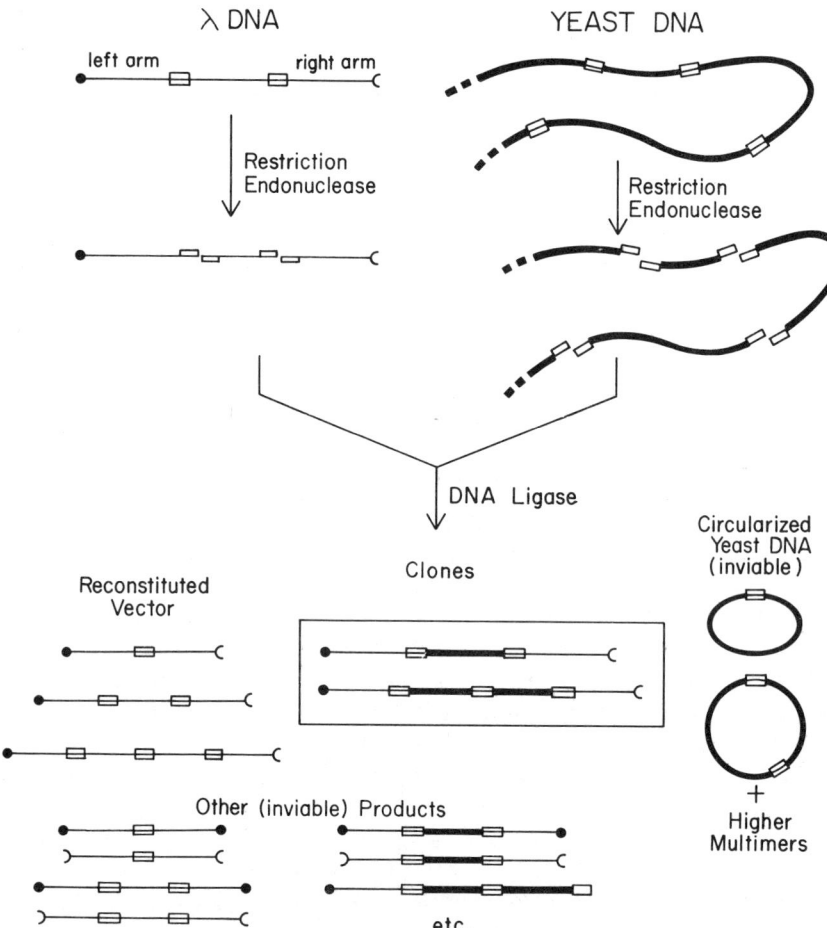

Figure 1 A schematic summary of the cloning of random restriction fragments into a bacteriophage λ cloning vector. It should be noted that more than one fragment can be cloned at once, resulting in joints not found in yeast DNA. The primary difficulty with this form of cloning is the recovery of viable reconstituted (or uncleaved) vector DNA molecules. Methods used to avoid this include purification of the arms after cleavage and before ligation (e.g., Maniatis et al. 1978), selection against a gene in the central fragment of the vector (e.g., Karn et al. 1980), screening against a gene in the central fragment (e.g., Murray and Murray 1975), and cleavage of the vector with an additional restriction enzyme that cleaves the central fragment but not either arm.

ucts is increased at high-input DNA concentrations. Another relevant point is that the normal substrate for encapsidation of λ DNA is a long end-to-end polymer of λ genomes called a concatemer (Feiss and Margulies 1973); such a product is favored by very high-input DNA concentrations as well. For these reasons, DNA ligations intended to generate hybrid clones are conducted at a DNA concentration as high as practicable. This is in contrast to the more complicated considerations for ligations in which plasmid clones are generated (see below).

Ligation mixtures can be converted into viable phages by one of two methods: (1) DNA transfection of suitably treated *E. coli* (Mandell and Higa 1970) or (2) packaging in vitro using cell-free extracts (Hohn and Murray 1977; Sternberg et al. 1977).

An important virtue of this cloning method is the automatic selection for molecules in which DNA is inserted. With suitable modifications (mostly aimed at avoiding the reincorporation of the central fragment of the vector; see the legend to Fig. 1), as many as 99% of all the phages resulting from such a cloning experiment will contain inserts of the added foreign DNA. Another major advantage of this method is its efficiency: With in vitro packaging, as many as 10^7 recombinants per microgram of foreign DNA can be recovered. Finally, hybridization screening methods (see below) are feasible, which can screen as many as 25,000 λ recombinant phages on a single petri plate.

λ phage vectors are available that will accept fragments made with endonucleases *Eco*RI, *Hin*dIII, *Bam*HI, *Sst*I, *Sal*I, and *Xho*I. Various variant vectors are available that will accept inserts up to 25 kb in length, and others are capable of accepting fragments as small as a few base pairs. The structures of popular vectors and directions for their use are given by Davis et al. (1980); more complete descriptions and lists have been published elsewhere (Klein and Murray 1979; deWet et al. 1980; Karn et al. 1980) A useful scheme for inserting very large (~ 20 kb) random fragments into an *Eco*RI vector has been devised by Maniatis et al. (1978); with such a system, the screening of mammalian genomes for single-copy DNA has become routine.

Cloning with Plasmids

Plasmid cloning vectors are small circular DNA molecules that are capable of autonomous replication in *E. coli*. These vectors contain unique targets for restriction enzymes that produce cohesive ends 4 bases or more in length and that carry one gene or more whose activity can be selected in *E. coli*. Insertion of foreign DNA into such plasmids can be accomplished in several ways, but the principal one is simple ligation of plasmid DNA cut with the appropriate restriction enzyme in the presence of excess foreign DNA cut with the same enzyme (Cohen et al. 1973).

The chemical reactions (Fig. 2) that produce viable plasmid DNA are more difficult to manipulate than in the case of λ cloning (above), where one simply uses the highest concentrations achievable. For plasmids, hybrid DNA molecules are produced by an initial bimolecular reaction between the cleaved vector DNA and the cleaved foreign DNA. This must then be

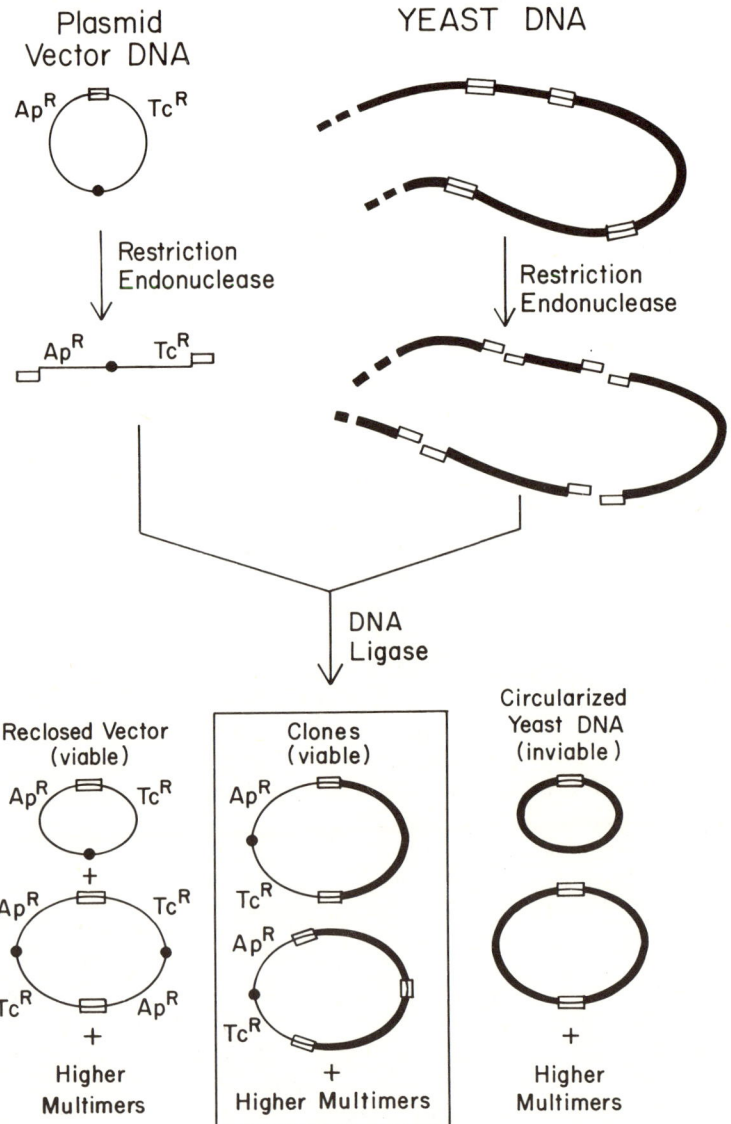

Figure 2 A schematic summary of cloning random restriction fragments into a typical plasmid vector such as pBR322 (Bolivar et al. 1977). (Ap^R, Tc^R) Genes specifying resistance to ampicillin and tetracycline, respectively; (●) the origin of plasmid DNA replication, As with λ, more than one fragment can be cloned, resulting in joints not found in yeast. Methods for avoiding the recovery of reclosed or (uncleaved) vector plasmids without inserts are described in the text.

followed by a unimolecular circularization of the hybrid. The first reaction is favored at high DNA end concentration, whereas the second requires a low DNA end concentration. The accompanying nonproductive reactions are the unimolecular circularization of the vector or foreign DNA and the bimolecular oligomerization of the vector and foreign DNAs. An additional consideration in predicting (and then manipulating) the outcome of these reactions is that the longer the DNA molecule, the farther apart are its ends, with the consequence that the effective end concentration for the unimolecular reaction (but not the bimolecular reactions) is lower for larger fragments (Jacobson and Stockmayer 1950; Wang and Davidson 1966). The optimization of conditions is thus not always straightforward.

In simple cases where the length of the foreign DNA fragments is substantial (comparable to or greater than the legnth of the vector), the DNA concentration is kept fairly low (to avoid extensive bimolecular oligomerization, and the major undesirable and viable product of the ligation is circularized vector containing no foreign DNA. As summarized below, most plasmid cloning schemes deal with this problem either biochemically (by interfering with the plasmid circularization reaction) or genetically (by allowing the identification or selection of plasmids containing foreign DNA segments).

One biochemical method that avoids the vector recircularization problem makes use of terminal deoxynucleotide transferase to produce homopolymer "tails" at the 3' ends of the plasmid and complementary homopolymer tails at the ends of the DNA to be inserted (Lobban and Kaiser 1973; Wensink et al. 1974; Petes et al. 1978). This method has the additional virtues of allowing, simply, the insertion of random fragments of foreign DNA of virtually any size and of reducing to a minimum the attachment together of yeast DNA sequences not contiguous in the yeast genome. A second biochemical method is the use of a phosphatase to remove the 5' phosphate from the cleaved vector DNA before ligation (Seeburg et al. 1977). Since ligase will not join 5'-hydroxyl and 3'-hydroxyl ends, recircularization of the vector is prevented. However, the insertion of a piece of foreign DNA with intact 5'-phosphate ends allows ligation of one strand at each side of the inserted DNA. The resultant molecule is stable and can be used to transform bacteria normally. A third biochemical method that serves to prevent recircularization of the vector is to cut the vector and the DNA to be inserted each with two different restriction endonucleases. Although both enzymes will produce cohesive ends, the two ends of the vector will not cohere with each other, and circular forms can only be obtained after ligation bimolecularly. This method is particularly advantageous for subcloning of small fragments and has the virtue that the orientation of a given fragment with respect to the vector is determined in advance.

The simplest and best genetic method for assuring the recovery of inserted

DNA is to select directly for it, as was done in the cloning of the yeast genes *LEU2, URA3, ARG4,* and so on, and in subcloning the *HIS3* gene. Selection of this kind is not generally available, however, and an alternate method is to select or screen for loss (because of insertion of foreign DNA) of a gene in the vector. For example, the commonly used vector pBR322 specifies resistance to tetracycline, and the gene for this characteristic contains several usable restriction targets unique to the plasmid. Since the plasmid also carries a gene specifying resistance to ampicillin, transformants can be selected by their resistance to ampicillin and, also (simultaneously or subsequently), tetracycline sensitivity can be selected (B. Bochner, pers. comm.) or assessed.

There are several important virtues to the plasmid cloning systems. First is the high purification afforded by these vectors: A typical popular plasmid, pBR322 (Bolivar et al. 1977), is only 4362 nucleotide pairs in length (Sutcliffe 1979). Therefore, a typical gene (1000 bp) comprises about 20% of the total DNA in the plasmid. In contrast, the same gene will comprise only about 2% of the DNA when inserted in a λ cloning vehicle. This advantage of plasmid clones is particularly important in applications requiring large amounts of pure DNA, such as nucleotide sequence analysis. Another advantage is the simplicity of manipulation of plasimd DNA. The small size and limited number of restriction endonuclease targets simplifies the insertion of additional DNA fragments as well as the deliberate formation of deletion, insertion, and point mutations in the cloned gene of interest.

Many alternative systems of cloning DNA into plasmids are available. Some vehicles already contain segments of DNA that can be selected for in *E. coli* or yeast; others contain, in addition, segments that allow independent replication in yeast as well as *E. coli;* and yet others are arranged to allow efficient transcription and/or translation of the foreign DNA in *E. coli* and/or yeast. Vectors exist that will accept fragments made with virtually any restriction endonuclease producing cohesive ends. Using terminal deoxynucleotide transferase or the blunt-end ligation activity of T4 DNA ligase, random fragments of DNA can be cloned into plasmids. In addition, systems have been worked out (and used with yeast vectors; Nasmyth and Reed 1980) for cloning genome segments at random using partial digests with enzymes of relatively low specificity (e.g., *Sau*3A, whose target is ↓GATC) that produce cohesive ends homologous to those produced by an enzyme, which cleaves the plasmid vector just once (e.g., *Bam*HI, whose target is G↓GATCC).

The particular purpose of cloning experiments should dictate the use of the vector system. Random segments to be screened by physical methods (hybridization, heteroduplex), in which completeness of representation and large-size inserts are desired, are usually best done in λ; although these ends can be achieved in plasmids, the experiments are more cumbersome. On the other hand, if a small segment of DNA is to be purified or if selection in yeast is ultimately to be carried out, plasmid systems are advantageous.

Isolation of Particular Genes

In the following sections, screening methods are reviewed for finding DNA segments corresponding to particular yeast genes in a pool of in vitro recombinants. Successful isolation of a clone by most such methods should not, by itself, be taken as proof that the gene desired has actually been cloned; in almost every case, further evidence is required for such an assurance. This issue is discussed further, below, in Identifying Cloned DNA Segments with Yeast Genes.

Hybridization Methods for Recognizing Genes

One of the most common ways in which particular genes can be recognized is through the availability of a hybridization probe. Often, such a probe can be acquired by the prior isolation of an RNA (a straightforward matter for rRNA or tRNA, more complex for an mRNA). If the RNA can be made radioactive (either in vivo or in vitro), the RNA can be used as a probe for hybridization to the DNA of random clones. Alternatively, a DNA copy of the RNA can be made with reverse transcriptase and cloned in a plasmid. The lack of homology between plasmids such as pBR322 and λ allows the use of such a plasmid clone to screen a λ pool, or vice versa.

The method of Benton and Davis (1977) is the simplest of this kind. As many as 10,000 plaques on a single petri plate can be tested for homology to a probe simply by making a replica of the plaques on nitrocellulose paper, denaturing and fixing the DNA to the paper, hybridizing with the probe, and comparing the autoradiogram of the paper with the original petri plate. The probe need not be particularly pure: Enrichments for particular kinds of molecules (e.g., tRNAs) suffice to narrow down the number of candidate clones to a manageable number. It is even possible to do "differential plaque hybridization" in which cDNA copies of mRNA are made from RNA preparations from cells induced or not induced for a particular enzyme system (e.g., galactose utilization). The plaques in a completely representative pool from the genome that hybridize to the probe from induced but not from uninduced cells frequently contain genes that are transcribed preferentially after induction (St. John and Davis 1977). Plasmid clones can also be screened in a similar way (Grunstein and Hogness 1975; Hanahan and Meselson 1980). The method works best on smaller numbers of clones per petri dish and requires the use of a plasmid vector that replicates to a high number of copies per cell.

For yeast, both plasmid and phage libraries have been screened by hybridization for clones of genes that specify rRNA (Kramer et al. 1976; Petes et al. 1978), tRNA (Beckman et al. 1977; Goodman et al. 1977; Petes et al. 1978; Valenzuela et al. 1978; Oslon et al. 1979b), and abundant mRNAs (Petes et al. 1978). The probes in these cases were usually the RNA species themselves. In addition, cDNA copies of mRNA have been used to find particular genes. Genes for glyceraldehyde-3-phosphate dehydrogenase (Holland and Holland 1979), histones (Hereford et al. 1979), and ribosomal proteins (Woolford et al. 1979) were found in yeast by partially purifying mRNA and using it, or a cDNA copy, as probes. In these cases, the mRNA was isolated on the basis of its ability to direct synthesis of the product of the desired gene in a cell-free translation system.

Yet another source of probes for particular genes is cloned DNA from other eukaryotes. The yeast actin gene was cloned by using a cDNA clone of an actin gene from *Dictyostelium discoideum* (Ng and Ableson 1980); a similar approach seems to be feasible for other highly conserved genes such as the tubulins (N. Neff and D. Botstein, unpubl.). Of course, if there is evidence of repeated genes, a clone of one of them suffices to isolate the rest.

In certain circumstances, synthetic DNA probes can be used as probes. The feasibility of such an approach was first described for the special case of yeast cytochrome *c*, where extensive genetic and protein sequence analysis had made possible the correct prediction of about 44 bases in the *CYC1* gene (Szostak et al. 1977; Montgomery et al. 1978).

Function of Yeast Genes in E. coli

The first isolation of a gene encoding a protein in yeast was done not by any hybridization scheme but through the ability of some yeast genes to function in *E. coli* (Struhl et al. 1976; Ratzkin and Carbon 1977; Struhl and Davis 1977; Bach et al. 1979). Such genes could be isolated on the basis of the yeast gene's ability to complement for an auxotrophic mutation in *E. coli*. It has since become clear that many yeast genes will function to some degree in *E. coli*, whereas others will not. When the yeast genes are functional in *E. coli*, the yeast DNA sometimes provides both a promoter and start of translation recognized by *E. coli*, and sometimes the promoter clearly must be provided by the vector. In one case (*URA3*), the yeast strain from which the gene originates makes a difference: The gene is always expressed in *E. coli* when the DNA originates from one yeast strain, whereas the expression in *E. coli* depends on a promoter in the vector when the gene comes from a different yeast strain (M. Rose and D. Botstein, unpubl.).

The ability of some yeast genes to function in *E. coli* has made possible the unambiguous identification of cloned DNA segments with the corresponding yeast gene. This was accomplished in two ways, each of which has general application. In the first instance, two mutant alleles of the yeast *HIS3* gene were cloned on the basis of homology with the wild-type gene (which functions in *E. coli*). The two mutant DNA fragments carried on λ were allowed to recombine in *E. coli* and produced recombinants that had regained function in *E. coli* (Struhl and Davis 1977). In the second instance, a nonsense mutation of the *HIS3* gene was cloned and found to function in *E. coli* only if the cell carried a nonsense suppressr (Struhl et al. 1979a). The latter method has also been used in the case of the *URA3* gene of yeast (M. Rose and D. Botstein, unpubl.).

Return of Cloned Genes to Yeast: DNA Transformation

Given that the primary purpose of cloning and analyzing genes is to understand what they do in vivo, it is essential to be able to return cloned genes to the organism of origin. Among eukaryotes, yeast is still the only organism in which it is routinely possible to return a cloned gene and observe its normal function by DNA transformation. It is the simplicity of DNA transformation in yeast, as well as the particularly advantageous way in which the organism responds to the introduction of cloned DNA, that has made yeast such a favorable system for the study of gene structure and function using recombinant DNA methods.

Basic Features of Transformation in Yeast

Transformation of yeast with cloned yeast DNA was first accomplished by Hinnen et al. (1978), who used the *LEU2* gene (which had already been isolated on the basis of its ability to function in *E. coli*) as a marker readily selectable in yeast. Having such a selectable marker is essential, since DNA transformation is (not surprisingly) a rare event. To achieve high sensitivity in their assays, Hinnen et al. constructed a yeast strain carrying two frameshift mutations in *LEU2*; this strain does not revert to leucine independence at a detectable frequency. This precaution also turned out to be important, because simple transformation of yeast is still rare enough that reversion of simple point mutations produces a disturbing background in transformation experiments. The cloned *LEU2* gene was introduced as part of a circular plasmid (the plasmid vector was ColE1) that could be amplified in *E. coli*. Such amplification of DNA also turns out to be essential, since the transformation of yeast with DNA is very inefficient.

With these tools, it was possible to devise a procedure (production of spheroplasts by enzymatic digestion of the cell wall, incubation with DNA in the presence of calcium ions and polyethylene glycol, followed by regeneration of the treated spheroplasts in selective medium) that allows the stable introduction of the functional *LEU2* gene. It was immediately shown that this event was usually accompanied by the stable introduction of the vector (ColE1) sequences as well.

Further analysis of transformants obtained in this way led to a very important conclusion. Virtually all of the transformation events could be

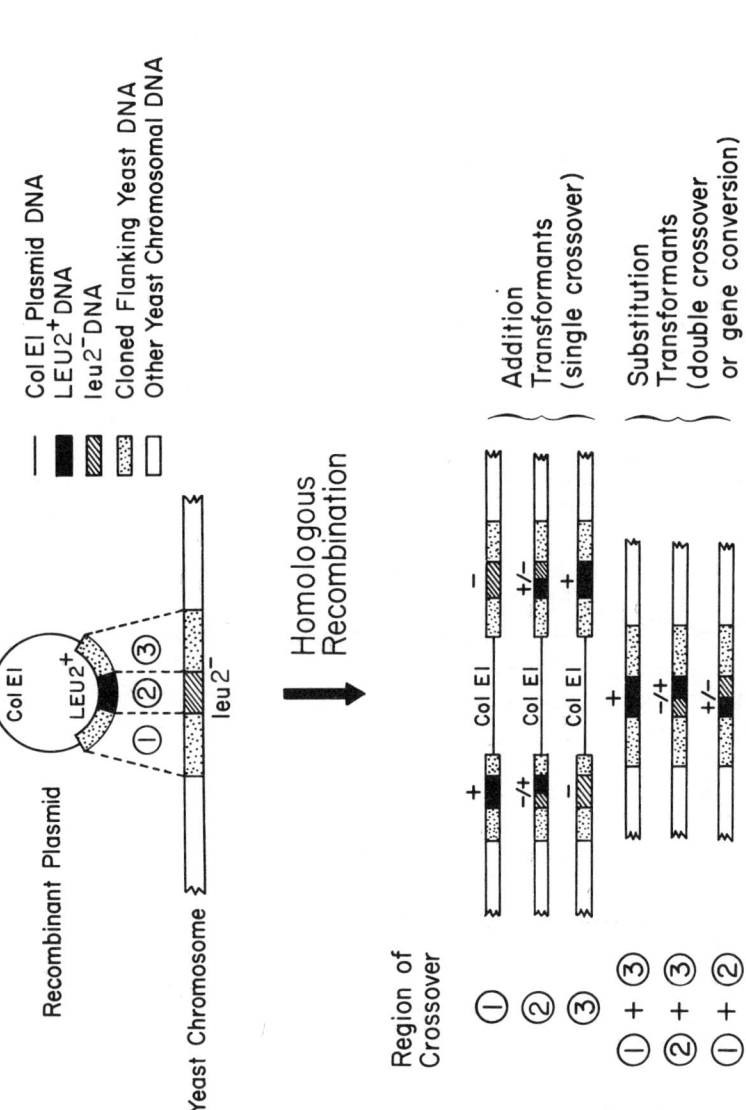

Figure 3 Schematic summary of the events resulting in integrative transformation of yeast (data from Hinnen et al. 1978). The crossover that results in integration can occur at many different points in the homologous region between the cloned sequence on the plasmid and the copy resident in the yeast genome. Likewise, the crossover resulting in excision can occur anywhere in the homologous region. Different combinations of crossover positions are shown to illustrate the possibility that mutant information can be transferred from plasmid to genome and vice versa.

accounted for if one assumed that the transforming DNA was recognized by yeast recombination systems that integrated them into homologous chromosomal sequences (Fig. 3). Thus, Hinnen et al. (1978) frequently observed events envisioned as a single reciprocal recombination between the incoming circular DNA and homologous chromosomal sequences, which we call "addition" transformants. These transformants have the property that the plasmid vector sequences are also integrated, forming a duplication of yeast DNA flanking whatever foreign DNA (usually bacterial in origin) was in the vector. They also observed, less frequently, simple substitutions of the resident doubly mutant *LEU2* gene, which are simply interpreted as double crossover (or gene-conversion) events and which we call "substitution" transformants. Both of these classes of transformants seem to involve recombination at the *LEU2* locus and nowhere else. Hinnen et al. also observed additional tranformants at loci other than *LEU2*; these later turned out to be the result of the presence, in the clone carrying the *LEU2* gene, of another DNA sequence that is repeated several times in the yeast genome. This third class of transformants is not observed with the cloned *HIS3* or *URA3* genes and, for that matter, is not observed with the *LEU2* gene when the repeated sequences are removed from the transforming cloned DNA.

The success of DNA transformation in yeast has led to many studies of transformation with many different yeast genes. Most of these behave in transformation like the *LEU2* gene: The transformation frequency is low, and stable transformants are usually of the addition type, producing a duplication of the selected gene flanking the vector sequence (Fig. 3). Cotransformation studies with two plasmids carrying different selectable markers were done, which showed that a small fraction of the yeast cells put through the transformation procedure are "competent" to take up and integrate DNA so that transformants selected for acquisition of one of the plasmids frequently turn out to have integrated the second plasmid as well (G.R. Fink, pers. comm.). Plasmids containing two or more yeast genes (e.g., both *LEU2* and *URA3*) were used in transformation and behave as anticipated: Transformants generally contain all of the DNA in the plasmid integrated at one of the loci represented on the plasmid (S.C. Falco and D. Botstein, unpubl.; G.R. Fink, pers. comm.). The integration of plasmids containing two genes was soon used to good effect in mapping the locus of the rDNA genes (Petes 1980) by taking advantage of the integration of a plasmid containing both the *LEU2* gene and rDNA into the many tandem copies of the latter. Ordinary genetic analysis of the location of the integrated *LEU2* gene sufficed to map the rDNA.

It was soon discovered, however, that there are special sequences in yeast which, when cloned into plasmids and reintroduced into yeast by DNA transformation, yield a very much higher frequency of transformation. The first of these special sequences is the DNA of the so-called 2μ plasmid (Broach 1981). It was shown that the addition of a fragment of 2μ plasmid

DNA to a plasmid containing a selectable gene (such as the *LEU2*, *HIS3*, or *URA3* gene) increased the transformation frequency many thousandfold (Beggs 1978; Gerbaud et al. 1979; Struhl et al. 1979b). This increase was the result of a difference in mechanism by which the transformed DNA is maintained in yeast: Unlike the low-frequency transformants, the plasmids in 2μ-plasmid-promoted transformants appear primarily not to be integrated into the yeast chromosomes but instead appear to replicate autonomously. More recently, it has been shown that some fragments of the 2μ plasmid DNA can indeed allow a plasmid containing them to replicate autonomously, whereas others result in high-frequency transformation because the incoming plasmid integrates into the established autonomous 2μ plasmid (Broach and Hicks 1980).

Another form of autonomous replication was discovered when the *TRP1* and *ARG4* genes (both of which were identified by their function in *E. coli*) were used in transformation experiments. In both of these cases, the transformation frequency was elevated many thousandfold over the frequency observed for most genes and, as in the case of plasmids containing 2μ plasmid DNA, most of the plasmid DNA appeared to be replicating autonomously (Hsiao and Carbon 1979; Stinchcomb et al. 1979; Struhl et al. 1979b). Further analysis of the *TRP1* case showed that it was a unique neighboring DNA sequence, and not the *TRP1* gene itself, that confers upon the plasmid the property of automomous replication (Stinchcomb et al. 1979). Many such autonomously replicating sequences (ARSs) have since been found (Chan and Tye 1980; Stinchcomb et al. 1980).

In summary, the introduction of circular plasmid DNA containing selectable yeast DNA sequences back into yeast cells results in the detectable acquisition of characteristics encoded in the plasmid DNA in different ways, depending upon the presence or absence of special sequences allowing autonomous replication. If no such sequence is present, maintenance of the introduced DNA apparently depends upon integration of the DNA by recombination into homologous sequences in the yeast genome. If 2μ DNA is present on the introduced plasmid, the plasmid can (either by itself or by integration into the resident 2μ plasmid) be maintained autonomously; this results in a greatly elevated frequency of transformation. If the introduced plasmid contains an ARS sequence, then again a much larger number of transformants are recovered in which the plasmid appears to be replicated autonomously.

Properties of Plasmid Vectors Useful in Yeast

Understanding the different modes by which DNA can be maintained in yeast has made possible the construction of four different kinds of plasmid vector systems useful to the yeast molecular biologist. The four kinds have been named after the way in which they are maintained in yeast after transformation. The plasmids that can only be maintained by integration into homologous chromosomal DNA are designated YIp (*y*east *i*ntegrating *p*lasmid); those that use a fragment of the 2μ plasmid for maintenance are call YEp (*y*east *e*pisomal *p*lasmid); the plasmids that are maintained because they contain an autonomously replicating sequence are called YRp (*y*east *r*eplicating *p*lasmid); and the plasmids that contain a functional centromere are called YCp (*y*east *c*entromere *p*lasmid). The structures of popular examples of the YIp, YEp, and YRp type are given in Figure 4; YCp plasmids are YRp plasmids that contain functional centromeric DNA in addition. Each of these systems has particular advantages and disadvantages, which will be discussed in turn.

All the systems, however, have principles that are common and deserve emphasis. First, all employ selectable markers (most commonly *LEU2, HIS3, URA3,* and *TRP1*) that can be selected in yeast strains mutant for that gene. In the cases of the four aforementioned genes, selection can be carried out in *E. coli* as well as in yeast. For all of the four except *TRP1*, stable (i.e., nonreverting) mutant alleles are available (Botstein et al. 1979). In addition to these selectable markers, most of the plasmids in general use also carry genes specifying resistance to antibiotics; selection for the presence of these genes can be carried out in any *E. coli* strain. Second, all of the systems carry some kind of origin of replication that allows maintenance of the plasmid (often at high copy number) in *E. coli*. Virtually all experiments require amplification of at least the vector DNA in *E. coli* before transformation of yeast. Third, all of these vectors contain restriction targets that occur just once or twice in the plasmid, which facilitates their use in cloning foreign DNA.

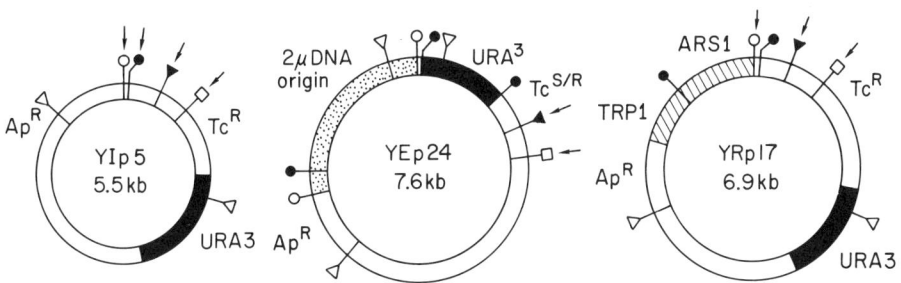

INTEGRATING VECTOR   EPISOMAL VECTOR   REPLICATING VECTOR

Figure 4 Three typical yeast cloning vectors are shown to scale with a partial restriction map. (→) Single targets for particular enzymes that are suitable for cloning. The different restriction targets are symbolized as follows: *Eco*RI (○); *Hind*III ●; *Pst*I (▽); *Bam*HI (▼); *Sal*I (□). All of the vectors contain only yeast DNA (as indicated) and pBR322 DNA. (Ap^R, Tc^R) Genes specifying resistance to ampicillin and tetracycline, respectively; (Tc$^{S/R}$) gene specifying low-level tetracyline resistance. The plasmid YIp5 is now being used as the basis for construction of many YIp, YEp, and YCp vectors by the addition of the appropriate DNA segment(s). The advantage of YIp5 as a base is that the *URA3* gene is short, has very few sites for useful restriction endonucleases, and is located so that both the Tc^R and Ap^R genes of pBR322 are still intact.

YIp Plasmid Vectors: Properties and Uses. The integrating plasmids are probably the most generally useful systems for genetic analysis. Although the frequency of transformation is extremely low, the transformants one obtains by integration are very stable when compared to transformants of YRp or YEp plasmids. The rate of segregation of YIp plasmids is usually much less than 1% per generation of growth in a nonselective medium. An integrated YIp plasmid can be handled as an ordinary genetic marker; a diploid heterozygous for such an integrated plasmid segregates the plasmid as an ordinary Mendelian trait. Since YIp plasmids integrate at positions of homology between DNA on the plasmid and DNA on the chromosome, it has been possible to map the genetic origin of cloned DNA by classical genetic methods.

The integration of YIp plasmids as a formal genetic system very closely resembles the genetics of bacterial episomes (Campbell 1962). Both systems allow the formation, by homologous recombination events, of addition (duplication) recombinants as well as substitution recombinants. It follows that the yeast integrating plasmid systems and the bacterial episomes should be comparably useful in genetic analysis and manipulation. Indeed, the YIp plasmids have begun to be exploited in this way, as discussed in greater detail below. Methods of introducing mutations using YIp plasmids have been developed. Fine-structure mapping using cloned mutations has begun. Complementation studies using cloned mutations appear to be feasible. Integration of plasmids containing two segments of yeast DNA deriving from different chromosomal locations should, in principle, make possible the construction of deletions, inversions, and transpositions as was done in *E.coli* with the specialized transducing phages and episomes.

YEp Plasmid Vectors: Properties and Uses. Plasmids that include fragments of the yeast 2μ plasmid have two important advantages. First, they transform yeast very efficiently (typically 10,000 times better than YIp vectors), and yet the transformants are reasonably stable. Segregation of the plasmid typically occurs at a rate of about 1% per generation of growth in a nonselective medium. Second, the copy number of the plasmid after transformation in yeast is quite high. Estimates (from hybridization studies and from production of an enzyme encoded in the transforming plasmid) range from 25 to 100 copies per cell (Clark-Walker and Miklos 1974; Gerbaud et al. 1979; Chevallier et al. 1980).

YRp Plasmid Vectors: Properties and Uses. Plasmids containing ARSs have been studied, largely because of their intrinsic biological interest. They transform yeast very efficiently, but the transformants are exceedingly unstable, segregating at rates well above 1% per generation of growth in nonselective medium. Occasional stable variants can be found among transformants; these appear to be cases in which the plasmid, including the ARS segment, has integrated into a homologous region in the same way that YIp plasmids do (Stinchcomb et al. 1979; Nasmyth and Reed 1980). This observation can be used to increase the frequency with which one can achieve integration. The gene copy number per cell found with YRp systems is generally lower than that found with YEp systems. This may be an advantage in cloning genes that are deleterious when present in many copies. The properties of YRp plasmid systems suggest that they ultimately may represent the most useful systems, genetically, since they offer high frequencies of transformation (like YEp systems) coupled with the possibility of integration through regions of homology (like YIp systems).

YCp Plasmid Vectors: Properties and Uses. The recent description (Clarke and Carbon 1980; Hsiao and Carbon 1981; Stinchcomb et al. 1982) of sequences apparently deriving from yeast centromeres, which stabilize the copy number, limit the replication, and regularize the segregation of plasmids carrying an ARS sequence, offers obvious ways to improve even further the usefulness of YRp vectors. A few YCp vectors have been constructed; these behave as independent chromosomes, segregating mitotically at a rate of less than 1% per generation. In meiosis, they generally show Mendelian segregation.

Since YEp, YRp, and YCp plasmids replicate in yeast autonomously, it is easy to recover YEp and YRp plasmids from yeast by a simple DNA extraction followed by transformation of *E. coli*. Thus, with these plasmid systems it is now routinely possible to transfer information back and forth between yeast and *E. coli* by transformation.

These properties have made the YEp, YRp, and YCp vectors the systems of choice for experiments in which yeast genes are to be isolated by their ability to complement mutations in yeast. In a typical scheme, fragments of yeast DNA are inserted into a vector that carries a marker (*LEU2, TRP1,* or *URA3*) selectable in yeast, as well as drug markers selectable in *E. coli*. Ligation mixtures are first amplified by transformation of *E. coli* or, in some cases, used directly to transform yeast that is defective in the gene of interest as well as the marker gene. This method has been very successful, because the high transformation frequency in yeast and the ease of recovery of the plasmids facilitate the screening of adequate numbers of independent plasmids.

USES OF RECOMBINANT DNA TECHNOLOGY IN YEAST MOLECULAR BIOLOGY RESEARCH

Characterization of Cloned Genes

Most research using cloned genes aims to understand the function and regulation of the gene in yeast. The first experiments, in virtually every case, are simple deletion experiments (usually in the form of serial subclonings) intended to determine the extent of the gene(s) under study.

Generally, such experiments result in the definition of a small (1–3 kb) restriction fragment that complements the yeast gene after transformation into a mutant yeast. The nucleotide sequence of many fragments containing yeast genes has been determined and, thereby, the coding regions have been defined more exactly; in no case, however, is it yet entirely clear how much material beyond the coding sequences is involved in normal expression of any gene. Nucleotide sequence analysis of yeast tRNA genes resulted in the discovery that many (but not all) tRNA genes contain intervening sequences in the neighborhood of the anticodon (Abelson 1979; Olson 1981). In one case (yeast actin; Gallwitz and Sures 1980; Ng and Abelson 1980), an intervening sequence within a protein-coding sequence was discovered. By using the RNA gel-transfer hybridization method (so-called "Northern technique"; Alwine et al. 1979), it is possible to determine what DNA sequences are transcribed when genes are expressed. The positions at which transcripts begin are spliced, and end can be mapped with great precision (within just a few nucleotides) by the hybridization-protection method of Berk and Sharp (1978). Recently, intervening sequences have been discovered using this technology in mRNAs encoding ribosomal proteins (Rosbash et al. 1981). The RNA gel-transfer hybridization method is semiquantitative with respect to the numbers of transcripts and has been used to determine that some regulated genes are regulated at the level of mRNA synthesis and/or degradation (St. John and Davis 1979; Kramer and Andersen 1980; Denis et al. 1981). The direction of transcription in vivo can also be determined by using DNA strand-separation techniques (these are conveniently done with λ vectors), by subcloning into filamentous phage vectors, or by R-looping methods using the electron microscope (Thomas et al. 1976). Such investigations have yielded interesting and unexpected results, including divergent transcription in the *MAT* (Klar et al. 1981; Nasmyth et al. 1981) and *GAL* (St. John and Davis 1979) loci, the histone genes (Hereford et al. 1979), and the rDNA (Philippsen et al. 1978). When genes are expressed in *E. coli* using bacterial transcription signals, the direction of reading can be deduced by observing the orientation of a cloned fragment required to allow the use of a particular promoter. The direction of reading of the *HIS3* gene was first determined in this way (Struhl and Davis 1980; Struhl et al. 1980). This method is particularly useful in λ, where the promoters and their regulation are well documented, and less useful in plasmid vectors, where the possible promoters (especially secondary ones) are incompletely characterized.

Having a cloned piece of DNA encoding only a single transcribed gene makes possible quantitative solution hybridization analysis of the regulation of transcription. Such methods have been used to show regulation at the level of transcription for the *URA3* (Bach et al. 1979) and *CYC1* (Zitomer et al. 1979) genes. These methods also made possible estimates of the half-life of these mRNAs.

Uses of Cloned Genes in Genetic Analysis

Characterization of variant or mutant forms of the cloned gene has been exceedingly fruitful in several cases. Such an analysis led to the definition of DNA sequence polymorphisms within rDNA, which could be recognized by differences in the length of restriction fragments (Petes and Botstein 1977). The restriction-fragment-length polymorphism turned out to be a simple Mendelian marker that could be followed in standard mitotic or meiotic genetic analysis, which showed that yeast contains only a single cluster of tandemly repeated rDNA and that meiotic crossing-over within the cluster is very rare. By using the polymorphism, the rDNA cluster was mapped by standard methods to chromosome XII (Petes 1979a,b).

Gel-transfer hybridization (Southern 1975) using a cloned gene to probe the entire yeast genome can often define linked restriction-fragment-length polymorphisms between different yeast strains, which can then be mapped. The eight genes encoding tyrosine tRNA were mapped to the eight tyrosine-inserting nonsense suppressor loci by the use of restriction-fragment-length polymorphisms linked to each of the cloned tRNA genes (Olson et al. 1979a).

Gel-transfer hybridization studies using cloned yeast DNA also resulted in the identification of the translocatable yeast element Ty1 (Cameron et al. 1979) and were used in conjunction with the cloning of mutants to demonstrate that Ty1 can alter regulation when inserted in front of structural gene-coding sequences (Chaleff and Fink 1980; Errede et al. 1980; Roeder and Fink 1980; Roeder et al. 1980; Scherer and Davis 1981; Ciriacy and Williamson 1981; Williamson et al. 1981).

A cloned gene can be used to correlate a genetic deletion map with the physical map. The clone is used to probe the genomes of the deletions in gel-transfer experiments. Changes in the sizes of homologous fragments allow the deduction of the physical size of the deletion. Such a correlated map of the *HIS4* gene has been constructed by Roeder and Fink (1980). A limitation of this approach is the requirement that deletion mutants in yeast be available. This problem can be overcome by using subcloned restriction fragments or deletions on plasmid vectors that can be returned to yeast by DNA transformation. If such plasmids are used to transform point mutations in yeast, recombination between the plasmid and the chromosome can be used to construct a physically correlated deletion map of the gene. Preliminary results using deletions of the *URA3* gene on a YEp (2μ) vector and several mapped *ura3* mutations indicate that such a procedure can yield a deletion map of considerable resolution (S.C. Falco and D. Botstein, unpubl.).

A straightforward method for analyzing mutants is through direct cloning of mutant (or variant) forms: Recovery of mutants or variants directly from the yeast genome can easily be accomplished by applying plaque or colony-hybridization methods to libraries constructed in *E. coli* from the DNA of the mutant or variant yeast strain (Struhl and Davis 1977). The mutants can then be analyzed physically or biochemically. The practicality and power of this approach was recently demonstrated in the analysis of mutations affecting the *SUP4* tRNA^{Tyr} locus (Kurjan et al. 1980). Mutants at the *SUP4* tRNA^{Tyr} locus were isolated and mapped in yeast; representatives from each region of the map were then cloned and sequenced, allowing a structure-function analysis and the definition of a site in the tRNA's intervening sequence that affects synthesis or function of the tRNA.

A selective method for cloning specifically mutant forms of a cloned gene was devised by Roeder and Fink (1980) and by Scherer and Davis (1980). This method has general applicability and depends upon homologous integration into the genome after transformation of yeast strains with YIp plasmids. The wild-type cloned gene on a YIp vector, which also contains a marker (such as ampicillin resistance) selectable in *E. coli*, is used to transform a haploid yeast strain carrying the mutant allele. Integration produces a heteroallelic duplication flanking the vector sequence. Digestion of the total DNA of yeast carrying this duplication with a restriction enzyme that cleaves in vector sequences, but not in the gene of interest or in the marker selectable in *E. coli*, results in a fragment that, when circularized (or cloned in a λ vector) and introduced (by transformation or infection) into *E. coli*, will often (depending on where the crossover during integration originally took place; see Fig. 3) carry the mutant form of the gene.

It seems also worth noting that integrative transformation makes possible the transposition of yeast genes to virtually any position on the yeast genome. All that is required is the construction of a YIp plasmid containing the gene to be transposed and a piece of DNA homologous to the position to which one desires to transpose the gene. If one desires to know the origin on the genetic map of a piece of cloned DNA, e.g., one simply places the cloned DNA into a YIp vector carrying one or more selectable markers (*LEU2* and/or*URA3*), transforms the appropriately marked (*leu2⁻* and/or *ura3⁻*) strains, and maps genetically the new position of the selectable marker by conventional genetic analysis. Petes (1980) used the integration of *LEU2* into the rDNA cluster to verify its map position as well as to detect unequal meiotic crossing-over events.

If one were to use random fragments of the yeast genome in a YIp vector, then one would be able to transpose genes to random positions in the genome. In principle, this approach should make possible many of the genetic manipulations currently done in bacteria using transposons (Kleckner et al. 1977). By using a piece of DNA as a "portable region of homology," which can be placed at many points in the genome, one could anticipate the construction of deletions, duplications, inversions, and translocations with endpoints specified in advance. It should be noted, however, that these methods assume that the recombination systems of yeast carry out the required crossovers at a frequency high enough to allow detection above the background of conversion events.

Manipulation of small regions or large chromosomal segments by homologous recombination, as described above, is sometimes complicated by vectors containing multiple regions of homology with different chromosomal locations. These complications can be circumvented in part by two recently developed techniques. Some regions of homology, e.g., the selectable marker, can be deleted from the yeast genome (see below), thereby reducing the number of recombination sites. The second approach originates from the observation (Orr-Weaver et al. 1981) that linear DNA with free ends located in DNA sequences homologous with yeast chromosomal DNA generally transforms yeast with higher efficiency (100–1000-fold), and the transforming DNA is generally found integrated in sequences homologous to the site of the cleavage. Therefore, by simply cleaving the DNA one can target the site of integration.

Alteration of Cloned Genes

One of the most promising opportunities offered by the ability to isolate virtually any gene for which a function is known in yeast is the possibility of manipulating these genes specifically with the aim of studying their function and regulation. In general, there are two aspects of this problem. The first is the introduction of alterations in the gene, and the second is the return of the altered gene to yeast. In many cases, the manipulation is really only useful if the unaltered gene in yeast can be replaced by the altered form.

Several kinds of mutagenesis can be applied to cloned genes on plasmids of phages. Among the simplest is the production of deletions. λ phages carrying deletions can be selected simply by application of chelating agents (Parkinson and Huskey 1971). This procedure, when applied to a λ vector carrying yeast DNA, results in the deletion of yeast sequences primarily, since most vectors contain very little nonessential λ DNA. Such deletions were made in *E. coli* and used to study the function of the yeast *HIS3*-gene function in both *E. coli* and in yeast (Struhl and Davis 1980; Struhl et al. 1980). It was found that transcription of the gene can begin in yeast DNA sequences both in *E. coli* and in yeast, and that in both cases the promoter sequences appear to lie within about 100 bases of the amino terminus of the protein-coding sequence. In this case, it was not essential to replace the chromosomal *HIS3* gene with the deleted forms; transformation of yeast with integrating and replicating plasmid vectors sufficed when the resident chromosomal gene was mutant.

Deletions can be produced by subcloning internal restriction fragments or by removing internal restriction fragments. Scherer and Davis (1979) made a deletion of the latter type in the *HIS3* gene and used this deletion to replace the normal gene in the yeast genome. Their procedure, which has very general application, was to place the deletion in an integrating (YIp) vector also containing the *URA3* gene. A haploid *HIS3$^+$ ura3$^-$* strain was transformed with the plasmid, and Ura$^+$ transformants integrated at the *HIS3* locus (producing a duplication with the deleted and wild-type forms of the gene flanking the vector and *URA3* DNA) were selected (see Fig. 3). Segregants lacking the vector (i.e., phenotypically again Ura$^-$) were found that were also His$^-$, representing excision of the vector and *URA3* sequences by a crossover that left the deletion in the yeast genome (diagrammed for a similar case at the bottom of Fig. 3). The structure of the deletion in the genome was confirmed by gel-transfer experiments showing that the restriction fragment in the genome containing the *HIS3* gene was shortened by the expected length.

Another convenient kind of mutation easily made in cloned genes is a frameshift due to the insertion of synthetic pieces of DNA (usually containing the sequence of a target for a restriction enzyme) whose length is not a multiple of 3 bases (Heffron et al. 1979). Such mutations were made in one of the histone structural genes (Rykowski et al. 1981) and inserted into the genome of a diploid yeast strain according to the procedure of Scherer and Davis (1979). The construction was confirmed by gel-transfer experiments showing the existence of the new restriction target within the gene. In this case, the fact that all four spores in tetrads produced by sporulation of the deletion-bearing diploid are viable constitutes strong evidence that the particular histone gene mutated is not essential for viability of yeast.

Integrative transformation can be used in another way to produce mutations in the yeast genome. If one subclones a restriction fragment totally internal to a gene of interest into a YIp vector containing a marker selectable in yeast, additional transformants that have integrated into that gene will produce a duplication, as shown in Figure 3. However, in this case, both of the duplicated copies will be defective, because one flanking copy will be missing the beginning of the gene and the other will be lacking the end: The gene will have been split by the integration event. This method has been used recently to show that the single yeast actin gene is essential for viability (Shortle et al. 1982). When a YIp vector carrying a fragment entirely internal to the actin-coding sequence and also carrying the *URA3* gene was used to transform a diploid *ura3$^-$* strain, the transformants in which the plasmid had integrated at the actin locus (verified by gel-transfer hybridization) segregated two viable Ura$^-$ spores and two dead spores in every tetrad. This constitutes another general (and relatively simple) method for constructing mutations in genes whose phenotypes are not known but for which well-characterized clones are available. It is worth noting that the presence of the selectable marker allows mapping of the locus by ordinary genetic methods.

A variety of methods can be applied to a cloned gene to produce point mutations, ranging from simple treatment of the plasmid in vitro or in *E. coli* with standard mutagens to specialized methods involving chemical synthesis of altered DNA sequences (see, e.g., Hutchinson et al. 1978; Muller et al. 1978; Itakura and Riggs 1980; Shortle et al. 1980). To return these mutations to the yeast genome, adaptations of the method of Scherer and Davis (1979) can be used, in principle, to carry out efficient localized mutagenesis of arbitrarily small regions of the yeast genome. Alternatively, in cases where the phenotypes of mutations can be assessed while in duplicated or episomal form, simple transformation with the mutagenized plasmid may suffice. The use of YCp vectors may be particularly important in this application since they maintain and limit copy number per cell. Aberrant regulation (or even lethality) may be encountered when gene dosage is high through the use of YRp or YEp vectors. A rather specialized example of this kind of experiment has been published recently by Wallace et al. (1980), who removed the intervening sequence from tRNA^{Tyr} in its ochre-suppressing mutant (*SUP6*) form by a method involving chemical DNA synthesis and found that the intervening sequence is not essential for the expression of the tRNA in the plasmid as judged from suppression of ochre mutations in yeast.

A form of gene alteration particularly useful for the study of gene expression and regulation is the fusion of genes to one another. Gene fusions to the gene of *E. coli* (*lacZ*), which specifies β-galactosidase, have been notably useful in all kinds of studies of regulation of other bacterial operons and genes (including autogenously regulated ones; for reviews, see Bassford et al. [1978]; Ptashne et al. [1980]), because β-galactosidase retains its activity even when as many as about 30 residues are removed from its aminoterminal end and replaced with other proteins. Recently, the range of usefulness of such fusions to the *E. coli* β-galactosidase gene has been extended to yeast. The aminoterminal end of the yeast *URA3* gene (Rose et al. 1981) or the yeast *CYC1* gene (Guarente and Ptashne 1981) was fused to most of the *lacZ*-coding sequence by different methods; the fusions were constructed in *E. coli* on YEp plasmid vectors. In both cases, the fused genes made a hybrid protein with β-galactosidase activity in yeast. The synthesis of the hybrid proteins in yeast was regulated as expected from the regulation of the genes to which *lacZ* had been fused. This fusion technology should extend to yeast most of the possibilities for the study of gene function and regulation already realized in bacterial systems.

Identifying Cloned DNA Segments with Yeast Genes

There are two phases in any in vitro recombination endeavor aimed at isolating particular genes. The first phase is the isolation of a hybrid phage or plasmid carrying part or all of the gene of interest; the second phase is the accumulation of evidence showing that the DNA isolated is indeed the gene desired. A complete proof of the identity of DNA in a clone with the yeast gene from which it is derived is not always feasible; nevertheless, some standards can be set for yeast, which make use of both genetic and physical analysis along the lines summarized in this paper.

The most direct and powerful association between a cloned DNA segment and the yeast gene that it represents is function; i.e., the ability of the cloned DNA, when returned to yeast, to complement a yeast mutant defective in the gene of interest. Many yeast genes have been cloned using complementation as the initial means for identifying the DNA segment of interest. However, complementation is by itself an inadequate criterion for identity between the gene and the cloned DNA, because the phenotype of a mutation in a gene could be suppressed by means other than the introduction of an intact copy of the same gene. A completely different gene could, when introduced as a cloned DNA segment (especially when introduced in plasmid form on a YEp or YRp vector), cause such suppression. Furthermore, some genes are duplicated in the genome (*SUC, MAL,* and *MAT* are good examples), and further analysis is required to show that the cloned gene is not an activated form of a known or hitherto unsuspected extra copy of the yeast gene sought.

A good adjunct to complementation is map position. Several methods for associating a cloned DNA segment with a genetic locus in yeast are available. One of these is to use a YIp vector to integrate (by homology) the cloned segment as well as another yeast marker (e.g., *URA3* if one uses YIp5). The integrated *URA3* function can then be followed in tetrad analysis, and appropriate crosses can be used to show that the cloned segment directs the plasmid to integrate at the correct locus, indicating that the DNA is homologous to DNA closely linked to the site of the mutation that it complements. If the DNA segment that was cloned is not repeated in the yeast genome (which can be determined by gel hybridization; see below), then the double association (complementation and map position) suffices to identify the cloned segment with the gene whose function it exhibits.

A method of particular usefulness when repeated genes are involved is the identification of a particular mutation with a DNA polymorphism. A cloned DNA segment homologous to the repeated gene is used in gel-transfer hybridization experiments to probe the DNA of several yeast strains; differences in the pattern represent restriction-fragment-length polymorphisms. If one carries out tetrad analysis of crosses in which both a mutation in the gene of interest and restriction-fragment-length polymorphism can be followed, then one can determine linkage between the polymorphism and the mutation and, in this way, associate the gene carrying the mutation with its restriction map. This method was used to identify the several RNA^{Tyr} genes (represented by suppressor mutations) with cloned DNA segments originally obtained by homology with the tRNA

(Olson et al. 1979a) and to identify cloned DNA segments with the *MAT, HML,* and *HMR* loci (Hicks et al. 1979).

Another kind of evidence associating a gene with cloned DNA entails the recovery from the yeast genome of mutant alleles of the gene by homology with the cloned DNA. In some of the circumstances, the demonstration that mutants have been recovered is by itself convincing evidence of identity between the DNA segment and the gene. Struhl et al. (1979a) cloned a nonsense mutant of *HIS3* that could be shown to function in a suppressor-dependent way in *E. coli;* Carlson and Botstein (1982) showed that they had cloned the *SUC2* gene (as opposed to any of the other *SUC* genes) by cloning a *suc⁻* amber mutation and showing that it functions in yeast only in the presence of an amber suppressor.

Finally, a direct way of associating a cloned DNA segment with a gene in yeast is to make a mutation in the cloned segment and introduce this mutation into the yeast genome by using one of the methods described above. This approach has the particular virtue that it can be used to investigate the function of segments of DNA for which no corresponding gene or mutation has yet been described. It is sometimes the case, however, that genetic tests are not relevant or possible. In such cases, the best confirmation that one has cloned the gene of interest is hybridization to restriction digests by the gel-transfer method of Southern (1975). This method allows the comparison, to a resolution limited only by the number of restriction endonuclease targets and the patience of the investigator, of the restriction maps of the cloned DNA and the homologous regions of the yeast genome. Such experiments are essential in the characterization of every cloned segment, if only to rule out the adventitious cloning of similiar DNA introduced by contamination. The sensitivity of the cloning methods (especially selection methods) is substantially greater than the purity achievable in biochemical procedures such as DNA isolation. It is therefore unfortunately commonplace to find a rare clone that answers some selection or screening scheme that does not derive from the organism of interest. Hybridization back to the donor strain also often leads to the realization that particular genes are repeated which, as described above, further complicates the identification of a cloned segment with a gene of interest. Another important reason for checking clones by hybridization to the donor genome is the possibility of rearrangements during the in vitro recombination process. In some systems, such rearrangements are expected. For example, cloning into λ will produce phages that have several fragments taken at random from the donor genome but are now linked together. Sometimes this is not obvious because, during the manipulations, one of the sites is damaged and is no longer cut by the enzyme used. Another example is cloning of direct or inverted repetitions: Some plasmid systems are extremely intolerant of such DNAs, and deletions occur that can only be detected by some sort of hybridization experiment.

CONCLUSION

We hope that the summary of experimental methodologies and approaches given above will stimulate further development and exploitation of yeast to answer basic questions in eukaryotic molecular biology. It seems appropriate, however, to emphasize those elements that we believe to be the key advantages of the yeast system for such studies. First and foremost, it seems to us, is the opportunity that the yeast system provides for the blending of the ideas and methods of classical genetic analysis and modern biochemical and biophysical ideas and methods. Simple cloning of genes can now be done in almost any organism and, by itself, is only of limited value. It is the ability to return cloned genes (intact or suitably altered) to yeast, either as extra copies or as replacements for the normal form in the normal position in the genome, which makes yeast uniquely suited to this blending of ideas and methods that has been so successful in prokaryotic molecular biology.

REFERENCES

Abelson, J. 1979. RNA processing. *Annu. Rev. Biochem.* **48**: 1035.

Alwine, J.C., D.J. Kemp, and G.R. Stark. 1977. A method for the detection of specific RNAs in agarose gels by transfer to diazobenzyloxymethyl paper and hybridization with DNA probes. *Proc. Natl. Acad. Sci.* **74**: 5350.

Bach, M.L., F. Lacroute, and D. Botstein. 1979. Evidence for transcriptional regulation of OMP-decarboxylase in yeast by hybridization of mRNA to the yeast structural gene cloned in *E. coli. Proc. Natl. Acad. Sci.* **76**: 386.

Bassford, P., J. Beckwith, M. Berman, E. Brickman, M. Casadaban, L. Guarente, I. Saint-Girons, A. Sarthy, M. Schwartz, H. Shuman, and T. Silhavy. 1978. Genetic fusions of the *lac* operon: A new approach to the study of biological processes. In *The operon* (ed. J.H. Miller and W.S. Reznikoff), p. 245. Cold Spring Harbor Laboratory, Cold Spring Harbor, New York.

Beckman, J.S., P.F. Johnson, and J. Abelson. 1977. Cloning of yeast transfer RNA genes in *Escherichia coli. Science* **196**: 205.

Beggs, J.D. 1978. Transformation of yeast by a replicating hybrid plasmid. *Nature* **275**: 104.

Benton, W.D. and R.W. Davis. 1977. Screening gt recombinant clones by hybridization to single plaques in situ. *Science* **196**: 180.

Berk, A.J. and P.A. Sharp. 1978. Spliced early mRNAs of simian virus 40. *Proc. Natl. Acad. Sci.* **75**: 1274.

Bolivar, R., R.L. Rodriquez, P.J. Greene, M.C. Betlach, H.L. Heyneker, H.W. Boyer, J.H. Crosa, and S. Falkow. 1977. Construction and characterization of new cloning vehicles. II. A multipurpose cloning system. *Gene* **2**: 95.

Botstein, D., S.C. Falco, S. Stewart, M. Brennan, S. Scherer, D.T. Stinchcomb, K. Struhl, and R. Davis. 1979. Sterile host yeast (SHY): A eukaryotic system of biological containment for recombinant DNA experiments. *Gene* **8**: 17.

Broach, J.R. 1981. The yeast plasmid 2µ circle. In *The molecular biology of the yeast* Saccharomyces. I. *Life cycle and inheritance* (ed. J. Strathern et al.), p. 445. Cold Spring Harbor Laboratory, Cold Spring Harbor, New York.

Broach, J.R. and J.B. Hicks. 1980. Replication and recombination functions associated with the yeast plasmid, 2µ circle. *Cell* **21**: 501.

Cameron, J.R., E.Y. Loh, and R.W. Davis. 1979. Evidence for transposition of dispersed repetitive DNA families in yeast. *Cell* **16**: 739.
Campbell, A. 1962. Episomes. *Adv. Genet.* **11**: 110.
Carlson, M. and D. Botstein. 1982. Two differentially regulated mRNAs with different 5' ends encode secreted and intracellular forms of yeast invertase. *Cell* **28**: 145.
Chaleff, D.T. and G.R. Fink. 1980. Genetic events associated with an insertion mutation in yeast. *Cell* **21**: 227.
Chan, C.S.M. and B.-K. Tye. 1980. Autonomously replicating sequences in *Saccharomyces cerevisiae*. *Proc. Natl. Acad. Sci.* **77**: 6329.
Chevallier, M.-R., J.-C. Bloch, and F. Lacroute. 1980. Transcriptional and translational expression of a chimeric bacterial-yeast plasmid in yeast. *Gene* **11**: 11.
Ciriacy, M. and V.M. Williamson. 1981. Analysis of mutations affecting Ty-mediated gene expression in *Saccharomyces cerevisiae*. *Mol. Gen. Genet.* **181**: 556.
Clarke, L. and J. Carbon. 1980. Isolation of a yeast centromere and construction of functional small circular chromosomes. *Nature* **287**: 504.
Clark-Walker, G.D. and G.L.G. Miklos. 1974. Localization and quantification of circular DNA in yeast. *Eur. J. Biochem.* **41**: 359.
Cohen, S.N., A.C.Y. Chang, H.W. Boyer, and R.B. Helling. 1973. Construction of biologically functional plasmids *in vitro*. *Proc. Natl. Acad. Sci.* **70**: 3240.
Davis, R.W., D. Botstein, and J.R. Roth, eds. 1980. *Advanced bacterial genetics*. Cold Spring Harbor Laboratory, Cold Spring Harbor, New York.
Denis, C.L., M. Ciriacy, and E.T. Young. 1981. A positive regulatory gene is required for accumulation of the functional messenger RNA for the glucose-repressible alcohol dehydrogenase from *Saccharomyces cerevisiae*. *J. Mol. Biol.* **148**: 355.
deWet, J.R., D.L. Daniels, J.L. Schroeder, B.G. Williams, K. Denniston-Thompson, D.D. Moore, and F.R. Blattner. 1980. Restriction maps for twenty-one Charon vector phages. *J. Virol.* **33**: 401.
Errede, B., T.S. Cardillo, F. Sherman, E. Dubois, J. Deschamps, and J.-M. Wiame. 1980. Mating signals control expression of mutations resulting from insertion of a transposable repetitive element adjacent to diverse yeast genes. *Cell* **22**: 427.
Feiss, M. and J. Margulies. 1973. On maturation of the bacteriophage λ chromosome. *Mol. Gen. Genet.* **127**: 285.
Gallwitz, D. and I. Sures. 1980. Structure of a split yeast gene: Complete nucleotide sequence of the actin gene in *Saccharomyces cerevisiae*. *Proc. Natl. Acad. Sci.* **77**: 2546.
Gerbaud, C., P. Fournier, H. Blanc, M. Aigle, H. Heslot, and M. Guerineau. 1979. High frequency yeast transformation by plasmids carrying part or entire 2μm yeast plasmid. *Gene* **5**: 233.
Goodman, H.M., M.V. Olson, and B. Hall. 1977. Nucleotide sequence of a mutant eukaryotic gene: The yeast tyrosine-inserting ochre suppressor *SUP4-o*. *Proc. Natl. Acad. Sci.* **74**: 5453.
Grunstein, M. and D.S. Hogness. 1975. Colony hybridization: A method for the isolation of cloned DNAs that contain a specific gene. *Proc. Natl. Acad. Sci.* **72**: 3961.
Guarente, L. and M. Ptashne. 1981. Fusion of *E. coli lac*Z to the cytochrome *c* gene *Saccharomyces cerevisiae*. *Proc. Natl. Acad. Sci.* **78**: 2199.
Hakura K. and A.D. Riggs. 1980. Chemical DNA synthesis and recombinant DNA studies. *Science* **209**: 1401.
Hanahan, D. and M. Meselson. 1980. Plasmid screening at high colony density. *Gene* **10**: 63.
Heffron, F., M. So, and B.J. McCarthy. 1979. Insertion mutations affecting transposition of Tn*3* and replication of ColE1 derivative. *Cold Spring Harbor Symp. Quant. Biol.* **43**: 1279.
Hereford, L., K. Fahrner, J.L. Woolford, Jr., M. Rosbash, and D.B. Kaback. 1979. Isolation of yeast histone genes H2A and H2B. *Cell* **18**: 1261.
Hicks, J., J.N. Strathern, and A.J.S. Klar. 1979. Transposable mating type genes in *Saccharomyces cerevisiae*. *Nature* **2**: 478.

Hinnen, A., J.B. Hicks, and G.R. Fink. 1978. Transformation of yeast. *Proc. Natl. Acad. Sci.* **75:** 1929.

Hohn, B. and K. Murray. 1977. Packaging recombinant DNA molecules into bacteriophage particles *in vitro. Proc. Natl. Acad. Sci.* **74:** 3259.

Holland, M.J. and J.P. Holland. 1979. Isolation and characterization of a gene coding for glyceraldehyde-3-phosphate dehydrogenase from *Saccharomyces cerevisiae. J. Biol. Chem.* **254:** 5466.

Hsiao, C.-L. and J. Carbon. 1979. High-frequency transformation of yeast by plasmids containing the cloned yeast *ARG4* gene. *Proc. Natl. Acad. Sci.* **76:** 3829.

———. 1981. Characterization of a yeast replication origin (*ars2*) and construction of stable minichromosomes containing cloned yeast centromere *DNA (CEN3). Gene* **15:** 157.

Hutchinson, C.A., S. Phillips, M.H. Edgell, S. Gillam, P. Jahnki, and M. Smith. 1978. Mutagenesis at a specific position in a DNA sequence. *J. Biol. Chem.* **253:** 6551.

Itakura, K. and A.D. Riggs. 1980. Chemical DNA synthesis and recombinant DNA studies. *Science* **209:** 1401.

Jacobson, H. and W.H. Stockmayer. 1950. Intramolecular reaction in polycondensations. I. The theory of linear systems. *J. Chem. Phys.* **18:** 1600.

Karn, J., S. Brenner, L. Barnett, and G. Cesareni. 1980. Novel bacteriophage λ cloning vector. *Proc. Natl. Acad. Sci.* **77:** 5172.

Klar, A.J.S., J.N. Strathern, J.R. Broach, and J.B. Hicks. 1981. Regulation of transcription in expressed and unexpressed mating type cassettes of yeast. *Nature* **289:** 239.

Kleckner, N., J. Roth, and D. Botstein. 1977. Genetic engineering *in vivo* using translocatable drug-resistance elements: New methods in bacterial genetics. *J. Mol. Biol.* **116:** 125.

Klein, B. and K. Murray. 1979. Phage λ receptor chromosomes for DNA fragments made with restriction endonuclease I of *Bacillus amylodiquefaciens* H. *J. Mol. Biol.* **133:** 289.

Kramer, R.A. and N. Andersen. 1980. Isolation of yeast genes with mRNA levels controlled by phosphate concentration. *Proc. Natl. Acad. Sci.* **77:** 6541.

Kramer, R.A., J.R. Cameron, and R.W. Davis. 1976. Isolation of bacteriophage λ containing yeast ribosomal RNA genes: Screening by *in situ* hybridization to plaques. *Cell* **8:** 227.

Kurjan, J., B.D. Hall, S. Gillam, and M. Smith. 1980. Mutations at the yeast *SUP4* TRNA^{Tyr} locus: DNA sequence changes in mutations lacking suppressor activity. *Cell* **20:** 701.

Lobban, P.E. and A.D. Kaiser. 1973. Enzymatic end-to-end joining of DNA molecules. *J. Mol. Biol.* **78:** 453.

Mandell, M. and A. Higa. 1970. Calcium-dependent bacteriophage DNA infection. *J. Mol. Biol.* **53:** 159.

Maniatis, T., R.C. Hardison, E. Lacy, J. Lauer, C. O'Connel, and D. Quon. 1978. The isolation of structural genes from libraries of eukaryotic DNA. *Cell* **15:** 687.

Montgomery, D.L., B.D. Hall, S. Gillam, and M. Smith. 1978. Identification and isolation of the yeast cytochrome *c* gene. *Cell* **14:** 673.

Muller, W., H. Weber, F. Meyer, and C. Weissmann. 1978. Site-directed mutagenesis in DNA: Generation of point mutations in cloned β-globulin complementary DNA at the positions corresponding to amino acids 121 to 123. *J. Mol. Biol.* **124:** 343.

Murray, K. and N.E. Murray. 1975. Phage lambda receptor chromosomes for DNA fragments made with restriction endonuclease III of *Haemophilus influenzae* and restriction endonuclease I of *Escherechia coli. J. Mol. Biol.* **98:** 551.

Murray, N.E. and K. Murray. 1974. Manipulation of restriction targets in phage lambda to form receptor chromosomes for DNA fragments. *Nature* **251:** 476.

Nasmyth, K.A. and S.I. Reed. 1980. Isolation of genes by complementation in yeast: Molecular cloning of a cell-cycle gene. *Proc. Natl. Acad. Sci.* **77:** 2119.

Nasmyth, K.A., K. Tatchell, B.D. Hall, C. Astell, and M. Smith. 1981. A position effect in the control of transcription at yeast mating type loci. *Nature* **289:** 244.

Ng, R., and J. Abelson. 1980. Isolation and sequence of the gene for actin in *Saccharomyces cerevisiae. Proc. Natl. Acad. Sci.* **77:** 3912.

Olson, M.V. 1981. Applications of molecular cloning to *Saccharomyces*. In *Genetic engineering* (ed. T.K. Setlow and A. Hollaender), vol. 3, p. 57. Plenum Press, New York.

Olson, M.V., K. Loughney, and B.D. Hall. 1979a. Identification of the yeast DNA sequences that correspond to specific tyrosine-inserting nonsense suppressor loci. *J. Mol. Biol.* **132:** 387.

Olson, M.V., B.D. Hall, J.R. Cameron, and R.W. Davis. 1979b. Cloning of the yeast tyrosine transfer RNA genes in bacteriophage lambda. *J. Mol. Biol.* **127:** 285.

Orr-Weaver, T.L., J.W. Szostak, and R.L. Rothstein. 1981. Yeast transformation: A model system for the study of recombination. *Proc. Natl. Acad. Sci.* **78:** 6354.

Parkinson, J.S. and R.J. Huskey. 1971. Deletion mutants of bacteriophage lambda. I. Isolation and initial characterization. *J. Mol. Biol.* **56:** 369.

Petes, T.D. 1979a. Meiotic mapping of yeast ribosomal DNA on chromosome XII. *J. Bacteriol.* **138:** 185.

———. 1979b. Yeast ribosomal DNA genes are located on chromosome XII. *Proc. Natl. Acad. Sci.* **74:** 5091.

———. 1980. Unequal meiotic recombination within tandem arrays of yeast ribosomal DNA genes. *Cell* **19:** 765.

Petes, T.D. and D. Bostein. 1977. Simple Mendelian inheritance of the reiterated ribosomal DNA of yeast. *Proc. Natl. Acad. Sci.* **74:** 5091.

Petes, T.D., J.R. Broach, P.C. Wensink, L.M. Hereford, G.R. Fink, and D. Botstein. 1978. Isolation and analysis of recombinant DNA molecules containing yeast DNA. *Gene* **4:** 37.

Philippsen, P., M. Thomas, R.A. Kramer, and R.W. Davis. 1978. Unique arrangement of coding sequences for 5S, 5.8S, 18S, and 25S ribosomal RNA in *Saccharomyces cerevisiae* as determined by R-loop and hybridization analysis. *J. Mol. Biol.* **123:** 387.

Ptashne, M., A. Jeffrey, A.D. Johnson, R. Maurer, B.J. Meyer, C.O. Pabo, T.M. Roberts, and R.T. Sauer. 1980. How the λ repressor and cro work. *Cell* **19:** 1.

Ratzkin, B. and J. Carbon. 1977. Functional expression of cloned yeast DNA in *E. coli*. *Proc. Natl. Acad. Sci.* **74:** 487.

Roeder, G.S. and G.R. Fink. 1980. DNA rearrangements associated with a transposable element in yeast. *Cell* **21:** 239.

Roeder, G.S., P.J. Farabaugh, D.T. Chaleff, and G.R. Fink. 1980. The origins of gene instability in yeast. *Science* **209:** 1375.

Rosbash, M., P.K. Harris, J.L. Woolford, Jr., and J.L. Teem. 1981. The effect of temperature-sensitive RNA mutants on the transcription products from cloned ribosomal protein genes of yeast. *Cell* **24:** 679.

Rose, M., M.J. Casadaban, and D. Botstein. 1981. Yeast genes fused to beta-galactosidase in *E. coli* can be expressed normally in yeast. *Proc. Natl. Acad. Sci.* **78:** 2460.

Rykowski, M.C., J.R. Wallis, J. Choe, and M. Grunstein. 1981. Histone H2B subtypes are dispensable during the yeast cell cycle. *Cell* **25:** 477.

St. John, T. and R.W. Davis. 1979. Isolation of galactose-inducible DNA sequences from *Saccharomyces cerevisiae* by differential plaque filter hybridization. *Cell* **16:** 443.

Scherer, S. and R.W. Davis. 1979. Replacement of chromosome segments with altered DNA sequences constructed *in vitro*. *Proc. Natl. Acad. Sci.* **76:** 4951.

———. 1981. Studies on the transposable element Ty1 of yeast. II. Recombination and expression of Ty1 and adjacent sequences. *Cold Spring Harbor Symp. Quant. Biol.* **45:** 581.

Seeburg, P.H., J. Shine, S.A. Martial, J.D. Baxter, and H.M. Goodman. 1977. Nucleotide sequence and amplification of structural gene for rat growth hormone. *Nature* **270:** 486.

Shortle, D., J. Haber, and D. Bostein. 1982. Lethal disruption of the yeast actin gene by integrative DNA transformation. *Science* **217:** 371.

Shortle, D., D. Koshland, G.M. Weinstock, and D. Botstein. 1980. Segment-directed mutagenesis: Construction *in vitro* of point mutations limited to a small predetermined region of a circular DNA molecule. *Proc. Natl. Acad. Sci.* **77:** 5375.

Southern, E.M. 1975. Detection of specific sequences among DNA fragments separated by gel electrophoresis. *J. Mol. Biol.* **98:** 503.

Sternberg, N., D. Tiemeier, and L. Enquist. 1977. In vitro packaging of a lambda D *am* vector containing *Eco*RI fragments of *E. coli* and phage P1. *Gene* **1:** 255.

Stinchcomb, D., C. Mann, and R.W. Davis. 1982. Centromeric DNA from *Saccharomyces cerevisiae*. *J. Mol. Biol.* **158:** 157.

Stinchcomb, D.T., K. Struhl, and R.W. Davis. 1979. Isolation and characterization of a yeast chromosomal replicator. *Nature* **282:** 39.

Stinchcomb, D.T., M. Thomas, J. Kelly, E. Selkev, and R.W. Davis. 1980. Eukaryotic DNA segments capable of autonomous replication in yeast. *Proc. Natl. Acad. Sci.* **77:** 4559.

Struhl, K. and R.W. Davis. 1977. Production of a functional eukaryotic enzyme in *Escherichia coli*: Cloning and expression of the yeast structural gene for imidazoleglycerol phosphate dehydratase (*his3*). *Proc. Natl. Acad. Sci.* **74:** 5255.

———. 1980. A physical, genetic and transcriptional map of the cloned *his3* gene region of *Saccharomyces cerevisiae*. *J. Mol. Biol.* **136:** 309.

Struhl, K., J.R. Cameron, and R.W. Davis. 1976. Functional genetic expression of eukaryotic DNA in *E. coli. Proc. Natl. Acad. Sci.* **73:** 1471.

Struhl, K., R.W. Davis, and G.R. Fink. 1979a. Suppression of a yeast amber mutation in *Escherichia coli. Nature* **279:** 78.

Struhl, K., D.T. Stinchcomb, and R.W. Davis. 1980. A physiological study of functional expression in *Escherichia coli* of the cloned yeast imidazoleglycerol phosphate dehydratase gene. *J. Mol. Biol.* **136:** 291.

Struhl, K. D.T. Stinchcomb, S. Scherer, and R.W. Davis. 1979b. High-frequency transformation of yeast: Autonomous replication of hybrid DNA molecules. *Proc. Natl. Acad. Sci.* **76:** 1035.

Sutcliffe, J.G. 1979. Complete nucleotide sequence of the *Escherichia coli* plasmid pBR322. *Cold Spring Harbor Symp. Quant. Biol.* **43:** 77.

Szostak, J.W., J.I. Stiles, C.P. Bahl, and R. Wu. 1977. Specific binding of a synthetic oligodeoxynucleotide to yeast cytochrome *c* mRNA. *Nature* **265:** 61.

Thomas, M., J. Cameron, and R.W. Davis. 1974. Viable molecular hybrids of bacteriophage lambda and eukaryotic DNA. *Proc. Natl. Acad. Sci.* **71:** 4579.

Thomas, M., R.L. White, and R.W. Davis. 1976 Hybridization of RNA to double-stranded DNA: Formation of R-loops. *Proc. Natl. Acad. Sci.* **73:** 2294.

Valenzuela, P., A. Venegas, F. Weinberg, R. Bishop, and W.J. Rutte. 1978. Structure of yeast phenylalanine tRNA genes: An intervening DNA segment within the region coding for the tRNA. *Proc. Natl. Acad. Sci.* **75:** 190.

Wallace, R.B., P.F. Johnson, S. Tanaka, M. Schold, K. Itakura, and J. Abelson. 1980. Directed deletion of a yeast transfer RNA intervening sequence. *Science* **209:** 1396.

Wang, J.C. and N. Davidson. 1966. Thermodynamic and kinetic studies on the interconversion between the linear and circular forms of phage lambda DNA. *J. Mol. Biol.* **15:** 111.

Wensink, P.C., D.J. Finnegan, J.E. Donelson, and D.S. Hogness. 1974. A system for mapping DNA sequences in the chromosomes of *Drosophila melanogaster*. *Cell* **3:** 315.

Williamson, V.M., E.T. Young, and M. Ciriacy. 1981. Transposable elements associated with constitutive expression of yeast alcohol dehydrogenase II. *Cell* **23:** 605.

Woolford, J.L., Jr., L.M. Hereford, and M. Rosbash. 1979. Isolation of cloned DNA sequences containing ribosomal protein genes from *Saccharomyces cerevisiae*. *Cell* **18:** 1247.

Wu, R. and E. Taylor. 1971. Nucleotide sequence analysis of DNA. II. Complete nucleotide sequence of the cohesive ends of bacteriophage λ DNA. *J. Mol. Biol.* **57:** 491.

Zitomer, R.S., D. Montgomery, D.L. Nichols, and B.D. Hall. 1979. Transcriptional regulation of the yeast cytochrome *c* gene. *Proc. Natl. Acad. Sci.* **76:** 3627.

Appendices

APPENDIX I
Genetic Map of *Saccharomyces cerevisiae*

Compiled by Robert K. Mortimer and David Schild
Department of Biophysics and Medical Physics and
Donner Laboratory, University of California
Berkeley, California 94720

The compilation of genetic mapping data that appeared in the companion volume of this series[1] and in more detail as a complete review (Mortimer and Schild 1980) was completed in January 1980. That map described the locations of 317 genes on 17 chomosomes and 3 fragments. In the succeeding 2½ years there have been many changes in and additions to that map; because of this, we were asked to prepare an updated map for inclusion in this volume. Ninety-eight new genes have been located on the map, and these genes, their locations, and the sources of the new data are described in the accompanying list of mapped genes. The group of genes located on chromosome XVII of the 1980 map have been found not to define a separate chromosome but instead to be located on the left arm of chromosome XIV (Klapholz and Easton-Esposito 1982). A centromere-linked marker, *KRB1*, has been excluded from chromosomes I-XVI and now serves to identify a new chromosome XVII (Wickner et al. 1982a). Several linkages based only on mitotic crossing-over or trisomic analysis have been confirmed by tetrad analysis. Some sequence ambiguities have been cleared up and others have been introduced. Two genes (*lys10* and *SAD1*) that appeared on the 1980 map have been deleted, and the lengths of several intervals have been changed by the new data. The newly mapped genes have mostly fallen in already established intervals, although five chromosome arms have been extended by new linkages. The total genetic map length is now at least 5000 centiMorgans (cM), and 413 genes are located on this map.

Some of the specific changes in the map are discussed below.

Chromosome I

The gene *ts11* has been shown to be an allele of *cdc24* and to be located distal rather than proximal to *cyc3* (P. Oeller, D. Kaback, and J. Pringle, pers. comm.). *FLO4* is an allele of *FLO1* and has been removed from the map.

---
[1]Mortimer, R.K. and D. Schild. 1981. Appendix II. Genetic map of *Saccharomyces cerevisiae*. In *The molecular biology of the yeast* Saccharomyces. *Life cycle and inheritance* (ed. J. Strathern et al.), p. 641. Cold Spring Harbor Laboratory, Cold Spring Harbor, New York.

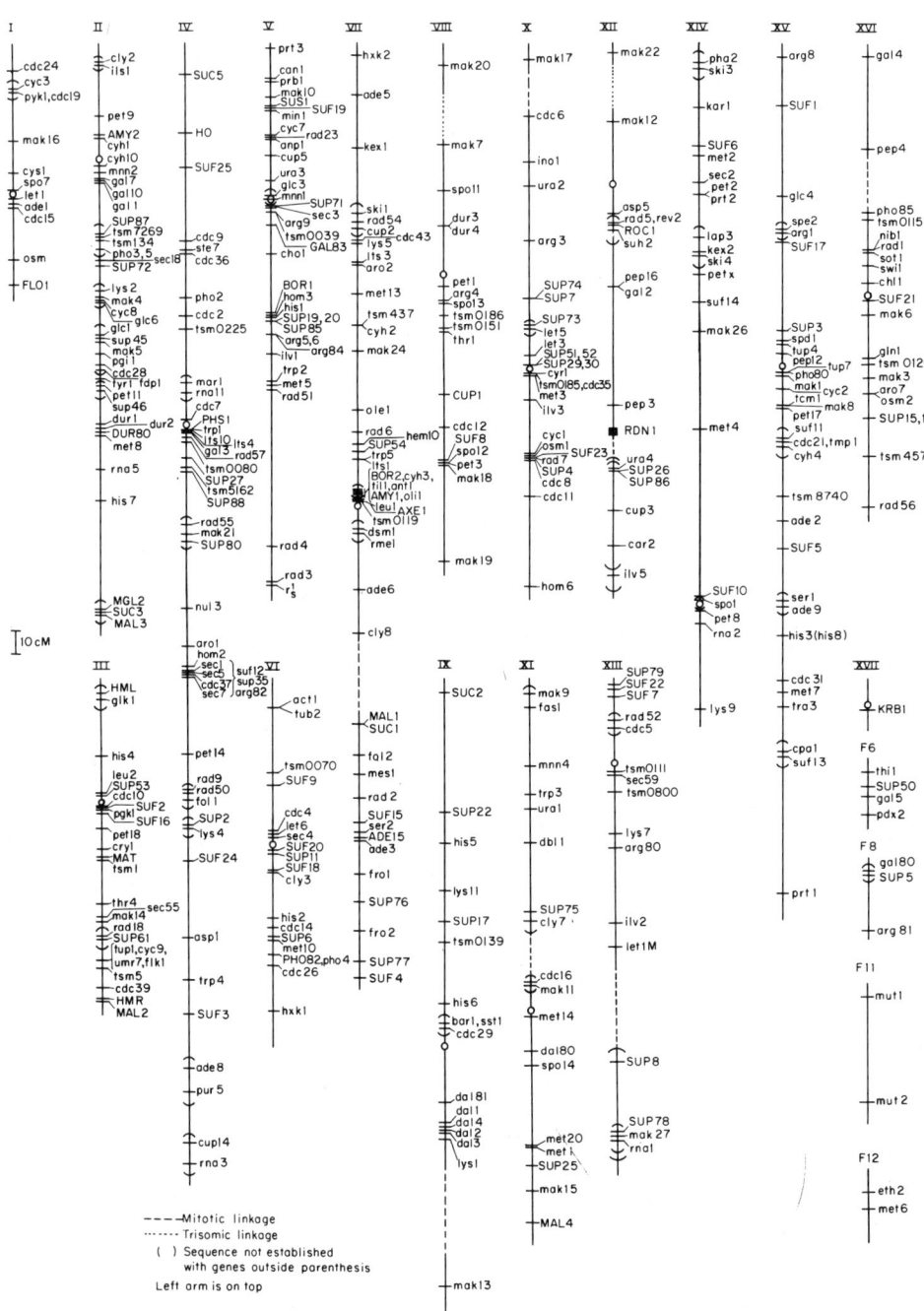

GENETIC MAP OF *SACCHAROMYCES CEREVISIAE*

List of Mapped Genes

Gene	Map position	Reference[a]
act1	6L	S.C. Falco, J. Thomas, and D. Botstein (pers. comm.)
AMY2	2L	30, 19
anp1	5L	23
arg84	5R	15
cdc10	3R	3, 24
cdc24	1L	P. Oeller, D. Kaback, and J. Pringle (pers. comm.)
cdc31	15R	D. Schild (pers. comm.);
cdc35		see *tsm0185*
cdc37	4R	S. Reed (pers. comm.)
cdc43	7L	A. Adams and J. Pringle (pers. comm.)
cly2	2L	30, 24
cup2	7L	J.W. Welch and S. Fogel (pers. comm.)
cup3	12R	J.W. Welch and S. Fogel (pers. comm.)
cup5	5L	J.W. Welch and S. Fogel (pers. comm.)
cup14	4R	J.W. Welch and S. Fogel (pers. comm.)
cyr1	10R	22
dal3	9R	T.G. Cooper and H.-S. Yoo (pers. comm.)
dal80	11R	T.G. Cooper and G. Chisholm (pers. comm.)
dal81	9R	T.G. Cooper and V. Turoscy (pers. comm.)
dsm1	7R	9
dur3	8L	T.G. Cooper and M. Mojumdar (pers. comm.)
dur4	8L	T.G. Cooper and M. Mojumdar (pers. comm.)
DUR80	2R	T.G. Cooper and G. Chisholm (pers. comm.)
gal83	5R	21
glc3	5L	J. Pringle (pers. comm.)
glc4	15L	P. Oeller and J. Pringle (pers. comm.)
glc6	2R	J. Pringle (pers. comm.)
gln1	16R	A.P. Mitchell (pers. comm.)
hem10	7L	29
hom6	10R	D. Schild (pers. comm.)
ilv2	13R	J. Petersen, M. Kielland-Brandt, G. Bank, S. Holmberg, and T. Nilsson-Tillgren (pers. comm.)
ilv5	12R	J. Petersen, M. Kielland-Brandt, G. Bank, S. Holmberg, and T. Nilsson-Tillgren (pers. comm.)
ino1	10L	5
kar1	14L	7, 18
KRB1	17R	30
lap3	14L	18; R. Trumbly (pers. comm.)
mak8	15R	31
met2	14L	18, 24
met4	14L	18, 24, 30
nib1	16L	14

List of Mapped Genes (*Continued*)

Gene	Map position	Reference[a]
met20	11R	20
osm	1R	27
pep3	12R	32
pep4	16L	E. Jones (pers. comm.)
pep12	15L	E. Jones and G. Fabian (pers. comm.)
pet2	14L	18, 24
pha2	14L	18, 24
PHS1	4R	28
prt2	14L	18, 24
rad9	4R	6
rad23	5L	23
rad54	7L	6; J. Game (pers. comm.);
rna6		see *tsm7269*
sec1	4R	C. Fields and R. Schekman (pers. comm.)
sec2	14L	C. Fields and R. Schekman (pers. comm.)
sec3	5R	C. Fields and R. Schekman (pers. comm.)
sec4	6L	C. Fields and R. Schekman (pers. comm.)
sec5	4R	C. Fields and R. Schekman (pers. comm.)
sec7	4R	C. Fields and R. Schekman (pers. comm.)
sec18	2R	C. Fields and R. Schekman (pers. comm.)
sec55	3R	C. Fields and R. Schekman (pers. comm.)
sec59	13R	C. Fields and R. Schekman (pers. comm.)
spo1	14L	B. DiDomenico and R.E. Easton pers. comm.)
spo7	1L	R.E. Esposito and C. Waddell (pers. comm.)
spo12	8R	17
spo14	11R	M. Townsend, B. DiDomenico, S. Klapholz, and R.E. Esposito (pers. comm.)
SUC5	4L	2, 16
SUF1	15L	11
SUF3	4R	11
SUF4	7R	11
SUF6	14L	11
suf12	4R	4
suf13	15R	4

List of Mapped Genes (*Continued*)

Gene	Map position	Reference[a]
suf14	14L	4
SUF15	7R	10
SUF16	3R	10
SUF17	15L	10
SUF18	6R	10
SUF19	5L	10
SUF20	6R	10
SUF21	16R	10
SUF22	13L	10
SUF23	10R	10
SUF24	4R	10
SUF25	4L	10, 11
SUP72	2R	12
SUP73	10L	12
SUP74	10L	12
SUP75	11L	D. Hawthorne (pers. comm.)
SUP76	7R	12
SUP77	7R	12
SUP78	13R	12
SUP79	13L	12
SUP80	4R	12
SUP85	5R	12
SUP86	12R	12
SUP87	2R	12
SUP88	4R	12
tcm1	15R	31
tsm0039	5R	1
tsm0070	6L	1
tsm0080	4R	1
tsm0111	13R	1
tsm0115	16L	30
tsm0119	7L	1
tsm0120	16R	1
tsm0139	9L	1
tsm0151	8R	1; F. Boutelet and F. Hilger (pers. comm.)
tsm0185 (*cdc35*)	10R	1; F. Boutelet and F. Hilger (pers. comm.)
tsm0186	8R	1; F. Boutelet and F. Hilger (pers. comm.)
tsm0800	13R	1
tsm4572	16R	13
tsm5162	4R	1
tsm7269 (*rna6*)	2R	13
tsm8740	15R	13
tub2	6L	J. Thomas, S.C. Falco, and D. Botstein (pers. comm.)

[a] References are listed numerically at the end of this paper.

Chromosome II

The genes *AMY2* and *cly2*, which provisionally had been excluded from chromosomes I–XVI (Mortimer and Hawthorne 1973; Lucchini et al. 1979) have been shown to be linked and to be located on the left arm of chromosome II (Wickner et al. 1982a). *DUR80* has been shown to be tightly linked to *dur1* and *dur2*, and these three genes are close to *met8* (T.G. Cooper and G. Chisholm, pers. comm.). The temperature-sensitive (ts) gene *tsm7269* is an allele of *rna6* (Hilger et al. 1982). Several ambiguities described previously (Mortimer and Schild 1980) still remain concerning the relative order of genes on the right arm of this chromosome. *glc6* is distal to *lys2*, but it has not been positioned relative to *mak4* and *cyc8* (J. Pringle, pers. comm.).

Chromosome III

The cell-cycle gene *cdc10* has been shown by molecular techniques to be on 3R rather than on 3L (Clarke and Carbon 1980). The marker *SAD1* was shown to be due to a structural alteration and has been removed from the map (J. Strathern, pers. comm.).

Chromosome IV

The most distal gene on the left arm of this chromosome has been identified as *SUC5* (Kawasaki 1979; Carlson et al. 1981). Fifteen new genes have been positioned on this chromosome, which now has a total length of approximately 500 cM. The marker *PHS1* fails to recombine with *trp1* (Takahashi et al. 1980); six markers are now tightly linked to *trp1* and the centromere. The genes *sec1*, *sec5*, *cdc37*, and *sec7* are located in that order between 2.5 and 6.2 cM distal to *hom2* (C. Fields and R. Schekman, pers. comm.). Three other genes, *suf12* (Culbertson et al. 1982), *sup35*, and *arg82* have been mapped in the same region, but these genes have not been tested against one another or against the *sec1-sec5-cdc37-sec7* group.

Chromosome V

Eleven new genes have been located on this chromosome. *SUF19* falls between *min1* and *mak10* (Gaber and Culbertson 1982), but the position of *SUS1* relative to these three genes is unknown. *tsm0039* fails to recombine with *arg9* (Boutelet and Hilger 1980); *sec3* is located between *arg9* and the centromere (C. Fields and R. Schekman, pers. comm.). The UGA suppressor *SUP85* fails to recombine with the ochre suppressor *SUP20*, and these two suppressors may be alleles (Hawthorne 1981). *arg84* defines a regulatory region for the *arg5 arg6* genes (Jacobs et al. 1980).

Genetic Map of S. cerevisiae   645

Chromosome VI

The actin and β-tubulin genes are tightly linked and are the most distal markers on 6L (S.C. Falco, J. Thomas, and D. Botstein, pers comm.); *sec4* is located between *cdc4* and *cen6* (C. Fields and R. Schekman, pers. comm.); *let6* (ochre-suppressible lethal) falls near the temperature-sensitive lethals *cdc4* and *sec4* and may be an allele of one of them. The frameshift suppressors *SUF20* and *SUF18* flank the ochre suppressor *SUP11* (Gaber and Culbertson 1982).

Chromosome VII

cup2 and *rad54* are located, respectively, 6 cM and 10 cM distal to *lys5* (Dowling 1982; J. Game; J.W. Welch and S. Fogel; both pers. comm.), but the order of these genes relative to *ski1* is unknown. *cdc43* is about 1 cM from *lys5*, but it has not been established whether it is proximal or distal to *lys5* (A. Adams and J. Pringle, pers. comm.). The temperature-sensitive lethal *tsm0119* is in the *leu1-cen7* interval (Boutelet and Hilger 1980). *dsm1*, which affects premeiotic DNA synthesis, has a phenotype similar to that of *rme1*. Since both genes map in the same region, they are probably alleles (Fast 1978; R.E. Esposito, pers. comm.). Four suppressors have been located near or distal to *ade3* (Hawthorne 1981; Gaber and Culbertson 1982; Gaber et al. 1982).

Chromosome VIII

dur3 and *dur4* fail to recombine and are located on 8L (T.G. Cooper and M. Mojumdar, pers. comm.). Two temperature-sensitive lethals, *tsm0186* and *tsm0151* are located in the *arg4-thr1* interval (Boutelet and Hilger 1980, and pers. comm.). *spo12* is in the *SUF8 pet3 mak18* cluster, tightly linked to *pet3* (Klapholz and Esposito 1982).

Chromosome IX

Only *tsm0139, dal81,* and *dal3* have been added to this chromosome (Boutelet and Hilger 1980; T.G. Cooper, H.-S. Yoo, and V. Turoscy, pers. comm.). *dal3* was positioned relative to *dal1, dal4,* and *dal2* only by cloning and restriction analyses.

Chromosome X

ino1 is located between *ura2* and *cdc6* on 10L (Donahue and Henry 1981). The UGA suppressor *SUP74* fails to recombine with the ochre suppressor *SUP7*, and these two suppressors may be allelic (Hawthorne 1981). *cyr1* defines the adenylate cyclase gene (Matsumoto et al. 1982), and *tsm0185,* which maps in the same region and is allelic with *cdc35*, is deficient in adenylate cyclase activity (Boutelet and Hilger 1980; F. Boutelet and F. Hilger, pers. comm.). *SUF23* fails to recombine with *rad7* and represents the sixth gene in the cluster around *cyc1* (Gaber and Culbertson 1982). *hom6* is now the distal marker on 10R (D. Schild, pers. comm.).

Chromosome XI

SUP75, dal80, and *spo14* are the new genes on this chromosome (Hawthorne 1981; T.G. Cooper and G. Chisholm; M. Townsend, B. DiDomenico, S. Klapholz, and R.E. Esposito; both pers. comm.).

Chromosome XII

pep3 is approximately 10 cM proximal to the rDNA genes (Zamb and Petes 1982). *SUP86* does not recombine with *SUP26,* and these two suppressors may be allelic (Hawthorne 1981). *cup3* is located between *ura4* and *car2* (J. Welch and S. Fogel, pers. comm.). *ilv5* is 47 cM from *ura4,* but its position relative to *car2* is unknown (J. Petersen, M. Kielland-Brandt, G. Bank, S. Holmberg, and T. Nilsson-Tillgren, pers. comm.).

Chromosome XIII

SUF22 has been shown to be 2.6 cM distal to *SUF7* on 13L (Gaber and Culbertson 1982), and *SUP79* has been shown to map near these two frameshift suppressors (Hawthorne 1981). New data place *rad52* and *cdc5* close to each other, and their order relative to the centromere is now ambiguous (C. Fields and R. Schekman; F. Boutelet and F. Hilger; both pers. comm.) Three temperature-sensitive lethals map near the centromere on 13R. *tsm0111* and *tsm0800* are separated by about 9 cM (Boutelet and Hilger 1980, and pers. comm.), and *sec59* falls near *tsm0111* (C. Fields and R. Schekman, pers. comm.); it is possible that *sec59* is an allele of one of the other two temperature-sensitive lethals. *ilv2* is located 38 cM distal to *lys7* but fails to show linkage to *met6* (J. Petersen, M. Kielland-Brandt, G. Bank, S. Holmberg, and T. Nilsson-Tillgren, pers. comm.), which had been positioned previously in the same region. Both *eth2* and *met6* have been removed from XIII and placed on fragment 12. *SUP78* is about 30 cM from *SUP8,* but it has not been tested against *rna1* or *mak27* (Hawthorne 1981). It is possible that *rna1* and *SUP78* are located on opposite sides of *SUP8.*

Chromosome XIV

Seven of the eight genes formerly located on chromosome XVII have been found to be interspersed with the five genes that had defined the left arm of chromosome XIV (Klapholz and Easton-Esposito 1982). *SUF6, suf14,* and *sec4* have also been added to 14L (Culbertson et al. 1982; Gaber et al. 1982; C. Fields and R. Schekman, pers. comm.). The most proximal marker, *met4,* shows significant linkage to *pet8* (Klapholz and Easton-Esposito 1982; Wickner et al. 1982a). The relative order of the 15 genes on 14L are based partly on recombination studies between the two sets of genes (Culbertson et al. 1982; Gaber et al. 1982; Klapholz and Easton-Esposito 1982; Wickner et al. 1982a) and partly on combining linkage data from the former chromosomes XIV and XVII. The marker *lys10* had been placed proximal to *met4* by mitotic crossover analysis (Mortimer and Hawthorne 1973). Since this marker is no longer available, it has been removed from the map. *spo1* is about 4 cM from *pet8* and provisionally is placed across the centromere from this gene, although the sequence of *spo1, cen14,* and *pet8* is still uncertain (B. DiDomenico and R.E. Esposito, pers. comm.).

Chromosome XV

SUF1 and *glc4* fall in the *arg1-arg8* interval on 15L, and *SUF17* is 3 cM proximal to *arg1* (Gaber et al. 1982; P. Oeller and J. Pringle, pers. comm.). *pep12* is between *tup4* and *cen15,* and the relative order of *tup4, cen15, tup7,* and *mak1* has been changed (E. Jones and G. Fabian, pers. comm.). *cdc31* is close to *met7* (D. Schild, pers. comm.), and *suf13* is 20 cM distal to *tra3;* this frameshift suppressor shows meiotic linkage to *prt1* (Culbertson et al. 1982). *prt1* had been located previously on 15R only by mitotic analyses (Mortimer and Hawthorne 1973).

Chromosome XVI

pep4 is 40 cM proximal to *gal4* and not tightly linked to *rad1* (E. Jones, pers. comm.), but this gene has not yet been tested against other 16L markers. *nib1* fails to recombine with *rad1* (Holm 1982), and *tsm0115* is distal to *rad1* (F. Boutelet and F. Hilger, pers. comm.). *SUF21* is 1 cM from *cen16* on the right arm (Gaber and Culbertson 1982), and *gln1* and *tsm0120* fall in the *mak6-mak3* interval (Boutelet and Hilger 1980; A. Mitchell, pers. comm.).

Chromosome XVII

The centromere-linked killer bypass gene *KRB1* has been convincingly eliminated from centromere markers defining chromosomes I–XVI and now defines a new chromosome XVII (Wickner et al. 1982a).

Fragments

No changes in fragments 6, 8, or 11 have been reported. Fragment 12 contains *eth2* and *met6,* which had formerly been located on chromosome

XIII. *met6* is not where it had previously been located (J. Petersen, M. Kielland-Brandt, G. Bank, S. Holmberg, and T. Nilsson-Tillgren, pers. comm.) and, independently, it has been excluded from chromosome XIII (D. Schild, pers. comm.). Preliminary data place fragment 12 on chromosome X (D. Schild, pers. comm.).

MAPPING METHODS

In the companion volume of this series we reviewed several genetic mapping procedures applicable to yeast. Since the appearance of that paper, several new approaches have been described. We have shown that diploids homozygous for *rad52* lose chromosomes at a high rate (Mortimer et al. 1981) and have developed *rad52* strains suitable for chromosome-loss mapping (D. Schild and R.K. Mortimer, unpubl.). Klapholz and Esposito (1982) have shown that *spo11/spo11* diploids have almost no meiotic recombination and have developed a set of strains suitable for mapping with this method. Falco et al. (1982) have found that Cir^+ diploids that contain an integrated 2μ plasmid express heterozygous markers at a high frequency if these markers are in repulsion to the site of integration. This method can be used to map any cloned gene (S.C. Falco and D. Botstein, pers. comm.) and is extremely efficient. The above three methods only localize the gene to a chromosome or chromosome arm. Subsequent crosses are needed to position the gene to a site on the chromosome. Gaber et al. (1982) have developed a set of strains that collectively carry 50 markers dispersed over the genome. These strains have been used to map 23 frameshift suppressors. A similar set of strains have been developed by R. Contopoulou and R. Mortimer (pers. comm.) and are available from the Yeast Genetics Stock Center.

ACKNOWLEDGMENTS

This work was supported by the Office of Health and Environmental Research, Office of Energy Research, U.S. Department of Energy, under contract no. DE-AC03-76F00098.

REFERENCES

1. Boutelet, F. and F. Hilger. 1980. A mapping study on fourteen centromere-linked temperature-sensitive mutations. In *10th International Conference of Yeast Genetics and Molecular Biology*, p. 177. Louvain-la-Neuve, Belgium.
2. Carlson, M., B.C. Osmond, and D. Botstein. 1981. Genetic evidence for a silent *SUC* gene in yeast. *Genetics* **98**:41.
3. Clarke, L. and J. Carbon. 1980. Isolation of a yeast centromere and construction of functional small circular chromosomes. *Nature* **287**:504.
4. Culbertson, M.R., R.F. Gaber, and C.M. Cummins. 1982. Frameshift suppression in *Saccharomyces cerevisiae*. V. Isolation and genetic properties of non-group specific suppressors. *Genetics* **102**:361.

5. Donahue, T.F. and S.A. Henry. 1981. Inositol mutants of *Saccharomyces cerevisiae*: Mapping the *ino1* locus and characterizing alleles of the *ino1*, *ino2* and *ino4* loci. *Genetics* **98**:491.
6. Dowling, E.L. 1982. "Meiotic recombination and sporulation in repair-deficient strains of yeast." Ph.D. thesis, University of California, Berkeley.
7. Dutcher, S.K. and L.H. Hartwell. 1982. The role of *S. cerevisiae* cell division cycle genes in nuclear fusion. *Genetics* **100**: 175.
8. Falco, S.C., Y. Li, J.R. Broach, and D. Botstein. 1982. Genetic properties of chromosomally integrated 2μ plasmid DNA in yeast. *Cell* **19**:573.
9. Fast, D. 1978. "Premeiotic DNA synthesis and recombination in the yeast *Saccharomyces*." Ph.D. thesis, University of Chicago.
10. Gaber, R.F. and M.R. Culbertson. 1982. Frameshift suppression in *Saccharomyces cerevisiae*. IV. New suppressors among spontaneous co-revertants of the group II *his4-206* and *leu2-3* frameshift mutations. *Genetics* **101**:345.
11. Gaber, R.F., L. Mathison, I. Edelman, and M.R. Culbertson. 1982. Frameshift suppression in *Saccharomyces cerevisiae*. VI. Complete genetic map of twenty-five suppressor genes. *Genetics* (in press).
12. Hawthorne, D.C. 1981. UGA suppressors in yeast. In *Alfred Benzon Symp. 16, Molecular genetics in yeast* (ed. D. von Wettstein et al.), p. 291. Munksgaard, Copenhagen.
13. Hilger, F., M. Prevot, and R. Mortimer. 1982. Genetic mapping of *arg*, *cpa*, *car* and *tsm* genes in *Saccharomyces cerevisiae* by trisomic analysis. *Curr. Genet.* **6**:93.
14. Holm, C. 1982. Sensitivity to the yeast 2μ DNA is conferred by the nuclear allele *nib1*. *Mol. Cell. Biol.* **2**:985.
15. Jacobs, P., J.C. Juniaux, and M. Grenson. 1980. A *cis*-dominant regulatory mutation linked to the *argB-argC* cluster in *Saccharomyces cerevisiae*. *J. Mol. Biol.* **139**:691.
16. Kawasaki, G. 1979. "Karyotypic instability and carbon source effects in cell cycle mutants of *Saccharomyces cerevisiae*." Ph.D. thesis, University of Washington, Seattle.
17. Klapholz, S. and R. Esposito. 1982. A new mapping method employing a meiotic rec⁻ mutant of yeast. *Genetics* **100**:387.
18. Klapholz, S. and R. Easton-Esposito. 1982. Chromosome XIV and XVII of *Saccharomyces cerevisiae* constitute a single linkage group. *Mol. Cell. Biol.* **2**:1399.
19. Lucchini, G., M.L. Carbone, M. Cocucci, and M.L. Sensi. 1979. Nuclear inheritance of resistance to antimycin A in *Saccharomyces cerevisiae*. *Mol. Gen. Genet.* **177**:139.
20. Masselot, M. and Y. Surdin-Kerjan. 1977. Methionine biosynthesis II. Gene-enzyme relationships in the sulfate assimilation pathway. *Mol. Gen. Genet.* **154**:23.
21. Matsumoto, K., A. Toh-e, and Y. Oshima. 1981. Isolation and characterization of dominant mutations resistant to carbon catabolite repression of galactokinase synthesis in *Saccharomyces cerevisiae*. *Mol. Cell. Biol.* **1**:83.
22. Matsumoto, K., I. Uno, Y. Oshima, and T. Ishikawa. 1982. Isolation and characterization of yeast mutants deficient in adenylate cyclase and cAMP-dependent protein kinase. *Proc. Natl. Acad. Sci.* **79**:2355.
23. McKnight, G.L., T.S. Cardillo, and F. Sherman. 1981. An extensive deletion causing overproduction of yeast iso-2-cytochrome *c*. *Cell* **25**:409.
24. Mortimer, R.K. and D.C. Hawthorne. 1973. Genetic mapping in *Saccharomyces*. IV. Mapping of temperature-sensitive genes and use of disomic strains in localizing genes. *Genetics* **74**:33.
25. Mortimer, R.K. and D. Schild. 1980. Genetic map of *Saccharomyces cerevisiae*. *Microbiol. Rev.* **44**:519.
26. Mortimer, R.K., R. Contopoulou, and D. Schild. 1981. Mitotic chromosome loss in a radiation-sensitive strain of the yeast *Saccharomyces cerevisiae*. *Proc. Natl. Acad. Sci.* **78**:5778.
27. Stateva, L. and P. Venkov. 1981. Meiotic mapping of the nuclear determinant of cell lysis of the osmotic dependent *Saccharomyces cerevisiae* mutant VY1160. *Mol. Gen. Genet.* **181**:414.

28. Takahashi, T., M. Hojito, and K. Sakai. 1980. Genes controlling hydrogen-sulfide production in *Saccharomyces cerevisiae*. *Bull. Brew. Sci.* **26**:29.
29. Urban-Grimal, D. and R. Labbe-Bois. 1981. Genetic and biochemical characterization of mutants of *Saccharomyces cerevisiae* blocked in six different steps of heme biosynthesis. *Mol. Gen. Genet.* **183**:85.
30. Wickner, R.B., F. Boutelet, and F. Hilger. 1982a. Evidence for a new chromosome in *S. cerevisiae*. *Mol. Cell. Biol.* (in press).
31. Wickner, R.B., S.P. Ridley, H.M. Fried, and S.M. Ball. 1982b. Ribosomal protein L3 is involved in replication and maintenance of the killer double-stranded RNA genome of *Saccharomyces cerevisiae*. *Proc. Natl. Acad. Sci.* **79**:4706.
32. Zamb, T.J. and T.D. Petes. 1982. Analysis of the junction between ribosomal RNA genes and single-copy chromosomal sequences in the yeast *Saccharomyces cerevisiae*. *Cell* **28**:355.

APPENDIX II
Biochemical Markers for Yeast Organelles

Compiled by Randy Schekman
Department of Biochemistry
University of California
Berkeley, California 94720

The following enzymes have convenient assays that can be used to follow yeast organelles during cell fractionation.

Compartment	Marker enzyme	Reference
Cell surface	invertase, acid phosphatase	2, 8
Plasma membrane	chitin synthetase, vanadate-sensitive Mg^{++}-ATPase	4, 18
Secretory vesicles	invertase, acid phosphatase	8, 14
Mitochondria:		
outer membrane	kynurenine hydroxylase, porin (29-Kd protein)	3 6
intermembrane space	L-lactate ferricyanide reductase (cytochrome b_2)	1
inner membrane	cytochrome oxidase	11
matrix	fumarase	15
Vacuole:		
membrane	α-mannosidase	16
soluble contents	alkaline phosphatase, carboxypeptidase Y, proteases A and B, ribonuclease	17
Golgi	mannoprotein outer chain oligosaccharide mannosyltransferase	13, 5
Endoplasmic reticulum	NADPH cytochrome c oxidoreductase, UDP-GlcNAc-N-actylglucosaminyl dolichyl-phosphate transferase	9 10
Nucleus:		
nucleoplasm	aspartate transcarbamoylase	12
nuclear envelope	same as endoplasmic reticulum (no distinctive markers yet)	
Cytoplasm	glyceraldehyde-3-phosphate dehydrogenase	7

REFERENCES

1. Appelby, C.A. and R.K. Morton. 1959. Lactic dehydrogenase and cytochrome b_2 of baker's yeast. *Biochem. J.* **71**: 492.
2. Arnold, W.N. 1972. Location of acid phosphatase and β-fructofuranosidase within yeast cell envelopes. *J. Bacteriol.* **112**: 1346.
3. Bandlow, W. 1972. Membrane separation and biogenesis of the outer membrane of yeast mitochondria. *Biochim. Biophys. Acta* **282**: 105.
4. Durán, A., B. Bowers, and E. Cabib. 1975. Chitin synthetase zymogen is attached to yeast plasma membrane. *Proc. Natl. Acad. Sci.* **72**: 3952.
5. Esmon, B., P. Novick, and R. Schekman. 1981. Compartmentalized assembly of oligosaccharide chains on exported glycoproteins in yeast. *Cell* **25**: 451.
6. Freitag, H., W. Neupert, and R. Benz. 1982. Purification and characterization of a pore protein of the outer mitochondrial membrane from *Neurospora crassa*. *Eur. J. Biochem.* **123**: 629.
7. Krebs, E.G. 1955. Glyceraldehyde-3-phosphate dehydrogenase from yeast. *Methods Enzymol.* **1**: 407.
8. Linnemans, W.A.M., P. Boer, and P.F. Elbers. 1977. Localization of acid phosphatase in *Saccharomyces cerevisiae*: A clue to cell wall formation. *J. Bacteriol.* **131**: 638.
9. Mahler, H. 1955. DPNH cytochrome *c* reductase. *Methods Enzymol.* **2**: 688.
10. Marriot, M. and W. Tanner. 1979. Localization of dolichyl phosphate- and pyrophosphate-dependent glycosyl transfer reactions in *Saccharomyces cerevisiae*. *J. Bacteriol.* **139**: 565.
11. Mason, T.L., R.O. Poyton, D.C. Wharton, and G. Schatz. 1973. Cytochrome *c* oxidase from baker's yeast. *J. Biol. Chem.* **248**: 1346.
12. Nagy, M., J. Laporte, B. Penverne, and G. Hervé. 1982. Nuclear localization of aspartate transcarbamoylase in *Saccharomyces cerevisiae*. *J. Cell Biol.* **92**: 790.
13. Nakajima, T. and C.E. Ballou. 1975. Yeast mannoprotein biosynthesis: Solubilization and selective assay of four mannosyltransferases. *Proc. Natl. Acad. Sci.* **72**: 3912.
14. Novick, P. and R. Schekman. 1979. Secretion and cell-surface growth are blocked in a temperature-sensitive mutant of *Saccharomyces cerevisiae*. *Proc. Natl. Acad. Sci.* **76**: 1858.
15. Racker, E. 1950. Spectrophotometric measurements of the enzymatic formation of fumaric and *cis*-aconitic acids. *Biochim. Biophys. Acta* **4**: 211.
16. Van der Wilden, W., P. Matile, M. Schellenberg, J. Meyer, and A. Wiemken. 1973. Vacuolar membranes: Isolation from yeast cells. *Z. Naturforsch.* **28c**: 416.
17. Wiemken, A., M. Schellenberg, and K. Urech. 1979. Vacuoles: The sole compartment of digestive enzymes in yeast (*Saccharomyces cerevisiae*)? *Arch. Microbiol.* **123**: 23.
18. Willsky, G.R. 1979. Characterization of the plasma membrane Mg^{2+}-ATPase from the yeast, *Saccharomyces cerevisiae*. *J. Biol. Chem.* **254**: 3326.

Subject Index

a cell(s)
  agglutination factors in, 351-352
  pheromone-induced changes in, 352-354
Acetaldehyde dehydrogenase, 4, 20
Acetate transport, 404, 409
Acetate utilization, 19-20
Acetohydroxy acid isomeroreductase pathway, 232, 233
  regulation of levels of, 235-236, 262
Acetohydroxy acid synthase (AHAS), 230, 232
  control of activity of, 234-235
  regulation of levels of, 235-236, 251, 263
Acetyl-CoA, 19-20
  in arginine biosynthesis, 188, 189, 258-259
  carboxylation, in fatty acid synthesis, 127-128
  in cysteine biosynthesis, 208, 209, 258
  in leucine biosynthesis, 231, 233, 258-259
  in lysine biosynthesis, 184, 196-198, 258
  in methionine biosynthesis, 221, 258
  in sterol biosynthesis, 112
  synthesis of, 20
Acetyl-CoA carboxylase
  biotin-dependent enzyme, 123
  mutants in, 117, 125
  regulation of, 127-128
Acetyl-CoA synthetase, 4, 20
Acetylglutamate kinase, 186, 189
  and acetylglutamylphosphate reductase, 191
  cross-pathway control of, 250
  mutations affecting, 190-192, 245
  repression by arginine, 190, 260
Acetylglutamate synthase, 186, 188
  cross-pathway control of, 250
  glucose repression of, 193-194, 258, 260
  inhibition by acetyl-CoA, 193-194, 258
  regulation of, 189
    by arginine, 190, 260
    by glucose, 193-194, 258, 260
Acetylglutamylphosphate reductase, 186
  and acetylglutamate kinase, 191
  cross-pathway control of, 250
  mutations affecting, 190-192, 245
  repression by arginine, 190, 260
Aconitase, 4, 20
Actin, 371, 628
Active transport. See Transport
Acyclic polyol transport, 404, 409
ADC1 gene, 591
ade1 phenocopy, in asp5 mutants, 215
Adenine transport, 425-427
  inhibitors of, 425, 427
  mutations affecting, 425, 427
S-Adenosylmethionine (SAM)
  compartmentation in vacuole, 439
  in phospholipid biosynthesis, 135
  in polyamine biosynthesis, 194
  pools of, 227-228
  in regulation of methionine biosynthesis, 225-229
    effect of lomofungin and cycloheximide on, 228
    posttranscriptional mechanism for, 228
  in sterol biosynthesis, 123
  use in radioactive labeling, 444
S-Adenosylmethionine synthesis, 213, 225, 227
S-Adenosylmethionine synthetase, 218, 225, 227
  mutations affecting, 227
  regulation of, 262
S-Adenosylmethionine transport
  active mechanism for, 421, 423-424

653

Subject Index

S-Adenosylmethionine transport (*continued*)
  inhibitors of, 424
  mutations affecting, 424
  into vacuoles, 441
Adenylate biosynthesis. *See* Purine biosynthesis
Adenylosuccinate lyase, 273, 275
Adenylosuccinate synthetase, 273, 275
  effect of mutations in, 276–277, 278
  feedback inhibition of, 277
  gene coding for, 276, 278
  involvement in regulation of purine biosynthesis, 278
Agglutination
  antigen-antibody
    as probe for RNA polymerase subunit function, 578
    as probe for RNA polymerase subunit homology, 572–574
    use in isolation of RNA polymerases, 567
    use in mannoprotein mutagenesis, 340–341
  factors involved in mating, 351–352
Alanine biosynthesis, 231, 239
Alcian blue, 339, 340
Alcohol, dehydrogenase, 4, 13–14, 20, 591
Aldolase, 4, 13
Alkaline phosphatase. *See* Phosphatase, alkaline
Allantoate
  cellular localization of, 439
  sequestration by vacuoles, 69, 71
Allantoate transport
  by active mechanism, 78, 411, 412
  gene coding for, 46
  inhibitors of, 412
  mutations affecting, 412
  regulation of, 46, 411, 412
  system, role in ureidosuccinate transport, 429
Allantoicase, 41
  gene coding for, 44
  induction of, 46
  nitrogen catabolite repression of, 82
Allantoin, 40–49
  cellular localization of, 439
  as inducer in allantoin transport, 410, 411
  as nitrogen source, 80–81
  sequestration by vacuole, 69, 71, 445–446
Allantoinase, 41
  gene coding for, 43
  induction of, 46
  nitrogen catabolite repression of, 82
Allantoin degradation, 40–49, 78–80. *See also* DUR1,2 gene
  genetics of, 43–44
  induction of enzymes in, 44, 46–49
  mutants in, 46–49
  nitrogen catabolite repression of, 80–82
  pathway, 41–43
  regulation of, 44, 46–49
  transcriptional control of, 46–47
Allantoin permease, 44
  gene coding for, 44
  nitrogen catabolite repression of, 82
  transcriptional control of, 49
Allantoin transport, 78, 400–402, 410–412
  by active mechanism, 410, 411
  induction of, by allantoin, 46, 410, 411
  inhibitors of, 410
  nitrogen repression of, 411, 412
  transinhibition of, 412, 446–447
Allophanate, as inducer, 43
  of allantoin degradation, 44, 46, 78–80
  of urea transport, 78–80, 412
Allophanate hydrolase, 43
  gene coding for, 44
  induction of, 46, 48, 49
  mRNA for
    half-life, 49
    transcription, 49–50
    transport, 48
  as multifunctional protein, 44
  mutants in, 47–48
  nitrogen catabolite repression, 82
  transcriptional control of, 49
  transport, 400–401
Allosuppressors, 475–477
  cold sensitivity of strains containing, 476
  viability of strains containing, 476
Allyl alcohol resistance, 13–14
α cells
  agglutination factors in, 351–352
  pheromone-induced changes in, 352–354
α-Amanitin
  differential sensitivity of RNA polymerase A and A* to, 577
  differential sensitivity of RNA polymerases to, 563, 566, 585
  mechanism of inhibition by, 582
  as probe for RNA polymerase function, 585
  resistance of tRNA transcription to, 514
Amidophosphoribosyltransferase, 271, 274
  effect of mutations in, 276
  feedback inhibition of, 277
  gene coding for, 276
  repression, by adenine, 278
Amino acid(s)
  cellular localization, 182, 183, 439
  pools, effect of nitrogen source on, 67, 442
  salvage pathway during nitrogen starvation, 65–66
  starvation, effect on RNA precursor uptake, 443
Amino acid biosynthesis, 181–264
  control systems for (general), 249–259
    cross-pathway control, 249–257
    glucose repression, 257–259
  control systems for (specific)
    cis-acting, 244–249
    trans-acting, 241, 244
  magnitude of responses to regulation in,

Subject Index   655

259-264
pathways, 183-241
Amino acid transport
anabolic systems, 420-425
basic amino acid permease, 420-423
catabolic systems, 411, 416-420
during nitrogen starvation, 65
pleiotropic mutations affecting, 418-419
Amino acid transport, general (GAP), 411, 416-418
inhibition by nitrogen, 417-418
mutations affecting
pleiotropic, 418-419
specific, 416-418
regulation of, 411, 416-418
substrate specificity of, 417
into vacuoles, 440-441
α-Aminoadipate, 63-64, 197
α-Aminoadipate aminotransferase, 186, 196
cellular localization, 196
glucose repression of, 198
regulation, 198
α-Aminoadipate reductase, 186, 196, 260
cross-pathway control of, 198, 251
mutations affecting, 196
α-Aminoisobutyric acid, compartmentalization of, 445-446
Ammonia transport
by active mechanism, 411, 413-416
effect of ionic strength on, 414
multiple system for, 411, 414-416
mutations affecting
pleiotropic, 419
specific, 414-416
regulation, 415-416
use of methylamine in studying, 411, 413-416
Ampicillin resistance, 612, 614
Anthranilate synthase, 201, 203
complex, with indoleglycerolphosphate, synthese, 201, 206-207
cross-pathway control of, 206, 250
feedback inhibition of, 204
regulation of levels, 206-207, 261, 264
Antibiotics, 540-541
Antisuppressors, 475-476. *See also SUP4* locus
analysis of tRNA transcription using, 594-596
undermodification of tRNA in, 509-510
Apocarboxylase ligase mutants, 117, 125
ARG5,6 gene, monocistronic vs. polycistronic, 285-286
Arginase, 58-60, 187, 189
cytosolic localization of, 60
gene coding for, 61
inactivation of OTCase by, 62-63, 190
induction of, 60-63, 190
mechanism of induction by lysine, 71
mRNA, 62

mutations affecting, 61-62
nitrogen catabolite repression of, 82
transcriptional control of, 49
Arginine
cellular localization, 439
as inducer in arginine degradation, 190
sequestration by vacuole, 69, 71
transport, 78
Arginine biosynthesis
cross-pathway control in, 193, 249, 250
genes and enzymes in, 186, 187
glucose repression of, 193-194
localization of enzymes in, 189
magnitude of regulatory response in, 259, 260, 264
metabolic flow in, 189-190
mutations affecting, 190-193, 244
pathway, 184, 188-189
regulation of enzyme levels in, 190-194
Arginine degradation, 58-63, 78-80
mutants in, 60-62
pathway, 58-60
regulation of, 60-62
sequential induction in, 78-80
transcriptional control of, 62
Arginine permease, and canavanine sensitivity, 503
Arginine transport
by active mechanism, 420-422
mutations affecting, 420, 422
into vacuoles, 440-441
ARO1 locus, 201, 202
ARS sequences, use in gene cloning, 620
L-Asparaginase, 364
Asparaginase I, 49-50
Asparaginase II, 50-52
gene coding for, 50
mutations affecting, 50
nitrogen repression, 51-52
regulation, 50-52
Asparagine biosynthesis
genes and enzymes for, 216
pathway, 215
Asparagine degradation, 49-52
Asparagine synthetase, 215, 216
Asparagine transport, 411, 420
Aspartate aminotransferase, 215, 216
Aspartate biosynthesis
genes and enzymes for, 216
pathway, 213, 215
Aspartate carbamoyltransferase, 265, 266
and carbamoylphosphate transferase, 269-270
feedback inhibition of, 270
nuclear localization, 265
as nuclear marker, 651
regulation of levels, 268, 270
Aspartate semialdehyde dehydrogenase, 215, 216, 219

Aspartate semialdehyde dehydrogenase (*continued*)
  regulation of levels, 214, 219–220, 244, 261
  role of tRNA^{Thr} in regulation, 220, 229, 244
Aspartokinase, 215, 216
  cross-pathway control, 219, 249, 252
  inhibition of, 219
  regulation of levels, 214, 219, 220, 244, 261
  role of tRNA^{Thr} in regulation, 220, 229, 244
ATPase, plasma membrane, 449–450
  as biochemical marker, 651
  role in transport, 450
AUG codon. *See* Initiation codon(s)
Azasterols, 122–123

Bacteriophage λ
  as a cloning vehicle, 609–611
  deletion formation, by chelating agents, 627
Berkeley body, 383–385, 397
Beryllium, 240
Biotin
  as cofactor for
    acetyl-CoA carboxylase, 123
    apocarboxylase ligase, 125
    pyruvate carboxylase, 20
  in urea degradation, 43
Biotin transport
  by active mechanism, 430–432
  inhibitors of, 430–432
  repression of, 431, 432
Borrelidin, 219
BSI mutant, 146
Bud
  formation, effect on cell wall, 345, 347
  growth
    mechanism, 370–372
    in secretion mutants, 383
Budding
  interruption of, during mating, 367
  role in localized secretion, 366–367
Bud scar, 345

Canavanine resistance
  in *arg80*, *arg81*, and *arg82* mutants, 191
  for isolation of amino acid permease
    mutants, 420
  for isolation of *gap* mutants, 416
  for isolation of tRNA mutants, 503
Canavanine transport, 420–422
Carbamoylphosphate synthetase, 186, 189, 265, 266
  and aspartate carbamoyltransferase, 269–270
  cross-pathway control, 192, 193, 249, 250, 252–253
  feedback inhibition of, 270
  localization, 189
  multiple species, 189
  mutations affecting, 244
  regulation, 189, 192–194, 260, 264, 268–270
  repression by arginine, 190, 244

  two-subunit enzyme, 192
Carbohydrate metabolism, 1–27
  catabolite repression in, 22–23
  enzymes and genes of, 4–7
  localization of enzymes in, 24
  oxygen induction in, 22
  pathway, 3
  Pasteur effect in, 25
Carboxypeptidase S, 73–75
Carboxypeptidase Y (proteinase C), 73–75
  gene coding for, 75
  as marker for vacuoles, 651
  mutations affecting, 75
  posttranslational modification of, 74
Carboxypeptidase Y synthesis
  localization, 386–387
  localization of precursors in, 377
  role of *PEP4*-gene product, 377
Cardiolipin, 108–112, 134
Catabolic synergism, in regulation of arginase, 63
Catabolite inactivation
  mechanism, 23–24
  mutants, 24
Catabolite repression, 22–23
  effect on mRNA levels, 22
  effector metabolite, 22–23
  mechanism, 22
  mutants, 23
Catabolite repression, nitrogen. *See* Nitrogen catabolite repression
Cell-cycle
  control
    of chitin synthesis, 369
    of ornithine decarboxylase levels, 194
    of ribosomal protein synthesis, 547
    of secretion, 367
    of synthesis of RNA polymerase, 585
  G_1 arrest mechanisms, 71–73
  vacuole morphology during, 68
Cell growth
  effect on cell wall, 344–345
  inhibition, by suppressors, 478–479
  in secretion mutants, 382
Cell surface. *See also* Cell wall
  biochemical marker for, 651
  bud-limited growth,
    during mating, 367
    mechanism, 367
  structure, 363
Cell wall, 335–357. *See also* Cell surface
  changes in, during
    bud formation, 345
    cell growth, 344–345
    septation, 345, 348
    spore germination, 350
    sporulation, 349–350
  composition, 336–342, 362–364
  effect of inositol-deprivation on, 142

Subject Index   657

effect of mating factors on, 352–354
immunochemical determinants in, 343, 344
localization of components in, 343–344, 346–347
during protoplast formation, 348
during protoplast regeneration, 348–349
role of disulfide bonds in, 355
role in mating, 351–354
role in transport, 448
and spore coat formation, 349–350
Centromeres, use in cloning vectors, 621, 622
Chemostat, glucose-limited growth, 2
Chitin
  effect of mating factors on, 353
  localization, 346–347
Chitin synthesis
  cell-cycle control, 369
  localization, 368–369
  during septation, 345
Chitin synthetase, 102
  localization, 368, 376
  as marker for plasma membrane, 651
  as marker for secretory vesicles, 651
  protease activation, 369, 376–377
Chitosomes, 376
Chlorolactate resistance, 322
Cholesterol, 104
Choline-requiring mutants, 118, 143–144
Choline transport
  by active mechanism, 431, 432–433
  inhibitors, 431, 432–433
  mutations affecting, 433
Chorismate mutase, 200, 202
  control of activity, 204
  mutations affecting, 201, 205, 245
  regulation of levels, 205, 261
Citrate synthase, 4, 20, 258
Cloning, 607–632. *See also* Transformation
  analysis of gene expression using, 624
  analysis of transcription using, 589–592
  avoidance of vector recircularization by differential restriction enzyme digestion, 623
    homopolymer tailing, 623
    phosphatase treatment, 623
  into bacteriophage λ, 609–611
    advantages, 611
    products formed, 609–610
    transfection of *E. coli*, 611
    vectors available, 611
    in vitro packaging, 611
  characterization of cloned gene, 623–624
  function of heterologous genes, 598
  function of yeast genes in *E. coli*, 616–617
  identification of cloned gene, 629–631
    by functional complementation, 630
    by homology with mutant alleles, 631
    by mapping, 630
    by restriction-fragment polymorphisms, 630
    by Southern analysis, 631
  lacZ gene fusions, 629
  of mutant genes, 625–626
  into plasmids, 611–615
    advantages, 614–615
    avoidance of vector recircularization, 613
    effect of DNA concentration, 613
    methods for selecting recombinant, 613–614
    products formed, 611–612
  screening methods, 615–617
    by functional complementation in *E. coli*, 616–617
    by hybridization with cDNA probe, 616
    by hybridization with DNA from other eukaryotes, 616
    by hybridization with RNA probe, 615–616
    by hybridization with synthetic DNA probe, 615
  of transport-associated genes, 438
  use in determining *URA3* mRNA levels, 271
  use in isolating *cis*-dominant mutations, 246 247
  use in mutant analysis, 625–627
  use of restriction-fragment polymorphisms in, 625
  use in studying cross-pathway control, 255
  in vitro mutagenesis, 627–629
  yeast DNA bank, 608
  yeast DNA library, 608
  yeast DNA transformation, 617–620
  yeast plasmid vectors, 620–623
    E. coli origin of replication, 621
    selectable markers, 621
Codon
  -anticodon wobble, 489–492
  usage, 490–491, 520–523
  correlation with number of tRNA genes, 490–491, 496, 520–523
  role in translational control, 520–523
Coenzyme A (CoA)
  inactivation of α-IPM synthase by, 235
  regulation of amino acid biosynthesis by, 230
Colchicine
  effect on secretion, 371
  effect on vesicle transport, 371
Compartmentation, 438–440
  of amino acids, 182, 183
  in mitochondrion, 183
  in vacuole, 182–183
  effect on cellular transport, 447
  of histidine, 422
Competitive inhibition, of transport systems, 402–403
Concanavalin A, 102
Copper transport, 437
Cordycepin, RNA polymerase inhibition by, 583

Cordycepin sensitivity, and adenosine
   utilization, 273
Cross-pathway control, 249-257
   in arginine biosynthesis, 207, 249, 250
   characteristics, 249, 252-253
   DNA sequences involved in, 255-257
   genes and enzymes examined for, 250-251
   in histidine biosynthesis, 241, 249, 250
   in homoserine biosynthesis, 220-221
   interaction of, with pathway specific control, 252-253
   involvement of tRNAs in, 228, 252
   in isoleucine-valine biosynthesis, 239, 249, 251
   in leucine biosynthesis, 239, 249, 251
   level of effect, 252, 255
   in lysine biosynthesis, 198, 249, 251
   mutations affecting, 253-255
   in tryptophan biosynthesis, 207, 249, 250
Cryptopleurine
   as inhibitor of translational elongation, 540
   resistance, 545
CYC1 gene. *See also* Initiation codon(s); Iso-1-cytochrome *c*; Translation initiation, *CYC1* gene
   cis-acting mutations in, 246-247, 616
   mRNA initiation in, 315, 320, 591
   mutations in, 301-330, 591
      chlorolactate selection for, 322
      isolation procedures for, 322-330
      revertants of, 302-323, 308-309, 322-323, 326-330
      using in vitro mutagenesis for, 591
      nonsense suppressors examined using, 465-469
   sequences required for transcription of, 591
   Ty element insertion in, 246-248
Cycloheximide
   effect of
      on enzyme repression by SAM, 228
      during heat shock, 549
      on rRNA synthesis, 548
      during sporulation, 539
      on translation elongation, 540
Cycloheximide resistance, ribosomal protein involved in, 535, 545
Cysteine biosynthesis, 207-212
   genes and enzymes for, 210-211
   mutations affecting, 209, 212
   pathway, 208
   by transulfuration from methionine, 212, 225
Cysteine synthase
   mutations affecting, 209, 223-224
   reaction, 209, 210, 216, 223
Cysteine transport, 421, 423
Cytochrome *b*, 127
Cytochrome coenzymes, 121
Cytidine transport, 427
Cytidylate biosynthesis. *See* Pyrimidine biosynthesis
Cytochrome *c* oxidoreductase (NADPH-), 651
Cytochrome oxidase, 651
Cytokinesis, mutants in, 372
Cytoplasm, biochemical markers for, 651
Cytosine transport, 425-427
   inhibitors of, 425, 427
   mutations affecting, 425, 427

Degradation of
   allantoin, 40-49
   arginine, 58-63
   asparagine, 49-52
   proline, 55-58
   proteins, 71-75
δ sequences, in *SUP4* locus, 505
Deoxy-*arabino*-heptulosonate-phosphate (DAHP) synthase, 198, 202
   phenylalanine sensitive, 203, 204
   regulation, 201, 204, 261
   two isozymes for, 201
   tyrosine-sensitive, 203, 204
2-Deoxyglucose
   induction of glycolytic enzymes by, 14
   inhibition of glucan synthesis by, 345, 369-370
   use in selection of hexokinase mutants, 10
6-Deoxyglucose, 14
Diffusion. *See* Transport
Digitonin, 369
Dihydrofolate reductase, 279, 282, 283
Dihydroorotase,
   induction of, 268, 270-271
   reaction, 266, 269
Dihydroorotase dehydrogenase, 266, 269
   induction of, 268, 270-271
   mutations affecting, 271
Dihydroxyacetone-phosphate reductase, 4, 14
Dihydroxy acid dehydratase
   reaction, 232, 233
   regulation of levels, 235-236, 263
Doubling time, in amino acids, 41
Douglas-Hawthorne model, 162-166
DUR1,2 gene. *See also* Allantoin degradation; Allophanate hydrolase; Urea carboxylase
   effect of *DAL80*- and *DAL81*-gene products on, 46-49
   induction of, by allophanate, 44, 46
   insertion mutations in, 247
   monocistronic gene, 44
   mRNA in, 48-49, 82
   mutations affecting expression of, 46-49
   nitrogen repression of, 46

Escherichia coli
   expression of yeast genes in, 616-617
   host-vector systems, 608-615
Elongation factors, 540
Emetine, 540

Subject Index   659

EMS mutagenesis, of *CYC1* gene, 329
Endoplasmic reticulum (ER), 394
  biochemical markers for, 651
  characterization of, 377-378
  formation of secretory vesicles from, 364
  mannoprotein glycosylation in, 372-374
  in secretion mutants, 383-384, 396
Enolase, 4, 13
Epiarginasic regulation, 62-63, 190
Epimerase, 15-16
  regulatory genes for, 161
  structural genes for, 161
  transcriptional control, 167-168
Ergosterol. *See also* Sterol(s)
  biosynthesis, 105, 112
  localization, 122
  mutations affecting, 114-115, 120-121
  structure, 104
Ethanolamine-requiring mutants, 118, 143-144
Ethanol transport, 404, 409
Ethanol utilization, 19-20

Fatty acid(s), 123-132
  deprivation, 128-129, 131-132
  replacement, 128-132
  structures, 124
Fatty acid biosynthesis, 123-124
  cellular localization, 106
  mutants in, 117, 124-132
  regulation, 127-128
Fatty acid desaturase, 123
  experimental manipulation of, 129-132
  mutants, 117, 126
Fatty acid synthetase, 123-124
  experimental manipulation of, 128-132
  mutants, 117, 123-124
Feedback inhibition
  of histidine transport, 445
  universality of, in transport systems, 445-446
Fermentation, 13-14
Fluoroacetate resistance, 21
Formate, 279
Fractionation (cellular)
  for analysis of
    cell surface, 348
    lipid distribution, 102
    RNA polymerases, 565
    secretory components, 362, 377-381
    transport components, 448-452
  biochemical markers used in, 651
Fructose, 15
Fructose-biphosphate aldolase, 4, 13
Fructose diphosphatase, 24
Fructose transport. *See* Hexose transport
Fumarase
  as marker for mitochondrial matrix, 651
  reaction, 4, 21

Galactokinase, 15-16

gene coding for, 161
regulatory genes affecting, 161
transcriptional control, 164, 167-168
Galactose metabolism, 158-168, 176-177
α-D-Galactose-1-phosphate uridyltransferase.
  See Transferase
Galactose transport
  by facilitated diffusion, 404, 406
  induction, by galactose, 404, 406
Galactose transport protein, 161
Galactose utilization, 15-16
  regulation of, 162-166
    comparison with phosphatase synthesis, 173, 175-177
    Douglas-Hawthorne model for, 162
    revised model for, 166-168
  regulatory genes for, 161
  structural genes for, 161
GAL3 gene, mutations in, 161
GAL4 gene
  constitutive expression of, 164-165
  effect of mutations in, 161, 163-165
  fine-structure mapping in, 165-166
  function of
    Douglas-Hawthorne model for, 162
    genetic characterization of, 163-166
    revised model for, 166-168
  nonsense mutations in, 165-166
GAL80 gene
  effect of mutations in, 161, 163-165
  function of
    Douglas-Hawthorne model for, 162
    genetic characterization for, 163-166
    revised model, 166-168
gal81 locus
  effect of mutations in, 161, 163-165
  fine-structure mapping in, 165-166
  function of
    Douglas-Hawthorne model for, 162
    genetic characterization for, 163-16
    revised model for, 166-168
  nonsense mutations in, 165-166
GAP system. *See* Amino acid transport, general
gdh1 mutation, effect on nitrogen catabolite repression, 83-86
Gene cloning. *See* Cloning
Gene dosage, effect on regulation, 629
Gene fusions, 629
Genome (yeast), size of, 608
Glucan, 336-337
  effect of mating factors on, 353-354
  localization, 347, 363
Glucan synthesis
  inhibition by 2-deoxyglucose, 369-370
  localization, 368-370
Glucan synthetase, 370
Glucokinase, 4, 10-11
Gluconase, activity during mating, 353-354
Gluconeogenesis, 21

Glucosamine, 15
  auxotrophs
    effect on septation, 345
    effect on spore coat formation, 348, 355
  pathway into glycolysis, 15
Glucose
  in catabolite inactivation, 23-24
  in catabolite repression, 22-23
  and the Pasteur effect, 25
  phosphorylation, 10-11
  repression of arginine biosynthesis, 193-194, 258-259
  repression of leucine biosynthesis, 244, 258-259
Glucose-6-phosphate dehydrogenase, 7, 12
Glucose-6-phosphate epimerase, 13
Glucose transport, 10, 403-406. *See also* Hexose transport
  by facilitated diffusion, 403-405
  role of phosphorylation in, 405
Glusulase, 348, 362
Glutamate biosynthesis
  genes and enzymes in, 186
  pathway, 184, 185
Glutamate dehydrogenase, NAD-, 52
  cross-pathway control in, 251, 252
  mutations affecting, 55
  regulation, 53
Glutamate dehydrogenase, NADP-, 52, 186
  cellular localization, 185
  as effector of nitrogen catabolite repression, 83-86
  mutations affecting, 54-55
  mutations in, effect on GAP, 417-418
  nitrogen catabolite repression of, 53
  regulation, 53, 185
  role in glutamate biosynthesis, 184, 185, 186
Glutamate kinase, 55
Glutamate synthase, 55
  reaction, 184, 185, 186
  regulation, 185
Glutamate transport, 421, 424
Glutamime biosynthesis
  genes and enzymes in, 186
  magnitude of regulatory responses in, 259, 260
  pathway, 184, 185
  regulation, 260
Glutamine synthetase, 55, 184, 185, 186
  effect of nitrogen source on, 185
  as effector of nitrogen catabolite repression, 87-88
  mutants in, 188
  regulation, 260
Glutamyl phosphate reductase, 55
Glyceraldehyde-3-phosphate dehydrogenase, 4, 13
  gene coding for, 10
  as marker for cytoplasm, 651

Glycerol kinase, 5, 18
Glycerol-1-phosphatase, 5, 14
Glycerol-3-phosphate, 133
Glycerol-3-phosphate dehydrogenase, 5, 18
Glycerol transport, 404, 409
Glycerol utilization, 18
Glycine, in single carbon metabolism, 209, 279
Glycine biosynthesis, 207-209, 210, 212
  in acetate medium, 209
  genes and enzymes for, 210
  pathway, 208
Glycine decarboxylase, 209, 210
Glycogen degradation, 25-26
Glycogen phosphorylase, 26
Glycogen synthesis, 25-26
Glycogen synthetase, 25-26
Glycolysis, 2, 9-15
  enzymes and genes of, 4-7
  mutants in, 9-15
  pathway, 3
  regulation, 14-15
Glycosides, 16-18
Glycoside utilization, 16-18
Golgi apparatus
  accumulation in secretory mutants, 397
  biochemical markers for, 651
  characterization, 378
  mannoprotein synthesis in, 373-374
Growth rate
  on amino acids, 41
  effect on ribosome concentration, 546-547
  effect on RNA polymerase content, 573
  in minimal medium, 8
Guanine transport, 425-428
  inhibitors of, 425, 427-428
  mutations affecting, 427
Guanylate biosynthesis. *See* Purine biosynthesis

Harden-Young effect, 27
Heat shock, effect on ribosome synthesis, 549-551
Heme mutants
  effect on sterol biosynthesis, 121-122
  of fatty acid desaturase, 116, 127
  in lipid biosynthesis, 115-116
Hexokinase A (P-I), 5, 10-11
Hexokinase B (P-II), 5, 10-11
Hexose-monophosphate oxidation, 18-19
Hexose-monophosphate shunt, 18-19
Hexose transport, 403-406
HIS3 gene
  sequences involved in cross-pathway control, 256-257
  sequences required for transcription, 591
  in vitro mutagenesis, 246, 247
HIS4 gene
  cis-dominant mutations in, 246
  encoding of multifunctional protein by, 240, 244-245

Subject Index   661

fusions, to study cross-pathway control of, 255
sequences involved in cross-pathway control, 256–257
Ty element insertions in, 246
in vitro isolation of mutations in, 247
Histidine biosynthesis, 240–243
  in *ade3* mutants, 240, 283
  cross-pathway control, 241, 249, 250
  genes and enzymes, 243
  magnitude of regulatory response in, 259, 263
  metabolic flow in, 240–241
  pathway, 240, 242
  regulation of enzyme levels in, 241
Histidine transport
  effect of histidine sequestration on, 422
  feedback inhibition of, 445
  multiple systems for, 421, 422–423
  mutations affecting, 423
Hogness box
  in *CYC1*, 256
  in *GAP491*, 256
  in *HIS3*, 256
  in *HIS4*, 256
Homocitrate synthase, 186, 196
  feedback inhibition of, by lysine, 197
  inactivation by CoA, 197, 258
  mutations affecting, 197
  regulation, 198, 260
Homocysteine synthase, 209, 211, 216, 223
  mutations affecting, 209, 223–224
  regulation, 212, 262
Homocysteine synthesis, 213, 221–224
  mutations affecting, 223–224
  two pathways for, 223–224
Homoserine acetyltransferase
  reaction, 216, 221, 258–259
  regulation of, 225–226, 262
Homoserine biosynthesis
  metabolic flow in, 219
  mutations affecting, 219–221
  pathway, 213, 215, 219
  regulation of enzyme levels, 214, 219–221, 244
Homoserine dehydrogenase, 216, 219
  inhibition of, 219
  regulation of levels of, 214, 220–221, 229, 261
Homoserine kinase, 216, 221, 262
Hypoxanthine transport, 425–427
  inhibitors of, 425, 427
  mutations affecting, 425, 427

ICR-170, 470
IMP dehydrogenase, 273, 275, 278
Indoleglycerolphosphate synthase, 201, 203
  complex with anthranilate synthase, 201, 206–207
  cross-pathway control of, 206, 249, 250
  regulation of levels, 205–207, 261
Initiation codon(s). *See also CYC1* gene; Iso-1-cytochrome *c*; Translation initiation, *CYC1* gene
  comparison to prokaryotic, 310
  in *CYC1* gene, 302–304, 309–314, 316–330
  mutations of, 302–314
  positions in mRNA, 285, 310–312
Inorganic phosphate
  in phosphate synthesis, 168, 170–173
  transport of, 172
Inosinate biosynthesis. *See* Purine biosynthesis
Inositol
  -requiring mutants, 117, 137–143
  -secreting mutants, 118
Inositol biosynthesis, 133
  in mutants in phosphatidylethanolamine methylation, 144–147
  regulation, 138–140
Inositol deprivation, 141–143
Inositolless death, 141
Inositol-1-phosphate synthase, 137, 139–140
  mutations affecting, 138–141, 145–146
  regulation
    mRNA half-life during, 140
    by proteolysis, 140
Inositol starvation, in selection of secretion mutants, 382
Inositol transport
  by active mechanism, 431, 433
  mutations affecting, 433
  repression, 431, 433
Integration, chromosomal, of plasmid DNA, 617–620
Intervening sequences. *See* Introns, tRNA
Introns, tRNA, 490–491. *See also* Splicing; tRNA
  comparison of, with other eukaryotes, 499–501
  enzymology of splicing of, 510–513
  location within gene, 497, 500–501
  mutations in, effect on splicing, 505–508
  Northern analysis of, 624
  role of splicing in gene expression, 523
  splicing of
    in *rna1* mutants, 497–502
    in vitro systems for, 499
    use of in vitro mutagenesis to study, 507–508
Invertase, 16
  cytoplasmic, 375
  external, 337, 342
  gene coding for, 375
  glycosylation, 366
  localization, 363–364
  in mannan structural mutants, 374
  in mannoprotein mutants, 341–342
  as marker for cell surface, 651

Invertase (*continued*)
mRNA of, in repressed cells, 375
mutants in secretion of, 382-385
role in sucrose transport, 407
secretion of, 365
transcriptional control of, 366
Invertase synthesis
localization of intermediates in, 375-376
stages, 384-385
In vitro mutagenesis. *See* Mutagenesis, in vitro
Ion transport, 433-438
Isocitrate dehydrogenase, 5, 20
Isocitrate lyase
in glycine biosynthesis, 209, 210
in growth on ethanol, 5, 20, 21
Iso-1-cytochrome *c*. *See also CYC1* gene;
Initiation codon(s); Translation initiation;
CYC1 gene
aminoterminal region, requirement for, 309
levels in *cyc1* mutants, 311, 312, 313
methionine aminopeptidase specificity in, 304-306
translation initiation of, 302-330
Isoleucine biosynthesis
cross-pathway control, 239, 249, 251
genes and enzymes in, 232
localization of enzymes in, 233
magnitude of response to regulation in, 259, 263, 264
metabolic flow in, 234-235
pathway, 230-233
regulation of enzymes levels, 235-238
Isomaltose, 17
α-Isopropylmalate (α-IPM) isomerase, 232, 233
cellular localization, 234
mutations affecting, 233-234, 238-239
regulation of levels of, 238-239, 263
α-Isopropylmalate synthase, 232, 233
cellular localization 234
control of activity of, 235
cross-pathway control of, 239, 249, 250
effect of carbon source on, 239, 258-259
inactivation of, by CoA, 258-259
mutations affecting, 238-239
proteolysis of, 239
regulation of levels, 238-239, 263
two genes coding for, 233-234
β-Isopropylmalate (β-IPM) dehydrogenase, 232, 233
cellular localization of, 234
mutations affecting, 233-234, 238-239
regulation of levels, 238-239, 263

α-Ketoglutarate dehydrogenase, 5, 20
Killer toxin, 364
Kluyveromyces lactis
mannoprotein mutants of, 356
mannoprotein structure in, 343
Kynurenine hydroxylase, 651

Labeling, radioactive
of nucleic acids, 443-444
of proteins, 442-443
L-Lactate ferricyanide reductase, 651
Lactate transport, 404, 409
Lactate utilization, 19
Lactose utilization, 17
lacZ fusions, to *HIS4* gene, 255
Leloir pathway, 160
Leucine biosynthesis
cross-pathway control in, 239, 249, 251
genes and enzymes in, 232
magnitude of regulatory response in, 259, 263, 264
metabolic flow in, 234-235
pathway, 230-234
regulation of enzyme levels, 238-239
Leucine transport
inhibitors of, 425
multiple systems for, 421, 424-425
Ligase, tRNA. *See* Splicing,
Lipid(s), 101-148. *See also* Phospholipid(s);
Sterol(s)
composition in membrane fractions, 108-112
depletion, effect on mitochondria, 119
Lipid biosynthesis
cellular localization, 103-108
mutants in, 113-122
Lomofungin, 48, 228, 569
Lysine biosynthesis, 196-198
cross-pathway control in, 198, 249, 251
genes and enzymes in, 187
magnitude of regulatory response in, 259, 260
metabolic flow in, 197-198
mutations affecting, 196-197
pathway, 196-197
regulation of enzyme levels, 187
Lysine transport
multiple systems for, 421, 422
mutations affecting, 422

Magic Spot, ppGp, 550
Magnesium transport, 434, 436-437
Malate dehydrogenase, 5, 21
Malate synthase, 5, 20
Malic dehydrogenase, 5, 21
Maltase, 16, 17
Maltose transport, 404, 407, 408
by active mechanism, 407
induction of, 407-408
inhibitors of, 407-408
mutants in, 407
Maltose utilization, 16
Manganese transport, 434, 436-437
Mannitol transport, 404, 409
Mannoprotein, 337-343
effect of mating factors on, 353
genus variation, 343

localization, 346–347, 363
mutants
  invertase in, 341–342
  isolation of, 338–341
  physiological defects of, 341
  secretion in, 342
  mutations affecting, 338–342
  secretion by exocrine process, 364–365
  structure, 337–343
Mannoprotein synthesis
  localization, 372–376
  in secretion mutants, 382–385
  stages, 372–376, 384–385
Mannose, 15
Mannose transport. See Hexose transport
α-Mannosidase
  localization, 374
  as marker for vacuolar membrane, 651
Mannosyltransferase, 338–342, 651
Mating
  pheromone-induced changes during, 352–354
  role of cell wall in, 351–354
  -type control, and Ty elements, 247
Melibiase, 16
Melibiose utilization, 16, 17
Melizitose utilization, 17–18
Membrane(s)
  composition and synthesis, 101–148
  effect of fatty acid deprivation on, 128–129, 131–132
  effect of fatty acid replacement on, 128–132
  effect of inositol deprivation on, 142–143
  lipid composition, 108–112
  phospholipid synthesis in, 102–108
  sterol requirements of, 112–123
Membrane fluidity, 127
Methionine, inhibitor in single carbon metabolism, 283
Methionine aminopeptidase, specificity of, 302, 304–306
Methionine biosynthesis. See also Homoserine biosynthesis
  cross-pathway control in, 228–229, 249, 251
  magnitude of regulatory response in, 259, 261–262, 264
  metabolic flow in, 225–226
  mutations affecting, 222–229
  pathway, 213, 215, 219, 221–225
  regulation of enzyme levels, 226–230
  role of SAM in, 225–229
  role of THF in, 225, 229, 279
  role of tRNA in, 228
  sulfate reduction in, 222
Methionine transport
  multiple systems for, 421, 423
  mutations affecting, 423
Methylamine, in study of ammonia transport, 411, 413–416
Methylamine transport. See Ammonia transport

Methylation
  of rRNA, 543
  in sterol biosynthesis, 123
α-Methylglucoside transport, 404, 408
  by active mechanism, 404, 408
  by facilitated diffusion, 404, 408
  regulation, 404, 408
α-Methylglucoside utilization, 17
Methyl mercury resistance, 226
5-Methyltryptophan (5MT)-sensitive mutants, 253
Michaelis-Menton equation, for analysis of transport
  of ammonia, 414
  of glutamate, 424
  of histidine, 422
  of proline, 419
  of sulfate, 433
  of uracil and uridine, 428
Microtubules, in secretory vesicle transport, 371
Mitochondria
  during anaerobic growth, 135–136
  biochemical markers for, 651
  effect of fatty acid deprivation on, 131–132
  effect of fatty acid substitution on, 131–132
  effect of lipid depletion on, 119
  effect of sterol deprivation on, 122
  effect of sterol inhibitors on, 122–123
  effect of sterol mutations on, 120–122
mRNA
  of acid phosphatase, 366
  capping of, 444
  comparison of, between eukaryotes and prokaryotes, 316
  5′ end, in CYC1 gene, 315, 320
  of galactokinase, in galactose induction, 164
  half-life, 49
  of arginase, 62
  determination by hybridization to cloned genes, 624
  of inositol-1-phosphate synthase, 140
  of invertase, 366
  of ribosomal proteins, 550
  monocistronic vs. polycistronic, 284–286, 317, 321, 539
  polyadenylation of, 592
  regulation of transcription, 589–592
  mutations affecting, 589–592
  structures and sequences involved, 589–592
  ribosomal binding site in, 317
  transcription,
    by RNA polymerase B, 586, 589–592
    in vitro systems for, 587–588
  of transferase, during galactose induction, 164
  translation of, 302–321
  transport, of allophanate hydrolase, 48

Multifunctional proteins, loci encoding, 284–286
  ADE3, 284–285
  ADE5,7, 285–286
  ARG5,6, 285–286
  ARO1, 285
  DUR1,2, 44, 45
  HIS4, 284–285
  TRP5, 284–285
  URA2, 286
Mutagenesis, of *CYC1* gene, 302–304, 306, 322–330
Mutagenesis, in vitro, 627–629
  analysis of transcription initiation using, 591–592
  analysis of tRNA introns using, 507–508
  deletion formation by
    chelating agents, 627
    integrative transformation, 627–628
    restriction fragment removal, 627–628
  frameshift formation, 628
  point mutation formation, 628–629
  use of integrative transformation for, 628
Mycotoxin, 582

NAD-glutamate dehydrogenase. *See* Glutamate dehydrogenase, NAD-
NADP-glutamate dehydrogenase. *See* Glutamate dehydrogenase, NADP-
Neomycin, 475
Neurospora crassa, phosphatase synthesis in, 173, 175–177
Nitrogen catabolite repression, 80–90
  of allantoin degradation, 80–82
  effector metabolites of, 83–89
    glutamate dehydrogenase, NADP-, 83–86
    glutamine synthetase, 87–88
    URE2-gene product, 86–87
    URE3-gene product, 88–89
  and enzyme induction, 82
  of glutamate dehydrogenase, NAD-, 53
  level of effect of, 82–83
Nitrogen metabolism, 39–90
Nitrogen metabolism, transport of compounds involved in, 410–420
Nitrogen repression
  of allantoate transport, 411, 412
  in allantoin degradation, 46
  of allantoin transport, 412
  in asparagine degradation, 51
  of general amino acid transport, 411, 416
  in proline degradation, 58
  of proline transport, 58
  of transport, 444–445
  of urea transport, 411, 412–413
  of ureidosuccinate transport, 426, 429
Nitrogen starvation
  amino acid metabolism during, 65–66
  amino acid permease activity during, 65
  effect on vacuolar constituents, 66–67
  mobilization of sequestered metabolites during, 69, 71
  protein degradation during, 71–75
  sporulation during, 75–76
Nonsense suppressors. *See* Suppressors
Northern hybridization, 624
Nuclease S1, as probe for pre-tRNA structure, 501–502
Nucleic acids, radioactive labeling of, 443–444
Nucleolus, 544
Nucleotide biosynthesis, 264–278
Nucleus, biochemical markers for, 651
Nystatin
  in isolation of sterol mutants, 120
  in study of allantoin transport, 410

One carbon metabolism. *See* Single carbon metabolism
Ornithine decarboxylase, 194, 195
  cell-cycle control of, 194
  mutations affecting, 194
Ornithine transaminase, 58–60
  cytosolic localization, 60
  gene coding for, 61
  induction of, 60–62
  mutations affecting, 61–62, 245
Ornithine carbamoyltransferase (OTCase), 187, 189, 191, 192, 193, 264
  cross-pathway control of, 250
  inactivation, in arginase complex, 62–63, 190
Ornithine transport, 420–422
Orotidine-5′-phosphate (OMP) decarboxylase, 266, 269, 270
  induction of, 268, 270–271
  mutations affecting, 270, 271
  transcriptional control of, 271
Osmotic shock
  loss of transport activity during, 451
  loss of transport proteins during, 451
Osmotic stabilization of
  mutant phenotypes, 475
  spheroplasts, 409
Ouabain resistance, 449
Oxalacetate, 19–20
Oxalurate, in allantoin degradation, 412, 413
Oxalurate transport
  by active mechanism, 411, 413
  regulation of, 411, 413
Oxygen induction, 22

PAPS reductase, 209, 212, 217, 220
Paromomycin
  alleles suppressible by, 477–478
  effect on translation, 540–541
  phenotypic suppression by, 475, 477, 481
  sensitivity, in suppressor mutants, 472
Pasteur effect, 25
pBR322, 611–615

Subject Index   **665**

Periplasmic space, 448
Petite mutants, 131-132
Phenylalanine biosynthesis, 198-207
  genes and enzymes for, 202-203
  magnitude of regulatory response in, 259, 261
  metabolic flow in, 204-205
  mutations affecting, 201
  pathway, 198-201
  regulation of enzyme levels, 205-206
PHO4 gene
  effect of mutations in, 169
  product, as activator of phosphatase gene expression, 171-173
PHO80 gene
  effect of mutations in, 169
  product, as repressor of *PHO4*, 171-173
PHO81 gene
  effect of mutations in, 169
  as mediator of *PHO80* and *PHO85* expression, 171-173
  phosphate activation of, 171
pho82 locus
  effect of mutations in, 169-173
  mapping of, 172
PHO85 gene
  effect of mutations in, 169
  product, as repressor of *PHO4*, 171-173
Phosphatase, acid
  constitutive, 168-177
    localization of, 363-364
    mutations affecting, 169-170
  glycosylation, 168, 366
  localization, 168, 376
  as marker for cell surface, 651
  as marker for secretory vesicles, 651
  mutants in secretion of, 382-384
  repressible, 168-177
    derepression of, in phosphate transport mutants, 436
    localization, 363-364
    mutations affecting, 169-171
    regulatory model for, 171-177
    secretion of, 394
    transcriptional control of, 169, 366
Phosphatase, alkaline
  intracellular localization of, 168
  as marker for vacuoles, 651
  nonspecific, 168-177
  and glycosylation, 366
  mutations affecting, 170-171
  regulatory model for, 171-177
  transcriptional control of, 171
  specific, 168-177
Phosphatase synthesis. *See also PHO4* gene; *PHO80* gene; *PHO81* gene; *pho82* locus; *PHO85* gene; Phosphatase, acid; Phosphatase, alkaline
  genetic loci involved, 168-172
  regulation of, 171-177

comparison to galactose utilization, 173, 175-177
  in *Neurospora crassa*, 173, 175-177
  role of phosphate in, 162-163
Phosphate metabolism, 168-177
Phosphate transport, 172
  by active mechanism, 434-436
  mutations affecting, 436
  role of cations in, 435-436
Phosphatidylcholine
  regulation of, during choline/ethanolamine starvation, 144
  structure of, 134
  synthesis
    coordinate control with inositol biosynthesis, 144-147
    mutants in, 117-118, 144-147
    regulation of, 136-137, 138-139
Phosphatidylethanolamine
  in mitochondrial membranes, 109-112
  mutants in methylation of, 144-147
  regulation of biosynthesis of, 136-137
  structure of, 134
Phosphatidylinositol, 133-134
  biosynthesis
    coordinate control with phosphatidylcholine biosynthesis, 144-147
    mutants in, 117-118, 144-147
    regulation of, 138-139
    during inositol deprivation, 141
    regulation of, 136-137, 143-144
  structure of, 134
Phosphatidylserine, 133-135
  biosynthesis, mutants in, 118, 143
  structure, 134
Phosphatidylserine synthase, 134-135, 143-144
Phosphoenolpyruvate carboxykinase, 5, 21, 24
Phosphofructokinase, 11-12
Phosphoglucomutase, 6, 16, 160, 161
Phosphoglucose isomerase, 6, 11, 12
Phosphoglycerate kinase, 6, 13
Phosphoglycerate mutase, 5, 13
Phospholipid(s), 132-147. *See also* Lipid(s)
  in mitochondrial membranes, 108-112
  in yeast membrane fractions, 108-112
Phospholipid biosynthesis, 132-135. *See also* Lipid biosynthesis
  cellular localization, 106-107
  mutants in, 117-118, 137-147
  pathways, 126
  regulation of, 135-137
Phosphomannose isomerase, 6, 15
Phosphoribosyltransferase, 240, 243, 250, 263
Phosphorylation
  of eukaryotic ribosomal proteins, 533
  of RNA polymerases, 571, 574
  role in regulation of RNA polymerases, 584-585

Phosphorylation (*continued*)
  role in transport, 405
Plasmalemma, 350
Plasma membrane, 384
  assembly, 386
  comparison to vacuolar membrane, 448-449
  disruption by basic proteins, 439
  purification of, 380-381
    biochemical marker for, 651
    ER contamination, 381
  role in transport, 448-449
Plasmids, 611-615
Polarity, 285
Poly(A) addition, sequences for, 592
Polyamine biosynthesis, 194, 195
Polyamines
  cellular roles for, 194
  pool size of, 194
Polycistronic operons. *See* mRNA
Polyphosphate, 26, 440
Polyphosphate kinase, 26
Porin, 651
Posttranslational modification, of proteinases, 73-74
Prephenate dehydratase, 200, 203
  control of activity of, 204-205
  mutations affecting, 201
  regulation of levels, 205, 261
Prephenate dehydrogenase, 201, 203
  control of activity of, 204-205
  mutations affecting, 201
  regulation of levels, 205, 261
Proline degradation, 56-58
  control of, 57-58
  mitochondrial localization of, 57
  mutations affecting, 57
  pathway, 56, 57
Proline oxidase, 57
  gene coding for, 57
  mitochondrial localization of, 60
Proline synthesis
  genes and enzymes in, 186
  mutations affecting, 57
  pathway, 55, 56, 184, 188
Proline transport, 78, 411, 419-420
  inhibition, by histidine, 411, 420
  multiple systems for, 411, 419-420
  mutations affecting, 419-420
  nitrogen repression of, 58
Promoters, sequences involved in, 589-592
Protease(s). *See* Proteinase(s)
Proteinase(s), 73-75
  activation of chitin synthetase by, 369
  in carboxypeptidase Y processing, 377
  and RNA polymerase B, 565, 570
Proteinase A, 73-75
  as marker for vacuoles, 651
  mutations affecting, 74-75
  posttranslational modification of, 73-74

Proteinase B, 73-75
  effect of, on carbamoylphosphate synthetase and aspartate carbamoyltransferase, 270, 286
  gene coding for, 74
  as marker for vacuoles, 651
  mutations affecting, 74-75
Protein degradation, 71-75
  genetics of, 74-75
  during nitrogen starvation, 71-73
Protein kinase, phosphorylation of RNA polymerase by, 584-585
Proteins, radioactive labeling of, 442-443
Protein synthesis. *See* Translation
Protein, transport
  cellular localization, 448
  isolation and purification, 450-452
  mutations affecting, 451
Proteolysis
  in asparagine degradation, 51
  of inositol-1-phosphate synthase, 140
Protoplast
  formation
    effect on distribution of plasma-membrane components, 369
    stabilization by concanavalin A, 102
    stabilization by sorbitol, 409
    use in study of cell wall, 348
    use in study of secretion, 362
    for yeast DNA transformation, 617
  regeneration, 348-349
ψ^+, 473-474
  effect on frameshift suppression, 474
  effect on nonsense suppression, 473-474
  effect on phenotypic suppression, 474
  nonsense suppression by, 473-474
Purine biosynthesis, 271-278
  genes and enzymes of, 274-275
  genetic analysis of, 276-279
  linkage to pyrimidine biosynthesis, 277
  metabolic flow in, 277-278
  pathway, 271-273, 276-277
  regulation of enzyme levels in, 278
  role of THF in, 279
Purine interconversion, 273
Purine transport, 425-428, 441
Purine utilization, 273
Pyrimidine biosynthesis, 264-271
  genes and enzymes in, 266-267
  genetic analysis of, 269-271
  magnitude of regulatory responses in, 268
  metabolic flow in, 270
  pathway, 265-270
  regulation of enzyme levels in, 270-271
Pyrimidine transport, 426, 428-429
Pyrimidine utilization, 269
Pyrroline-5-carboxylate (P5C) dehydrogenase, 57, 188
  cellular localization, 57, 188

Subject Index 667

gene coding for, 57
Pyrroline-5-carboxylate reductase, 55, 186, 188
  cellular localization, 188
  gene coding for, 57
  role in proline biosynthesis, 188
Pyruvate, in TCA cycle, 19-21
Pyruvate carboxylase, 6, 20, 24
Pyruvate decarboxylase, 6, 13
Pyruvate dehydrogenase, 6, 20
Pyruvate kinase, 6, 13
Pyruvate utilization, 19-20

Raffinose utilization, 17
rDNA genes
  mapping by cotransformation, 619
  restriction-fragment-length polymorphism in, 625
Recombinant DNA. *See* Cloning
Recombination
  with linear DNA, 627
  manipulation of chromosome using, 626-627
  during transformation, 617-620
Replication origins, use in cloning, 621-623
Respiration, 19-27
  catabolite inactivation in, 23-24
  catabolite repression in, 22-23
  localization of enzymes in, 24
  mutants in, 20-21
  oxygen induction of enzymes in, 22
  TCA cycle in, 19-21
ϱ^-, effect on suppression, 475-476
Riboflavin transport, 430, 432
Ribosomal proteins
  assembly of, 544
  defective, in ribosomal suppressors, 472-473
  gene isolation, 616
  intervening sequences in mRNAs of, 624
  localization of genes for, 545
  methylation of, 533
  mutants in assembly of, 546
  numbering system for, 531-532, 534-537
  phosphorylation of, 533
  properties of, 532-538
  and rRNA, coordinate control of synthesis, 552-553
  synthesis of, 544
  synthesis, during heat shock, 549
    effect of sporulation on, 551
    in presence of cyclohexamide, 549
    in *rna* mutants, 550-553
  synthesis of mRNA for, mutations affecting, 550
Ribosomal RNA. *See* rRNA
Ribosome(s), 529-553
  defects, in ribosomal suppressor mutants, 472-473, 481-482
  as limiting factor for translation, 539
Ribosome synthesis, 542-545

effect of growth rate on, 546-547
effect of heat shock on, 549-551
mutations affecting, 545-546, 550-553
regulation of, 546-553
role in cell cycle, 547
rna1 mutants, 496
  effect on ribosomal protein synthesis, 551
  effect on tRNA maturation, 497-502, 510-513
RNA polymerase(s), 561-599. *See also* Transcription
  catalytic properties, 566, 580-581
  chromatographic properties, 566
  common subunits among, 572-576
  electrophoretic determination of, 573-574
  fingerprinting determination of, 574
  immunological determination of, 574-575
  effect of growth rate on, 564-565, 583-585
  function of, 585-586
  immunological relationships among, 572-576
  inhibition of
    by α-amanitin, 582
    by chelators, 583
    by polyanions, 583
    by substrate analogs, 578, 583
    by tRNA, 582-583
  molecular properties of, 567-571
  mutants in, 578-579
  phosphorylation of, 571
  polypeptides in
    biochemical properties of, 570-571
    functional role of, 576-580
    molar ratios of, 570
    organization of genes for, 571-572
  proteolysis of, during purification, 565
  purification procedures for, 563-567
  regulation of, 583-585
    during cell-cycle, 584
    by phosphorylation, 584-585
  subunit dissociation during purification, 565
RNA polymerase I, 586-587. *See also* RNA polymerase A
RNA polymerase II. *See also* RNA polymerase B
  function in heterologous systems, 597-598
  sequence requirements for initiation by, 589-592
  sequence requirements for termination by, 592
  in vitro template specificity of, 587-588
RNA polymerase III, 593-594. *See also* RNA polymerase C
  sequence requirements for initiation by
    role of coding sequences, 593-594
    role of 5'-flanking sequences, 593-594
    in *Xenopus* systems, 593-594
  in vitro transcription by
    protein components required for, 588
    specificity of, 588

RNA polymerase A. *See also* RNA polymerase I
  association with RNase-H activity, 579-580
  dissociation into A*, 569
  in DNA replication, 586
  effect of growth rate on, 564-565
  immunological relationship
    with other yeast species, 575-576
    with RNA polymerases B and C, 574-575
  immunoprecipitation of, 567, 568
  independent regulation of, 584
  phosphorylation of, 571
  polypeptide composition of, 568-569
  polypeptide function in, 576-577
  pyridoxal 5′-phosphate inhibition of, 578
  in rRNA synthesis, 586-587
  species variation of, 564
  substrates and catalytic properties, 566, 581
RNA polymerase B. *See also* RNA polymerase II
  dissociation into B*, 570
  effect of growth rate on, 564-565
  immunological relationship
    with other eukaryotic polymerases, 576
    with RNA polymerase A, 574-575
  in mRNA synthesis, 586-588
  mutant in, 578
  phosphorylation of, 571
  polypeptide composition of, 569-570
  polypeptide function in, 576-577
  proteolysis of, 564, 565, 570
  substrates and catalytic properties, 566, 580-581
RNA polymerase C. *See also* RNA polymerase III
  catalytic properties of, 566, 581
  effect of growth rate on, 565
  in 5S and tRNA synthesis, 585-586
  phosphorylation of, 571
  polypeptide content, 569-570
  properties, 581
RNase H, 579-580
RNase P, 514
rRNA
  gene isolation, 616
  methylation, 543
  mutant in processing of, 545
  posttranscriptional modifications of, 542-543
  processing of primary transcript, 543
  properties, 530
  and ribosomal proteins, coordinate control of synthesis of, 552-553
  synthesis
    during heat shock, 549
    mutations affecting, 550-551
  transcription
    during amino acid starvation, 548
    inhibition by cyclohexamide, 548
    role in regulation of ribosomal synthesis, 552-553
  rRNA genes
    chromosomal organization of, 542
    transcription of, 542-543

Saccharopine dehydrogenase, 186, 196
  cross-pathway control of, 198, 250
  mutations affecting, 197
  regulation of, 261
Saccharopine reductase, 186, 196
  mutations affecting, 197
  regulation of, 198, 261
sec mutants, 381-385
Secreted proteins
  localization of, 362-364
  vacuolar, 364
Secretion, 361-387, 394-397. *See also* Invertase; Mannoproteins; Phosphatase, acid; Secreted proteins, Secretory vesicles
  cell-cycle regulation of, 367
  localization of, 366-368
  in mannoprotein mutants, 342
  of mannoproteins, 364-365
  during mating, 367
  mechanism of, 364, 367, 384-385
  mutants
    accumulation of secretory vesicles in, 382-383, 396
    Berkeley body formation in, 383
    bud growth in, 383
    cell growth in, 382
    class-A, 382-383
    class-B, 382-384
    ER in, 383-384, 396
    isolation and characterization of, 382
    mutations affecting, 381-387
    order of events in, 384-385
    organelles involved in
      ER, 377-378
      Golgi, 378
      plasma membrane, 380-381
      secretory vesicles, 380
      vacuoles, 378-380
    role of bud in, 366-367
    role of secretory vesicles in, 364-365
    and surface growth, 366-368
    transcriptional control of, 366
Secretory vesicles, 394
  accumulation in secretion mutants, 382-383, 396
  biochemical markers for, 651
  characterization of, 380
  fusion with plasma membrane, 371-372
  microtubule-mediated transport of, 371
  role in bud-limited growth, 367
  role in mannoprotein secretion, 364-365
Septation
  effect on cell wall, 345, 348
  in glucosamine auxotrophs, 345

Subject Index **669**

Sequential induction
 in arginine degradation, 78–80
 during nitrogen starvation, 77–79
Serine, as source of single carbons, 279
Serine biosynthesis, 207–209, 210, 212
 in acetate medium, 209
 genes and enzymes for, 210
 pathway, 208
Serine hydroxymethyltransferase
 in glycine biosynthesis, 209–210
 cellular localization, 209
 mutations affecting, 209
 in single carbon metabolism, 279, 282
 inhibition of, 283
 mutations affecting, 283
 regulation of, 284
 two species of, 209
Shine-Dalgarno sequence, 316–317
σ element, role in tRNA transcription, 594
Sinefungin, 123
Single carbon metabolisn, 279–284
 effect of mutations in, 283
 genes and enzymes in, 282
 metabolic flow in, 283
 pathway, 279–283
 role of mitochondria and cytoplasm in, 283
Sorbitol transport, 404, 409
Southern hybridization
 in analysis of *his4* mutations, 246
 in genetic analysis, 625, 631
Spermine synthase, 194, 195
Spheroplast. *See* Protoplast
Splicing
 in mitochondrial rRNA, 543–544
 of tRNA introns
  effect of intron mutations on, 505–508
  endonuclease activity in, 512, 523–524
  intermediates of, 511–512, 522–523
  ligase specificity in, 512, 522–523
  mutational analysis of, 503–508
  in *rna1* mutants, 497–502
  role of ATP in, 511–512
  role of nuclear membrane in, 513
  role of pre-tRNA structure in, 500–502
  time course of, 510–511
  in vitro system for, 499, 510–511
Spore coat, 349–350, 355
Spore germination, 350
Sporulation
 effect on cell wall, 349–350
 effect of nitrogen starvation on, 75–76
 effect on ribosomal protein synthesis, 551
 inhibition by suppressors, 479
 spore coat formation during, 349–350
 translation during, 539
Squalene, 105, 112
Sterol(s). *See also* Ergosterol; Lipid(s)
 in mitochondrial membranes, 108
 -requiring mutants, 115–117, 119–122

 structure requirements of membranes for, 113–123
 in yeast membrane fractions, 108
Sterol biosynthesis, 112
 cellular localization, 106
 inhibitors of, 122–123
 mutants in, 114–115, 119–122
 nystatin-resistant, 120–121
 temperature-sensitive, 114–115, 121
Sterol deprivation, effect on mitochondria, 122
Sterol esters, 119
Streptomycin sensitivity, 546
Strepzyme, 348
Stringent response, 548
Succinate dehydrogenase, 6, 21
Succinyl-CoA synthase, 6, 20–21
Sucrose transport, 404, 407
Sucrose utilization, 16
Sulfate reduction
 cross-pathway control of, 251
 in methionine biosynthesis, 222
 regulation of, 226–227, 244
Sulfate transport
 feedback inhibition of, 435
 and homocysteine synthase, coordinate control of, 435
 inhibitors of, 433, 435
 multiple systems for, 433–435
 mutations affecting, 433, 435
 repression of, 434, 435
 role of cations in, 433
Sulfite reductase, 217, 222
 cross-pathway control of, 251
 mutations affecting, 222–223
 regulation of, 227, 262
SUP4 locus
 analysis of, by gene cloning, 626
 analysis of tRNA transcription using, 594–597
 δ sequences in, 505
 fine-structure analysis of, 503–505
 isolation of mutants, with canavanine, 594
 mutations in, 503–508
Suppressible alleles, 477–478
 discriminating, 477–478
 highly effective, 477
Suppression, 463–482. *See also* Allosuppressors; Antisuppressors, Suppressible alleles; Suppression, phenotypic; Suppressors
Suppression, phenotypic, 475
 effect of ψ⁺ on, 474
 by hypertonic media, 475
 by paromomycin, 475
Suppressors. *See also* Allosuppressors; Antisuppressors, Suppression; Suppression, phenotypic
 effect on translation, 541
 frameshift, 470, 471, 480–481

Suppressors (*continued*)
  effect of ψ^+ on, 474
  number of loci producing, 470, 471, 480–481
  inhibition of growth by, 478–479
  inhibition of sporulation by, 479
  instability, 478–479
  missense, 470–471
  nonsense
    chromosomal distribution of, 495–497
    effect of ψ^+ on, 473–474
    ribosomal, 471–473, 481–482
    UAA, 465–469, 477–482
      amino acids inserted by, 466–469
      and codon-anticodon wobble, 492
      efficiency of, 467–468
      identification of mutant tRNA, 480
      number of loci producing, 467–468
    UAG, 465–469, 477–482
      amino acids inserted by, 464–469
      efficiency of, 467–468
      number of loci producing, 467–468
    UGA, 469–470
      number of loci producing, 469–470
      in in vitro systems, 469

TATAAA element
  location of, 589–591
  role in transcription initiation
    in eukaryotes, 589
    in yeast, 589–591
Tetracycline, 162, 614
Tetrahydrofolate (THF), in methionine biosynthesis, 225, 229
Tetrahydrofolate derivatives. *See also* Single carbon metabolism
  in methionine synthesis, 279
  in purine biosynthesis, 279
  synthesis and interconversion of, 279–283
    localization of enzymes for, 280–281
    mutations affecting, 283
    role of mitochondria in, 283
Thiamine transport
  by active mechanism, 430, 432
  inhibitors of, 430, 432
  mutations affecting, 430
  repression of, 430, 432
Thiolutin, 578, 583
Thioredoxin and thioredoxin reductase, 269
Threonine biosynthesis. *See also* Homoserine biosynthesis
  control of enzyme activity of, 214, 221
  magnitude of regulatory response in, 259, 261
  pathway, 213, 215, 219, 221
Threonine deaminase, 230, 232
  control of activity of, 234
  mutations affecting, 238, 245
  regulation of levels of, 235–236, 263
  role in regulation of isoleucine-valine

enzymes, 236–238
Threonine synthase, 215, 221
Thymidine kinase, 443
Thymidine transport, mutants capable of, 443
Thymidylate biosynthesis. *See* Pyrimidine biosynthesis
Thymidylate synthase, 267, 269
  cellular localization, 269
  mutations affecting, 269
Tonoplast
  isolation, 439
  morphology, 69, 70
Transcription. *See also CYC1* gene; mRNA; RNA polymerase(s)
  analysis of, by recombinant DNA techniques, 624
  comparison, with other eukaryotes, of
    initiation by RNA polymerase II, 589, 597–598
    initiation by RNA polymerase III, 592–594, 597
    termination by RNA polymerase II, 592, 598
    termination by RNA polymerase III, 595, 597–598
  direction of, determination by cloning, 624
  inhibitors of, 581–583
  regulation of, 583–598
    by RNA polymerase II
      analysis of, by in vitro mutagenesis, 591–592
      signals involved in termination, 592
      structures and sequences required for, 589–592
    by RNA polymerase III
      mutational analysis of, 592–597
      sequences involved in initiation, 593–594
      sequences involved in termination, 595, 597
  in vitro systems for, 586–589
Transferase
  gene coding for, 161
  regulatory genes for, 161
  transcriptional control of, 164, 167–168
Transfer RNA. *See* tRNA
Transformation, yeast DNA, 617–620
  analysis of *CYC1* gene transcription using, 591
  cotransformation
    study of, 619
    use in rDNA mapping, 619
  efficiency of, 617
  integration into chromosome, 617–620
    double crossover events, 619
    frequency of, 619
    single crossover event, 619
  use in cloning of mutations, 626
  use in creating deletions, 628
  use in gene transposition, 626–627

Subject Index   **671**

procedure for, 617
Transinhibition of, 446–447
  allantoate transport, 412
  allantoin transport, 412, 446–447
  ureidosuccinate transport, 429
Translation. *See also CYC1* gene; Initiation codon(s); Iso-1-cytochrome *c*; Translation initiation, *CYC1* gene
  cell-free systems for, 539–540
  effects of antibiotics, 540–541
  factors affecting rate of, 539
  identification of factors required for, 540
  of monocistronic vs. polycistronic RNAs, 284–286, 317, 321, 539
  mutations affecting, 541
  during sporulation, 531
Translation initiation, *CYC1* gene, 301–330. *See also CYC1* gene; Initiation codon(s); Iso-1-cytochrome *c*,
  comparison with prokaryotes, 310, 316–321
  effect of mRNA leader region on, 315–316
  effect of mRNA secondary structure on, 314
  mutational analysis, 302–330
  reinitiation after termination codon, 310, 316–320
  relocation of site of, 302–304, 306–309
  scanning hypothesis, 317, 319, 321
  sequence specificity of, 306–309, 312–315
Transport, 399–453
  active, 401–402
    energy dependence, 402
    involvement of carrier protein, 402
    source of energy for, 441
  activity in cells of multiple ploidy, 448
  of allantoate, 46
  of allantoin, 46
  of allophanate hydrolase, 48
  of amino acids, 78
  of anabolic amino acids, 411–425
  of anabolic nitrogenous compounds, 420–429
  of carbohydrates, 403–409
  of catabolic amino acids, 411, 416–420
  of catabolic nitrogenous compounds, 409–420
  by diffusion, 401
    effect of inositol starvation on, 142
  by facilitated diffusion, 401–402
    energy independence, 401
    involvement of carrier protein, 401–402
  genes cloned, 438
  genetic loci for, 438
  of glucose, 10, 403–406
  of ions, 433–438
  of peptides, 421, 425
  of purines, 425–428
  of pyrimidines, 426, 428–429
  regulation of, 444–447
    by feedback inhibition, 445–446
    by nitrogen repression, 444–445
    by transinhibition, 446–447
    by vacuolar compartmentation, 447
  types of, 401–402
  of urea, 44
  use of competitive inhibition to analyze, 402–403
  use of Michaelis-Menton equation to analyze, 402
  into vacuoles, 440–441
  of vitamins, 429–433
Transport components, isolation and purification, 448–452
Transposable elements. *See* Ty element insertions
Trehalase, 26
Trehalose degradation, 25–26
Trehalose-phosphate phosphatase, 26
Trehalose-6-phosphate synthetase, 26
Trehalose synthesis, 25–26
Trehalose transport, 404, 408–409
Trehalose utilization, 17
Triazolealanine resistance, 240, 253, 254
Tricarboxylic-acid (TCA) cycle, 19–21, 196
  localization of enzymes in, 21, 24
  mutants in, 20–21
  pathway, 20
Trichodermin
  as inhibitor of translation termination, 540
  as inhibitor of translocation, 48
  resistance, 534, 545
Trinitrobenzene sulfonate, in analysis of secretion, 386
Triose-phosphate isomerase, 7, 12
tRNA, 487–524. *See also* Introns, tRNA; Suppression; Suppressor(s); tRNA genes; tRNA synthesis
  analysis of role of intervening sequences in, 629
  codon-anticodon wobble, 489–492
  codon recognition by, 489–492
  gene isolation, 472
  as inhibitor of RNA polymerase, 582–583
  initiator
    eukaryotic and prokaryotic, 310, 316
    role of hypermodified base in recognition by, 310, 316
  methionyl, role in methionine biosynthesis, 228
  minor species, 490–491, 520–523
  mutations affecting, 503–510
    canavanine-resistance in selection for, 503
    rna1, 497–502
    SUP4 locus, 503–508
  number of species, 489–492
  precursors
    accumulation in *los1* mutants, 510
    nuclease sensitivity, 485–486
    structural models for, 500–502
  regulation of transcription of, 594
  role in cross-pathway control, 252

tRNA (continued)
role of σ element in transcription of, 594
suppressor mutations in, 463–471
synthesis, as regulatory mechanism, 522–526
undermodification of
effect of, 509–510
in SAM mutants, 228
tRNA genes, 487–524. *See also* Suppression; Suppressor(s); tRNA
analysis of intervening sequences in, 624
chromosomal distribution of, 492–496
chromosomal juxtaposition of, 495
copy number, 490–491
number of, 490–496
Southern analysis of, 505
tRNA introns. *See* Introns, tRNA
tRNA synthesis. *See also* Introns; tRNA; RNA polymerase I; RNA polymerase A; Splicing
addition of 3' terminus, 513–520
initiation and termination sites, 515, 593–594
maturation of 5' terminus, 513–520
modification of bases, 509–510, 513–520
regulation of transcription in, 516–519
role of RNA polymerase III, 514, 585, 586, 593–594
splicing of introns, 510–513, 523–524
stages in, 515–518
in vitro heterologous systems for, 514–520, 593–594
Tryptophan biosynthesis, 198–207
cross-pathway control in, 207, 249, 250
genes and enzymes for, 202–203
magnitude of regulatory response in, 259, 261, 264
metabolic flow in, 204–205
mutations affecting, 201, 204, 206, 207
pathway, 198–204
regulation of enzyme levels, 205–207
Tryptophan synthase, 201, 203, 250
Tubulosine, 540
Tunicamycin
inhibition of invertase and acid phosphatase synthesis, 374
inhibition of mannoprotein synthesis, 366
2µ plasmid, use in cloning, 619–620
Ty element insertions, 47, 246–248
in *ADR2* gene, 246–248
in *CAR1* gene 83, 89
in *CYC1* gene, 246–248
in *DUR1,2* gene, 47, 89
effect of, on gene expression, 589
in *HIS4* gene, 246, 248
and mating-type control, 247
properties of, 247, 248
Southern analysis of, 625
in urea degradation, 82
Tyrosine biosynthesis, 198–207
genes and enzymes of, 202–203
magnitude of regulatory response in, 259, 261

metabolic flow in, 204–205
mutations affecting, 201
pathway, 198–201
regulation of enzyme levels, 205–206

UDP-glucose-4-epimerase. *See* Epimerase
UDP-glucose pyrophosphorylase, 16
Ultraviolet light, in *CYC1* mutagenesis, 303, 329
Uracil, rate of efflux, 428
Uracil transport, 426, 428
multiple systems for, 426, 428
mutations affecting, 428
URA3 gene
mRNA levels, 271
regulation of, on multicopy plasmid, 271
Urea
in allantoin metabolism, 78–80
degradation of, 41, 43
as nitrogen source, 80–81
Urea carboxylase, 43
gene coding for, 44
induction of, 46
as multifunctional protein, 44
nitrogen catabolite repression of, 82
transcriptional control of, 49
Urease, lack of, in *S. cerevisiae*, 41
Urea transport, 78, 411, 412–413
active mechanism, 411, 412–413
gene coding for, 44
induction of, 46, 412
inhibitors of, 412
mutations affecting, 413
facilitated diffusion, 411, 413
gene coding for, 44
regulation of, 46
URE2-gene product, in nitrogen catabolite repression, 86–87
URE3-gene product, in nitrogen catabolite repression, 88–89
Ureidoglycollate hydrolase, 41
gene coding for, 44
induction of, 46
nitrogen catabolite repression of, 82
transcriptional control of, 49
Ureidosuccinate transport, 426, 428–429
by active mechanism, 426, 428–429
by allantoate transport system, 429
inhibitors of, 429
mutations affecting, 428–429
regulation of, 426, 429
Uridine diphosphoglucose-4-epimerase. *See* Epimerase
Uridine transport, 426, 428
multiple systems for, 426, 428
mutations affecting, 428
Uridylate biosynthesis. *See* Pyrimidine biosynthesis
Uridyl transferase, 15

Vacuole, 394
  biochemical markers for, 651
  contents of, 439–440
    enzymes, 67
    metabolites, 67
    during nitrogen starvation, 66–67
    polyphosphate, 26
    proteins, 364, 379–380
  isolation of, 439, 440
  membrane, 379, 448–449
  morphology, cell-cycle changes in, 68
  role in metabolite compartmentation, 69, 71, 438–447
    of amino acids, 182, 183
    effect of, on transport, 447
  role in secretion, 378–380
  transport systems for, 440–441
Valine biosynthesis
  cross-pathway control of, 249, 251
  genes and enzymes in, 232
  localization of enzymes in, 233
  magnitude of regulatory response in, 259, 263, 264
  metabolic flow in, 234–235
  pathway, 231–233
  regulation of enzyme levels, 235–238
Vesicles, plasma membrane
  ATPase activity in, 449–450
  use in transport studies, 449–450
Vitamin transport, 429–433

Wobble, codon-anticodon, 489–492

YCp vectors, 621, 623
YEp vectors, 620–623
YIp vectors, 620–622
YRp vectors, 620–623

Zinc, in RNA polymerase, 568
Zymolase, 348, 362

Gene Index

aap, 415, 418–419, 438
aas1, 229, 253–255
aas2, 229, 253–255
aas3, 229, 253–255
acc1, 117, 125
acc2, 117, 125
aco1 (*glu1*), 3, 4, 20, 185
ACR3, 23
adc1, 3, 4, 13–14, 590, 591, 598
ade1, 215, 271, 272, 274, 276–278
ade2, 271, 272, 274, 276–277, 504, 594
ade3, 240, 276, 280, 281, 282, 283–285
ade4 (*pur6*), 272, 274, 276–278
ade5, 272, 274, 276, 285, 469, 478
ade6, 272, 274, 276
ade7, 272, 274, 276, 285, 469, 478
ade8, 272, 274, 276, 280, 282
ade12, 272, 275, 276–278
ade13, 272, 275, 276
adm1, 3, 4, 13
adr1, 14, 598
adr2, 3, 4, 13–14, 246–248, 590
ADR3, 14
ama, 54
amt, 414–415, 419, 438
apf, 418–419
app1, 425, 426, 427, 438
arg1, 184, 187, 190, 244, 250, 260
arg2, 184, 186, 190, 244, 250, 260
arg3, 184, 187, 190–191, 244–245, 250, 252, 255, 260
arg4, 184, 187, 190, 250, 252, 260, 620
arg5, 184, 186, 190–192, 244–245, 250, 260, 285–286
arg6, 184, 186, 190–192, 244–245, 250, 260, 285–286
arg7, 184, 186, 190, 250, 260
arg8, 184, 186, 190, 244, 250, 260
arg10, 184, 187, 190, 244, 260
arg80, 61–63, 82–83, 190–192, 244–245
arg81, 61–63, 82–83, 190–192, 244–245, 254–255
arg82, 61–63, 82–83, 190–192, 244–245
aro1A, 199, 201, 202, 285
aro1B, 199, 201, 202, 285
aro1C, 199, 201, 202, 285
aro1D, 199, 201, 202, 285
aro1E, 199, 201, 202, 261, 285
aro2, 199, 202
aro3, 199, 201, 202, 204, 261
aro4, 199, 201, 202, 204, 261
aro7, 199, 201, 202, 245, 261, 477
asn1, 213, 215, 216
asn2, 213, 215, 216
asp1, 50
asp2, 50–51
asp3, 50–51, 364
asp4, 51
asp5, 50, 213, 215
asu1, 476
asu2, 476
asu3, 476
asu4, 476
asu5, 476
asu6, 476
asu7, 476
asu8, 476
asu9, 476, 481
ASU10, 481
asu11, 481

bin1, 64

675

Gene Index

bin2, 64
bin3, 64
BOR1, 219

can1 (*argp1*), 416–417, 420, 421, 422–423, 438, 451, 503, 504, 594
car1, 42, 49, 61–63, 77, 82–83, 87, 89, 184, 187
car2, 56, 59, 61–62, 77
car80 (*cargR*), 49, 61–62, 82–83
cargA, 247
cargB, 247
cat1, 23
cat2, 23
cat80, 23
ccr1, 23
ccr2, 23
ccr3, 23
CCR80, 23
cdc3, 372
cdc4, 65
cdc7, 65
cdc10, 372
cdc11, 372
cdc12, 372
cdc19 (*pyk1*), 3, 6, 9, 13
cdc21 (*tmp1*), 265, 267, 269, 280, 281, 282
cdc24, 346, 367, 369, 371–372
cdc25, 76
cdc28, 65
cdc35, 76
cho1, 111, 118, 143–147
chr1, 217, 434
cif, 24
cit1 (*glu3*), 3, 4, 20, 185
cpa1, 184, 186, 190, 192–193, 244–245, 250, 252–253, 260, 264
cpa2, 184, 186, 190, 192–193, 244, 250, 252–253, 260, 264
CPA80, 192, 245
cpa81, 192, 244
cps1, 74
cry1, 541
cyc1, 255–257, 301–330, 464, 468, 473–474, 479, 590, 591–592, 616, 624, 629
cyc4, 115, 121
cyc7, 246–248
cyh2, 541
cys1, 208, 210, 212, 213
cys2, 208, 210, 212, 213
cytp, 427

dal1, 42, 43–44, 46, 77, 82
dal2, 42, 44, 46, 77, 82
dal3, 42, 44, 46, 49, 77, 82–83
dal4, 42, 44, 49, 77, 82–83, 411, 412, 438
dal5 (*uep1*), 42, 44, 46, 411, 412, 426, 429, 438
dal80, 46–47, 49, 62, 82–83

dal81, 46–47, 49, 62, 82–83
dap, 276
dhu1, 428
DHU2, 428
dip1, 545
dsf, 17
dsf6, 407, 438
dsf7, 407, 438
dsf17, 407, 438
dsf21, 407, 438
dur1,2, 42, 44, 45, 46, 49, 59, 71, 77, 82–83, 89, 247
dur80, 47–48

erg1, 114, 121
erg2, 114, 120
erg3, 114, 120
erg5, 114
erg6, 114, 120
erg7, 114, 121
erg8, 115, 121
erg9, 115, 121
erg10, 115, 121
erg11 (*SG1*), 115, 121
erg12 (*GL7*), 113, 115, 119, 121
eth2 (*sam1*), 208, 211, 213, 218, 221, 225, 227, 229, 244, 262
eth3, 228, 244
eth10 (*sam2*), 208, 211, 213, 225, 227–228, 244
exb1, 364

fas1, 117, 124–125, 128–129, 130, 131, 132
fas2, 117, 124–125, 128–129, 130, 131
fcy1, 265, 267
fcy2, 265, 267, 427, 438
fdp1, 24
for1, 270
fui1, 265, 267
fur1, 265, 267, 270
fur4, 265, 267, 426, 428, 438

gal1, 15, 161, 164–165, 167, 168, 176, 286, 406
gal2, 161–162, 167, 404, 406, 438
gal3, 16, 161–162, 406
gal4, 161–168, 173, 175, 176, 406
gal5 (*pgm2*), 3, 6, 16, 161
gal7, 15, 161, 164, 167, 168, 286, 318, 406
gal10, 15, 161, 167, 168, 286, 318
gal80, 161–165, 167–168, 173, 175, 176, 406
GAL81, 161–168, 173, 175, 176
gap1, 411, 415–420, 422, 424–425, 429, 438, 444, 451
gap491, 255–257
gcr1, 15
gdc1, 208, 209, 210, 280, 281, 282
gdh1 (*gdha*, *ure1*), 51, 54–55, 75, 83–88, 90, 184, 185, 186, 250, 417–418

Gene Index   **677**

gdhCR (*glu80, ure2*), 54–55, 85–89, 418
gen^c, 254–255
gl7 (*erg12*), 113, 115, 119, 121
glc1, 26
glk1, 3, 4, 10
gln1, 87–88, 185, 186, 251, 260
gln3, 188
glu1 (*aco1*), 3, 4, 20, 185
glu3 (*cit1*), 3, 4, 20, 185
glu80 (*gdhCR, ure2*), 54–55, 85–89, 418
gnrR, 88
gpm1 (*pgm1*), 3, 5, 6, 12, 16
gua1, 272, 275, 276, 278
gua2, 276
gua3, 276
gut1, 3, 5, 18
gut2, 3, 5, 18

hem1 (*ole3*), 113, 116, 121–122, 127
hem2 (*ole4, olerg4*), 113, 116, 121, 127
hem3 (*ole2, olerg2*), 113, 115, 116, 121, 127
hem4, 116, 121
hem5, 116, 121
hex2, 23
hip1, 421, 423, 438, 445
his1, 240–243, 250, 252, 263
his2, 240, 242, 243, 250, 252, 263, 471
his3, 242, 243, 246–247, 250, 252, 255–257, 263, 590, 591, 614, 616–617, 619–621, 624, 626, 628, 631
his4, 240, 246–248, 252, 255–257, 284–285, 470, 477, 481, 625
his4A, 242, 243, 250, 252, 263
his4B, 242, 243, 250
his4C, 242, 243, 250, 263
his5, 242, 243, 263
his6, 242, 243
his7, 242, 243
HML, 631
HMR, 631
hom2, 213, 216, 261
hom3, 213, 216, 219, 261
hom6, 213, 216, 261
hxk1, 3, 5, 10, 14
hxk2 (*glr1, hex1*), 3, 5, 10, 14, 23

icl1, 3, 5, 21
ils1, 252
ilv1, 230, 231, 232, 236–238, 245, 263, 471, 477
ilv2, 230, 231, 232, 234, 250, 263
ilv3, 231, 232, 233, 263
ilv4, 231, 232, 263
ilv5, 231, 232, 263
ino1, 110, 111, 117, 138, 140–141, 146
ino2, 111, 117, 138, 139, 140–141, 145–147
ino3, 117
ino4, 111, 117, 138, 139, 140–141, 145–147
ino5, 117

ino6, 117
ino7, 117
ino8, 117
ino9, 117
ino10, 117

kgd1, 3, 5, 20

leu1, 231–234, 250, 263, 477
leu2, 231–234, 250, 263, 477, 478, 614, 617, 619, 621, 626
leu3, 233
leu4, 232, 234, 250, 263
leu5, 232, 234, 250, 263, 504
los1, 510, 520
lyp1, 417, 421, 422–423, 438
lys1, 63–64, 184, 187, 197, 251, 261, 504
lys2, 63–64, 184, 187, 196–197, 251, 260, 473, 477–479
lys3, 184, 187, 196
lys4, 64, 184, 196
lys5, 63–64, 184, 187, 196, 251, 260
lys6, 64, 197
lys7, 64, 184, 187, 196
lys8, 64, 197
lys9, 64, 184, 187, 197, 261
lys10, 64
lys12, 64, 184, 196
lys13, 184, 187, 197, 261
lys14, 184, 187, 197, 261

mal1, 17
mal2, 17
mal3, 17
mal4, 17
mal6, 17, 22
MAT, 624, 631
mata, 247
mata1, 590
matα, 247
matα1, 590, 592
mdh1, 3, 5, 21
mdh2, 21
mel1, 364
mep1, 411, 414, 438
mep2, 411, 414, 438
mes1, 219, 228–229, 252
met1, 213, 217, 223
met2, 208, 211, 213, 216, 221, 224–226, 262
met3, 213, 217, 222, 225, 262
met4, 213, 217, 223
met5, 213, 217, 222–223, 251, 262
met6, 208, 211, 213, 217, 222, 225, 262, 280, 281, 282
met7, 225, 283
met8, 213, 217, 223, 477
met10, 213, 217, 222–223, 251, 262
met13, 225
met14, 213, 217, 222–223, 225

678 Gene Index

met15, 225–226
met16, 213, 217, 222, 225
met17, 208, 209, 210, 211, 213, 216, 217, 222–224
met18, 213, 217, 222–223, 250, 262
met19, 213, 217, 222–223, 250, 262
met20, 213, 217, 222–223, 250, 262
met21, 225
met22, 213, 217, 222
met23, 225
met24, 225
met25, 208, 209, 210, 211, 212, 213, 216, 218, 223–224
metp1, 421, 423, 438
mg2, 438
mgl1, 17
mgl2, 17, 404, 408, 413
mgl3, 17
mgl4, 17, 407
mnn1, 338–340
mnn2, 338–340, 355–356
mnn3, 340
mnn4, 339–340, 344
mnn5, 339–340
mnn6, 339–340
mnn7, 341, 374
mnn8, 341, 374
mnn9, 339, 341, 374
mnn10, 341, 374
mod1, 476, 510
mod2, 476, 510
mod3, 476, 510
mod4, 476, 510
mod5, 476, 509–510
mod6, 476, 510

ndr1, 253–254
ndr2, 253
ngl3, 54–55

ole1, 113, 117, 126, 128–129, 130, 131–132
ole2 (*hem3, olerg2*), 113, 115, 116, 121, 127
ole3 (*hem1*), 113, 116, 121–122, 127
ole4 (*hem2, olerg4*), 113, 116, 121, 127
olerg1, 116, 121
olerg2 (*hem3, ole2*), 113, 115, 116, 121, 127
olerg3, 116, 121
olerg4 (*hem2, ole4*), 113, 116, 121, 127
olerg5, 116, 121
olerg6, 116, 121
opi1, 118, 141
opi2, 118, 141
opi3, 111, 118, 139, 141, 145–147
opi4, 118, 141

pdc1, 3, 6, 13
pep4 (*pho9*), 75, 170–173, 174, 176–177, 377, 386
pep17, 377, 386

pfk1, 3, 6, 9, 12
pgi1, 3, 6, 11
pgk1, 3, 6, 9, 12, 590
pgm2 (*gal5*), 3, 6, 16, 161
pha2, 199, 201, 203, 261
pho1, 170
pho2, 169–173, 174, 176–177
pho3, 169–172, 364
pho4, 169–173, 174
pho5, 169–173, 174, 177, 364, 590
pho6, 169–170, 172
pho7, 169–170, 172
pho8, 170–173, 174, 177
pho9 (*pep4*), 75, 170–173, 174, 176–177, 377, 386
pho80, 169–173, 174, 175, 396
pho81, 169–173, 174, 175–176
PHO82, 169–173, 174, 175–176
PHO83, 170–172, 174
pho84, 170–172, 434, 436, 438
pho85, 169–173, 174, 175
pmi1, 3, 6, 15
ppr1, 271
pra2, 74
prb1, 74, 369
prb2, 74
prb3, 74
prb4, 74
prc1, 75
pro1, 57, 184, 186, 188
pro2, 57, 184, 186, 188
pro3, 56, 57–58, 59, 184, 186, 188
pt-r162, 438
pur1, 277–278
pur5, 275–276
pur6 (*ade4*), 272, 274, 276–278
put1, 56, 57–58, 59, 60
put2, 56, 57–58, 59, 60
put3, 58
put4, 411, 420, 438
pyk1, 3, 6, 9, 13, 590

rna1, 48, 497, 498, 499, 510, 513, 520, 551
rna2, 550–553
rna3, 550–553
rna4, 550–553
rna5, 550–553
rna6, 550–553
rna7, 550–553
rna8, 550–553
rna9, 550–553
rna10, 550–553
rna11, 550–553
rpoB1, 578–579, 585

sal1, 476
sal2, 476
sal3, 476
sal4, 476

Gene Index **679**

sal5, 476
samp1, 424, 438
samp2, 424
samp3, 424
sap3, 421
sdh1, 3, 6, 21
sec1, 142, 381-385, 396
sec2, 382, 384-385
sec3, 384-385
sec4, 384-385
sec5, 384-385
sec6, 384-385
sec7, 384-385
sec8, 384-385
sec9, 384-385
sec10, 384-385
sec12, 385
sec13, 385
sec14, 384-385
sec15, 384-385
sec16, 385
sec17, 385
sec18, 383-385, 396
sec20, 385
sec21, 385
sec22, 385
sec23, 385
sel1, 217, 434
ser1, 208, 209, 210, 283
ser2, 207, 208, 210
SG1 (*erg11*), 115, 121
sin1, 476
sin2, 476
spd1, 76
spe1, 194-195
spe2, 194-195
spe3, 194-195
spe4, 194-195
spe10, 194-195
SPE40, 194-195
SRN1, 551
srp1, 551, 553
suc1, 17, 364, 375
suc2, 364, 366, 375, 631
suc3, 364
suc4, 364
suc5, 364
suc6, 364
SUF1, 470, 471, 474, 477, 478, 480
SUF2, 471, 481
SUF3, 471, 477, 478, 480
SUF4, 470, 471, 474, 477, 478, 480
SUF5, 470, 471, 477, 478, 480
SUF6, 470, 471, 474, 477, 478, 480
SUF7, 471, 481
SUF8, 471, 481
SUF9, 471, 481
SUF10, 471, 481
suf11, 471, 481

suf12, 481
suf13, 481
suf14, 481
SUF15, 480
SUF16, 481
SUF17, 480
SUF18, 481
SUF19, 481
SUF20, 481
SUF21, 481
SUF22, 481
SUF23, 481
SUF24, 481
SUF25, 481
SUH1, 471, 476
SUH2, 471, 476
SUP2, 466, 467, 476, 494
SUP3, 466, 494
SUP4, 465, 466, 468, 494, 503, 504, 505-506, 520, 594-597, 626
SUP5, 465, 466, 468, 476, 494
SUP6, 465, 466, 494, 629
SUP7, 465, 466, 474, 494, 509
SUP8, 466, 476, 494
SUP11, 466, 468, 476, 494, 496
SUP15, 466
SUP16, 465, 466, 467, 476, 480, 494
SUP17, 465, 466, 467, 477, 480, 494
SUP19, 465, 466, 467, 469, 480, 494
SUP20, 466, 476
SUP22, 466, 481
SUP25, 469
SUP26, 466, 467, 470, 494
SUP27, 466, 470, 494
SUP28, 466
SUP29, 466, 494
SUP30, 466
SUP32, 466
SUP33, 466
sup35, 471-473, 477, 481
sup45, 471-473, 477, 481
SUP46, 472-473, 475, 481, 536, 541
SUP50, 469
SUP51, 466
SUP52, 466, 467, 468, 495
SUP53, 466, 495
SUP54, 466, 494
SUP55, 466
SUP56, 466
SUP61 (*SUP-RL1*), 465, 466, 467, 468, 484, 494, 495, 502
SUP71, 470
SUP72, 470
SUP72, 470
SUP73, 470
SUP74, 470
SUP75, 470
SUP76, 470
SUP77, 470

SUP78, 470
SUP79, 470
SUP80, 470
SUP85, 470
SUP86, 470
SUP87, 470
SUP88, 470
SUP90, 470
SUP101, 470
SUP-RL1 (SUP61), 465, 466, 467, 468, 484, 494, 495, 502
SUQ5, 466
tcm1, 541, 546
tf1, 253
thp1, 432
thr1, 213, 216, 221, 262
thr2, 213, 216, 221
tmp3, 208, 209, 210, 217, 265, 267, 269, 281, 282, 283
tpi1, 3, 7, 12, 590
tra1, 229, 240
tra2, 255
tra3, 207, 253, 254
tra4, 253, 254
trp1, 199, 203, 250, 261, 477, 620–621
trp2, 199, 201, 203, 204, 206–207, 250, 252, 261
trp3, 199, 201, 203, 206–207, 250, 252, 261
trp4, 199, 203, 207, 250, 252, 261
trp5, 203, 250, 252, 261, 284–285, 471, 473, 477, 480
tyr1, 19, 201, 203, 261

uep1 (dal5), 42, 44, 46, 411, 412, 426, 429, 438
upf1, 477
upf2, 477
ura1, 265, 266, 268, 269, 271
ura2B, 86, 192, 265, 266, 268, 269–270, 285–286, 428–429
ura2C, 86, 192, 265, 266, 268, 269–270, 285–286, 428–429
ura3, 265, 266, 268, 269, 271, 614, 616–617, 619–621, 624–626, 628–630
ura4, 265, 266, 268, 478
ura5, 265, 266, 268, 269–270
urap, 428
urd1, 426
ure1 (gdh1, gdha), 51, 54–55, 75, 83–88, 90, 184, 185, 186, 250, 417–418
ure2 (gdhCR, glu80), 54–55, 85–89, 418
URE3, 88–89
urid-k 265, 267
uridp, 438
urid-rh, 265, 267